Flavins and Flavoproteins

Proceedings of the
Eighth International Symposium
Brighton, England, July 9 – 13, 1984

Editors
R. C. Bray · P. C. Engel · S. G. Mayhew

W
DE
G

Walter de Gruyter · Berlin · New York 1984

Editors

Robert C. Bray, B. A., Ph. D., Sc. D.
The University of Sussex
School of Molecular Sciences
Brighton BN1 9QJ
Great Britain

Paul C. Engel, M. A., D. Phil.
Department of Biochemistry
University of Sheffield
Sheffield S10 2TN
Great Britain

Stephen G. Mayhew, B. Sc., Ph. D.
Department of Biochemistry
University College Dublin
Belfield, Dublin 4
Ireland

Library of Congress Cataloging in Publication Data

Flavins and flavoproteins.
Includes bibliographies and indexes.
1. Flavins--Congresses. 2. Flavoproteins--Congresses.
I. Bray, R. C. (Robert C.), 1928- . II. Engel, Paul C. III. Mayhew, S. G. (Stephen G.), 1940- . IV. International Symposium on Flavins and Flavoproteins (8th : 1984 : Brighton, East Sussex)
QP67I.F5F53 1984 574.19'218 84-23034
ISBN 3-11-009879-2

CIP-Kurztitelaufnahme der Deutschen Bibliothek

Flavins and flavoproteins : proceedings of the internat. symposium. – Berlin ; New York : de Gruyter
8. Brighton, England, July 9 – 13. 1984. – 1984.
ISBN 3-11-009879-2

 Printing: Gerike GmbH, Berlin. Binding: Dieter Mikolai, Berlin. – Printed in Germany.

Flavins and Flavoproteins

PREFACE

The Eighth International Symposium on Flavins and Flavoproteins was held at the University of Sussex (Brighton, England) from the 8-13 July 1984. This coincided with an important anniversary for the flavin world, the 50th anniversary of the synthesis of lumiflavin by Richard Kuhn in 1934.

The organisers are indebted to a small but select group of bodies listed elsewhere who contributed financially to the anniversary symposium. In particular substantial help (though at an alarmingly late date) from the International Union of Biochemistry (through the Interest Group on Oxidases and Redox Enzymes) made it possible to provide travel support, if on a somewhat nominal scale, to all speakers who required it and to a few others. Support from other sources made it possible to subsidise registration fees, particularly for an encouragingly large contingent of younger participants; it is in their hands that the future development of flavin research will rest. It has to be said that many firms, especially in the pharmaceutical industry, to whom we turned for support were unable to help, and this was disappointing. These firms are among the most obvious beneficiaries of research into a group of enzymes that not only includes prime drug targets but also is intimately involved in drug metabolism. In view of the minimal support we were able to offer to senior scientists we had serious doubts about likely attendance until a month or two before the meeting. Under these circumstances it was especially gratifying that, in the event, so many travelled from all over the world to take part in the symposium.

In planning the programme the Organising Committee decided to continue the pattern adopted at the Ann Arbor Meeting, of morning lectures, afternoon poster sessions and evening discussions. This stamina-taxing regime again proved successful and the evening sessions were well attended throughout. The usefulness of the evening sessions owed much to the efforts of the discussion leaders, who receive our special thanks : Martha Ludwig, Franz Muller, Steve Mayhew, Bruno Curti, Graham Palmer, Dale Edmondson. The rigours of the British licensing laws meant that the discussions tended to finish after "closing time" in the outside world, but the bar extension allowed opportunities for continued discussion well into the night.

This was the first time that the 'Flavins Symposium' had been held in Britain. Our famed weather lived up to its reputation and had overseas participants in total confusion by the end of the week. On the whole, however, we were lucky and we were certainly spared the alleged divine anger that led to York Minster being struck by lightning in the same week! It rained while we talked science, but the sun always managed to come out for our various excursions.

The Chairman would like to thank his colleagues on the Organising Committee for their help and support. Thanks particularly are due to Vince Massey for his guidance which helped to ensure that the meeting followed faithfully in the traditions of earlier ones of this series. The international character of the Organising Committee made it inevitable that there should also be a Local Committee charged with much of the immediate decision-making. However, in this context, "local" encompassed divers parts of the British Isles. Thus it came about that Barry Smith (who has never worked on flavins) was cajoled into joining the Local Committee, as the representative of the Molecular Enzymology Group of the Biochemical Society, and then into shouldering a substantial part of the organisational work during the meeting. The success of the symposium as a whole was due in no small measure to his efforts and those of his helpers, not to mention the generous hospitality of the School of Chemistry and Molecular Sciences. Especial thanks are also due to the long-suffering Conference Secretary, Mrs. Margaret Stovold.

Each symposium seems to witness the application of a new crop of techniques to the flavin field. This time the impact of recombinant DNA techniques was in evidence for the first time, with the cloning of a number of flavoproteins being reported and detailed structural comparisons based upon DNA sequencing being presented. The advances in the application of high-resolution n.m.r. techniques were also prominent. The use of flavin analogues with chemically reactive substituent groups formed the subject of numerous communications and has justified a separate section in our division of the contents of this book.

The task of sorting the diverse contributions into categories has been extremely difficult since, inevitably, so many articles belong in two or more categories. The editors have had to make some very arbitrary decisions but hope that some logic may be discerned in the groupings at which they have finally arrived. This is the first time that the symposium volume has been produced from camera-ready copy. This has entailed some problems for authors and editors alike. Only time will show whether these were justified by unprecedentedly rapid publication. We hope so!

One of our intended session chairmen, Henry Kamin, was prevented by illness from attending the Symposium. His thoughtful comments were missed, and we wish him a speedy recovery.

The next symposium is to be held at Emory University, Atlanta Georgia in 1987. We look forward to many exciting developments over the intervening three years. The flavoprotein field is as vigorous as ever, and we feel that the pioneers of 1934 would be pleased to see where their researches have led.

R.C. Bray Brighton
P.C. Engel Sheffield
S.G. Mayhew Dublin

August 1984

ORGANISING COMMITTEE

R.C. Bray (Brighton) Chairman
H. Beinert (Madison)
B. Curti (Milano)
S. Ghisla (Konstanz)
J.-M. Lhoste (Orsay)
V. Massey (Ann Arbor)
D.B. McCormick (Atlanta)
G.E. Schulz (Heidelberg)
C. Veeger (Wageningen)
C. Walsh (Cambridge, MA)
C. H. Williams, Jr. (Ann Arbor)
K. Yagi (Nagoya)

LOCAL COMMITTEE

P.C. Engel (Sheffield)
S.G. Mayhew (Dublin)
B.E. Smith* (Brighton)

*representing the Molecular Enzymology Group of the Biochemical Society.

ACKNOWLEDGEMENTS

The organisers gratefully acknowledge the financial support received from the following organisations:

International Union of Biochemistry
Biochemical Society
The Royal Society
Roche Products
The Wellcome Foundation
I.C.I. Pharmaceuticals Division

CONTENTS

FLAVIN CHEMISTRY

ENZYME REACTION MECHANISMS

Modified Flavins in Flavoproteins

BIOMEDICAL ASPECTS

Practical Applications of Flavoprotein Studies

PARTICIPANTS

AUSTRALIA

ENTSCH, B.

AUSTRIA

WESSIAK, A.

CANADA

ARMSTRONG, D.A.

STEVENSON, K.J.

SURDHAR, P.S.

TSAI, C.S.

EIRE

MAYHEW, S.G.

Ó NUALLAIN, E.M.

FRANCE

CAPEILLÈRE-BLANDIN, C.

JANOT, J.M.

LABEYRIE, F.

LEDERER, F.

URBAN, P.

GERMANY

BACHER, A.

BEINERT, W.-D.

BIEMANN, M.

BOCK, M.

BRANDSCH, R.

BRETZ, N.H.

DECKER, K.

ELSNER, M.

GEISSLER, J.

GHISLA, S.

HINKKANEN, A.

HORN, E.

KRAUTH-SIEGEL, R.L.

KURFÜRST, M.

LUDWIG, H.

MACHEROUX, P.

MEYER, B.

NIELSEN, P.

OTT, U.

PAI, E.F.

SCHIRMER, R.H.

SCHULZ, G.E.

SIMON, H.

ZELLER, H.-D.

ZIPPLIES, M.F.

ITALY

CURTI, B.

NEGRI, A.

PAGANI, S.

RATTI, S.

RONCHI, S.

SIMONETTA, M.

TEGONI, M.

VANONI, M.A.

ZANETTI, G.

JAPAN

IKEGAMI, T.

ITAGAKI, E.

KOIKE, K.

KOIKE, M.

JAPAN (continued)

MATSUI, K.
MIURA, R.
NAKANO, H.
NONAKA, Y.
NISHINA, Y.
NISHINO, TAKESHI
NISHINO, TOMOKO
SHIN, M.
TOJO, H.
TSUGE, H.
WADA, K.
YAGI, K.
YAMANO, T.
YOSHIKAWA, S.
YUBISUI, T.

NETHERLANDS

ADDINK, R.
DE KOK, A.
EWEG, J.K.
LAANE, C.
MAGER, H.I.X.
MOONEN, C.T.W.
MÜLLER, F.
RUITENBEEK, W.
SCHIERBEEK, A.J.
SCHREUDER, H.A.
VAN BERKEL, W.J.H.
VAN DEN BERG, W.A.M.
VAN DER LAAN, J.M.
VEEGER, C.
VERVOORT, J.
VISSER, C.M.
WIJNANDS, R.A.

POLAND

KOZIOŁ, J.

SWEDEN

BRÄNDÉN, C.-I.
LINDQVIST, Y.
LINDSKOG, S.
MANNERVIK, B.
NEUJAHR, H.
SAHLMAN, L.
SEJLITZ, C.T.

SWITZERLAND

FENDRICH, G.

UNITED KINGDOM

ANDERSON, R.F.
BRAY, R.C.
BROWN, N.L.
CAMMACK, R.
CROSS, A.R.
DEISTUNG, J.
EADY, R.
ELLISON, P.A.
ENGEL, P.C.
GEORGE, G.N.
GREER, S.
GUEST, J.R.
HAWKES, T.R.
HEELIS, P.F.
HILL, S.
HOLMES, C.F.B.
JONES, G.
JONES, O.T.G.
LAMBERT, N.
McCAPRA, F.
MORPETH, F.F.

UNITED KINGDOM (continued)

NI' BHRIAIN, N.

PERHAM, R.N.

RICE, D.W.

RYLEY, J.G.

SHAW, L.

SMITH. B.E.

THORNELEY, R.N.F.

TROWER, M.K.

TRUDGILL. P.W.

TURNER, N.

VENTOM, A.

WALKER, C.J.

WALPOLE, C.

WATERS, B.W.

WOODLAND, M.P.

UNITED STATES

ABELES, R.

ACKRELL, B.A.C.

ALBARELLA, J.P.

ARSCOTT, L.D.

AYLING, J.E.

BAILEY, J.W.

BALDWIN, T.O.

BALLOU, D.

BARBER, M.J.

BRUICE, P.Y.

BRUICE, T.C.

CHURCHICH, J.E.

CLAIBORNE, H.A.

COOK, R.J.

DETMER, K.

EDMONDSON, D.E.

FLOSS, H.G.

GENNIS, R.B.

GIEGEL, D.A.

GORELICK, R.

GUSTAFSON, W.G.

HALL, C.L.

HAMZAH, R.

HASTINGS, J.W.

HILLE, R.

HUSAIN, M.

IKEDA, Y.

JONES, K.

JORNS, M.S.

KARPLUS, A.

KEARNEY, E.B.

KELLER, P.J.

KLAPPER, M.H.

LEE, J.S.

LUDWIG, M.L.

McCORMICK, D.B.

McINTIRE, W.S.

MASSEY, V.

MATHEWS, F.S.

MATTHEWS, R.G.

MILLER, S.M.

MORRIS, D.L.

PALMER, G.

PORTER, D.J.

POULSEN, L.L.

REEVES, H.C.

ROKITA, S.

SCHLOSS, J.V.

SHERRY, B.

SOLOMONSON, L.P.

STANKOVICH, M.T.

UNITED STATES (continued)

STEENKAMP, D.J.

STEWART, R.C.

SWAFFORD, J.R.

SWENSON, R.P.

TANAKA, K.

THORPE, C.

TU, S.-C.

WALSH, C.

WEYLER, W.

WILLIAMS, C.H. Jr.

WILLIAMSON, G.

WINGARD, L.B. Jr.

ZIEGLER. M.M.

FLAVIN CHEMISTRY

THE INFLUENCE OF HYDROGEN BOND FORMATION WITH THE N(1) ATOM ON THE ORBITAL STRUCTURE OF FLAVIN

Franz Müller, Jan Kees Eweg, and Viola Szczesna
Department of Biochemistry, Agricultural University, De Dreijen 11, 6703 BC Wageningen, The Netherlands

Ben Hesper
Central Computer Center, Leiden State University, Leiden, The Netherlands

Summary

Vapour phase spectroscopy on various flavins revealed a strong tendency of the molecule to form an intramolecular hydrogen bonded complex between the hydroxyl group of the side chain and the N(1) atom when a suitable solvent is lacking to interact with. The hydrogen bond affects, at the same time, the molecule's π and σ framework. The σ interaction, hitherto not taken into account in a description of the mechanisms of flavin catalysis, is very peculiar. The hydrogen bond causes a breakdown of existing through-bond interactions between the N(1) and O(2) and between the O(4) and N(5) lone pairs. As a consequence, a new interaction between O(2) and O(4) is created, increasing the lone pair character of N(5). This strongly indicates that the redox and bonding properties of flavin can be 'tuned' by hydrogen bond formation.

Introduction

In spite of the wealth of chemical and physical data on flavins and flavoproteins currently available, we still do not understand the mechanism by which a specific chemical reaction of the prosthetic group in a particular flavoprotein is 'tuned'. Already a decade ago, it was suggested that hydrogen bond formation between flavin and the apoflavoprotein is one of the major tuning mechanisms (1). This suggestion did not receive much attention until it was rediscovered by others (2,3) and its application broadened.

Flavins and Flavoproteins

Additionally, the so-called 'butterfly movement of reduced flavin was supposed to be another regulation mechanism, regulating the reactivity of the reduced prosthetic group towards molecular oxygen (4). Previously we have shown by ultra violet photoelectron spectroscopy (5) that reduced flavin is coplanar or almost coplanar in the vapour phase. Theoretical calculations alone seem not yet to be able to describe the conformation of reduced flavin adequately (6). More recently we could demonstrate that reduced flavin in solution is also almost coplanar (7). Consequently there is no intrinsic electronic property which can be held responsible for the bended conformation in the solid state (8,9). Hence the actual conformation in the solid state must be ascribed either to the bulky substituents at the N(5) and/or N(10) atom (cf. discussion in ref. 5 and 7) or to the fact that the crystals investigated contain hydrogen-bonded dimers (9). Undoubtedly bending of reduced flavin will affect the redox potential of the molecule since it affects orbitals and their energies (5). In flavoproteins several other factors play a role in the regulation of the redox potential. A new concept has been recently developed taking into account the integrety of the apoflavoprotein-flavin complex in a much broader sense than hitherto done (see C.T.W. Moonen, J. Vervoort, F. Müller, this symposium).
At any rate it is desirable to study the influence of the formation of a hydrogen bond to N(1) of flavin by model compounds in order to evaluate its influence on the chemical and physical properties of free flavin.
Such an evaluation will then allow a caution extrapolation to flavoprotein and allow to estimate possible beneficial contributions to flavin-dependent biocatalysis.
The possibility to sublime free flavin opened the way to study these molecules by photoelectron spectroscopy and to obtain some insight into the orbital structure of flavin, i.e. the basic knowledge required for a thorough interpretation of the chemical reactivity of flavin. In this contribution we focus the attention on the influence of hydrogen bond formation on the orbital structure of flavin.

Results and Discussion

Ultraviolet photoelectron spectroscopy is very powerful. It provides spectra

TABLE 1

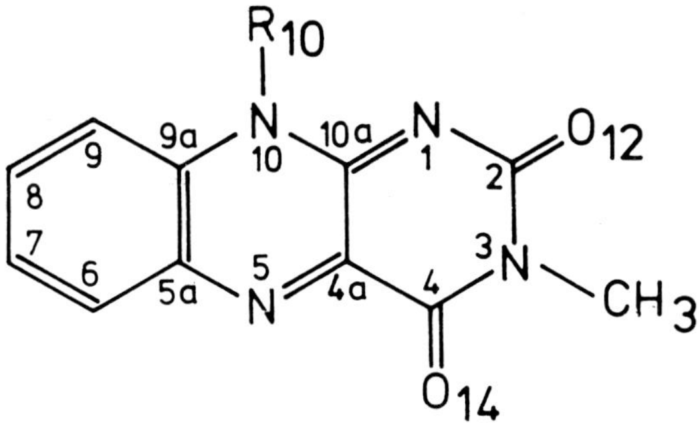

FLUORESCENCE LIFETIMES (NS) OF DERIVATIVES, MEASURED WITH A MODE-LOCKED AR ION LASER, A XP2020 PHOTOMULTIPLIER AND SINGLE PHOTON COUNTING.

DERIVATIVE:	VAPOUR PHASE (493 K):	SOLUTION*(77 K):
R_{10} = METHYL	1.7	10.1
R_{10} = N-BUTYL		9.9
R_{10} = N-$C_{18}H_{37}$	1.9	9.5
R_{10} = CH_2-CH_2-CH_2-OH	6.1 AND 15.0 (7.2 AND 18.1 AT 423 K)	9.7

*SOLVENT 2-METHYL-TETRAHYDROFURAN

EXCITATION AT 476.5 NM, 50 MW AVERAGE MODE-LOCKED POWER; DETECTION AT 526 NM.

which are a plot of the number of electrons, photoionized per unit time from flavin vapour, versus their binding energy. The latter quantity can directly be compared to calculated orbital energies, whereas the intensity of the ultraviolet photoelectron bands, particularly its dependence on the energy of the ionizing photons, provides information about the orbital character. The advantage of our ultraviolet photoelectron spectra over the X-ray photoelectron spectra presented at a previous flavin symposium

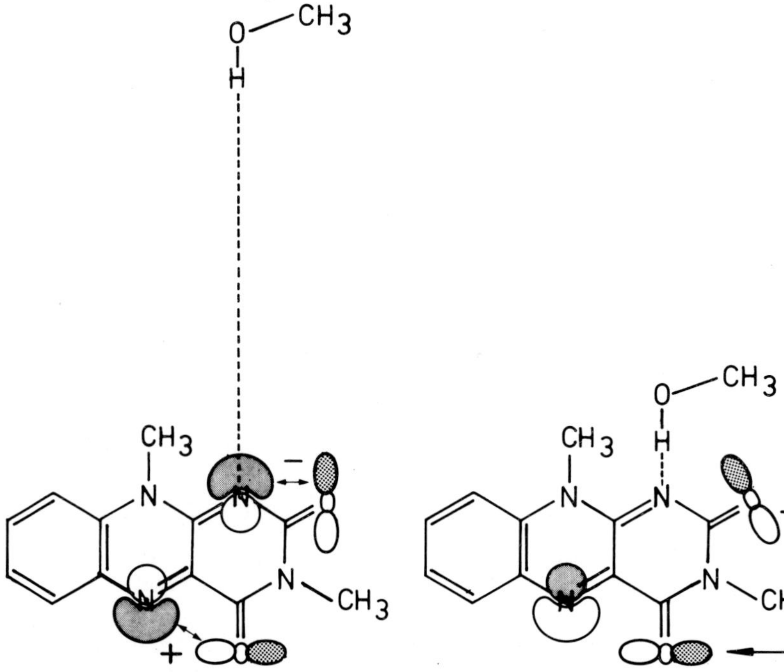

43 : LC LP(N_1)-LP(O_2)
41 : LC LP(O_4)+LP(N_5)

Figure 1. The theoretical model used to account for hydrogen bond formation towards N(1) of flavins. Left: through-bond interactions of the highest occupied n orbitals in free flavin; Right: ibid. for a hydrogen bond towards N(1).

(10) is the fact that our spectra give a picture of the flavin valence shell. The correlation of the ultraviolet photoelectron spectra with MO calculations using the CNDO/S scheme with Nishimoto-Matage integrals (6 and refs. therein) is very satisfactory, allowing to draw firm conclusions from the experimental data. The experiments were run with both He(I) (21.21 eV = 58.46 nm) and He (II) (40.82 eV = 30.38 nm) ionizing photons.
When the N(10) substituent of flavin contains a hydroxyal group, the fluorescence lifetime in the vapour phase increases by almost one order of magnitude (11). This behaviour is not observed in a condensed medium (Table 1). This observation was explained by intramolecular hydrogen bond formation of the hydroxyl group at the N(10) substituent towards N(1), the

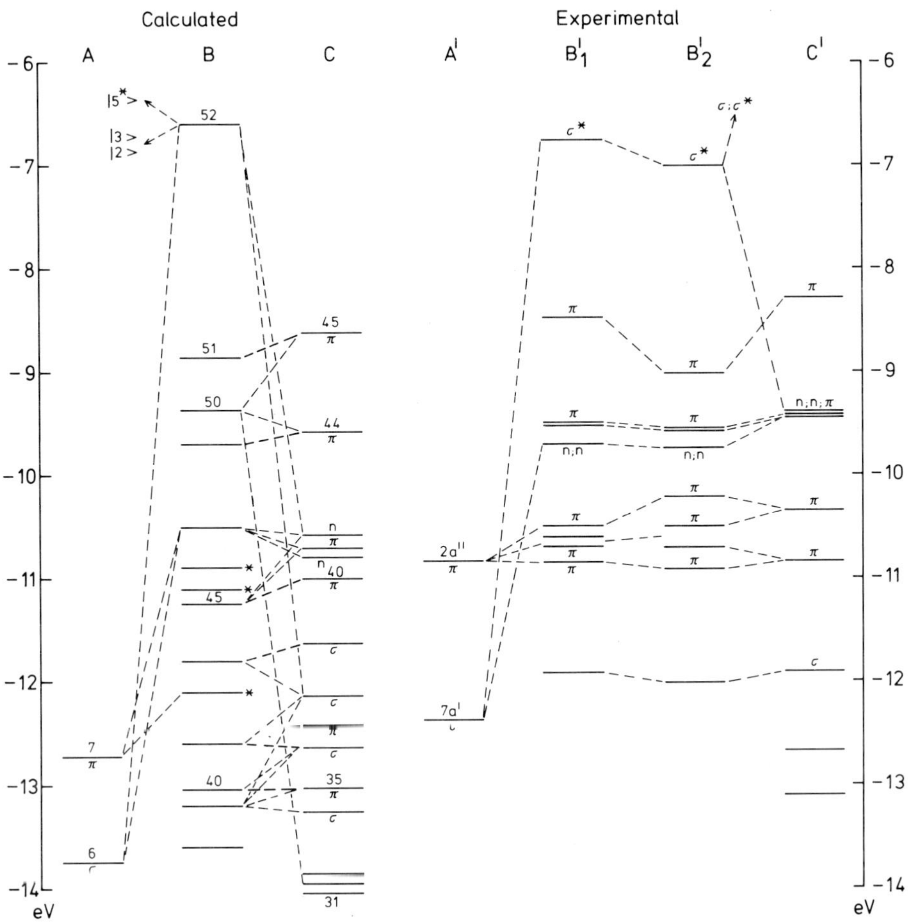

Figure 2. Correlation diagram of the orbitals of a hydrogen-bonded isoalloxazine-methanol complex with those of the parent compounds. Orbital energies and correlations from ref. 7. A and A': Methanol; B: Isoalloxazine-methanol complex (cf. Fig. 1); B_1': 3-methyl-10-(γ-hydroxypropyl)isoalloxazine; B_2': 3-methyl-10-(β-hydroxyethyl)isoalloxazine; C: 3-10-dimethyl-isoalloxazine; C': e-methyl-10-(n-butyl)isoalloxazine.

only plausible mechanism to account for the inhibition of vibronic coupling in the emissive excited state, necessary to increase its lifetime (11). The reduction of vibronic coupling must be quite considerable since its tendency to lengthen the decay rate of the excited state has to counteract the shortening of the decay rate induced by the process of the so-called sequence congestion (11). This is an effect due to thermal population of

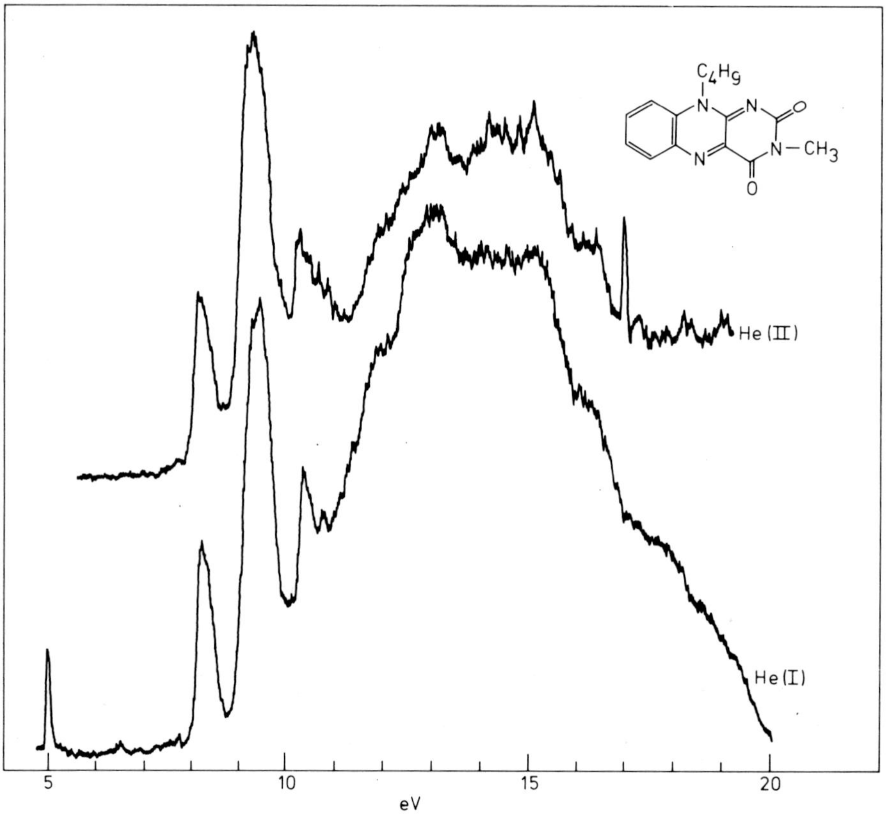

Figure 3. Ultraviolet photoelectron spectrum of 3-methyl-10-(n-butyl)iso-alloxazine, 510 K.

vibronic states (T > 400 K), also responsible for the inhomogeneity of some decay constants given in Table 1 (11). In condensed media, there is no internal complexation because solvation of the flavin molecules apparently prevails.

The theoretical model to account for the hydrogen bond formation towards N(1) was a supermolecule description of a methanol molecule approaching 3,10-dimethylisoalloxazine as pictured in Fig. 1. The consequences for the molecular orbital (MO) level diagram are given in Fig. 2 ('calculated'). Details on the calculation procedure and the values obtained for the correlations of the hydrogen bonded orbitals of the complex with the orbitals of the constituents can be found elsewhere (12). Important is the behaviour

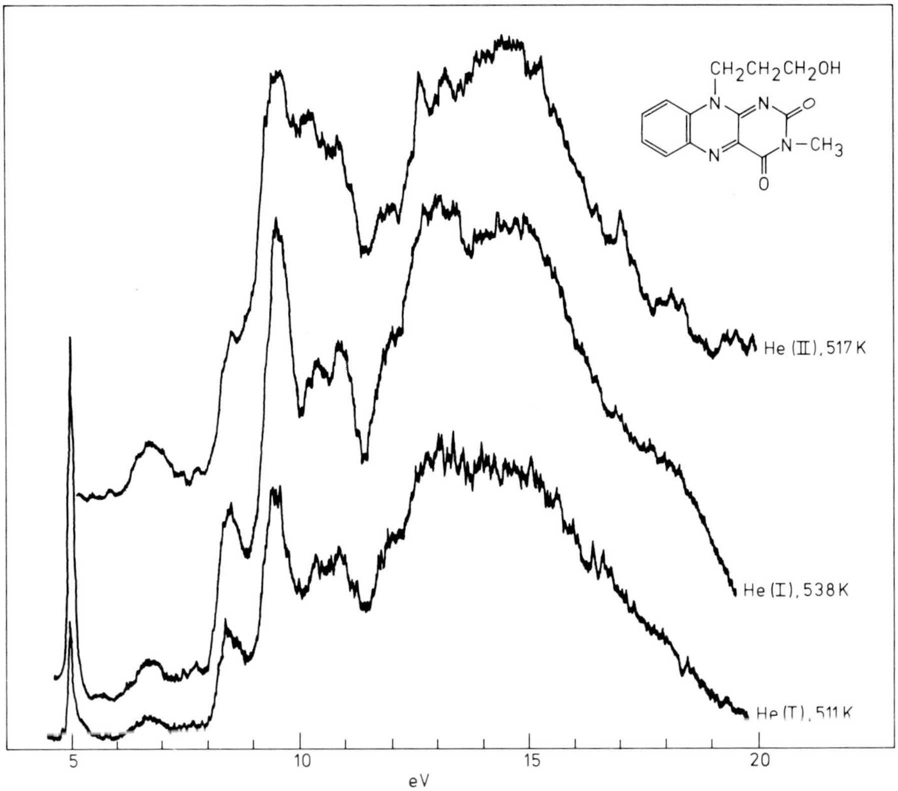

Figure 1. Ultraviolet photoelectron spectrum of 3-methyl-10-(γ-hydroxypropyl)isoalloxazine.

of the orbitals of the complex labeled 50 and 52, both destabilized by basically antibonding interaction. In the case of orbital 50, this effect could be proved to originate from a typical CNDO/S artefact. It was caused by the incapability of the calculation method to cope with the steric effects connected to the short distance between the N(10) methyl substituent and the methyl group of methanol. Orbital 50 disappeared completely from the MO diagram by using an artifical isoalloxazine in which the N(3) and N(10) methyl groups were replaced by protons in a new calculation. Orbital 52, however, could not be made to disappear from the level scheme by this method. It occurred in every calculated geometry with an energy which seemed to depend only on the hydrogen bond length. The character was for

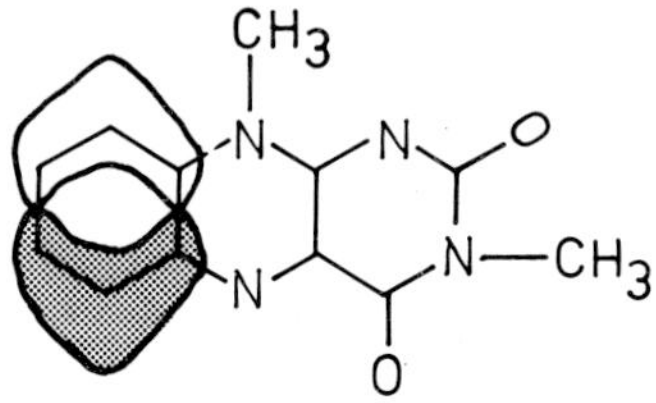

44: approx. Benzene e_{1gS}

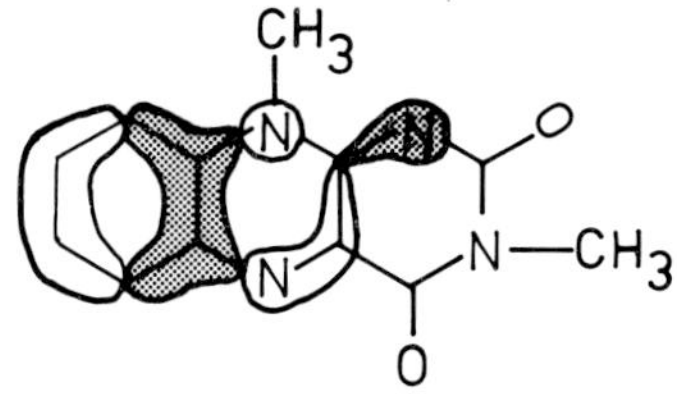

45: approx. Benzene e_{1gA}

Figure 5. Shape of the highest occupied π orbitals in isoalloxazine.

81% an antibonding linear combination of the methanol O(2py), the N(1)(2s) and the N(1)(2py) atomic orbitals. The orbital contained hardly any methanol hydrogen 1s character, indicative that it could hardly be involved in an energetical stabilization of the hydrogen bonded complex. At first sight it seemed an audacious assumption to accept this result as a basis for further investigation, knowing so many characteristics of a typical CNDO failure (5,12). Yet we did, because of two arguments. First, the calculated character described above was physically not completely unrealistic. Moreover the net interaction between the constituents of the complex was calculated to be bonding, the total energy decreased (more negative!) upon complex formation and similarly did the total dipole moment (12). Secondly, a destabilisation of an n-orbital on N(1) upon hydrogen bond formation could finally explain the well-known red shift of the flavin absorption band at ∿360 nm on going to protic solvents (12), a property hitherto not satisvactorily interpreted (5,12). The experimental confirmation of this behaviour came from ultraviolet photoelectron spectra. The photoelectron spectrum of 3-methyl-10-(n-butyl)isoalloxazine, unable to form an intramolecular hydrogen bonded complex, is given in Fig. 3. It has all the characteristics of the normal flavin photoelectron spectrum and can be considered as typical for a flavin (5,12). Note that the sharp peaks at 5 eV in the He(I) spectrum and at ∿ 17 eV in the He (II) spectrum are ghosts caused by autoionization of Helium. The spectrum of 3-methyl-10-(γ-hydroxypropoyl)-isoalloxazine (Fig. 4) clearly has an additional band at ∿6.5 - 7.0 eV, the same energy region where orbital 52 was calculated. The compound 3-methyl-10-(β-hydroxyethyl)isoalloxazine shows a similar feature in its photoelectron spectrum (12). The appearance of this low-energy band in

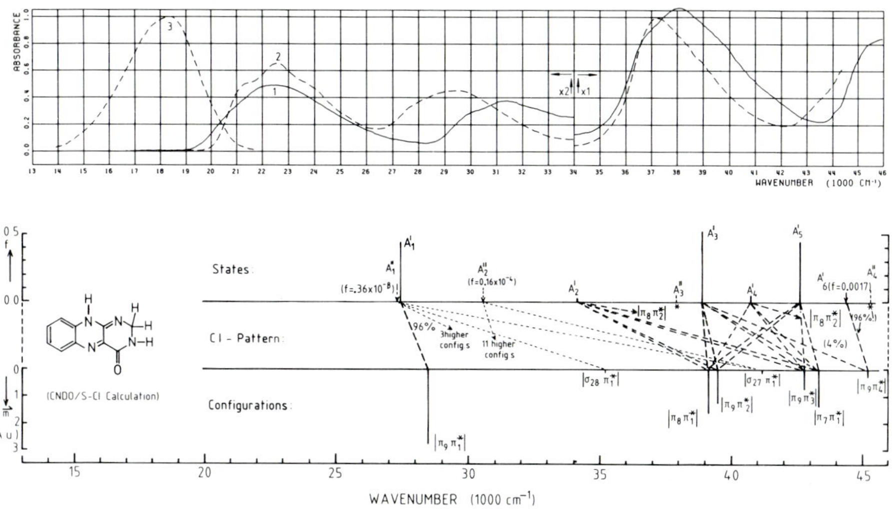

Figure 6. Experimental and theoretical data on the spectrum of 3,7,8,10-tetramethyl-2-desoxy-2-dihydroisoalloxazine. Upper panel: absorption and fluorescence spectra (1 cm cuvette, solvent acetonitrile) 1: absorption spectrum of 2-dihydroflavin (c=33.3 μM), 2: absorption spectrum of 3-methyl-lumiflavin (c=26.4 μM), 3: fluorescence spectrum of 2-dihydroflavin (arbitrary units). Note the expansion of the vertical scale in the low wavenumber region). Lower panel: calculated spectrum of the indicated model compound (f=oscillator strength, m transition moment in atomic units, a': ππ* state, A'': nπ* state-stick bar expanded by a factor of 1000 in the plot).

both the He(I) and He(II) spectra is conclusive evidence that it is not a ghost caused by spurious radiation of ionizing power. This is extensively discussed elsewhere (12). Hence, we finally assigned the band at ∿ 7.0 eV to the calculated orbital 52 of the complex.

The complete interpretation of the flavin photoelectron spectrum now provides the insight, at least on a qualitative basis, into the effects on the orbital structure of flavin upon hydrogen bond formation towards N(1). The first detailed analysis of the photoelectron spectra of flavins (5) showed that the two highest occupied orbitals in oxidized flavin originate from the benzene e_{1g} level. Their original degeneracy is lifted in the condensed aromatic system of the flavin nucleus (5), their calculated energy in 3,10-dimethylisoalloxazine is given in Fig. 2 and their shape (note the corres-

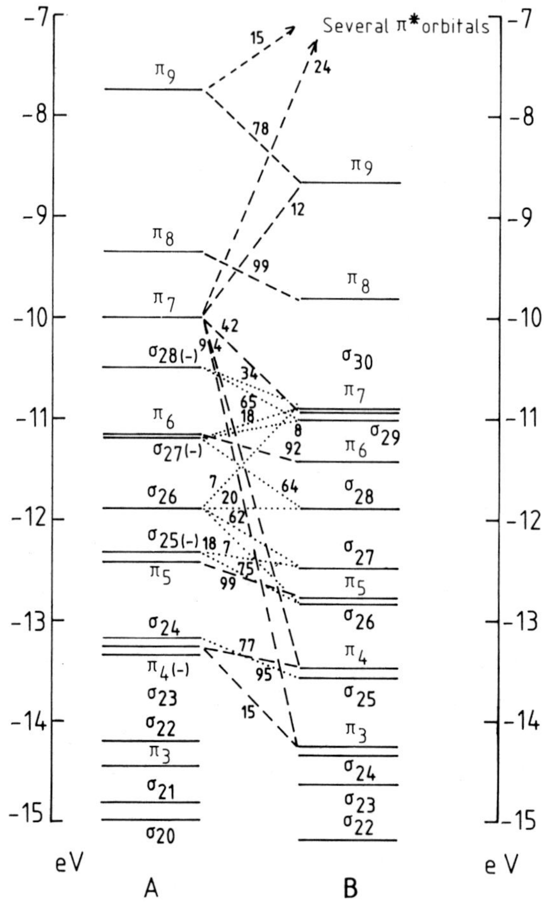

Figure 7. Orbital energies and orbital correlations (CNDO/S method). A: 2-desoxy-2-dihydroisoalloxazine (cf. Fig. 6), B: 'ideal' lumiflavin (cf. ref. 5). Details of calculation procedure in refs. 5 and 12. Correlations are indicated by their percentages.

ponding labeling) is pictured in Fig. 5. Orbital 45 is localized for 56% in the benzene subnucleus of flavin, the 'tail' extending towards N(1) represents ∿ 36% of its density (12). Orbital 44 is localized for 85% in the benzene subnucleus. The two highest occupied n orbitals, on the other hand, indicated by 43 and 41 in Figs. 1 and 2, are through-bond interacting linear

combinations of the N(1) and O(2), and the O(4) and N(5) lone pairs as pictured in Fig. 1. From substituent effects on photoelectron spectra it was established that their calculated energy contained an error of $\sim$1.2 eV (5) and that they are (nearly) degenerated with orbital 44 in the benzene subnucleus. Conclusive evidence for this follows especially from the spectrum of 5-deazaflavin whereas the calculational error can be explained from the MO-characters (5). This is the reason for the appearance of the intense second band in the flavin spectrum at $\sim$ 9.6 eV (Fig. 3). The degeneracy of orbitals 41, 43 and 44 can exist by virtue of their spatial separation (Figs. 1 and 5). The combined results yielded finally the effect of hydrogen bond formation towards N(1).
The hydrogen bond affects, at the same time, both the π and σ systems of the flavin molecule. The effect of the proton at N(1) on the π system is predominantly an electrostatic one. It can easily be visualized that orbital 45, having a correlation of 93% with orbital 51 of the complex (cf. Figs. 2 and 5) stabilizes, owing to the electrostatic effect on the 'tail' of orbital 45.
Similarly, the lowest unoccupied flavin orbital (localized for 50% on the 10, 10a, 1, 4a and 5 position according to the calculations) is stabilized by electrostatic interaction. Such a mechanism obviously affects the flavin redox properties which reside in the concerned orbitals. The interaction with the through-bond system 43 (Figs. 1,2) is the reason why orbital 52 of the complex (cf. Fig. 1) is destabilized: the hydrogen bond causes breakdown of the energetically favourable, linear combination of the N(1) and O(2) lone pairs. The result is a rearrangement of through-bond interactions as shown in Fig. 1, leading to an increase of the lone pair character on N(5). Thus, the hydrogen bond affects the binding properties of N(5) also, an important catalytical site of flavins, by a 'domino-effect' on through-bond interactions. A highly interesting result was obtained by a recalculation of the whole process using the INDO approximation (13,14) with Pariser-Parr integrals. This method is known to give more accurate results on charge densities (13,14) than CNDO/S. First, the results were carefully checked against the existing data to make sure that no significant discrepancies did occur. It was found that INDO, albeit with less accurate quantitative results on spectroscopic data, predicted exactly the

same behaviour as discussed above. Secondly, the effect on charge densities was analysed yielding the following salient changes in electron densities upon hydrogen bond formation:

Center	Electron density without H bond	Electron density with H bond	Difference
N(5)	0.085	0.070	0.015
O(2)	0.422	0.399	0.023
N(10)	0.101	0.067	0.034
C(10a)	-0.289	-0.303	0.014

The 0.086 electron lost by these centers is roughly half of the electron transfer (0.171 electron) from isoalloxazine to methanol. The other half originates from all other centers. Surprisingly the N(1) atom does not occur in this summary: it looses only 0.008 electron. Closer examination of the INDO results, however, shows this to be caused by a net loss of 0.147 σ electron accompanied by a net gain of 0.139 π electron. This suggests a hydrogen bond stabilization by a similar mechanism as encountered sometimes in transition metal complexes, i.e. a synergetic cooperation between a direct bond and a backbonding mechanism.

The spectral properties of the newly synthesized derivative 3,7,8,10-tetramethyl-2-desoxy-2-dihydroisoalloxazine (in the following 2-dihydroflavin) lacking the carbonyl function in position 2 and, therewith the corresponding through-bond coupling, seem to fit into the model discussed here. By assigning the flavin S_2 state to a mixed state of both $\pi\pi^*$ and $n\pi^*$ character (7) (the $n\pi^*$ character predominantly originating from the N(1)-O(2) through-bond combination of lone pairs) a satisfactory explanation can be given for the solvatochromy of the second absorption band of flavins (12) (cf. above). Removal of carbonyl function is expected to produce an increase of the imine-like character on N(1) accompanied by a change in energy of the N(1) lone pair. This will reduce the admixture of $n\pi^*$ character into the S_2 state. Such behaviour is indeed suggested by the spectrum of 2-dihydroflavin. The second absorption band is significantly blue-shifted and displays a little more vibrational structure when compared to ordinary flavin (Fig. 6). The first absorption band displays exactly the opposite behaviour.

Apparently some excited state character moves from the S_2 to the S_1 region upon formal replacement of O(2) by two protons. The fluorescence of 2-dihydroflavin is also red shifted and shows a large solvent shift ($\sim$1400 cm^{-1} to the red) on going to aqueous solution. At the same time the quantum yield reduces by almost one order of magnitude (0.03 in acetonitrile, 0.004 in water) at room temperature. A possible explanation could be an emissive $n\pi^*$ state (hidden in absorption) becoming nearly isoenergetic with the triplet $\pi\pi^*$ owing to the solvent shift. Presumable the first triplet $\pi\pi^*$ state will not be affected seriously in 2-dihydroflavin since it contains one electron configurations involving orbitals with low density in the neighbourhood of the 2-position (12).

Obviously, close lying excited states favour radiationless decay and a low quantum yield. Probably, the emissive state does not correspond to the first absorption band of 2-dihydroflavin since the latter shifts only $\sim$ 240 cm^{-1} to the red on going to aqueous solution. The band is probably the envelope of more than one transition. The solvatochromy of the second absorption band is strongly reduced in 2-dihydroflavin. Its red shift is approximately 1280 cm^{-1}, roughly half the value of ordinary flavin (12). However, for an $n\pi^*$ transition from a pure imine-like nitrogen lone pair, the shifts are in the wrong direction. A CNDO/S calculation surprisingly shows a pure imine lone pair on N(1) to be non-existent in 2-dihydroflavin (Fig. 7). The highest occupied σ orbital in this molecule is calculated as a linear combination of orbitals 43 and 41 (σ_{30} and σ_{29} in Fig. 7B) in such a way that the O(2) contribution effectively cancels and produces a tripartite through-bond coupled system between the N(1), O(4) and N(5) lone pairs. As a consequence there is poor agreement between the experimental and calculated spectrum. Clearly this highly interesting molecule needs further investigation.

Acknowledgements

We thank Miss Lyda Verstege for typing the manuscript and Mr. M.M. Bouwmans for preparing the figures. This study was carried out under the auspices of the Dutch Foundation for Chemical Research (SON) with financial aid from the Dutch Organisation for the Advancement of Pure Research (ZWO).

References

1. Müller, F.: Naturforsch. 276, 1023-1026 (1972).

2. Hemmerich, P., Massey, V.: in Oxidases and Related Redox Systems (T.E. King, H.S. Mason, M. Morrison,eds.), pp. 379-403, Oxford) Pergamon Press 1982.

3. Massey, V., Hemmerich, P.: Biochem.Soc.Trans. 8, 246-257 (1980).

4. Hemmerich, P., Massey, V., Michel, H., Schug, C.: Structure and Bonding 48, 93-123 (1982).

5. Eweg, J.K., Müller, F., van Dam, H., Terpstra, A., Oskam, A.: J.Am.Chem.Soc. 102, 51-61 (1980).

6. Palmer, M.H., Simpson, I., Platenkamp, R.J.: J.Mol.Struct. 66, 243-263 (1980).

7. Moonen, C.T.W., Vervoort, J., Müller, F.: Biochemistry 23 (1984) in press.

8. Kierkegaard, P., Norrestam, R., Werner, P., Csöregh, I., von Glehn, M., Karlsson, R., Leijonmarck, M., Rönnquist, O., Stensland, B., Tillberg, O., Torbjörnsson, L.: in Flavins and Flavoproteins (H. Kamin ed.), pp. 1-22, Baltimore, University Park Press 1971.

9. Norrestam, R., von Glehn, M.: Acta Cryst. B28, 434-440 (1972).

10. Palmer, M.H., Wheeler, J.R., Platenkamp, R.J., Visser, A.J.W.G.: Flavins and Flavoproteins, (V. Massey and C.H. Williams jr. eds.), New York, Elsevier North Holland pp. 584-589, 1982.

11. Eweg, J.K. Müller, F., Bebelaar, D., van Voorst, J.D.W.: Photochem. Photobiol. 31, 435-443 (1980).

12. Eweg, J.K., Müller, F., van Dam, H., Terpstra, A., Oskam, A.: J.Phys.Chem. 86, 1642-1651 (1982).

13. Pople, J.A., Beveridge, D.L.: Approximate Molecular Orbital Theory, Berlin, McGraw-Hill, 1970.

14. Dewar, M.J.S., Worley, S.D.: J.Chem.Phys. 50, 654-667 (1969).

SPECTRAL AND PHOTOCHEMICAL PROPERTIES OF ALLOXAZINES

Jacek Kozioł, Anna Koziołowa, Wojciech Babś, Jan Dawidowski, Danuta Panek-Janc, Małgorzata Stroińska, Violetta Szczęsna, Henryk Szymusiak and Bożena Tyrakowska

Institute of Commodity Sciences, Academy of Economics
Marchlewskiego 146, Poznań, Poland

Introduction

Alloxazine derivatives are studied in our group because they are: the isomers of isoalloxazines, products of their chemical, photochemical and biochemical decomposition, despite of similarities they exhibit distincly different properties, they undergo interesting chemical and photochemical reactions under relatively mild conditions.

Results

7-cyanoalloxazine. To compare with the relatively well recognized properties of methyl derivatives of alloxazines (1,2) it was decided to check the impact of electron accepting substituents in the benzene ring of the molecule. As first the 7-cyanoalloxazine was synthesized and its spectral properties studied. This compound exhibits well resolved absorption spectra with maxima at 380.5, 320.5, 264.5, 247.5 and 218.5 nm (buffer pH 6.0), or 377.5, 317.0, 263.0 and 252 (shoulder) in dioxane (insoluble in less polar solvents). Along with the shifts due to change in solvent polarity most prominent in the region of minimum at 335 nm (shifts to 343 in dioxane) and a shoulder is appearing at about 307 nm. Maximum at longest wavelengths shows no distinct vibronic fine structure. Intensity

Flavins and Flavoproteins

ratio of the two maxima at longest wavelengths (E 9.2 and 6.0 $x10^3$ in dioxane) resembles spectra of isoalloxazines. Fluorescence emission maximum appears at about 435 nm. In presence of pyridine strong phototautomerism occurs (second emission maximum at about 536 nm) and negligible in presence of acetic acid. Mass spectral fragmentation shows characteristic for alloxazines sequence (m/e: 239, 196, 168, 141, 115, 114...) leaving the cyano group intact.

Photoreduction of alloxazines in viscous media. To estimate the angles between transition moments of the two low energy electronic transitions of alloxazine and its derivatives monomethylated in the benzene ring polarization of fluorescence was measured. The results obtained for solutions of alloxazines in glycerol at -10°C in different series were inconsistent and of poor reproducibility. Accidentaly it has been found that solutions of alloxazines in glycerol undergo efficient photoreduction even under dim day light or during measurements in a spectrofluorometer. More detailed studies on deaerated solutions have shown that photoreduction is equally efficient also for alloxazines methylated at N-1 and N-3 and 3-methyllumiflavin, the reaction is of pseudofirst order and fully reversible, its temperature dependence is low (decrease by about 20% when going from 20°C to -10°C), less efficient in 1,3-propanediol and 1,2-propanediol (10 times and 20 times, respectively), does not proceed in ethylene glycol. Most probably the non hydroxylic hydrogen at C-2 of glycerol molecule is the reactive one as in lecithine(3). Thus, use of glycerol should be avoided the more so due to low permeability of gases reoxidation even in vigorously shaked solution is very slow.

Phototautomerism. Studies on alloxazine-isoalloxazine phototautomerism (4) were continued. To avoid possible formation of anionic species solutions of 3-methyllumichrome containing different concentrations of acetic acid in mixture with methanol or dioxane were used. Reaction rate constants estimated using Stern-Volmer equation and the diffussion controlled reaction approach differ by more than order of magnitude ($2.9x10^8$ to $9.4x10^9 l.mol^{-1}.s$ at 20^oC).Activation energies estimated by different methods also do not coincide. Inspite of distinct influence of solvent viscosity (in glycerol) equal phototautomeric efficiency (ratio of isoalloxazinic vs. alloxazinic fluorescence intensity) was observed for 3-undecyllumichrome in 1% acetic acid in cyclohexane and paraffine. In pure acetic acid the hiqhest efficiency is usually reached at 190K but in some experiments it remains constant down to 77K. It is assumed that spatial arrangement of solute and solvent molecules are responsible for the observed phenomena.

Band shape analysis. The solvent polarity and substitution site in the benzene ring dependency of alloxazine absorption and emission spectra is very significant (1,4). Attempts were continued to find reliable and objective method for band shape analysis. To this aim an interface was designed for Cary 118C-spectrophotometer allowing for direct recording of spectra in digitalized form by a computer. Using 3-methyllumiflavin as a model compound and its solutions in dioxane-water mixtures the obtained spectra were smoothed, 2-nd and 4-th derivative calculated and used to control the fit. Both long walength maxima were resolved using Gaussian curves located initially at frequencies indicated by the fine structure, keeping decreasing spacing, and also broadening of half-widths with increasing frequency. Calculations were made using least squares method until the fit was better than 0.2%. Reproducible results give resolution of both lowest energy maxima into 5 subbands sets of different progression. The assumed 0-1 transition of the second maximum shifts by about 2000 $cm^{-1}$when going from water to dioxane (the other subbands of this progression less). Around 28 kK an irreqularity appears which might be ascribed to excitation of a nonbonding electron (appropriate solvent polarity dependence). At frequencies from 29 to 34 kK regular discrepancies were found in more polar solvents, probably indicating the presence of differently solvated solute molecules.

Photodecomposition. Alloxazines are practically photostable in aqueous solutions but undergo slow photodecomposition under prolonged irradiation in alcohols, acetic acid and pyridine (5). Alloxazine and its monomethyl derivatives were irradiated using polychromatic light (250W high pressure Hg lamp, filter cutting of light below 350 nm).In alcohols all alloxazines decompose slowly to many products. In pyridine only 9-monomethyl alloxazine decomposes into 3 major fluorescent products (under anaerobic conditions much faster). All alloxazines undergo photoreaction in acetic acid (under anaerobic conditions photoreduction) but only 6-monomethylalloxazine yields a single photoproduct. This compound was isolated, and work on elucidation of its structure is in progress.

Quarternary flavinium salts. Quarternary flavinium salts and their reduced derivatives show considerable photoreactivity. Among the photoproducts formed during photolysis in different organic solvents no alloxazines were found.

References

1. Kozioł, J., Koziołowa, A., Konarski, J., Panek-Janc, D., Dawidowski, J.: Flavins and Flavoproteins, K.Yagi and T. Yamano ed., Japan Sci.Soc. Press, Tokyo, University Park Press, Baltimore, 475-484 (1980).
2. Kozioł, J., Tyrakowska, B., Müller, F.: Helv. Chim. Acta 64, 1812-1917 (1981).
3. Schmidt, W., Hemmerich, P.: J. Membrane Biol. 60, 129-141 (1981).
4. Koziołowa, A.: Photochem. Photobiol. 29, 459-471 (1978).
5. Kozioł, J.: Photochem. Photobiol. 5, 55-62 (1966).

π...π-INTERACTIONS OF FLAVINS : NOVEL COENZYME-MODELS OF THE CYCLOPHANE - TYPE

Matthias F. Zipplies and Heinz A. Staab*

Abteilung Organische Chemie, Max-Planck-Institut für medizinische Forschung, D-6900 Heidelberg

The exact mechanism of redox-equivalent-transfer in NAD(P)H-dependent flavoproteins, as well as the possible relevance of charge-transfer complexation in the course of this reaction, still remain a controversial issue [1]. The results of the X-ray structure analysis of the flavoenzyme glutathione reductase [2] suggest that intermolecular π...π-interactions between the nicotinamide- and flavin-systems play a significant role in their redox-catalysis. More information concerning the exact steric requirements and physico-chemical properties of flavin-π-complexes seems desirable. As in the case of other weak interactions (charge-transfer complexes [3], excimers [4]) the concept of fixing the interacting units in definite orientation has been applied by us to this problem.

One target molecule for the investigation of flavin-nicotinamide π...π-interactions is the phane molecule 1, which contains a nicotinamide- and a flavin-moiety in similar geometrical arrangement as in the enzyme glutathione reductase. As preparatory step in the approach to 1, we synthesized

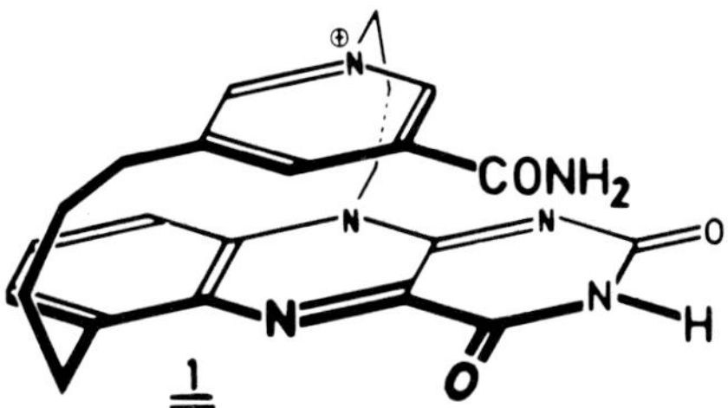

1

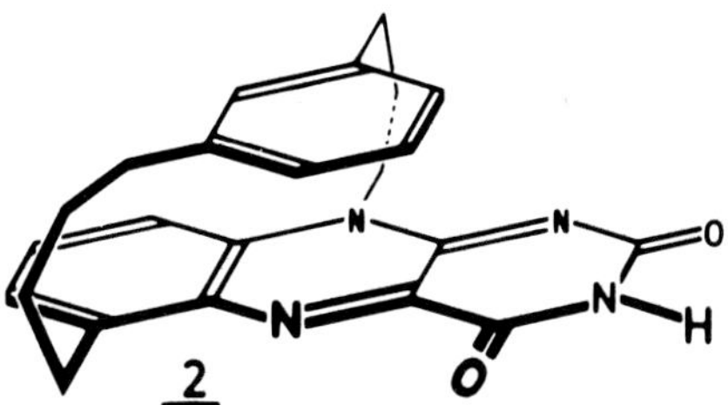

2

Flavins and Flavoproteins

[4]metacyclo[3](10,6)isoalloxazinophane (2). An X-ray structure investigation of 2 revealed, that the aromatic ring-systems are indeed arranged in the expected way [5].

By further continuing our efforts for the synthesis of 1, we succeeded in the synthesis of [4](3,1)pyridinio[3](10,6)=isoalloxazinophane (3) [6], which in comparison to 1 is lacking the carboxamide-substituent in position 3 of the pyridinium ring. When the flavin portion of 3 is photoreduced in the presence of EDTA, the disappearance of the 450 nm maximum in its uv/vis-spectrum is accompanied by the development of a new long wavelength absorption, most likely originating from a charge-transfer interaction between the 1,5-dihydro-iso=alloxazine-(donor-) and the pyridinium-(acceptor-) subunits of semi-reduced 3 (fig. 1).

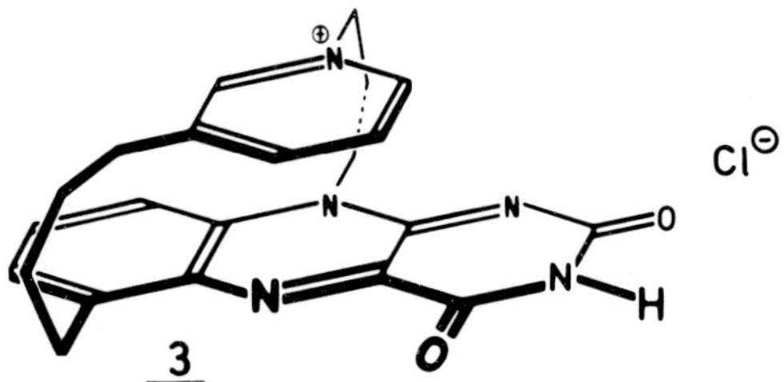

3

The new absorption is strongly pH-dependent: at pH 4.0 it possesses a maximum at 410 nm (ε = 2000); deprotonation of the flavin N-1-atom in buffer pH 9.0 lowers the maximum extinction of the absorption band, while broadening it considerably at the same time (cf. fig. 2,3). This observation is in agreement with the expected better donor-properties of the dihydro=flavin-anion in zwitterionic $3\text{-H}]^-$.

The comparison of the spectra of reduced 2 and reduced 6,10-dimethyl-isoalloxazine (4) at different pH-values shows, that the integration of a flavin-ring in a cyclophane system with concomitant enforced flattening of the dihydroflavin leads to an observable shift of the lowest energy transition of the 1,5-dihydroflavin-chromophore to longer wavelengths.

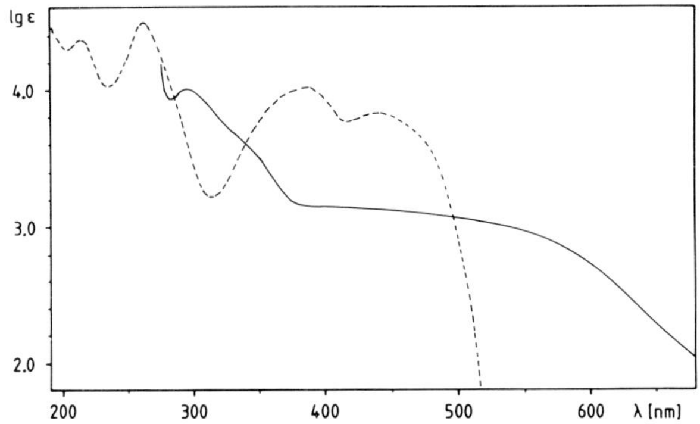

Fig. 1: Absorption spectra of 3 :
--- in water, pH 7.0; —— reduced, pH 9.0.

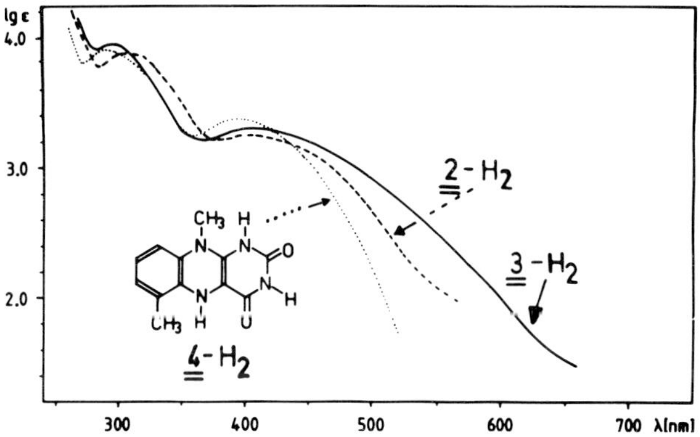

Fig. 2: Absorption spectra of photo-
reduced 2, 3, 4 (citrate, pH 4.0).

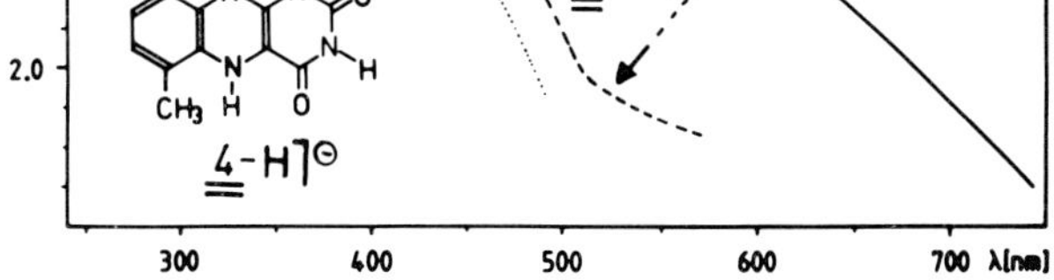

Fig. 3: Absorption spectra of photo-
reduced 2, 3, 4 (borate, pH 9.0).

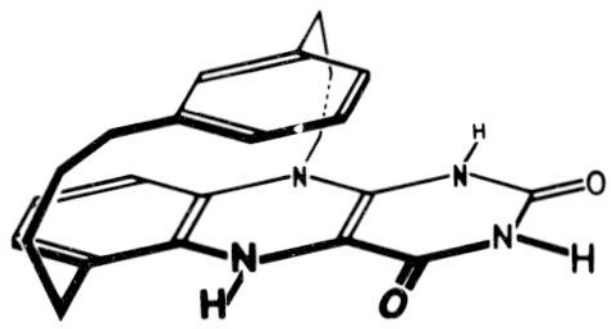

2-H2

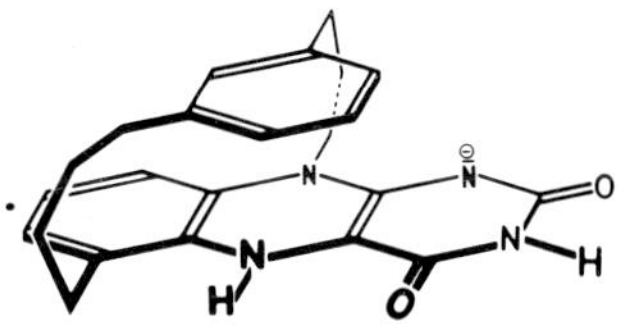

2-H]⊖

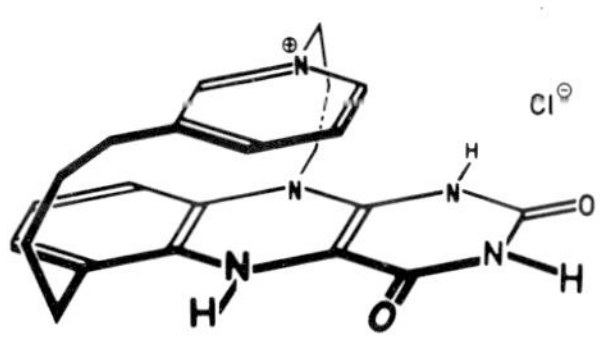

3-H2

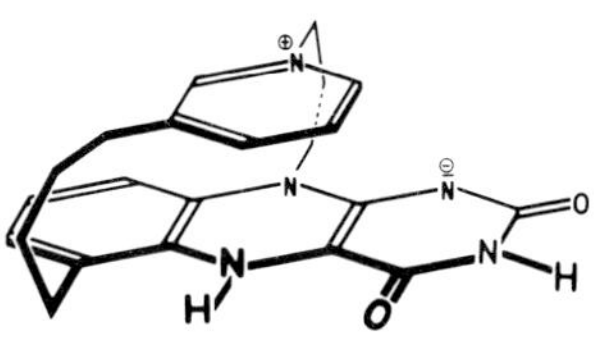

3-H]⊖

However, the effect of the transanular π-interaction of the pyridinium-ring in photoreduced 3 is by far larger. The absorption found for 3-H]⁻ (pH 9.0) between 350 and 700 nm is presumably a superposition of the 1,5-dihydro-isoalloxazine-$\pi\pi^*$- with the charge-transfer-transition, which is centered at approximately 520 nm (ε = 1000).

These results show, that in semi-reduced 3 there exists considerable π-orbital overlap between a reduced isoalloxazine- and an oxidized pyridinium-ring, which are held in a specific geometrical orientation. The same set of orbitals, which is involved in the observed charge-transfer interaction, might also play a role in thermal electron transfer in flavin-nicotinamide redox-reactions.

Further investigations on transanular interactions in these and related flavinophane-systems are under way [7].

References

1) cf. Walsh, C.: Acc. Chem. Res. 13, 148 (1980); Massey, V., Hemmerich, P.: Biochem. Soc. Trans. Biochem. Rev. 8, 246 (1980); Bruice, T.C.: Acc. Chem. Res. 13, 256 (1980).

2) Pai, E.F., Schulz, G.E.: J. Biol. Chem. 258, 1752 (1983).

3) Schanne, L., Staab, H.A.: Tetrahedron Lett. 1984, 1721 and references cited therein.

4) Staab, H.A., Riegler, N., Diederich, F., Krieger, C., Schweitzer, D.: Chem. Ber. 117, 246 (1984) and references cited therein.

5) Zipplies, M.F., Krieger, C. Staab, H.A.: Tetrahedron Lett. 1983, 1925.

6) Staab, H.A., Zipplies, M.F., to be published elsewhere.

7) Zipplies, M.F., Staab, H.A.: Tetrahedron Lett. 1984, 1035.

A LASER FLASH PHOTOLYSIS STUDY OF THE TRIPLET STATES OF LUMICHROMES.

Paul F. Heelis, G. O. Phillips.

Research Division, North East Wales Institute,
Kelsterton College, Connah's Quay, Deeside,
Clwyd, CH5 4BR, U.K.

Introduction

Lumichromes (I) or 7, 8-dimethylalloxazines represent a class of nitrogen heterocycles related to lumazines and the biologically

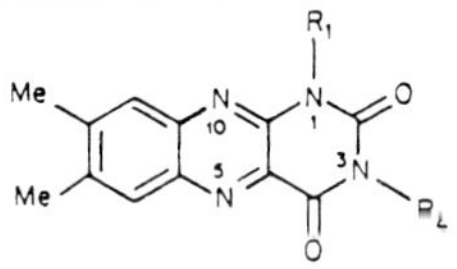

I

important flavins. Previously a large body of data on the photochemical properties of isoalloxazines has been published detailing a wide range of photochemical reactions. In contrast comparatively little is known concerning the reactions and intermediates of alloxazines.

The present study consists of a laser flash photolysis study of lumichrome triplet states. Absorption, spectra and acid-base properties are presented in aqueous solution and in pyridine or glacial acetic acid.

Flavins and Flavoproteins

Results

1) Flash photolysis in aqueous solution.

Flash photolysis of 1,3-dimethyllumichrome (DMLC) in the pH range 2-10 revealed the existence of two different transient species, one (λ max 285 375nm) predominent in the pH range 4-10 and the second (λ max 290, 430 and 590nm) in the pH range < 4. From previous studies (1) the transient found in the pH range 4-10 can be identified as the neutral triplet state (^3LCH).

From a study of the pH dependence of the transient spectra, the second transient species is clearly formed from and is in equilibrium with ^3LCH and can hence be identified as the protonated triplet $^3LCH_2^+$. A pka of 3.75 was thus determined for the equilibria in equation (1)

$$^3LCH + H^+ \underset{pka\ 3.75}{\rightleftharpoons} {}^3LCH_2^+ \qquad (1)$$

The absolute extinction coefficients of ^3LCH and $^3LCH_2^+$ were determined by utilising their one-electron reduction by added substrates (RH) e.g. EDTA, according to equation 2.

$$^3LCH + RH \longrightarrow LCH_2^{\cdot} + R^{\cdot} \qquad (2)$$

For complete quenching of ^3LCH and using the known extinction coefficients of $LCH_2^{\cdot}$ (2) the corresponding extinctions and absolute spectra of ^3LCH and $^3LCH_2^+$ were calculated and are shown in figure 1.

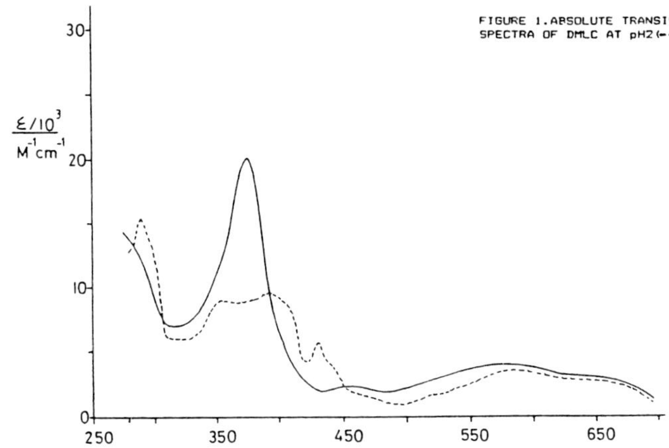

Flash photolysis of N(1) methyllumichrome (N(1)MeLc) in the pH range 2-8 produced essentially identical results to DMLC in terms of the spectra and pka of $^3LCH_2^+$ and ^3LCH. Of some interest is that excitation at pH > 9 involves the monoanion of N(1) MeLc, however, the same transient spectrum (i.e. that assigned to ^3LCH) is observed. Hence it is suggested that although the anionic triplet $^3LC^-$ is formed initially rapid protonation occurs eq.3.

$$^3LC^- + H^{\oplus} \rightleftharpoons {}^3LCH \qquad \text{pka } 12 \qquad (3)$$

Flash photolysis of N(3) methyllumichrome (N(3)MeLC) in the pH range 2-8 produced the characteristic transient spectra of ^3LCH and $^3LCH_2^+$. However, in this case, the pka was found to be ~ 8, cf 3.7 for DMLC and N(1)MeLC. Excitation of the monoanion (pH > 9) produced a new transient spectrum which can be assigned to the triplet anion ($^3LC^-$). An accurate value for the pka value for the equilibrium in eq. 3 could not be obtained due to the closeness of the protonation equilibria (eq. 1).

Flash photolysis of lumichrome itself gave results closely similar to those of N(3)MeLc.

2) Flash Photolysis in glacial acetic acid or pyridine.

Flash photolysis of DMLC or N(1)MeLC produced the transient absorption (λ max 380,440 and 590nm) shown in figure 3. In contrast, N(3)MeLc or LC itself produced a transient with λ max 450nm (also shown in figure 3). While no firm assignments can be made, it is suggested that the transient spectra for N(3)MeLC or LC represent that of the isoalloxazine or tautomeric triplet state as glacid acetic acid is known to promote alloxazine ⟶ isoalloxazine tautemerism in the singlet state.

FIGURE 3. TRANSIENT DIFFERENCE SPECTRA OF DIMETHYL-LUMICHROME (●-●) AND N(3)MELC (O-O) IN GLACIAL ACETIC ACID.

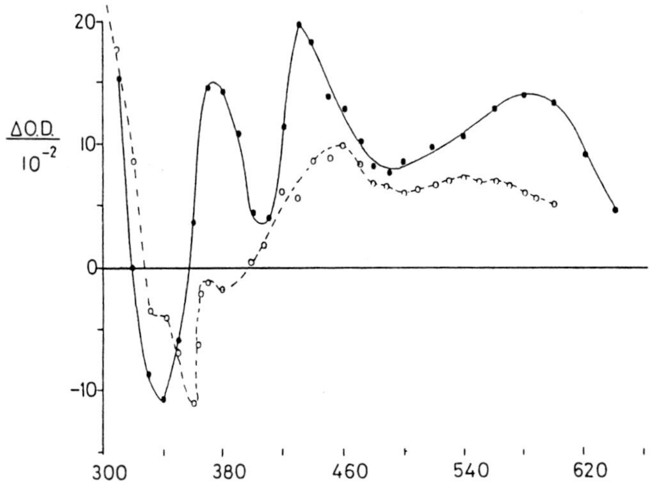

In contrast, all four lumichrome derivatives in pyridine gave the transient spectrum of ^{3}LCH (essentially as observed in aqueous solution) suggesting that this solvent promotes tautomerism in the singlet state but not the triplet state.

References

1. Gradowski, M.S., Veyret, B., Weiss, K. : Photochem. Photobiol. 26, 341 (1977).

2. Heelis, P.F., Parsons, B.J., Phillips, G.O., Land, E.J., Swallow, A.J., : J. Phys.Chem. 86, 5169 (1982).

THE DARK FORMATION OF RADICALS IN FLAVINIUM CATION/ACID SYSTEMS

R. Addink and H.I.X. Mager

Biochemical and Biophysical Laboratory of the Delft University of Technology, 67 Julianalaan, 2628 BC DELFT, The Netherlands.

N^1- and N^5-alkylflavinium salts, dissolved in low polar solvents, may show the formation of covalent adducts (eqs. 1a; 1b). The adducts reduce the flavinium cations at room temperature in the dark to give the flavin radicals and counter-radical cations (eqs. 2a; 2b)[1,2]:

$$\text{1-RFl}^{+}_{ox},\text{A}^{-} \rightleftharpoons \text{1-RFl-10}^{a}\text{-A} \xrightarrow{\text{1-RFl}^{+}_{ox}} \text{1-RFl}\bullet + (\text{1-RFl-10}^{a}\text{-A})^{\overset{+}{\bullet}} \quad \text{--(1a; 2a)}$$

$$\text{5-RFl}^{+}_{ox},\text{A}^{-} \rightleftharpoons \text{5-RFl-4}^{a}\text{-A} \xrightarrow{\text{5-RFl}^{+}_{ox}} \text{5-RFl}\bullet + (\text{5-RFl-4}^{a}\text{-A})^{\overset{+}{\bullet}} \quad \text{---(1b; 2b)}$$

Unprotonated 1-RFl• radicals are unstable. Their formation was not directly observed, but revealed by sequential reactions like an anaerobic N^{10}-dealkylation or a reaction with O_2 (eq. 3). $O_2^{\overset{-}{\bullet}}$ and/or $\text{1-RFl-10}^{a}\text{-OO}\bullet$

$$\text{1-RFl}\bullet + O_2 \longrightarrow \text{1-RFl}^{+}_{ox} + O_2^{\overset{-}{\bullet}} \rightleftharpoons \text{1-RFl-10}^{a}\text{-OO}\bullet \quad \text{---(3)}$$

probably act as one-electron donors in a reaction with $(\text{1-RFl-10}^{a}\text{-A})^{\overset{+}{\bullet}}$ or its product of hydrolysis (eq. 4). The back transfer of an electron to

$$(\text{1-RFl-10}^{a}\text{-A})^{\overset{+}{\bullet}} + H_2O \longrightarrow (\text{1-RFl-10}^{a}\text{-OH})^{\overset{+}{\bullet}} + \text{HA} \quad \text{--(4)}$$

$(\text{1-RFl-10}^{a}\text{-OH})^{\overset{+}{\bullet}}$ gives the blue-coloured $10,10^{a}$-ring opened product 2 (λ_{max} $\sim$ 600 nm) as exemplified by eq. 5 (cf. note).[3] It is noticed that a flavinium salt system (eqs. 1a+2a) can become an O_2-activation model (eq.3) in which, however, the O_2-uptake may be nullified by an O_2-generation in a secondary electron transfer step (eq. 5).

In competition, other sequential reactions may occur like a decomposition of the counter-radical cation (eq. 6a) followed by a decarboxylation (eq.7) and, finally, by a secondary electron transfer (eq. 8).

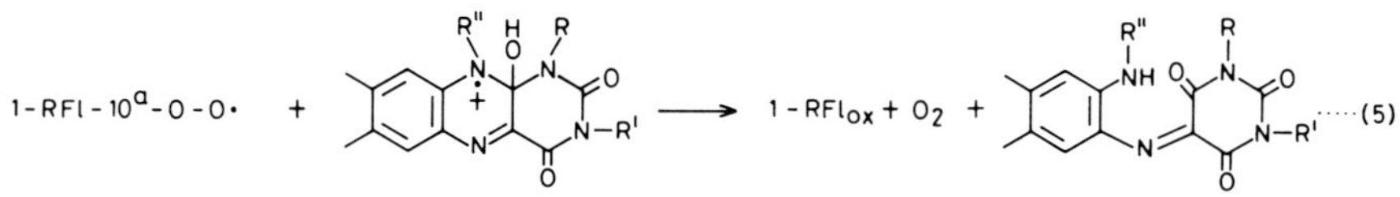

1 2

Flavins and Flavoproteins

$$(1\text{-RFl-}10^{a}\text{-A})^{+\bullet} \longrightarrow 1\text{-RFl}^{+}_{ox} + A\bullet \quad \text{--(6a)} \qquad A\bullet(= R_{a}COO\bullet) \longrightarrow R_{a}\bullet + CO_2 \text{--(7)}$$

$$1\text{-RFl}\bullet + R_{a}\bullet \longrightarrow 1\text{-RFl}^{+}_{ox} + R^{-}_{a} \quad \text{----(8)}$$

The recycling of both primary radical species to 1-RFl^{+}_{ox} (eqs. 6a; 8) appeared from experiments in which TCA was catalytically decarboxylated by the flavinium cation. In contrast, TCA can be regenerated (eq. 4), which opens the possibility to accomplish the conversion of a $4^{a},10^{a}$-ethylenedioxyflavin adduct into 2 by the use of a limited amount of TCA:

$$1\text{-RFl-}4^{a},10^{a}\text{-}(OCH_2)_2 \xrightarrow{HA} 1\text{-RFl}^{+}_{ox} \longrightarrow \text{eqs. 1a+2a+4} \xrightarrow[HA]{} 2$$

The choice of the solvent is important as is illustrated by the fate of TCA in comparative experiments: a decarboxylation in acetonitrile versus the use of limited amounts of TCA in the $10,10^{a}$-ring opening reaction in water-saturated benzene or chloroform.

Due to a greater stability, 5-RFl• radicals can be accumulated in low polar solutions of N^{5}-alkylflavinium salts (eqs. 1b+2b). In a general procedure, 5-ethyl-3-methyllumiflavinium salts ($5\text{-EtFl}^{+}_{ox},A^{-}$) were prepared *in situ* by adding an acid to $5\text{-EtFl-}4^{a}\text{-OR}_{b}$ adducts (eq. 9; R_{b}= H; Me).

$$5\text{-EtFl-}4^{a}\text{-OR}_{b} + HA \longrightarrow 5\text{-EtFl}^{+}_{ox},A^{-} + R_{b}OH \quad \text{---(9)}$$

The spectral changes showed the spontaneous conversions of a 4^{a}-flavin adduct or of 5-EtFl^{+}_{ox} (Fig. 1) into 5-EtFl• or $5\text{-EtFlH}^{+\bullet}$ (Fig. 2). In some cases, the occurrence of $5\text{-EtFl-}4^{a}\text{-A}$ transients was indicated by the appearance of spectra similar to curve a. Under the conditions that an organic acid ($5\text{x}10^{-4} \sim 10^{-1}$M) was present in a 17-3000 fold excess, the counter-radical cations were unstable giving reactions like:

a) a *decomposition* to 5-EtFl^{+}_{ox} or 5-EtFl• and products derived from free acyloxy radicals A• or from species represented as A^{+} (eqs. 10; 11):

$$(5\text{-EtFl-}4^{a}\text{-A})^{+\bullet} \longrightarrow 5\text{-EtFl}^{+}_{ox} + A\bullet \;\text{-(10)} \qquad (5\text{-EtFl-}4^{a}\text{-A})^{+\bullet} \longrightarrow 5\text{-EtFl}\bullet + A^{+} \;\text{-(11)}$$

b) a *secondary electron transfer* in which the solvent, the acid or a product might act as the secondary electron donor X (eq. 12):

$$(5\text{-EtFl-}4^{a}\text{-A})^{+\bullet} + X \longrightarrow 5\text{-EtFl-}4^{a}\text{-A} + X^{+\bullet} \quad \text{---(12)}$$

c) a *disproportionation* giving the starting adduct (or a product derived from it) and, formally, a di-cation:

$$2\ (5\text{-EtFl-}4^{a}\text{-A})^{+\bullet} \longrightarrow 5\text{-EtFl-}4^{a}\text{-A} + (5\text{-EtFl-}4^{a}\text{-A})^{++} \quad \text{---(13)}$$

A disproportionation is reflected by an anaerobic N^{5}-dealkylation being a final result of a degradation of an intermediate like 3 which is on the same level of oxidation as the di-cation. The reaction of 3 with H_2O is

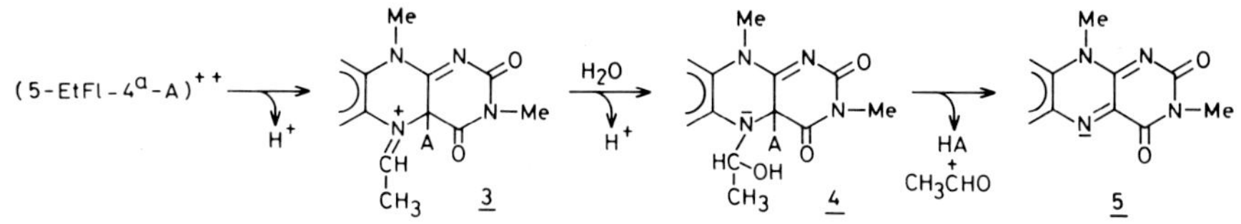

considered to be crucial for the formation of 3-methyllumiflavin 5.

The yield of 5-EtFl• (50% based on the sum of the eqs. 1b+2b) is increased by the sequential reactions. It is 100% based on the sum of the eqs. 1b+2b+10; 1b+2b+11; or 1b+2b+12; and, 66.7% based on the sum of the eqs. 1b+2b+13.

The use of HA= TFA, TCA, AcOH, α-ketoglutaric and salicylic acid gave 5-EtFl•/5-EtFlH^{+}• and 3-methyllumiflavin in yields of 60-90% and 6-21%, respectively. The time to obtain maximal radical formation (t_{max}) varied from 3 min - 24 h in dependence on the nature of the solvent and, on the nature and the concentration of the acid. The experimental results indicated the

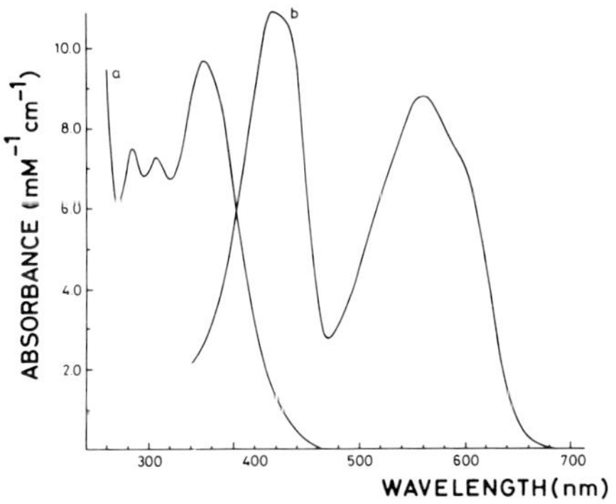

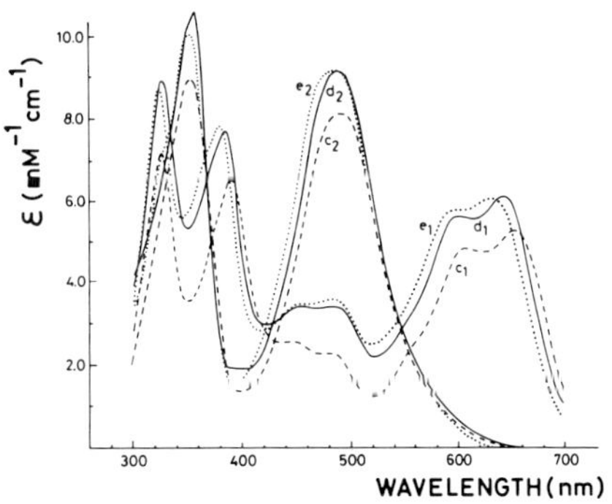

Fig. 1. (a) Molar absorbance of 5-EtFl-4^{a}-OMe, *freshly* dissolved in dried $CHCl_3$. (b) Molar absorbance of 5-EtFl$^{+}_{ox}$ in $CHCl_3$ containing TFA (1.3×10^{-1}M).

Fig. 2. Quantitative formation of 5-EtFl• obtained on adding HCOOH to 2.7 ∿ 3.0×10^{-5} M solutions of 5-EtFl-4a-OMe under N_2 at 25° in the dark:
(c_1) in *benzene* (1.3×10^{-3}M HCOOH; t_{max}= 3h), λ_{max} in nm (ε): 652 (5200); 606 (4800); 480 (2300); 446 (2500); 389 (6400); 325 (7100).
(d_1) in *$CHCl_3$* (6.0×10^{-4}M HCOOH; t_{max}= 4.5 h), λ_{max} in nm (ε): 642 (6100); 599 (5500); 484 (3400); 452 (3400); 383 (7500); 325 (8900).
(e_1) in *MeCN* (1.3×10^{-2}M HCOOH; t_{max}= 40 min), λ_{max} in nm (ε): 631 (6000); 600 (5700); 486 (3500); 456 (3400); 379 (7700); 324 (8600).

(d_2-e_2) Molar absorbances of 5-EtFlH^{+}• in benzene, $CHCl_3$ and MeCN, respectively, obtained on adding 1 volume % of TFA to the solutions of 5-EtFl• (c_1-e_1); the final concentration of TFA was 1.3×10^{-1}M. λ_{max} in nm (ε); *benzene*: 352 (9000); 490 (8200); *$CHCl_3$*: 356 (10,600); 489 (9200); *MeCN*: 353 (10,100); 486 (9200).

simultaneous occurrence of a disproportionation (eq. 13) and one or more reactions as represented by eqs. 10-12. The competitive nature of these sequential reactions was shown. The disproportionation (eq. 13) which limits the yield of 5-EtFl• can be suppressed in favour of a decomposition (eq. 10). This will also result in an increased accumulation of 5-EtFl•, if the radicals which are concomitantly produced (eq. 7) will not trap 5-EtFl• to give a *stabilized* 4^a-flavin adduct (eq. 14; e.g. $R_a^\bullet = Cl_3C^\bullet$).

$$5\text{-EtFl}^\bullet + R_a^\bullet \longrightarrow 5\text{-EtFl-}4^a\text{-}R_a \quad \text{---(14)}$$

Apart from its role in the formation of a dihydroflavin ester (eq. 1b), HCOOH was expected to advance the occurrence of some sequential reactions by providing a radical cation more liable to a decomposition (eq. 10; $A^\bullet =$ $HCOO^\bullet$) or by being or providing a secondary electron donor X in the secondary electron transfer (eq. 12). Moreover, a radical termination (eq. 14; $R_a^\bullet = H^\bullet$) would not produce a relatively, stabilized adduct as with TCA, but the (4^a-unsubstituted) dihydroflavin 5-EtFlH. This would also lead to 5-EtFl• in a comproportionation (eq. 15). Therefore, the use of HCOOH

$$5\text{-EtFlH} + 5\text{-EtFl}_{ox}^{+} \longrightarrow 2\ 5\text{-EtFl}^\bullet + H^+ \quad \text{---(15)}$$

promised the quantitative accumulation of 5-EtFl•. This promise was indeed fulfilled (Fig. 2).

REFERENCES.

1. Mager, H.I.X., Addink, R.: *Flavins and Flavoproteins* (edited by V. Massey and Ch. Williams), pp. 284-293. Elsevier, Amsterdam (1982); Tetrahedron 39, 3359-3366 (1983).
2. Mager, H.I.X., Addink, R.: Tetrahedron (1984, in the press).
3. *Note*: the back transfer of an electron to a radical cation may result in a $10,10^a$-ring opening. Based on the same principle, disproportionations (eqs. 16; 17) are supposed to be other possibilities to give $10,10^a$-ring opened transients.

$$2\ 1\text{-RFl-}10^a\text{-OO}^\bullet + H^+ \longrightarrow 1\text{-RFl}_{ox}^{+} + O_2 + 10,10^a\text{-ring opened hydroperoxide or carbonyloxide} \quad \text{----(16)}$$

$$2\ (1\text{-RFl-}10^a\text{-OH})^{\overset{+}{\bullet}} \longrightarrow \underline{2} + (1\text{-RFl-}10^a\text{-OH})^{++} \quad \text{----(17)}$$

ENDOR STUDIES ON FLAVIN RADICALS

Michael Bock, Martin Elsner, Harry Kurreck
Institut für Organische Chemie der Freien Universität Berlin
D-1000 Berlin

Introduction

In a pioneering work Massey and Palmer were the first to identify paramagnetic species in flavin containing enzymes generated by photoreduction (1). Since the radical state of the isoalloxazine ring system exhibits a very large hyperfine anisotropy - essentially arising from the nitrogens in positions 5 and 10 - and the Brownian motion at room temperature is slow the EPR spectra of protein bound flavin radicals exhibit only broad lines. In contrast, the EPR spectra of the radicals of protein free coenzymes and some model compounds give moderately resolved EPR signals. In fact, proton and nitrogen hyperfine coupling constants ("hfc") of the lumiflavin radicals could be extracted from these spectra, e. g. by Müller et al (2 - 5).
Taking advantage of the much better resolution power of Electron Nuclear DOuble Resonance ("ENDOR") spectroscopy often complete sets of hfc can unambiguously be determined (6, 7). Moreover, by applying the "general TRIPLE" technique relative signs are accessible. ENDOR-induced EPR allows different radical species present in the sample under study to be recorded selectively.
In the present paper fluid solution ENDOR studies of a variety of flavin and thiaflavin radicals are reported (8, 9). Moreover, in order to obtain information about anisotropic contributions of the hyperfine interactions ENDOR spectra of flavin radicals in disordered solids were taken.

Flavins and Flavoproteins

Results

Isotropic ^{1}H, ^{2}H, and ^{14}N hfc of a variety of flavin and thiaflavin radicals were determined by performing ENDOR experiments in fluid solutions. Assignments to molecular positions were achieved by selective deuteration and/or substitution. It is demonstrated that previous assignments of some of the hfc from conventional EPR have to be revised (2 - 5). For the first time a small spin density could be detected in the pyrimidin moiety of the flavin skeleton. Not unexpectedly, the spin density distribution of the flavin system is significantly affected by the replacement of the nitrogen in position 5 by sulphur (10, 11). In the flavin series hfc of the different protonation states were determined. Whereas the cation and the neutral radical have a quite similar spin density distribution that of the anionic species turned out to be remarkably different. Comparison of the radicals from lumiflavin, riboflavin, FMN, and FAD allows it to be concluded that in the case of increasing residues at position 10 the β-protons of the corresponding methylene group are no longer equivalent. In the thiaflavin series it could be shown by ENDOR-induced EPR (Figure 1) that the synthetic route leads to isomers of the diamagnetic precursors and thereby to different radicals (12).

In agreement with previous studies of, e.g., J.S. Hyde and coworkers (7, 13, 14) it is shown that the most prominent features of the ENDOR spectra in disordered solids are the matrix signal and those of the protons of freely rotating methyl groups. These couplings could be assigned to molecular positions by replacement of methyl groups by α-protons and/or by introduction of larger substituents. Since the methyl proton ENDOR signals are additionally split different field settings allow an orientational selection and thereby determination of different anisotropic hyperfine components. Knowledge of these results is no doubt a prerequisite for the studies of native flavoenzyme free radicals by means of ENDOR.

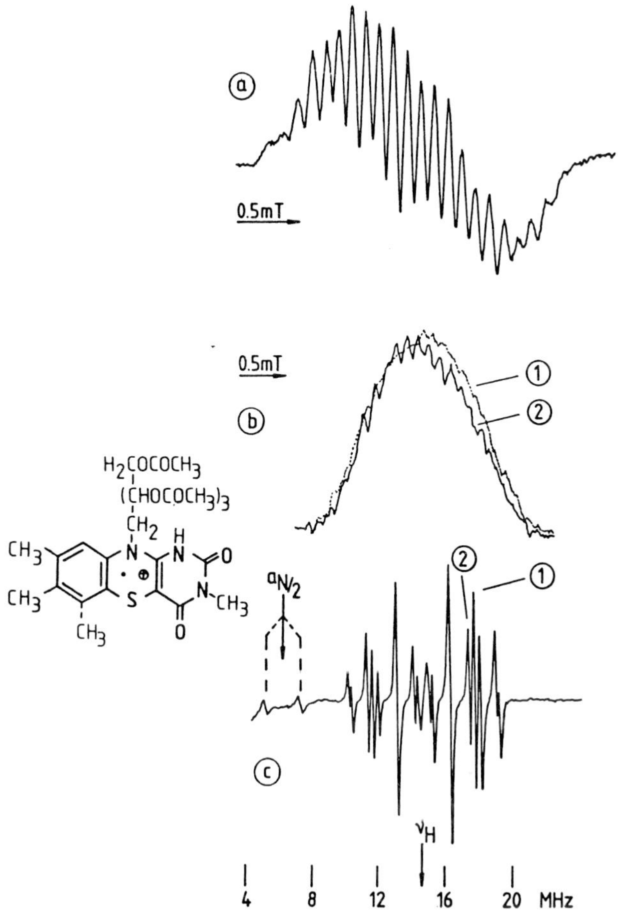

Fig. 1. EPR and ENDOR spectra of a mixture of the two isomers of 7,8- and 6,7-dimethyl-tetraacetyl-thiariboflavin radical cation in toluene/CF_3COOH at 280 K.

ⓐ EPR spectrum (superposition).

ⓑ ENDOR-induced EPR spectra of the 7,8-isomer ① and the 6,7-isomer ② (① and ② indicate the rf pump frequency).

ⓒ ENDOR spectrum (superposition); mw-power 30 mW; rf-power 0,4 mT (14 MHz) in the rotating frame; field modulation of the NMR field (10 kHz), amplitude 70 kHz; 128 sweeps (30 s/sweep); time constant 40 ms. Instrumentation see ref. 15.

Acknowledgement

We thank Prof. H. Fenner and Dr. R. Grauert (thiaflavins) and Prof. F. Müller (labelled flavins) for their help.

References

1. Massey, V., Palmer, G.: Biochem. 5, 3181 (1966)
2. Westerling, J., Mager, H.I.X., Berends, W.:Tetrahedron 31, 437 (1975)
3. Ehrenberg, A., Müller, F., Hemmerich, P.: Eur. J. Biochem. 2, 286 (1967)
4. Müller, F., Hemmerich, P. Ehrenberg, A., Palmer, G., Massey, V.: Eur. J. Biochem. 14, 185 (1970)
5. Müller, F., Grande, H.J., Harding, L.J., Dunham, W.R., Visser, A.J.W.G., Reinders, J.H., Hemmerich, P., Ehrenberg, A.: Eur. J. Biochem. 116, 17 (1981)
6. Dorio, M.M., Freed, J.H.: Multiple Electron Resonance Spectroscopy, Plenum, New York 1979
7. Eriksson, L.E.G., Hyde, J.S., Ehrenberg, A.: Biochim. Biophys. Acta 192, 211 (1969)
8. Bock, M., Lubitz, W., Kurreck, H., Fenner, H., Grauert, R.: J. Am. Chem. Soc. 103, 5567 (1981)
9. Kurreck, H., Bock, M., Bretz, N., Elsner, M., Kraus, H., Lubitz, W., Müller, F., Geissler, J., Kroneck, P.M.H.: J. Am. Chem. Soc. 106, 737 (1984)
10. Fenner, H.,Grauert, R., Hemmerich, P., Michel, H., Massey, V.: Eur. J. Biochem. 95, 183 (1979)
11. Fenner, H., Grauert, R., Hemmerich, P.: Liebigs Ann. Chem. 193 (1978)
12. Fenner, H, Grauert, R., Tessendorf, L.: Arch. Pharm. 314, 874 (1981)
13. Hyde, J.S., Rist, G.H., Eriksson, L.E.G.: J. Phys. Chem. 72, 4269 (1968)
14. Kevan, L., in Kevan, L., Kispert, L.D.: Electron Spin Double Resonance Spectroscopy, Wiley, New York 1976
15. Kurreck, H., Kirste, B., Lubitz, W.: Angew. Chem. 96, 171 (1984); Angew. Chem. Int. Ed. Engl. 23, 173 (1984)

ON THE ROLE OF SOME FLAVIN ADDUCTS AS ONE-ELECTRON DONORS

H.I.X. Mager and R. Addink

Biochemical and Biophysical Laboratory of the Delft University of Technology, 67 Julianalaan, 2628 BC DELFT, The Netherlands.

The analysis, the purification or the use of particular flavin derivatives is sometimes hampered by the spontaneous formation of radicals of which the mechanism was not understood. Concerning the problems caused by the unexpected appearance of radicals only some casual remarks have been made in the literature. Ghisla et al[1,2] noticed that they could not record the NMR-spectra of 5-ethyl-3-methyllumiflavinium perchlorate ($5\text{-EtFl}^{+}_{ox}, A^{-}$; $A^{-} = ClO_4^{-}$) and the derived 4^a-pseudobase ($5\text{-EtFl-}4^a\text{-OH}$) due to the formation of radicals. In attempts to recrystallize this pseudobase from aqueous and organic solvents, Mager[3] observed the occurrence of some ring transformations preceded by an excessive production of 5-EtFl• radicals.

In this laboratory, the spontaneous transformations of a variety of N^1- and N^5-alkylflavinium salts have been extensively studied.[4-6] It was found that these processes start with the formation of a $1,10^a$- or $4^a,5$-dihydroflavin ester ($1\text{-RFl-}10^a\text{-A}$; $5\text{-RFl-}4^a\text{-A}$). These adducts reduce the flavinium cations e.g. 5-EtFl^{+}_{ox} to give the flavin radical 1 and a counter-radical cation 2 in a 1:1 ratio. Under the experimental conditions applied, the yield of 5-EtFl• increased as a result of subsequent conversions of 2 by the large excess of acid used.

The electron transfer property is not limited to the dihydroflavin ester e.g. $5\text{-EtFl-}4^a\text{-X}$ ($X^{-} = A^{-}$), but it is also shown by other adducts having the same dihydroflavin structure ($X^{-} = HO^{-}$; HOO^{-}; RO^{-}). It is demonstrated by the accumulation of 5-EtFl• occurring in dilute solutions of the adducts in e.g. benzene or chloroform which were kept under N_2 at room temperature in

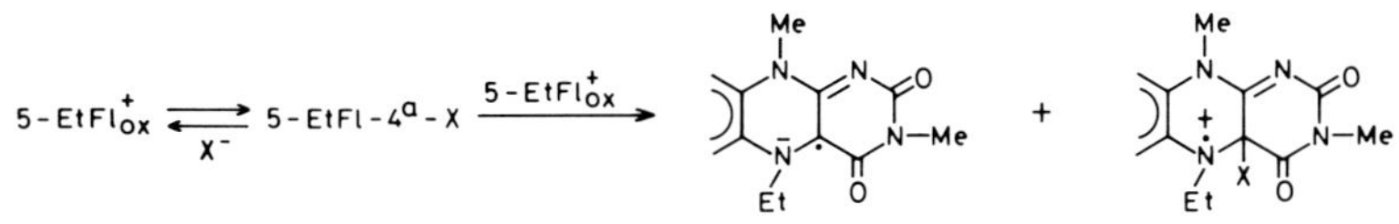

1 2 ·····(1;2)

the dark. As compared with $5\text{-EtFl}^{+}_{ox},A^{-}$ salts, the conversions of $5\text{-EtFl-4}^{a}\text{-X}$ (X= OH; OMe), dissolved in purified solvents, took place very slowly requiring several days. The overall spectral changes in solvents like benzene and toluene cover two features: (a) the yield of the accumulation of 5-EtFl• does not exceed the calculated value of 50%; (b) an additional accumulation of a new intermediate. This is a radical cation 2 or a derivative from it as it immediately provides an additional amount of $5\text{-EtFlH}^{+}\cdot$ (curves d-c, Fig. 1) on adding a large excess of an acid (eq. 3a).

$$(5\text{-EtFl-4}^{a}\text{-X})^{+}\cdot + H^{+} \longrightarrow 5\text{-EtFlH}^{+}\cdot + X^{+} \quad ---(3a)$$

Concerning the conversions of $5\text{-EtFl-4}^{a}\text{-X}$ (X= OH; OMe; eqs. 1+2) comparable cxperiments showed a spread of t_{max}-values. A possible cause for this might be the presence of traces of acidic or anionic impurities. The addition of a limited amount of an acid gives a good reproducibility and an acceleration of the process without influencing the overall spectral change. The acid can be a carboxylic acid (e.g. TFA; TCA), a sulfonic acid and even, but less recommendable, a mineral acid like H_2SO_4. For example, the addition of β-naphtalenesulfonic acid in relative amounts of 33-50% with reference to the concentration of $5\text{-EtFl-4}^{a}\text{-OH}$ (∿ 3×10^{-5}M) resulted in a complete conversion of the adduct giving a mixture of 5-EtFl• (30-35%; cf. curve b, Fig. 1), 3-methyllumiflavin (4-8%) and $(5\text{-EtFl-4}^{a}\text{-OH})^{+}\cdot$ in yields of 45-50% (cf. curve d) in 2-3 h. The use of TFA in relative concentrations of 16-190% gave 5-EtFl• (39-45%), 3-methyllumiflavin (2-4%) and $(5\text{-EtFl-4}^{a}\text{-OH})^{+}\cdot$

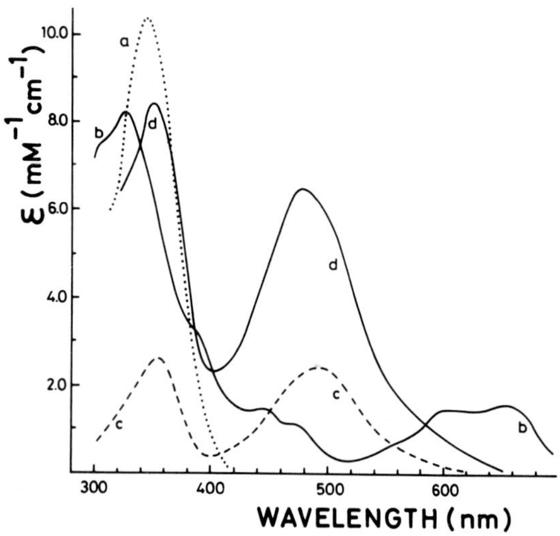

Fig. 1. (a) Molar absorbance of $5\text{-EtFl-4}^{a}\text{-OH}$, *freshly* dissolved in benzene. (b) Spectral change after 3 h, effected by β-naphtalenesulfonic acid in a relative amount of 33.3%; the conversion occurred at 25° in the dark. (c) the spectrum of the amount of $5\text{-EtFlH}^{+}\cdot$ corresponding to the amount of 5-EtFl• represented by the 600-700 nm region of b. (d) The actual spectral change obtained on acidification of b (1.3×10^{-1}M TFA). The additional amount of $5\text{-EtFlH}^{+}\cdot$ is formed in the decomposition (eqs. 3a; 3b) of $(5\text{-EtFl-4}^{a}\text{-OH})^{+}\cdot$.

in yields of 25-35% in a considerably shorter reaction time (1h - 4 min).

The accelerating effect of an acid is due to the formation and the consumption of a dihydroflavin ester (eqs. 4+5+6). The fact that a limited amount of an acid does not influence the overall spectral change, implies that the counter-radical cations mainly or completely consist of the hydroxy-derivative 2 (X= OH) arisen by hydrolysis of the primary radical cation (eq. 7).

$$\text{5-EtFl-4}^{a}\text{-X} + \text{HA} \xrightarrow[\text{HX}]{} \text{5-EtFl}^{+}_{ox}, \text{A}^{-} \rightleftharpoons \text{5-EtFl-4}^{a}\text{-A} \text{ -------(4; 5)}$$

$$\text{5-EtFl-4}^{a}\text{-A} + \text{5-EtFl}^{+}_{ox} \longrightarrow \text{5-EtFl}\bullet + (\text{5-EtFl-4}^{a}\text{-A})^{\overset{+}{\bullet}} \text{ -------(6)}$$

$$(\text{5-EtFl-4}^{a}\text{-A})^{\overset{+}{\bullet}} + H_2O \longrightarrow (\text{5-EtFl-4}^{a}\text{-OH})^{\overset{+}{\bullet}} + \text{HA} \text{ --------------(7)}$$

Concerning the identification of the products derived from the species $X^{+}= HO^{+}$ it could be established that no hydroxylation of e.g. the aromatic solvent had occurred. This suggests that the acidic decomposition of the radical cation 2 (X= OH) is accompanied by a generation of O_2 (eq. 3b).

$$(\text{5-EtFl-4}^{a}\text{-OH})^{\overset{+}{\bullet}} \xrightarrow{(H^{+})} \text{5-EtFlH}^{\overset{+}{\bullet}} + \tfrac{1}{2}\, O_2 \text{ -----(3b)}$$

We have already concluded that the counter-radical cations 2 may show various sequential reactions.[5,6] In our current studies, we are trying to establish the conditions which bring about specifically the back transfer of an electron to a radical cation. In the N^{1}-alkylflavin series, we already found[5] that the back transfer of an electron to $(\text{1-RFl-10}^{a}\text{-OH})^{\overset{+}{\bullet}}$ led to a (dark) $10,10^{a}$-ring opening while, basically, a similar one-electron return has been claimed by McCapra[7] to be a chemiluminescent reaction. The nature of the electron-donor might be one of the causes producing different effects. In connection with the still unknown mechanism of bacterial bioluminescence, Schuster[8] suggested the occurrence of an intramolecular CIEEL pathway. In the line of this suggestion, we have formulated the transients derived from 3: the flavin moiety donates an electron to the peroxide bond resulting in the formation of a radical pair 4a. Loss of a proton

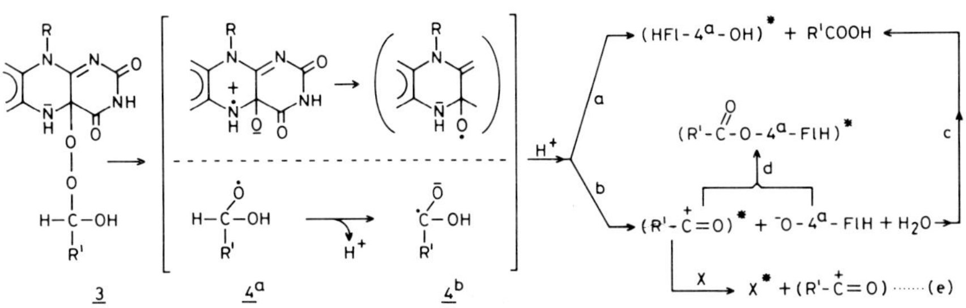

(4b) followed by back transfer of an electron to the flavin radical might result in the excited pseudobase (pathway a). Alternatively, we propose the primary formation of an excited acylium cation $(R'\overset{+}{C}{=}O)^*$ by pathway b, followed by c, d or e producing by *energy transfer* the excited pseudobase, dihydroflavin ester or fluorescer X, respectively. These pathways might be of relevance to some particular bio- and chemi-luminescent systems.[9-11]

In relation to the radical pair 4a, 4b the accumulation of the model compound 2 (X= OH) is considered to be an important step forward, since it will enable us to check the validity of the hypothesis on an intramolecular CIEEL mechanism for peroxyflavin adducts 3, while it might consequently lead to the development of new chemiluminescent model systems.

REFERENCES

1. Ghisla, S., Hartmann, U., Hemmerich, P., Müller, F.: Ann. Chem. 1388-1415 (1973).
2. Ghisla, S., Hastings, J.W., Favaudon, V., Lhoste, J.M.: Proc. Natl. Acad. Sci. USA 75, 5860-5863 (1978).
3. Mager, H.I.X.: Tetrahedron Letters 3549-3552 (1979).
4. Mager, H.I.X., Addink, R.: *Flavins and Flavoproteins* (edited by V. Massey and Ch. Williams), pp. 284-293. Elsevier, Amsterdam (1982).
5. Mager, H.I.X., Addink, R.: Tetrahedron 39, 3359-3366 (1983).
6. Mager, H.I.X., Addink, R.: Tetrahedron, in the press (1984).
7. McCapra, F.: Proc. R. Soc. Lond. B215, 247-272 (1982).
8. Schuster, G.B.: Acc. Chem. Res. 12, 366-373 (1979).
9. Lee, J., Carreira, L.A., Gast, R., Irwin, R.M., Koka, P., Daune Small, E., Visser, A.J.W.G.: *2nd Symp. on Bio- and Chemi-luminescence* (edited by M. DeLuca and W.D. McElroy), pp. 103-113, Academic Press (1981).
10. Matheson, I.B.C., Lee, L., Müller, F.: Proc. Natl. Acad. Sci. USA 78, 948-952 (1981).
11. Shepherd, P.T., Bruice, T.C.: J. Am. Chem. Soc. 102, 7774-7776 (1980).

PHOTOINACTIVATION OF FLAVIN REDOX CATALYSIS: THE REDUCTIVE FLAVIN PHOTOADDUCT FORMATION

Ulrich Ott, Rainer Traber, Horst E.A. Kramer
Institut für Physikalische Chemie der Universität Stuttgart, Pfaffenwaldring 55, D-7000 Stuttgart 80, W-Germany

Introduction

The chemical processes involved in flavin dependent dehydrogenation are too complex as to describe them simply as an "abstraction of molecular hydrogen from an organic compound". Within the efforts that have been made to reveal the mechanisms of these redox reactions, a lot of model compounds have been employed and in this connection the photochemistry of lumiflavins has achieved great importance. Previous investigations dealt with photoinduced reduction reactions which showed radical character as well as ionic one, sometimes including alkylated intermediates or stable photoadducts. The adduct formation stops redox turnover and therefore has been used as a model for inhibition of flavin enzymatic activity (1,2).

Mechanistic Considerations

The possible reaction pathways of flavin photoadduct formation are described in Fig.1. Hemmerich and coworkers (3,4,5) discussed a nucleophilic attack of a carbanionic intermediate (steps a+b) and a possible protolysis of the formed adduct (step i). This reaction pathway has been revealed by Traber et al. (6), while flavin alkylation with preceding hydride transfer (steps a+c+h) never has been reported. Concerning

Flavins and Flavoproteins

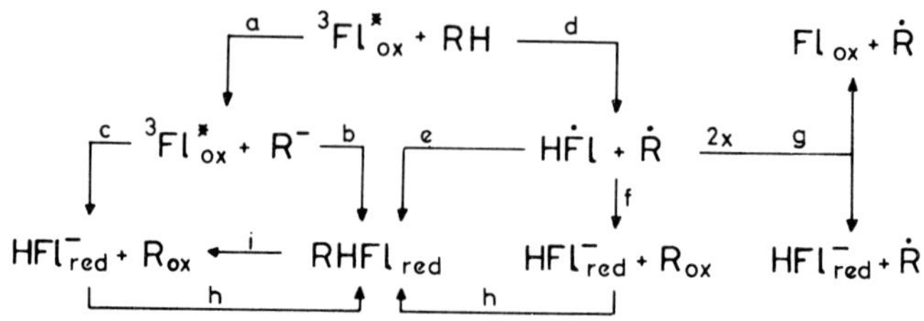

Figure 1. Reaction pathways of photolytic reduction and adduct formation ($RHFl_{red}$) of excited flavin triplet ($^3Fl^*_{ox}$) with organic substrates (RH).

the formation of radicals (step d) and subsequent reactions, such as radical combination (step e), dismutation (step g), or a second electron transfer with a following dark reaction (steps f+h), many investigations have been made (2,7-11).

Results

In order to get further informations about the adduct formation mechanisms and the structural requirements of the substrates, we have investigated the photoreactions with various thioureas and a propargylamine. In this paper, we discuss the results, obtained with N-allylthiourea (ATU) and N,N-dimethylpropargylamine (DMPA). In case of the substrate ATU, continuous illumination of flavin yields a 4a,5-adduct (Fig.2b) with a typical absorption band at 365 nm, whereas with DMPA a mixture of 4a,5-adduct and 5-adduct (cyanine structure) is obtained (1,2,12). In both reactions, only a small amount of the reduced solution is reoxidisable by oxygen.

Primary processes of the photoreactions. The reactive photoexcited state was proven by flash photolysis to be the flavin triplet state. Within the triplet quenching, an electron transfer reaction is predominant, yielding the neutral flavosemiquinone with 63 % (ATU) or 72 % (DMPA) relative to the quenched flavin triplet. As to the ATU-reaction, the remaining 27 % flavin triplet were deactivated to the singlet ground state. This agrees with the initial absorption of the intermediates at 440 nm (Fig.2a). The extrapolation of the 483 nm curve (the isosbestic point between flavin and neutral flavosemiquinone) to the end of the flash leads back to the starting absorbance (Fig. 2a) which also indicates, that only flavosemiquinone and flavin in its ground state are present at this time. With the substrate DMPA, the non-radical triplet quenching is more complex because of the high 440 nm bleaching just after the flash. Here a further, less absorbing species, possibly the 4a,5-adduct or the unsubstituted dihydroflavin, has to be produced directly from the triplet state. <u>Fate of the intermediates</u>. In case of ATU as substrate, the analysis of the flavosemiquinone decay ($k = 5.15\times10^{9}\ M^{-1}\ s^{-1}$) and the absorption curves in Fig.2a reveals three reactions: electron back donation, dismutation of the flavin radicals and mainly adduct formation by radical combination (see step e, Fig.1). In the photoreaction with DMPA the kinetic of flavin radical decay also shows a clear second order reaction with the rate constant of $k = 2.7\times10^{9}\ M^{-1}\ s^{-1}$ which points to a dismutation reaction (13). This, however, is not consistent

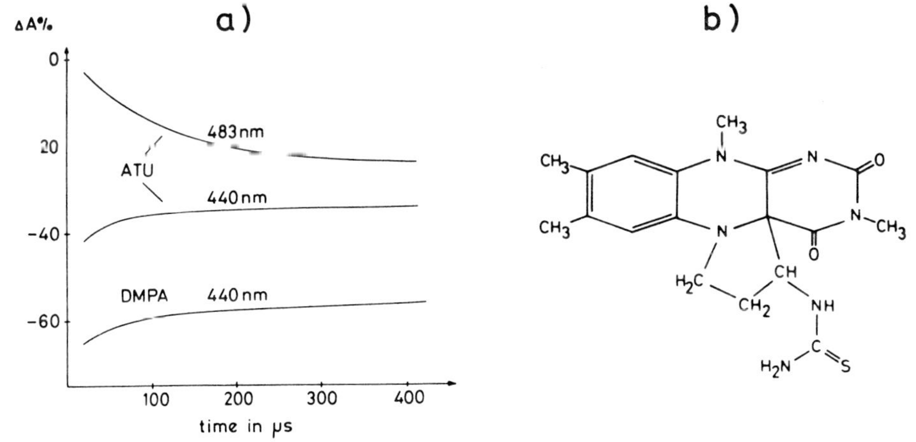

Figure 2. Flash photolysis of 3-methyllumiflavin (7 μM) with ATU (2 mM) and DMPA (10 mM) in anaerobic, aqueous phosphate buffer, pH 7. Time dependence of the change of absorbance after flash decay, related to the initial Fl_{ox} concentration. Insert: Structure of flavin-ATU-adduct.

with the slow increase of the 440 nm curve in Fig.2a, which requires at least a competition reaction, yielding a weakly absorbant species, i.e. the 4a,5-adduct (Fig.1, step e) or unsubstituted dihydroflavin (step f). In contrast to the ATU-reaction, a further change of absorbance far into the range of seconds is observed which gives rise to the interpretation of adduct formation (5-adduct, cyanine structure) after flavin reduction (step h).

References

1. Gärtner, B., Hemmerich, P., Zeller, E.A.: Eur. J. Biochem. 63, 211-221 (1976)

2. Simpson, J.T., Krantz, A., Lewis, F.D., Kokel, B.: J. Am. Chem. Soc. 104, 7155-7161 (1982)

3. Hemmerich, P.: Prog. Chem. Org. Nat. Prod. 33, 451-527 (1976)

4. Knappe, W.-R., Hemmerich, P.: Liebigs Ann. Chem. 1976, 2037-2057 (1976)

5. Haas, W., Hemmerich, P.: Biochem. J. 181, 95-105 (1979)

6. Traber R., Kramer, H.E.A., Knappe, W.-R., Hemmerich, P.: Photochem. Photobiol. 33, 808-814 (1981)

7. Traber, R., Kramer, H.E.A., Hemmerich, P.: Pure & Appl. Chem. 54, 1651-1665 (1982)

8. Hore, P.J., Volbeda, A., Dijkstra, K., Kaptein, R.: J. Am. Chem. Soc. 104, 6262-6267 (1982)

9. Novak, M., Miller, A., Bruice, T.C., Tollin, G.: J. Am. Chem. Soc. 102, 1465-1467 (1980)

10. Heelis, P.F.: Chem. Soc. Rev. 1982, 15-39 (1982)

11. Traber, R., Kramer, H.E.A., Hemmerich, P.: Biochemistry 21, 1687-1693 (1982)

12. Maycock, A.L.: J. Am. Chem. Soc. 97, 2270-2272 (1975)

13. Hemmerich, P., Knappe, W.-R., Kramer, H.E.A., Traber, R.: Eur. J. Biochem. 104 , 511-520 (1980)

FLAVIN OXYGEN CHEMISTRY BROUGHT TO DATE

Thomas C. Bruice

Department of Chemistry
University of California, Santa Barbara
Santa Barbara, California USA 93106

In this contribution the author provides a summary of the research accomplishments of his laboratory which deal with chemistry related to the flavoenzyme mixed function oxidases. Studies which are discussed are those which have come to fruition since the 7th Symposium on Flavins and flavoenzymes. When added to the chapter published in the volume edited by Williams and Massey (1) the salient features of our contributions to this area of research are reviewed. The reader is also directed to a recent review in the Israeli Journal of Chemistry (2).

For the reader who may wish to gain an understanding of the important aspects of our investigations, with little investment of time, the following statements can be made:

1) 1,5-Dihydroflavins (Fl_{red}) react with molecular oxygen by way of a $1e^-$ transfer to yield as intermediate a radical pair consisting of flavin radical (Fl_{sem}) and superoxide ($O_2^{\cdot-}$). This radical pair can partition back to 1,5-dihydroflavin and O_2 or forward to a 4a,5-dihydroflavin substituted by a hydroperoxy anion at the 4a-position. This flavin hydroperoxy anion becomes protonated to yield 4a-hydroperoxy flavin which may then eliminate hydrogen peroxide to yield flavin (Fl_{ox}). This step-wise mechanism is shown in eq. 1. (Note: Scheme II in ref. 1 is an alternate mechanism which is now known to be

Flavins and Flavoproteins

incorrect, ref. 3). The 4a-Hydroperoxy flavin moiety

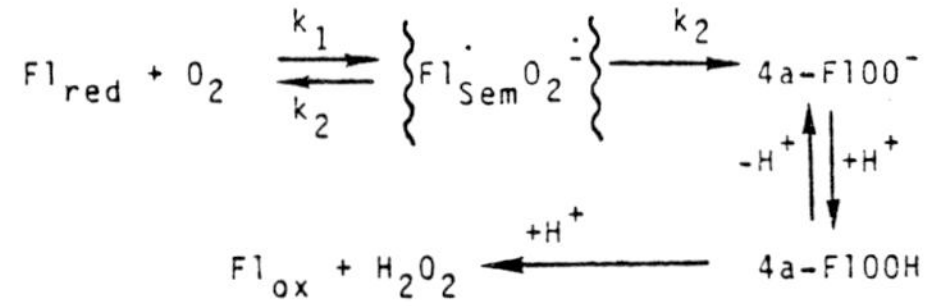

(1)

when enzyme bound, is the oxygen transfer agent of flavoenzyme mixed function oxidases. When Fl_{red} is alkylated at (N)5 the 4a-hydroperoxy flavin can be isolated.

2) There is no evidence which can be offered in support of the much quoted carbonyl oxide mechanism (Scheme I) for the flavoenzyme mixed function oxidases nor for a concerted mechanism which leads to 4a,5-ring opening (Scheme II).

Scheme I

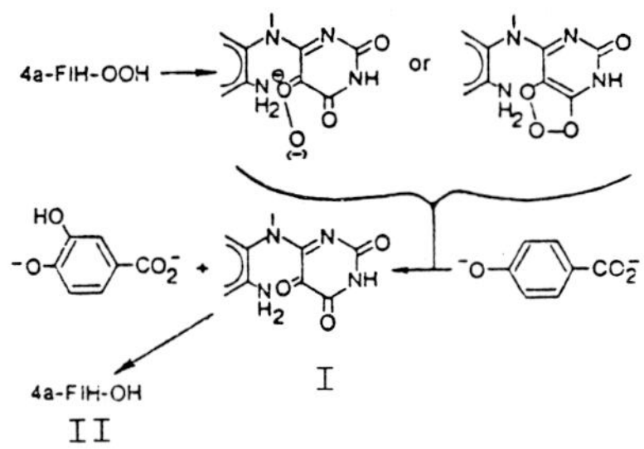

Scheme II

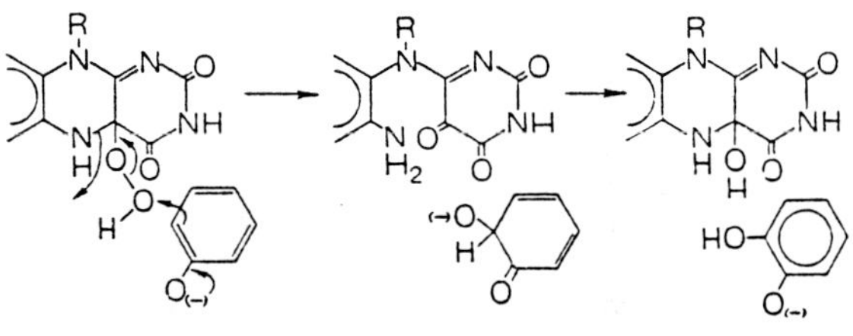

(II)

The syntheses of III and IV has been accomplished and from their spectra, in a variety of solvents and at varied basicities, it can be concluded that there is no support for the proposal of I (Schemes I and II) as the spectrally observed intermediate in the reaction of phenol mono oxygeneses. No evidence for the intermediacy of I has been offered for any other flavin mixed function oxidase.

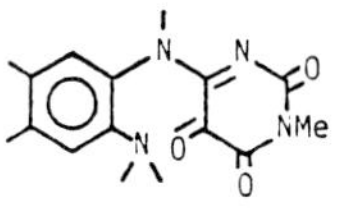

III

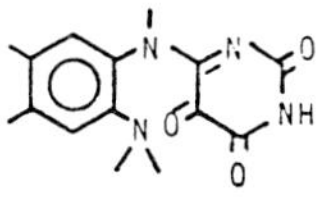

IV

3) The mechanism for the microsomal flavin mixed function oxidase involves the direct conversion of the flavin hydroperoxide to its pseudo-base (eq. 2).

(2)

The reaction occurs by nucleophilic attack of the substrate (:X) upon the terminal oxygen of the 4a-hydroperoxy flavin. The second order rate constant for reaction of nucleophiles (:X) with hydroperoxides and percarboxylic acids depends upon the pK_a of the generated hydroxyl compound (ROH in eq. 2). The large oxygen donation potential of 4a-flavin hydroperoxide is due to the low (for an alcohol) pK_a of II.

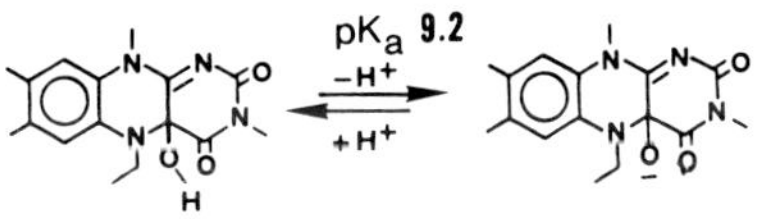

(3)

II

4) There are at present no model systems that duplicate the spectral observations made with the various phenolate hydroxylases. Perhaps suitable mechanisms will evolve for the enzymatic reactions which will be based on our observations that numerous phenolate and indole anion substrates are dioxygenated by the anion of 4a-hydroperoxyflavin. A suitable example is provided in Scheme III.

Scheme III

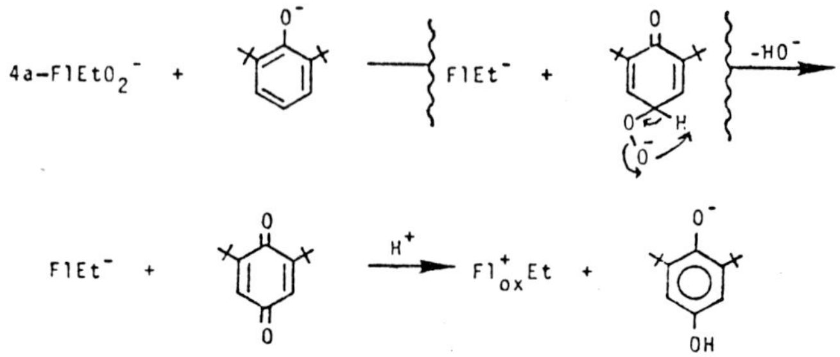

In what follows there is provided in greater detail a description of our recent contributions which have led to a portion of the conclusions just presented. The mechanism of eq. 1 for the reduction of oxygen to hydrogen peroxide by 1,5-dihydroflavins was reached by a combination of electrochemical and kinetic experiments (3). From the one-electron reduction potential for both oxidized flavin $\longrightarrow$ flavin radical and flavin radical $\longrightarrow$ 1,5-reduced flavin plus the one-electron reduction potentials for $O_2 \longrightarrow O_2^{\overline{\cdot}}$ and $O_2^{\overline{\cdot}} \longrightarrow H_2O_2$ there were calculated the standard free energies for the reactions:

$$\text{1,5-reduced flavin} + O_2 \longrightarrow \text{flavin radical} + O_2^{\overline{\cdot}} (\Delta G_1^0) \qquad (4)$$

$$\text{Flavin radical} + O_2^{\overline{\cdot}} \longrightarrow \text{flavin} + H_2O_2 (\Delta G_2^0) \qquad (5)$$

Plots of ΔG_1^0 vs. pH were found to parallel in shape plots of $\Delta G_f^{\ddagger}$ vs. pH for the bimolecular O_2 oxidation of 1,5-reduced-5-ethyl-3-methyllumiflavin (Figure 1). In the plot of Figure 1 the free energy content of the transition state ($\Delta G_f^{\ddagger}$) exceeds the free energy content of the $O_2^{\overline{\cdot}}$ + flavin radical pairs (ΔG_1^0) by 15 kJM^{-1}. In eq.1, $\Delta G_f^{\ddagger}$ pertains to k_1. The values of $\Delta G_f^{\ddagger} - \Delta G_1^0$ equals the free energy of activation for reduction of flavin radicals by $O_2^{\overline{\cdot}}$ ($\Delta G_r^{\ddagger}$). The free energy of activation $\Delta G_r^{\ddagger}$ pertains to k_{-1} of eq. 1. Since $\Delta G_f^{\ddagger} >> \Delta G_r^{\ddagger}$, the critical transition state for the reaction of O_2 with 1,5-dihydro-5-ethyl-3-methyllumiflavin must closely resemble an ($Fl_{sem}O_2^{\overline{\cdot}}$) pair (pH range 3 to 10.5). A reaction

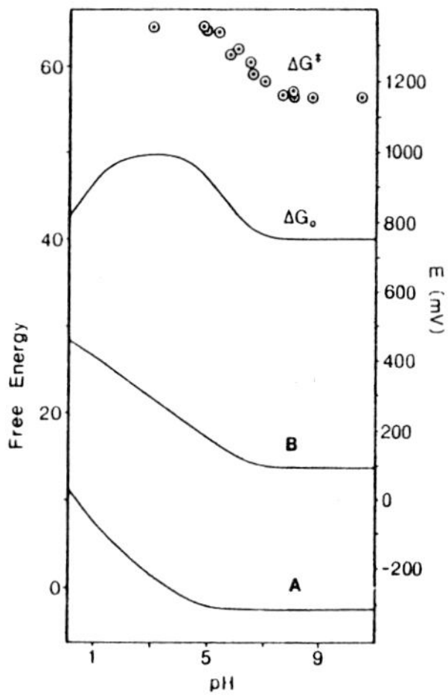

Figure 1. Plots of the experimentally determined free energies of activation ($\Delta G^{\ddagger}$) and calculated standard free energies (ΔG^0), for formation of flavin radical plus superoxide on reaction of 1,5-dihydro-5-ethyl-3-methylluminflavin with O_2 (30 ^{0}C; H_2O) vs. pH. Lines a and b represent the pH dependencies of the E_m for $O_2 + 1e^- \longrightarrow O_2^{\overline{\cdot}}$ and $FlEt_{sem} + 1e^- \longrightarrow FlEt_{red}$ respectively.

coordinate cartoon is shown in Figure 2. The pH dependence of the apparent second-order rate constant for reaction of 1,5-dihydro-5-ethyl-3-methyllumiflavin with O_2 is quantitatively explained by the assumption of second-order reactions of neutral (k_a) and anionic (k_b) dihydroflavin species with O_2.

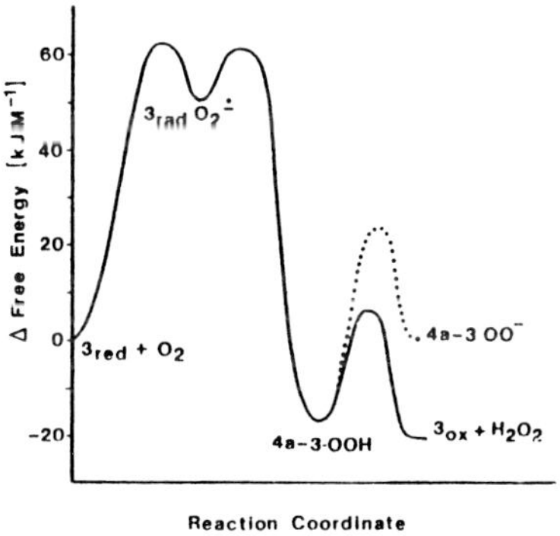

Figure 2. Reaction coordinate cartoon for the oxidation of N(5)-ethyl-1,5-dihydro-3-methyllumiflavin by O_2 at 30 ^{0}C in H_2O at pH 4.6.

The numerical values of k_a and k_b are such that the fraction $HO_2^{\cdot}/O_2^{\overline{\cdot}}$ produced as intermediates are as predicted from the pK_a of $HO_2^{\cdot}$ (Scheme IV).

Scheme IV

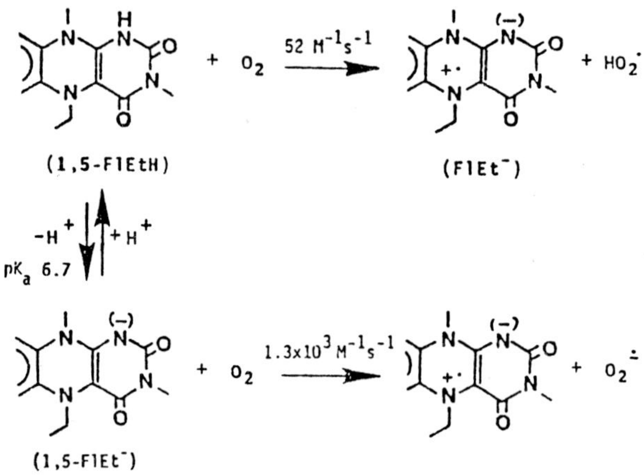

The mechanisms of the reaction of a series of 1,5-dihydroflavins, possessing various substituents, with O_2 was studied at pH 4.6 (pK_a $HO_2^{\cdot}$). As in the instance of N^5-ethyl-3-methyllumiflavins, values of ΔG_1° and ΔG_2° were calculated from one-electron reduction potentials of oxygen and flavin species, $\Delta G_f^\ddagger$ by kinetic studies, and $\Delta G_r^\ddagger$ as the difference $\Delta G_f^\ddagger - \Delta G_1^\circ$. In all instances $\Delta G_f^\ddagger >> \Delta G_r^\circ$ so that $\{Fl_{sem}O_2^{\dot{-}}\}$ pairs serve as reasonable intermediates. These oxidations are neither general nor specific-acid catalyzed and since the rate constant k_r can be as large as $10^{11} s^{-1}$ it follows that further reactions of $\{Fl_{sem}O_2^{\dot{-}}\}$ must occur by radical coupling (k_2 in eq. 1) to yield flavin 4a-hydroperoxide anion, which may then be protonated to yield the 4a-hydroperoxyflavin. These results have been obtained from studies of 1-blocked, 5-blocked and 1,5-diblocked flavins. From the kinetics for reaction of O_2 with normal unblocked flavins we show that the second-order reaction of 1,5-dihydroflavin with O_2 cannot be derived from initial rates due to the very early onset of autocatalytic oxidation.

The synthesis of III and IV were carried out by percarboxylic acid oxidation of the appropriate N(5)-diethylated lumiflavin as shown in eq. 6 (4,5).

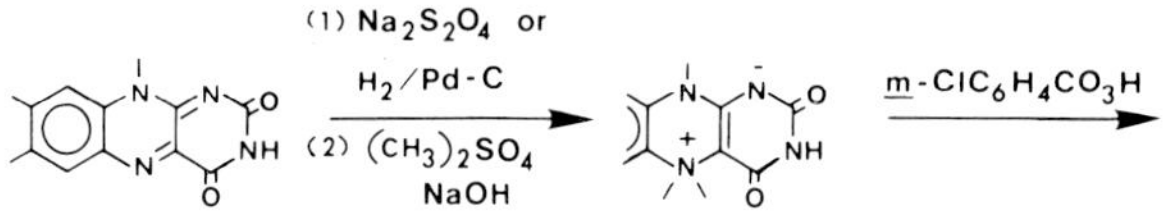

IV (6)

In Figure 3 there is compared the UV-VIS spectrum of III to that of an absorbing species formed, in time, between the 4a-hydroperoxyflavin (4a-FlHOOH) and 4a-hydroxyflavin (II) with p-hydroxybenzoate hydroxylase. This absorbing species was assumed (6) to be the 6-amino-5-oxo-3H, 5H-uricil (Species I in Schemes I and II).

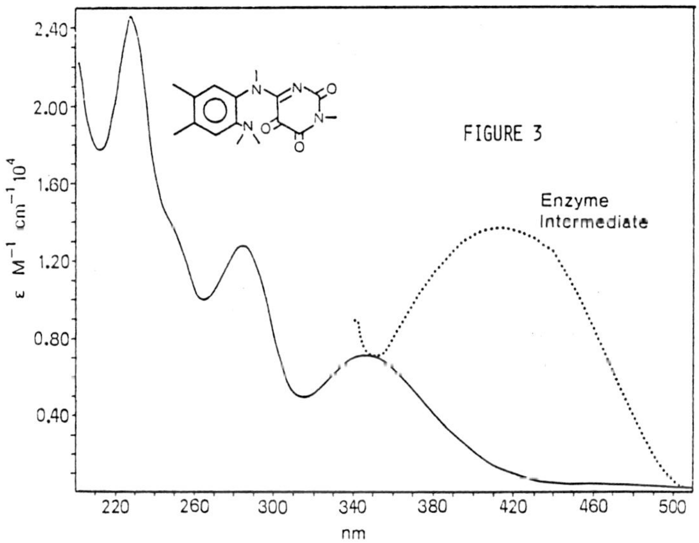

Examination of Figure 3 shows that the spectrum of III does not resemble at all that of the enzyme bound intermediate (both in λ_{max} and most importantly in ε, 7,000 vs. 15,000 cm^{-1} M^{-1}). Compound IV shows the same spectral characteristics as III. Further, the spectrum of IV is not changed by the dielectric constant of the media or upon solution in t-BuOH containing t-BuO^-K^+. The latter observation shows that the long-wave length absorbance of the enzyme intermediate cannot be due to formation of a quinoid species by isomerization (eq. 7) or ionization of the N(3)-H. The steric influence of the dimethyl substituents of the dimethylamino moiety of III and IV

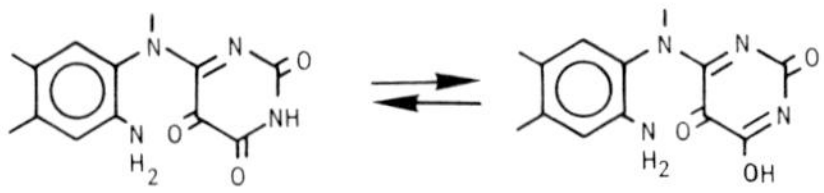
(7)

in preventing sufficient planarity, to allow internal charge transfer to occur and thereby change the spectra of the synthetic 6-amino-5-oxo-3H,5H-uracil to that of the enzymatic intermediate, has been considered (7). However, there is no evidence that such structures are subject to charge transfer nor could structure I (Scheme I and II) become coplanar in the coiled conformation shown.

The microsomal flavomonooxygenase of animals is responsible for the N-oxygenation of tertiary- and secondary amines as well as hydroxylamines and the S-oxygenation of organic sulfides (eq. 2). That this mechanism involves a direct nucleophilic displacement by substrate at the terminal oxygen of the hydroperoxyflavin (eq. 2) is shown by the following observations. Displacements of this nature are common in hydroperoxide chemistry where there is no possibility of special effects as ring opening to give a carbonyl oxide (as postulated in Scheme I and II). In the reaction of eq. 2 there is expected to be a relationship between the basicity of RO^- and the second order rate constant for O-O bond scission accompanying attack of X:. In Figure 4 there is plotted the log of the second order rate constant for reaction of the organic sulfide thioxane k_S vs. the pK_a of ROH (8). Figure 4 represents a correlation between the free energy of activation for thioxane sulfur displacement upon hydroperoxides (ROOH) and the standard free energy for proton dissociation from ROH. Inspection of Figure 4 shows that a linear free energy correlation exists for thirteen powers of ten in the proton bicicity of RO^- and eight and a half powers of ten in rate.

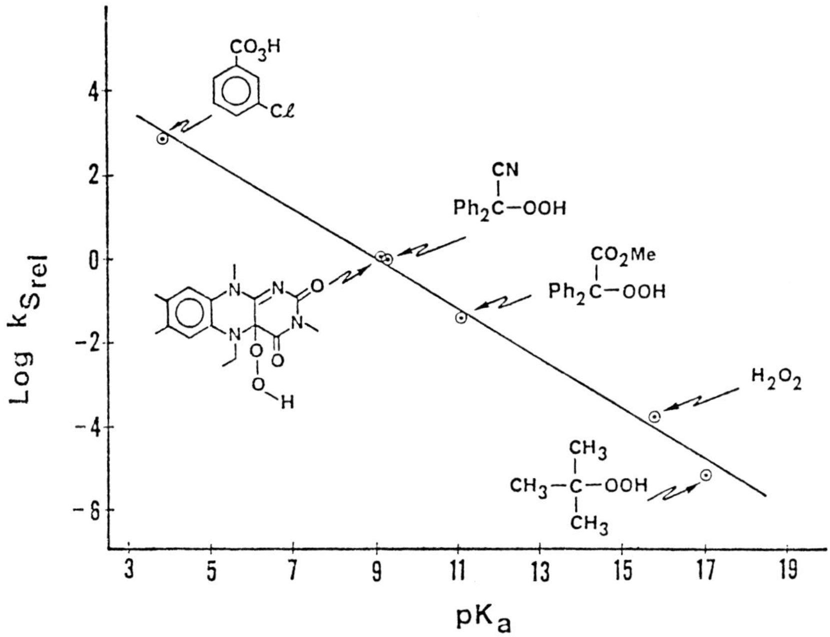

Figure 4. Plot of the log of the second-order rate constants (relative to the rate constant for 4a-FlEtOOH) for sulfoxidation of thioxane (k_S) by the ROOH species vs. the pK_a of ROH species (solvent absolute t-BuOH for rate constants and H_2O for pK_a values, 30 °C).

The slope of the plot is -0.6. Very similar plots may be made for the N-oxidation of N,N-dimethylbenzylamine and the oxidation of I^- by the same series of hydroperoxides and percarboxylic acid. The slope for these plots is also -0.6. The slope of these plots tell us that O-O bond breaking has proceeded to about 60% in the transition states. In an earlier study (9) we found that, for the reaction of secondary-, and tertiary amines and hydroxylamines, a plot of amine pK_a vs. the log of the second order rate constant for attack of these nitrogen bases upon peroxides was of slope 0.2. A general picture emerges for the reaction of equation 2 where bond making has not proceeded as far as bond breaking. Note that in figure 4 the point for the 4a-hydroperoxyflavin fits the correlation line.

So there it is, 4a-hydroperoxyflavin possesses a rather remarkable monooxygen donation potential because the pK_a of 4a-hydroxyflavin (II) is about 9.2. This is low for an alcohol and is attributable to the electronegativity of the nitrogen substituents at the 1, 5 and 10 positions and the carbonyl group at the 4-position of the isoaloxazine ring structure. The pK_a of 9.2 for 4a-hydroxyflavin was calculated employing the $\rho_I\sigma_I$ method of Charton (10). A kinetic pK_{aap} value for the dissociation of the hydroxylic proton has been experimentally determined as 9.9 in a study of the solvolysis of 4a-hydroxy-5-ethyl-4a,5-dihydrolumiflavin (11).

Acknowledgement: The author thanks the National Institutes of Health and the National Science Foundation for continuing support of his laboratory in its studies of flavin chemistry.

References

1. Bruice, T.C. in "Flavins and Flavoproteins" (Massey, V., Williams, C.H., eds.), Elsivier/North-Holland, 1982, ch.44.
2. Bruice, T.C.: Is. J. Chem., (1984).
3. Ebertein, G., Bruice, T.C.: J. Am. Chem. Soc., 105, 6685 (1983).
4. Wessiak, A., Bruice, T.C.: J. Am. Chem. Soc., 105, 4809-4825 (1983).
5. Wessiak, A., Noar, J.B., Bruice, T.C.: Proc. Natl. Acad. Sci. (USA), 81, 332-336 (1984).
6. Entsch, B., Ballou, D.P., Massey, V.: J. Biol. Chem., 251, 2550, 7369 (1976).
7. Wessiak, A., Schoyfer, L.M., Yuan, L.-C, Bruice, T.C., Massey, V.: Proc. Natl. Acad. Sci. (USA), 81, July (1984).

8. Bruice, T.C.: J. Chem. Soc. Chem. Comm., 14 (1983). Bruice, T.C., Noar, J.B., Ball, S.S., Vankataram, U.V.: J. Am. Chem. Soc., 105, 2452 (1983).

9. Ball, S.S., Bruice, T.C.: J. Amer. Chem. Soc., 102, 6498-6503 (1980).

10. Charton, M.: J. Org. Chem., 29, 1222 (1964). See also Fox, J.F., Jencks, W.P.: J. Am. Chem. Soc., 96, 1436 (1974)

11. Venkataram, U.V., Bruice, T.C.: J. Chem. Soc. Chem. Comm. In Press.

PULSE RADIOLYSIS STUDIES ON THE EQUILIBRIA BETWEEN REDUCED AND OXIDIZED FREE FLAVIN SPECIES AND THE EFFECT OF MOLECULAR OXYGEN

Robert F. Anderson

Cancer Research Campaign Gray Laboratory, Mount Vernon Hospital, Northwood, Middlesex HA6 2RN, England

Introduction

Oxidized flavin (F), dihydroflavin (FH_2) and flavosemiquinone radicals ($FH^{\bullet}$) are known to be in equilibrium with a charge-transfer complex ($FH_2 \rightarrow F$) (1,2).

$$FH^{\bullet} + FH^{\bullet} \underset{k_{-1}}{\overset{k_1}{\rightleftharpoons}} FH_2 \rightarrow F \underset{k_{-2}}{\overset{k_2}{\rightleftharpoons}} FH_2 + F$$

Estimates of K_1 (k_1/k_{-1}) and K_2 (k_2/k_{-2}) have been made using the temperature jump technique following mixing of F and FH_2 (3). Pulse radiolysis is a convenient method for producing $FH^{\bullet}$ (4,5) and we have reinvestigated K_1 and K_2 using various initial [$FH^{\bullet}$] and [F], and the perturbation of the equilibria by molecular oxygen.

Results and Discussion

1) Determination of equilibrium constants. $FH^{\bullet}$ were produced within *ca.* 20 μs by pulse radiolysis (55 or 110 Gy in 1 μs) of deaerated solutions (pH 5.9) containing FMN (0.25-10 mmol dm^{-3}) and formate ions (0.1 mol dm^{-3}) (4,5). The bimolecular decay of $FH^{\bullet}$ in solutions of low [F] (0.1 mmol dm^{-3}) proceeds through a dismutation reaction (4,6,7). However, with increasing [F] it is seen that less of the $FH^{\bullet}$ absorption band measured at 560 nm decays and the charge-transfer band measured at 880 nm increases, figure 1. (Absorptions are presented as the product of the radiation chemical yield, G and extinction coefficient, ε relative to the aerated

Flavins and Flavoproteins

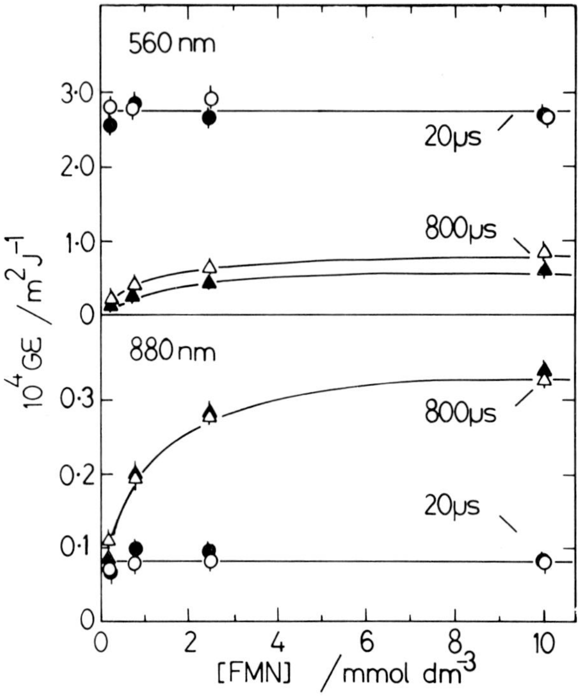

Figure 1. Absorption changes following pulse radiolysis (open symbols 55 Gy, closed 110 Gy) of deaerated solutions (pH 5.9) of FMN.

KSCN dosimeter where $G((SCN)_2^-)$ = 0.29 μmol J^{-1} and ε (472 nm) = 758 m^2 mol^{-1} (8)). At high [F] (⩾ 10 mmol dm^{-3}) a plateau in Gε measured at 800 μs is seen. This is interpreted as k_{-2} [F] >>k_2 which permits K_1 to be studied. The initial $[FH^\bullet]_i$ (at 20 μs) for both 55 Gy and 110 Gy can be calculated using ε(560 nm) $FH^\bullet$ = 510 m^2 mol^{-1} (4,9). (It is calculated that ε(880 nm) $FH^\bullet$ = 15 m^2 mol^{-1} and ε(880 nm) $FH_2 \rightarrow F$ = 152.5 m^2 mol^{-1}). At equilibrium $[FH^\bullet]_i$ = 2 $[FH_2 \rightarrow F]_e$ + $[FH^\bullet]_e$ and assuming ε(560 nm) $FH^\bullet$ >>ε(560 nm) $FH_2 \rightarrow F$ calculations for both doses yield K_1 = 1.3 x 10^5 dm^3 mol^{-1}. Since at I = 0.1 k_1 = 9.6 x 10^8 dm^3 mol^{-1} s^{-1} (7), k_{-1}= 7.4 x 10^3 s^{-1}. Our value for k_{-1} is slightly higher than that previously published (3) whereas k_1 (and K_1) are considerably higher.

At [F] <10 mmol dm^{-3} all species are present at equilibrium. The $[FH^\bullet]_e$ is measured at 560 nm and the $[FH_2 \rightarrow F]_e$ then calculated from measurements

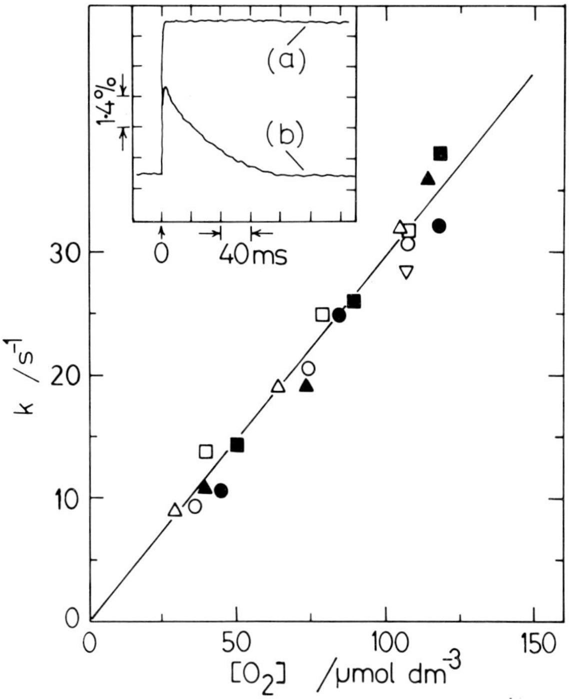

Figure 2. Decay rates of transients observed at △480, ○560, □880 nm following pulse radiolysis (80 Gy, closed symbols 40 Gy) of FMN (0.5 mmol dm^{-3}) in solution (pH 5.9) containing O_2 (▽2.5 mmol dm^{-3}). $[O_2]$ corrected for loss resulting from dose. Insert: Oscilloscope traces (absorbance vs. time) at 880 nm (a) deaerated (b) $[O_2]$ = 120 μmol dm^{-3}.

at 880 nm(FH_2 does not absorb significantly at 560 or 880 nm). The $[FH_2]_e$ is then deduced from $[FH^{\bullet}]_i = [FH^{\bullet}]_e + 2[FH_2{\rightarrow}F]_e + 2[FH_2]_e$. $K_1K_2 = [FH_2][F]/[FH^{\bullet}]^2$, and calculations for $[F] \approx 0.75$ and 2.5 mmol dm^{-3} at both 55 and 110 Gy yield $K_1K_2 = 281 \pm 101$ and hence $K_2 = 2 \pm 1 \times 10^{-3}$ mol dm^{-3}. This value is higher than the previously estimated maximum (3).

2) Effect of added oxygen. Experiments using FMN (0.5 mmol dm^{-3}) and low $[O_2]$ (50-120 μmol dm^{-3}) showed that all long-lived species detected at 480, 560 and 880 nm decayed to F at a rate proportional to $[O_2]$, figure 2. The second order rate constant is 3×10^5 dm^3 mol^{-1} s^{-1}. The observed rate was uninfluenced by (i) the size of the radiation dose (ii) an

increase in [F]. As the rate constant is greater than that for the upper limit of $k(FH^{\cdot} + O_2)$ $<3.8 \times 10^4\ dm^3\ mol^{-1}\ s^{-1}$ (10) (for lumiflavin), we are presumably observing the rate determining step in the oxidation of FH_2, reaction (3).

$$FH_2 + O_2 \longrightarrow FHOOH \qquad (3)$$

$$FHOOH \longrightarrow F + H_2O_2 \qquad (4)$$

Under our experimental conditions reaction (3) proceeds at a rate slower than the establishment of K_1 and K_2 to be followed by the relatively fast breakdown of the FHOOH species ($t_{\frac{1}{2}} \sim 2.5$ ms (11)), reaction (4).

Acknowledgments

This work is financially supported by the Cancer Research Campaign.

References

1. Beinert, H.: J. Amer. Chem. Soc. 78, 5323-5328 (1956).
2. Gibson, Q.H., Massey, V.A., Atherton, N.M.: Biochem. J. 85, 369-383 (1962).
3. Swinehart, J.H.: J. Amer. Chem. Soc. 88, 1056-1058 (1966).
4. Land, E.J., Swallow, A.J.: Biochemistry 8, 2117-2125 (1969).
5. Anderson, R.F.: Ber. Bunsenges. physik. Chem. 80, 969-972 (1976).
6. Holmstrom, B.: Photochem. Photobiol. 3, 97-114 (1964).
7. Anderson, R.F.: Biochim. Biophys. Acta 723, 78-82 (1983).
8. Schuler, R.H., Patterson, L.K., Janata, E.: J. Phys. Chem. 84, 2088-2089 (1980).
9. Anderson, R.F.: Biochim. Biophys. Acta 722, 158-162 (1983).
10. Vaish, S.P., Tollin, G.: Bioenergetics 2, 61-72 (1971).
11. Anderson, R.F.: In Flavins and Flavoproteins, Massey, V., Williams, C.H. (eds.) pp 278-283, Elsevier North Holland, Amsterdam 1982.

EFFECT OF pH ON THE OXIDATION-REDUCTION PROPERTIES OF 8α-IMIDAZOLE FLAVINS

Gary Williamson and Dale E. Edmondson
Dept. of Biochemistry, Emory University, Atlanta, GA 30322 USA

Introduction

Since the original elucidation of the structure of the covalent flavin of succinate dehydrogenase as 8α-N(3)-histidyl FAD (1,2), the number of flavoenzymes known to contain covalently-bound flavin coenzymes has been steadily increasing (3). To date, there is no known rationale to explain why some flavoenzymes require a covalent flavin while others contain normal flavin cofactors bound by non-covalent interactions. As a first step towards gaining an insight into this question, we report here on the effect of imidazole ionization on the redox properties of the flavin ring of 8α-imidazole-flavins and, conversely, the effect of flavin reduction on the ionization properties of the 8α-imidazole ring. The results show a rather strong inter-relation and also show that the ionization of the 8α-imidazole ring may play a role in oxido-reduction reactions involved in catalysis by those enzymes containing a covalent 8α-histidyl flavin.

Results and Discussion

8α-(N)-Imidazolylriboflavin (ImRf) was synthesized by the condensation of imidazole with 8α-bromo-TARF as described previously (1,4). Oxidation-reduction potentials were determined spectrocoulometrically using an apparatus described by Stankovich (5). ^{1}H-NMR studies were performed using a Nicolet 360MHz spectrometer. The structure of ImRf and the designation of the relevant ionization sites is given in Fig. 1.

Flavins and Flavoproteins

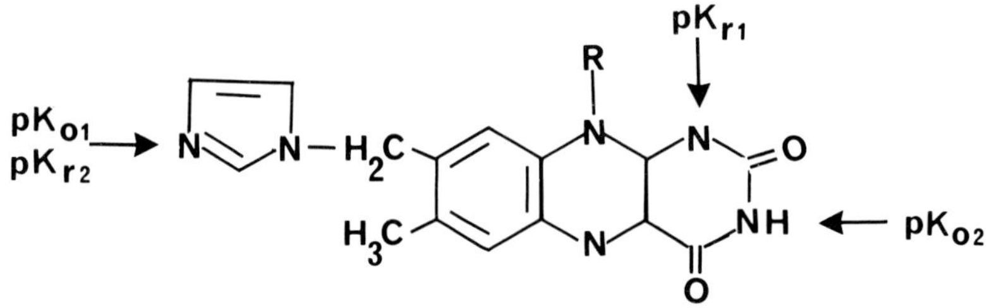

Fig. 1. Structure of 8α-(N)-imidazolylriboflavin. R designates the ribityl side chain. The pK_o and pK_r designations are for those ionizations in the oxidized and hydroquinone forms, respectively. The ring skeleton of the isoalloxazine ring is presented without designation of the redox level.

The effect of flavin reduction on the ^1H resonances of both the imidazole and flavin rings are shown in Fig. 2. Peak assignments are based on those previously reported at 300 MHz for 8α-[N(1) and N(3) histidyl]-riboflavin (6). In addition, decoupling experiments were done to verify previous assignments as well as FlH_2 assignments. The spectra in Fig. 2 show that reduction of the flavin results in significant upfield shifts (1.5 ppm) in the Fl(6)-H and Fl(9)-H resonances and smaller upfield shifts (0.5-0.6 ppm) in the Fl(7)-CH_3 and Fl(8)-CH_2 resonances. The effect of reduction on the imidazole resonances were smaller (0.1-0.2 ppm upfield).

It is well-known that the chemical shift value of the Im(2)-H is sensitive to the protonation of the adjacent N(3) in mono-N-alkylated imidazoles. The pH-dependence of the Im(2)-H chemical shift for either the oxidized or hydroquinone forms of ImRf results in an upfield shift of 1.0 ppm on deprotonation of the Im N(3). As shown in Table 1, oxidized ImRf exhibits an imidazole pK_a of 6.0-6.2 while the pK_a of the hydroquinone form increases to a value of 7.1. Thus, reduction of the electron-deficient isoalloxazine ring results in a pK_a shift of the 8α-imidazole to a value similar to that of free imidazole or 1-methyl imidazole.

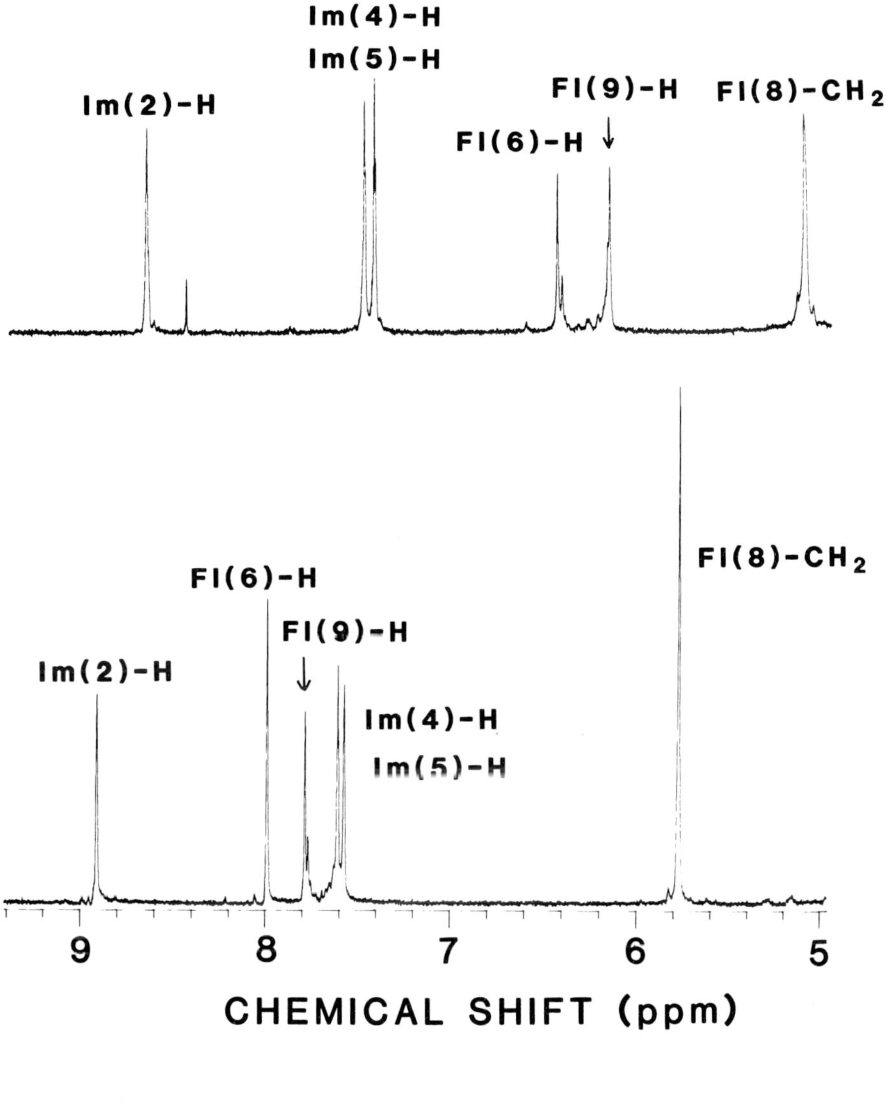

Fig. 2. ^1H NMR spectra at 360MHz on (top) hydroquinone and (bottom) oxidized forms of ImRf in ^2H_2O, pH 2.5. The flavin concentration was 10 mg/ml. The reduced spectrum was obtained after reduction by an excess of dithionite in an anaerobic NMR tube under an Argon atmosphere.

Table 1. Comparison of pK_a Values For the Oxidized and Reduced Forms of 8α-Imidazolyl riboflavin With Those of Riboflavin

Position	Method	ImRf	Riboflavin
Imidazole N(3) (pK_{o1})	^{1}H-NMR	6.2	
	Fluorescence	6.0	
	Absorption	6.0	
	Potentiometric	6.0	
	Redox Potential	6.0	
Flavin N(3) (pK_{o2})	Fluorescence	9.7	
	Absorption	9.5	
	Potentiometric	9.5	
	Redox Potential	9.7	10.0[a]
Flavin N(1) (pK_{r1})	Redox Potential	5.5	6.3[a]
Imidazole N(3) (pK_{r2})	^{1}H-NMR	7.1	
	Redox Potential	7.1	

[a]Values taken from reference 7.

The data in Table 1 show that the imidazole pK_a for oxidized ImRf determined by absorption, fluorescence and potentiometric techniques agrees quite well with that determined by NMR. The value obtained (6.0-6.2), is much higher than that determined previously (pK_a ∿3.8) for 3-methyl-8α-(N)-imidazolyl-lumiflavin (2).

To determine whether the ionization of the 8α-imidazole ring could influence the redox potential of the flavin ring, the effect of pH on E_m was determined for ImRf (Fig. 3). E_m values were determined at each pH from Nernst plots of Em *versus* Log OX/RED (Figure 3, *inset*). Both oxidative and reductive titrations were in excellent agreement. Error analysis showed the precision to be better than 0.8mV and slopes ranging from 34 to 38mV (depending on pH). A computer fit of the data showed the imidazole pK_a of 6.0 for the oxidized form to increase to 7.1 on flavin reduction (Fig. 3 and Table 1). The pK_a of the N(1) position of the hydroquinone was found to be

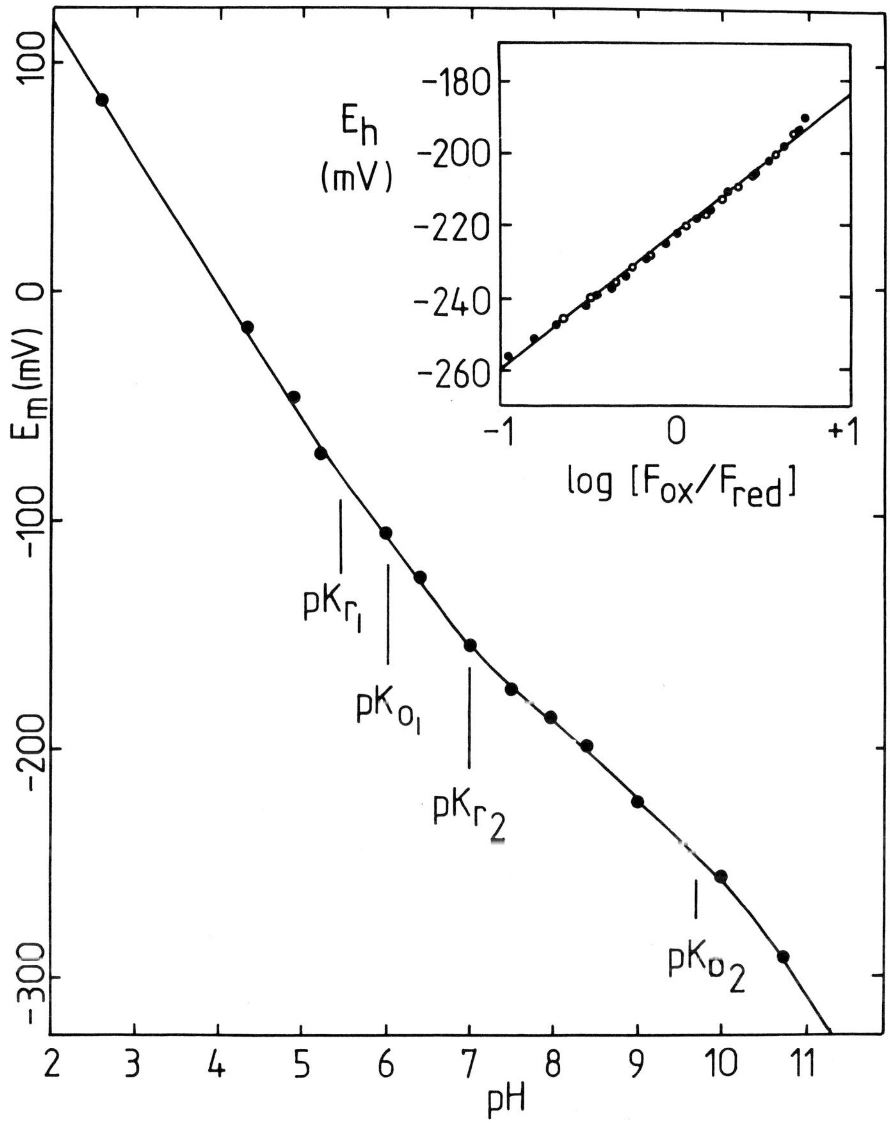

Fig. 3. Effect of pH on the midpoint potentials of ImRf. The closed circles are the experimental data points while the solid line is a computer-simulation using the indicated pKa values (see Table 1). The inset is a typical Nernst plot of the spectrocoulometric titration data in which the closed circles represent the reductive titration and the open circles represent the oxidative titration.

5.5 which is 0.8 units lower than that determined for riboflavin (7). The pK_a of the N(3) position of the oxidized ImRf was found to be 9.7, in good agreement with the values determined spectrally and potentiometrically (Table 1) and only slightly less than the value of 10.0 determined for riboflavin (6). Similar experiments with 8α-[N-methyl-imidazolyl]-tetra-o-acetyl-riboflavin showed only pK_a values corresponding to the Fl N(1) ionization of the hydroquinone and the Fl N(3) ionization of the oxidized form (data not shown).

The results presented above show that the ionization of the imidazole ring of ImRf is a) altered on flavin reduction and b) is observable in redox potential measurements. Thus, future experiments on the redox and catalytic properties of flavoenzymes containing 8α-histidyl flavins must take into account the ionic form of the 8α-histidyl substituent as well as the flavin ring in the interpretation of pH-dependent experiments.

This work was supported by NIH grant GM-29433.

References

1. Walker, W.H., Singer, T.P., Ghisla, S., Hemmerich, P.: Eur. J. Biochem. 26, 279-289 (1972).
2. Salach, J., Walker, W.H., Singer, T.P., Ehrenberg, A., Hemmerich, P., Ghisla, S., Hartmann, U.: Eur. J. Biochem. 26, 267-278 (1972).
3. Merrill, A.H., Lambeth, J.D., Edmondson, D.E., McCormick, D.B.: Ann. Rev. Nutr. 1, 281-317 (1981).
4. Singer, T.P., Edmondson, D.E.: Meth. Enzymol. 66, 253-264 (1980).
5. Stankovich, M.T.: Anal. Biochem. 109, 295-308 (1980).
6. Edmondson, D.E., Kenney, W.C., Singer, T.P.: Biochemistry 15, 2937-2945 (1976).
7. Draper, R.D., Ingraham, L.L.: Arch. Biochem. Biophys. 125, 802-808 (1968).

STUDIES OF INTERMEDIATES IN REACTIONS OF FLAVINS AND SULPHYDRYL COMPOUNDS

P.S. Surdhar and D.A. Armstrong

Chemistry Department
University of Calgary
Calgary Alberta
T2N 1N4

Introduction

Glutathione reductase, lipoamide dehydrogenase and thioredoxin reductase are three enzymes which possess protein-bound disulphides in the active site region (1). The two-electron-transfer pathways for the direction of substrate oxidation can be described as in reaction [1]:

$$\text{Substrate} \xrightarrow{1} \text{S-S} \xrightarrow{2} \text{Fl} \xrightarrow{3} \text{NAD}^+ \qquad [1]$$

Here Fl represents a strongly bound flavin molecule, S-S the protein disulphide, and NAD^+ the oxidised nicotinamide co-factor. Step 1 leaves the enzyme disulphide doubly reduced (1), while step 2 transfers the electrons to the flavin. The structures existing before and after step 2 are thus taken to be I and III below. Structure II is one

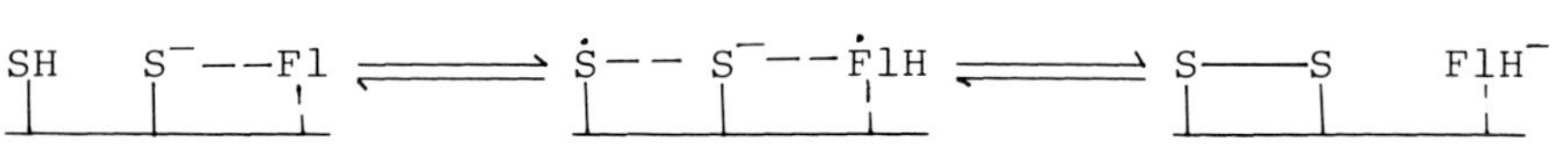 [2]

I II III

type of short-lived intermediate, which should be considered in the overall mechanism. Complex formation between thiolate and •FlH would contribute to the stabilisation of II, and an earlier spectroscopic study yielded evidence for that at pH 9

Flavins and Flavoproteins

and 10 (2). This paper is concerned with spectra obtained at pH 7 by the method of pulse radiolysis as described earlier (2-5). Flavin radicals (·FlH) were produced quantitatively by reduction of the flavins, flavin adenine dinucleotide (FAD) and lumiflavin (LFl), with $\cdot CO_2^-$ radicals or disulphide anion radicals ($D\dot{S}_2^-$) formed from oxidised or reduced dithiothreitol (DS_2 or $D(SH)_2$). The reactions were carried out in nitrous oxide or argon purged buffered solutions, containing known concentrations of $D(SH)_2$ or β-mecaptoethanol (RSH).

Results and Discussion

Spectra of the flavin radicals of LFl and FAD from present and earlier work (4, 5) are shown by the solid lines in Figure 1 (a) and (b) respectively. The $D\dot{S}_2^-$ radical produced the same spectrum as $\cdot CO_2^-$ under those conditions (circles in Figure 1a). As shown by the dashed lines, in the presence of 10-20 mM $D(SH)_2$ the absorbance in the 350 to 500 nm region was enhanced. The spectral changes, which are similar to those at pH 9 and 10 (2), are best explained by complex formation:

$$\cdot FlH + RSH \rightleftharpoons \cdot FlH - - SR^- + H^+ \quad [3]$$

The product of reaction [3] is known to be converted to FlH^- by further reactions (6).

The squares in Figure 1b show the spectrum of ·FlH in the presence of 10^{-4}M RSH. As reported earlier, for these conditions there was no observable change in the radical spectrum (2). More extensive work has shown that the concentration of sulphydryl, at which perturbation of the ·FlH spectrum occurs, is dependent on the nature of the flavin and the RSH molecule. The radical spectrum of LFl in the presence of 10^{-2}M RSH at pH 7 is demonstrated by the solid line in

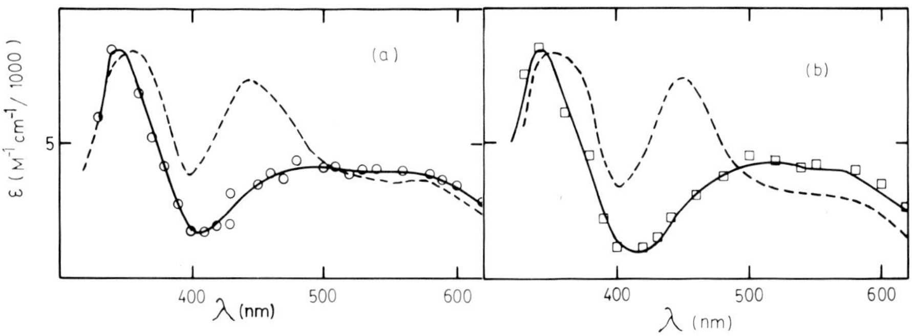

Figure 1. Absorption spectra at pH 7 of flavin radicals of FAD (a) and LFl (b) produced in N_2O-saturated formate solutions in the absence of sulphydryls by $\cdot CO_2^-$ (——), $D\dot{S}_2^-$ (O), and by $\cdot CO_2^-$ in the presence of 10^{-4}M RSH (□); and by $D\dot{S}_2^-$ in the presence of 10-20 mM $D(SH)_2$ (- - -).

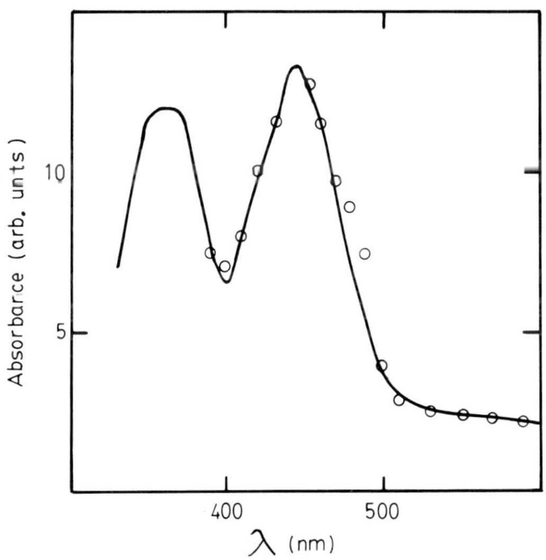

Figure 2. LFl radical spectrum 80 μs after formation in N_2O-saturated 49 μM LFl, 10^{-3}M $D(SH)_2$ and 10^{-2}M RSH at pH 7 normalised to circles at 450 nm (——); Spectrum from ref. (7) of intermediates 12.5 ms after mixing 57 μM lipoamide dehydrogenase with 10 mM NHDH (O).

Figure 2. The circles depict the spectrum of an intermediate observed in the thermal reduction of lipoamide dehydrogenase in a stopped flow experiment of Matthews et al (7) normalised to fit the present absorption curve at 450 nm. Both curves exhibit maxima near 450 nm and a flat absorbance in the range 500 to 600 nm. However, more detailed studies will be required to determine whether the species in the two systems are in fact similar.

References

1. Williams, C.H., Jr.: Enzyme, 13, 89 (1976); Wilkinson, K.D., Williams, C.H., Jr.: J. Biol. Chem. 254, 852 (1976); Arscott, L.D., Thorpe, C., Williams, C.H., Jr.: Biochemistry 20, 1514 (1981).
2. Surdhar, P.S., Ahmad, R., Armstrong, D.A.: Can J. Chem. 62, 580 (1984).
3. Elliot, A.J., Wilkinson, F., Armstrong, D.A.: Int. J. Radiat. Biol. 38, 1 (1980).
4. Elliot, A.J., Ahmad, R., McIntosh, L., Burke, D., Stevenson, K.J., Armstrong, D.A.: Flavins and Flavoproteins. Proceedings of the Seventh Symposium, Elsevier, New York. Amsterdam. Oxford 1981.
5. Ahmad, R., Wu, Z., Armstrong, D.A.: Biochemistry 22, 1806 (1983).
6. Ahmad, R., Armstrong, D.A.: Can. J. Chem. 62, 171 (1984).
7. Matthews, R.D., Ballou, D.P., Williams, C.H. Jr.: J. Biol. Chem. 254, 4974 (1979).

A KINETIC STUDY ON THE ACID-CATALYZED PHOSPHATE MIGRATION IN RIBOFLAVIN PHOSPHATES

Peter Nielsen, Peter Rauschenbach, and Adelbert Bacher
Lehrstuhl für Organische Chemie und Biochemie der Technischen Universität München, 8046 Garching, German Federal Republic

It has been recognized for some time that synthetic preparations of riboflavin 5'-phosphate (5'-FMN) contain substantial amounts of other riboflavin phosphates (for review see Ref. 1). Highly pure 5'-FMN could only be obtained by affinity chromatography on immobilized apoflavodoxin (2). A more detailed study of other phosphoric esters of riboflavin and their interconversion was hampered by the lack of appropriate analytical methods. Only the acid-catalyzed conversion of 5'-FMN to 4'-FMN had been studied to some extent (1).

Results

The various monophosphates and diphosphates of riboflavin can be conveniently separated by reverse-phase HPLC (3). A typical separation of a commercial FMN sample is shown in Fig. 1. By combined application of conventional ion exchange chromatography, partial hydrolysis with phosphatase, and preparative reverse-phase HPLC, we could prepare samples of all four monophosphates and five diphosphates of riboflavin in pure form. Their structures were unequivocally established by chemical methods. 2'-FMN occurs only in trace amounts in commmercial FMN. However, the compound could be obtained by acid-catalyzed isomerization of 3'-FMN.

We have performed a detailed kinetic study of the phosphate

Flavins and Flavoproteins

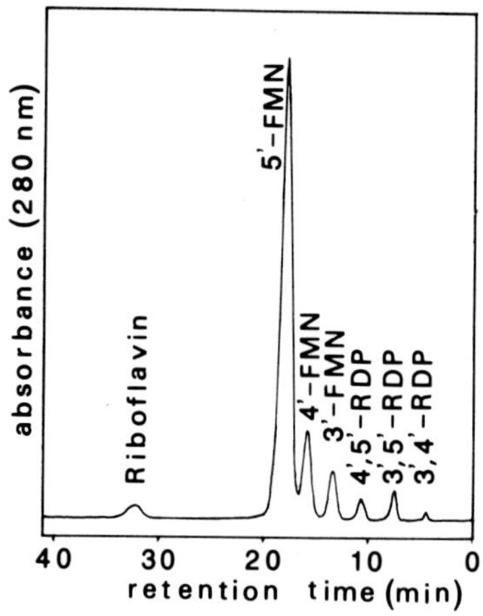

Fig.1: Analytical reverse-phase HPLC of commercial FMN. Column, Nucleosil 10 C_{18}; eluant, 0.1 M ammonium formate/ 0.1 M formic acid, 17 % methanol. RDP, riboflavin diphosphate.

migration in the riboflavin phosphates. Each of the four pure riboflavin monophosphates (i.e. 5'-FMN, 4'-FMN, 3'-FMN and 2'-FMN) was incubated under a variety of conditions (pH, ionic strength, temperature) and the isomerization reaction was followed by reverse-phase HPLC. Fig. 2 shows the concentration profile for isomerization of 5'-FMN and 3'-FMN at 50°C and pH 1. We could subsequently show that all experimental data can be described by the following sequence of first order reactions:

$$\text{5'-FMN} \underset{k_{-1}}{\overset{k_1}{\rightleftharpoons}} \text{4'-FMN} \underset{k_{-2}}{\overset{k_2}{\rightleftharpoons}} \text{3'-FMN} \underset{k_{-3}}{\overset{k_3}{\rightleftharpoons}} \text{2'-FMN}$$

This kinetic system is characterized by a set of four differential equations. The analytical integration leads to algorithms which allow the calculation of theoretical concentra-

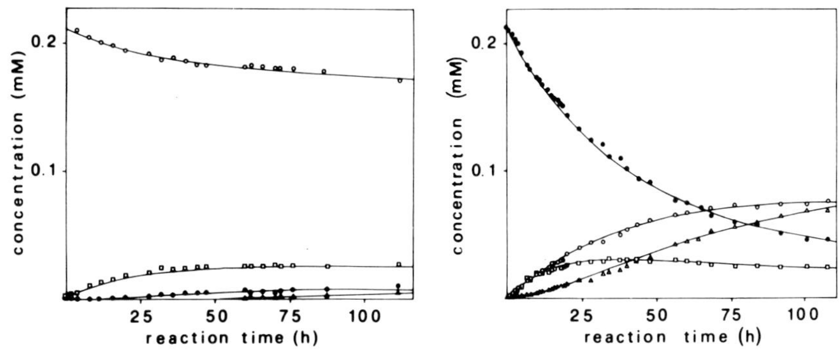

Fig. 2: Isomerization of 5'-FMN (right) and 3'-FMN (left) at pH 1.0 and 50°C. Closed lines, were calculated with the parameters shown in Table 1; symbols, experimental data: O 5'-FMN, □ 4'-FMN, ● 3'-FMN, Δ 2'-FMN.

tion curves. In order to determine the velocity constants, we developed a computer program which fits the theoretical curves with the experimental data (Fig 2). This procedure yields the kinetic constants shown in Table 1 which are in excellent agreement with the experimental data.

Table 1: Velocity constants for the acid-catalyzed isomerization of riboflavin phosphates.
* IS, ionic strenght.

reaction conditions			velocity constants (10^{-6} s^{-1})					
pH	IS*	T(°C)	k_1	k_{-1}	k_2	k_{-2}	k_3	k_{-3}
1.0	0.1	50	1.60	9.41	2.21	3.06	2.92	1.55
		60	5.53	27.9	6.34	9.02	8.74	4.68
		70	13.9	69.4	19.1	23.2	17.5	8.53
		80	33.3	152.7	39.5	52.7	44.7	25.3
0.3	0.5	50	7.02	48.3	19.9	23.7	17.6	8.23
0	1.0	50	13.4	86.1	38.7	48.2	32.0	17.8
2.0	0.01	80	1.66	6.09	4.30	2.04	1.73	2.31
1.0	1.0	50	1.76	10.3	6.10	4.14	3.31	2.31
1.0	3.0	50	2.99	15.1	4.24	5.81	4.86	3.15

At thermodynamic equilibrium, the reaction mixture contains about 65 % 5'-FMN, 11 % 4'-FMN, 8 % 3'-FMN and 15 % 2'-FMN (pH 1.0, 50°C).

The activation enthalpies (pH 1.0, 65°C) are in the range of 80-90 kJ/Mol, the activation entropies in the range of -70 to -100 J K^{-1} Mol^{-1}.

In summary, our data show that the acid-catalyzed phosphate migration in riboflavin phosphates is a relatively fast reaction in the pH range < 2 and at temperatures above room temperature.

A similar sequence of first order reactions was found for the acid-catalyzed phosphate migration of riboflavin diphosphates:

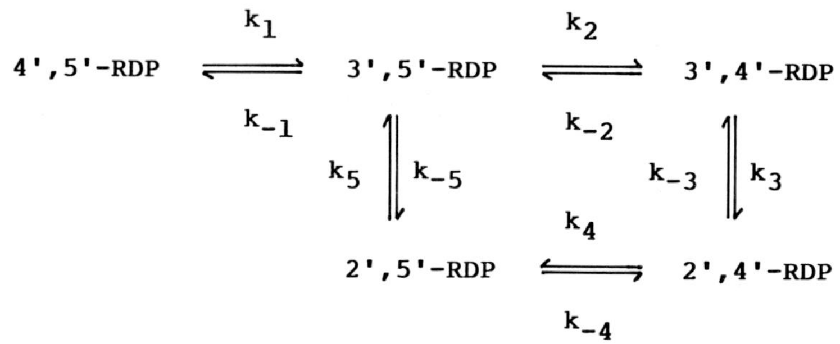

We thank the Deutsche Forschungsgemeinschaft and the Fonds der Chemischen Industrie for support.

References

1. Scola-Nagelschneider, G., Hemmerich, P.: Eur. J. Biochem. 66, 567-577 (1976).

2. Mayhew, S.G., Strating, M.J.J.: Eur. J. Biochem. 59, 539-544 (1975).

3. Nielsen, P., Rauschenbach, P., Bacher, A.: Anal. Biochem. 130, 359-368 (1983).

CHEMICAL STRUCTURE OF NEKOFLAVIN

Kunio Matsui, Sabu Kasai
Division of Biology, Research Institute for Atomic Energy,
Osaka City University
Sumiyoshi-ku, Osaka 558, Japan

Introduction

Choroids or retinae of some animals contain high concentration of unphosphorylated riboflavin (RF)(1,2,3,4). One of the authors found an unidentified flavin, nekoflavin (NF)(5), in addition to a fairly high concentration of RF in cat's choroids (6,7). The compound was found to be a flavin and to have very similar UV spectra to RF, but to differ in Rf value from any of the known flavins at that time, i.e., RF, FMN, FAD, and RF glucoside. The Rf values of the periodate oxidation product of NF were also different from those of RF. It was named "nekoflavin" from the Japanese name of cat "neko". The content of NF in the choroids was very small, and was estimated as about 0.2-0.3 μg a piece of the choroid from the absorbance. Finally, we could obtain about 8 mg of NF acetate from about 25,000 of cat eye. From the molecular weight of the acetate, NF was assumed as either 7α- or 8α-hydroxyriboflavin (7αHORF, 8αHORF). The PMR spectrum of NF acetate did not coincide with that of 7αHORF acetate. Then, NF was assumed as 8αHORF, which was found in human urine by Ohkawa *et al.* (8).

Results and Discussion

NF was extracted from cat choroids and purified by column

chromatographies as described previously (9), and finally by chromatography on Biogel-P2. The NF preparation was acetylated with acetic anhydride and pyridine by the usual method, and the acetate was recrystallized from ethanol. The molecular weight of the acetate was estimated as 602 by FD mass spectrometry. The difference of molecular weight, 58, from that of RF acetate, 544, suggested a composition difference, $C_2H_2O_2$, as one of the most probable, i.e., NF acetate may have an additional group, CH_3-CO-O-, to RF acetate. The PMR spectrum of NF acetate suggested the presence of CH_3 and CH_2 group linked to an aromatic ring (see Table I). Therefore, NF was assumed as either 7αHORF or 8αHORF.

7αHORF acetate

7αHORF was prepared from 4-nitro-o-xylene according to Scheme 1. The product was acetylate as described above, and the acetate was recrystallized from ethanol.

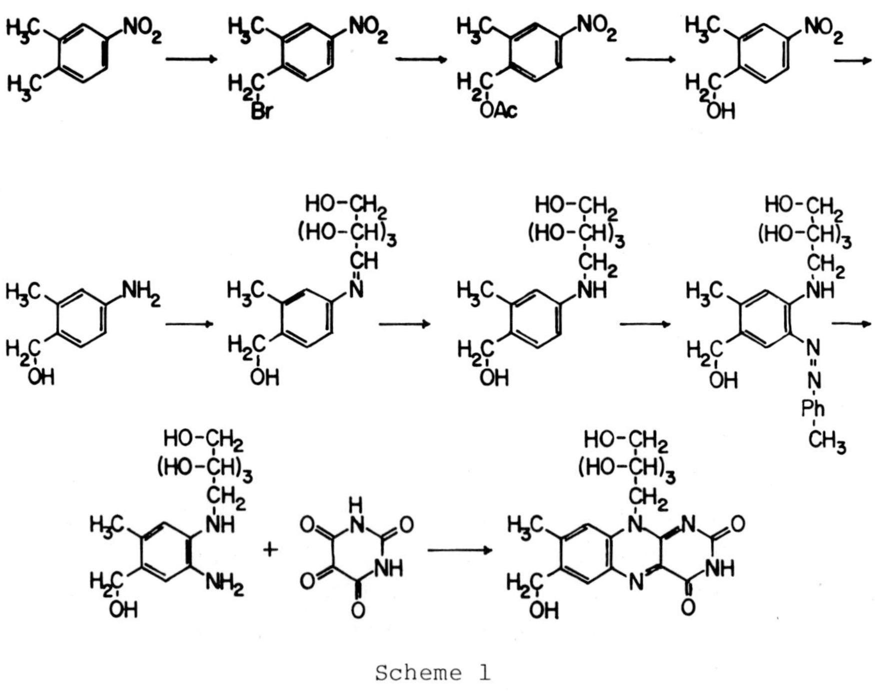

Scheme 1

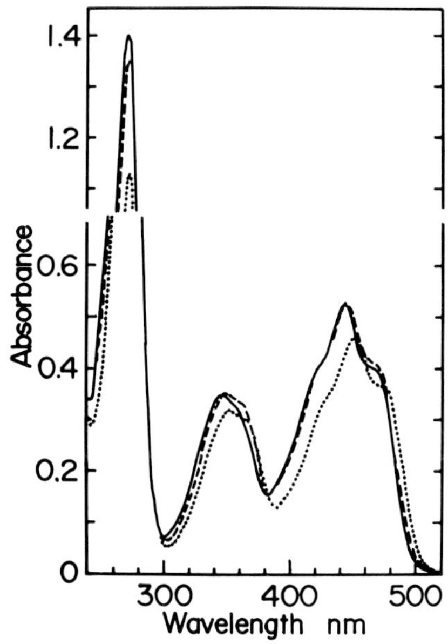

Fig. 1. UV Absorption spectra of flavin acetates in $CHCl_3$.
——, NF acetate; ----, 7αHORF acetate; ·······, RF acetate.

Comparison of the acetates of NF, 7αHORF, and RF

The Rf values of the 3 flavin acetates were identical (0.63) in a TLC using silica gel plate and the solvent $CHCl_3$-methanol (20 : 1 v/v). But that of 7αHORF acetate (0.68) differed from that of NF and RF acetates (both 0.75) in a TLC using solvent isoamyl alcohol-methyl ethyl ketone-acetic acid-water (40 : 40 : 7 : 13 v/v/v/v). The UV absorption spectra of the three acetates are shown in Fig. 1. The spectrum of NF acetate is very similar to that of 7αHORF acetate, and a little different from that of RF acetate. Their absorption peaks are in 272, 345, 443; 272, 348, 444, and 272, 351, 450nm, respectively. The most definitive differentiation is PMR spectrum (Table I). NF and 7αHORF acetates have each CH_2 and CH_3 group linked to aromatic nucleus, and δ-values of each CH_2 group are different. They have also each extra acetyl-CH_3 group in addition to four acetyl-CH_3 linked to ribityl chain. These facts clearly

Table I. Chemical shift of protons of flavin acetates. Solvent, $CDCl_3$

Acetate	δ-Value ppm										
	CH_3-CO-						CH_3	CH_2	aryl-H		NH
NF	1.78		2.08	2.19	2.21	2.28	2.60	5.22	7.58	8.22	8.65
7αHORF	1.75	1.85	2.06		2.20	2.27	2.59	4.54	7.54	8.17	8.85
RF	1.73		2.05		2.21	2.27	2.42 2.55		7.54	7.89	9.57

differentiate NF from 7αHORF, and strongly support the assumption that NF is 8αHORF.

References

1. von Euler, H., Adler, E.: Hoppe-Seyler's Z. Physiol. Chem. 228, 1-12 (1934).
2. Manabe, M.: Acta Soc. Ophthalmol. Jap. 44, 1149-1170 (1940).
3. Yagi, K.: Jap. J. Physiol. 2, 27-33 (1951).
4. Pirie, A.: Nature 183, 985-986 (1959).
5. Matsui, K.: J. Biochem. Tokyo 57, 201-206 (1965).
6. Elliot, J. H., Futterman, S.: Arch. Ophthalmol. 70, 531-534 (1963).
7. Matsui, K., Yanagiya, K.: J. Nutr. Sci. Vitaminol. 19, 187-191 (1973).
8. Ohkawa, H., Ohishi, N., Yagi, K.: J. Biol. Chem. 258, 5623-5628 (1983).
9. Matsui, K.: in Methods in Enzymology Vol. 18B (McCormick, D. B., Wright, L. D. eds.) pp. 433-437 Academic Press, New York & London 1971.

Flavoprotein Structure

BINDING MODE AND ACTION OF FAD IN GLUTATHIONE REDUCTASE

G.E. Schulz

Institut für Organische Chemie und Biochemie der
Universität Freiburg
7800 Freiburg, Albertstr. 21, F.R.Germany

Summary

The contacts between the prosthetic group FAD and the apoprotein are described. The conformation of bound FAD is compared with other protein-bound dinucleotides. The interaction between FAD and the substrates NADPH and GSSG are discussed as well as the present view on the catalytic mechanism. Comparison with p-hydroxybenzoate hydroxylase indicate an electronic intermediate of the isoalloxazine. General features of FAD-binding domains are emerging, which show an evolutionary connection between many FAD-binding enzymes.

Introduction

The catalytic reaction of glutathione reductase (EC 1.6.4.2) can be subdivided into 2 steps

$$H^+ + NADPH + E \rightleftharpoons NADP^+ + EH_2$$

$$EH_2 + GSSG \rightleftharpoons 2\ GSH$$

where the intermediate EH_2 is the stable 2-electron reduced form of the enzyme (1). The enzyme function is to keep the cellular concentration of the reduced form of glutathione (GSH) high and that of the oxidized form (GSSG) low (2). Glutathione reductase is a member of the family of disulfide reductases, which also includes lipoamide dehydrogenase and thioredoxin reductase (3). Moreover, the enzyme shows structural similarities with other FAD-binding enzymes (4,5,6,7). Human erythrocyte glutathione reductase has been crystallized

Flavins and Flavoproteins

(8) and structurally elucidated by X-ray methods (9) and amino acid sequence analysis (10). The crystalline enzyme is catalytically competent. Therefore, all structural data obtained with these crystals are relevant for the enzyme´s behaviour in solution.

The placement of FAD in the enzyme

Glutathione reductase is a dimer with M_r = 104800 containing 1 molecule FAD per subunit. Each subunit can be subdivided

Fig.1 FAD as positioned in the chain fold of glutathione reductase; a) FAD-binding domain; b) Central domain.

into 4 domains and a flexible N-terminal arm of 18 residues. Domains are defined as continuous segments of the polypeptide chain with numerous internal and fewer external contacts (11). FAD binds to the domain of residues 19 through 157, which contains a very long excursion tying both subunits together (Fig. 1a). Apart from the excursion the chain fold of this domain may turn out as a general FAD-binding fold, which is structurally related to other nucleotide-binding folds (12). Moreover, FAD contacts the central-domain (residues 294 through 364) at a number of positions as shown in Fig.1b. This domain has no structural relationship to other nucleotide-binding proteins.

The polar contacts between FAD and the polypeptide chains are shown in Fig.2a. The majority of the contacts are from residues of domain 19 through 157, which is consequently named FAD-binding domain. The flavin portion of FAD is surrounded by residues from all 4 domains of one subunit as well as from the C-terminal Interface-domain (residues 365 through 478) of the other subunit. Thus, as many as 5 domains contribute to the catalytic center of the enzyme. The FMN moiety of FAD is completely engulfed by the polypeptide, whereas the AMP moiety of FAD has some contact with the solvent. Such a binding mode is only possible if the protein opens up before or during the binding process. Appropriate entrance slits for FAD can be produced by a movement of domains relative to each other.

A rather intriguing feature of glutathione reductase is the absence of any positively charged residue around the pyrophosphate portion of FAD, which is to be expected to carry 2 negative charges. The only stabilization of these charges is provided by the dipole field of helix 29 through 42. In the light of this problem it seems likely that one of the isolated globular features of the electron density map which have been interpreted as water molecules, is a positively charged ion instead. This is presumably the case for the larger globule W4 located between the phosphate and carbonyl groups as shown in Fig.2a. Here a Mg^{2+} ion is the candidate of choice. With FAD carrying Mg^{2+} the apo-protein has a good chance of being stable on its own, because there are no uncompensated positive charges at the FAD-binding site repelling each other. This feature is of physiological significance, as intermittent riboflavin deficiency in the diet would not cause the enzyme to be destroyed but put it on standby.

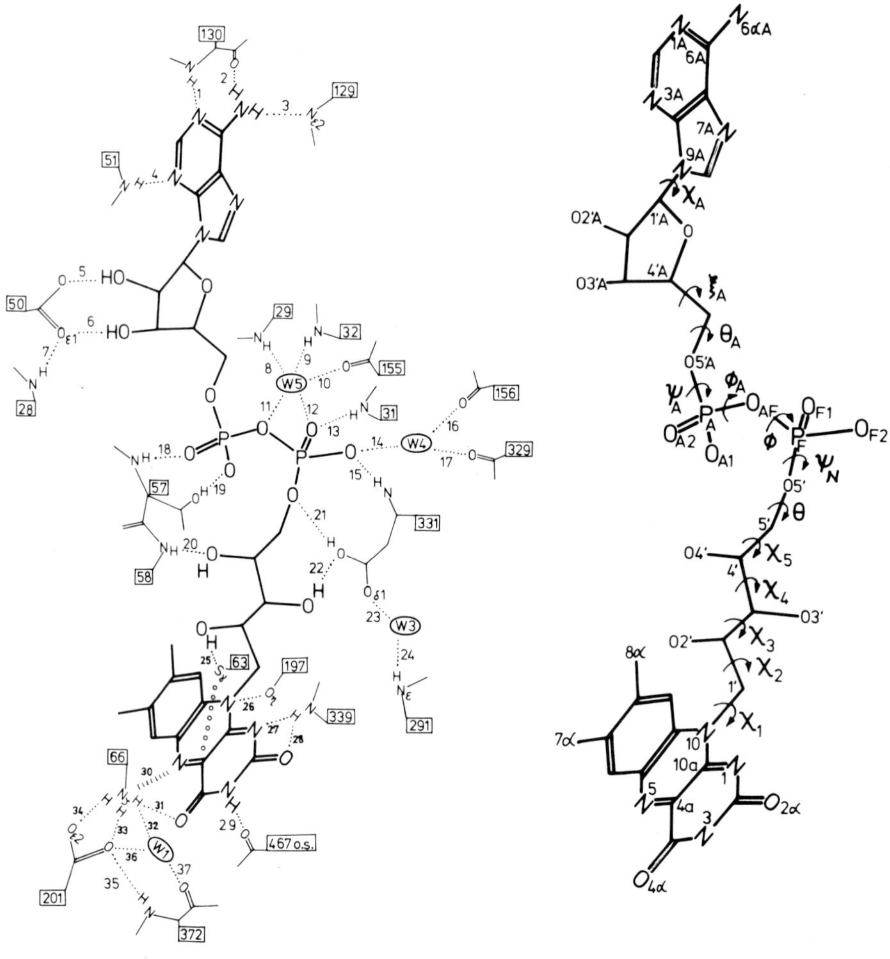

a) b)

Fig.2 FAD as bound glutathione reductase;
a) polar contacts; b) dihedral angles.

The conformation of FAD

FAD is rigidly bound to the apo-protein as demonstrated by its clear outlines in the electron density map. Therefore, it can be related to other protein-bound dinucleotides. A comparison with p-hydroxybenzoate hydroxylase, the other structurally known FAD containing protein, shows that all torsion angles are either in identical or neighbouring ranges (7). In both enzymes, the FMN moieties of FAD are very extended, the assumed conformations are close to all-trans. In contrast, flavodoxin which contains the mononucleotide FMN has a ribitol with a quite different, more curled conformation. The AMP moiety of FAD can be compared with eight other structures. The observed C-2´endo pucker (Table 1) is rather exceptional. The only other known example occurs in glyceraldehyde-3-phosphate dihydrogenase. The C-2´endo pucker tends to shorten the nucleotide. It is more commonly observed in the nicotinamide riboses of NAD(P) and in mono- and polynucleotides.

Table 1 Ribose puckers and distances in protein-bound dinucleotides.

	ribose pucker in adenosine	nicotinamide	distance C4´-C4´ (%)
M_4 lactate dehydrogenase (13)	C3´-endo	C3´-endo	71
H_4 lactate dehydrogenase (13)	3´	2´	85
malate dehydrogenase (14)	3´	2´	83
alcohol dehydrogenase (15)	3´	3´	95
glyceraldehyde-3-P dehydrogenase "red" subunit (16)	2´	3´	84
glyceraldehyde-3-P dehydrogenase "green" subunit (16)	3´	2´	91
6-phosphogluconate dehydrogenase (17)	3´	3´	90
dihydrofolate reductase (18)	3´	2´	94
glutathione reductase, NADPH (19)	3´	3´	92
glutathione reductase, FAD (19)	2´		85
p-hydroxybenzoate hydroxylase FAD (7)			97

FAD is bound in a very elongated manner as indicated by the observation that the distance between the C-4´ atoms of the sugars is 85% of its maximal possible length. As shown in Table 1 this is a quite general observation: elongated dinucleotides form larger contact areas with the apo-protein than compact dinucleotides. Therefore, they exert more influence. The bound FAD was also compared at the more detailled level of dihedral angles (Fig.2b). The adenine orientation is in the anti-range of χ_A as observed usually. The other torsion angles, except ψ_A (-15°) and ψ_N (-111°), are also in their commonly observed ranges shown in Fig.3. The ψ_A, ψ_N angles in glutathione reductase cause a distinct kink at the pyrophosphate.

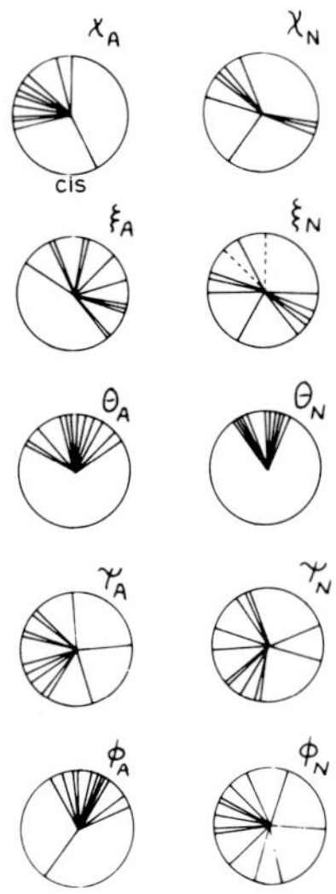

Fig.3 Observed dihedral angles of protein-bound NAD(P) and FAD; the angles are defined clockwise with zero in cis position.

The interaction between FAD and substrates

During catalysis FAD is engaged in a direct contact between its isoalloxazine and the nicotinamide of NADPH. However, there is no direct contact between FAD and the other substrate, GSSG. In the second part of the catalytic reaction formulated above, FAD operates on GSSG via the redox-active disulfide bridge Cys-58 : Cys-63 of the apo-protein, which touches the isoalloxazine (20).

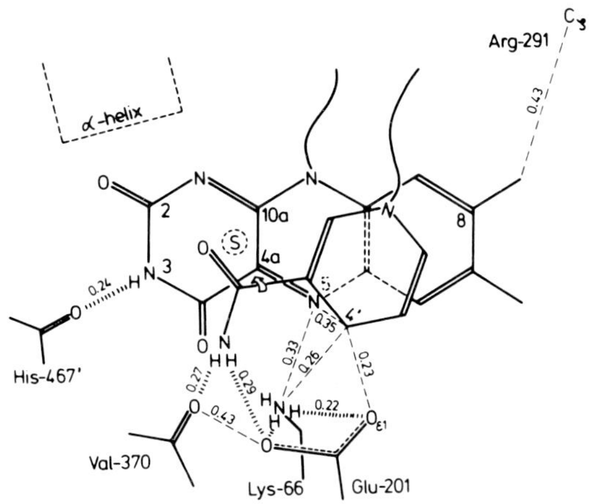

Fig.4 The complex between FAD and NADPH in glutathione reductase.

The interaction between nicotinamide and isoalloxazine is shown in Fig.4. The isoalloxazine ring system is planar and assumed to lie in the paper plane. Also in the plane is the carbonyl of His-467´, and the α-helix. The sulfur of Cys-63 is located 0.34 nm below the paper plane. The nicotinamide ring lies 0.34 above the isoalloxazine, its carboxamid group is twisted as marked. The N-atom is hydrogen-bonded to the polypeptide environment, the oxygen is pushed onto the pteridin ring. Since hydrogen atoms of proteins cannot be visualized in an X-ray analysis, none of the given H-atoms has a direct experimental base.

The transferred H_S-atom of nicotinamide (21) is close to N5 of isoalloxazine and N_ζ of Lys-66. Since N5 is tightly covered by the alkyl part of Lys-66, it is rather unlikely that

H_s is bound to N5 in the flavin plane. Moreover, there is no other reasonable pathway for H_s from the re- to the si-face of isoalloxazine. As a consequence, one has to conclude that H_s does not proceed to the redoxactive disulfide but eventually remains as a proton in the nicotinamide-binding pocket. The nature and the sequence of the reduction equivalents transferred from NADPH to FAD and from FAD to the redoxactive disulfide cannot directly be determined by X-ray structure analysis.

In the first part of the reaction the enzyme picks up 2 protons when proceeding from E to EH_2 (22). This happens most likely at the glutathione binding side of isoalloxazine. Presumably, on breaking the protein disulfide bridge, the distal sulfur of Cys-58 is protonated together with His-467´, which is tightly hold by the carboxyl group of Glu-472´. As indicated in Fig.5, it is likely that both these protons are donated to the sulfurs of glutathione. First, the proton of His-467´is

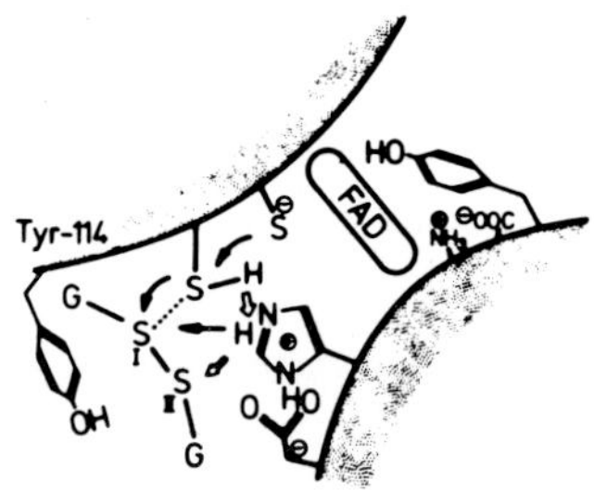

Fig.5 A likely reaction scheme for the oxidation of EH_2

released together with glutathione-II. Then, the proton of Cys-58 is donated via His-467´to the leaving glutathione-I.

An isoalloxazine intermediate revealed by convergent evolution

When comparing the environments of isoalloxazine in p-hydroxybenzoate hydroxylase and glutathione reductase there emerges an exciting feature, a common α-helix pointing with its N-terminal end at the N1, O2α-region, as shown in Fig.6.

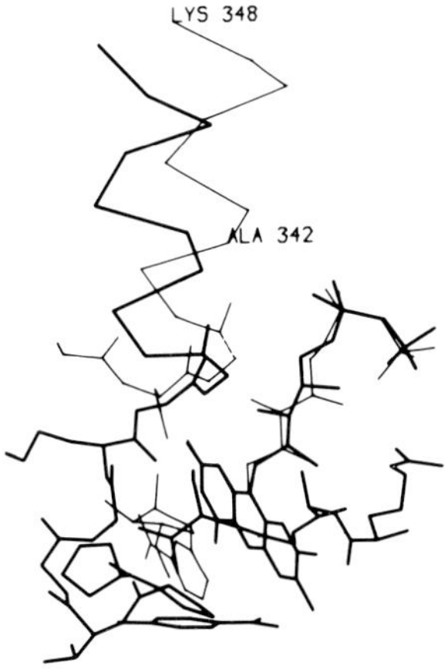

Fig.6 Superposition of the isoalloxazines of p-hydroxybenzoate hydroxylase (thick lines) and glutathione reductase

These α-helixes are located in clearly non-homologuous parts of the polypeptide chains and thus cannot be related by divergent but only by convergent evolution. The dipole fields of these helixes stabilize a negative charge (7) around N1, O2α. This would suggest a transition state of the isoalloxazine which is negatively charged in this region, as for instance a semiquinone anion. In both enzymes the si-face of isoalloxazine is covered by the polypeptide chain and NADPH approaches from the re-face. However, because of symmetry this should not be significant. Either it is a consequence of divergent evolution or it is just chance.

FAD-binding domain structures

A superposition of the FAD-binding domains of glutathione reductase and p-hydroxybenzoate hydroxylase is given in Fig.7.

These enzymes have no obvious relationship and there is no detectable amino acid sequence homology. Still, there are 2 domains which can be superimposed with a residual average devia-

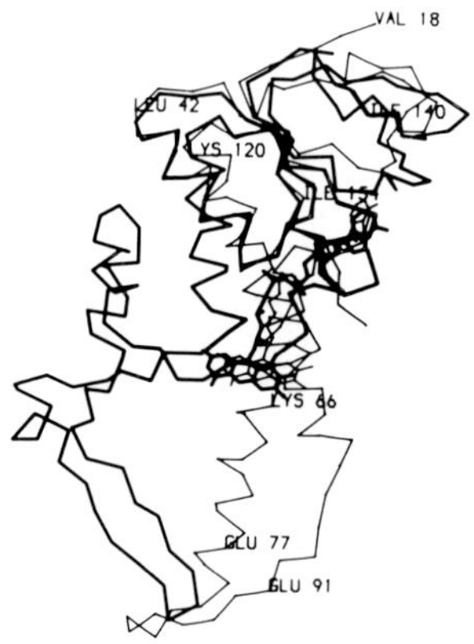

Fig.7 Superposition of the FAD-binding domains of p-hydroxybenzoate hydroxylase (thick lines) and glutathione reductase

tion between equivalenced C_α-atoms of only 1.2Å. These are the FAD-binding domains shown in Fig.7. Moreover, after equivalencing these domains, also the bound FAD molecules are well adjusted with a residual error of only 1.9Å. Therefore, if the location of FAD in one of these enzymes were not known, it could have been predicted from the other. The resembling chain folds consist of a 4-stranded parallel ß-pleated sheet with 2 α-helices at one side and a 3-stranded antiparallel ß-sheet at the other. FAD binds at the C-terminal ends of the parallel sheet strands. This corresponds well to the situation in a number of NAD(P)-binding domains, which, however, have only the parallel sheet and the α-helices, but not the antiparallel sheet in common with the FAD-binding domains. A further conspicuous feature of the 2 proteins are the long excursions which in both cases are located between the second ß-strand and the second α-helix, as shown in Fig.7. These excursions are not related with each other. There are no equivalent features in other known nucleotide-binding proteins.

Considering these similarities the question arises, whether there exists a general characteristic FAD-binding chain fold. With only 2 examples at hand, this question cannot yet be answered. However, there seems to be a universal feature, which applies to all dinucleotide-binding proteins: It consists of the first 2 ß-strands and the intervening α-helix (24). This feature is exactly superimposable in the example of Fig.7. With the same geometry and dimensions and with the same geometric relation to the dinucleotides it is also present in the other proteins concerned.

There is a good chance, that this ßαß-unit can be recognized in the amino acid sequence. Invariantly, it is at the N-terminus of the domain, which in turn is in most cases at the N-terminus of the protein. Almost exclusively, glycines occur at the end of the first ß-strand, at the helix start, and after the first helix turn at relative positions 6,8,11 of the ßαß-unit. Glu or Asp are always located at the end of the second ß-strand (position 28±1), except for the NADP-binding domains, which contain an Arg one position further down the chain. In all cases, the first ß-strand consists of hydrophobic residues with a high ß-forming propensity (11). Ala and Gly also occur at the second helix turn (relative position 14) and at the helix end. Glu or Asp bind to the 2′and/or 3′hydroxyls of the adenine ribose, Arg binds to the respective 2′phosphate. Glycines and alanines allow tight turns between ß-strands and helix and close packing of these secondary structures. Taken together, these markers suffice to recognize the special ßαß-feature in an amino acid sequence, as for instance in the enzyme D-amino acid oxidase (23). However, only in some cases, where proteins have a rather close evolutionary relationship, it seems possible to derive the complete FAD-binding chain fold from the amino acid sequence, as has been done with mercuric reductase (6) and lipoamide dehydrogenase (5).

Acknowledgement

The contributions of Drs. R.H. Schirmer, E.F. Pai, J. Drenth, R. Wierenga to this paper are highly appreciated.

References

1. Williams, C.H.,Jr.: The Enzymes (Boyer,P.D., ed.) 3rd edit., vol.13, pp.89-172, Academic Press, New York 1976

2. Akerboom, T.P.M., Bilzer,M., Sies,H.: J.Biol.Chem. 257, 4248-4252 (1982)

3. Holmgren,A.: Experientia (Basel) Suppl.36, 149-180 (1980)

4. Williams,C.H.,Jr., Arscott,L.D., Schulz,G.E.: Proc.Natl. Acad.Sci. USA 79, 2199-2207 (1982)

5. Rice,D.W., Schulz,G.E., Guest,J.R.: J.Mol.Biol. 174, 483-496 (1984)

6. Brown,N.L., Ford,S.J., Pridmore,R.D., Fritzinger,D.A.: Biochemistry 22, 4089-4094 (1983)

7. Wierenga,R.K., Drenth,J., Schulz,G.E.: J.Mol.Biol. 167, 725-739 (1983)

8. Schulz,G.E., Zappe,H., Worthington,D.J., Rosemeyer,M.A.: FEBS-Letters 54, 86-88 (1975)

9. Thieme,R., Pai,E.F., Schirmer,R.H., Schulz,G.E.: J.Mol. Biol. 152, 763-782 (1981)

10. Krauth-Siegel,R.L., Blatterspiel,R., Saleh,M., Schiltz,E., Schirmer,R.H., Untucht-Grau,R.: Eur.J.Biochem. 121, 259-267 (1982)

11. Schulz,G.E., Schirmer,R.H.: Principles of Protein Structure Springer Verlag New York 1979

12. Hofsteenge,J., Vereijken,J.M., Weijer,W.J., Beintema,J.J., Wierenga,R.K., Drenth,J.: Eur.J.Biochem. 113, 141-150(1980)

13. Grau,U.M., Trommer,W.E., Rossmann,M.G.: J.Mol.Biol. 151, 289-307 (1981)

14. Webb,L.E., Hill,E.J., Banaszak,L.J.: Biochemistry 12, 5101-5109 (1973)

15. Abdallah,M.A., Biellmann,J.-F., Nordström,B., Brändén, C.-I.: Eur.J.Biochem. 50, 475-481 (1975)

16. Moras,D., Olsen,K.W., Sabesan,M.N., Buehner,M., Ford,G.C., Rossmann,M.G.: J.Biol.Chem. 250, 9137-9162 (1975)

17. Abdallah,M.A., Adams,M.J., Archibald,I.G., Biellmann,J.-F., Helliwell,J.R., Jenkins,S.E.: Eur.J.Biochem. 98, 121-130 (1979)

18. Matthews,D.A., Alden,R.A., Freer,S.T., Xuong,N., Kraut,J.: J.Biol.Chem. 254, 4144-4151 (1979)

19. Schulz,G.E., Schirmer,R.H., Pai,E.F.: J.Mol.Biol. 160, 287-308 (1982)

20. Pai,E.F., Schulz,G.E.: J.Biol.Chem. 258, 1752-1757 (1983)

21. Stern,B.K., Vennesland,B.: J.Biol.Chem. 235, 209-212(1960)

22. Massey,V., Williams,C.H.,Jr.: J.Biol.Chem. 240, 4470-4480 (1965)

23. Ronchi,S., Minchiotti,L., Curti,B., Zapponi,M.C., Bridgen, J.: Biochem.Biophys.Acta 427, 634-643 (1976)

24. Wierenga,R.K., Hol,W.G.J.: Nature, 302, 842-844 (1983)

ACTIVE SITE CHEMICAL MODIFICATION AND SEQUENCING OF FLAVOPROTEINS

Charles H. Williams, Jr., L. David Arscott and Richard P. Swenson
Veterans Administration Medical Center and Dept. of Biological Chemistry, University of Michigan, Ann Arbor 48105, Michigan, U.S.A.

Fascination with the flavin has rightly dominated the study of flavoproteins over the years. However, amino acid residues must take active roles in the catalytic process as they do in other proteins. Several observations in the mid-fifties suggested roles for the protein in flavin binding (1,2). The protein role was confirmed in succinate dehydrogenase by Kearney (3) a few years later by isolation of the FAD with a peptide attached. We have always known subliminally, that it is the protein which confers on the flavin its specific reactivity, and this is being elegantly demonstrated in the flavin replacement studies of Massey and Ghisla (4) and their colleagues, not forgetting Peter Hemmerich up until his death. Although there are scattered allusions in the fifties to the chemical modification of protein groups in flavoenzymes and even the 1926 report by Szent-Gyorgyi (5) of the inhibition of xanthine oxidase by cyanide, these have either not been followed through or the follow up has not indicated that the protein *per se* is the site of action. Perhaps the first clear catalytic role ascribed to an amino acid residue in a flavoenzyme was that of the protein disulfide in lipoamide dehydrogenase by Massey and Veeger (6), and Searls and Sanadi (7) in 1960.

There has been a steady increase in the attention given to the structure of the protein of flavoenzymes and to the function of specific amino acid residues at the past five flavin symposia. At the meeting in Durham in 1969, only six papers addressed protein structural problems. At the last meeting this was one of the dominant themes. In the twelve years between Durham and Ann Arbor, the amino acid sequences and X-ray crystal structures of several flavodoxins were determined, mechanism-based inhibitors were introduced, homology among *E. coli* and pig heart lipoamide dehydrogenase, and yeast glutathione reductase was discovered, functions were postulated for several amino acid residues and many extensive sequences were reported.

Table I illustrates amino acid residues identified by chemical modification which may play specific roles in flavoproteins. Proposed

Flavins and Flavoproteins

functions have been included even where they are speculative. We must be careful as usual to add the appropriate caveats. It is sad but true that the result of any chemical modification may be due to steric hindrance by the bulk of the modifying group, to a conformational change in the enzyme, or, as we would hope, to the fact that the modified residue has been altered in its function. The flavoprotein literature is commendably free

Enzyme	Residue	Possible Function
Succinate dehydrogenase	His (8)	covalent FAD binding
Monoamine oxidase	Cys (9)	covalent FAD binding
Xanthine oxidase	Lys (10)	reductive half-reaction
D-amino acid oxidase	His-217 (11-14)	reductive half-reaction
	Arg (15,15a)	N1 anion stabilization
	Tyr-55 (11,16)	substrate binding
	Lys-204 (11,16)	substrate binding
	Tyr-224 (17,18)	reductive half-reaction
	His-307 (19)	-
	Tyr-228 (a)	-
L-lactate oxidase	His (20)	reductive half-reaction
Cyclopiazonate oxidocyclase	His-a (21)	reductive half-reaction
	His-b (21)	substrate binding
Flavocytochrome b_2	Cys-a,b,c (22)	FMN binding pocket
Pyridoxamine-5P oxidase	His (23)	substrate binding pocket
Hydroxynitrile lyase	Cys (24)	FAD reduction
Lipoamide dehydrogenase	Cys-45 (25-27)	interchange with substrate
	Cys-50 (25-27)	electron transfer to FAD
	His-444 (28)	proton abstraction from substrate
Glutathione reductase	Cys-58 (29)	interchange with substrate
	Cys-63 (29)	electron transfer from FAD
	His-467 (30)	protonation of substrate
Mercuric reductase	Cys-135 (31,32)	interchange with substrate
	Cys-140 (31,32)	electron transfer from FAD
p-Hydroxybenzoate hydroxylase	Glu-49 (33)	-
D-Lactate dehydrogenase	Cys (34)	-
Bacterial Luciferase	Cys (35)	substrates protect
	His (35)	$FMNH_2$ binding
	amino (35)	-

Table I. Flavoprotein Amino Acid Residues Identified by Chemical Modification. (a) S. Ronchi, L. Minchiotti, M. Galliano and B. Curti, unpublished results as quoted in (14).

of claims that an important residue has been modified based only on the presumed specificity of the reagent, when in fact the effect was probably due to the modification of an indeterminant number and type of residues.

The fact that the half-reactions of flavoproteins can be observed separately, means that the effects of chemical modification can, in principle, be assigned to one or the other. Porter and Bright (17) working with Ronchi and Curti (18) have shown that the modification of Tyr-224 by N-chloroleucine leads to a 1000 fold decrease in the rate of the reductive half reaction of D-amino acid oxidase.

Pyridine Nucleotide-Disulfide Oxidoreductases

The group of enzymes for which lipoamide dehydrogenase is the prototype contains an active center cystine and a base (36-38). In the stable 2-electron reduced intermediate, the FAD is oxidized and the cystine is reduced (37). Properties of these enzymes in which the nascent thiol nearer the amino-terminus has been alkylated indicate that interchange between this thiol and the substrate is prevented, while the nascent thiol nearer the carboxyl-terminus is still able to interact with the flavin (25-27,29). The base must be so positioned that it can accept or donate a proton from/to the substrate and can stabilize either thiol as the thiolate depending on the stage of catalysis. These residues have been identified by x-ray crystallography in glutathione reductase (39,39a). The histidine residue first postulated on the basis of redox potentials and rates of the reductive half reaction (37,38), and identified in the glutathione reductase sequence (39a), has only recently been chemically modified in <u>E. coli</u> lipoamide dehydrogenase (28). Mercuric reductase, the newest member of this group, shows very similar catalytic properties (31).

The sequence of <u>E. coli</u> and pig heart lipoamide dehydrogenase, and of yeast glutathione reductase demonstrated that at least in the active site

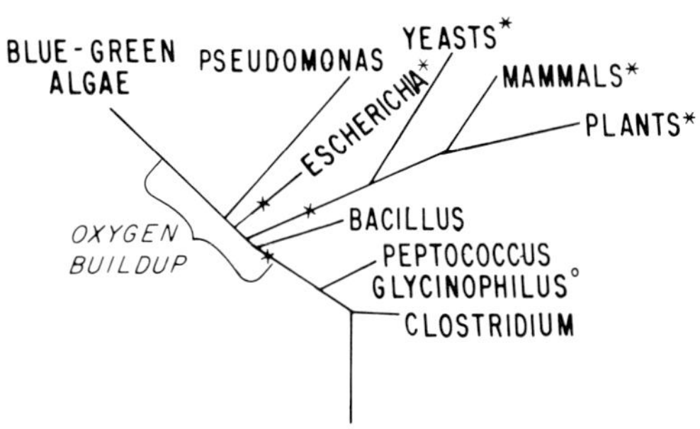

Figure 1. Phylogenetic Tree of Prokaryotes and Eukaryotes. Possible points of divergence of glutathione reductase from lipoamide dehydrogenase, x. Contains both lipoamide dehydrogenase and glutathione reductase, *. Glutathione reductase not reported, o.

region these two enzymes were homologous and that thioredoxin reductase probably was not (40). The homology was shown to extend to all parts of the molecule (41). The evolution of two enzymes from a common ancestor and their differentiation to distinct functions is a common occurrence in nature. It has been postulated that glutathione reductase evolved during the oxygen buildup era as organisms needed to protect their protein thiols against oxidation (42). Since lipoamide dehydrogenase is found in several anaerobic organisms, it is possible that it is the progenitor of glutathione reductase as shown in Figure 1. The divergence could have occurred either before or after the divergence of the eukaryotes. The homology between glutathione reductase and lipoamide dehydrogenase has now been fully documented from the gene sequence of lipoamide dehydrogenase and has been extended to mercuric reductase.

The sequence of pig heart lipoamide dehydrogenase is incomplete (see Arscott et al., this volume). The available portions are compared to the complete sequences of human erythrocyte glutathione reductase (43), E. coli lipoamide dehydrogenase (44) and Pseudomonas aeruginosa transposon mercuric reductase (32) in Figures 2, 3, 5 and 6. Glutathione reductase and E. coli lipoamide dehydrogenase have been compared previously (45) and very few changes were required to align the other two sequences. Figure 2 shows the FAD-1 domain. The degree of homology is highest in the regions of the FAD adenine (position 23-50, see below) and the redox active disulfide (positions 53-67). The x-ray crystal structure shows that the side chain of Lys-66 extends from the polypeptide back bone on the si side of the isoalloxazine ring to the re side or the side on which the pyridine nucleotide binds in glutathione reductase. The next lysine residue extends into the glutathione binding space (46). The question now becomes, can the lysines in the other three enzymes homologous with the second lysine in glutathione reductase take up potentially crucial positions for assisting in electron transfer between the flavin and the pyridine nucleotide? This residue is a prime candidate for site-directed mutagenesis which should be quite possible in the E. coli enzymes.

Figure 3 shows the sequences in the pyridine nucleotide domain. A highly characteristic pattern, also observed in the adenine section of the FAD binding site (Figure 2), is impressively conserved at the junction between the first beta strand and the first helix as observed for p-hydroxybenzoate hydroxylase, D-amino acid oxidase and glutathione reductase (47). Starting at residue 192, it goes apolar, apolar, glycine, dash, glycine, dash, dash, glycine, apolar, dash, dash, alanine and finally a glutamate which contacts the two prime oxygen of the ribose.

```
                                            @   * @ *   *
A C R Q E P Q P Q G P P P A A G A V A S Y D Y L V I G G G S         30
                            A D Q P I D A D V T V I G S G P
                              S T E I K T Q V V V L G A G P
80R K V R G W M A A A E K H S G N E P P V Q V A V I G S G G

  @     @ *     @ @   @ *             *       @ @ * @   *   *
G G L A S A R R A A E L G A R A A V V E S H - K L G G T C V N       60
G G Y V A A I K A A Q L G F K T V C I E K N E T L G G T C L N
A G Y S A A F R C A D L G L E T V I V E R Y N T L G G V C L N
A A M A A A L K A V E Q G A Q V T L I E R - G T I G G T C V N

* * *   * @ *                                     @
V G C V P K K V M W N T A V H S E F M H D H A D Y G F P S C         90
V G C I P S K A L L N N S H Y Y H M A H G K D F A S R G I E
V G C I P S K A L L H V A K V I E E A K A L A E H G I V F G
V G C V P S K I M I R A A H I A H L R R E S P F D G G I A A

                                        *
E G K F N W R V I K E K R D A - Y V S R L N A I Y Q N N L T - - K 120
M S E V R L N L E K M                 A L T G G I A H L F K
E P K T D I D K I R T W K E K - V I N Q L T G G L A G M A K - - G
T V P T I D R S K L L A Q Q Q A R V D E L R H A K Y E G I L G G N

          @   *       @             @         @
S H I E I I R G H A A F T S D P K P T I E V S G - K K Y T A P       150
          V N G Y G K                   A D - G S T E V I N T
R K V K V V N G L G K F T - - G A N T L E V E G E N G K T V I
P A I T V V H G E A R F K D D Q S L T V R L N - E G G E R V V

          @ @ * @ *
- - - H I L I A T G      HRBC GLURED
- - K N I L I A T G      PIG H LIPDH       FAD-1
N F D N A I I A A G      E.COLI LIPDH      DOMAIN
M F D R C L V A T G      PSEUD MERRED
```

```
                  @               @ @                   @
HRBC GLURED   G M P S T P H E S Q I P G A - S L G - I T S D G F   180
PIG H LIPDH   S E
E.COLI LIPDH  S R P I Q L - - P F I P H E - D P R I W D S T D A
PSEUD MERRED  A S P A V P - - - P I P G L K E S P Y W T S T E A

            @   @     @   *   @   @     *               @ @
F Q L E E L P G R S V I V G A G Y I A V E M A G I L S A L G         210
                K M V V I G A G V I G V E
L E L K E V P E R L L V M G G G I I G L E M G T V Y H A L G
L A S D T I P E R L A V I G S S V V A L E L A Q A F A R L G

                                                @
S K T S L M I R H D K V L R S F D S M I S T N C T E E L E N         240
-                                                         -
S Q I D V V E M F D Q V I P A A D K D I V K V F T K R I S K
S K V T V L A R N T L F F R E - D P A I G E A V T A A F R A

          @               @           @             @
A G V E V L K F S Q V K E V K K T L S G L E V S M V T A V P         270
                                                  V T P F P
K F N L M L E - T K V T A V E A K E D G I Y V T M E G K - -
E G I E V L E H T Q A S Q V - A H M D G - E F V - L T - - -

                        @     * @ @ @ *   *
G R L P V M T M I P D V D C L L W A I G R V P                       300
G - I T I D E                 L V C I G R E P    PYRIDINE
- K A P A E P Q - - R Y D A V L V A I G R V P    NUCLEOTIDE
- T T H G E L - - - R A D K L L V A T G R T P    DOMAIN
```

Figure 2, left; Figure 3, right. Sequence Comparisons. The residue numbering and domain designation are based on glutathione reductase. The 88 apparently non-homologous residues at the amino-terminus of mercuric reductase have been omitted. Positions at which all four enzymes are identical have been designated with an *, and those at which three of the enzymes are identical with an @.

```
                          * * *   *     * *     *                                 *
pLipDH        - D A D V T V I G S G P G G Y V A A I K A A Q L G - - - F K T V C I E K
pLipDH-PN         - K M V V I G A G V I G V E L
cLipDH        - K T Q V V V L G A G P A G Y S A A F R C A D L G - - - L E T V I V E R
cLipDH-PN       - E R L L V M G G G I I G L E M G T V Y H A L G S Q I D V V E M F D Q
GluRed        - S Y D Y L V I G G G S G G L A S A R R A A E L G - - - A R A A V V E S
GluRed-PN       - G R S V I V G A G Y I A V E M A G I L S A L G S K T S L M I R H D K
TRRed       - K H S K L L I L G S G P A G
MerRed        - P V Q V A V I G S G G A A M A A A L K A V E Q G - - - A Q V T L I E R
MerRed-PN       - E R L A V I G S S V V A L E L A Q A F A R L G S K V T V L A R N T L
DAAO                M R V V V I G A G V I G L S T A L C I H E R Y - - H S V L Q P L D V
FumRed        - Q A D L A I V G A G G A G L R A A I A A A Q A N P N A K I A L I S K V
NADH-DH       - L K K I V I V G G G A G G L E M A T Q L G H K L G R K K K A K I T L V
NADH-DH-PN        - N I A I V G G G A T G V E L S E L H N A V K Q L H S Y G Y K G L T
PHBH          M K T Q V A I I G A G P S G L L L G Q L L H K A G - - - I N D V I L E R
LacDH         - N N K I T V V G V G A V G M A C A I S I L M K D - L A D E V A L V D V
                          * * *   *     * *     *                                 *
```

Figure 4. AMP Binding Site Homology. * Highly conserved positions. pLipDH, pig heart lipoamide dehydrogenase; cLipDH, E. coli K12 lipoamide dehydrogenase (44); GluRed, glutathione reductase (43); TRRed, thioredoxin reductase (O'Donnell and Williams, unpublished); DAAO, D-amino acid oxidase (13,14); MerRed, Mercuric Reductase (32); NADH-DH, NADH dehydrogenase (I. G. Young et al. (1981) Eur. J. Biochem. 116, 165-170); FumRed, fumarate reductase; (S. T. Cole (1982) Eur. J. Biochem. 122, 479-484): PHBH, p-hydroxybenzoate hydroxylase (47); LacDH, lactate dehydrogenase (W. Eventoff et al. (1977) P.N.A.S. 74, 2677-2681).

The final homologous position in the pyridine nucleotide binding site cannot be important in NADP-linked enzymes since the two-prime position is phosphorylated. This pattern is remarkably constant in all FAD-enzymes and these sequences have been summarized in Figure 4. The fruits of recombinant DNA technology become obvious here since almost half of these sequences are based on the gene sequence. They are all compared to the lactate dehydrogenase which is not a flavoprotein. The FAD-2 domains are compared in Figure 5.

Finally, Figure 6 compares the sequences in the interface domain. We have to face the surprising fact that the mercuric reductase lacks a histidine at the homologous postion. The mechanistic work of Fox and Walsh has shown clearly that mercuric reductase is very similar to the other two enzymes - in fact virtually identical by all criteria tested (31). So far as we have been able to detect, none of the histidines in mercuric reductase are homologous with this position in the other three enzymes.

Veeger and Massey (48) demonstrated the remarkable sensitivity of pig heart lipoamide dehydrogenase to cupric ion. The activity in the physiological reaction drops rapidly to zero while the diaphorase activity increases some 30 fold. In later work Casola and Massey (49) showed that two cysteine residues disappear upon copper treatment and Matthews et al.

```
                                          @ @     *   @
                     HRBC GLURED          N T K D L S L  300
                     PIG H LIPDH          F T Q N L G L
                     E.COLI LIPDH         N G K N L D A
                     PSEUD MERRED         N T R S L A L

      *             @   @   @       @   @   *   *   *
N K L G I Q T D D K G H I I V D E F Q N T N V K G I Y A V G  330
E E L G I E L R                           I P N I A A I G
G K A G V E V D D R G F I R V D K Q L R T N V P H I F A I G
D A A G V T V N A Q G A I V I D Q G M R T S N P N I Y A A G

@       @   @     @       *     @       @
D V C G K A L L T P V A I A A G R K L A H R L F E Y K E D S  360
D V V A G P M L A H K A E D E G I I(C)V E G M A G C - - A V
D I V G Q P M L A H K G V H E G H V A A E V I A G H - - K H
C - T D Q P Q F V Y V A A A A G T R A A I N M T G G - - D A

    *
K L D Y
H I D Y                                 FAD-2
Y F D P                                 DOMAIN
A L D L
```

```
        *   @     @   *   @ @   * *     *   @ @
    N N I P T V V F S H P P I G T V G L T E D E A I H K      390
    N C V P S V I Y T H P E V A W V G K S E E Q - L K E
    K V I P S I A Y T E P E V A W V G L T E K E A - K E
    T A M P A V V F T D P Q V A T V G Y S E A E A H H D

* * @                         @ *         @   @     *
Y G I E N V K T Y S T S F T P M Y H A V T K R K T K C V M K  420
E G I E Y K V G K F P F A A N S - R A K T N A D T D G M V K
K G I S Y E T A T F P W A A - S G R A I A S D C A D G M T K
- G I E T D S R T L T L D - N V P R A L A N F D T R G F I K

        @ @   *       @     @ *         @ @ *
M V C A N K E E K V V G I H M Q G L G C D E M L Q G F A V A  450
I L G Q K S T D R V L G A H I I G P G A G E M I N E A A L A
L I F D K E S H R V I G G A I V G T N G G E L L G E I G L A
L V I E E G S H R L I G V Q A V A P E A G E L I Q T A A L A

      @         @   @         @ @ *     *         @
V K M G A T K A D F D N T V A I H P T S S E E L V T L R  478 HRBC GR
L E Y G A S C E D I A R V C H A H P T L S E A F R E A N L A
I E M G C D A E D I A L T I H A H P T L H E S V G L A A E V
I R N R M T V Q E L A D Q L F P Y L T M V E G L K L A A Q T

A S F G K A I N F                478   PIG H LIPDH     INTERFACE
F E G S I T D L P N K A K K K    473   E.COLI LIPDH    DOMAIN
F N K D V K Q L S C C A G        561   PSEUD MERRED
```

Figure 5, left; Figure 6, right. Sequence Comparisons. The residue numbering and domain designations are based on glutathione reductase.

(50,51) identified the sequences around these residues. They are Cys-457 and Cys-464. Homologous cysteines are not found in any of the other three enzymes. While E. coli lipoamide dehydrogenase is somewhat sensitive to copper, both the rate and the extent of the effects are markedly less. It would be reasonable to assume that the effects in the pig heart lipoamide dehydrogenase are due to the formation of an intrachain disulfide which repositions the base to prevent its stabilization of the thiolate which normally interacts with the FAD. This would be expected to interfere with the physiological reaction while increasing the diaphorase reaction. In the native enzyme, the diaphorase reaction (which does not involve the nascent thiols) must be inhibited by the potential for interaction between the FAD and the dithiol. With the base repositioned by disulfide bond formation, the FAD would be free to interact exclusively with the pyridine nucleotide and the dye. That the disulfide bond is most likely intrachain can be tested by imposing the lipoamide dehydrogenase sequence on the glutathione reductase structure in order to test if either of the cysteine side chains can reach across the interface boundary to make an interchain bond. It seems highly likely that this is the case in the E. coli enzyme where at least in two dimensions this cysteine is very close to the two fold axis. The fact that the reaction is much slower in the E. coli enzyme must mean that the positioning is not ideal.

Methylation of D-Amino Acid Oxidase

D-amino acid oxidase is the prototype enzyme for the dehydrogenase/oxidase group in which the production of a carbanion is a presumed early step in the reductive half reaction (52-54 and refs. in 11). It is important to identify the base responsible for the abstraction of the proton. Figure 7 shows the binding of the substrate and reviews the roles proposed for several residues. Tyr-55 and Lys-204 in D-amino acid oxidase have been suggested as substrate binding groups on the basis that their mutually exclusive modification is prevented by the competitive inhibitor benzoate, and since the enzyme is stereospecific, its zwitterionic substrates reasonably would be positioned by potentially cationic and anionic groups (11,16). His-217 is the putative base. The removal of a proton from an alkyl carbon can only be envisioned if the proton is activated. It has been suggested that binding through ion pairs would serve to distort the Sp^3 alpha carbon into a more planar Sp^2-like conformation in which orbital overlap should weaken the C-H bond (11). His-217 is the site of reaction with dansyl chloride resulting in the loss of activity (12). While this

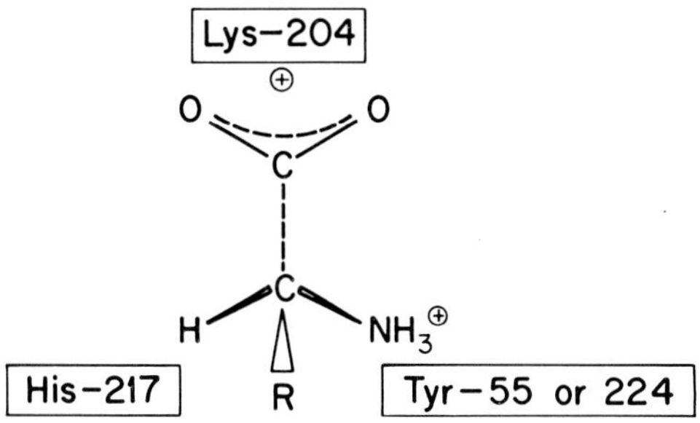

Figure 7. Binding of D-Amino Acids Through Ion Pairs Positioning the C-Alpha Hydrogen.

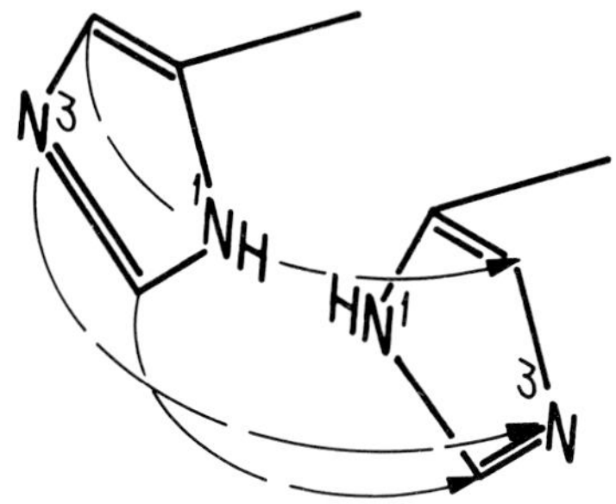

Figure 8. Rotation of the Imidazole Side Chain about the C-beta--C-gamma bond.

is of interest, the dansyl group is very large and we sought a much smaller moiety.

Recently we reported on a modification of D-amino acid oxidase which may contribute to our understanding of the reductive half reaction (55). Further studies on the methylated enzyme are reported by Swenson et al. in this volume. It was shown that methyl-p-nitrobenzene sulfonate bound to the enzyme prior to reacting with His-217. The competitive inhibitor benzoate protected against the loss of activity, and prevented the modification of His-217. The marked effect of benzoate on the flavin spectrum of the native enzyme indicates that its binding site is quite close and it has been assumed to be at least partially coincident with that of the substrate (56). The modification of chymotrypsin by methyl-p-nitrobenzene sulfonate resulted in the exclusive methylation of the N1 of a single histidine residue (57), while the reaction with phospholipase A2 yielded a specific histidine residue methylated at N3 (58). Given that this reagent binds to D-amino acid oxidase prior to reaction, it was somewhat surprising to find that both the one and three nitrogens of histidine 217 were modified in a ratio of two to one under all conditions tried (55). This result implies that the imidazole side chain of His-217 is free to rotate in the active site assuming that the reagent is tightly bound. Figure 8 shows a complete flip of an imidazole side chain. In the enzyme any motion would probably be limited to a few degrees of wobble. Attempts to separate the two modified forms of D-amino acid oxidase have thus far failed and all of our studies (Swenson et al., this volume) have been carried out on the mixture.

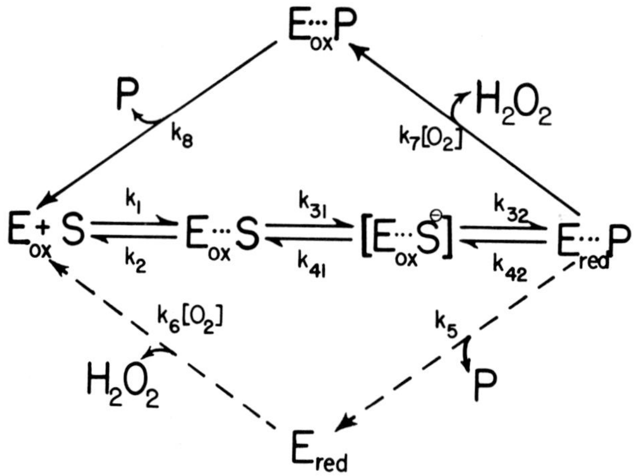

Figure 9. Kinetic Scheme for D-Amino Acid Oxidase.

The methylation of His-217 profoundly affects the reductive half reaction making it at least partially rate limiting. Figure 9 shows a slightly expanded version of the kinetic mechanism of D-amino acid oxidase based on that of Gibson and Massey (59) and of Porter and Bright (60). The reductive half reaction measured as the rate of production of the reduced enzyme-product charge transfer complex is fast in the native enzyme. Breakdown of this complex, measured in the absence of oxygen, k_5, is very slow. Catalysis of the oxidation of most substrates such as alanine proceeds through the upper loop with dissociation of the product being rate-limiting. Table II summarizes the steady state and rapid reaction data for native enzyme and methylated enzyme. The overall rate of the reductive half reaction is decreased from 1200 to 20 per second while the steady state rate remains essentially unchanged. These data suggest that the reductive half reaction must now be at least partially rate limiting.

	DAAO	Me-DAAO
k_3	1200 s^{-1}	20 s^{-1}
k_5	0.017 s^{-1}	0.015, 0.36 s^{-1}
T. N.	10-17 s^{-1}	26 s^{-1}

Table II. Kinetic data for Native and Methylated D-Amino Acid Oxidase with Alanine.

This is born out by the fact that the methylated enzyme shows a two-fold deuterium isotope effect with alanine as the substrate, in contrast to the native enzyme which shows no isotope effect with alanine. This may indicate that proton abstraction, k_{31}, has been altered by the methylation of His-217 rather than reduction to the charge transfer complex per se, k_{32}, but it cannot yet be said with any certainty which substep is affected by the modification. Clearly if His-217 is the base, it is still functional, albeit at a much reduced rate.

The steps in the reaction cycle are shown in Figure 10 which is adapted from (4). Substrate binds most reasonably through ion pairs with its ammonium and carboxylate groups, and for the sake of discussion we have shown the ion pair partners as Lys-204 and Tyr-55, or alternatively Tyr-224. Proton abstraction of the alpha hydrogen by His-217 is pictured as the first step following binding. The covalent bond between substrate and flavin is formed and breaks down to give the reduced enzyme-product charge transfer complex. The anionic flavin is stabilized by a positive charge in the protein (4,15a). Formation of a carbanion through proton abstraction by a base in the enzyme must be facilitated by a combination of factors. If indeed ion pairs are involved in substrate binding they would serve to decrease the strength of the alpha-carbon-hydrogen bond making the proton extractable by a base such as imidazole. The identity

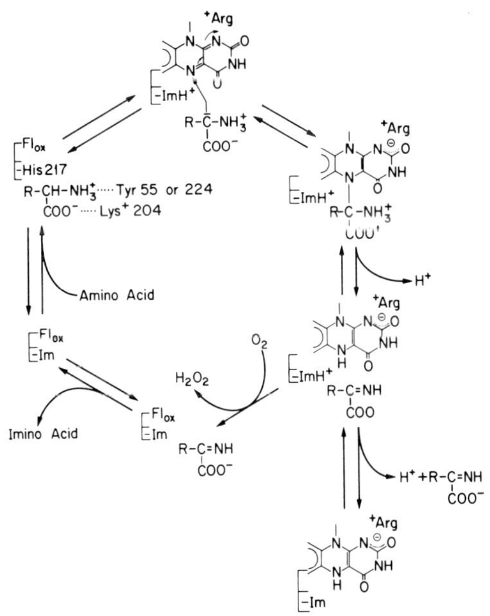

Figure 10. Reaction Scheme for D-Amino Acid Oxidase. Adapted from (4).

of this base is not clear at this point and our suggestion that it is His-217 is very tentative. We have not studied the properties of the enzyme modified at Tyr-55 in the reductive half reaction, since this involves its separation from enzyme modified at Lys-204. On the other hand, if Tyr-224 is the ion pair partner of the substrate amino group, then the observation of Porter and Bright that the rate of the reductive half reaction is reduced by 1000 fold would be quite reasonable. The absence of this ion pair would abstraction of the alpha-carbon more difficult. While the chemical modification studies described here have not allowed a definitive assignment of residue roles, they have, suggested a number of further experiments, especially if the methyl-D-amino acid oxidase can be resolved.

Acknowledgement

This work has been supported by the Medical Research Service of the Veterans Administration and by Grant GM21444 from the National Institutes of Health.

References

1. Nygaard, A. P. and Theorell, H.: Acta Chem. Scand. 8, 1489 (1954).
2. Green, D. E., Mii, S. and Kohout, P.: J. Biol. Chem. 217, 551 (1955).
3. Kearney, E. B.: J. Biol. Chem. 235, 865 (1960).
4. Massey, V. and Ghisla, S.: Colloquium - Mosbach, Biological Oxidations, Springer-Verlag, Berlin 1983.
5. Szent-Gyorgyi, A.: Biochem. Z. 173, 275-278 (1926).
6. Massey, V. and Veeger, C.: Biochim. Biophys. Acta, 40, 184-185 (1960).
7. Searls, R. L. and Sanadi, D. R.: Biochem. Biophys. Res. Comm. 2, 189-192 (1960).
8. Walker, W. H. and Singer, T. P.: J. Biol. Chem. 245, 4224 (1970).
9. Walker, W. H., Kearney, E. B., Seng, R. and Singer, T. P.: Biochem. Biophys. Res. Comm. 44, 287 (1971).

10. Nishino, T., Tsushima, R. H. and Massey, V.: J. Biol. Chem. 257, 7348-7353 (1982).

11. Nishino, T., Massey, V. and Williams, C. H., Jr.: J. Biol. Chem. 255, 3610-3616 (1980).

12. Swenson, R. P., Williams, C. H., Jr. and Massey, V.: J. Biol. Chem. 258, 497-502 (1983).

13. Swenson, R. P. et al.: J. Biol. Chem. 257, 8817-8823 (1982).

14. Ronchi, S. et al.: J. Biol. Chem. 257 8824-8834 (1982).

15. Ferti, C., Curti, B., Simonetta, M. P., Ronchi, S., Galliano, M. and Minchiotti, L.: Eur. J. Biochem. 119, 553-557 (1981).

15a. Fitzpatrick, P. F. and Massey, V.: J. Biol. Chem. 258, 9700-9705 (1983).

16. Swenson, R. P., Williams, C. H., Jr. and Massey, V.: J. Biol. Chem. 257, 1937-1944 (1982).

17. Porter, D. J. T. and Bright, H. J.: J. Biol. Chem. 251, 6150-6153 (1976).

18. Ronchi, S., Galliano, M., Minchiotti, L., Curti, B., Rudie, N. G., Porter, D. J. T. and Bright, H. J.: J. Biol. Chem. 255, 6044-6046 (1980).

19. Ronchi, S., Minchiotti, L., Galliano, M, Curti, B. and Ghisla, S.: Flavins and Flavoproteins (Yagi, K. and Yamano, T., eds.) 511-515 (1980).

20. Murphy, C. J., Shepherd, M. G. and Sullivan, P. A.: Biochemistry 22, 1665-1669 (1983).

21. Steenkamp, D. J., Schabort, J. C., Holzapfel, C. W. and Ferreira, N. P.: Biochim. Biophys. Acta 358, 126-143 (1974).

22. Pompon, D. and Lederer, F.: Eur. J. Biochem. 131, 359-365 (1983).

23. Horiike, K., Tsuge, H. and McCormick, D. B.: J. Biol. Chem. 254, 6638-6643 (1979).

24. Jaenicke, L. and Preun, J.: Eur. J. Biochem. 138, 319-325 (1984).

25. Thorpe, C. and Williams, C. H., Jr.: J. Biol. Chem. 251, 3553-3557 (1976).

26. Thorpe, C. and Williams, C. H., Jr.: J. Biol. Chem. 251, 7726-7728 (1976).

27. Williams, C. H., Jr., Thorpe, C. and Arscott, L. D.: Developments in Biochemistry, 1, (Singer, T. P. and Ondarza, R. N., eds.) Elsevier/North-Holland, New York 1978

28. Adamson, S. R., Robinson, J. A. and Stevenson, K. J.: Biochemistry 23, 1269 (1984).

29. Arscott, L. D., Thorpe, C. and Williams, C. H., Jr.: Biochemistry 20, 1513 (1981).

30. Boggaram, V. and Mannervik, B.: Biochem. Biophys. Res. Comm. 83, 558-564 (1978).

31. Fox, B. S. and Walsh, C. T.: Biochemistry 22, 4082-4088 (1983).

32. Brown, N. L., Ford, S. J., Pridmore, R. D. and Fritzinger, D. C.: Biochemistry 22, 4095-4102 (1983).

33. Claiborne, A., Hemmerich, P., Massey, V. and Lawton, R.: J. Biol. Chem. 258, 5433-5439 (1983).

34. Olson, S. T. and Massey, V.: Biochemistry 19, 3137-3144 (1980).

35. Ziegler, M. M. and Baldwin, T. O. Curr. Top. Bioenerg. 12, 65-113 (1981).

36. Massey, V. and Veeger, C.: Biochim. Biophys. Acta 48, 33-47 (1961).

37. Matthews, R. G. and Williams, C. H., Jr.: J. Biol. Chem. 251, 3956-3964 (1976).

38. Matthews, R. G., Ballou, D. P. and Williams, C. H., Jr.: J. Biol. Chem. 254, 4974-4981 (1979).

39. Schulz, G. E., Schirmer, R. H., Sachsenheimer, W. and Pai, E. F.: Nature 273, 120-124.

39a. Untucht-Grau, R., Schulz, G. E. and Schirmer, R. H.: FEBS Letters 105, 244-248 (1979).

40. Jones, E. T. and Williams, C. H., Jr.: J. Biol. Chem. 250, 3779-3784 (1975).

41. Williams, C. H., Jr., Arscott, L. D. and Schulz, G. E.: Proc. Natl. Acad. Sci. USA 79, 2199-2201 (1982).

42. Fahey, R. C.: Adv. Exp. Med. Biol. 86A, 1-30 (1977).

43. Krauth-Siegel, R. L., Blatterspiel, R., Saleh, M., Schiltz, E., Schirmer, R. H. and Untucht-Grau, R.: Eur. J. Biochem. 121, 259-267 (1982).

44. Stephens, P. E., Lewis, H. M., Darlison, M. G. and Guest, J. R.: Eur. J. Biochem. 135, 519-527 (1983).

45. Rice, D. W., Schulz, G. E. and Guest, J. R.: J. Mol. Biol. 174, 483-496 (1984).

46. Thieme, R., Pai, E. F., Schirmer, R. H. and Schulz, G. E.: J. Mol. Biol. 152, 763-782 (1981).

47. Hofsteenge, J., Vereijken, J. M., Weijer, W. J., Beintema, J. J., Wierenga, R. K. and Drenth, J.: Eur. J. Biochem. 113, 141-150 (1980).

48. Veeger, C. and Massey, V.: Biochim. Biophys. Acta 64, 83-100 (1962).

49. Casola, L., Brumby, P. E. and Massey, V.: J. Biol. Chem. 241, 4977-4984 (1966).

50. Matthews, R. G., Arscott, L. D. and Williams, C. H., Jr.: Biochim. Biophys. Acta 370, 26-38 (1974).

51. Matthews, R. G. and Williams, C. H., Jr.: Biochim. Biophys. Acta 370, 39-48 (1974).

52. Walsh, C. T., Schonbrunn, A. and Abeles, R. H.: J. Biol. Chem. 246, 6855-6856 (1971).

53. Porter, D. J. T., Voet, J. B. and Bright, H. J.: J. Biol. Chem. 248, 440-4416 (1973).

54. Massey, V. Ghisla, S., Ballou, D. P., Walsh, C. T., Cheung, Y. T. and Abeles, R. H.: Flavins and Flavoproteins (Singer, T. P., ed.) Elsevier Sci. Pub. Co., N. Y., 199-212 (1976).

55. Swenson, R. P., Williams, C. H., Jr. and Massey, V.: J. Biol. Chem. 259, 5585-5590 (1984).

56. Quay, S. and Massey, V.: Biochemistry 16, 3348-3353 (1977).

57. Swenson, R. P., Williams, C. H., Jr. and Massey, V.: J. Biol. Chem. 259, 5585-5590 (1984).

58. Swenson, R. P., Williams, C. H., Jr. and Massey, V.: J. Biol. Chem. 259, 5585-5590 (1984).

59. Massey, V. and Gibson, Q. H.: Fed. Proc. 23, 18-29 (1964).

60. Porter, D. J. T., Voet, J. G., and Bright, H. J.: J. Biol. Chem. 252, 4464-4473 (1977).

MOLECULAR GENETIC APPROACHES TO THE STUDY OF *E. coli* FLAVOPROTEINS

J. R. Guest† and D. W. Rice*

Departments of Microbiology† and Biochemistry*, Sheffield University, Sheffield S10 2TN, U.K.

Summary

The emergence of molecular genetic techniques such as gene manipulation and nucleotide sequencing has provided new tools for analysing the structure-function relationships of enzymes. Their application to the study of *E. coli* flavoproteins has yielded primary structures for NADH dehydrogenase, lipoamide dehydrogenase, succinate dehydrogenase and fumarate reductase. Each possess some common features and the flavoprotein subunits of succinate dehydrogenase and fumarate reductase exhibit a remarkable degree of mutual homology. Comparisons between the lipoamide dehydrogenase sequences and the previously-defined structure of human glutathione reductase have been particularly useful for interpreting structure-function relationships in the bacterial enzyme.

Introduction

The advent of gene cloning and nucleotide sequence analysis, and its application to protein biochemistry, has not only accelerated the rate at which structural information has accumulated, but it has opened up new avenues of study that could not even have been contemplated a few years ago. At first the main advantage seemed to be in the provision of highly-enriched sources of specific proteins, achieved by amplifying the number of copies of the corresponding genes or by increasing the expression

Flavins and Flavoproteins

of the genes by coupling to powerful promoter sequences. However, the development of rapid methods for nucleotide sequence analysis has meant that primary structures can be derived directly from the genes, without even the need for purifying the proteins. This is particularly valuable for proteins of very low abundance or for membrane-bound proteins, which are often difficult to isolate. The resulting sequences fall into two groups with respect to their immediate value for identifying the detailed structure-function relationships of the corresponding enzymes. Most useful are the sequences of enzymes that resemble proteins of known 3D-structure. Then there are novel sequences for which there is no relevant 3D-structure, although some may belong to families of related proteins. Examples of each type will be discussed below. There are now other applications of molecular genetics and gene manipulation technology that are potentially even more important for investigating protein structure-function relationships. These involve studying the consequences of structural modification of the genes. Thus, macrolesions of gene and protein can be engineered by specific deletion, rearrangement or hybrid formation, at the level of the gene, and microlesions involving single codons can be introduced by site-directed and other methods of *in vitro* mutagenesis. These offer far-reaching possibilities for probing function, altering stabilities and specificities, and redesigning or creating new enzymes.

Cloning and Sequence Analysis of Some Flavoprotein Genes of *E. coli*

The first *E. coli* flavoprotein gene to be cloned and sequenced was NADH dehydrogenase (NADHDH), closely-followed by fumarate reductase (FRD), and more recently by lipoamide dehydrogenase (LPDH) and succinate dehydrogenase (SDH). These are listed in Table 1 together with some other flavoprotein genes that have cloned but not sequenced. The genes have been isolated by screening a variety of gene-banks containing segments of *E. coli* DNA, for hybrid phages or plasmids that complement the lesions of mutants with defects in the flavoprotein genes or neighbouring genes. Other flavoprotein genes that have been identified and would be amenable to the same cloning strategy are listed in Table 1.

Table 1 Cloning and Sequence Analysis of Some Flavoprotein Genes of *E. coli*

Enzyme	Gene	Status	(Reference)
NADH dehydrogenase	*ndh*	Sequenced	(1)
Fumarate reductase	*frdABCD*	"	(2-4)
Succinate dehydrogenase	*sdhABCD*	"	(5,6)
Lipoamide dehydrogenase	*lpd*	"	(7)
D-Lactate dehydrogenase	*dld*	Cloned	(8)
Glutathione reductase	*gor*	"	(10)
Glycerol-3-phosphate dehydrogenase (ana)	*glpA*	"	(11)
Glycerol-3-phosphate dehydrogenase (aer)	*glpD*	Mapped	(12)
Pyruvate oxidase	*poxB*	"	(12)
D-Amino acid dehydrogenase	*dadA*	"	(12)

In Sheffield there has been a longstanding interest in the gene-protein relationships of the citric acid cycle enzymes, especially the 2-oxoglutarate dehydrogenase complex (OGDHC) and succinate dehydrogenase (SDH). This is not only because they contain flavoprotein subunits but also because they each have an enzymatic counterpart in the pyruvate dehydrogenase complex (PDHC) or fumarate reductase (FRD). Thus it was envisaged that a detailed structural analysis of the corresponding genes and proteins may provide insights into their evolution and the evolution of the citric acid cycle.

The Pyruvate and 2-Oxoglutarate Dehydrogenase Complexes

The PDH and OGDH complexes are multienzyme complexes that catalyse the oxidative decarboxylation of pyruvate or 2-oxoglutarate to the substrate (acetyl-CoA) or an intermediate (succinyl-CoA) of the citric acid cycle, CO_2 and NADH. Their dehydrogenase (E1) and acetyltransferase (E2) components are encoded by the *ace* and *suc* operons at 3 and 17 min in the *E. coli* chromosome, and studies with mutants have established that the lipoamide dehydrogenase (LPDH or E3) components of both complexes are encoded by a single gene (Fig. 1). This gene (*lpd*) is linked to the *ace* operon and is now known to be expressed as the distal gene of the *ace* operon and as an independent gene (M. E. Spencer and J. R. Guest, unpublished).

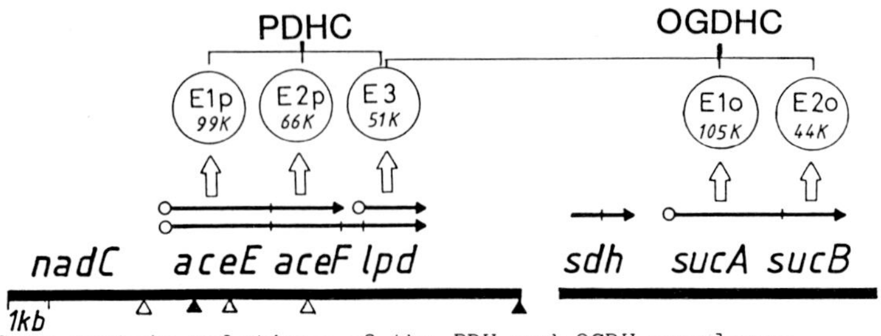

Fig. 1. Gene-protein relations of the PDH and OGDH complexes. Proteins (kd), and HindIII (Δ) and EcoRI (▲) sites are indicated.

The ace, suc and lpd genes have been isolated by cloning neighbouring genes (nadC and gltA), and extending the cloned inserts by prophage integration, aberrant excision, and selection for λnadC-aceEF-lpd and λgltA-sdh-sucAB transducing phages (13, 14). The nucleotide sequences of an 8000bp segment of DNA containing the aceE, aceF and lpd genes (7, 15, 16), and a 10000bp segment containing the gltA (citrate synthase, CS), sdh and sucA,B genes (5, 6, 17-19), have now been defined by the M13/dideoxy method. Transcriptional promoters and terminators have been identified and the coding regions have been translated into primary structures for the corresponding proteins.

The dehydrogenase components (E1p and E1o) exhibit very little sequence homology (comparison matrix A; Fig. 2), so it would appear that despite their analogous catalytic functions, molecular interactions and genetic organization, these components do not share a close evolutionary ancestor. By contrast, the acyltransferase components (E2p and E2o) are remarkably homologous. They possess lipoyl domains comprising three (E2p) or one (E2o) 100-residue lipoyl-binding regions and these are linked to each other and to analogous 300-residue subunit-binding domains by some proline- and alanine-rich segments (Fig. 3). The homologies are indicated by the dotted diagonals in the comparison matrices (Fig. 2B and C) and they suggest that the evolution of the E2 components has involved gene duplication. New E2p components having only one or two hybrid lipoyl-binding regions have recently been created by selective in vitro deletion within the aceF gene (H. M. Lewis and J. R. Guest, unpublished). These engineered proteins are active and it is not clear why E2p normally contains three such regions.

The E3 component (LPDH) resembles other disulphide oxido-reductases e.g. glutathione reductase (GR; Fig. 2D), see below.

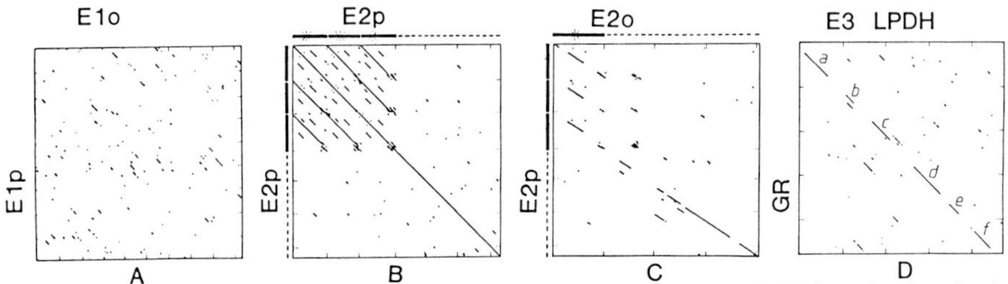

Fig. 2. Comparison matrices for components of PDHC and OGDHC. Analysis by DIAGON (20): probability <0.001; span = 21; scale = 100 residues.

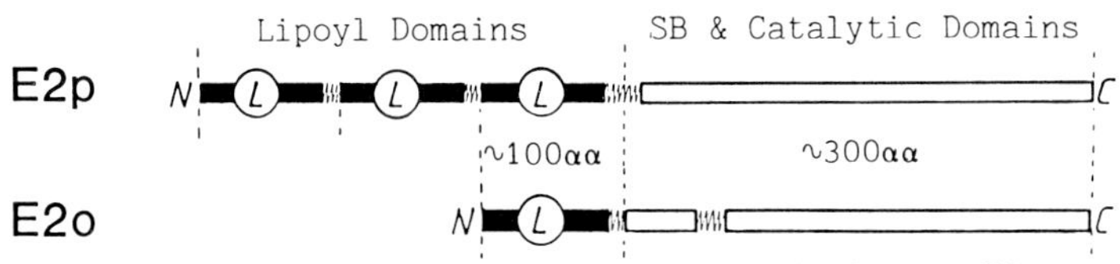

Fig. 3. Diagram showing the domain structures of E2p and E2o.

Succinate Dehydrogenase and Fumarate Reductase

SDH and FRD are membrane-bound enzymes catalysing the interconversion of succinate and fumarase. SDH functions in the citric acid cycle as a primary dehydrogenase donating electrons to the aerobic respiratory chain. Its synthesis is repressed under anaerobic conditions and by glucose via cAMP and the catabolite gene activator protein, CRP. In contrast, FRD is the terminal reductase of an anaerobic energy-generating respiratory chain, which can support growth on non-fermentable substrates with fumarate as the ultimate electron acceptor. Its synthesis is repressed by oxygen or nitrate and derepressed anaerobically by mechanisms involving an anaerobic gene activator protein, Fnr. The existence of two such enzymes raises questions of whether they are related enzymes, whose functional diversity is imposed by the mechanisms controlling their synthesis, or whether they are of entirely separate origin.

The corresponding sdh and frd genes are located at 17 and 94 min and they were isolated by extending the transducing range of gltA phages (14) or by screening gene banks for frd mutant complementation (21). The sequence of a 4500bp segment of DNA containing the sdh genes has now been defined (5, 6) and this has enabled direct comparisons between SDH and FRD, which had been characterized earlier by sequencing the frd genes (2-4).

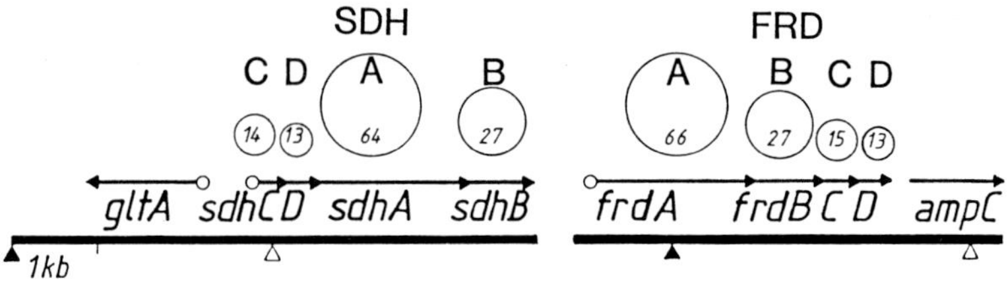

Fig. 4. Gene-protein relationships of SDH and FRD. Subunit sizes (kd), and HindIII (Δ) and EcoRI (▲) sites are indicated.

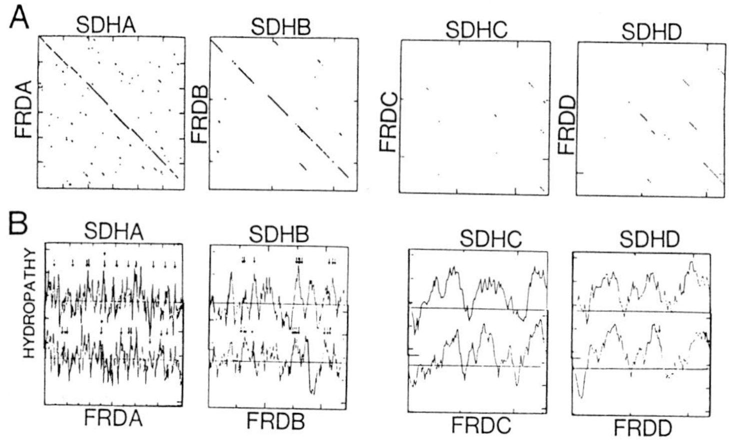

Fig. 5. Comparison matrices(A) and hydropathy profiles(B) of SDH and FRD. Analysis by DIAGON (20): probability <0.001; span = 21. Hydropathy averages at span = 7 (28); arrows = Cys; scale = 100 residues.

The gene-protein relationships for the two operons are shown in Fig. 4. Each contains four genes encoding analogous flavoprotein (A), iron-sulphur protein (B), and two hydrophobic protein (C and D) subunits, although the gene orders are different. The comparison matrices in Fig. 5 reveal striking sequence homologies between the A and B subunits (detailed elsewhere, 22), but these do not extend to C and D, which are similar in composition and hydropathy profile (Fig. 5) but not sequence.

The flavoprotein subunits contain characteristic AMP-binding sites (Table2) and their flavin-attachment sites closely resemble the flavopeptide of beef heart SDH (23).

		*	
SDHA (Beef Heart)		SHTVAAZGGIBLAAGBMDEBZWR	
SDHA (*E. coli*)	(43)	SHTVSAQGGITVALGNTHEDNWE	(65)
FRDA (*E. coli*)	(43)	SHTVAAEGG-SAAVAQDH-DSFE	(63)

These alignments illustrate the high degree of sequence conservation at the flavin-attachment sites of the three enzymes. There is no comparable region in NADH dehydrogenase, which uses non-covalent FAD (5, 8), but it is interesting to note that this region occupies the same position in the primary structure as the cystine peptides of glutathione reductase (GR) and LPDH, which interact with the isoalloxazine ring of FAD. Most of the flavoprotein cysteine residues are scattered and only one occupies the same position. This could correspond to the active-site cysteines detected in beef heart SDH and Vibrio succinogenes (24, 25). Close to this cysteine is a histidine residue (in a sequence, His-Pro-Thr, that occurs at the active-sites of GR and LPDH) that could have a proton donor-acceptor function at the active site (26).

```
                                   *                   *
SDHA (E. coli)  (229) GVPPQDMEMWQFHPTGIAGAGVLVTEGCRGEGGYLLNKHGER (270)
FRDA (E. coli)  (220) GVPLRDMEFVQYHPTGLPGSGILMTEGCRGEGGILVNKNGYR (261)
```

Ten of the eleven cysteine residues in the iron-sulphur protein subunits are conserved in three clusters of 3 or 4 cysteine residues (Fig. 5) but it would be premature to assign these to specific iron-sulphur centres (6). Despite their similar sizes and compositions the hydrophobic subunits show very little sequence homology (Fig. 5). However, they have hydropathy profiles suggesting the presence of three transmembrane α-helices per subunit (Fig. 5) and this could indicate comparable functions as anchors for the A and B subunits, although other functions in proton or electron transport have not been excluded.

Despite the differences in gene orders the SDH and FRD of E. coli are remarkably similar, they could certainly have evolved from common ancestors by processes involving gene duplication and even operon duplication and rearrangement. The greater divergence between the C and D subunits compared with A and B could reflect differences in the selection pressures operating to maintain hydrophobicity rather than catalytic properties. Their functional compartmentation in aerobic and anaerobic respiration would appear to be largely due to the mechanisms regulating their expression, and this is to some extent confirmed by the partial replacement of SDH function as a consequence of the aerobic derepression of FRD synthesis that accompanies amplification of the frd genes (27).

Disulphide Oxido-Reductase Flavoproteins

Structural comparison: The structural and mechanistic similarities in the family of flavin-containing pyridine nucleotide-disulphide oxido-reductases that include GR and LPDH, have been recognized for several years (29). In particular, they contain a highly conserved cystine peptide which undergoes reduction and oxidation during catalysis. The availability of the primary structure of LPDH, derived from the lpd gene, has enabled direct comparison with human erythrocyte GR for which the three dimensional structure is known (30). A complete alignment of GR and LPDH based on the DIAGON analysis has been made, and 6 major regions of homology (a to f of Fig. 2D) have been identified (31). Some 50% of the 463 equivalenced residues are either identical or conservatively substituted, and the complete sequence alignment is shown in Fig. 6.

The polypeptide chain of GR comprises 5 consecutive sections: a flexible N-terminal segment; the FAD binding domain; the NADPH domain; the central domain; and the dimer interface domain (30). All but the flexible N-terminal segment have regions of unambiguous homology between the two sequences and it is concluded that the overall domain organization of both enzymes is identical. It is also apparent that the regions of weaker homology and the positions of insertions/deletions between the two enzymes are largely associated with the molecular surface. Large structural differences occur in only two places. The first is the N-terminus of LPDH which is located on the surface of the FAD domain and differs from GR which possess an extra 15 residues comprising a flexible N-terminal segment. The second occurs at the C-terminal region of GR that lines a broad cleft which binds the substrate. LPDH has 18 additional residues here which may be associated with substrate recognition.

FAD binding: A layout of the catalytic site of GR illustrating the relative orientations of GSSG, the redox disulphide, FAD and NADP, is presented in Fig. 7. The FAD prosthetic group is deeply buried within the protein in a domain whose structure consists of a central parallel β-sheet flanked on one face by 3 α-helices and on the other by an anti-parallel β-meander (32). The two sequences indicate a similar structure for this

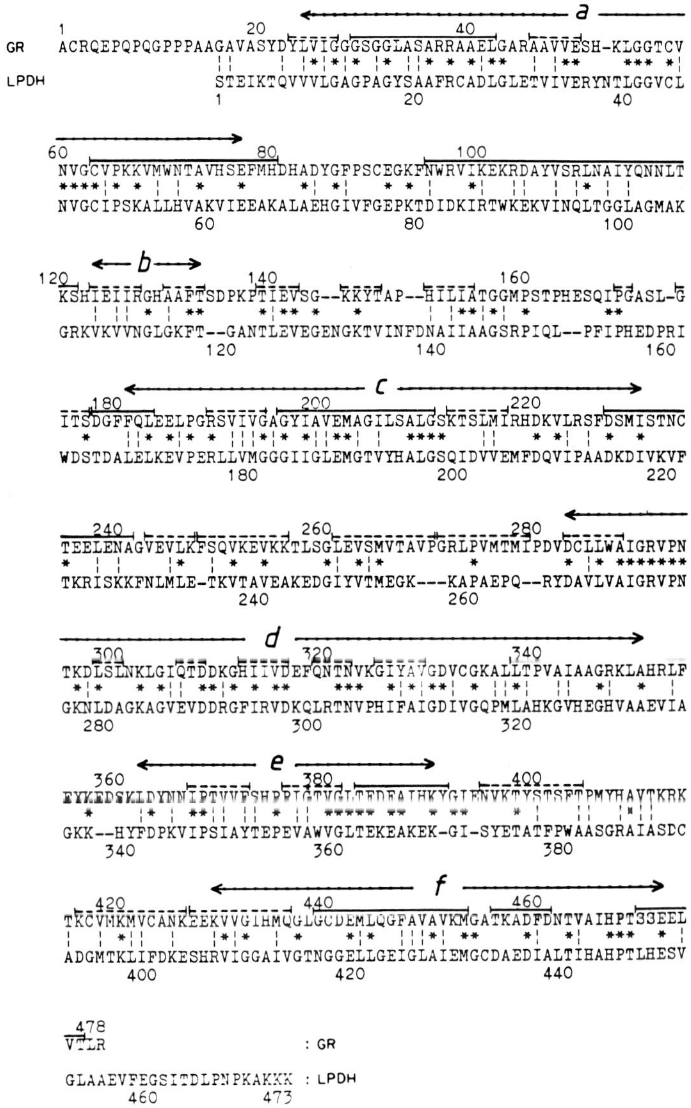

Fig. 6. Alignment of the sequences for GR (upper) and LPDH (lower). Conserved residues are indicated by asterisks and substitutions with scores > 0.10 in MDM_{78} by vertical bars. α-helices and β-sheets in GR are represented by continuous and broken lines respectively.

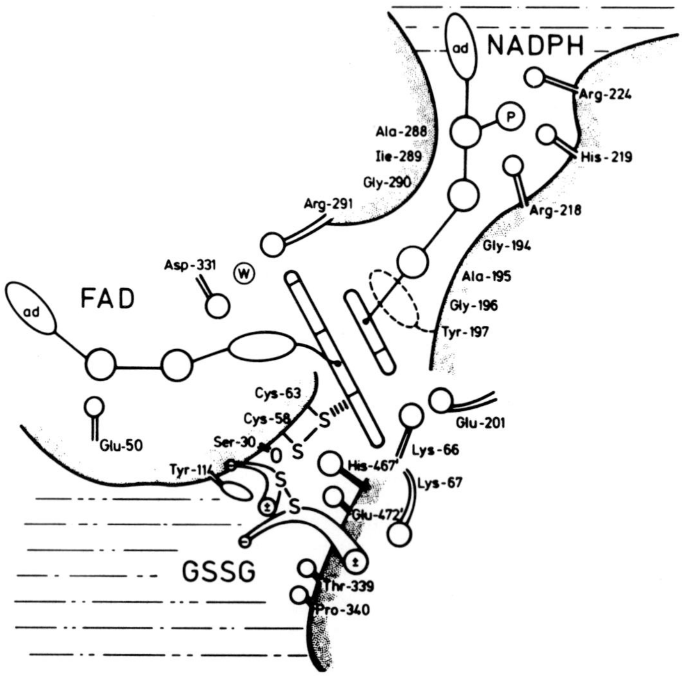

Fig. 7. Schematic diagram of the catalytic centre of GR including FAD, NADP and glutathione. The redox disulphide is formed between Cys58 and Cys63. The broken line denotes the position that Tyr197 adopts when NADPH is absent.

Table 2 Sequence conservation in a segment of the FAD-binding folds of LPDH and GR with part of the FAD-binding domains of SDHA and FRDA

Enzyme	Domain	
SDHA	FAD	(8-31)...DAVVIGAGGAGIARLAQISQS--GQTCALLS...
FRDA	FAD	(6-36)...DLAIVGAGGAGLRAAIAAAQANPNAKIALIS...
LPDH	FAD	(7-35)...QVVVLGAGPAGYSAAFRCADL--GLETVIVE...
GR	FAD	(22-50)...DYLVIGGGSGGLASARRAAEL--GARAAVVE...
LPDH	NAD	(176-211)...RLLVMGGGIIGLEMGTVYHAL--GSQIDVVEMFDQVIP...
GR	NADP	(189-224)...RSVIVGAGYIAVEMAGILSAL--GSKTSLMIRHDKVLR...

Residues important in cofactor recognition are underlined.

domain in LPDH and GR, and in addition many of the specific contacts between the FAD moiety and GR are retained in LPDH, signifying a similar geometrical arrangement for FAD in both enzymes (31).

Protection from auto-oxidation: The reduced and half-reduced forms of GR are relatively stable to oxygen (33). It has been suggested that this may in part stem from a conformational flexibility induced by NADPH binding (32). In GR the side chain of Tyr197 shields the re-face of the isoalloxazine ring of FAD from contact with solvent and moves away by rotation around the $C\alpha - C\beta$ bond on binding NADPH. This allows the aromatic rings of FAD and NADPH to stack against one another (Fig. 7) (34). On reduction of FAD by NADPH, $NADP^+$ is expelled from the active site and the tyrosyl side chain moves back possibly protecting the isoalloxazine ring from attack by oxygen. In LPDH the polypeptide chain around Tyr197 is strongly conserved but the tyrosine residue itself is replaced by an isoleucine, which cannot undergo a comparable large conformational change. Thus, if LPDH does indeed possess a protection mechanism, it must be different from that of GR and this may account for the greater oxygen sensitivity of the EH_2 form of LPDH.

NAD(P) binding: Strong homology between GR and LPDH occurs throughout the NAD(P) binding domain, and particularly in the region interacting with the pyrophosphoryl group of NADP: residues 194-196 (Gly-Ala-Gly in GR, Gly-Gly-Gly in LPDH) and 288-291 (Ala-Ile-Gly-Arg in GR and LPDH). The NADP-binding domain of GR has a similar chain topology to that of the FAD-binding domain (35) and the sequence conservation in segments of the FAD and NAD(P) domains of several flavoproteins is shown in Table 2. In all cases, a similar pattern of residues emerges with relatively strict conservation at some positions and conservative substitutions at others. This is consistent with their overall structural equivalence but there is also evidence of specific substrate recognition. The 2' phosphate group of NADP is bound to GR by multiple interactions with Arg218, His219 and Arg224 (36), and the substitution of these residues to Met, Pro and Phe in LPDH is consistent with its requirement for NAD rather than NADP as the pyridine nucleotide cofactor. The substitution Ile217 (GR) to Glu204 (LPDH) can also be related to nucleotide specificity and to the structural equivalence of the FAD and NADP domain of GR. When the co-ordinates of the FAD and NADP domains of GR are superimposed Ile217 (NADP domain) is structurally equivalent to Glu50 (FAD domain). The function of Glu50 is to form hydrogen bonds to the 2' and 3' hydroxyl groups of the FAD ribose and consequently the substitution of Ile217 by Glu in LPDH is likely to provide a residue to bind the ribose hydroxyl groups of the NAD moiety in this enzyme.

Substrate binding and the catalytic site: Around the glutathione binding site non-conservative substitutions occur including Ser30, Tyr114, Thr339 and Pro340 (Pro Leu Ala His in LPDH), as well as the possible addition in this region of 18 residues at the C-terminus of LPDH. It is assumed that these substitutions are related to lipoamide binding. However, the region around the catalytic site shows strong homologies, with residues contributed by many parts of the polypeptide chain conserved (31). Thus in LPDH the catalytic disulphide Cys58-Cys63 and the polar couple (His467-Glu472) identified as playing a role in proton donation during the catalytic cycle are retained. Therefore a catalytic reaction analogous to that in GR with His467 as a proton donor/acceptor can be envisaged for LPDH.

Conclusion

The successful completion of the primary structure of the lpd gene coupled to the availability of the 3-D structure for a related enzyme, has enabled detailed inferences to be made concerning the molecular structure of LPDH and emphasizes the value of this type of approach.

Acknowledgements

The authors wish to acknowledge the contribution of G. Schulz to the GR/LPDH comparison and the SERC and MRC for financial support.

References

1. Young, I. G., Rogers, B. L., Campbell, H. D., Jaworowski, A., Shaw, D.C.: Eur. J. Biochem. 116, 165-170 (1981).
2. Cole, S. T.: Eur. J. Biochem. 122, 479-484 (1982).
3. Cole, S. T., Grundström, T., Jaurin, B., Robinson, J. J., Weiner, J. H.: Eur. J. Biochem. 126, 211-216 (1982).
4. Grundström, T., Jaurin, B.: Proc. Natl Acad. Sci. U.S.A. 79, 1111-1115 (1982).
5. Wood, D., Darlison, M. G., Wilde, R. J., Guest, J. R.: Biochem. J. in press.
6. Darlison, M. G., Guest, J. R.: Biochem. J. in press.
7. Stephens, P. E., Lewis, H. M., Darlison, M. G., Guest, J. R.: Eur. J. Biochem. 135, 519-527 (1983).
8. Young, I. G., Jaworowski, A., Poulis, M.: Biochemistry 21, 2092-2095 (1982).
9. Neidhardt, F. C., Vaughn, V., Phillips, T. A., Bloch, P. L.: Microbiol. Rev. 47, 231-284 (1983).
10. Greer, S., Perham, R. N.: this volume.
11. Schryvers, A., Weiner, J. H.: Can. J. Biochem. 60, 224-231 (1982).
12. Bachmann, B. J.: Microbiol. Rev. 47, 231-284 (1983).
13. Guest, J. R., Stephens, P. E.: J. Gen. Microbiol. 121, 277-292 (1980).
14. Spencer, M. E., Guest, J. R.: J. Bacteriol. 151, 542-552 (1982).
15. Stephens, P. E., Darlison, M. G., Lewis, H. M., Guest, J. R.: Eur. J. Biochem. 133, 155-162 (1983).
16. Stephens, P. E., Darlison, M. G., Lewis, H. M., Guest, J. R.: Eur. J. Biochem. 133, 481-489 (1983).
17. Ner, S. S., Bhayana, V., Bell, A. W., Giles, I. G., Duckworth, H. W., Bloxham, D. P. Biochemistry, 22, 5243-5249 (1983).
18. Darlison, M. G., Spencer, M. E., Guest, J. R.: Eur. J. Biochem. 141, 351-359 (1984).
19. Spencer, M. E., Darlison, M. G., Stephens, P. E., Duckenfield, I. K., Guest, J. R.: Eur. J. Biochem. 141, 361-374 (1984).
20. Staden, R.: Nucleic Acid. Res. 10, 2951-2961 (1982).
21. Cole, S. T., Guest, J. R.: Mol. Gen. Genet. 178, 409-418 (1980).
22. Guest, J. R., Darlison, M. G., Wilde, R. J., Wood, D.: this volume.
23. Kenney, W. C., Walker, W. H., Singer, T. P.: Biol. Chem. 247, 4510-4513 (1972).
24. Kenney, W. C., Mowery, P. C., Seng, R. L., Singer, T. P.: J. Biol. Chem. 251, 2369-2373 (1976).
25. Unden, G., Kröger, A.: FEBS Lett. 323-326 (1980).
26. Vik, S. B., Hatefi, Y.: Proc. Natl Acad. Sci. U.S.A. 78, 6749-6753 (1981).
27. Guest, J. R.: J. Gen. Microbiol. 122, 171-179 (1981).
28. Kyte, J., Doolittle, R. F. J. Mol. Biol. 157, 105-132 (1982).
29. Williams, C. H.: In The Enzymes (Boyer, P. D., ed.) Academic Press New York 13, 89-173 (1976).
30. Thieme, R., Pai, E. F., Schirmer, R. H., Schulz, G. E.: J. Mol. Biol. 152, 763-782 (1981).
31. Rice, D. W., Schulz, G. E., Guest, J. R.: J. Mol. Biol. 174, 483-496 (1984).

32. Schulz, G. E., Schirmer, R. H., Pai, E. F.: J. Mol. Biol. 160, 287-308 (1982).
33. Coleman, R. F., Black, S.: J. Biol. Chem. 240, 1796-1803 (1965).
34. Pai, E. F., Schulz, G. E.: J. Biol. Chem. 258, 1752-1757 (1983).
35. Schulz, G. E.: J. Mol. Biol. 138, 335-347 (1980).
36. Pai, E. F., Schirmer, R. H., Schulz, G. E.: In Structures of Complexes Between Biopolymers and Low Molecular Weight Molecules (Bartmann, W., Snatzke, G., eds). Wiley, Chichester, 39-52 (1982).

GLUTATHIONE REDUCTASE: MUTATION, CLONING AND SEQUENCE ANALYSIS OF THE GENE IN *E. coli*

Shaun Greer and Richard N. Perham

Department of Biochemistry, University of Cambridge,
Tennis Court Road, Cambridge CB2 1QW, England

Summary

A mutation in the *Escherichia coli* gene *(gor)* for glutathione reductase (EC 1.6.4.2) has been identified and mapped. Mutants deficient in glutathione reductase are not impaired in growth, showing that *E. coli* has no absolute need for this enzyme. A double mutant *(gor trx*B) of *E. coli* has been constructed that is simultaneously deficient in glutathione reductase and thioredoxin reductase (EC 1.6.4.5). The double mutant is severely crippled in growth, implying that these enzymes provide the major, but not necessarily the only routes for coupling NADPH with the reduction of ribonucleoside diphosphates required for DNA biosynthesis. The *gor* gene has been cloned and partially sequenced as a prelude to the study of the enzyme by means of protein engineering.

Introduction

Glutathione reductase (EC 1.6.4.2) is a widespread enzyme which catalyses the reduction of oxidized glutathione by NADPH:

$$\text{GSSG} + \text{NADPH} + \text{H}^+ \longrightarrow \text{NADP}^+ + 2\text{GSH}$$

It is a member of an important family of flavoprotein oxidoreductases that includes dihydrolipoamide dehydrogenase (EC 1.6.4.3), thioredoxin reductase (EC 1.6.4.5) and, a recent addition, mercuric reductase. Dihydrolipoamide dehydrogenase is well-known and widely studied as a component of the 2-oxo-acid dehydrogenase multienzyme complexes (1,2,3) in which it acts to re-

Flavins and Flavoproteins

oxidize the dihydrolipoic acid residues of the lipoate acyltransferase components in an NAD^+-dependent reaction:

$$Lip(SH)_2 + NAD^+ \longrightarrow LipS_2 + NADH + H^+$$

Thioredoxin reductase catalyses the $NADP^+$-linked reduction of a disulphide bridge in the small protein, thioredoxin, that acts in turn as a source of reducing power in the conversion of ribonucleotides to deoxyribonucleotides (4):

$$TrS_2 + NADPH + H^+ \longrightarrow Tr(SH)_2 + NADP^+$$

And mercuric reductase in bacteria is part of a plasmid-encoded system for the detoxification of mercuric ions, catalysing the following reaction (5):

$$Hg^{2+} + NADPH \longrightarrow Hg^\circ + NADP^+ + H^+$$

All these enzymes are dimers with an M_r of about 105 000 and all possess a disulphide bridge in each subunit which is alternately oxidized and reduced as part of the catalytic mechanism (for review see 6). This common feature has served as a focal point for amino acid sequence studies. As shown in Table 1, the sequences around the redox-active disulphide bridges of dihydrolipoamide dehydrogenase, glutathione reductase and mercuric reductase are highly homologous, implying that they have arisen by divergent evolution from a common ancestor (6,7,8,9,10,11). On the other hand, the bridge size and the amino acid sequence surrounding it in thioredoxin reductase are sufficiently different for it to be likely that this enzyme has arisen by convergent evolution towards a common mechanism (7).

The three-dimensional structure of human glutathione reductase is known from its complete amino acid sequence (10) and X-ray crystallographic analysis at 2Å resolution (13), and this has significantly sharpened our appreciation of the reaction mechanism (14). From other lines of enquiry, the *lpd* gene of *Escherichia coli*, encoding dihydrolipoamide dehydrogenase (15), and the *mer*A gene of transposon Tn 501 from *Pseudomonas aeruginosa* encoding mercuric reductase (16), have been cloned and their nucleotide sequences determined. This has enabled the complete primary structures

Table 1: Amino acid sequences around redox-active disulphide bridges

Dihydrolipoamide dehydrogenase[a]	
(a) Pig heart(6,7,8)	T V C I E K B Z T L G G T C L N V G C I P S K
(b) *E. coli*(6,7)	Y N T L G G V C L N V G C I P S K
(c) *B. stearothermophilus*(9)	G N L G G V C L N V G C I P S K
Glutathione reductase[a]	
(a) Human erythrocytes(10)	A A V V E S H K L G G T C V N V G C V P K K
(b) Yeast(6)	L G G T C V N V G C V P K
Mercuric reductase[a]	
P. aeruginosa(11)	G T I G G T C V N V G C V P S K
Thioredoxin reductase	
E. coli(6,12)	A C A T C D G F

[a]Sequences of tryptic peptides

of the corresponding enzymes to be inferred, but further work is somewhat hampered by the lack of complementary X-ray diffraction studies.

In the present paper we describe our own recent efforts to identify mutations in the gene *(gor)* for *g*lutathione *o*xido*r*eductase of *E. coli*, and to clone and sequence the gene as a preamble to a deeper analysis of the mechanism of flavoprotein oxidoreductases by means of the newer techniques of protein engineering.

Isolation and mapping of the *gor* gene of *E. coli*

Glutathione is not an essential metabolite in *E. coli*, for mutants unable to synthesize it are unaffected in growth (17,18). Thus we suspected that mutants unable to reduce oxidized glutathione might also not be impaired in growth. A method for screening *E. coli* colonies for glutathione reductase *in situ* on agar plates was therefore developed (19).

This is based on the reaction of almost colourless 5,5'-dithiobis-(2-nitrobenzoic acid)(Nbs_2) with reduced glutathione to give the strongly yellow-coloured Nbs^- anion, and consisted of pouring a soft agar overlay containing NADPH, Nbs_2 and oxidized glutathione onto agar plates of *E.coli* colonies previously lysed by exposure to chloroform vapour:

$$GSSG + NADPH + H^+ \longrightarrow NADP^+ + 2GSH$$

$$2GSH + Nbs_2 \longrightarrow GSSG + 2Nbs^- + 2H^+$$

Only colonies having glutathione reductase can produce reduced glutathione, so only those colonies turn yellow.

This method was used to screen colonies of *E. coli* mutated by bacteriophage Mu insertion (19). Mu is a bacteriophage which can lysogenize *E. coli*; it does so by inserting itself pseudo-randomly into the chromosome of the cell. If the insertion point is within a gene, a complete block of normal transcription and translation ensues, i.e. a complete block of gene function. The Mu phage used was Mu*cts*Ap. Stocks of this Mu phage are easily produced because it has a temperature-sensitive repressor and thus at elevated temperatures it is forced to enter the lytic phase of its life-cycle. *E. coli* cells containing inserted phage are easy to select because they are ampicillin-resistant; and the ampicillin-resistance also makes the insertions easy to map because it is a readily-scorable trait.

Of 15 000 colonies screened in two independent experiments, eight were obtained that could not reduce glutathione. Analysis of cell-free extracts showed that these mutants truly lacked glutathione reductase.

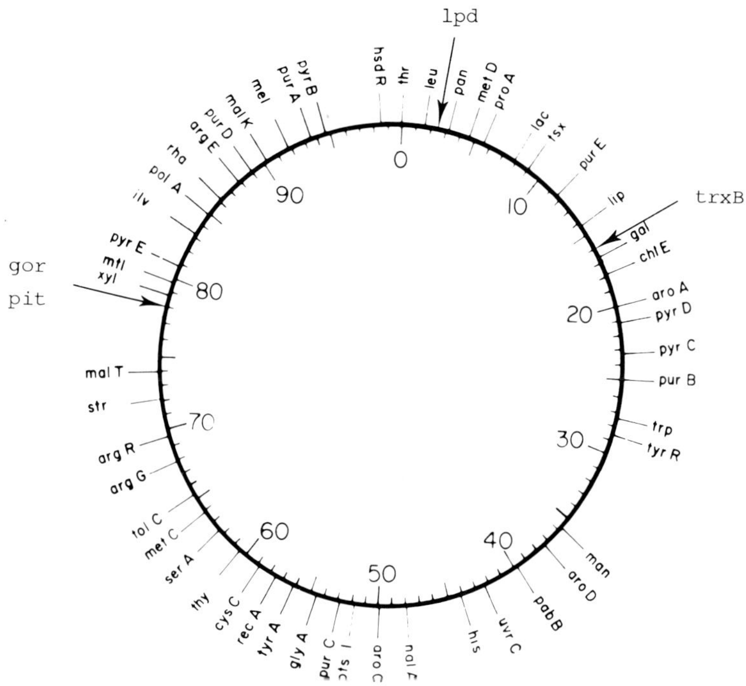

Fig. 1. Linkage map of *E. coli* K-12. The position of the genes for glutathione reductase *(gor)*, dihydrolipoamide dehydrogenase *(lpd)*, thioredoxin reductase (*trx*B) and an inorganic phosphate transport system *(pit)* are shown. The map positions are those listed by Bachmann *et al.* (20,21) except for *trx*B which has been mapped only approximately (22) and was not included.

One of these strains was selected and the *gor* lesion was mapped by transduction with bacteriophages P1 and T4GT7 as vectors: it was located between 77 and 78 min on the *E. coli* genome (Fig. 1). As suspected initially, the *gor* mutation caused no defect in growth (19).

Properties of a *gor*-*trx*B double mutant

Deoxyribonucleotides in *E. coli* are synthesized by reduction of the corresponding ribonucleotides in the presence of the enzyme ribonucleotide diphosphate reductase (23). An essential cofactor for ribonucleotide reductase is either thioredoxin or glutaredoxin, small proteins (M_r 12 000) containing an intrachain disulphide bridge (4). Thioredoxin is reduced by NADPH in a reaction catalysed by thioredoxin reductase, an enzyme encoded by gene *trx*B (24); glutaredoxin requires glutathione/glutathione reductase for its reduction (Fig. 2).

The thioredoxin and glutaredoxin systems function in parallel (4): the simultaneous presence of both thioredoxin reductase and glutathione reductase is not required since mutants defective in thioredoxin reductase (*trx*B) (24) and glutathione reductase (*gor*) (19) are individually viable. The obvious question is whether a *gor*-*trx*B double mutant is also viable - that is, whether reducing power must reach ribonucleotide reductase *via* one or other of these enzymes, or whether yet another pathway exists.

Combination of the *gor* and *trx*B mutations in the same *E. coli* cell is not a trivial operation because neither of these mutations is itself directly selectable. Fortunately, the *gor* mutation we describe above is selectable because of the Mu*cts*Ap used to block gene function. The obvious approach is to transduce the *gor* :: Mu*cts*Ap construct from this mutant into a strain already known to contain the *trx*B mutation. This recipient strain must be lysogenic for bacteriophage Mu so that there is Mu repressor in the cell - otherwise the incoming Mu in the *gor* :: Mu*cts*Ap construct will simply enter

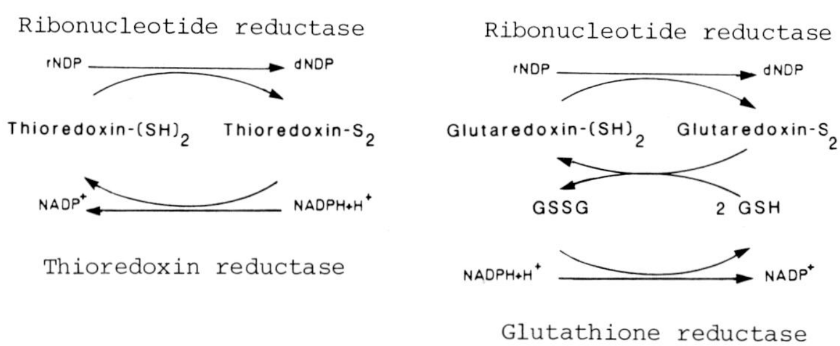

Fig. 2. Accepted pathways for the biosynthesis of deoxyribonucleotides

the lytic phase and kill the potential double mutant even before it can form. The *trx*B strain described by Fuchs (24) was used as the starting point for the manipulations. It was first made lysogenic for Mu*cts*Km, a Mu phage bearing a gene for kanamycin-resistance. This new strain was then used as the recipient in a bacteriophage P1-mediated transduction using a *gor* :: Mu*cts*Ap donor and selecting for ampicillin-resistant progeny. There are two possible outcomes of such a transduction; the incoming Mu may recombine with the Mu already inserted in the chromosome of the recipient, leaving the *gor* gene of the recipient intact, or the incoming *gor* gene may recombine with the *gor* gene of the recipient, rendering the recipient glutathione reductase-negative. Thus two classes of ampicillin-resistant recombinants are expected in roughly a one-to-one ratio. This in fact is what was found.

One class of recombinants grew much like the recipient strain whereas the second class grew much more slowly. As expected, the first class still possessed glutathione reductase but the second class lacked it. They also still lacked thioredoxin reductase, as shown in Table 2. These *gor*-*trx*B double mutants grow poorly relative to the parent strains, but they do grow. The requirement for either of these two enzymes is therefore not, in *E. coli*, an absolute one.

Table 2: Levels of glutathione reductase, thioredoxin reductase and dihydrolipoamide dehydrogenase activities in mutated *E. coli* cells

Strain	Mutations	Cell-free extract			
			Enzyme activities (nmol/min/mg prot)		
		Prot.concn (mg/ml)	GR	TR	DLD
JF202	gor^{+}, $trxB^{-}$, lpd^{+}	4.1	104	n.d.	600
SG2	gor^{-}, $trxB^{+}$, lpd^{+}	5.4	n.d.	39	800
14a	gor^{-}, $trxB^{-}$, lpd^{+}	3.4	n.d.	n.d.	790

n.d. = not detectable

Cloning the *E. coli* gene for glutathione reductase

The *gor* mutation described above is not ideal for cloning work because of the presence of the Mu*cts*Ap, which makes the strain both temperature-sensitive and ampicillin-resistant. This makes difficult the use of normal transformation methods and common plasmids. A successful attempt was made to remove this impediment by deleting the *gor* gene and the Mu phage inserted in it from the *E. coli* chromosome. The strategy adopted for this was as follows. The *gor* gene is known to map close to the *pit* locus (Fig. 1) which encodes one of the two *E. coli* transport systems for inorganic phosphate and in particular is the system largely responsible for the uptake of arsenate ion (25). Strains in which the *pit* locus is mutated are thus relatively insensitive to arsenate in the growth medium. Growth of the *gor* :: Mu*cts*Ap mutant on arsenate-containing medium at elevated temperature will presumably select for cells which are not temperature-sensitive or arsenate-sensitive. Among these cells should be *pit*-*gor* deletion mutants.

The arsenate medium used for the initial screen provided only a weak counter-selection against cells with a functional *pit* locus. Strains isolated from this test were checked on ampicillin-containing medium for loss of the Mu phage. One strain was ampicillin-sensitive, though when tested on the original selection medium it proved much more arsenate-sensitive than its parent. However, it was sensitive to bacteriophage Mu and, when a cell-free extract was assayed, it was found to be glutathione reductase-negative. It presumably contains a deletion of the *gor* gene and neighbouring DNA which also renders the cells arsenate-sensitive. The isolation of this strain off a selective medium to which it is extremely sensitive must be regarded as highly fortuitous.

As we have seen, the *gor* mutation in *E. coli* has no known phenotype and there is no growth test that can be used to select for glutathione reductase-containing cells, an otherwise obvious route for cloning the gene from a gene bank. We could of course apply in reverse the direct colony assay used to detect the original *gor* mutants (see above) but that was judged to be too tedious. However, the arsenate-sensitivity of the

Δ*gor* mutant is probably due to the deletion of a portion of the genome close to the *gor* gene. We argued that if it were possible to clone sections of DNA conferring arsenate-resistance back into this strain, the *gor* gene might be cloned simultaneously if the map distance between it and and arsenate-resistance locus was small enough. For these experiments a counter-selection medium for the Δ*gor* mutant was developed on which it was possible to select against the mutant in favour of its *gor* :: Mu*cts*Ap parent or a wild-type *E. coli*. Reversion of the arsenate-sensitivity of the Δ*gor* mutant to arsenate-resistance was undetectable and therefore did not interfere with the selection.

By this means, the Δ*gor* recipient strain was used to screen the Clarke and Carbon (26) bank of recombinant plasmids containing *E. coli* DNA for arsenate-resistance. Of the arsenate-resistant clones isolated during the screen, ten were tested by colony assay for the presence of glutathione reductase. All of the strains appeared glutathione reductase-positive. After screening the cells for plasmid, the strain containing the smallest plasmid was used for the isolation of plasmid DNA by a rapid boiling method (27). This DNA preparation transformed the Δ*gor* mutant to arsenate-resistance; and when a cell-free extract was assayed it was clear that the cells had also been rendered glutathione reductase-positive.

The Clarke and Carbon bank strain carrying this plasmid was isolated and plasmid DNA prepared (28). On agarose gel electrophoresis this DNA preparation appeared homogeneous and, as expected, it could transform the Δ*gor* mutant to arsenate-resistance. Assay of cell-free extracts showed that the arsenate-resistant transformants were also glutathione reductase-positive. Analysis of linearized plasmid DNA on agarose gel electrophoresis showed the plasmid to be about 16.5 kilobase pairs in length. It proved difficult to remove the inserted *E. coli* chromosomal DNA either whole or piecemeal by digestion with restriction enzymes, so random-sheared pieces of whole plasmid DNA were cloned into bacteriophage m13mp9 and these clones were then sequenced using the dideoxy-chain termination method (29,30). Because of the strong homology expected between human and *E. coli* glutathione reductase, it was hoped that some of these clones would be recognized as

E. coli	LysMetGlyAlaThrLysLysAspPheAspAsnThrValAlaIle AAGATGGGGGCAACCAAAAAAGACTTCGACAATACCGTCGCCATT
Human	LysMetGlyAlaThrLysAlaAspPheAspAsnThrValAlaIle
	452 460
E. coli	HisProThrAlaAlaGluGluPheValThrMetArgTER CACCCAACGGCGGCAGAAGAGTTCGTGACAATGCGTTAAATGTTA
Human	HisProThrSerSerGluGluLeuValThrLeuArgTER
	467 470 478

Fig. 3. DNA sequence (tentative) of the 3'-terminal region of the *E. coli gor* gene. The amino acid residues overscored are identical with those of human erythrocyte glutathione reductase, the numbering of which is taken from ref. 10.

part of the *gor* gene. This turned out to be the case. For example, a clone containing DNA encoding the carboxy-terminus of glutathione reductase was isolated from among those sequenced (Fig. 3). From this clone, a radioactive probe to this portion of the gene was made and used to probe Southern blots of plasmid DNA cut with various restriction enzymes. In this way, the *gor* gene was localized to one end of the chromosomal DNA insert carried in the plasmid (Fig. 4). This section of the DNA was isolated and the complete DNA sequence of the *E. coli gor* gene should soon be available.

P : Pst 1

S : Sph 1

T : Tth III-1

Scale: ⊢—⊣ = 1 kbp

Fig. 4. Restriction map of the Col E1 plasmid carrying the *E.coli gor* gene

Discussion

Several interesting conclusions can be drawn from the results we have obtained thus far. First, glutathione reductase is not itself an essential enzyme in *E. coli*, though a cell devoid of glutathione reductase and thioredoxin reductase is crippled in growth. The existence of twin pathways for ribonucleotide reduction via glutaredoxin and thioredoxin can be viewed as valuable protection for this essential metabolic pathway. The crippled growth of the *gor*, *trx*B double mutant would imply that glutathione reductase and thioredoxin reductase provide the major routes for coupling NADPH with the reduction of ribonucleoside diphosphates. In the absence of these enzymes, thioredoxin and glutaredoxin may be entering a more general network of disulphide oxidoreductions in the cell. An entertaining possibility in the context of flavoprotein dehydrogenases is the involvement of the NAD^+-linked 2-oxo acid dehydrogenase complexes and dihydrolipoamide dehydrogenase, since free dihydrolipoic acid can readily reduce thioredoxin (A. Holmgren and R.N. Perham, unpublished work).

It is clear from the C-terminal sequence of the *E. coli* enzyme (Fig. 3) that the histidine residue assigned a specific role as proton donor in the mechanism of the human enzyme (14) is present as part of a highly-conserved amino acid sequence. A comparable histidine residue has been implicated in the mechanism of dihydrolipoamide dehydrogenase (31) but is lacking in the related enzyme mercuric reductase (16). The cloning and complete nucleotide sequence of the *gor* gene will open the way to a sophisticated analysis of the role of this histidine and other residues in the mechanism by means of protein engineering.

Acknowledgements

We thank the Department of Education for Northern Ireland for the award of a Studentship to S.G. We are grateful to Dr. M. Jones-Mortimer and Mr. N. Davis for their help with Mu-phage mutagenesis and mapping experiments, and to Dr. J. Fuchs and Professor J. Guest for gifts of bacterial strains.

References

1. Reed, L.J.: Acc. Chem. Res. 1, 40-46 (1974).
2. Lowe, P.N., Hodgson, J.A., Perham, R.N.: Biochem. J. 215, 133-140 (1983).
3. Packman, L.C., Hale, G., Perham, R.N.: EMBO Journal 3, 1315-1319 (1984).
4. Holmgren, A.: Experientia (suppl.) 36, 149-180 (1980).
5. Fox, B.S., Walsh, C.T.: J. Biol. Chem. 257, 2498-2503 (1982).
6. Williams, C.H. Jr.: The Enzymes, 3rd Edition 13, 89-173 (1976).
7. Perham, R.N., Harrison, R.A., Brown, J.P.: Biochem. Soc. Trans. 6, 47-50 (1978).
8. Williams, C.H. Jr., Arscott, L.D., Schulz, G.E.: Proc. Natl. Acad. Sci. USA 79, 2199-2202 (1982).
9. Packman, L.C., Perham R.N.: FEBS Lett. 139, 155-158 (1982).
10. Krauth-Siegel, R.L., Blatterspiel, R., Saleh, M., Schiltz, E., Schirmer, R.H., Untucht-Grau, R.: Eur. J. Biochem. 121, 259-267 (1982).
11. Fox, B.S., Walsh, C.T.: Biochemistry 22, 4082-4088 (1983).
12. Ronchi, S., Williams, C.H. Jr.: J. Biol. Chem. 247, 2083-2086 (1972).
13. Thieme, R., Pai, E.F., Schirmer, R.H., Schulz, G.E.: J. Mol. Biol. 152, 763-782 (1981).
14. Pai, E.F., Schulz, G.E.: J. Biol. Chem. 258, 1752-1757 (1983).
15. Stephens, P.E., Lewis, H.M., Darlison, M.G., Guest, J.R.: Eur. J. Biochem. 135, 519-527 (1983).
16. Brown, N.L., Ford, S.J., Pridmore, R.D., Fritzinger, D.C.: Biochemistry 22, 4089-4094 (1983).
17. Apontoweil, P., Berends, W.: Biochim. Biophys. Acta 399, 10-22 (1975).
18. Fuchs, J.A., Warner, H.R.: J. Bacteriol. 124, 140-148 (1975).
19. Davis, N.K., Greer, S., Jones-Mortimer, M.C., Perham R.N.: J. Gen Microbiol. 128, 1631-1634 (1982).
20. Bachmann, B.J., Low, K.B., Taylor, A.L.: Bacteriol. Rev. 40, 116-167 (1976).
21. Bachmann, B.J.: Microbiol. Rev. 47, 180-230 (1983).
22. Fuchs, J., Haller, B., Tuggle, C.K.: Personal communication.
23. Thelander, L., Reichard, P.: Annu. Rev. Biochem. 48, 133-158 (1979).
24. Fuchs, J.A.: J. Bacteriol. 129, 967-972 (1977).
25. Sprague, G.F.Jr., Bell, R.M., Cronan, J.E.Jr.: Molec. gen. Genet. 143, 71-77 (1975).
26. Clarke, L., Carbon, J.: Cell 9, 91-99 (1976).

27. Holmes, D.S., Quigley, M.: Analyt. Biochem. 114, 193-197 (1981).

28. Maniatis, T., Fritsch, E.F., Sambrook, J.: Molecular Cloning, A Laboratory Manual. Cold Spring Harbor Laboratory 1982.

29. Sanger, F., Coulson, A.R., Barrell, B.G., Smith, A.J.H., Roe, B.A.: J. Mol. Biol. 143, 161-178 (1980).

30. Biggin, M.D., Gibson, T.J., Hong, G.F.: Proc. Natl. Acad. Sci. USA 80, 3963-3965 (1983).

31. Adamson, S.R., Robinson, J.A., Stevenson, K.J.: Biochemistry 23, 1269-1274 (1984).

THE COENZYME BINDING SITE OF GLUTATHIONE REDUCTASE

Correlation of X-ray Studies with Kinetic Data

Emil F. Pai, Elke Horn
Max-Planck-Institut für Medizinische Forschung, Abt. Biophysik, Jahnstr. 29, D-6900 Heidelberg, FRG

Georg E. Schulz
Institut für Organische Chemie und Biochemie der Universität, Albertstr. 21, D-7800 Freiburg, FRG

Introduction

Glutathione reductase is a ubiquitous flavoenzyme catalyzing the reduction of oxidized glutathione by NADPH (1). Its sequence and three-dimensional structure at high resolution are known (2, 3). The stereochemistry of its catalysis has been deduced from X-ray analyses of several reaction intermediates and inhibitor complexes (4). During this study numerous analogues as well as fragments of NADPH and GSSG had been soaked into crystals of the enzyme.

In a first attempt to relate the results of the "static" X-ray method to the behaviour of the enzyme in solution the coenzyme binding site was probed. At first, inhibition constants of substances were measured that had been shown to bind to the NADPH-site of the crystalline enzyme (5). These analogues either had a modified purine or the substituent of the pyridine base had been changed. Then fragments of the NADPH-molecule were tested. For comparison, compounds that had not been used in the crystallographic work were included, too. Finally, Michaelis constants of several modified substrates were determined.

Flavins and Flavoproteins

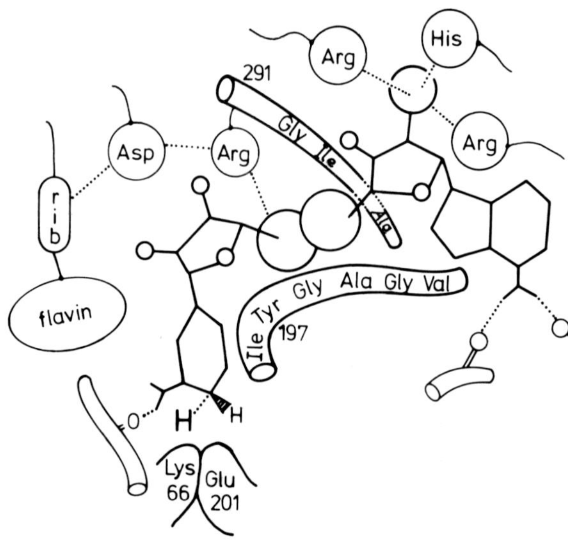

Fig. 1: Sketch of NADPH bound to the coenzyme site of glutathione reductase. Some of the dipolar interactions are indicated by dotted lines.

Results

The results of the kinetic experiments and the x-ray data are clearly consistent (e.g. Fig. 2). There are four regions in the coenzyme which have been studied by modifying the native structure.

The positive charge of an oxidized nicotinamide ring, even when balanced in part by electronic effects as in $AADP^+$, results in repulsion of this ring from its binding site near the isoalloxazine ring of FAD, at least in the oxidized enzyme, E .
The EH_2-form of the enzyme, however, is able to accomodate both reduced and oxidized coenzymes there. Enzyme irreversibly kept in its EH_2-state by reacting the sulfur of Cys-58 with iodoacetamide does bind the entire $NADP^+$ molecule except for the nicotinamide ring .

While 2'-P-5'-ADP carrying three negative charges at its pyro-

KINETIC CONSTANTS

SUBSTRATES:

	K_M (µM)		K_M (µM)
*NADPH	7	NHDPH	50
*NADH	500	3′-NADPH	170

α-NADPH, α-NADH AND NMNH WERE NO SUBSTRATES FOR GLUTATHIONE REDUCTASE.

INHIBITORS:

	K_I (µM)		K_I (µM)
*NεADP$^+$	2000	§NAD$^+$	2900
NHDP$^+$	3000	*AcPADP$^+$	400
ITP-RIBOSE	8000	*THIONADP$^+$	200
3′-NADP$^+$	3000	*AADP$^+$	50
2′,3′-cNADP$^+$	3400	*NADP$^+$	190
α-NADP$^+$	300	*2′-P-5′-ADPRIBOSE	75
§NMN$^+$	45000	*2′-P-5′-ADP-CH_3	150
*5′-ATP	6400	*2′-P-5′-ADP	5000
*5′-ADP	5400	*2′-P-5′-AMP	30
*5′-AMP	12000	*2′-AMP	500

TAB. 1: KINETIC CONSTANTS OF SUBSTRATES (K_M) AND INHIBITORS (K_I) OF GLUTATHIONE REDUCTASE. *ANALYZED BY CRYSTALLOGRAPHIC METHODS. § DID NOT BIND TO THE CRYSTALLINE ENZYME.

phosphates is only weakly bound to the enzyme (K_I = 5000 µM) its methylester, 2'-P-5'-ADP-CH_3, is fixed at its appropriate place (K_I = 150 µM).

The 2'-phosphate group which distinguishes NADPH from NADH provides a major part of the binding energy, as can be seen by comparing 5'-AMP (K_I=12mM) vs. 2'-P-5'-AMP (K_I = 30 µM) or 2'-AMP (K_I = 500 µM). 2'-P-5'-AMP certainly is the ligand of choice when purifying glutathione reductase by affinity chromatography (6). When NADH is soaked into crystals of the enzyme the phosphate/sulfate ion at the 2'-site is not removed but only pushed aside, thus mimicking an NADPH molecule.

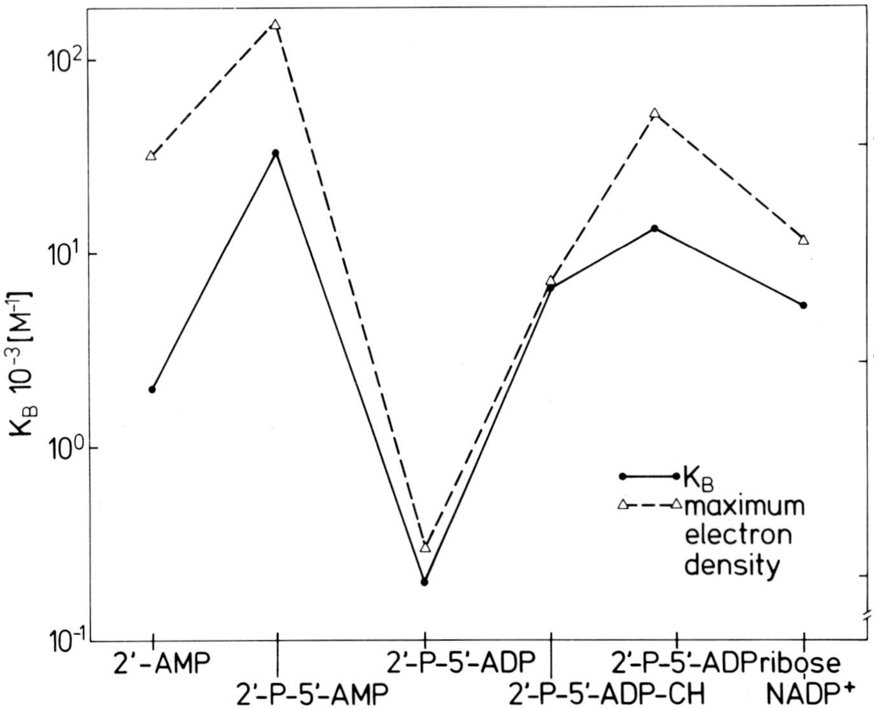

Fig. 2: Comparison between log K_B (●——●) and the normalized maximum electron density of the corresponding difference-Fourier map (△--△) for a series of compounds soaked into crystalline glutathione reductase under very similar conditions.

References

1. Williams, C.H., Jr.: The Enzymes (Boyer, P.D., ed.), 3rd ed. vol. 13, pp. 89-172, Academic Press, New York (1976).
2. Krauth-Siegel, R.L., Blatterspiel, R., Saleh, M., Schiltz, E., Schirmer, R.H., Untucht-Grau, R.: Eur. J. Biochem. 121, 259-267 (1982).
3. Thieme, R., Pai, E.F., Schirmer, R.H., Schulz, G.E.: J. Mol. Biol. 152, 763-782 (1981).
4. Pai, E.F., Schulz, G.E.: J. Biol. Chem. 258, 1752-1757 (1983).
5. Pai, E.F., Schulz, G.E.: Flavins and Flavoproteins (Massey,V., Williams, C.H., eds.) pp. 3-10, Elsevier, Amsterdam (1982).
6. Krohne-Ehrich, G., Schirmer, R.H., Untucht-Grau, R.: Eur. J. Biochem. 80, 65-71 (1977).

^{13}C-NMR STUDY ON THE ACTIVE SITES OF LIPOAMIDE DEHYDROGENASE AND GLUTATHIONE REDUCTASE

Willy A.M. van den Berg, Jacques Vervoort, Chrit T.W. Moonen and Franz Müller

Department of Biochemistry, Agricultural University, De Dreijen 11, 6703 BC Wageningen, The Netherlands

Inger Carlberg and Bengt Mannervik, Department of Biochemistry, Arrhenius Laboratory, University of Stockholm, S-106 91 Stockholm, Sweden

Introduction

Lipoamide dehydrogenase from A.vinelandii and glutatione reductase from human erythrocytes have been investigated. Both enzymes exist in three distinct redox states, i.e. oxidized (E), two-electron (EH_2) and four-electron (EH_4) reduced states (1-4). EH_2 is a catalytically important intermediate in both enzymes. The stability of EH_2 differs for both enzymes (1-4). The natural prosthetic group in both enzymes has been replaced by selectively ^{13}C enriched FAD derivatives. The ^{13}C-NMR spectra should reveal the oxidation state and the microenvironment of the prosthetic group in the various redox states of the two proteins. The three-dimensional structure of glutathione reductase has been solved at a high resolution (5). These data form a solid base for a NMR study on this enzyme. The results presented in the following have to be considered as tentative. In order to arrive at a detailed interpretation of the ^{13}C-NMR results additional experiments and ^{15}N-NMR data are needed.

Results and discussion

The ^{13}C-NMR spectra of lipoamide dehydrogenase and glutatione reductase are shown in figure 1 and figure 2, respectively. The spectra demonstrate that the ^{13}C resonances of FAD can be easily distinguished from the natural abundance ^{13}C resonances of the proteins. The latter resonances represent the carbonyl and carboxyl groups (170-180 ppm), arginine C_ζ (158.2 ppm) and

Flavins and Flavoproteins

tyrosine, tryptophan, phenylalanine and histidine residues (110-140 ppm) (6). The natural abundance ^{13}C resonance line at about 158 ppm exhibits an intensity very similar to those of the ^{13}C atoms of protein-bound FAD and can therefore be used as an internal standard to estimate roughly the concentration of the prosthetic group in the different redox states.

^{13}C chemical shifts of lipoamide dehydrogenase and glutathione reductase in the three redox states are compared with those of free FMN (7) in Table 1. The results can be summarized as follows. The ^{13}C chemical shifts due to C(4) and C(10a) of both enzymes in the oxidized state resemble very much those of FMN. In contrast C(2) in glutathione reductase and C(4a) in lipoamide dehydrogenase are shifted downfield whereas C(2) in lipoamide dehydrogenase and C(4a) in glutathione reductase are shifted upfield. These results indicate that the protein-bound flavin in both enzymes is strongly polarized (7), but that C(4a) in glutathione reductase possesses a higher π electron density than that in lipoamide dehydrogenase. Furthermore C(2) in glutatione reductase is more polarized than that in lipoamide dehydrogenase. A comparison of the ^{13}C chemical shifts in the EH_2 redox state of both enzymes

Table 1. Comparison of the ^{13}C chemical shifts of protein-bound FAD in lipoamide dehydrogenase from A.vinelandii (pH 7.0) and in glutathione reductase (pH 7.0) with those of free FMN.

^{13}C-Atom	^{13}C Chemical Shifts (ppm) in							
	Lipoamide Dehydrogenase			Glutathione Reductase			Free FMN	
	E	EH_2	EH_4	E	EH_2	EH_4	FMN	$FMNH^-$
C(2)	158.3	158.6	159.4	160.4	161.6	161.0	159.8	158.2
C(4)	163.2	163.8	159.7	163.3	163.9	162.0	163.7	157.7
C(4a)	137.6	140.3	98.9	135.7	139.9	99.7	136.2	101.4
C(10a)	152.2	153.0	157.2	152.3	153.8	159.1	152.1	155.5

shows that all carbon atoms are shifted downfield as compared to those in E state. This means that the π electron density is apparently decreased in EH_2 as compared to E. The EH_2 state of the enzymes consists of oxidized

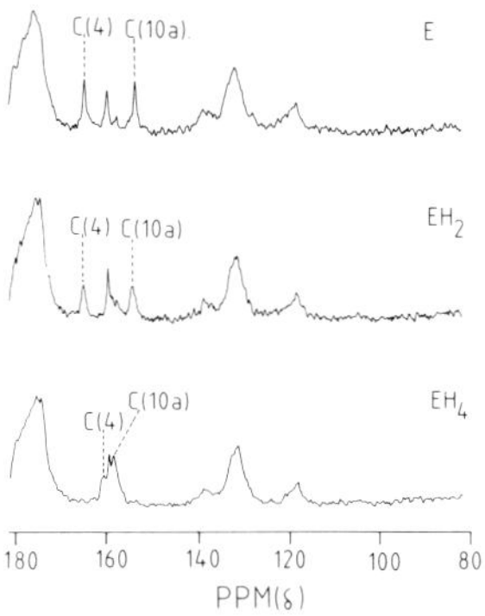

Figure 1. ^{13}C-NMR spectra of lipoamide dehydrogenase from A.vinelandii in 0.1 M phosphate buffer, pH 7.0, in the three redox states.

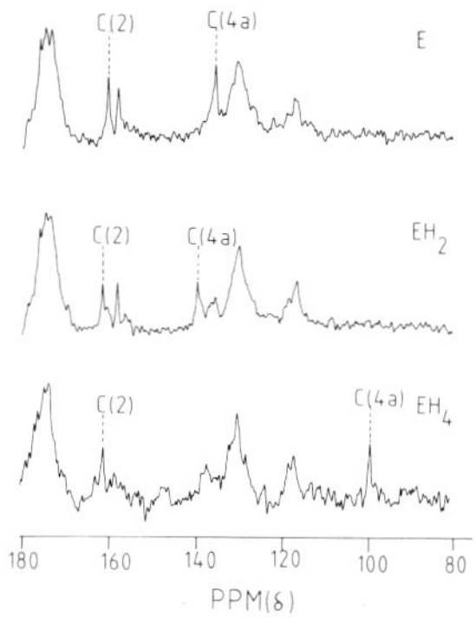

Figure 2. ^{13}C-NMR spectra of glutathione reductase in 0.1 M phosphate buffer, pH 7.0, in the three redox states.

flavin and two thiol groups in the active center, forming a charge transfer complex (1). The negative charge on the thiolate group, which is located close to the prosthetic group, (5) could induce the downfield shifts. The existence of a charge transfer complex should manifest itself by an increase of charge density in the prosthetic group, as is shown by results obtained on old yellow enzyme (see Beinert, Rüterjans and Müller, this symposium). An increase in charge density in the prosthetic group should lead to an upfield shift of the resonances. This is not observed for the four carbon atoms under consideration. To clarify this point further experiments must be awaited. Just a few additional observations on the EH_2 states will be discussed here. A rough estimation of the concentration of EH_2 indicates that glutathione reductase contains more EH_2 than lipoamide dehydrogenase. This is in agreement with the observed stability of the intermediate of different enzymes (1). Addition of NAD^+ to EH_2 of lipoamide dehydrogenase does not influence the chemical shifts but the intensity of the resonance lines is slightly reduced, indicating disproportionation of EH_2 (3). On the other hand addition of $NADP^+$ to EH_2 of glutathione reductase leads to a great loss of intensity of the resonance lines. It is not yet clear if this is either due to the formation of flavin radical or due to an exchange reaction of protein-bound $NADP^+$. In the former case a fast electron exchange between the flavin molecules would strongly broaden the ^{13}C resonances (8). In the latter case a rough calculation shows that the nicotinamide moiety of $NADP^+$ exchanges with a rate of about 100-1000 s^{-1}. In the four electron reduced state (EH_4) the carbonyl functions at position 2 and 4 are strongly polarized, in glutathione reductase to a greater extent than in lipoamide dehydrogenase. In addition there is no doubt that the protein-bound FAD in EH_4 is ionized in both enzymes, as is deduced by a comparison of the chemical shifts of the protein-bound FAD with those of reduced ionized ($FMNH^-$) flavin (see Table 1).

Acknowledgements

We thank Miss Lyda Verstege for typing the manuscript, Dr. C.A.H. Rasmussen for carefully reading the manuscript and Mr. M.M. Bouwmands for preparing the figures. This study was supported in part by the Dutch Foundation for Chemical Research (SON) with financial aid from the Dutch Organization for the Advancement of Pure Research and the Swedish Natural Science Research Council.

References

1. Williams, C.H.: Enzymes, 3rd Ed. 13, 89-173 (1976).
2. Mathews, R.G., Ballou, D.P., Williams, C.H.: J.Biol.Chem. 254, 4974-4981 (1979).
3. Wilkinson, K.D., Williams, C.H.: J.Biol.Chem. 254, 852-862 (1979).
4. Ascott, L.D., Thorpe, C., Williams, C.H.: Biochemistry, 20, 1513-1520 (1981).
5. Pai, E.F., Schulz, G.E.: J.Biol.Chem. 258, 1752-1757 (1983).
6. Allerhand, A.: Methods Enzymol. 61, 458-549 (1979).
7. Moonen, C.T.W., Vervoort, J., Müller, F.: Biochemistry 23, (1984), in press.
8. Moonen, C.T.W., Müller, F.: Eur.J.Bichem. 140, 303-309 (1984).

X-RAY CRYSTALLOGRAPHIC STUDIES ON LIPOAMIDE DEHYDROGENASE FROM *AZOTOBACTER VINELANDII*

A.J. Schierbeek, J. Drenth & W.G.J. Hol
Lab. of Chemical Physics, Nijenborgh 16, 9747 AG Groningen, The Netherlands

Lipoamide dehydrogenase is an FAD containing NAD^+-dependent enzyme which exists in A.vinelandii as a dimer of two identical subunits with Mr = 51,000 per polypeptide chain. It is part of the pyruvate dehydrogenase and α-ketoglutaric acid dehydrogenase complex and catalyses the reoxidation of dihydrolipoic acid which is covalently bound in amide linkage to the core enzyme of the complex.

$$NAD^+ + \text{dihydrolipoamide} \rightleftharpoons NADH + H^+ + \text{lipoamide}$$

It crystallizes as thin plates from solutions containing 10 mg/ml protein in 0,1 M K-phosphate pH 7.5, 0,1 mM EDTA and 20% (w/v) PEG 4000 in space group $P2_12_12_1$ with a=64,0, b=83.8 and c=193 Å and one dimer per asymmetric unit. The crystals are quite sensitive to radiation damage. The effective lifetime of the crystals could be prolonged with synchrotron radiation. In combination with cooling down to -10°C the resolution was increased up to 2.4 Å. Therefore a native dataset will be collected at DESY, EMBL at Hamburg using synchrotron radiation.

Lipoamide dehydrogenase displays extensive homology with Glutathione reductase. The three-dimensional structure of the latter is known. Elucidation of the structure of lipoamide dehydrogenase will be essential to explain the functional differences between this enzyme and Glutathione reductase.

Flavins and Flavoproteins

MOBILITY OF LIPOAMIDE DEHYDROGENASE IN AND OUT OF THE PYRUVATE DEHYDROGENASE COMPLEX FROM *AZOTOBACTER VINELANDII*

Arie de Kok and Antonie J.W.G. Visser

Department of Biochemistry, Agricultural University, De Dreijen 11, 6703 BC Wageningen, The Netherlands

Lipoamide dehydrogenase is a dimer of M_r = 104 ± 5 kDa, as determined by sedimentation equilibrium experiments. Up to four dimers can be bound to the tetrameric acetyltransferase component of the pyruvate dehydrogenase complex, $K_d < 0.5$ nM (1,2). Time dependent anisotropy measurements of the flavin fluorescence indicate two rotational correlation times (τ_c); a short one in the order of several picoseconds, ascribed to rapid internal motion of the coenzyme and a longer one, 17-20 ns, corresponding to the rotation of a sphere of 40-50 kDa and ascribed to a hinged subunit movement. No change in the longer τ_c is observed when the enzyme is bound to the pyruvate dehydrogenase complex (containing one dimer of lipoamide dehydrogenase) or to the transacetylase tetramer (containing one to four dimers of the enzyme). This indicates that the binding domains for lipoamide dehydrogenase (and probably also for pyruvate dehydrogenase) consist of flexible polypeptide chains. Movements of the peripheral components of the complex may therefore play a role in catalysis comparable to the flexibility of the polypeptide chains to which the lipoylgroups are bound (3).

Upon dilution to catalytic concentrations the longer τ_c decreases (fig. 1). A more extensive decrease is observed upon dilution of the monoalkylated enzyme in the micromolar range. In the latter case sedimentation equilibrium experiments indicate a monomer-dimer equilibrium with a K_d in the order of 5-10 µM. To explain the decrease in τ_c it is assumed that dissociation causes an increase in the internal mobility of the FAD containing protein segment which could correspond with the flavin binding domain. In all cases FAD remains bound to the protein as indicated by the fluorescence lifetime and anisotropy parameters.

Flavins and Flavoproteins

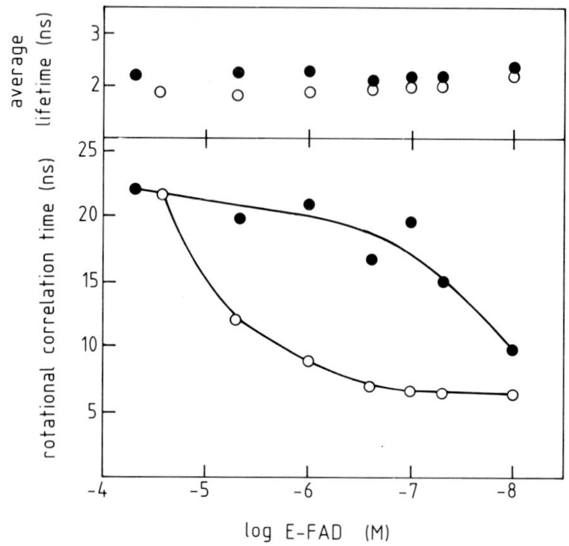

Figure 1. Concentration dependence of rotational correlation times of native (●) and monoalkylated (○) lipoamide dehydrogenase. Temp 20°C, 50 mM phosphate buffer, pH 7.0. The experimental conditions are described in ref. 4.

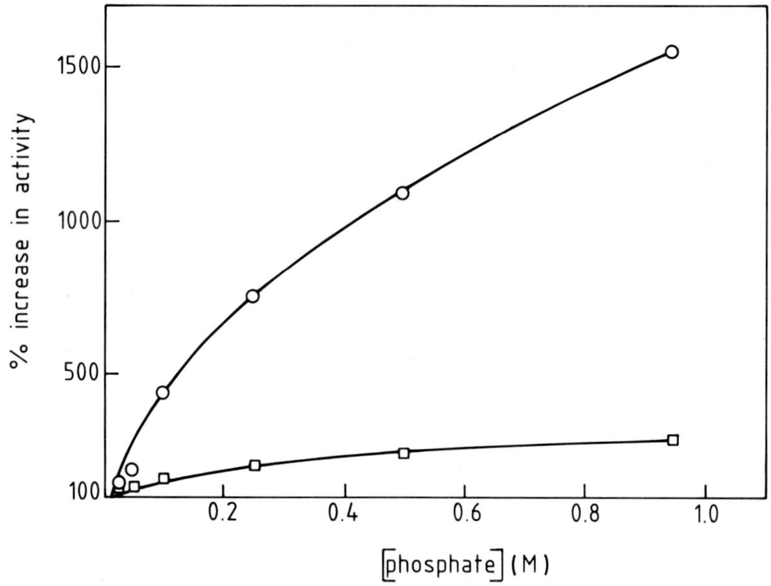

Figure 2. Effect of phosphate concentration on the NADH → lipoamide activity of the isolated (o) and complex bound lipoamide dehydrogenase (□). Potassium phosphate pH 7.0 containing 1 mM EDTA was used. temp. 25°C. Cuvettes contained 0.7 mM lipoamide and 0.1 mM NADH.

The NADH → lipoamide activity of the native enzyme is strongly dependent on the ionic strength (fig. 2). No specific ionic effects were observed. These effects are largely abolished when the enzyme is bound to the complex (fig. 2), when NAD^+ (1mM) is added or when the NADH → DCPIP reaction is studied (not shown). It is concluded that high salt concentrations, NAD^+ and binding to the complex prevent dissociation of the enzyme at catalytic conditions. The differences observed between the lipoate and the DCPIP activity are probably due to the fact that the monomer is inactive with respect to reduction of lipoate (cf. glutathion reductase), whereas DCPIP reduction is catalyzed by the monomer and the dimer.

As a further indication of dissociation we have used the sensitivity of the monomer to tryptic digestion. The oxidized, native enzyme is very stable to trypsin. No degradation is observed after 24 hrs of incubation at 0^oC with 2% trypsin. Both the monoalkylated and the NAD(P)H reduced enzyme are rapidly degraded (fig. 3). In the latter case the NADH → lipoamide activity is lost in a first order process. The NADH → DCPIP activity remains

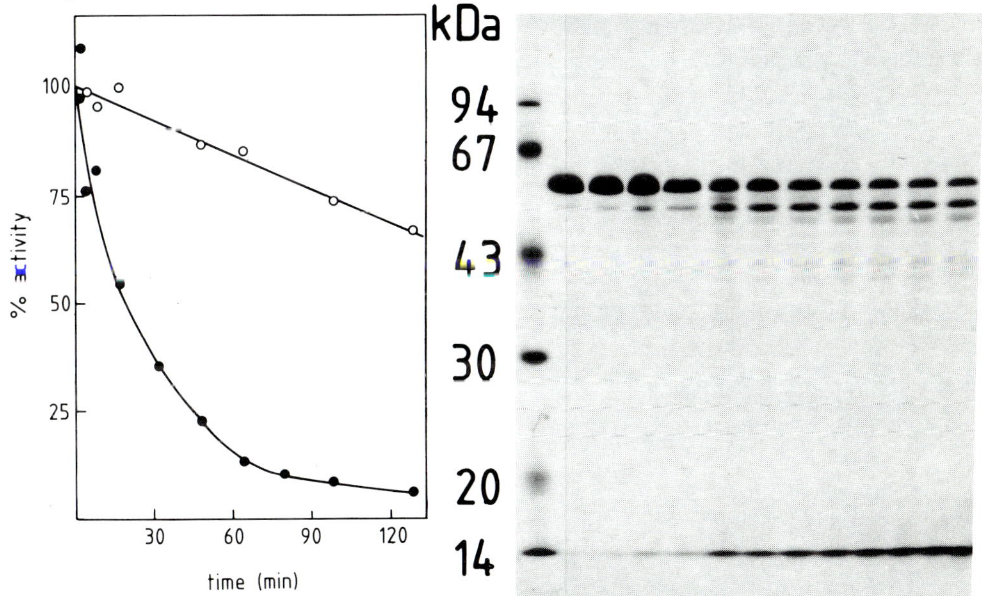

Figure 3. Trypsinolysis of lipoamide dehydrogenase.
15 μM of the enzyme in 50 mM potassium phosphate pH 7.0 containing 0.5 mM EDTA was reduced with 2 mM NADH and incubated with 0.2% TPCK treated trypsin. Reactions were stopped with trypsin inhibitor and analysed for NADH → lipoamide activity (•), NADH → DCPIP activity (o) or applied to a SDS gel.

high. The degradation pattern indicates that proteolysis starts with the loss of a 8 kDa peptide. The alkylated redox active cysteine is in position 44 in the E.coli enzyme (5). Incorporation of N-ethyl [2,3-^{14}C]maleimide and subsequent radioautography of the proteolysed samples after separation on SDS-gels yielded the same pattern as shown in fig. 3 for protein staining. Therefore it is concluded that degradation starts at the C-terminus. In the homologous enzyme glutathion reductase the C-terminus is part of the interphase domain (6). Dissociation may expose lysine or arginine residues of this domain. In the E.coli enzyme a cluster of these residues is found near position 400; cleavage at this position will yield a fragment of the observed size.

It is concluded that the isolated enzyme dissociates at catalytic concentrations and that dissociation is enhanced in the presence of NAD(P)H and in the monoalkylated state. NAD^+ seems to shift the monomer-dimer equilibrium to the dimeric state. A similar conclusion has been drawn by Van Muiswinkel Voetberg (7) with the pigheart enzyme. The question arises whether cleavage of the redox active S-S bridge could function as a trigger for changes in subunit interactions and whether such changes could play a role in catalysis, especially because binding to the complex prevents dissociation but does apparently not restrict the motional freedom of the subunits.

References

1. Bosma, H.J., De Kok, A., van Markwijk, B.W., Veeger, C.: Eur.J.Biochem. 140, 273-280 (1984).
2. Bosma, H.J., de Kok, A., Westphal, A.H., Veeger, C.: Eur.J.Biochem. in press.
3. Wawrzynczak, E.S., Perham, R.N., Roberts, G.C.K.: FEBS Lett. 131, 151-154 (1981).
4. Van Hoek, A., Vervoort, J., Visser, A.J.W.G.: Biochem.Biophys.Meth. 7, 243-254 (1983).
5. Stephens, P.E., Lewis, H.M., Darlison, M.G., Guest, J.R.: Eur.J.Biochem. 135, 519-527 (1983).
6. Pai, E.F. and Schultz, G.E.: Flavins and Flavoproteins, Massey, V. and Williams, C.H. eds. 3-10, Elsevier, North Holland Amsterdam, 1982.
7. Van Muiswinkel-Voetberg, H., Veeger, C.: Eur.J.Biochem. 33, 285-291 (1973).

PARTIAL AMINO ACID SEQUENCE OF PIG HEART LIPOAMIDE DEHYDROGENASE

L. David Arscott, Martin Berman and Charles H. Williams, Jr.

Veterans Administration Medical Center and Department of Biological Chemistry, The University of Michigan, Ann Arbor, Michigan 48105, U.S.A.

Preliminary studies of the amino acid sequence of pig heart lipoamide dehydrogenase showed that it was homologous with human erythrocyte glutathione reductase. The homology extended over all domains of glutathione reductase (1). These enzymes have subsequently been shown to be homologous with the prokaryote enzymes E. coli lipoamide dehydrogenase (2), and Pseudomonas aeruginosa transposon mercuric reductase (3). We now present the sequence of the amino-terminal 88 residues and the carboxyl-terminal 156 residues of pig heart lipoamide dehydrogenase.

Sequencing studies have been carried out on reduced and carboxymethylated enzyme. Pig heart lipoamide dehydrogenase is a mixture of an indeterminate number (2 - 12) of isozymes. No attempt was made to separate these prior to sequencing. The amino-terminus of lipoamide dehydrogenase was established by sequencing the whole protein. The first five residues agreed with sequences of peptides isolated subsequently from digests with trypsin (T) and Staphylococcus aureus V8 protease (S) and from cyanogen bromide (CB) reaction mixtures.

Four of the eight tryptic peptides in the 88 residue amino-terminal portion were placed by homology with human erythrocyte glutathione reductase constituting the first 54 residues (1). Chemical overlaps for the tryptic peptides are given in Figure 1 with the exception of the T1 T2 connection which is supported by amino acid analysis of S2 which includes S3. S1 did not contain the expected carboxyl-terminal D either by sequence or by amino acid analysis.

Flavins and Flavoproteins

SEQUENCE OF THE C-TERMINAL 156 RESIDUES OF
PIG HEART LIPOAMIDE DEHYDROGENASE

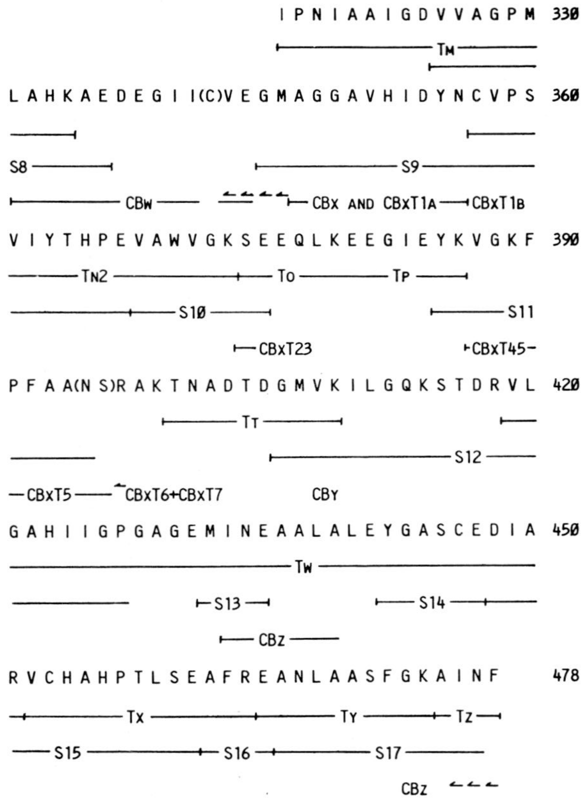

FIGURE 2. PEPTIDE IDENTIFICATION AS IN THE FOOTNOTE TO FIGURE 1.

SEQUENCE OF THE N-TERMINAL 88 RESIDUES OF
PIG HEART LIPOAMIDE DEHYDROGENASE

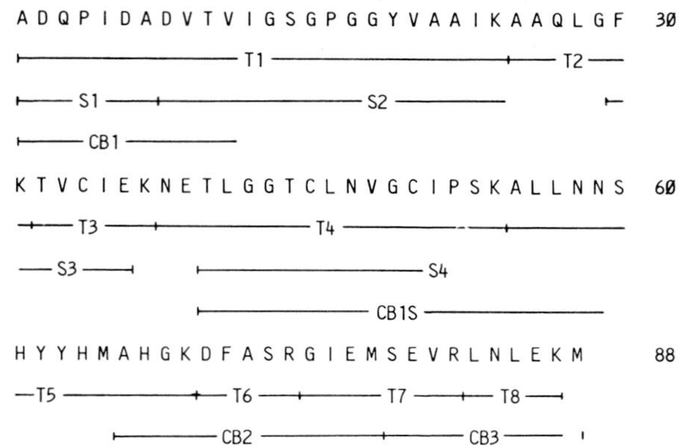

FIGURE 1. PEPTIDES WERE DERIVED FROM DIGESTION WITH TRYPSIN (T) AND <u>STAPHYLOCOCCUS AUREUS</u> PEPTIDASE (S), AND FROM TREATMENT WITH CYANOGEN BROMIDE (CB). THE TRYPTIC PEPTIDES WERE NUMBERED IN ORDER (1 TO 8 IN THIS FIGURE AND M TO Z IN FIGURE 2). PEPTIDES FROM THE <u>STAPHYLOCOCCUS AUREUS</u> PROTEASE DIGEST WERE NUMBERED IN ORDER WITHOUT REGARD TO MISSING PEPTIDES (1 TO 4 IN THIS FIGURE AND 8 TO 17 IN FIGURE 2.)

Tryptic peptides Tm, Tn2 and Tw within the 156 residue carboxyl-terminal portion of pig heart lipoamide dehydrogenase were originally placed through homology with human erythrocyte glutathione reductase (1) (Figure 2). The residue numbering in the figure is arbitrary, deriving from comparisons with glutathione reductase. Amino acid analysis indicates 474 residues (4). The T peptide set for this region was not complete and must be supplemented by T peptides derived from CB fragments, and in one segment of nine residues, the sequence depends on an S peptide. Two sections within this region gave difficulty in sequencing which will be discussed more fully elsewhere. While we feel confident of the sequence around C-342 it can not be considered firm. The sequence following A-394 is less certain and this region may reflect heterogeneity due to isozymes.

Difficulties encountered in the isolation of the cyanogen bromide peptides representing the last third of the FAD-1 domain and all of the pyridine nucleotide domain of lipoamide dehydrogenase may also reflect the potential problem of isozymic heterogeneity. Figures 2, 3, 4 and 6 in the Chapter by Williams et al. in this volume give our best estimate of the homology between pig heart lipoamide dehydrogenase and human erythrocyte glutathione reductase (5), E. coli lipoamide dehydrogenase (2) and Pseudomonas aeruginosa transposon mercuric reductase (3). The residue numbering is based on glutathione reductase. Placements in the pyridine nucleotide domain are tentative since chemical overlaps have not been determined. One peptide placed tentatively in this domain (Figure 3, position 189) presumably represents the beta-alpha-beta region which contacts the adenine of NAD by analogy to glutathione reductase. Matthews and Williams isolated a cysteine-containing peptide from enzyme treated with a very high concentration of iodoacetamide. The thiol was protected by NAD (6). This peptide can now be placed, by homology with the E. coli enzyme, at the end of the pyridine nucleotide domain (Figure 3). Placing this peptide into a model of glutathione reductase reveals several potential contacts with the pyridine nucleotide.

The sequence and alignment of 244 residues of the 474 indicated by amino acid analysis of pig heart lipoamide dehydrogenase has been presented.

The 88 residues constituting the amino-terminus of the protein are satisfactorily documented and can be considered firm. This region includes the oxidation-reduction disulfide (4) and the FAD binding site. The 156 residues at the carboxyl-terminus are likewise documented with two exceptions involving one and one or two residues. This region includes the base implicated in the mechanism of lipoamide dehydrogenase (7) as well as the two cysteine residues oxidized by cupric ion (8). Assignments in the pyridine nucleotide domain are based primarily on homology with three other pyridine nucleotide disulfide oxidoreductases.

Acknowledgement

This work has been supported by the Medical Research Service of the Veterans Administration and by a grant (GM21444) from the National Institute of General Medical Sciences.

References

1. Williams, C. H., Jr., Arscott, L. D. and Schulz, G. E.: Proc. Nat. Acad. Sci. (USA) 79, 2199-2201 (1982).

2. Stephens, P. E., Lewis, H. M., Darlison, M. G. and Guest, J. R.: Eur. J. Biochem. 135, 519-527 (1983).

3. Brown, N. L., Ford, S. J., Pridmore, R. D. and Fritzinger, D. C.: Biochem. 22, 4089-4095 (1983).

4. Williams, C. H., Jr. in The Enzymes (P. D. Boyer, ed.) 3rd Ed., vol. XIII, Academic Press, New York, 1976, pp. 89-173.

5. Untucht-Grau, R., Blatterspiel, R., Krauth-Siegel, R. L., Saleh, M., Schiltz, E., Schirmer, R. H. and Whittmann-Liebold, B. in Flavins and Flavoproteins (V. Massey and C. Williams, Jr., eds.) Elsevier/North Holland, New York, 1982, pp. 38-43.

6. Matthews, R. G. and Williams, C. H., Jr.: Biochim. Biophys. Acta 370, 39-48 (1974).

7. Matthews, R. G. and Williams, C. H., Jr.: J. Biol. Chem. 251, 3956-3964 (1976).

8. Thorpe, C. and Williams, C. H., Jr.: Biochem. 14, 2419-2424 (1975).

THE AMINO ACID SEQUENCE ENCOMPASSING THE ACTIVE-SITE HISTIDINE RESIDUE OF LIPOAMIDE DEHYDROGENASE FROM ESCHERICHIA COLI LABELLED WITH A BIFUNCTIONAL ARSENOXIDE.

Charles F.B. Holmes and Kenneth J. Stevenson, Division of Biochemistry, Department of Chemistry, University of Calgary, Calgary T2N 1N4, Alberta, Canada.

Introduction

Trivalent organoarsenicals have been used to investigate the structure and functions of proteins possessing vicinal thiol groups (1) and in particular the 2-keto acid dehydrogenase multienzyme complexes (E. coli) that utilize lipoic acid (2). The radiolabelled bifunctional reagent p-[(bromo-acetyl)amino]phenyl arsenoxide ($BrCH_2{}^{14}CONHPhAsO$) in the presence of substrates, irreversibly inhibited the overall activity of pyruvate dehydrogenase (PD) complex and the lipoamide dehydrogenase (E3) component (2). Treatment of the inhibited PD complex with 2,3-dithiopropanol (to break dithioarsinite bonds) increased the PD complex activity only to the level of E3 inhibition. The reagent is an interesting form of an active-site directed irreversibly inhibitor. The site of alkylation in E3 was established as the nitrogen-3 position of an imidazole ring through the isolation of N^3-[^{14}C] carboxymethyl histidine (2). This paper establishes the sequential location of the alkylated histidine in the active site of lipoamide dehydrogenase.

Results

PD complex was inhibited with $BrCH_2{}^{14}CONHPhAsO$ to only 50% and the radio-labelled E3 component was isolated (2). The extent of incorporation of the reagent was indicated with the recovery (after acid hydrolysis) of 0.37 residues of N^3[^{14}C]-carboxymethyl histidine. The purified, inhibited E3

Flavins and Flavoproteins

was further reduced and carboxymethylated. Subsequent fragmentation with cyanogen bromide led to the purification of a radiolabelled peptide (peptide B) using high performance liquid chromatography (HPLC) techniques (Figure 1). The amino acid composition of Peptide B (∿34 residues) was S-CM-Cys, 0.8, Asp, 3.1; Ser, 1.6; Thr, 1.9; Glu, 3.2; Pro, 1.7; N^3-[^{14}C]-CM-His, 0.2; Gly, 2.9; Ala, 4.1; Val, 1.9; Ile, 1.8; Leu, 2.6; Phe, 1.1; His, 1.5; Lys, 3.1; Arg, 0.5, Homoserine, 0; Homoserine lactone, 0. (CM = carboxymethyl). The tryptophan content was not determined. Peptide B was subjected to Edman degradation in a Beckman 890C Sequencer equipped with a cold-trap and a P6 Sequemat Autoconverter. A Quadrol (0.1M, pH 9) buffer was used and polybrene and glycylglycine were used in a 4 cycle precycling stage. Phenylthiohydanton (PTH) amino acids (Figure 2) were quantitated using a Varian 5000 HPLC equipped with an Altrex 5μ Ultrasphere:ODS reverse phase column. Acidic amino acids were identified as their methyl

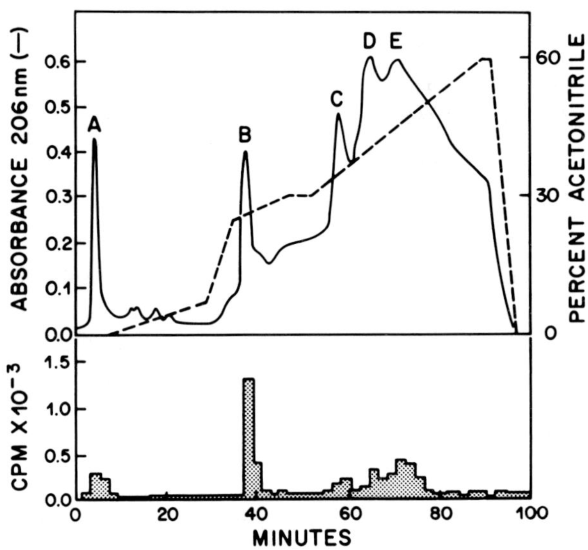

Figure 1. Reverse phase HPLC analysis of CNBr peptides from C^{14}-labelled lipoamide dehydrogenase inhibited with $BrCH_2{}^{14}CONHPhAsO$. The mixture was dissolved in 0.1% trifluoroacetic acid and injected onto a Varian cyanopropyl column (30 cm x 4 mm) equipped with a Brownlee cyanopropyl guard column. The peptides were eluted with increasing concentrations of acetonitrile containing 0.1% trifluoroacetic acid and were monitored at 206 nm with peak collection being made by hand. The flow rate was 1.1 mL/min. Aliquots were subjected to liquid scintillation counting.

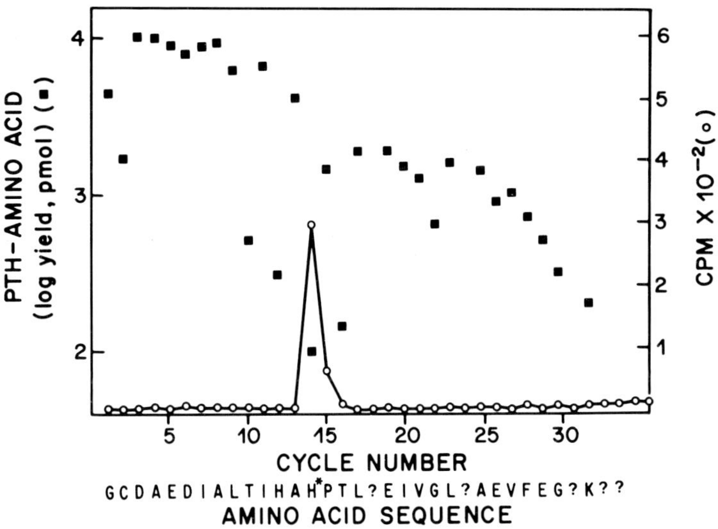

Figure 2. Recovery of PTH-amino acids and the placement of ^{14}C-labelled histidine from the Edman degradation of peptide B obtained from lipoamide dehydrogenase. (■) represents the log yield (pmol) of PTH-amino acids quantitated by HPLC. (○—○) represents the radioactivity. H* represents the histidine residue in peptide B modified with $BrCH_2{}^{14}CONHPhAsO$ as determined by high-voltage electrophoresis, amino acid and radiochemical analysis (2).

esters and in some cases PTH amino acids were back hydrolysed. The alkylated histidine residue is located in residue 14 of Peptide B.

Discussion

A comparison of the C-terminal regions of two dithiol reductases is presented in Table I. The histidine residue alkylated by $BrCH_2{}^{14}CONHPhAsO$ (when in the PD complex) can be assigned to residue 444 of E3 based on the amino acid sequence determined from the nucleotide sequence of the structural gene (3). The existence of this residue as the active site base (6) agrees well with structural studies on the homologous human erythrocyte glutathionine reductase (7) that clearly established a catalytic role for a histidine residue.

TABLE I

Comparison of the amino acid sequence of the C-terminal region of glutathionine reductase and lipoamide dehydrogenase.

Lipoamide Dehydrogenase	
E. coli (Protein Sequence)	430 ... * ... 450 MGCDAEDIALTIHAHPTL-EIVGLAAEVFEG-K-D-----
E. coli (Gene Sequence)[a]	MGCDAEDIALTIHAHPTLHESVGLAAEVFEGSITDLPNPKAKKK
Pig Heart (Protein Sequence)[b]	KGASCEDIARVCHAHPTLSE--------------------
Glutathionine Reductase (Protein Sequence)[c] (human erythrocyte)	MGATKADFDNTVAIHPTSSEELVTLR

a. Sequence from Stephens et al. (3), b. Sequence of Williams et al. (4), c. Sequence from Krauth-Siegel et al. (5). The numbering is based on amino acid sequence determined from the structural gene for lipoamide dehydrogenase (3). The histidine radiolabelled with $BrCH_2{}^{14}CONHPhAsO$ is identified with the asterisk and is residue 444.

References

1. Webb, J.L.: Enzyme and Metabolic Inhibitors, Vol. 3, p. 595, Academic Press, New York 1966.
2. Adamson, S.R., Robinson, J.A. and Stevenson, K.J.: Biochemistry 23, 1269-1274 (1984).
3. Stephens, P.E., Lewis, H.M., Darlison, M.G., Guest, J.R.: Eur. J. Biochem. 135, 519-527 (1983).
4. Williams, C.J. Jr., Arscott, L.D., Schulz, G.E.: Proc. Natl. Acad. Sci. USA 79, 2199-2207.
5. Krauth-Siegel, R.L., Blatterspiel, R., Saleh, M., Schiltz, E., Schirmer, R.H., Untucht-Grah, R.: Eur. J. Biochem. 121, 259-267 (1982).
6. Matthews, R.G., Ballow, D.P., Thorpe, C., Williams, C.H. Jr.: J. Biol. Chem. 252, 3199-3207 (1977).
7. Pai, E.F., Schulz, G.E.: J. Biol. Chem. 258, 1752-1757 (1983).

INACTIVATION OF PIG HEART LIPOAMIDE DEHYDROGENASE BY 1,3-DIBROMOACETONE

James S. Lee and Charles H. Williams Jr.

Veterans Administration Medical Center and Department of Biological Chemistry, University of Michigan, Ann Arbor, Michigan

Introduction

Lipoamide dehydrogenase (NADH:lipoamide oxidoreductase, EC 1.6.4.3) catalyzes the reversable reoxidation of dihydrolipoamide by NAD^+. The enzyme is composed of two subunits of 51,000 daltons, each containing an FAD and an oxidation-reduction active cystine (1). The mechanism of catalysis has been the subject of a great deal of study. It can be described as a Ping Pong Bi Bi mechanism in which the enzyme oscillates between an oxidized and two electron reduced form (2). Studies of the pH dependence of the oxidation-reduction potential of the enzyme and the reduction by dihydrolipoamide suggest the presence of a base, with a pK of 5.5 or below, the function of which is to deprotonate the dihydrolipoamide prior to nucleophillic attack on the active center disulfide (3,4).

1,3-Dibromoacetone has been used successfully as a bifunctional and monofunctional reagent in the investigation of nucleophiles at the active sites of papain and glyceraldehyde-3-phosphate dehydrogenase respectively (5,6). Here we report the initial investigation of the reactivity of dibromoacetone towards lipoamide dehydrogenase.

Experimental Procedure

Enzyme assays for lipoamide dehydrogenase activity contained 0.3 mM 3-acetylpyridine adenine dinucleotide, 0.08 mM dihydrolipoamide, and 0.3 mM EDTA in 60 mM phosphate, pH 7.6. The assays were initiated by addition of enzyme, and the reduction of 3-acetylpyridine adenine dinucleotide was followed at 363 nanometers. 1,3-Dibromo[2-^{14}C]acetone was synthesized by bromination of [2-^{14}C]acetone (New England Nuclear) by two equivalents of Br_2 in 48 percent HBr (7). The reaction mix was extracted twice with ether, the organic fractions pooled and evaporated under a stream of nitrogen. Dibromoacetone was isolated from the resulting mono-, di-, and

Flavins and Flavoproteins

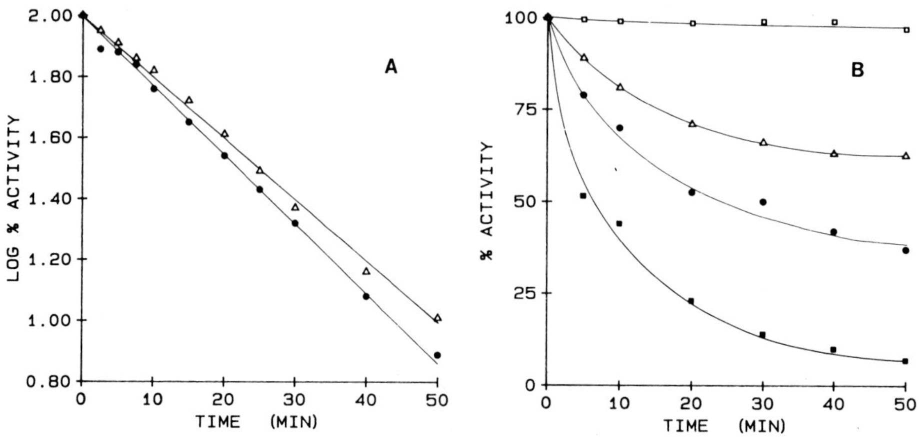

Figure 1. Incubation of lipoamide dehydrogenase in 20 mM phosphate pH 6.0 containing 0.3 mM EDTA, 25°. (A) 93 μM enzyme with (▲) 0.93 mM dibromoacetone; (●) 1.8 mM dibromoacetone. (B) 78 μM enzyme with (□) no additions; (△) 8% (v/v) methanol; (●) 8% methanol, 4 mM lipoic acid, and 2.0 mM dibromoacetone; (■) 2.0 mM dibromoacetone.

tribromoacetone by HPLC on a silica column (Waters, μPorasil) developed isocratically with 0.5% (v/v) tetrahydrofuran in hexane. The peak corresponding to dibromoacetone was collected and the solvent removed under reduced pressure. This synthesis resulted in colorless needles with strong lacrymatory properties and a melting point of 26-27° (lit. value is 26.5°(5)). The synthetic material co-eluted with recrystallized commercial dibromoacetone from HPLC. The specific activity of the radiolabeled compound was 1.95 mCi/mmole.

Results and Discussion

The inactivation of lipoamide dehydrogenase by 10 and 20 fold excess dibromoacetone exhibits pseudo-first order kinetics (Figure 1A). The half time of inactivation by 0.93 mM dibromoacetone was 13.9 minutes, and 14.9 minutes using 1.8 mM dibromoacetone. The inactivation remains pseudo-first order through approximately four half times. Including 4.0 mM lipoic acid in the incubation affords some protection from inactivation by dibromoacetone (Figure 1B). This result is complicated by the fact that methanol, required for solubilization of lipoic acid, causes a loss of activity alone, but the protection is at least 30%.

TABLE I

STEP	NMOLES ENZYME	DPM	MOLE [^{14}C]DIBROMOACETONE/ MOLE ENZYME
POST G-25	300	$14.2x10^5$	1.1
POST RED. ALKYLATION	250	$3.2x10^5$	0.3

Enzyme (132 μM) was inactivated with 1.3 mM 1,3-dibromo[2-^{14}C]acetone (specific activity of 4300 dpm/nmol), with addition of half as much reagent again at 14 minutes. After 50 minutes the reaction mix was applied to a Sephadex G-25 column to separate the enzyme from excess reagent. The resulting enzyme had 18% residual activity, and the visible spectrum indicated that the flavin was still bound. Quantitation of the radiolabeled material showed that 1.1 moles of [^{14}C] dibromoacetone had been incorporated per mole of lipoamide dehydrogenase monomer (Table I). The label, however, was unstable to the conditions of reductive alkylation (pH 8.0). Only 27% of the label remained following reductive alkylation and dialysis. Further studies indicated that the dibromoacetone modification of lipoamide dehydrogenase is stable at pH values below 7.0. Glutathione reductase is an enzyme very similar to lipoamide dehydrogenase in both mechanism and structure (1, 8). The base and oxidation-reduction active disulfide in glutathione reductase are contributed to the active site by different subunits (9). With this in mind, 73 μM lipoamide dehydrogenase was inactivated with 1.4 mM dibromoacetone with 1.4 mM NADH being added either coincident with, or 50 minutes after the addition of dibromoacetone. Crosslinking was then assessed by SDS polyacrylamide gel electrophoresis (Figure 2). The gels show two major bands for both coincident (lanes 3,4,8,and 9) and subsequent (lanes 5,6,10,and 11) addition of NADH. One band migrates to the same position as native lipoamide dehydrogenase (lane 1 and 12) at 51,000 daltons, the higher molecular weight band migrates at approximately 104,000 daltons. No high molecular weight band is seen in the absense of NADH addition. This higher molecular weight band is presumably a lipoamide dehydrogenase dimer in which the nucleophile which reacts with dibromoacetone resulting in inactivation of the enzyme has been crosslinked to one of the thiols formed upon reduction with NADH.

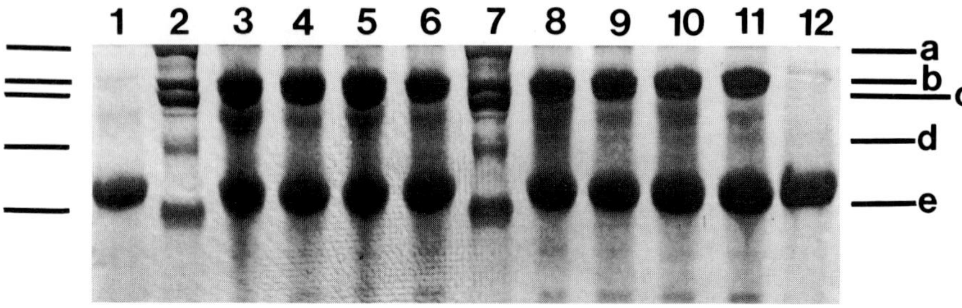

Figure 2. Segment of a polyacrylamide gel of lipoamide dehydrogenase inactivated by dibromoacetone in the presence of NADH, 7.5% polyacrylamide, 0.1% SDS, and 25 mM phosphate, pH 6.3. The portion of the gel not shown contained no bands. Lane 1 and 12, native lipoamide dehydrogenase; lane 2 and 7, standards; lane 3, 4, 8, and 9, NADH and dibromoacetone added at the same time; lane 5, 6, 10, and 11, NADH added 50 minutes after dibromoacetone. Lanes 8 through 12 were preincubated with β-mercaptoethanol prior to electrophoresis. Standards: (a) myosin, 200,000 daltons; (b) β-galactosidase, 116,250 daltons; (c) phosphorylase B, 92,500 daltons; bovine serum albumin, 66,200 daltons; ovalbumin, 45,000 daltons.

Acknowledgements This work has been supported by Medical Research Service of the Veterans Administration and by Grant GM21444 (C.H.W.) from the National Institute of General Medical Sciences.

References

1. Williams, C. H., Jr.: in The Enzymes (Boyer, P. D., ed) Vol. 13, 89-173, Academic Press, New York 1976
2. Massey, V., Gibson, Q. H., and Veeger, C.:Biochem. J. 77, 341-351 (1960)
3. Matthews, R. G.,Williams, C. H., Jr.:J. Biol. Chem. 251, 3956-3964 (1976)
4. Matthews, R. G., Ballou, D. P., Thorpe, C., and Williams, C. H., Jr.: J. Biol. Chem. 252, 3199-3207 (1977)
5. Husain, S. S., Lowe, G.:Biochem. J. 108, 855-859 (1968)
6. Moore, J., Jr., Fenselau, A.:Biochemistry 11, 3753-3762 (1972)
7. Rappe, C.:Acta Chem. Scand. 16, 2467-2468 (1962)
8. Williams, C. H., Jr.,Arscott, L. D., and Schulz, G. E.:Proc. Nat. Acad. Sci. USA 79, 2199-2201 (1982)
9. Schulz, G. E., Schirmer, R. H., Sachsenheimer, W., and Pai, E. F.: Nature 273, 120-124 (1978)

THE EVOLUTION OF MERCURIC REDUCTASE, AND A REDOX TRANSFER MODEL FOR MERCURIC ION DETOXIFICATION IN BACTERIA.

Nigel L. Brown

Department of Biochemistry, University of Bristol,
Bristol BS8 1TD, U.K.

and Dean Goddette

Department of Biochemistry, Washington University Medical School, St. Louis, MO 63110, U.S.A.

Introduction

The transposon Tn501 confers inducible resistance to mercuric ions in several eubacterial species. The resistance mechanism is catalytic, involving reduction of Hg(II) to Hg(0) by the enzyme mercuric reductase. This has been shown to be a flavin-containing enzyme and to be mechanistically similar to the dithioloxidoreductases glutathione reductase and lipoamide dehydrogenase (1,2). DNA sequence analysis of the merA gene encoding mercuric reductase has allowed us to predict the amino acid sequence of the polypeptide and to demonstrate sequence homology with human erythrocyte glutathione reductase (3). The full predicted amino acid sequence of E.coli lipoamide dehydrogenase has also been published (4). Mercuric reductase differs from these other enzymes in possessing an additional N-terminal region of approximately 80 amino acids. Further DNA sequence analysis of Tn501 (N.L. Brown, R.D. Pridmore & D.C. Fritzinger, unpublished data) has predicted the amino acid sequences of other gene products which are involved in the detoxification of mercuric ions. Examination of these sequences indicates that the evolution of mercuric reductase included a gene

Flavins and Flavoproteins

duplication event which gave rise to both the N-terminal region of mercuric reductase and to another gene product. Analysis of the primary structures of the gene products suggests a model for the detoxification of mercuric ions in prokaryotes.

Results

DNA sequence analysis of the mercury-resistance determinant of Tn501 showed that there are three open reading frames before the merA gene (5) and at least one after merA. This agrees with genetic data from the closely-related R-100 system (6,7) in which coding sequences are expected, in order, for the genes merR, merT, merC, merA and merD. The merR gene product regulates transcription of the other structural genes, of which only the functions of merT (transport of mercuric ions into the cell) and merA (mercuric reductase) are known. Analysis of the predicted primary sequence of the merT protein for membrane-spanning regions (8) indicates that there may be three trans-membrane helices. There are four cysteine residues in the merT protein; one pair would lie in

```
                           ↓
merC    MKKLFASLALAAVVAPVWAATQTVTLSVPGMTCSACPITVKKAISEVEGVSKVDVT
                                 :   ::::  :   ::     : ::    :
merA                           THLKITGMTCDSCAAHVKEALEKVPGVQSALVS

        FETRQAVVTFDDAKTSVQKLTKATADAGYPSSVKQ
             :         :   :: : :  ::
        YPKGTAQLAIVPGT-SPDALTAAVAGLGYKATLADAPLADNRVGLLDKVRGWMAAA
```

Figure 1. Alignment of the merC gene product with the amino-terminal end of mercuric reductase. Only identical aligned amino acids are marked, and the underline marks the start of the alignment of the glutathione reductase sequence (3). The arrow indicates a possible cleavage site for signal peptidase. The alignment of the sequences is merely illustrative, and has not been optimised.

the first helix near the outside face of the membrane, the other pair would be on the inside face of the membrane. The predicted sequence of the merC protein contains a potential signal peptide (9), but no other extensive hydrophobic region. The merC product may therefore be periplasmic. The region beyond the potential signal peptide shows homology to the N-terminal 'extra' region of mercuric reductase, including the pair of cysteines (marked in Fig. 1). The detailed structure of mercuric reductase, including the role of these cysteines, has been discussed previously (3). The predicted merD gene product contains two potential membrane-spanning regions.

Discussion

The homology between the merC gene product and mercuric reductase of Tn501 is entirely contained within the 'extra' N-terminal region of mercuric reductase. One evolutionary model is that mercuric reductase was elaborated from an ancestral lipoamide dehydrogenase or glutathione reductase lacking this 'extra' region. This region may have been acquired by duplication of the ancestral merC gene followed by a gene-fusion by a one copy with the ancestral mercuric reductase; or the merC gene may have evolved by duplication of the 'extra' region of mercuric reductase. In either case the evolutionary advantage would be more efficient mercury detoxification. The 'extra' region of mercuric reductase is not required for catalysis (2) but may be transiently involved in mercury binding.

The model for mercury detoxification in bacteria is based on structural predictions and enzymological studies (2). The components of the R-100 mer system are similar to those of Tn501 (5). We propose that Hg(II) is sequestered as a dithiol adduct of the merC gene product in the periplasmic space and transferred by a series of redox-exchange reactions, via the paired cysteines on the outside and inside regions of the merT protein, to the N-terminal cysteines of mercuric reductase and thence to the C-terminal cysteine pair. Hg(II) is reduced to Hg(O) and released to the cytoplasm. A possible role for the merD product is in facilitating release of Hg(O) from the cell. The model is satisfying in that the highly toxic Hg(II) is sequestered away from the cytoplasm. The model is now being tested and direct experimental evidence is being sought.

Acknowledgements. We are grateful to Drs. S. Silver and T.K. Misra for supplying information on the R-100 system and for discussion.

References

1. Fox, B.S., Walsh, C.T.: J. Biol. Chem. 257, 2498-2503 (1982).

2. Fox, B.S., Walsh, C.T.: Biochemistry 22, 4082-4088 (1983).

3. Brown, N.L., Ford , S.J., Pridmore, R.D., Fritzinger, D.C. : Biochemistry 22, 4089-4093 (1983).

4. Stephens, P.E., Lewis, H.M., Darlison, M.G., Guest, J.R.: Eur. J. Biochem. 135, 519-527 (1983).

5. Misra, T.K., Brown, N.L., Fritzinger, D.C., Banes, W.M., Haberstroh, L., Silver, S.: Proc. Nat. Acad. Sci. USA. In the press (1984).

6. Foster, T.J., Nakahara, H., Weiss, A.A., Silver, S.: J. Bacteriol. 140, 167-181 (1979).

7. Ni'Bhriain, N., Silver, S., Foster, T.J.: J. Bacteriol. 155, 690-703 (1983).

8. Kyte, J., Doolittle, R.F.: J. Mol. Biol. 157, 105-132 (1982).

9. Michaelis, S., Beckwith, J.: Ann. Rev. Microbiol. 36, 435-464 (1982).

THE STRUCTURAL STUDIES ON FERREDOXIN - $NADP^+$ OXIDOREDUCTASE FROM *Spirulina*, A BLUE - GREEN ALGA.

Keishiro Wada, Yoshio Yao, Toshiaki Tamura, Hiroshi Matsubara
Dept. of Biology, Fac. of Science, Osaka University, Toyonaka, Osaka 560, Japan

Keiun Kodo
Spirulina Biology Research Co., Yodogawa, Osaka 532, Japan

Introduction

Ferredoxin - $NADP^+$ oxidoreductase [EC.1.18.1.2., FNR] is a key enzyme of the electron transfer system in oxygenic photosynthesis. It has one FAD as a prosthetic group and catalyzes the electron transfer from reduced ferredoxin to $NADP^+$. Since the isolation of spinach enzyme (1), many reports have been published on the characterization of FNR from eukaryotic and prokaryotic algae in addition to some other higher plants. FNR has been known to form a one - to - one complex with ferredoxin or $NADP^+$ (2,3). Some chemical modification experiments (4,5,6) have revealed that a certain amino acid residue is important in the electron transfer and complex formation. Although a preliminary study on the three dimensional structure of FNR was reported (7), information on the primary structure of FNR was limited to only the amino-(N-) and carboxyl-(C-) terminal regions (8,9). We have tried to decide the complete sequence of *Spirulina* FNR which shows only a small heterogeneity in the N-terminal region as described previously (8).

Flavins and Flavoproteins

Methods

The purified FNR (actually cm - FNR, cysteine residues of which were previously carboxymethylated) was cleaved with BrCN in 70 % formic acid, and were subjected to Sephadex G-50 column chromatography in 10 % formic acid containing 8 M urea. The four fragments thus obtained were further purified with SP-Sephadex C-25 in the presence of urea. Two other batches of succinylated cm - FNR were digested with trypsin and staphylococcal protease, respectively, and the resulting peptides were purified by gel filtration, ion-exchange chromatography, high voltage electrophoresis and reverse phase HPLC. Amino acid analysis and sequence determination were carried out essentially as described in the previous paper (10).

Results and Discussion

The amino acid sequence of Spirulina FNR was determined as in Fig. 1.

```
          *         *         *         *         *
AKTDIPVNIYKPKNPYIGKCLSNEELVREGGTGTVRHLIFDISGGDLRYL
          *         *         *         *         *
EGQSIGIIPPGTDNNGKPHKLRLYSIASTRHGDHVDDKTVSLCVRQLEYK
          *         *         *         *         *
HPETGETVYGVCSTYLCNLEAGADVAITGPVGKEMLLPEDEDATIIMMAT
          *         *         *         *         *
GTGIAPFRAFLWRIFKEQHEDYKFKGLWALFFGIPYSPNILYQQELEELQ
          *         *         *         *         *
EEFPENFRLTLAISREQQNPEGGKMYIQDRIKENADQLWELIQKPNTHTY
          *         *         *         *
ICGLKGMEGGIDEGMSAAAGKFDVDWSDYQKELKKKHRWHVETY
```

Fig. 1 The complete sequence of Spirulina FNR. The N-terminus is heterogeneous. The preparation includes three other species of molecules starting with K, T, or D in addition to the molecule drawn here.

The total number of amino acid residues of FNR was 294 giving an M.W. of 34,135 including FAD. On the sequence analysis, we took much care of detecting heterogeneity. There was no heterogeneity except for the N-terminus. It is suggested that only one molecule of FNR is synthesized in Spirulina cell and the N-terminus is probably modified by protease(s) during the

enzyme preparation or even in living cells, resulting the N-terminal microheterogeneity.
The N-terminal sequence of Spirulina FNR was compared with spinach FNR reported by Hasumi et al. (9) as in Fig. 2.

```
Spirulina FNR                 A K T D I P V N I Y K P K N P Y I G
                                                  |           |
Spinach FNR  V E K H S R K M E E G I T V N K F K K T K K Y V G
```

Fig. 2 N-terminal sequences of Spirulina and spinach FNRs. Identical residues found in both FNRs are enclosed in boxes and similar residues are shown in vertical bars.

Among 18 residues compared, 6 residues were identical and 2 similar. The sequence of Spirulina FNR was also compared with these of other flavin-containing enzymes. Some similar parts were found in human erythrocyte glutathione reductase (11), p-hydroxybenzoate hydroxylase from Pseudomonas fluorescence (12) and others, all of which are FAD enzymes and NADPH - dependent enzymes. In particular, FAD - binding region of glutathione reductase, which had β-α-β folding containing -G-X-G-X-X-G-sequence was similar to the part from Pro-15 to Glu-51 in Spirulina FNR (Fig. 3). However, β-α-β folding could not be predicted in Spirulina FNR sequence by Chou and Fasman's secondary structure prediction method (13). This similarity on sequences in FAD containing enzymes seems to be ascribed to the convergent evolution.

```
                                        ┌─β or α─┐      ┌─β─┐
                      20        30        40         50
FNR                   PKNPYIGKCLSNEELVREGGTGTVRHLIFDIS-GGDLRYLEGQ
GRase                 GPPPAAGAVASYDYLVIGGGSGGLASARRAAELGARAAVVESH
LPDHase(E. coli)                 QVVVLGAGPAGYSAAFRCADLGLETVIVERY
LPDHase(pig heart)               DVTVIGSGPGGYVAAIKAAQLGFKTVCIEKN
HBHase(P. fluorescence)          QVAIIGAGPSGLLLGQLLHKAGIDNVILERQ
                                 └─β─┘ └────────α────────┘ └─β─┘
```

Fig. 3 Sequence comparison of FAD binding folds of several FAD enzymes. Identical residues found between FNR and GRase are shown by vertical bars.

References

1. Shin, M., Tagawa, K., Arnon, D. I.: Biochem. Z. 338, 84-96 (1963).

2. Shin, M., San Pietro, A.: Biochem. Biophys. Res. Commun. 33, 38-42 (1968).

3. Dykes, J., Davis, D. J.: Arch. Biochem. Biophys. 218, 46-50 (1982).

4. Valle, E. M., Carillo, N., Vallejos, R. H.: Biochim. Biophys. Acta 681, 412-418 (1982).

5. Bookjans, G., Böger, P.: Arch. Biochem. Biophys. 194, 387-393 (1979).

6. Zanetti, G., Gozzer, C., Sacchi, G., Curti, B.: Biochim. Biophys. Acta 569, 127-134 (1979)

7. Sheriff, S., Herriott, J. R.: J. Mol. Biol. 145, 441-451 (1981).

8. Wada, K., Tamura, T., Matsubara, H., Kodo, K.: J. Biochem. 94, 387-393 (1983).

9. Hasumi, H., Nagata, E., Nakamura, S.: Biochem. Biophys. Res. Commun. 110, 280-286 (1983).

10. Hase, T., Matsubara, H., Hutber, G. N., Rogers, L. J.: J. Biochem. 92, 1347-1355 (1982).

11. Krauth-Siegel, R. L., Blatterspiel, R., Saleh, M., Schiltz, E., Schirmer, R. H., Untucht-Grau, R.: Eur. J. Biochem. 121, 259-267 (1982).

12. Weifue, W. J., Hofsteenge, J., Vereijken, J. M., Jekel, P. A., Beintema, J. J.: Biochim. Biophys. Acta 704, 385-388 (1980).

13. Chou, P. Y., Fasman, G. D.: Adv. Enzymol. 47, 45-148 (1978).

THE AMINO ACID SEQUENCE AND PARTIAL TERTIARY STRUCTURE OF FERREDOXIN:NADP OXIDOREDUCTASE FROM SPINACH

P. Andrew Karplus, Jon R. Herriott, Kenneth A. Walsh
Department of Biochemistry SJ-70, University of Washington, Seattle, Washington 98195

Abstract

The primary structure of the redox protein ferredoxin:$NADP^+$ oxidoreductase has been investigated in order to further x-ray crystallographic studies aimed at elucidating its three-dimensional structure. We report here the amino acid sequence of the spinach enzyme and the fit of its carboxy-terminal half to the NADP binding domain as seen at 3.25 Å resolution. Most of the sequence was obtained by automated sequence analysis of peptides generated by digestion of carboxymethylated protein at arginyl and methionyl residues. The complete sequence comprises 314 amino acids, including a single disulfide. This work suggests that some of the multiple forms of FNR reported in the literature are due to partial digestion of the native protein at the amino-terminus. The sequence has extended identity with published sequences of FNR from other sources, but no striking similarity to other proteins of known sequence.

The crystallographic analysis of FNR has been extended to 3.25 Å resolution by the method of multiple isomorphous replacement. Attempts to obtain heavy atom derivatives with phasing power at higher resolution have been unsuccessful. The result at 3.25 Å shows similar features to the reported 3.7 Å resolution result (J. Mol. Biol. 145,441(1981)). The fit of residues 164 to 314 of the sequence to the NADP binding domain reveals a parallel sheet topology common to many nucleotide binding domains.

Flavins and Flavoproteins

ON THE NATURE OF FERREDOXIN:FERREDOXIN-NADP REDUCTASE COMPLEX

Masateru Shin and Naoko Sakihama
Department of Biology, Faculty of Science, Kobe University
Nada-ku, Kobe 657, Japan

Introduction

Ferredoxin-NADP reductase (FNR, EC 1.18.1.2) catalyses a reversible electron transport between ferredoxin and NADP (1, 2). Since FNR was found to form a stable complex with oxidized ferredoxin (3, 4, 5), it has been suggested that the complex played a functional role in the electron transport between the two proteins. The first proposal was made by Nelson and Neumann (5), who observed that both the complex formation and NADP photoreduction were similarly sensitive to ionic strength. Nakamura and Kimura (6) reported that the inhibition of NADPH-diaphorase activity of FNR by ferredoxin was ascribable to complex formation between ferredoxin and FNR. Furthermore, Davis and San Pietro (7) demonstrated that both complex formation and NADP photoreduction were simultaneously inhibited when ferredoxin was trinitrophenylated. The parallel influence on both complex formation and electron transport function was considered to be good evidence for actural involvement of the complex in electron transport between the two proteins. However, we present evidence here that complex formation between ferredoxin and FNR depends on the conformation of the ferredoxin molecule and is independent of its electron transport function.

Results and Discussion

We investigated in this work the effect of chemical modification of ferredoxin on complex formation and electron transport between ferredoxin and FNR. As shown in TABLE I, FNR formed a complex with trinitrophenylated ferredoxin, but did not react with this modified ferredoxin in the electron

Flavins and Flavoproteins

TABLE I. Effects of chemical modification of ferredoxin on its enzymatic properties and FNR-complexing ability

Trinitrophenylated (TNP-) and acetylated (acetyl-) ferredoxin were prepared as previously described (8, 9). Apo-ferredoxin was obtained according to Hong and Rabinowitz (10). Complex formation of FNR with proteins was determined by the immobilized FNR method (12). Electron transport function of ferredoxin was evaluated by measuring activities in the NADP photoreduction system of chloroplasts and NADPH-cytochrome c reduction system (3). Inhibition of NADPH-diaphorase activity by FNR was determined by the method described previously (8). Detection of the iron-sulfur center in ferredoxin was performed by measuring the absorption spectrum characteristic of the center or determining non-heme iron according to Kimura and Suzuki (13).

Protein	Complex formation	Electron transport	Iron-sulfur center	Inhibition of diaphorase
Ferredoxin				
native	+	+	+	+
apo-	-	-	-	-
TNP-	+	-	-	-
acetyl-	-	+	+	-
Apo-ferredoxin				
TNP-	+	-	-	
Ovalbumin				
native	-			
TNP-	-			

transport system. The lack of activity of the modified ferredoxin in the electron transport system was ascribable to elimination of the iron-sulfur center of ferredoxin by the treatment. This result is strikingly different from the result of Davis and San Pietro (7) that " The trinitrophenylated ferredoxin is unable to complex with the flavoprotein and, therefore, unable to participate in NADP photoreduction ". Our result shows that the ferredoxin-FNR complex should not be considered to be an "enzyme-substrate complex". Further evidence that complex formation and electron transport are independent was obtained with acetylated ferredoxin. Since acetylation left the iron-sulfur center of ferredoxin intact, the modified ferredoxin was active in the electron transport system but did not form a complex with FNR.

Inhibition of diaphorase activity of FNR by ferredoxin has been attributed to the complex formation between the two proteins (6). The interpretation was supported by the fact that acetylated ferredoxin, which did not complex with FNR, did not inhibit the activity. However, trinitrophenylated ferredoxin which did complex with FNR was inert in the inhibition. Further investigation will be necessary to clarify the mechanism of inhibition.

As previously reported, apo-ferredoxin did not form the complex, but this ability was recovered by reconstitution of iron-sulfur center (11). This suggested that the iron-sulfur center of ferredoxin is essential for the complex formation. However, trinitrophenylated ferredoxin, lacking the iron-sulfur center, could form the complex and acetylated ferredoxin, possessing the iron-sulfur center, could not. This indicates that the iron-sulfur center itself is not directly involved in the complex formation. Since removal of the iron-sulfur center or acetylation severely alter the conformation of ferredoxin molecule (14, 15), this suggests that the native conformation of ferredoxin molecule is the essential factor for complex formation. Thus, the iron-sulfur center would contribute to the complex formation indirectly by stabilizing the native conformation of ferredoxin. In trinitrophenylated ferredoxin, trinitrophenyl groups may act as a conformation stabilizer in place of the lost iron-sulfur center. Trinitrophenyl groups themselves are incapable of forming the complex with FNR, because trinitrophenylated ovalbumin failed to form a complex with FNR. The fact that trinitrophenylation of apo-ferredoxin restored the capability of complex formation is further evidence for this interpretation.

In conclusion, the complex between ferredoxin and FNR is formed independently of the electron transport function between the two proteins and depends on the specific protein conformation of ferredoxin. The complex should therefore be designated as a "STRUCTURAL COMPLEX".

References

1. Shin, M., Tagawa, K., Arnon, D.I.: Biochem. Z. 338, 84-96 (1963).
2. Shin, M., Arnon, D.I.: J. Biol. Chem. 240, 1405-1411 (1965).
3. Shin, M., San Pietro, A.: Biochem. Biophys. Res. Commun. 33, 38-42 (1968).
4. Foust, G.P., Mayhew, S.G., Massey, V.: J. Biol. Chem. 244, 964-970 (1969).
5. Nelson, N., Neumann, J.: J. Biol. Chem. 244, 1932-1936 (1969).
6. Nakamura, S., Kimura, T.: J. Biol. Chem. 246, 6235-6241 (1971).
7. Davis, D.J., San Pietro, A.: Biochem. Biophys. Res. Commun. 74, 33-40 (1977).
8. Sakihama, N., Niimi, K., Nakamura, M., Shin, M.: J. Biochem. 93, 1685-1690 (1983).
9. Sakihama, N., Shin, M.: Mem. Grad. School Sci. & Technol. Kobe Univ. 1-A, 13-18 (1983).
10. Hong, J.-S., Rabinowitz, J.C.: Biochem. Biophys. Res. Commun. 29, 246-249 (1967).
11. Shin, M., Sakihama, N., Niimi, K., Oshino, R.: Flavins and Flavoproteins, Massey, V. & Williams, C.H.,Jr., eds., pp. 672-674. Elsevier North Holland, New York 1982
12. Shin, M., Sakihama, N., Sugimoto, N., Miyazawa, T., Oshino, R.: J. Biochem. 91, 953-557 (1982).
13. Kimura, T., Suzuki, K.: J. Biol. Chem. 242, 485-491 (1967).
14. Hasumi, H.: J. Biochem. 92, 1049-1057 (1982).
15. Masaki, R., Wada, K., Matsubara, H.: J. Biochem. 81, 1-9 (1977).

PROPERTIES OF A CROSS-LINKED COMPLEX BETWEEN FERREDOXIN-NADP$^+$ REDUCTASE AND FERREDOXIN

Giuliana Zanetti, Bruno Curti
Dipartimento di Fisiologia e Biochimica Generali, Università di Milano, Milano, Italia

Introduction

As an approach to understanding the interactions between ferredoxin-NADP$^+$reductase (FNR) and its protein-substrate ferredoxin (Fd), the formation of a covalent complex between the two proteins (FNR-Fd) has been achieved by using a soluble carbodiimide. The cross-linked complex thus obtained, has been shown to possess many properties of the native complex: the 1:1 stoichiometry, the spectral perturbations, the cytochrome c reductase activity and the isoelectric point (1). In view of an utilization of the covalent complex for obtaining functional and structural information, a further characterization of such a complex was attempted by circular dichroism (CD) and fluorescence spectroscopy.

Results and Discussion

Several reports (2,3,4,5) have appeared on the alterations in the CD spectra of FNR and Fd following complex formation. Thus, we decided to examine and compare the CD spectra of the covalent and native complexes. Previously (2,3), it was shown that the difference CD spectra, in the visible region, of the native complex obtained either by subtraction of the spectra of the single proteins from the spectrum of the protein mixture or by

Flavins and Flavoproteins

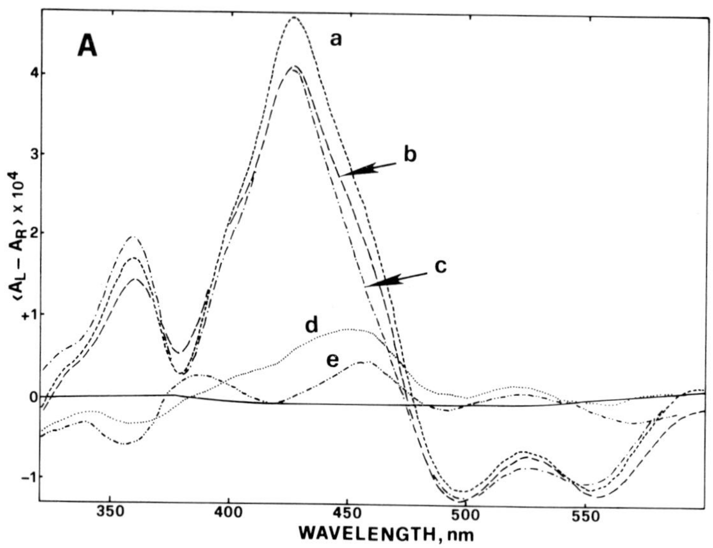

Fig. 1 - CD spectra of FNR-Fd (a) and of an equimolar mixture of FNR+Fd at low (b) and high ionic strength (c); difference CD spectra of the covalent (d) and native complex (e).
A) FNR-Fd was 18.6 μM in 0.1 M Tris-HCl, pH 7.4 (a, ---); FNR and Fd were 18.5 μM in 10 mM Tris-HCl, pH 7.4, before (b,— —) and after adding 0.3 M NaCl (c,— · —). Calculated difference spectra of the covalent (d,····) and native complex (e,—·· —). Light path 1 cm. B) Same conditions as in A), except that the protein concentration was ten times lower and the light path was 0.1 cm . A spectropolarimeter Jasco J-500 A was used.

calculating the difference between the spectra of the protein mixture at low and high ionic strength, were nearly identical. Thus, we made use of the latter technique and found it suitable also in the UV region. Namely, our near-UV difference CD spectrum of the native complex was very similar to that published by Shin (3); in the far-UV we confirmed the data of Knaff et al. (4) and Hasumi et al. (5), who found a difference CD spectrum for the native complex, even in this region. In Fig. 1, the CD spectra of FNR-Fd and of the FNR+Fd mixture at low and high ionic strength are shown, as well as the calculated dif-

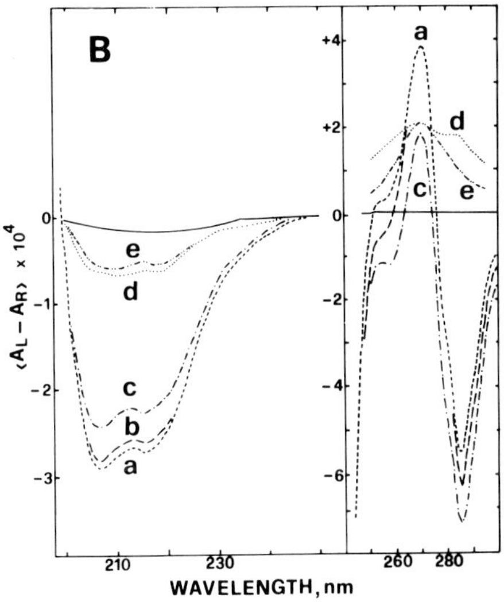

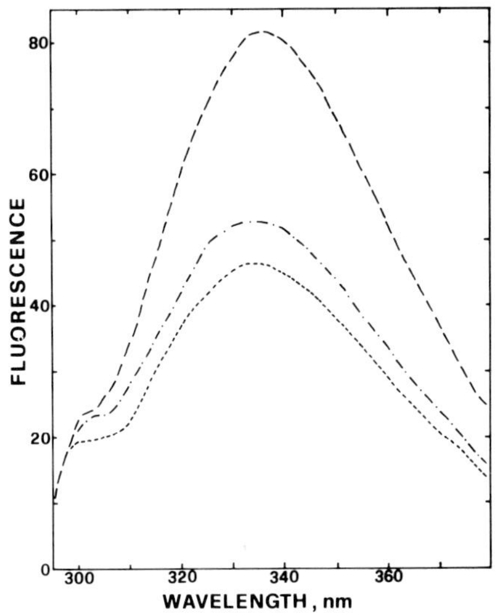

Fig. 2 - Fluorescence emission spectra of FNR, FNR+Fd and FNR-Fd. The proteins were dissolved in 5 mM phosphate pH 7 at 20° C. FNR (— —) and FNR-Fd (----) were 0.37 μM. The free reductase was titrated with Fd to maximal quenching (Fd 0.6 μM, —·—). Excitation wavelength was 270 nm.

ferences spectra for both the complexes. Whereas in the UV (Fig. 1,B) the complexes showed very similar spectra, in the visible region (Fig. 1,A) slight differences were apparent, especially in the 380-430 nm region. Curiously, analogous differences can be detected among the various published spectra for the native complex. Thus, our native complex has a difference CD spectrum very similar to that obtained by Cammack et al. (2), whereas the covalent complex has a difference spectrum more similar to that published by Hasumi et al. (5). Therefore, it is not clear if the differences in the 380-430 nm region depend on a slightly different conformation of the complexes.

An examination of the fluorescent properties of the cross-linked complex, revealed that the FAD fluorescence was quenched to the same extent as in the native flavoprotein; on the other

hand, as shown in Fig. 2, the protein fluorescence was greatly reduced in comparison to that exhibited by the free reductase. The same figure shows that addition of Fd to the flavoprotein resulted in a quenching of the protein fluorescence of about 40% at complete complex formation. The presence of 0.3 M NaCl abolished the fluorescence quenching only in the case of the native complex. The latter result suggests that cross-linking stabilizes the interactions between the two proteins towards even partial perturbation by high ionic strength.
All these data seem again to confirm the similarity between the cross-linked and the native complex, prompting us to use it for further investigation on the structure and function of the ferredoxin-reductase complex.

Acknowledgments

We would like to thank Dr. G. Vecchio for help in performing CD measurements. This work was supported by grants from the Consiglio Nazionale delle Ricerche and from the Ministero della Pubblica Istruzione of Italy.

References

1. Zanetti, G., Aliverti, A., Curti, B.: J. Biol. Chem. 259, 6153-6157 (1984).
2. Cammack, R., Neumann, J., Nelson, N., Hall, D.O.: Biochem. Biophys. Res. Commun. 42, 292-297 (1971).
3. Shin, M.: Biochim. Biophys. Acta 292, 13-19 (1973).
4. Knaff, D.B., Smith, J.M., Chain, R.K.: Arch. Biochem. Biophys. 199, 117-122 (1980).
5. Hasumi, H., Nakamura, S., Koga, K., Yoshizumi, H., Parcells, J.H., Kimura, T.: J. Biochem. (Tokyo) 91, 135-141 (1982).

ON THE ENIGMA OF OLD YELLOW ENZYME'S SPECTRAL PROPERTIES

Jan K. Eweg, Franz Müller, Willem J.H. van Berkel and Ben Hesper
Department of Biochemistry, Agricultural University, De Dreijen 11, 6703 BC Wageningen, The Netherlands

Introduction

Long wavelength absorbing species ($\lambda >$ 500 nm) of flavoproteins are often observed under catalytic turnover conditions, but also under non-catalytic conditions (1). In the former case the long wavelength absorbing species are real catalytic intermediates and are commonly observed by the oxidation or reduction of pyridine nucleotide-dependent flavoproteins (1). Other flavoproteins such as e.g. D-amino acid oxidase also form catalytically important coloured intermediates with the substrate (1). Deeply coloured species of no catalytic importance are formed by several flavoproteins with a variety of organic molecules. Old Yellow Enzyme belongs to this class of flavoproteins and is famous for the formation of complexes with a variety of phenolate derivatives. The coloured species are usually referred to as charge transfer complexes (2,3). For several flavoproteins, e.g. glutathione reductase (4), lipoamide dehydrogenase (5) and butyryl-CoA dehydrogenase (6) it has been shown that the interaction of a thiolate ion or a persulphide ion with the prosthetic group of the corresponding oxidized flavoprotein leads to charge transfer (CT) phenomena. The sulphur atom possesses electronic properties (weakly bound d-electrons) making it an excellent electron donor which, embedded in complexes with molecules possessing appropriate electron acceptor properties, may give rise to an electronically excited state in which charge is transferred from donor (D) to acceptor (A). The energy (in cm^{-1}) $\bar{\nu}_{CT}$ of such a CT-transition obeys approximately the equation (7-9):

$$hc\bar{\nu}_{CT} = I_D - E_A - W \qquad (1)$$

where I_D represents the ionization potential of D, E_A the electron affinity of A and W the difference in interaction between D and A in the ground

Flavins and Flavoproteins

and CT state, respectively. This is the most frequently observed CT interaction, denoted by Mulliken (7,8) as 'outer complex'. The CT states are inaccessible when the moieties are separated and the corresponding CT transitions may appear anywhere in the optical absorption spectrum. In case of large W or small $I_D - E_A$ (related to the energy difference between the highe st occupied orbital (HOMO) of D and the lowest vacant orbital (LUMO) of A) the CT band appears at lower energy (longer wavelength) than the individual transitions of D and A, respectively.

In rare cases, the value of W (basically related to changes in electron correlation upon electron transfer from D to A) may exceed $I_D - E_A$, leading to cross-over of the ground and CT states of the complex (if not forbidden by symmetry-derived 'no-crossing rules'). The 'inner complex' (7,8) thus formed is characterized by a strong change of polarization upon its formation and it exhibits electronic transitions which are totally different from those of its constituents which, in turn, are completely destroyed. The 'new' transitions are usually also red-shifted compared to those of the molecules which make-up the complex, simply because the complex as a whole is now a large 'supermolecule'. 'Inner complex' transitions, however, usually have larger oscillator strength than 'outer complex' CT-transitions. Regrettably, the literature is poisoned to some extent by indistinct use of the term 'charge transfer complex' for both the inner and outer complexes. In case of flavoproteins there is an additional complication since the flavin semiquinone also absorbs at long wavelength. The pure semiquinone state can, of course, easily be detected by its paramagnetism but, in case of an inner (closed shell) complex with flavin as acceptor, the flavin LUMO may contribute significantly to the ground state of the (diamagnetic) complex giving it some semiquinone-like optical properties. Clearly, precise characterization of each of the coloured intermediates of the flavoproteins in the outlined context is highly desirable for a better understanding of flavoproteins.

We decided to embark on the study of old yellow enzyme (OYE) since 'normal' CT interaction (outer complex) likely occurs in the sulfur-containing flavoproteins . OYE-phenolate complexes are very stable (2,3) but they appear to be the least understood of all. Our results, interpreted according to modern physical theories, are published elsewhere in great detail (9). In the scope of this contribution we limit ourselves to the most salient conclusions.

Results and discussion

The initial assignment of the OYE-phenolate complex absorbance at long wavelength to a CT transition (3) was based on the observed correlation of the Hammett σ_p constant of the phenol used to produce the transition and $\bar{\nu}_{max}$ of the absorption band. We prefer the use of eq. (1) instead (cf. above), since this equation adheres strictly to basic CT theory, whereas σ_p is essentially an empirical parameter describing the ground state charge distribution in an aromatic molecule (10,11). Occasionally σ_p is observed to correlate with excited state properties but the physical significance of this observation is doubted (10,11). For flavoproteins, W is expected to be constant (maybe even zero) since it is the protein which keeps the proposed D (phenolate) and A (FMN) together(3). Titrating OYE with various phenols (of which I_D is known (12)) should, therefore, yield direct proportionality between $\bar{\nu}_{CT}$ and I_D, since E_A is a constant in the experiments done (3). Our results (taking into account the character of the orbitals responsible for the actual value of I_D - important since it is the I_D of phenolate which is relevant) show the $\bar{\nu}_{CT}$ - I_D proportionality to be non-existent in OYE (9).

Close examination of various absorption and CD spectra of OYE allows for one firm conclusion: Upon binding to OYE, phenolate interferes with an already existing complex between FMN and the apoprotein; the induction of the long wavelength absorption band is likely a secondary effect (9). The crucial argument for this is provided by the CD and low-temperature absorption spectra (Fig. 1). It is highly remarkable that native OYE displays no CD in the region of the first FMN absorption band: every electronic transition in free FMN has an intrinsically non-zero rotational strength owing to the coupling between the aromatic system and the D-ribose side chain (9). Only the apoprotein can provide the necessary L center to compensate this coupling in such a delicately balanced manner that the net dichroism is zero. Obviously, this balance can only be reached for one electronic transition in a system with so many excited states of totally different character (9). Phenolate binding essentially disrupts the balanced inter-

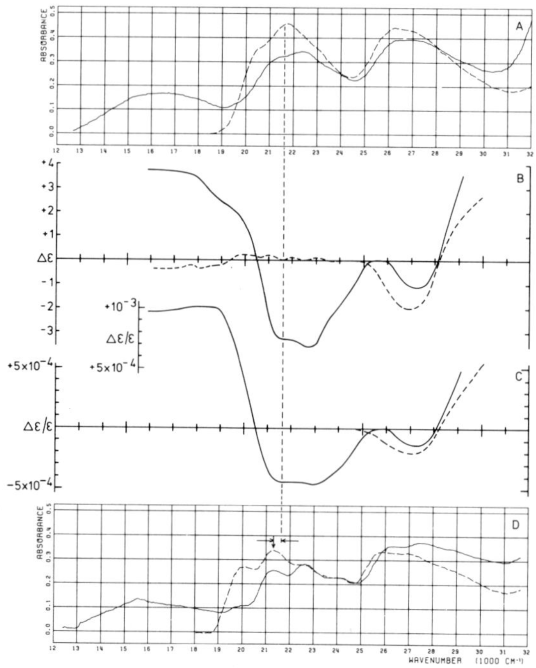

Figure 1. Absorption and circular dichroism spectra of free (---) and 4-hydroxy-N-n-butylbenzamide complexed (——) old yellow enzyme. A: Absorption spectra at 293K; B: circular dichroism spectra at 293K; C: Kuhn's anisotropy factor g = Δε/ε calculated from A and B; D: absorption spectra at 143K (From ref. 9).

action and produces the usual FMN CD characteristics: origin (dashed vertical line in Fig. 1) coincident with that of free FMN and a constant anisotropy-factor according to Kuhn (Δε/ε, Fig. 1 panel C) over the vibrational progression connected to this origin. The latter property is characteristic for a dipole-allowed electronic transition (9). Hence, the fine structure oserved in the low temperature spectrum of native OYE (Fig. 1, panel D) is not a progression originating at ∿ 20 000 cm^{-1} (500 nm), but the subbands at this wavenumber and that slightly shifted with respect to the vertical dashed line (Fig. 1) represent different electronic origins and, presumably, even belong to different species. Purely accidental, their separation is approximately one quantum of the vibration responsible for the progression in the first absorption band of FMN. It is also the 500 nm band which collapses upon complexation with phenolate, proving that it cannot be the ori-

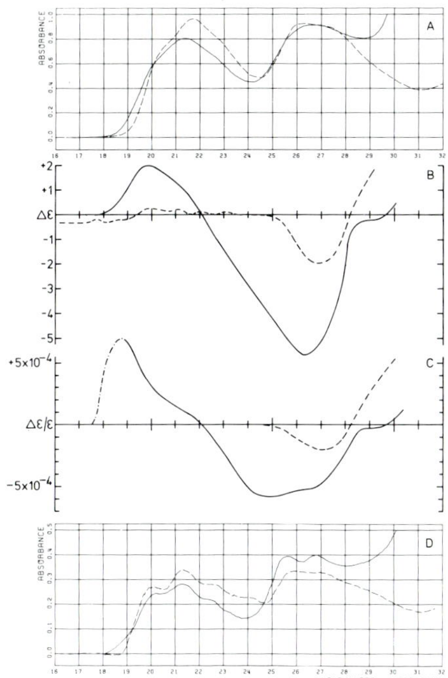

Figure 2. Absorption and dichroism spectra of free (——) and 4-methoxybenzaldehyde complexed (---) old yellow enzyme. For further details, see Fig. 1. (From ref. 9).

gin of a single progression extending to higher wavenumber since this would imply unrealistic changes and ratios of Franck-Condon factors (9). Additionally, it was claimed (3) that p-methoxybenzaldehyde should interact similarly with OYE as phenols, except for the ability to form the long-wavelength absorption band which can only be produced by the electron-rich phenolate, an argument in favour of the donor properties and consequent CT interaction of the latter. Repeating the foregoing experiments with p-methoxybenzaldehyde yielded less defined spectral characteristics but made immediately clear that the interactions of this compound with OYE differs completely from that of phenolate (Fig. 2).

Fluorescence and phosphorescence data (including life times), combined with MO calculations, provided several additional arguments that phenolate interacts with an existing FMN-apoenzyme complex of a yet unknown nature. The discovery of an additional fluorescence/phosphorescence emission in the blue and near uv and a very low content of unpaired spin will not be discussed here since there is no evidence whatsoever that the centers on which these properties reside are involved at the site of complex formation: they most likely are not! So there remains one conclusion: the charge of phenolate is responsible for the OYE phenomena observed as stated before (3). Direct influence on the spectra in the sense of a Stark-effect can be excluded since this is, contrary to the Zeeman-effect, not accompanied by circular polarization phenomena. This leaves one obvious preliminary suggestion for an alternative explanation: phenolate withdraws a proton from some network of hydrogen bonds in which possibly an FMN tautomeric form is involved (this may perhaps also explain the correlation with the Hammett σ_p). Which other chromophore (in a physical sense) in the network finally produces the long wavelength absorption is completely enigmatic (hence our title!), although the transition might finally prove to be of CT nature involving another donor than phenol. Recent NMR results (cf. Beinert, Rüterjans and Müller, this symposium) indicate strong polarization of the C(4) carbonyl group and the formation of a (strong) hydrogen bond to the N(5) atom of FMN in the native enzyme, contrary to the complexed enzyme where it appears to be normal. This behaviour fits into the proposed model outlined above.

It should also be considered that the proof or disproof of the existence of a CT should be supported by several physical techniques. Another good diagnostic tool is the measurement of the direction of polarization of the transition relative to that of the constituents. However, a polarization in the parallel direction of the constituents would not necessarily disprove the CT character. A polarization study has been done on the complex between oxidized flavin and anthranilic acid (13). The results clearly show the expected polarization in the complex and in this particular case support the CT character of the complex in question. This example can, however, not be used as a general support for the CT character of other coloured species between oxidized flavin and an organic molecule. In fact it has to be proven in each individual case. Measurement of the polarization in the OYE-phenolate complex in solution failed because no emissive state could be detected which could be identified with the long wavelength absorption band (9).

Acknowledgements

We thank Miss Lyda Verstege for typing the manuscript, Dr. C.A.H. Rasmussen for carefully reading the manuscript and Mr. M.M. Bouwmans for preparing the figures. This study was carried out under the auspices of the Dutch Foundation for Chemical Research (SON) with financial aid from the Dutch Organization for the Advancement of Pure Research (ZWO).

References

1. Massey, V., Ghisla, S.: Ann.N.Y.Acad.Sci. 227, 446-465 (1974).
2. Abramovitz, A.S., Massey, V.: J.Biol.Chem. 251, 5321-5326 (1976).
3. Abramovitz, A.S., Massey, V.: J.Biol.Chem. 251, 5327-5336 (1976).
4. Arscott, L.D., Thorpe, C., Williams, C.H. : Biochemistry 20, 1513-1520 (1981).
5. Wilkinson, K.D., Williams, C.H.: J.Biol.Chem. 254, 852-862 (1979).
6. Williamson, G., Engel, P.C.: Biochem.J. 211, 559-566 (1983).
7. Mulliken, R.S.: J.Phys.Chem. 56, 801-822 (1952).
8. Mulliken, R.S., Person, W.B.: Ann.Rev.Phys.Chem. 13, 107-126 (1962).
9. Eweg, J.K., Müller, F., van Berkel, W.J.H.: Eur.J.Biochem. 129, 303-316 (1982).
10. Jaffé, H.H.: Chem.Rev. 53, 191-261 (1953).
11. Hammond, P.R.: J.Chem.Soc. 471-479 (1964).
12. Turner, D.W., Baker, C., Baker, A.D., Brundle, C.R.:Molecular Photoelectron Spectroscopy, pp. 279-322, Wiley, London 1970.
13. Shieh, H.S., Ghisla, S., Hanson, L.K., Ludwig, M.L., Nordman, C.E.: Biochemistry 20, 4766-4774 (1981).

"ON THE ENIGMA OF OLD YELLOW ENZYME'S SPECTRAL PROPERTIES"

V. Massey, L. M. Schopfer, W. R. Dunham
Department of Biological Chemistry and Biophysics Research Division,
University of Michigan, Ann Arbor, Michigan, 48109

The Old Yellow Enzyme of brewers yeast was the first discovered flavoprotein, being isolated by Warburg and Christian in 1933 (1). In the intervening 50 years, its physiological function has remained enigmatic, as it would seem unlikely that the known NADPH-oxygen reductase activity would serve any useful function in the metabolism of the yeast. Despite this fundamental lack of knowledge of its catalytic function, a large body of information on the spectral characteristics of the enzyme has been accumulated. Starting from the isolation of the enzyme in 1969 in a green form (2), which was shown to be due to a complex between the yellow enzyme and p-hydroxybenzaldehyde (3), it was shown that a large variety of aromatic and heteroaromatic compounds would bind to the enzyme and result in long wavelength absorption bands (2-5). While some compounds, such as p-methoxybenzaldehyde, bind to the enzyme and cause pronounced perturbations in the flavin absorption spectrum without resulting in a long wavelength transition (5), a common feature to all compounds which cause the appearance of a long wavelength transition is the presence of an ionizable hydroxyl residue (5). A positive correlation between the energy (wavenumber) of the long wavelength transition and the Hammett para-constant for a series of p-substituted phenols, plus the pH-dependence of binding, provided the first experimental evidence that the long wavelength absorption bands were due to charge transfer transitions between the bound phenol in its ionized phenolate state as charge transfer donor and oxidized flavin as charge transfer acceptor. This ascription was supported very strongly by companion experiments where the native flavin, FMN, was replaced by artifical flavins of different redox potentials. Again, long wavelength transitions were observed on adding phenols, and in all cases the shift in the position of the long wavelength band was consistent with the concept of the oxidized

Flavins and Flavoproteins

flavin as charge transfer acceptor, ie. flavins of more negative potential than FMN resulted in higher energy transitions while flavins of more positive potential resulted in lower energy transitions (5).

Thus we had considered that the spectral ascription of charge transfer bands involving oxidized flavin and electron-rich ligands had been placed on a firm experimental base, at least with Old Yellow Enzyme, and that by extrapolation, similar bands found with other flavoenzymes could be interpreted in the same manner (see ref. 6 for a review). Further support for this concept came from resonance Raman studies with Old Yellow Enzyme (7,8), D-amino acid oxidase (8), fatty acyl CoA-dehydrogenase (9) and butyryl CoA-dehydrogenase (10). Recently the structure has been reported of a crystalline molecular complex between lumiflavin and o-aminobenzoic acid (11). The optical transition of these crystals at about 600 nm was shown to be a charge transfer transition because of the strength of the absorption perpendicular to the plane of the flavin and aminobenzoic acid molecules.

With this background we were therefore somewhat suprised to see a paper by Eweg et.al. (12) entitled "On the Enigma of Old Yellow Enzyme's Spectral Properties". This paper, according to its authors, presents data and arguments which destroy the conclusion from previous work about the charge transfer nature of OYE-phenolate complexes. Instead they attempt to explain the spectral properties of the enzyme in terms of an obscure network of hydrogen bonds which promote enol tautomerization at the flavin 4-position. On binding phenolate anion this network of hydrogen bonds is proposed to be disrupted, leading to the return of the flavin keto-tautomer, and to the appearance of a long wavelength transition. The purpose of this paper is to remove the confusion resulting from the work of Eweg et.al. by demonstrating the errors in their theory, their experiments and their interpretations, while at the same time extending our previous work with new results that even more fully document our previous conclusions.

Definition of Terms. A common source of confusion is the meaning of the term "charge transfer complex". This term has come to define those complexes that display charge transfer transitions in their optical spectra. A charge transfer transition usually refers to a change from an arbitrary ground state to an excited state where an electron has been transferred from the donor to the acceptor molecule. As stated by Eweg et.al. (12) the energy of this transition is given by

$$hc\nu_{ct} = I_d - E_a - W \qquad \text{(Eq 1)}$$

where h is Planck's constant, c the velocity of light, νct is the position of the charge transfer transition in wavenumbers, Id the ionization potential of the donor, Ea the electron affinity of the acceptor and W the binding energy in the charged-transferred state. The actual binding energy of the ground state of the complex is irrelevant to the above equation and to the concept of charge transfer. Indeed it is obvious from equation 1 that charge transfer itself does not contribute greatly to the overall bonding in the complex, because if it did, W would become larger than Id-Ea, and νct would become negative, ie. the ground and excited states would be reversed, and we would be describing a normal ionic bond with a charge-transferred ground state. Therefore, in most cases, and in OYE-phenolate complexes in particular, we are describing a complex where charge transfer is possible, but not a necessary result of the interaction of the donor and acceptor molecules. In the paper of Eweg et.al. (12) CNDO/S calculations showed that the molecular orbitals were not affected as long as the FMN-phenolate stacking distance was greater than 3.4 A^o. Since the x-ray crystallographic study on the lumiflavin-o-aminobenzoate complex showed that the stacking distance in this complex was at least 3.36-3.42 A^o, we can indeed assume that the W term in equation 1 is very small and that differences in W with different donor molecules can probably be ignored.

Therefore, the important parameters in equation 1 are the ionization potential of the donor molecule and the electron affinity of the acceptor molecule (the ionization potential of the FMN anion radical). Both the Id and the Ea terms of equation 1 can be written as ground state properties (ionization potentials), and contrary to the claim of Eweg et.al. one can expect to correlate ground state properties with the energy of the charge transfer transition. In particular, this type of correlation for both the donor and acceptor has been shown to be characteristic of charge transfer transitions using the Hammett para-constants (13). Such a correlation was shown for phenolate-OYE complexes by Abramovitz and Massey (5) and in the present case has been extended by the inclusion of several p-substituted phenolates not previously studied.

While the correlation with Hammett para-constants is impressive, an even more striking correlation is obtained when tha data are treated by the procedure described by Swain et.al. (14) which explicitly takes into account the contributions from both resonance and inductive effects. In Fig 1 the measured energy of the long wavelength transition is plotted against that calculated by the Swain equation

$$\nu_{ct} = h + Ff + Rr \qquad \text{(Eq 2)}$$

where F is the field constant and R is the resonance constant for each substituent, and h, f and r are the reaction constants calculated to fit the equation for the whole set of phenolates. The analysis of 12 phenolates yielded

$$\nu_{ct} = 16073 + 392F + 1360R \qquad s=275 \qquad \text{(Eq 3)}$$

with a correlation coefficient of 0.97, where s is the standard error in νct, and standard errors: 122, 206 and 74 respectively are for the three constants h, f and r. The large value of r indicates that in the series studied, the resonance effect of the substituent is dominant.

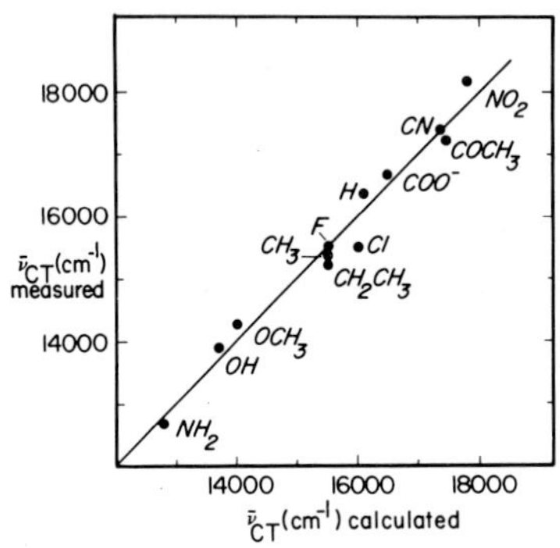

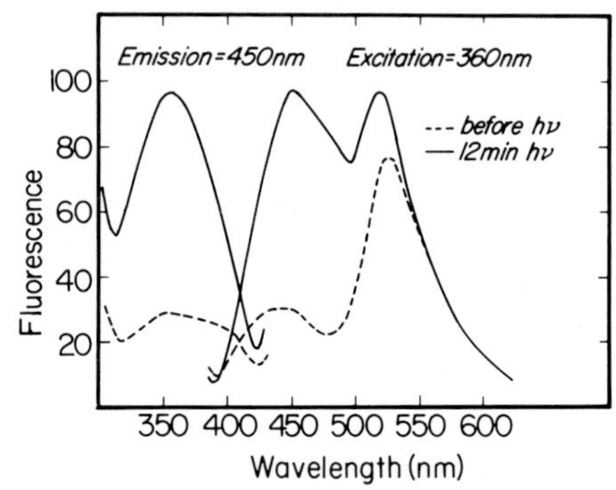

Fig. 1 (left) Comparison of observed and calculated energies for the long wavelength transitions of OYE-phenolate complexes. The Swain analysis yields $\nu ct=(16073+392F+1360R)\ cm^{-1}$, where F and R are the field and resonance constants for the para-substituted phenols used.
Fig.2 (right) Generation of a blue fluorophore in OYE by visible light. The dashed line shows the fluorescence excitation and emission spectra of a preparation of OYE (4) which had been stored at -20° for 2 years. It displays some FMN fluorescence, as seen from the emission maximum at 520 nm. The solid line shows the fluorescence after irradiation for 12 min at 0° with light from a commercial Sun gun (cut off ~330 nm) at an intensity of $\sim 8x10^{6}$ ergs $cm^{-2}s^{-1}$.

<u>On the Supposed Existence of Non-Flavin Chromophores in Old Yellow Enzyme</u>. One of the most remarkable features of Old Yellow Enzyme is its lack of fluorescence, a property used to advantage in the pioneering work of Theorell and his associates (15). Indeed freshly prepared enzyme is without

visible fluorescence, as noted by Theorell (16) and confirmed by work in this laboratory. However, on aging, a weak flavin fluorescence does develop. It appears to be this weak fluorescence which prompted some four pages of results and discussion by Eweg et.al. (12) and led to their admission that it is dominated by the emission of free FMN. In addition they detected a new fluorophore with an excitation maximum at 380 nm and an emission maximum at 444 nm, which they ascribe to the presence of a previously unrecognized chromophore. On examining five different preparations of Old Yellow Enzyme which had been stored at -20° for periods from 6 months to several years, we were able to confirm the presence of the same fluorescence. However, the intensity of this fluorescence varied significantly among the five samples indicating that it was probably an artifact. That this explanation is correct, and that the source of the artifact is from some photochemical modification of the protein, is supported by the finding that the intensity of the blue fluorescence is increased dramatically on exposure of the enzyme to visible light (Fig 2).

It was also claimed by Eweg et.al. that Old Yellow Enzyme exhibited an unpaired spin when examined by EPR at low temperature, with quantitation between 0.1-10% of the FMN concentration. Our attempts to reproduce these experiments were a complete failure. At 9° K, the 9 GHz EPR spectra at several microwave powers and field modulation amplitudes revealed no measurable signal at g=2, regardless of whether the OYE preparations were at pH 7 or pH 8.5, or whether they had been illuminated or not. We can only conclude that the EPR signals reported by Eweg et.al. were due to a contamination which is not present in pure preparations of the enzyme.

On the Origin of the S_0--S_1 and S_0--S_2 Electronic Transitions in Free Flavins, Old Yellow Enzyme and Other Flavoproteins. A fundamental error of interpretation by Eweg et.al. appears to have been in the assignment of the 0-0 transition of the first electronic absorption band S_0--S_1. With free flavin in aqueous solution the 0-0 band of the S_0--S_1 transition has been identified as occurring in the range 469-480 nm (21300-20830 cm^{-1}) by a number of workers employing a variety of techniques, including curve fitting of both absorbance and CD spectra (17), low temperature spectroscopy (18-20) and line shape calculation for coherent antistokes Raman spectroscopy (21). To these techniques, all of which are very time consuming or expensive, we wish to add the technique of second derivative optical absorption spectroscopy. The theoretical and practical abililty of this technique to resolve complex overlapping spectral bands has been described by Butler (22) and its application to tetraacetylriboflavin (TARF) in 95% CCl_4:5%CH_3CN is shown in Fig 3.

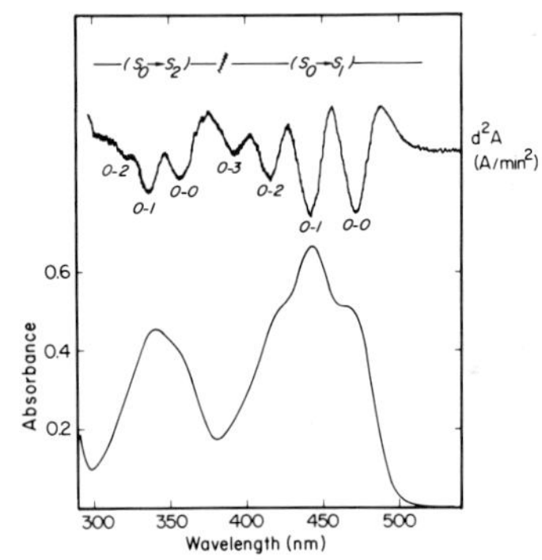

Fig. 3 Second derivative absorption spectrum of tetaacetylriboflavin in 95% CCl_4:5%CH_3CN, and the absorption spectrum of the same sample. Spectra were recorded with a Cary 219 spectrophotometer; the second derivative was recorded at a scan speed of 2nm per second.

According to Butler (22) the troughs in the second derivative spectra should correspond closely to the positions of maxima of components in the overall spectrum. The longest wavelength trough would then be expected to correspond to the S_0--S_1 0-0 transition. From Fig 3 it is clear that in this solvent the S_0--S_1 transition can be cleanly resolved into at least three and possibly four vibrational levels. The results of analyses such as that shown in Fig 3 for TARF in a variety of solvents, for FMN and FAD, and comparison with literature values are shown in Table 1. It is clear that the results from second derivative spectra agree well with those obtained by other methods. In particular, it should be noted that the value of the S_0--S_1 0-0 transition for FMN of 21280 cm^{-1} (470 nm) agrees very well with that, 21300 cm^{-1} or 469 nm, determined by Eweg et.al. (20) by low temperature spectroscopy. The results from the second derivative analyses of Table 1 give a vibrational progression in the S_0--S_1 transition which averages 1385 cm^{-1} for all the data listed. This compares reasonably with the average progression of 1340 cm^{-1} found by Edmondson and Tollin by curve fitting (17) and 1250 cm^{-1} quoted by Eweg et.al. (20) for results from low temperature spectra.

Table 1
Spectroscopic Properties of Flavins Derived from Second Derivative Spectroscopy and Comparison with Literature Values

Flavin	$\nu\ S_0$--S_1 0-0	0-1	0-2	0-3	$\nu\ S_0$--S_2 0-0	0-1	0-2
TARF in H_2O	21190	22620	24100		26880		
TARF in EtOH	21275	22725	24100	~25315	27470	29410	~31250
TARF in CH_3CN	21320	22780	24270	25500	27780	29850	~31250
TARF in 95% CCl_4	21230	22675	24100	25500	28090	29760	~31250
TARF in H_2O[a]	21050	22270	23580		27170	27030	28990
FMN	21280	22675	24040		26810	28170	~29850
FMN[a]	21140	22320	23920		26180	27780	29850
FMN[b]	21300						
FAD	21100	22420	23810		26595	27780	~29400
FAD[c]	20830						
N-3-undecyl lumiflavin[d]	20956						
N-3-methyl lumiflavin[e]	21190						

a) Ref 17 b) Ref 20 c) Ref 21 d) Ref 18 e) Ref 19

Table 2
Data from Second Derivative Spectra of OYE and Complexes for S_0--S_1 Transition

Flavin form	ν of vibronic transition (cm-1) 0-0	0-1	0-2	0-3	Δ 0-0 free flavin-enzyme bound
FMN	21280	22675	24040		
Native OYE	20240	21740	23200	~24400	1040
OYE + p-methoxy-benzaldehyde	19920	21320	22830	~23900	
OYE + p-cresol	21100	22620	23920		
OYE + p-fluorophenol	21190	22570	24040		
OYE + p-cyanophenol	21190	22620	24100		
OYE + p-aminophenol	20300	21740	23150	~24390	
3-methyl-FMN	21140	22520	23920		
3-methyl-FMN OYE	20160	21645	23040	~24100	980
7-Br-FMN	21230	22570	24100		
7-Br-FMN OYE	20120	21550	23040	24330	1110
4-$NHCH_3$-FMN	20450	21930	23360		
4-$NHCH_3$-FMN OYE	19230	20790	22220	23750	1220
4-$N(CH_3)_2$-FMN	20370	21835	23310		
4-$N(CH_3)_2$-FMN OYE	19160	20750	22170	23700	1210

A similar vibrational progression of 1385 cm^{-1} is obtained with OYE from second derivative spectra, except that the 0-0 transition is now moved from 470 nm (21280 cm^{-1}) in FMN to 494 nm (20240 cm^{-1}) in OYE (Table 2). A remarkably similar series of values can be estimated from the low temperature spectrum of the enzyme published by Eweg et.al. (12), with a vibrational progression of 1330 cm^{-1}. These authors considered the possibility that the 20000 cm^{-1} peak in their low temperature spectrum was due to the 0-0 transition, but rejected this on the basis of several experimental results which we consider to be incorrectly interpreted by them. The first of these involves circular dichroism spectra, which we will consider in a later section. The second involves the different effects which they observed with p-methoxybenzaldehyde, and p-hydroxy-N-n-butylbenzamide. The former, lacking an ionizable hydroxyl, binds to OYE but does not give a long wavelength transition, while the latter is a typical phenol giving rise to a long wavelength band (4,5). Eweg et.al. did not find any appreciable change in the low temperature spectrum of the enzyme on adding p-methoxybenzaldehyde, but did observe a shift back to that typical of free FMN in the presence of p-hydroxybutylbenzamide. On the basis of these results they stated "it may be concluded that the 20150 cm^{-1} (496 nm) and 21300 cm^{-1} (469 nm) bands represent different electronic origins and presumably, even belong to different species".

The fallacy of this conclusion results from a failure to saturate the enzyme with p-methoxybenzaldehyde. The Kd for this OYE complex has been reported to be $6x10^{-4}$ M (5). From the legend to their Fig 2, Eweg et.al. employed only equimolar quantities of OYE and p-methoxybenzaldehyde, at a concentration of $9.1x10^{-5}$ M. With a Kd of $6x10^{-4}$ M, this would correspond to 12% complex and 88% free enzyme! While the Kd may be lower under their low temperature conditions, it is clear that only partial complex formation had been achieved. Hence, it is not surprising that the low temperature spectrum was similar to that of free enzyme. When the enzyme is saturated with p-methoxybenzaldehyde, second derivative spectra reveal that the whole S_0--S_1 transition is in fact shifted to longer wavelengths, with the 0-0 transition at 502 nm (19920 cm^{-1}), and a vibrational progression of 4 bands separated by 1330 cm^{-1} (Table 2). Table 2 also shows the results with four typical phenolates which all give rise to long wavelength transitions. Three of these, p-methyl, p-fluoro and p-cyano phenols give results similar to those oberved by Eweg et.al. for p-hydroxybutylbenzamide, ie. the 0-0 transition is shifted back close to that of free FMN. With p-aminophenol, however, the 0-0 transition is close to that found with the uncomplexed enzyme. In all cases, the vibrational progression is the same, an average of 1410 cm^{-1}.

Table 3
Spectroscopic Parameters of S_0--S_1 Transition for Various Flavoproteins Obtained from Second Derivative Spectroscopy and Comparison with Literature Values[a]

Flavoprotein	ν of vibronic transition (cm^{-1}) 0-0	0-1	0-2	0-3
M. elsdenii flavodoxin	21230	22620	24040	
	(21230)	(22620)	(23640)	
Shethna flavoprotein	21050	22470	23980	~25000
	(21050)	(22420)	(23640)	
Lactate oxidase	20620	22025	23530	~24815
Lactate oxidase + D-lactate	20280	21785	23255	24570
Lactate oxidase + tartronate	21050	22470	23980	
Electron transfer flavoprotein	21600	23090	24500	
Lipoyl dehydrogenase	20660	22120	23580	24940
Ferredoxin-NADP reductase	20530	21930	23310	
	(20400)	(21690)	(22620)	
L-amino acid oxidase	20410	21830	23260	
	(20120)	(21510)	(22830)	
Butyryl-CoA dehydrogenase	20920	22370	23750	
" + acetoacetyl-CoA	21930	23365	24690	

a) Values in brackets from Edmondson and Tollin, ref 17.

Entirely analogous results are found with other flavoproteins, as listed in Table 3. These show, in agreement with results of Edmondson and Tollin (17), that the S_0--S_1 0-0 transition of flavins can be shifted considerably in different flavoproteins, ranging from 21600 cm^{-1} (463 nm) in pig kidney electron transfer flavoprotein to 20400 cm^{-1} (490 nm) in L-amino acid oxidase and 20240 cm^{-1} (494 nm) in OYE (cf. Table 2). From the 0-0 transition, the vibrational progression of 1414 cm^{-1} is found on average for all of the flavoproteins studied. From an inspection of Tables 1-3, it should be clear that the S_0--S_1 electronic transition of flavins is made up of four experimentally observed vibrational sub-bands: 0-0, 0-1, 0-2, 0-3, but that in many cases it is not possible to resolve the 0-3 vibrational level because of overlap of the S_0--S_1 and S_0--S_2 transitions.

Table 3 also lists the second derivative analysis of the yellow form of M. elsdenii butyryl-CoA dehydrogenase, and of its complex with acetoacetyl-CoA. The latter gives a long wavelength transition similar to those of OYE-phenolate complexes (23), and as with most of these complexes, the S_0-S_1 0-0 transition is changed considerably, to a value higher in energy (21930 cm^{-1}) than that of free FAD itself (21100 cm^{-1}). It is clear from the summary of results presented in Tables 1-3, that there is nothing in the spectral properties of OYE that cannot find its counterpart in other flavoproteins.

Circular Dichroism Spectra of Old Yellow Enzyme Much attention was paid by Eweg et.al. (12) to the CD properties of OYE, which is unusual among flavoproteins in having no CD bands at wavelengths greater than 400 nm, ie. the S_0--S_1 transition does not display CD absorption. Eweg et.al. state "This is a remarkable observation since optical activity of all transitions in flavins carrying a ribityl side chain or a derivative of it, is a rule rather than an exception, a fact well documented by experiment and theory." While we cannot offer any concrete explanation for the lack of CD absorption associated with the S_0--S_1 transition, we wish to point out that the phenomenon is not unique to OYE. The CD absorbance of this transition is missing in free FAD (24) and practically absent in lipoyl

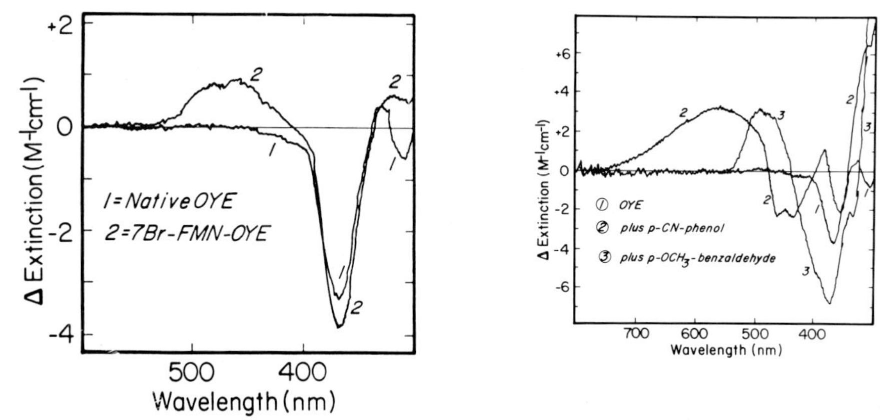

Fig. 4 Circular dichroism spectra of OYE. These were all recorded with a Durram-JASCO J-41 spectropolarimeter, in 0.1 M phosphate, pH 7.0, 6°. A: native OYE and 7-Br-FMN OYE. B: OYE and its complexes with p-cyanophenol and p-methoxybenzaldehyde.

dehydrogenase (25) and lactate oxidase (L. Schopfer, unpublished). Eweg <u>et.al.</u> interpret the CD spectrum as possibly indicating an interaction between the bound FMN and a protein tyrosine residue, as has been suggested before to explain the quenching of the flavin fluorescence in OYE (16). They use this as evidence in support of their explanation of the spectral properties of OYE. We will discuss this in a later section.

Whatever the explanation of the CD spectrum, it is clear that it is readily modified. Figure 4A shows the effect of replacing the native FMN with 7-Br-FMN. This modified flavin had been shown previously to bind tightly to the apoenzyme (Kd < 10^{-8} M) and to give a parallel set of long wavelength transitions with 9 different phenols in which the ν_{ct} was displaced 1000 cm^{-1} lower than that for the native enzyme (5). This important correlation was in fact completely ignored by Eweg <u>et.al.</u> (12). With 7-Br-FMN enzyme, a positive CD absorbance is now found in the 410-520 nm region. A similar positive band is induced on binding p-methoxybenzaldehyde to the native enzyme (Fig 4B). This should be contrasted to the results with a typical phenolate, p-cyanophenol (Fig 4B), where the CD absorbance of the S_0--S_1 transition is negative, and a positive CD absorbance associated with the charge transfer transition is clearly visible.

<u>Proposal of Eweg et.al. for Flavin-Protein Interactions in Old Yellow Enzyme</u> On the basis of the results and conclusions discussed in previous sections Eweg <u>et.al.</u> (12) concluded that their "results clearly demonstrated the inadequacy of a simple phenolate-FMN donor-acceptor charge transfer complex to explain the phenomena occurring on the addition of phenols to old yellow enzyme. Instead it was found that the phenolate anion interferes strongly with an existing tight complex between FMN and the apoprotein, probably an H-bonded structure in which FMN is tautomerized and interacts with an L-chiral center." Although they did not lay out their proposal in diagrammatic form, it is helpful for discussion purposes to be able to visualize their proposal. This has been done in Fig 5.

The shift in wavelength maximum from 445 nm to 462 nm when FMN is bound to the apoprotein is ascribed to an interaction with the apoprotein in which a "tyrosine donates a proton for an H-bond with a flavin carbonyl group and flavin donates its N(3) proton to an H-bond with the apoprotein. This may lead to a flavin tautomeric form". This is diagrammed in the top part of

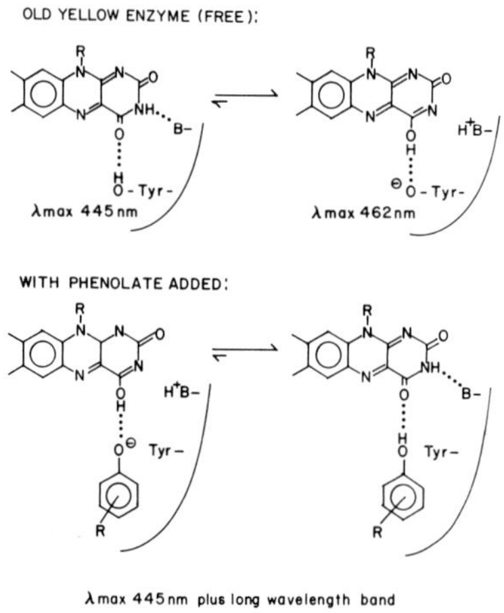

Fig. 5 Model of FMN-protein and FMN-phenolate interactions proposed by Eweg et.al. (12). This is a diagrammatic representation of their written statements.

Fig 5 for the free enzyme. In this explanation, the 20150 cm^{-1} (496 nm) absorption band is ascribed to the 0-0 transition of the flavin enol tautomer shown. They further state "Within this model, binding of the phenolate anion disrupts the existing FMN-apoprotein hydrogen-bonded structure, leading to the disappearance of the 20150 cm^{-1} (496 nm) absorption band, and formation of a new species absorbing at longer wavelength....(with) the recurrence of the protein-bound FMN to its keto form." This proposal, involving a proton transfer mechanism, is diagrammed in the lower half of Fig 5.

Quite apart from this model resting on questionable data and interpretations, as already discussed, it can be eliminated partly on the basis of already published information whose import was not taken into account by Eweg et.al. The first piece of information is that N(3)-methyl-FMN also binds to Old Yellow Enzyme apoprotein, and forms long wavelength absorption bands with added phenols in the same way as does native enzyme (5). We now report the spectral properties of this artifical enzyme, which are listed in Table 2, along with the properties of other artificial flavin forms of OYE. With native enzyme, the S_0--S_1 0-0 transition of FMN is shifted 1040 cm^{-1} to lower energy from 21280 cm^{-1} to 20240 cm^{-1}. An almost identical shift is observed (from 21140 cm^{-1} to 20160 cm^{-1}, delta 980 cm^{-1}) for 3-methyl-FMN on binding to apo-OYE, and with a very similar vibrational progression as found with FMN (see Table 2). Clearly 3-methyl-FMN cannot form an enol tautomer of the structure proposed by Eweg et.al., although a zwitterion with a negative charge on O(4) and a positive charge on N(3) cannot be ruled out. The conclusion that 3-methyl-FMN behaves like FMN is strengthened by the finding that with every phenolate tested (pentafluorophenol, p-cyanophenol, p-hydroxybenzaldehyde, p-chlorophenol, p-cresol and p-methoxyphenol) the long wavelength transitions have very similar extinction coefficients to those found with native enzyme and have wavelength maxima indistinguishable from those of native enzyme. Two of these are illustrated in Fig 6.

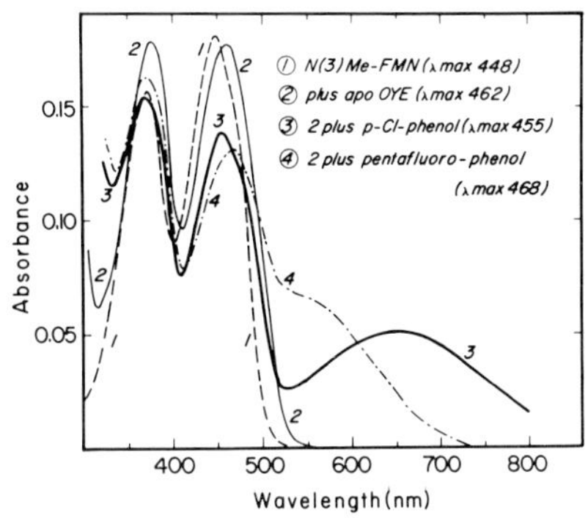

Fig. 6 Spectra of N(3)-methyl-FMN, free (curve 1) and bound to apoOYE (curve 2). Also shown are the effects of adding saturating concentrations of p-chlorophenol (2 mM; curve 3) and pentafluorophenol (1 mM; curve 4). Conditions, 0.1 M phosphate pH 7.0, 4°.

Table 2 also lists spectral comparisons of other artificial flavins both free and bound to OYE. It appears that a shift to lower energy of 1000-1200 cm^{-1} on binding to apoprotein is a common feature among all these flavins. Included in this list are two other flavins which are locked into a particular tautomeric form, in this case the form analogous to the native flavin enol tautomer (26); 4-$NHCH_3$-FMN and 4-$N(CH_3)_2$-FMN:

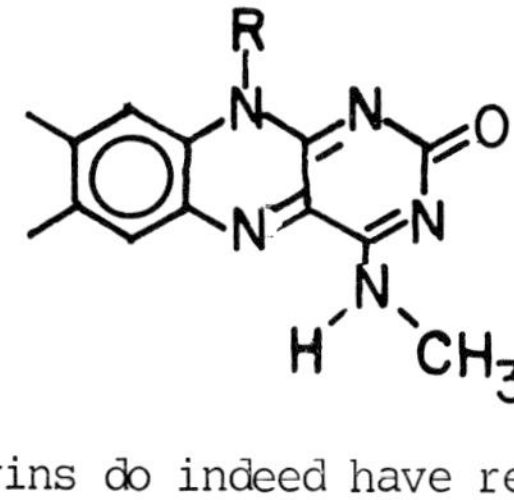

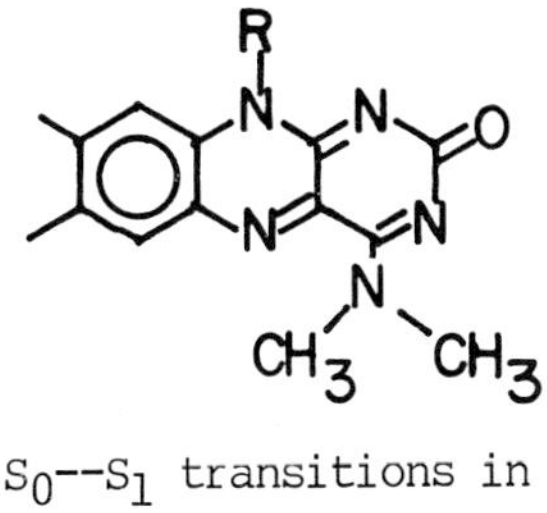

These flavins do indeed have red-shifted S_0--S_1 transitions in accord with the model of Eweg et.al. Therefore, on binding to the apoprotein, no change in the absorption envelope would be expected on the basis of their model. However, precisely the same type of shift is found as is observed with other flavins. This is illustrated in Figs 7A and 7B and Table 3. The long wavelength transitions induced on adding phenols are now less pronounced, particularly with 4-N$(CH_3)_2$-FMN enzyme, suggesting that the bulky flavin residue is interfering with the normal binding or orientation of the phenolate. This idea is consistent with the fact that 4-SCH_2CONH_2-FMN will not bind to the apoenzyme, whereas 4-thio-FMN binds quite tightly.

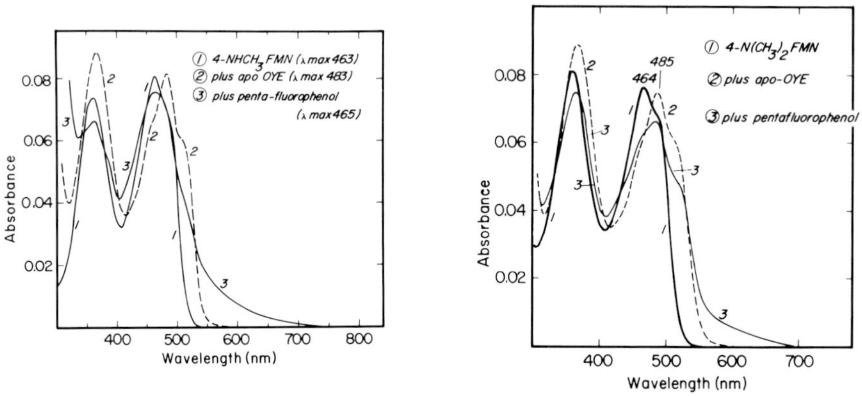

Fig. 7A Spectra of 4-$NHCH_3$-FMN, free (curve 1) and bound to apoOYE (curve 2), and plus a saturating concentration of pentafluorophenol (1 mM; curve 3).
Fig. 7B Results with 4-N$(CH_3)_2$-FMN, same conditions. (0.1 M phosphate, pH 7.0, 4°).

The State of the Bound Phenolate in Complexes with Old Yellow Enzyme
Another consequence of the Eweg et.al. model is the prediction that while the phenolate anion is bound by the protein, by the proposed proton transfer mechanism, it would become protonated in the complex (see Fig 5). This was the second occasion where already published results were at variance with the model, and simply not taken into account by Eweg et.al. Thus, in ref. 5, the titration of the enzyme by p-chlorophenol caused the appearance of a new absorbance band at about 315 nm which was noted to be coincident with the appearance of the long wavelength band, and which was ascribed to a red-shifted phenolate absorbance. This phenomenon is exhibited by every phenolate studied. A particularly dramatic

illustration of this is given in Fig 8, where OYE was titrated at pH 7.0 with p-nitrophenol. The dissociation constant for the complex is very low (< 10^{-8} M), so that there is a sharp endpoint in the development of the long wavelength band with stoichiometric addition of nitrophenol. This is accompanied by another intense transition in the region 400-460 nm, more intense than that of flavin, which can only be ascribed to the bound nitrophenolate anion. Figure 8B shows the difference spectrum between the complexed and the free enzyme in this wavelength region, to give an approximation of the spectral contribution of the bound nitrophenolate. For comparison, the absolute spectra of free nitrophenolate anion and the protonated nitrophenol are shown.

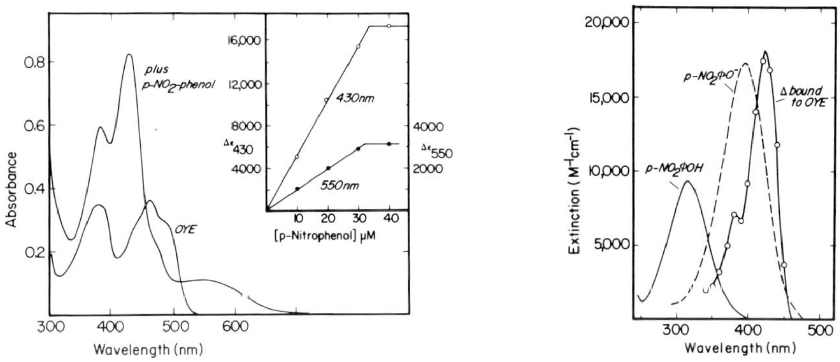

Fig. 8A Titration of OYE with p-nitrophenol. Enzyme (34 μM) in 0.1 M phosphate, pH 7.0, 4°, was titrated with p-nitrophenol, at the concentrations shown in the inset. Spectra were recorded after each addition. The spectrum "plus p-NO_2-phenol" is that calculated for the equivalence point, and was obtained by subtracting the small contribution of excess p-nitrophenol from the spectrum obtained with the highest concentration.

Fig. 8B Absorption spectra of p-nitrophenol at pH 1 (solid line) and pH 10 (dashed line). Also shown is the difference extinction calculated from the free and ligand-bound OYE spectra of Fig. 8A.

The same phenomenon is found with all phenols tested, and the results are summarized in Table 4. In each case, the extinction change observed with the enzyme-bound form matches roughly the absolute extinction of the free phenolate, and there is a relatively constant shift in the energy to lower values, a shift which varies from about 1900 cm^{-1} with nitrophenolate anion to 2500 cm^{-1} with phenolate and p-cresolate anions.

Table 4
Spectral Characteristics of Free and Enzyme-Bound Phenols

phenol	λ_{max} and Extinction Coefficients						$\Delta\nu$ free-bound phenolate
	Free phenol		Free phenolate		Enzyme-bound		
	nm	($M^{-1}cm^{-1}$)	nm	($M^{-1}cm^{-1}$)	nm	($M^{-1}cm^{-1}$)	cm^{-1}
phenol	268	(1600)	286	(2550)	308	(2400)	2500
p-cresol	275	(2400)	293	(3600)	316	(3400)	2480
p-chlorophenol	278	(1550)	296	(2400)	314	(3100)	1940
p-fluorophenol	276	(4500)	296	(5900)	316	(3400)	2140
p-methoxyphenol	286	(2300)	303	(2650)	324	(3300)	2140
p-aminophenol	~295	(~2000)	310	(~2600)	332	(~2300)	2140
p-nitrophenol	315	(9400)	396	(17300)	428	(18000)	1890

Equation 1 states that the energy of a charge transfer transition is due to three terms: the ionization potential of the donor (Id), the electron affinity of the acceptor (Ea) and the binding energy of the charge transferred (excited) state (W). We have argued that the latter energy is small and non-variable in our set of phenolate-OYE complexes. We will provide a rationale below for maintaining that the electron affinity of FMN, as expected, should remain relatively constant. Thus the shifts in energy of the charge transfer transition of the phenolate-OYE complexes should be accounted for totally by changes in the phenolate ionization potential. In ref. 5 and Fig. 1 we show that the energy of the charge transfer band can be correlated with the Hammett para-constants for our data set and with the Swain constants for field (F) and resonance (R) effects of substituents on a benzene ring. However, we have not made a direct connection (either theoretically or experimentally) between ionization potentials of the phenolate anions and the Swain or Hammett constants. This we shall do now.

Detailed justification of this connection does not seem to be available in the literature, although it is often stated (see 13 for review) that charge-transfer transitions show strong correlation with Hammett constants. In our treatment of this question, we shall correlate the Swain, rather than the Hammett, constants with various transition energies. We have chosen Swain constants since it has already been shown that Hammett constants can be well fit by the treatment of Swain et.al. and since the field and resonance effects are easily differentiated. Experimentally, ionization potentials of phenols can be accurately determined by high resolution photoelectron spectroscopy. Our best fit of the data of Palmer <u>et.al.</u> (27)

to the Swain equation gave the ionization potential of para-substituted phenols (in wavenumbers) as

$$\nu = 68520 + 4234F + 1350R \qquad s=462 \qquad \text{(Eq 4)}$$

where s is the standard error in ν. The standard errors (28) in the three coefficients are 350, 836 and 377, respectively. The F coefficient is less accurately determined than the R coefficient because F has one third the range of R for this data set. Correlation analysis is consistent with F being independent of R. Thus there is a clear correlation between the Swain constants and the ionization potentials for para-substituted phenols.

The ionization potentials of phenols are not necessarily analogous to the ionization potentials of phenolates. The only method which we could find for determination of phenolate anion ionization potentials was via the measurements of intrinsic acidities of substituted phenols. The use of these data is appropriate because relative acidities of phenols are chiefly determined by the ionization potential of the phenolate anion (29-31). Our best fit to the intrinsic acidity of phenols (31) (wavenumber) is

$$\nu = 11251 - 4094F - 1636R \qquad s=889 \qquad \text{(Eq 5)}$$

where the standard errors of the coefficients are 960, 1351 and 361, respectively. Comparing equations 4 and 5, we see that the F and R coefficients show remarkable agreement. Therefore, the ionization potentials of both the phenols and their phenolate anions have a very similar sensitivity to the para-substituent and there is a direct connection between phenolate ionization potentials and the Swain/Hammett constants for para-substituted phenolates. Correlation of the intrinsic acidity of phenols with Hammett constants was originally demonstrated by McMahon and Kebarle (28).

The theoretical basis for the phenol correlations is clearly set forth by Debies and Rabalais (32). The ionization potential is the energy of the highest occupied molecular orbital, HOMO. For phenols, the symmetry of this pi-orbital allows it to be delocalized easily into the 1 and 4 position substituents. Thus it is sensitive to both field and resonance effects. In contrast, the next lower orbital is also a pi-orbital, but with a node at the 1 and 4 positions. Therefore, the second ionization potential of phenols is much less subject to field and resonance effects.

Referring to equation 3, we find that the R coefficients from the phenol/phenolate correlations agree well with the R coefficient from the charge transfer transition energy correlation, but the field effect coefficient (F) is an order of magnitude higher for the phenols and phenolates than for the charge transfer energies. If the ionization potential is the only factor on the right hand side of equation 1 which is sensitive to the phenolate substituent, then both the field and resonance coefficients for the phenolate energies would be expected to be the same as these coefficients for the charge transfer energies.

To test for a dependence of the FMN electron affinity on the phenolate complexed to OYE, we performed a Swain analysis on the S_0--S_1 transition of FMN in the OYE-phenolate complexes. The energy in wavenumbers was

$$\nu = 21211 - 9F + 217R \qquad s=236 \qquad \text{(Eq 6)}$$

with standard errors: 173, 260 and 87 respectively. There is essentially no dependence of this flavin transition on the para-substituent of the phenolate in complex. Therefore, the changes in the charge transfer energy must be entirely due to the ionization potential of the phenolate. This implies that the differences in the field coefficient between the phenolates and the charge transfer transitions must be due to a reduction in the field effect on binding of phenolate to OYE.

In support of this proposal, we have found a similar effect on the F coefficient comparing the lowest energy optical transition of bound and free phenolates. A best fit for the free phenolates gave

$$\nu = 34720 - 736F + 839R \qquad s=209 \qquad \text{(Eq 7)}$$

where the standard errors of the coefficients were 154, 284 and 97, respectively. For phenolates bound to OYE, the fit was

$$\nu = 32220 - 194F + 800R \qquad s=195 \qquad \text{(Eq 8)}$$

with standard errors: 144, 265 and 91, respectively. Thus the binding of phenolates to OYE does not destroy the free energy correlation found for phenolates free in solution. It is however, accompanied by a 2500 cm^{-1} decrease in the energy of the lowest electronic transition and a significant decrease in the field term. This is in agreement with the previously mentioned observation that the field term for the ionization potential of free phenolates is an order of magnitude greater than that for the energy of the charge transfer transitions.

The pattern of energy shifts summarized in Table 2 suggests that on binding to apoprotein, FMN forms an interaction which causes the lowest energy optical transition to be decreased by about 1000 cm^{-1}, and that most phenolates substitute for FMN in this interaction when they bind to the protein. This interaction is not the dominant FMN-protein interaction because phenolate never displaces FMN from the protein. In this respect we agree with the conclusions of Eweg et.al. (12), but as detailed above, not in the other interpretations of their data. The loss of the field effect on phenolate binding suggests interaction of the phenolate anion with a positively charged protein residue. This residue presumably is able also to influence the spectral properties of the flavin, in a fashion common to FMN and all substituted flavins so far tested.

Supported by GM-11106 of the U.S. Public Health Service.

References

1. Warburg, O., Christian, W.: Biochem Z. 266, 377-414 (1933).

2. Matthews, R. G., Massey, V.: J. Biol. Chem. 244, 1779-1786 (1969).

3. Matthews, R. G., Massey, V., Sweeley, C. C.: J. Biol Chem 250, 9294-9298 (1975).

4. Abramovitz, A. S., Massey, V.: J. Biol. Chem. 251, 5321-5326 (1976).

5. Abramovitz, A. S., Massey, V.: J. Biol. Chem. 251, 5327-5336 (1976).

6. Massey, V., Ghisla, S.: Ann. N. Y. Acad. Sci. 227, 446-465 (1974).

7. Nishina, Y., Shiga, K., Tojo, H., Miura, R., Watari, H., Yamano, T.: J. Biochem. (Tokyo) 90, 1515-1520 (1981).

8. Kitagawa, T., Nishina, Y., Shiga, K., Watari, H., Matsumara, Y., Yamano, T.: J. Am. Chem. Soc. 101, 3376-3378 (1982).

9. Schmidt, J., Reinsch, J., McFarland, J. T.: J. Biol. Chem. 256, 11667-11670 (1981).

10. Williamson, G., Engel, P. C., Nishina, Y. Shiga, K.: FEBS Lett 138, 29-32 (1982).

11. Shieh, H-S., Ghisla, S., Hanson, L. K., Ludwig, M. L., Nordman, C. E.: Biochemistry 20, 4766-4774 (1981).

12. Eweg, J. K., Müller, F., van Berkel, W. J. H.: Eur. J. Biochem. 129, 303-316 (1982).

13. Foster, R.: Oganic Charge Transfer Complexes, Academic Press, New York, 1969 pp. 60-62.

14. Swain, G. G., Unger, S. F., Rosenquist, N. R., Swain, M. S.: J. Am. Chem. Soc. 105, 492-502 (1983).

15. Theorell, H., Nygaard, A. P.: Acta Chem Scand 8, 877-888 (1954).

16. Theorell, H.: Proc. 4th Intern. Congr. Biochem., Vienna 8, 167-174 (1959).

17. Edmondson, D. E., Tollin, G.: Biochemistry 10, 113-124 (1971).

18. Platenkamp, R. J., Van Osnabrugge, H. D., Visser, A. J. W. G.: Chem. Phys. Lett. 72, 104-111 (1980).

19. Sun, M. Moore, T. A., Song, P-S.: J. Am. Chem. Soc. 94, 1730-1740 (1972).

20. Eweg, J. K., Müller, F., Visser, A. J. W. G., Veeger, C., Bebelaar, D., VanVoorst, J. D. W.: Photochem Photobiol 30, 463-471 (1979).

21. Dutta, P. K., Spiro, T. G.: J. Chem. Phys. 69, 3119-3123 (1978).

22. Butler, W. L.: Methods in Enzymology 24, 3-25 (1972).

23. Engel, P. C., Massey, V.: Biochem. J. 125, 889-902 (1971).

24. Miles, D. W., Urry, D. W.: Biochemistry 7, 2791-2799 (1968).

25. Brady, A. H., Beychok, S.: J. Biol. Chem. 244, 4634-4637 (1969).

26. Massey, V., Claiborne, A., Biemann, M., Ghisla, S.: J. Biol. Chem. in press.

27. Palmer, M. H., Moyes, W., Spiers, M., Ridyard, J. N. A.: J. Mol. Struct. 52, 293-307 (1979).

28. Draper, N. R., Smith, H.: Applied Regression Analysis, John Wiley and Sons, New York, 1966 p 120.

29. Richardson, J. H., Stephenson, L. M., Brauman, J. I.: J. Am. Chem. Soc. 97, 2967-2970 (1975).

30. McMahon, T. B., Kebarle, P.: J. Am. Chem. Soc. 99, 2222-2230 (1977).

31. Catalàn, J., Macias, A.: J. Chem. Soc. Perkin II, 1632-1636 (1979).

32. Debies, T. P., Rabalais, J. W.: J. Electron. Spect. Relat. Phenom. 1, 355-370 (1972).

NMR STUDIES ON THE OLD YELLOW ENZYME

Wolf-Dieter Beinert, Heinz Rüterjans and Franz Müller

Institute of Biophysical Chemistry, University of Frankfurt,
D-6000 Frankfurt a.M. 70, Germany

Department of Biochemistry, Agricultural University,
6703 BC Wageningen, The Netherlands

Introduction

Old Yellow Enzyme apoenzyme (M=49,000) was recombined with different selectively ^{15}N- and ^{13}C-labeled flavin mononucleotides. The aim of the study was to investigate the interaction of FMN with the apoenzyme and the interaction of the protein with phenolic compounds by NMR techniques.

Results and Discussion

The ^{13}C resonances of the protein-bound FMN are shifted downfield as compared with those of free FMN, which indicates a strong polarisation of the protein-bound flavin. Especially the very strongly polarized carbonyl group at position 4 is remarkable. Complete reduction of the protein hardly affects the resonance positions of C(2) and C(4), suggesting that the two carbonyl functions are still polarized. The downfield shift of C(10a) in the reduced protein proves that the prosthetic group is ionized. However, with respect to the $FMNH^-$ the electron density on C(10a) is strongly decreased in the protein. According to recent interpretations of NMR results on flavins (1) the downfield shift of C(10a) in the reduced protein strongly suggests that the endocyclic angle of the N(5) atom is decreased as compared to that of the free $FMNH^-$.

Flavins and Flavoproteins

On the other hand C(4a) is extremely shifted upfield, which indicates a large π-electron density at this atom.
The most striking feature of the ^{15}N NMR spectra of the oxidized Old Yellow Enzyme (Fig.1) is the extreme high field position of the N(5) resonance, indicating a very strong hydrogen bond at the N(5) atom. This hydrogen bond is much stronger than that in free FMN in aqueous solution (2). Weak hydrogen bonds can

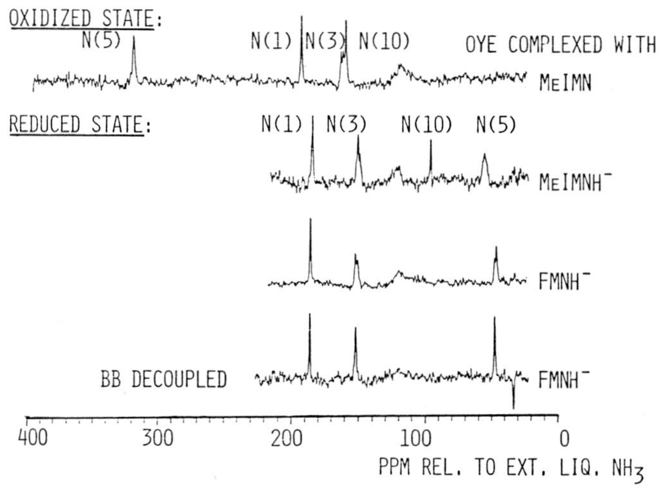

Fig.1. ^{15}N-NMR spectra of the Old Yellow Enzyme (OYE) complexed with different flavins in the oxidized and reduced state. (MeIMN: 7-methyl-10-ribityl-isoalloxazine-5'-phosphate, $MeIMNH^-$: anionic 1,5-dihydro-MeIMN, $FMNH^-$: anionic 1,5-dihydro-FMN)

also be derived for the N(1) and N(3) nuclei. The resonance due to N(10) is shifted upfield compared to that of free FMN. This suggests that the N(10) atom possesses a slightly bent configuration (1). In the reduced protein the high field position of the resonance due to N(5) indicates that the N(5) atom is strongly sp^3 hybridized. This is also supported by the $^{15}N-^{1}H$ coupling constant. Thus far this is the first flavoenzyme where an almost completely sp^3 hybridized N(5) atom is observed. The N(1) and N(10) resonance positions prove that the reduced flavin in the Old Yellow Enzyme is ionized.
Two different FMN derivatives have been used in this study.
The considerable difference between the ^{15}N NMR spectra of the

two compounds in the reduced Old Yellow Enzyme is remarkable. It is suggested that the two flavins are bound differently to the apoprotein. This interpretation needs however further experimental support.

Old Yellow Enzyme forms deeply coloured complexes by the interaction with phenolates. It still serves as a model for the physical study of similar complexes formed by many flavoproteins usually referred to as charge transfer complexes.

Addition of phenolate to Old Yellow Enzyme exerts some interesting effects on certain ^{13}C and ^{15}N resonances of the protein-bound flavin (Fig.2). With respect to the ^{13}C chemical shifts only those of C(4) and C(4a) are shifted upfield by

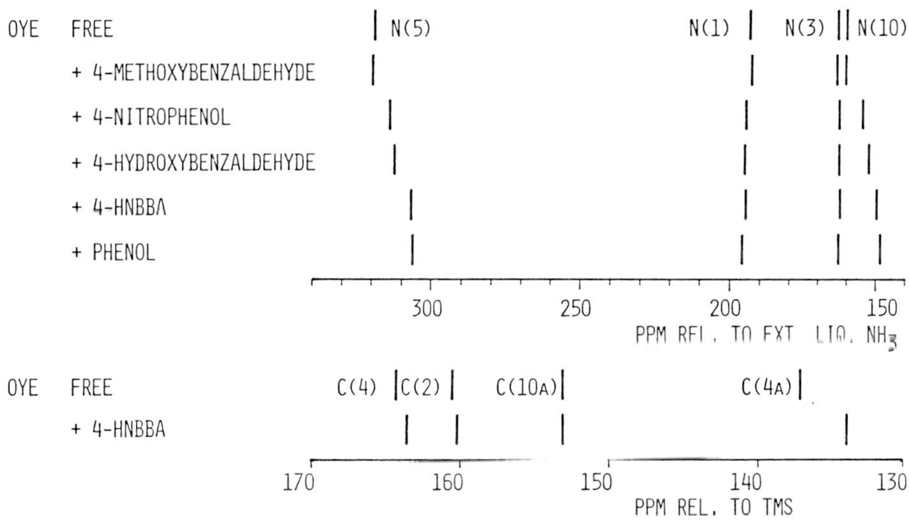

Fig.2. Interaction of the Old Yellow Enzyme with phenols: correlation diagrams of ^{15}N and ^{13}C chemical shifts of the protein complexed with oxidized labeled flavins ad different phenolates. (4-HNBBA: 4-hydroxy-N-n-butylbenzamide)

complexation. The same holds for the ^{15}N chemical shifts due to N(10) and N(5). The other carbon and nitrogen atoms, as far as investigated, are little or not at all affected by complexation.

An interesting observation is that the above mentioned upfield shifts of the N(10) and N(5) atoms show some parallelity with the pK_a value of the phenolic compounds. The upfield shifts can be interpreted as increase in π electron density at the corresponding carbon atoms and N(5) and an increase in sp^3 character of the N(10) atom. The ^{13}C resonances of the phenolate derivatives on the other hand show a downfield shift indicating loss of π electron density. The results thus clearly demonstrate that indeed charge is transferred from the phenolate to the flavin. Moreover, the polarization of the carbonyl group at position 4 is strongly affected by complexation of the protein by phenolate. The results also seem to be in support with the suggestions that binding of phenolate to Old Yellow Enzyme interferes with an existing hydrogen bond network in the free protein. This is strongly supported by the fact that binding of 4-methoxybenzaldehyde to Old Yellow Enzyme does not influence any of the chemical shifts of the protein-bound FMN. Therefore, the latter complex can safely be characterized as a molecular complex.

The ^{31}P resonance position of the FMN phosphorus atom bound to the Old Yellow Enzyme is extremely shifted upfield compared to the free or flavodoxin-bound FMN. This result apparently reflects a different binding mode of the FMN phosphate group to the Old Yellow Enzyme as compared with the flavodoxins.

Acknowledgements

This work was in part supported by the Deutsche Forschungsgemeinschaft (Ru 145/6-6) and the Dutch Foundation for Chemical Research (S.O.N.) with financial aid from the Dutch Organization for the Advancement of Pure Research (Z.W.O.).

References

1. Moonen, C.T.W., Vervoort, J., Müller, F.: Biochemistry 23 (1984) in press
2. Franken, H.-D., Rüterjans, H., Müller, F.: Eur.J.Biochem. 138, 481-489, (1984)

STRUCTURAL AND KINETIC CHARACTERISTICS OF DIMETHYLGLYCINE DEHYDROGENASE AND SARCOSINE DEHYDROGENASE

Robert J. Cook, David H. Porter, Kunio S. Misono, and Conrad Wagner

Vanderbilt University and VA Medical Center
Nashville, Tennessee, 37203, U.S.A.

The enzymes dimethylglycine dehydrogenase (DMGDH; EC 1.5.99.2 and sarcosine dehydrogenase (SDH; EC 1.5.99.1) catalyze the last two steps of choline degradation which results in the production of glycine. This takes place exclusively in the liver mitochondria. Both enzymes contain covalently bound flavin in addition to tightly, but not covalently bound H_4Pte-Glu_5. A mechanism has been proposed in which the oxidation of the N-methyl group of the substrate proceeds via electron transfer to the flavin followed by hydrolysis of the resulting Schiff base to produce formaldehyde. The latter reacts directly with the enzyme-bound $H_4PteGlu_5$ to yield 5,10-CH_2-$H_4PteGlu_5$.

Flavin-peptides (FP's) have been purified from DMGDH and SDH after proteolysis by trypsin or by trypsin/chymotrypsin (T/C). Treatment with nucleotide pyrophosphatase produced changes in the spectral and chromatographic properties of the T/C-FP's consistent with the conversion of FAD-peptides to FMN-peptides. The pKa of 4.5 for pH-dependent fluorescence quenching and lack of effect of borohydride reduction on the fluorescence of the purified T/C-FP's indicates an 8α-N(3)histidylflavin is present in both enzymes. N-terminal analysis of T/C-FP's suggested histidylflavin (HisF). Sequence analysis of the DMGDH-FP showed no detectable amino acid in the first cycle. Cycles 2 through 5 showed Ala, Ala, Gly and Leu which agreed with the amino acid composition of Ala_2, Gly, Leu plus HisF

Flavins and Flavoproteins

from the acid hydrolysate of DMGDH-FP. The composition of the T/C-FP of SDH was identical except for the substitution of Thr for an Ala. HisF was at the N-terminus and Leu was at the C-terminus suggesting a sequence of HisF-(Ala, Thr, Gly)-Leu. The sequences of the DMGDH-FP and SDH-FP produced by trypsin digestion are:

```
                               FAD
                               |
DMGDH-FP:  S E L T A G S T W H A A G L T T Y F H P G I N L K
SHD-FP:        L T S G T T W H T A G L G R
                               |
                               FAD
```

The amino acid sequences are very closely related but are not obviously similar to any of the known sequences for other enzymes with covalently bound flavin.

DMGDH is able to oxidatively demethylate either dimethylglycine (DMG) or sarcosine (SAR). DMGDH kinetics, with DMG as the substrate, are sharply biphasic with a Km of 0.05mM for the low concentration phase. With SAR as the substrate both DMGDH and SDH kinetics are Michaelis-Menten with Km's of 20mM and 0.05mM respectively. In the absence of folate, substrate and product determinations indicate that 1 mole of formaldehyde and of SAR or glycine are produced for each mole of DMG or SAR consumed, with the concommitant reduction of 1 mole of bound FAD. When DMG is the only substrate initially present for DMGDH, no glycine is produced by the enzymatic reaction. Methoxyacetic acid is a competitive inhibitor of DMGDH with Ki's of 0.19mM and 0.23mM for DMG and SAR respectively. Under anaerobic conditions, the substrate induced reduction of FAD in both DMGDH and SDH has the same rate in the presence or absence of H_4PteGlu. (This work was supported by NIH Grant #AM15289, Nutrition Training Grant #AM07083 and by the Veterans Administration.)

EVIDENCE FOR TWO SPATIALLY DISTINCT DOMAINS ON EACH SUBUNIT OF METHYLENETETRAHYDROFOLATE REDUCTASE

Rowena G. Matthews, Maria A. Vanoni and Shahrokh Khani

Biophysics Research Division and Department of Biological Chemistry, The University of Michigan, Ann Arbor, Michigan 48109

James F. Hainfeld and Joseph Wall

Department of Biology, Brookhaven National Laboratories, Upton, New York 11973

Methylenetetrahydrofolate reductase is a flavoprotein which catalyzes the NADPH-linked reduction of methylenetetrahydrofolate to methyltetrahydrofolate. This reaction proceeds with a standard free energy decrease of 9.4 kcal per mol at pH 7.2 and is effectively irreversible under cellular conditions. The enzyme is regulated by adenosylmethionine (AdoMet), which is the final common product of methyl transfer from methyltetrahydrofolate, and AdoMet has been shown to bind at an allosteric site and to initiate a hysteretic response (1,2). Because hysteretic (slow onset) inhibitors often cause aggregation/disaggregation of enzymes, we felt that it was valuable to examine the effect of AdoMet on the state of aggregation of native methylenetetrahydrofolate reductase. For this purpose we have used scanning transmission electron microscopy (STEM), which permits measurement of the masses of individual protein molecules under very mild conditions and using very small samples (10-50 μg) of protein (3). Such measurements give information on the mass distribution within protein samples as well as the mean mass. Specimens were prepared and imaged at the Brookhaven STEM Biotechnology Resource and were prepared as described by Mosesson et al. (4). Homogeneous preparations of methylenetetyrahydrofolate reductase were used. In each case, histograms of the mass distribution formed a bell shaped curve

Flavins and Flavoproteins

extending from 80-210 kDa. The mean masses and standard deviations were, 136 ± 29 kDa for enzyme which had been desalted by gel filtration using a Sephadex G 100 matrix; 145 ± 41 kDa, for enzyme which was not desalted prior to measurement, and remained in 50 mM potassium phosphate buffer, pH 7.2/0.3 mM EDTA/10% glycerol; 141 ± 36 kDa for enzyme which was lightly fixed with glutaraldehyde prior to application to the specimen grid; and 144 ± 29 kDa for enzyme which was not desalted and which had been preincubated with AdoMet for 10 minutes prior to application to the grid. Previous experiments had already established that the subunit molecular weight of the enzyme is 77 kDa as determined by gel electrophoresis in the presence of sodium dodecyl sulfate and by amino acid analysis of enzyme of known flavin content (5). Thus the native enzyme appears to be dimeric in both the presence and absence of AdoMet.

The same preparation of enzyme can also be used to prepare specimens which can be shadowed with uranyl sulfate in order to examine the molecular fine structure in more detail. In this case the negatively stained protein molecules appear to consist of 4 globular regions of roughly equal size (Figure 1).

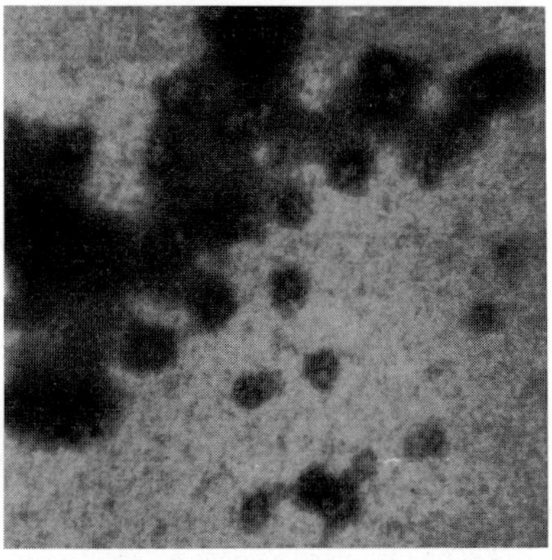

Figure 1. Fine structure of the dimeric protein, after negative staining with 2% uranyl acetate. x 20,000

Since these molecules are dimeric, each subunit must consist of two spatially distinct globular domains. We have provided confirmation of this observation by limited proteolysis of native enzyme with 1% (w/w) trypsin. The 77 kDa subunit is rapidly cleaved (with a half time of about 5 minutes) into 39 and 36 kDa fragments and the 36 kDa fragment is further degraded, via intermediates, into a 33 kDa fragment. The 39 and 33 kDa fragments are relatively stable to further digestion. Tryptic digestion causes no loss in catalytic activity, but the enzyme becomes completely desensitized to AdoMet. The half time for desensitization corresponds to the half time for the initial cleavage of the subunits into 39 and 36 kDa fragments. The two tryptic fragments remain closely associated after cleavage as evidenced by their co-chromatography on DEAE-Sephadex A-50 during elution with a 0-1 M NaCl gradient (the fragments elute at 0.37 M NaCl).

We have used 8-azidoAdoMet as a photoaffinity label for the AdoMet binding site on the protein. [^{3}H-methyl]-8-azidoAdoMet was prepared and purified as described previously (6). The compound induces hysteretic inhibition of methylenetetrahydrofolate reductase, with an apparent K_i which is comparable to that of AdoMet. Exposure of a mixture of enzyme (5 μM) and tritiated 8-azidoAdoMet (11 μM) to 254 nm light for 5 minutes resulted in covalent labelling of a small fraction of the protein (1.8%). Autoradiograms of samples taken before and during tryptic digestion of the labelled enzyme indicated that the tryptic label is associated with the 36 kDa fragment, but is released on further degradation of the 36 kDa fragment to 32 kDa.

We believe that the allosteric inhibition of methylenetetrahydrofolate reductase presents an interesting system for further study. The enzyme contains at least four types of binding sites, for FAD, NADPH, methylenetetrahydrofolate and AdoMet. It is tempting to postulate that we are looking at evidence of modular construction, with spatially localized catalytic and regulatory domains. It will be of interest to localize the binding sites associated with catalytic activity and see if they all lie within the 39 kDa fragment. Clearly, inhibition requires specific interactions of the two spatially distinct domains, and the positive charge induced by AdoMet binding may play a major role in inducing that interaction, since the structurally analogous thioether, adenosylhomocysteine, binds but does not lead to enzyme inhibition (1,2).

This work has been funded in part by National Institutes of Health Grant GM 24908 (RGM). Shahrokh Khani is supported as a fellow on an NIH Medical Scientist Training Grant, GM 07863. Scanning transmission electron microscopy was performed at the Brookhaven STEM, which is supported as a Biotechnology Resource (RR 01777) by the National Institutes of Health.

References

1. Kutzbach, C., and Stokstad, E.L.R.: Biochim. Biophys. Acta 250, 459-477 (1971).
2. Matthews, R. G., and Daubner, S. C.: in Advances in Enzyme Regulation 20, (G. Weber, ed.) pp 123-131, Pergamon Press, Oxford (1982).
3. Wall, J.: in Introduction to Analytical Electron Microscopy (Hren, J., Goldstein, J. I., and Joy, D. C., eds.) pp 333-342, Plenum, New York.
4. Mosesson, M. W., Hainfeld, J., Wall, J. and Haschemeyer, R. M.: J. Mol. Biol. 153, 695-718 (1981).
5. Daubner, S. C., and Matthews, R. G.: J. Biol. Chem. 257, 140-145 (1982).
6. Kaiser, I. I., Kladianos, D. M., VanKirk, E. A., and Haley, B.: J. Biol. Chem. 258, 1747-1751 (1983).

STRUCTURE OF NADH-CYTOCHROME $\underline{b}_5$ REDUCTASE OF HUMAN ERYTHROCYTES

Toshitsugu Yubisui, Toshiyuki Miyata*, Sadaaki Iwanaga*, Minoru Tamura, Satoshi Yoshida, Masazumi Takeshita

Department of Biochemistry, Medical College of Oita, Hazama-cho, Oita 879-56, and *Department of Biology, Faculty of Science, Kushu University 33, Fukuoka 812, Japan

Introduction

In hereditary methemoglobinemia, NADH-cytochrome $\underline{b}_5$ reductase is deficient in tissues of patients. The two forms of the enzyme deficiency are mainly now known; one of them is the "erythrocyte type" or "type I", and the other is the "generalized type" or "type II". At present the genetical mechanism leading to these two forms of the hereditary disease is not well explained, while the general properties of the enzyme from normal human erythrocytes as well as those from liver microsomes have been well characterized. In this study, to know the protein-chemical nature of the enzyme, and also as an approach to understand the hereditary deficiency of this enzyme, we determined the whole amino acid sequence of soluble NADH-cytochrome $\underline{b}_5$ reductase purified from normal human erythrocytes.

Materials and Methods

NADH-cytochrome $\underline{b}_5$ reductase was purified from normal human erythrocytes as described previously (1). The purified protein was reduced and carboxymethylated, and then cleaved by CNBr. The CN-peptide mixture was separated by gel filtration on a Sephadex G-50 column, and then purified by HPLC. Total of 9 CN-peptides were obtained, and amino acid compositions of these peptides were determined by amino acid analysis. The sequences of these peptides were determined by automated Edman-degrada-

Flavins and Flavoproteins

tion, directly or after further fragmentation with proteases. With the overlaps obtained by trypsin-peptides, whole amino acid sequence was determined (2).

Results and Discussion

From the sequence analysis the enzyme was found to be composed of 275 amino acid residues (Table I), and is hydrophilic as a whole, but two regions, from Phe-36 to Ile-71, and from Met-231 to Phe-275, were found to be very hydrophobic (Fig. 1). The sequence of the former hydrophobic region is similar to those of the AMP-binding sites of other flavoproteins (3-5) as shown in Table II. In the latter region, 3 of 4 total cysteines are

Table I. Amino acid composition of NADH-cytochrome $\underline{b}_5$ reductase

Amino acids	Amino acid analysis	Sequence analysis
	residues/molecule	
C	3.9[a]	4
D	25.4	19
N		6
T	12.7	13
S	12.7	13
E	24.5	15
Q		10
P	26.3	26
G	20.9	19
A	11.8	12
V	14.5	15
M	8.0[b]	8
I	20.9	22
L	26.3	26
Y	9.1	9
F	12.7	13
K	16.3	15
H	10.9	10
W	1.8	2
R	18.4	18
Total		275

a) Performic acid oxidation. b) Hydrolysis with methanesulfonic acid.

FQRSTPAITL ESPDIKYPLR LIDREIISHD TRRFRFALPS PQHILGLPVG 50
QHIYLSARID GNLVVRPYTP ISSDDDKGFV DLVIKVYFKD THPKFPAGGK 100
MSQYLESMQI GDTIEFRGPS GLLVYQGKGK FAIRPDKKSN PIIRTVKSVG 150
MIAGGTGITP MLQVIRAIMK DPDDHTVCHL LFANQTEKDI LLRPELEELR 200
NKHSARFKLW YTLDRAPEAW DYGQGFVNEE MIRDHLPPPE EEPLVLMCGP 250
PPMIQYACLP NLDHVGHPTE RCFVF

Figure 1. Amino acid sequence of NADH-cytochrome b_5 reductase

contained, proline content is high (20%), and unique sequences such as P-P-P or P-P-P-E-E-E are observed. Moreover, two tryptophan residues are in close position to the latter region. Indirect evidence shows that a reactive cysteine of this enzyme is participated in the NADH-binding (6, 7). So these two hydrophobic regions in this enzyme are very characteristic, and are considered to be important for the function. As there is another evidence which suggests that the flavin- and NADH-binding sites are in close proximity in the enzyme (6, 7), the two hydrophobic regions may come close by peptide-folding to interact to each other.

Recently, partial sequences of NADH-cytochrome b_5 reductases of pig and steer liver microsomes were reported (8, 9), and those sequences are identical to that of the erythrocyte enzyme we determined, except that Ser-22 in the erythrocyte enzyme is replaced by Asn in the liver microsomal enzymes as shown in Fig.2.

Table II. Sequence homology of NADH-cytochrome b_5 reductase with the AMP-binding sites of other flavoproteins.

b_5R	-PSPQ-HILGL-PVGQHIYLSARIDG-NL-VVRPI
GR	-ASYDYLVIGGGSGGLASARRAAELGARAAVVESH
PHBH	MKTQVAIIGAGPSGLLLGQLLHKAGINDVILERQ
DAO	MRVVVIGAGVIGLSTALCIHERY-HSV-LQPL

b_5R, NADH-cytochrome b_5 reductase of human erythrocytes (2); GR, glutathione reductase of human erythrocytes (3); PHBH, p-hydroxybenzoate hydroxylase of *Pseudomonas fluorescence* (4); DAO, D-amino acid oxidase of pig liver (5).

	Sequence	References
H.E.	FQRSTPAITLESPDIKYPLRLIDR--	(2)
P.L.	STPAITLENPDIKYPLRLID	(8)
S.L.	-WFLYDLIMKLFQRSTPAITLENPDIKYP	(9)

Figure 2. Sequence homology of NADH-cytochrome $\underline{b}_5$ reductase of human erythrocytes, and pig and steer liver microsomes. Identical residues are enclosed in a box. , membrane-binding domain (sequence has not been determined).

The sequence of steer liver microsomes suggests that the membrane-binding domain of this enzyme resides at the NH_2-terminal side. Common COOH-terminal amino acid, Phe, from pig and steer liver microsomes (8, 9), and human (2) and rabbit (10) erythrocytes support the suggestion described above.

References

1. Yubisui, T., Takeshita, M.: J. Biochem 91, 1467-1477 (1982)
2. Yubisui, T., Miyata, T., Iwanaga, S., Tamura, M., Yoshida, S., Takeshita, M., Nakajima, H.: J. Biochem. in press
3. Kraut-Siegel, R.L., Blatterspiel, R., Saleh, M., Schiltz, E., Schirmer, R.H., Untrecht-Grau, R.: Eur. J. Biochem. 121, 259-267 (1982)
4. Weijier, W.J., Hofsteenge, J., Vereijken, J.M., Jekel, P.A., Beintema, J.J.: Biochim. Biophys. Acta 704, 385-388 (1980)
5. Ronchi, S., Minchiott, L., Galliano, M., Curti, B., Swenson, R.P., Williams, C.H., Jr., Massey, V.: Flavins and Flavoproteins, pp.62-72, Elsevier North-Holland, New York 1982
6. Strittmatter, P.: Flavins and Flavoproteins, pp. 325-340, Elsevier Publishing Co., Amsterdam 1966
7. Yubisui, T., Takeshita, M.: Biochem. Internatl. 8, 319-327 (1984)
8. Crabb, J.W., Tarr, G.E., Yasunobu, K.T., Iyanagi, T., Coon, M.J.: Biochem. Biophys. Res. Commun. 95, 1650-1655 (1980)
9. Kensil, C.R., Hediger, M.A., Ozols, J., Strittmatter, P.,: J. Biol. Chem. 258, 14656-14663 (1984)
10. Yubisui, T.: unpublished data

STRUCTURAL COMPARISON OF THE SUCCINATE DEHYDROGENASE AND FUMARATE REDUCTASE OF *Escherichia coli*

J. R. Guest, M. G. Darlison, R. J. Wilde and D. Wood

Department of Microbiology, Sheffield University,
Sheffield S10 2TN, U.K.

E. coli contains two membrane-bound flavoenzyme complexes that catalyse the interconversion of succinate and fumarate: succinate dehydrogenase (SDH) - a primary dehydrogenase of *aerobic* respiration that is repressed anaerobically, and fumarate reductase (FRD) - the terminal reductase of an *anaerobic* respiratory chain that is repressed aerobically. The *sdh* genes encoding SDH have been cloned and sequenced along with the citrate synthase (*gltA*) and 2-oxoglutarate dehydrogenase complex (*sucAB*) genes in a 10000 base-pair segment of DNA (Guest *et al*. unpublished). This has enabled a direct structural comparison between SDH and FRD, which has already been characterized by sequencing the *frd* genes (1, 2, 3).

Organization of the *sdh* Genes

The nucleotide sequence has revealed an operon of four *sdh* genes that are analogous to the *frd* genes but the order differs in the *frd* operon.

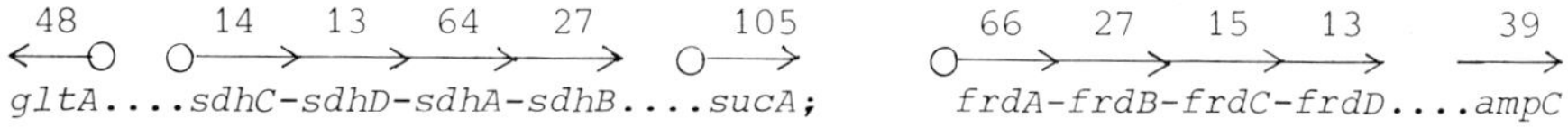

The genes encode comparable flavoprotein (A) and iron-sulphur protein (B) subunits, and two pairs of hydrophobic subunits (C & D): their sizes are shown in kilodaltons.

Flavins and Flavoproteins

Fig. 1. Alignments of Amino Acid Sequences of the Flavoprotein (A) and Iron-Sulphur Protein (B) Subunits of SDH and FRD. Identical residues (*) and conservative substitutions (¦) are indicated.

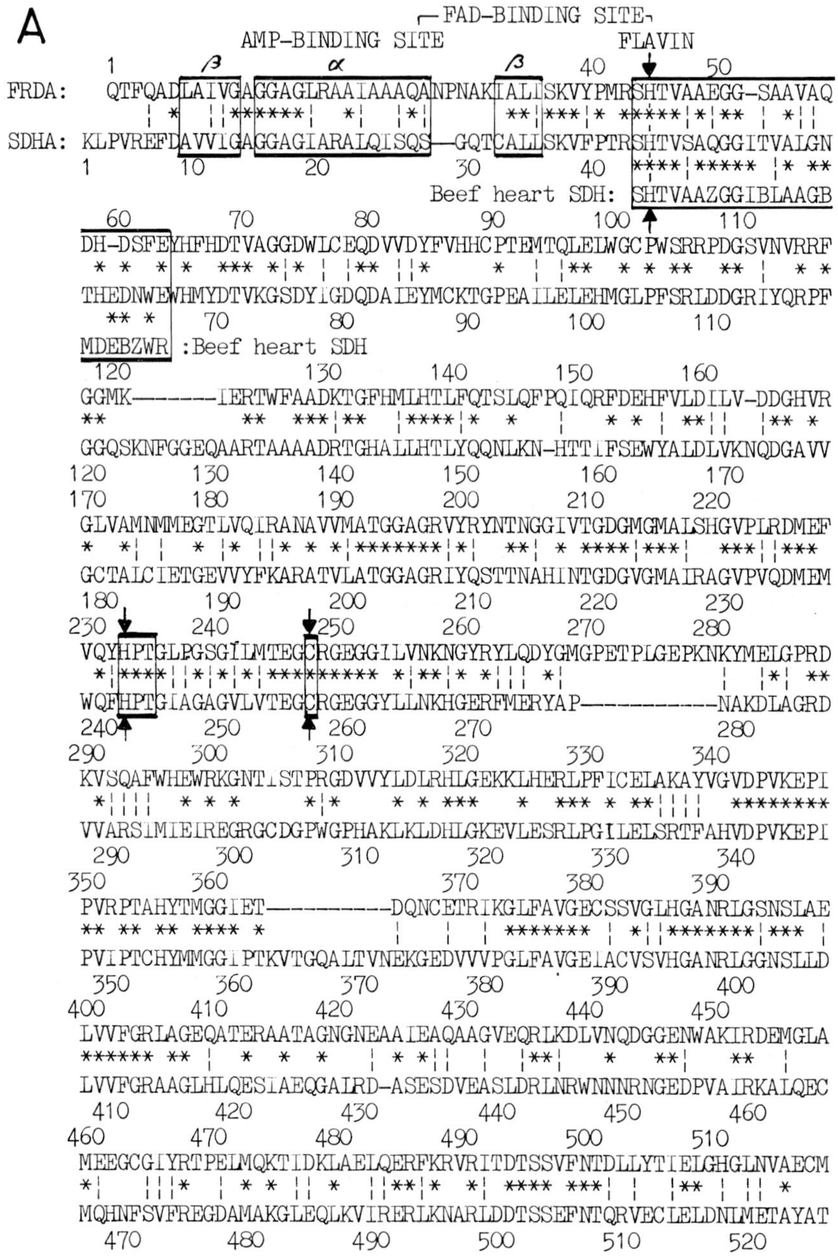

```
  520       530       540       550       560        570
AHSAMARKESRGAHQRLDEGCTERDDVNFLKHTLAFRDADGTTRLE-YSDVKI-TTLPPA
* **  * ****** *¦*    ¦*** * * *¦* ¦ ¦¦¦¦ **      ¦ *¦  ¦¦**
AVSANFRTESRGAHSRFDFP--DRDDENWLCHSLYLPESESMTRRSVNMEPKLRPAFPPK
 530       540         550       560       570       580
 580       590       601
KRVYGGEADAADKAEAANKKEKANG :FRDA
 * *
IRTY :SDHA
 587
```

B

```
       1    10       20         30        40        50      60  I
FRDB: AEMKNLKTEVVRYNPEVDTAPHSAFYEV-PYDA-TTSLLDALGYIKDNLAPDLSYRWSCRMAIC
          *  ¦ ¦ ****¦** **¦    *  ¦  ¦ ¦¦    *****  ¦*¦¦  * **¦*¦***  ¦¦*
SDHB:     RLEFSIYRYNPDVDDAPRMQDYTLEADEGRDMMLLDALIQLKEK-DPSLSFRRSCREGVC
          1        10        20        30        40        50
           70        80        90        100       110
GSCGMMVNNVPKLACKTFLRDYTD-GMKVE--ALANFPIERDLVVDMTHFIESLEAIKPY
** *¦ ¦*     ***  * ¦    ¦ * *¦    ¦*¦ ¦*¦ ******* ¦*     * ****
GSDGLNMNGKNGLACITPISALNQPGKKIVIRPLPGLPVIRDLVVDMGQFYAQYEKIKPY
60        70        80        90   II    100       110
120       130       140       150       160       170
TIGNSRTADQGTNIQTPAQMAKYHQFSGCINCGLCYAACPQFGLNPE-FIGPAATTLAHR
¦¦ *¦¦ ¦     ¦¦* * * *     ¦ ** *¦ * ¦¦** *  **¦ *****¦¦  * *
LLNNGQNPPAREHLQMPEQREKLDGLYECILCACCSTSCPSFWWNPDKFIGPAGLLAAYR
120       130       140       150  III  160       170
180       190       200       210       220       230          243
YNEDSRDHGKKERMAQLNSQNGVWSCTFVGYCSEVCPKHVDPAAAIQQGKVESSKDFLIATLKPR
¦  ****       *¦  *    ¦* *  ¦  *   ****  ¦¦*¦ **  ¦ *
FLIDSRDTETDSRLDGLSDAFSVFRCHSIMNCVSVCPKGLNPTRAIGHIKSMLLQRNA    :SDHB
180       190       200       210       220       230    237
```

Structural Comparisons of Subunits

The primary structures of analogous subunits have been compared by DIAGON analysis (4, 5) and alignments for the flavoprotein and iron-sulphur protein subunits are shown in Fig. 1.

1. Flavoprotein subunits. The SDHA and FRDA subunits exhibit a remarkable degree of sequence homology: 44% identity increasing to 64% when conservative changes are included (Fig. 1A). The FAD-binding sites are strongly conserved with respect to (i) the N-terminal AMP-binding sites, and (ii) the flavin-attachment sites, which closely resemble the flavopeptide of beef heart SDH (6). The cysteine residues are scattered and only one of the 10 (FRDA) or 11 (SDHA) residues is conserved. This could correspond to the active-site cysteine residue of beef heart SDH (7) and V. succinogenes FRD (8). It is close to the histidine residue of a

tripeptide that is conserved in glutathione reductase and lipoamide dehydrogenase (9). This could be the histidine residue associated with proton donor-acceptor function in SDH (10).

2. Iron-sulphur protein subunits. The SDHB and FRDB subunits show 38% identity and 58% homology when conservative changes are included (Fig. 1B). Ten of the eleven cysteine residues in each subunit are conserved in three clusters (I, II and III). These resemble ferredoxin clusters, but some, containing only 3 cysteine residues, may need residues from elsewhere to form functional clusters.

3. Hydrophobic subunits. The SDHC and SDHD subunits resemble the comparable FRD subunits in size, composition and hydrophobicity, but not sequence. They each contain three ~25-residue hydrophobic segments that could represent transmembrane α-helices. They probably anchor the A and B subunits to the membrane but could also function in proton translocation.

Conclusion

By gene cloning and sequence analysis the SDH and FRD of E. coli have been shown to contain similar flavoprotein and iron-sulphur protein subunits, and equivalent (but not homologous) pairs of hydrophobic subunits. It would appear that both enzymes have evolved from common ancestral precursors by gene duplication, functional diversification, and functional compartmentation in aerobic and anaerobic respiration by coupling to different regulators.

1. Cole, S. T.: Eur. J. Biochem. 122, 479-484 (1982).
2. Cole, S. T., Grundström, T., Jaurin, B., Robinson, J. J., Weiner, J. H.: Eur. J. Biochem. 126, 211-216 (1982).
3. Grundström, T., Jaurin, B.: Proc. Natl Acad. Sci. U.S.A. 79, 1111-1115 (1982).
4. Staden, R.: Nucleic Acid. Res. 10, 2951-2961 (1982).
5. Schwartz, R. M., Dayhoff, M. O.: Atlas of Protein Sequence and Structure, 5, 353-358 (1978).
6. Kenney, W. C., Walker, W. H., Singer, T. P.: J. Biol. Chem. 247, 4510-4513 (1972).
7. Kenney, W. C., Mowery, P. C., Seng, R. L., Singer, T. P.: J. Biol. Chem. 251, 2369-2373 (1976).
8. Unden, G., Kröger, A.: FEBS Lett. 117, 323-326 (1980).
9. Rice, D. W., Schulz, G. E., Guest, J. R.: J. Mol. Biol. 174, 483-496 (1984).
10. Vik, S. B., Hatefi, Y.: Proc. Natl Acad. Sci. U.S.A. 78, 6749-6753 (1981).

LACK OF ASSEMBLY OF SUCCINATE AND NADH-UBIQUINONE OXIDOREDUCTASES IN IRON-DEFICIENT RAT SKELETAL MUSCLE MITOCHONDRIA

Brian A. C. Ackrell, Bruce Cochran, Kent Larson, and Edna B. Kearney

Department of Biochemistry and Biophysics, University of California, San Francisco, CA 94143 and Molecular Biology Unit, Veterans Administration Medical Center, San Francisco, CA 94121

John J. Maguire

Membrane Bioenergetics Group, Lawrence Berkeley Laboratory and Department of Physiology-Anatomy, University of California, Berkeley, CA 94720

Peter R. Dallman

Department of Pediatrics, University of California, San Francisco, CA 94143

Introduction

Iron deficiency has been shown to decrease both the activities and EPR signals of the membrane-bound NADH and succinate dehydrogenases of rat skeletal muscle mitochondria (1). The effect is specific: cytochrome levels are less severely reduced and the activities of ATP'ase (1) fumarase, malate dehydrogenase, and 3-OH acyl CoA dehydrogenase, which contain no iron, increase slightly (2). We have examined, therefore, whether impaired forms of the NADH-and succinate-oxidizing complexes, lacking Fe-S clusters, are assembled in iron-deficient membranes, or whether fewer but fully competent complexes are present. A further point of interest is whether non-functional, incomplete complexes may be present where only iron-containing peptide components are affected.

Experimental procedures

Normal and iron-deficient inner membranes (SMP) of rat skeletal muscle mitochondria were obtained according to Maguire et al (1). NADH-Q (Complex I) and succinate-Q (Complex II) oxidoreductases were assayed with 2,3-dimeth-

Flavins and Flavoproteins

oxy-5-methyl-6-pentyl-1,4-benzoquinone (DPB) in lieu of Q as a measure of full complex integrity. Ferricyanide (rotenone-insensitive) and phenazine methosulfate (PMS) reduction were measured for detection of dehydrogenase activities of forms of Complexes I and II unable to complete electron transfer to Q. Analyses for histidyl FAD (Complex II) and non-covalently bound FMN (Complex I) and reconstitution experiments were carried out as in previous work or by published methods (3,4).

Results and Discussion

Both Complex I (12-15 subunits; 6 Fe-S clusters) and Complex II (4 subunits; 3 Fe-S clusters) of rat skeletal muscle mitochondria were similarly affected by iron deficiency: activities were about 30% of normal in both dehydrogenase and Q reductase assays, and the respective flavin prosthetic groups were similarly reduced (Table I). The catalytic turnovers, then, remain the same and one may conclude that the complexes present are fully competent with Fe-S clusters intact.

Decreased levels of Complexes I and II were also indicated by comparison of the relative staining intensities of constituent subunits following SDS-PAGE of iron-deficient and normal membranes. Bands sufficiently resolved for evaluation included the flavoprotein subunit of succinate

Table I

Effect of Iron Deficiency on Rat Skeletal Muscle Submitochondrial Particles

Analyses	Normal	Iron-deficient	Decrease %
FMN[a]	0.13	0.05	62
Histidyl FAD[a]	0.15	0.04	73
NADH-$Fe(CN)_6^{3+}$ reductase[b]	40.00	12.50	69
NADH-DPB reductase	3.00	1.05	65
Succinate-PMS reductase	2.10	0.60	72
Succinate-DPB reductase	2.10	0.60	72

[a] nmol/mg of protein

[b] activity at infinite acceptor concentration (30°) = μmol substrate oxidized/min/mg of protein

dehydrogenase and two Complex I subunits of 75kDa and 40kDa, respectively, neither of which is part of the low molecular weight form of NADH dehydrogenase. All three of these subunits showed diminished staining intensities in iron-deficient membranes, in accord with the lower concentrations of histidyl FAD and FMN determined by direct analyses (Table I). Of interest is that the 40kDa subunit may not be a non-heme iron protein (5,6); if so its lower concentration would suggest that the complex either may be assembled completely or not at all.

In further studies rat skeletal muscle SMP, after incubation at alkaline pH to dissociate endogenous succinate dehydrogenase and expose its binding sites, were able to bind succinate dehydrogenase from beef heart. The hybrid complexes formed were shown to have the same succinate-DPB reductase activity as the original rat complex (TN $\sim$ 15,000 min^{-1} at 30°) and the same inhibitor sensitivity. Not surprisingly, therefore, at least partial homology of rat muscle and beef heart succinate dehydrogenases was evident in reaction with antibodies generated in rabbits against the large and small subunits of the beef heart enzyme (Fig. 1). Antibody to the beef heart

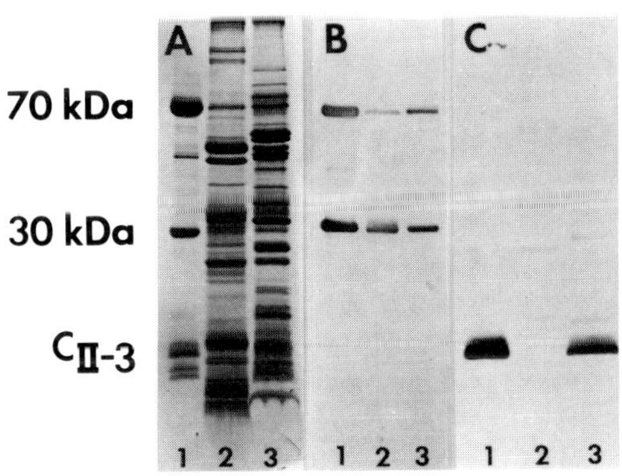

Fig. 1. Homology of rat and beef antigens. Lane 1, beef heart Complex II, lane 2, rat skeletal muscle succinate-cyt. *c* reductase, lane 3, beef heart SMP. After SDS-PAGE peptides were stained (panel A), or transblotted for identification by ELISA with primary antibodies to the 70- and 30-kDa subunits of beef heart succinate dehydrogenase (panel B), or to peptide C_{II-3} of beef heart Complex II (panel C).

peptide C_{II-3}, however, which functions in binding the enzyme to the membrane in beef heart mitochondria, failed to indicate comparable peptides in the rat skeletal muscle. A further difference is the sensitivity of the succinate dehydrogenase binding sites in rat membranes to prolonged alkaline treatment, not evident in beef heart SMP.

Iron-deficient and normal particles behaved alike in response to alkaline incubation, and in reconstitution tests. The greatly reduced initial activity of the iron-deficient membranes before alkaline incubation was not increased by addition of beef heart enzyme, in accord with the concept that the whole complex is missing, that there are no unfilled binding sites. Peptide C_{II-3}, it may be noted, is believed to be a *b*-type cytochrome and could itself suffer direct effects of iron limitation.

The data are most reasonably interpreted at this time as indicating that only fully competent dehydrogenase complexes are present in iron-deficient membranes, and no incomplete or impaired forms, and that iron-deficient complexes may not be assembled, or may be lost from the membrane after unsuccessful or incomplete assembly.

This work was supported by the Veterans Administration, the National Foundation for Cancer Research, and NIH Grants HL-16251, AG-04818, and AM-13897.

References

1. Maguire, J.J., Davies, K.J.A., Dallman, P.R., and Packer, L.: Biochim. Biophys. Acta 679, 210-220 (1982).
2. Cartier, L.-J., Cuddihee, R.W., Chen, M., Young, J.C., Garthwaite, S.M., and Holloszy, J.O.: Federation Proc. 41, 1738 (1982).
3. Ackrell, B.A.C., Ball, M.B., and Kearney, E.B.: J. Biol. Chem. 255, 2761-2769 (1980).
4. Burch, H., Bessey, O.A., and Lowry, O.H.: J. Biol. Chem. 175, 457-470 (1948).
5. Paech, C., Reynolds, J.G., Singer, T.P., and Holm, R.H.: J. Biol. Chem. 256 3167-3170 (1981).
6. Ragan, C.I., Galante, Y.M., and Hatefi, Y.: Biochemistry 21, 2518-2524 (1982).

CRYSTAL STRUCTURE STUDY OF TRIMETHYLAMINE DEHYDROGENASE

F. Scott Mathews, Louis W. Lim and N. Shamala

Department of Physiology and Biophysics, Washington University School of Medicine, St. Louis, MO 63110, USA

Summary

A wooden model of trimethylamine dehydrogenase has been constructed based on a 6.0A resolution electron density map. A 2.5A resolution map is currently being analyzed. The 166,000 dalton molecule contains two identical subunits each consisting of three distinct structural domains. The largest domain, ca. 40,000 daltons, contains the [4Fe-4S]* center and the covalently attached flavin. It is largely helical with only a limited amount of beta structure, the two smaller domains, ca. 24,000 and 19,000 daltons, are principally of the α/β type.

Introduction

Trimethylamine Dehydrogenase (EC 1.5.99.7) is produced by a variety of obligate or restricted facultative methylotrophic bacteria when grown with trimethylamine as their sole carbon source (1). It catalyses the oxidative demethylation of trimethylamine to dimethylamine and formaldehyde (2). Subsequent to the reaction, two electrons are transferred to an FAD-containing electron transfer flavoprotein *in situ* (3).

---

*Abbreviations used: [4Fe-4S], four-iron-four-sulfur; PMS, phenazine methosulfate; TMADH, trimethylamine dehydrogenase; MMC, methylmercuric chloride; PCMB, parachloromercury benzoate; HIPIP, high potential iron protein.

Flavins and Flavoproteins

When the artificial electron acceptor PMS is used, the overall reaction can be written

$$(CH_3)_3\ N + H_2O + PMS \rightleftharpoons (CH_3)_2\ NH + CH_2O + PMSH_2$$

TMADH from the bacterium W_3A_1 exists in solution as a tightly associated dimer consisting of two identical subunits of 83,000 daltons each (4). Each subunit contains one flavin, 6-S-cysteinyl-FMN covalently bound to the protein (5) and one iron-sulfur center of the $[4Fe\text{-}4S]^{2+}$ type (6). The amino acid sequence is unknown except for a 12 residue peptide following the point of covalent attachment of the flavin (7). When fully reduced by dithionite, the enzyme takes up 3 electron-equivalents per mole of subunit and displays an electron spin resonance spectrum similar to reduced ferredoxin (8). When reduced by the substrate, trimethylamine, or by dithionite in the presence of the inhibitor, tetramethylammonium chloride, only two electron equivalents are taken up and the electron spin resonance spectrum shows very strong coupling between the reduced [4Fe-4S] center and the flavin semiquinone.

Low Resolution X-ray Study

The quaternary structure of TMADH was established by a low resolution study of the native crystals (9). The crystals contain one dimer of 166,000 daltons per asymmetric unit and diffract to 1.8A resolution. Anomalous scattering at 5.0A resolution from the unresolved four-iron cluster showed the presence of two clusters per dimer. A rotation function analysis at 6.0A resolution showed the presence of a non-crystallographic molecular 2-fold axis relating the subunits perpendicular to the line connecting the two [4Fe-4S] clusters. The iron-sulfur and flavin content of the molecule and the probable identity of the subunits was subsequently confirmed by chemical studies (4).

Two heavy atom derivatives were prepared by growing crystals in the presence of MMC or PCMB. X-ray data were recorded to 6.0A resolution on a four-circle diffractometer. Six heavy atom binding sites were located. The sites formed three pairs which obeyed the molecular symmetry with respect to both their positions and their extent of binding. The pair of sites with highest affinity for MMC had the lowest affinity for PCMB and vice versa. Other mercurial reagents also bound at the same sites with varying extents of binding.

An electron density map was computed at 6.0A resolution using protein phases determined from the isomorphous and anomalous scattering differences. The mean figure of merit was 0.90. The protein density showed marked contrast to the solvent background and a high degree of 2-fold symmetry. The density map was averaged about the molecular 2-fold axis. After averaging, the intersubunit boundary could be clearly distinguished. Numerous rods of continuous density were visible but the course of the polypeptide chain could not be traced.

A model was constructed from the averaged 6.0A density map by tracing 2.0A thick sections of density onto 5/32 inch thick sheets of bass wood and gluing contiguous pieces of high density together to form a three-dimensional model.

Although the subunit interface was clearly defined in the model, the two subunits were connected by segments of high electron density in several places, a couple of which extended across the molecular 2-fold axis. The subunits were separated by cutting through the segments, after which each subunit could be easily split into three domains by breaking a small number of connecting points. The three domains of one subunit were painted three separate colors, those of the other subunit a fourth color and the domains were then reglued to form 2 separate subunits. Photographs of the

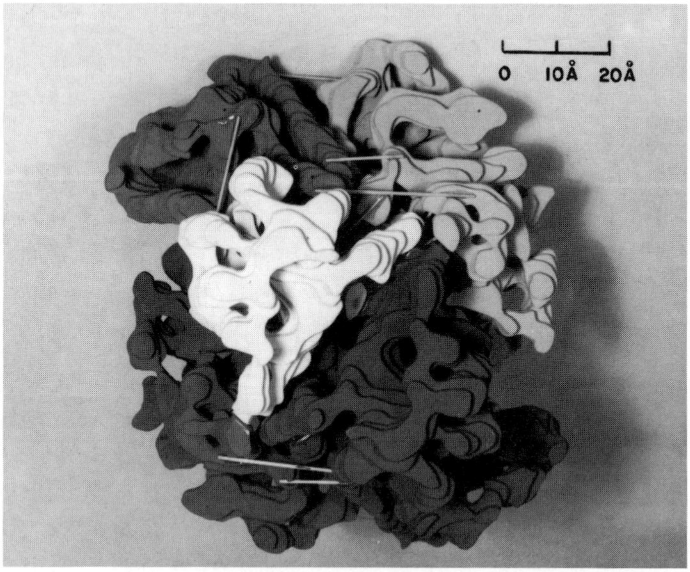

Fig. 1a. Bass wood model of TMADH, viewed from the top. One subunit is shown in a single dark shade and the other in 3 separate shades. The small domain is the lightest and the large domain in the upper right is the next lightest.

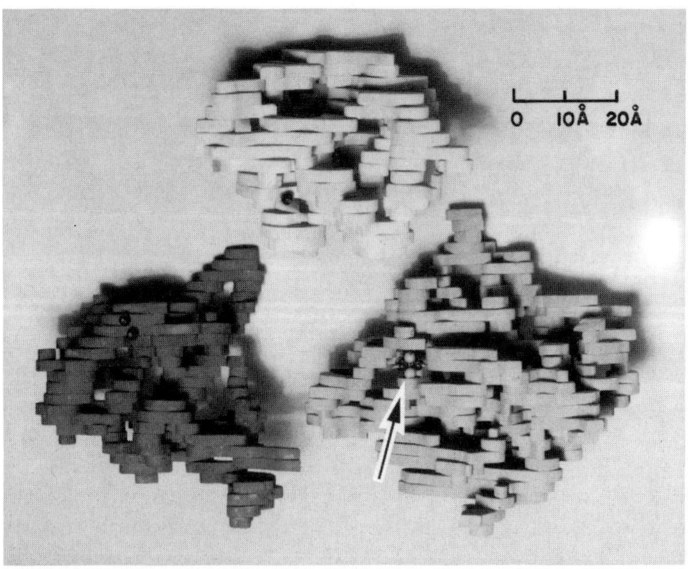

Fig. 1b. Separated domains of one subunit of TMADH viewed from the interior of the molecule. The position of the buried [4Fe-4S] cluster is indicated.

intact molecule and of the individual domains are shown in figures 1a and 1b, respectively.

The TMADH dimer forms a prolate ellipsoid of approximate dimensions 70 by 85 by 90A. The subunit interface is quite large and somewhat convoluted, accounting for the tight association of the subunits in solution. The three domains of each subunit are approximately 19,000, 24,000 and 40,000 daltons. They are packed with the small domain on top and the two larger domains side by side underneath. The domain and subunit packing is indicated schematically in figure 2. The iron-sulfur cluster is located in a pocket of the large domain adjacent to the medium domain and about 15A from the protein surface. The flavin could not be located in the low resolution map.

Both the small domain and the medium domain contain α/β-type supersecondary structures. Several superimposed sections of electron density for the small domain are shown in figure 3.

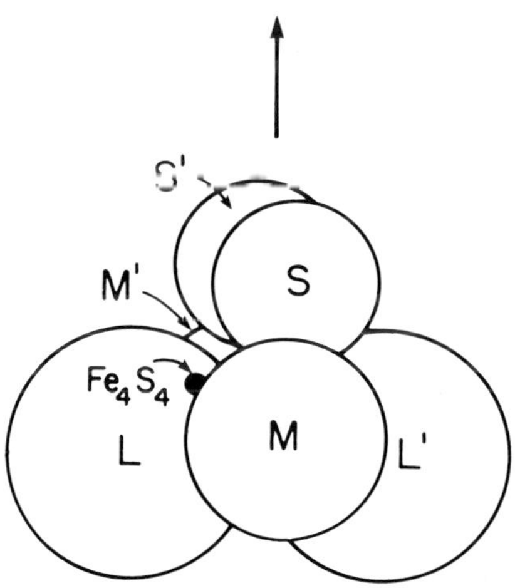

Fig. 2. Packing of domains in the TMADH dimer. S, M, and L refer to the small, medium and large domains described in the text. The primed symbols refer to the domains related by the 2-fold axis which is also shown. Although visible in this diagram, the [4Fe-4S] cluster is buried about 15A below the surface.

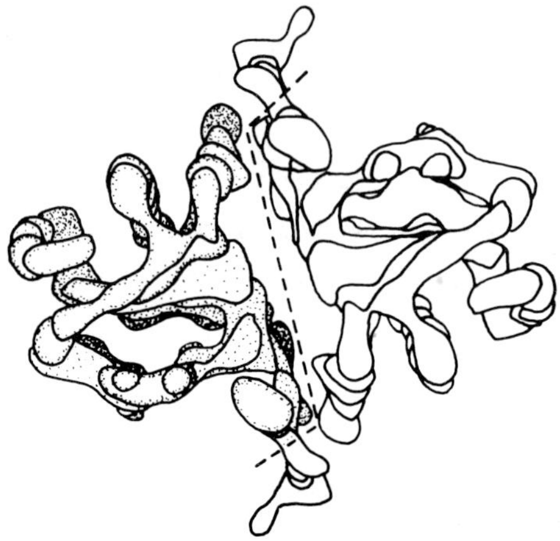

Fig. 3. Electron density at 6.0A resolution of the small domain comprising 6 sections. The view is perpendicular to the molecular 2-fold axis. The dashed line indicates the subunit and domain boundary. The central twisted beta sheet of each domain is surrounded by helix-like rods of density.

The left handed twist of the sheet was clearly visible as well as helix-like rods of density lying on either side. The α-carbon backbone of flavodoxin was superimposed on the density of the small domain on the MMS-X graphics system. Although the beta sheet was common to both structures the arrangement of connecting helices appeared to be different.

The α/β-type structure of the medium domain is also quite apparent. In addition, the medium domain contains several long helix-like rods of density adjacent to the large domain. These features of secondary structure in the medium domain (dark) can be seen in figure 1b. The large domain is mostly helical with little apparent β structure.

High Resolution X-ray Study

X-ray data to 2.5A resolution were recorded on a two-chamber multiwire area detector system at the University of

California, San Diego. Complete data, including anomalous scattering measurements, were recorded from one crystal each of the native, the MMC and the PCMB derivatives, plus a small amount of data from 2 additional native crystals, all in about 2 weeks. Each reflection was recorded several times and the agreement factors between the intensities of equivalent reflections were about 4% to 5%.

The heavy atom binding sites were verified in difference Patterson maps and then refined. The MMC derivative was well substituted. However, the substitution of PCMB was considerably weaker, probably because of variations in heavy atom concentrations. Only 4 of the 6 binding sites were occupied by PCMB, but the derivative still contributed significantly to the protein phases.

In order to utilize the anomalous scattering from the iron-sulfur cluster at high resolution it was necessary to locate the individual iron atoms. An anomalous difference Fourier for the native protein was computed at 2.5A resolution based

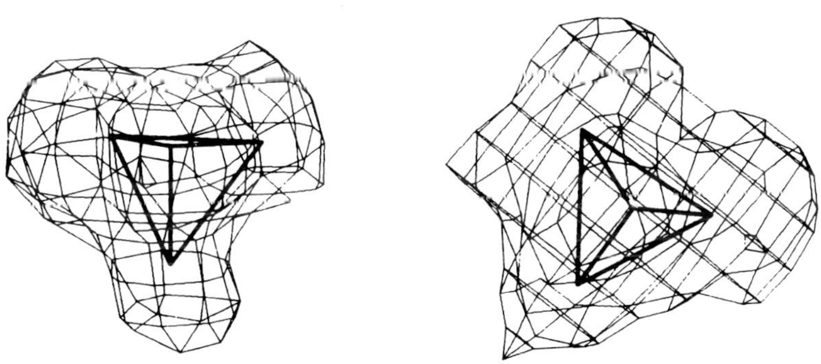

Fig. 4. Anomalous scattering difference Fourier at 2.5A resolution at the two four-iron positions. Tetrahedra representing resolved iron atoms are superimposed.

on single isomorphous phases derived from the MMC derivative. The anomalous density for each iron-sulfur cluster in the asymmetric unit is shown in figure 4. The individual iron atoms are nearly resolved, enabling the orientation of each cluster and the positions of the individual iron atoms to be defined.

The 2.5A electron density map based on isomorphous and anomalous scattering measurements was computed and averaged. The density of the iron-sulfur cluster is shown in figure 5, along with a skeletal model of the [4Fe-4S] cluster attached to 4 cysteinyl sulfur atoms. The cube-like nature of the cluster is evident indicating the marked contribution of the inorganic sulfur to the cluster density. Three of the four cysteine ligands are connected to the protein through high electron density while the fourth is connected through considerably weaker density.

Sections of the averaged 2.5A resolution electron density were plotted at 3 mm/A and transferred to Xerox transparencies, which were then stacked to form a minimap. The complete course of the polypeptide chain has been traced in the minimap with reasonable certainty, but approximate coordinates for only a few portions of the chain have been extracted and displayed on the graphics system. These include the covalently bound flavin with its attached peptide (about 22 residues) and the [4Fe-4S] cluster with its surrounding peptide (about 28 residues). The dodecapeptide sequence at the FMN group correlated well with the side-chain densities of the minimap and helped establish the direction of the polypeptide chain. The polypeptide chain begins in the large domain, passes through the medium and into the small domain and appears to return to the medium domain to finish.

The overall conclusions from the 6.0A model about the subunit and domain structure of TMADH are corroborated in the 2.5A

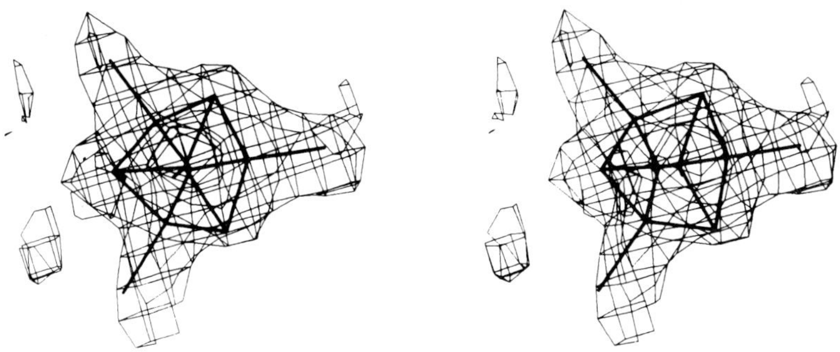

Fig. 5. Stereo diagram of [4Fe-4S] cluster plus 4 cysteinyl-S atoms superimposed on averaged 2.5A electron density map.

map. In a few places of the 6.0A model, portions of the density were assigned to the incorrect subunit or domain but these occurrences are rare. The numerous rods of density on the outside of the model in all three domains correspond quite well to segments of helix.

The presence of α/β structures in the medium and small domains has been verified. The small domain has a five stranded parallel beta sheet with well defined helices on one side only. On the other side the connecting peptides are extended and largely irregular in structure. The beta sheet topology is clearly different from that of flavodoxin. The medium domain also has a five stranded parallel beta sheet but with helices on both sides.

The tracing of the polypeptide chain of the large domain in the 2.5A map reveals many features which were not visible in the 6A model. In particular, it shows that it also contains an α/β supersecondary structure and that the FMN group is in the interior of this domain. A five stranded parallel beta sheet separates long helices on the surface from the FMN and the iron-sulfur center on the inside. The sheet was not recognized in the 6A model, probably because the interior of the large domain also contains substantial amounts of random coil segments. The flavin is situated in between the beta sheet and the [4Fe-4S] cluster but the polypeptide chain immediately surrounding the FMN does not appear to contain recognizable secondary structural features.

The [4Fe-4S] cluster is attached to a 21 residue peptide loop at the C-terminal end of the large domain. The cluster and an α-carbon diagram of the loop, plus three or four residues at either end, are shown in figure 6. There are three residues separating ligand 1 and 2, two separating ligands 2 and 3, and 16 residues separating ligands 3 and 4. The first half of the large loop is in a helical conformation, and passes near the molecular 2-fold axes. The loop then reverses direction and forms an extended chain, passing back to the iron-sulfur cluster. Several residues beyond the last connection to the cluster, the polypeptide chain proceeds into the medium domain.

The covalently bound flavin, superimposed on its electron density, is shown in figure 7. The isoalloxazine ring appears to be slightly bent. The center of the isoalloxazine

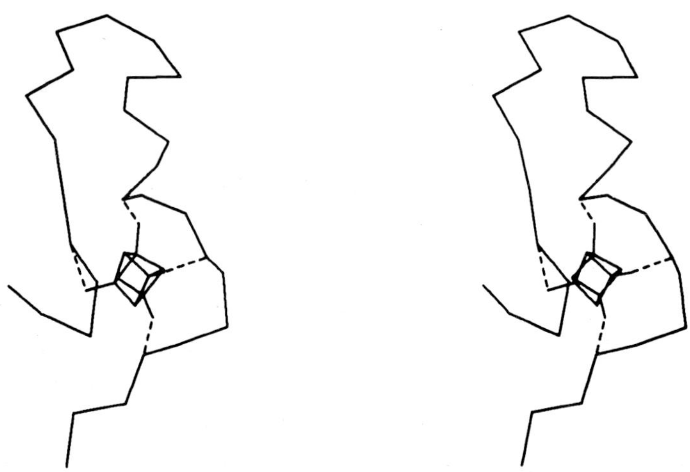

Fig. 6. Stereo diagram of the [4Fe-4S] cluster of TMADH attached to a 28-residue peptide loop through cysteine side chains.

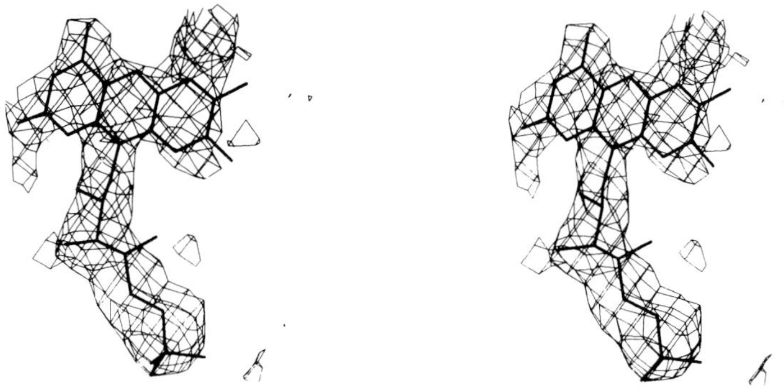

Fig. 7. Stereo diagram of the FMN group of TMADH superimposed on averaged electron density. The site of covalent attachment is at the C6 position at the upper right.

ring of FMN is about 12A from the center of the iron-sulfur cluster. Their closest contact distance is about 6A, from the 8-methyl group of FMN to one of the iron atoms. The relative positions and orientations of the two prosthetic groups and parts of their attached polypeptide chains are shown in figure 8.

Partially Reduced Protein

Preliminary studies at 6.0A resolution have been carried out on TMADH crystals which have been partially reduced by dithionite in the presence of tetramethylammonium chloride. Under these conditions, electronic interaction of the reduced iron-sulfur cluster and the flavin semiquinone have been observed in solution and the enzyme may exist in a different conformation from the fully oxidized form (8). An electron density difference map was computed and averaged. The most significant peaks occur at the iron-sulfur center and near the phenyl portion of the isoalloxazine ring. These peaks cannot yet be interpreted in structural terms but do suggest that a local conformational change at the active center has occurred.

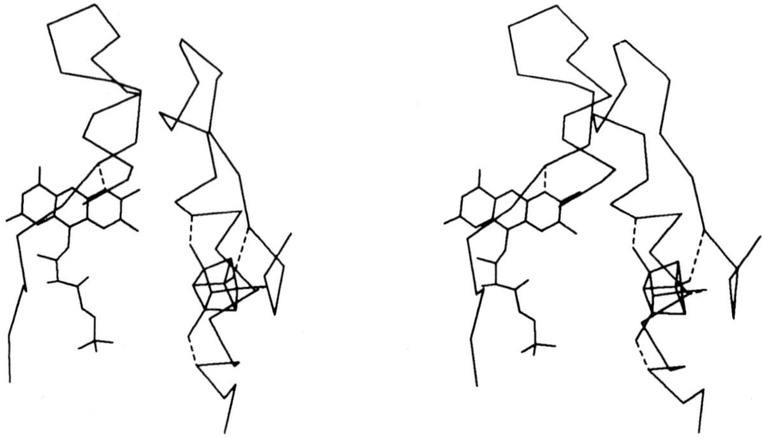

Fig. 8. Stereo diagram of the relative orientations of the FMN and [4Fe-4S] centers and portions of their attached polypeptide chains in TMADH.

Discussion

The division of the TMADH subunit into three distinct domains is the most notable feature of its structure. The functional role of the three domains is unknown at this time but it clearly is not to separate the prosthetic groups since both the flavin and the iron-sulfur cluster are contained within the large domain. This property of TMADH differs from other domain-containing proteins such as glutathione reductase (10) where the cofactors FAD and NADPH are located in separate domains.

The cysteine coordination patterns of TMADH and the 3 other known [4Fe-4S]-binding peptides are compared in figure 9. TMADH and the two ferredoxin structures contain three closely spaced cysteine ligands, but, the fourth ligand of only one of the ferredoxin [4Fe-4S] centers lies on the C-terminal side, 31 residues away. Also, the first pair of cysteines of TMADH are separated by 3 residues rather than 2, as in the ferredoxins. HIPIP contains only two cysteine ligands close together, the other two being separated by 16 and 13 residues. Only TMADH and P. aerogens ferredoxin contain significant amounts of helix within the [4Fe-4S] binding loops. Thus there does not seem to be any consistent structural homology among the [4Fe-4S] binding peptides other than the presence of 2 or 3 closely spaced cysteine residues.

The binding site for FMN differs considerably from that in flavodoxin (14) or that of the FMN fragment of FAD in glutathione reductase (10). In the last 3 structures the phosphate group is located at the N-terminus of an α-helix which flanks a parallel stranded beta sheet. In TMADH, no such helix exists near the phosphate site and the beta sheet is considerably further away. However, since FMN is covalently attached to TMADH such possible stabilization may not be required.

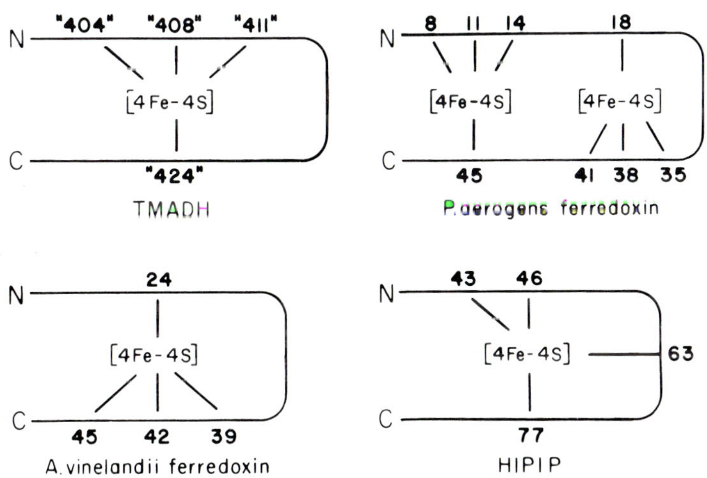

Fig. 9. Coordination pattern of cysteine side chains attachments to [4Fe-4S] clusters in TMADH, P. aeragens ferredoxin (11), A. vinelandii ferredoxin (12) and HIPIP (13). The attachment sites for TMADH are in quotations since they are arbitrarily positioned along the polypeptide chain.

Acknowledgement

This work has been supported by USPHS Grants GM20530 and GM31611. Thanks to Sue Eads for typing the manuscript.

References

1. Anthony, C.: The Biochemistry of Methylotrophs, Academic Press, N.Y. (1982).

2. Steenkamp, D.J. and Mallinson, J.: Biochim. Biophys. Acta 429, 705-719 (1976).

3. Steenkamp, D.J. and Gallup, M.: J. Biol. Chem. 253, 4086-4089.

4. Kasprzak, A., Papas, E.J. and Steenkamp, D.J.: Biochem. J. 211, 535-541 (1983).

5. Steenkamp, D.J., McIntire, W. and Kenney, W.C.: J. Biol. Chem. 253, 2818-2824 (1978).

6. Hill, C.L., Steenkamp, D.J., Holm, R.H. and Singer, T.P.: Proc. Natl. Acad. Sci., U.S.A. 74, 547-551 (1977).

7. Kenney, W.C., McIntire, W., Steenkamp, D.J. and Benisek, W.F.: FEBS Lett. 85, 137-140 (1978).

8. Steenkamp, D.J., Singer, T.P. and Beinert, H.: Biochem. J. 169, 361-369 (1978b).

9. Lim, L.W., Mathews, F.S. and Steenkamp, D.J.: J. Mol. Biol. 162, 869-876 (1982).

10. Thieme, R., Pai, E.F., Schirmer, R.H. and Schulz, G.E.: J. Mol. Biol. 152: 763-782 (1981).

11. Adman, E.T., Sieker, L.C. and Jensen, L.H.: J. Biol. Chem. 251: 3801-3806 (1976).

12. Ghosh, D., O'Donnell, S., Furey, Jr., W., Robbins, A.H. and Stout, C.D.: J. Mol. Biol. 158: 73-109 (1982).

13. Carter, C.W.: J. Biol. Chem. 252: 7802-7811 (1977).

14. Burnett, R.M., Darling, G.D., Kendall, D.S., LeQuesne, M.E., Mayhew, S.G., Smith, W.W. and Ludwig, M.L.: J. Biol. Chem. 249: 4383-4392 (1974).

THE FLAVIN DOMAIN OF ASSIMILATORY NADH:NITRATE REDUCTASE FROM *CHORELLA VULGARIS*.

Larry P. Solomonson and Michael J. Barber

Department of Biochemistry, University of South Florida
College of Medicine, Tampa, FL 33612, U.S.A.

Introduction

Assimilatory nitrate reductase (NR) from *Chlorella vulgaris* is a complex flavo-hemo-molybdoprotein that catalyses the rate limiting step, reduction of NO_3^- to NO_2^-, in nitrogen assimilation. The proposed structure for the enzyme is shown in Fig. 1. Native NR, molecular weight 380,000, is a homotetramer with dihedral cyclic symmetry (dimer of dimers) (1). The subunit molecular weight is approximately 95,000, and each contains 1 FAD, 1 Mo and 1 b_5-type cytochrome prosthetic groups (1,2). In addition to the physiological reduction of NO_3^- by NADH, NR also exhibits a variety of partial activities, classified as either NADH-dehydrogenase or NO_3^--reducing activities, which include the reduction of cytochrome c and ferricyanide by NADH (NADH:CR, NADH:FR) and NO_3^- reduction by reduced methyl viologen ($MV^{\cdot +}$:NR). FAD is required for the former activities whereas Mo is necessary for the latter (3). NADH:CR but not $MV^{\cdot +}$:NR activity is lost following incubation of native NR with N-ethylmaleimide (4) or a purified corn protease (5). $MV^{\cdot +}$:NR but not NADH:CR activity is lost following incubation of the native enzyme with CN^- in the presence of NADH via formation of an inactive NR-CN complex (4). We have examined a number of properties of the flavin domain of nitrate reductase to facilitate a description of the structure-function relationships between the various prosthetic groups and catalytic sites.

Flavins and Flavoproteins

Results

Spectral Characterization. Visible spectra of native NR in the oxidized and reduced state are shown in Fig. 2. In A, the enzyme was reduced with NADH while in B, dithionite was used as the reductant, both in the presence and absence of NAD^+. The spectra shown in the inserts demonstrate an increase in absorbance at long wavelengths when the enzyme is reduced by NADH or dithionite and NAD^+. This increased absorbance can be attributed to the formation of a reduced flavin-NAD charge transfer complex (6). No spectral changes were detected for the oxidized enzyme in the presence of NAD^+. The epr spectrum of NADH-reduced NR (pH 7) showed a free radical species (g=2.004, linewidth 1.4 mT). Identical spectra were obtained at pH 9 and following reduction of ^2H-exchanged NR. This radical was not detected during the course of potentiometric dithionite titrations within the potential range +150 to -350 mV. These results indicate the flavin species formed in the presence of NAD^+ is the anionic semiquinone, $FAD^{\cdot -}$, and suggests the mechanism of electron transfer between NADH and flavin of NR is similar to that described for NADH:cyt. b_5 reductase (6). NR pre-treated with N-ethylmaleimide or 4-maleimido-TEMPO spin label was not reduced by NADH, nor was there any increase in absorbance at long wavelengths when the enzyme was reduced with dithionite in the presence of NAD^+. The epr spectrum of spin-labeled NR exhibited only weak immobilization indicating the probe to be situated on the protein surface. Thus, a free sulfhydryl group(s) is necessary for binding of NADH to the enzyme and transfer of electrons to the flavin of NR.

Flavin Domain Polypeptide. Treatment of the native enzyme with a natural inactivation protein (protease) generates two fragments. The properties of these fragments are summarized in Table 1. A large fragment, devoid of FAD, contains Mo and heme, exhibits $MV^{\cdot +}$:NR activity and is composed of 4 equivalent polypeptides. A small fragment contains FAD and retains no quaternary structure. The NADH-binding site is associated with

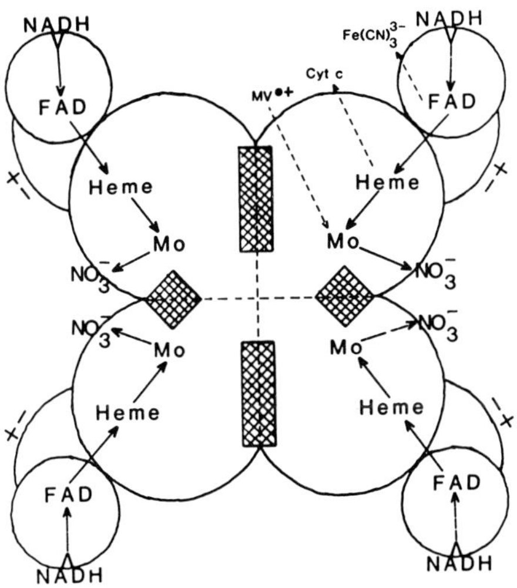

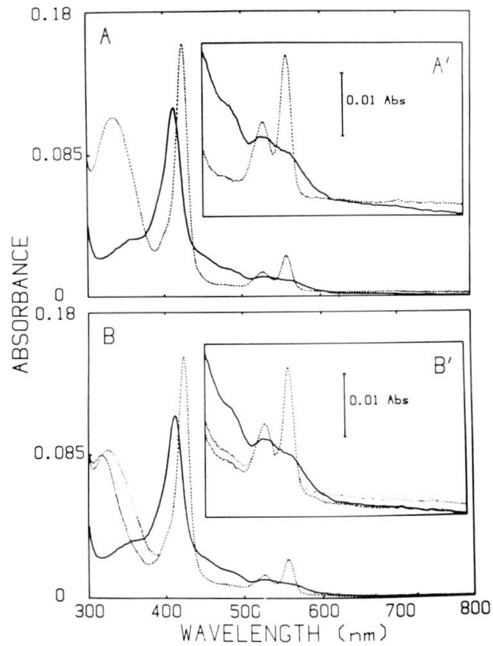

Fig. 1. Proposed structural model for nitrate reductase.

Fig. 2. Visible spectra of nitrate reductase. A. Oxidized NR in 50 mM Mops buffer, 0.1 mM EDTA, pH 7.0 (———). NADH-reduced enzyme (- - -). B. Oxidized NR (———). Dithionite-reduced enzyme (- - -). Dithionite-reduced enzyme plus NAD^+ (............).

Table 1. Properties of Proteolytic Fragments Generated by Action of Corn-Inactivator Protein on NR

Parameter	Native NR (Control)	Large Fragment	Small Fragment
Molecular Weight	376,000	283,000	28,000
Subunit Molecular Weight	99,000	69,000	28,000
FAD (nmol/nmol polypeptide)	1.0	0	0.9
Enzyme activity (% control)			
NADH → NO_3^-	100	0	0
NADH → Cyt. c	100	0	0
$MV^{\cdot +}$ → NO_3^-	100	105	0

this fragment as indicated by affinity labelling with [^{3}H]-fluorosulfonylbenzoyladenine and by NADH-reduction of FAD. The heme associated with the large fragment cannot be reduced by NADH suggesting a functional flavin domain is essential for electron transfer from NADH to heme. In addition, no increase in absorbance at long wavelengths was observed following reduction of this fragment by dithionite in the presence of NAD^+.

These results support the model for the structure of NR, shown in Fig. 1, in which the FAD/NADH-binding domains are exposed on the surface of the molecule while the quaternary structure is maintained via association sites on the heme/Mo domain.

This work was supported in part by grants from NSF (PCM-8214001) and USDA (GAM-8400528).

References

1. Howard, W.D., Solomonson, L.P.: J. Biol. Chem. 257, 10243-10250 (1982).

2. Giri, L., Ramadoss, C.S.: J. Biol. Chem. 254, 11703-11712 (1979).

3. Guerrero, M.G., Vega, J.M., Losada, M.: Ann. Rev. Plant Physiol. 32, 169-204 (1981).

4. Solomonson, L.P.: Biochim. Biophys. Acta 344, 297-308 (1974).

5. Yamaya, T., Solomonson, L.P., Oaks, A.: Plant Physiol. 65, 146-150 (1980).

6. Iyanagi, T., Watanabe, S., Anan, K.F.: Biochemistry 23, 1418-1425 (1984).

POLARIZED ABSORPTION SPECTRA OF FLAVOCYTOCHROME b_2 SINGLE CRYSTALS.

Andrea Mozzarelli, Mariella Tegoni* and Gian Luigi Rossi
Institute of Molecular Biology, University of Parma, Parma, Italy and
* Centre de Génétique Moléculaire du C.N.R.S., Gif-sur-Yvette, France.

Flavocytochrome b_2 is a tetrameric enzyme catalyzing the electron transfer from L-lactate to cytochrome c. Each monomer contains a FMN and a heme b. Trigonal crystals of S.cerevisiae flavocytochrome b_2 are under study by X-ray diffraction. The asymmetric unit cell contains 4 subunits related by 222 symmetry. Our previous studies have shown that flavocytochrome b_2 in the crystal is catalytically competent and can form a functional complex with cytochrome c.

In order to establish whether changes in the orientation of hemes occur in the course of electron transfer, polarized absorption spectra of oxidized and reduced flavocytochrome b_2 were obtained by single crystal microspectrophotometry. The incident light direction was normal to the c crystal axis and the electric vector of plane polarized light was either normal or parallel to the c crystal axis.

The polarization ratio (P.R.), λ-independent in the heme α, β and γ bands, was equal to 1.2 both for the reduced and oxidized state of the iron. This finding shows that the orientation of the heme plane does not appreciably change with the redox state of the iron.

Polarized absorption spectra of oxidized FMN in the 440-500 nm region could be calculated from the observed spectra by substracting the heme contributions. The P.R. was λ-dependent, probably due to vibronic components of different polarization, and it was found to be equal to 0.8 at $\lambda_{max} = 455$ nm.

Binding of sulfite to oxidized flavocytochrome b_2 abolishes the characteristic absorption spectrum of oxidized FMN, as in solution, and changes the P.R. value for the heme absorption bands that becomes equal to 1. This finding might be indicative of a change in the orientation of hemes, associated with the modification of FMN. No crystal damage was observed and the effect was readily reversed upon removal of sulfite.

Flavins and Flavoproteins

FMN:PROTEIN INTERACTIONS IN FLAVODOXIN FROM *A. nidulans*

Martha L. Ludwig, Katherine A. Pattridge, and George Tarr.
Biophysics Research Division and Department of Biological Chemistry, The University of Michigan, Ann Arbor, Michigan

Introduction

In contrast to most other flavodoxins, flavodoxins from several algae (1), including the cyanobacterium, *Synechococcus* PCC 6301 (*A. nidulans*) (2), have redox potentials for the oxidized/semiquinone equilibrium that are near the potential for free flavins. This unusual property of cyanobacterial flavodoxin prompted us to undertake an x-ray study of the protein. The structure of the oxidized form of algal flavodoxin has been determined at 2.0 Å resolution (3) and described by a molecular model which includes approximately half of the sidechains. Additional peptide sequences have now been included in the model of the oxidized form, using maps at 1.65 Å resolution. A study of the semiquinone form at 3.0 Å permits an initial comparison of the semiquinone and oxidized structures of algal flavodoxin.

Inclusion of New Sequence Data at 1.65 Å

The polypeptide chain was cleaved at Trp residues by reaction with skatole, using a modification of the method of Fontana (4). The peptides shown in Figure 1 were purified by HPLC and sequenced by manual Edman methods. They were placed in the chain by comparison with the electron density. In every case, location was simplified by the knowledge that the peptide N-terminus adjoined a Trp residue. Peptides 1 and 2 were

Flavins and Flavoproteins

found to overlap the peptide 62-67, isolated earlier from an S. aureus protease digest. The first nine residues of peptide 3 correspond to the published sequence (3) located at positions 120 through 128 of the model, and peptide 4 is the C-terminus of the chain. Peptide 3 completes identification of the residues that are inserted in the fifth strand of β sheet (3) and peptide 1 is a functionally important sequence as it corresponds to a chain reversal that contacts FMN.

```
[1]   (W)-N-V-G-E-L-Q-S
       57     60

[2]   (W)-E-G-I-Y-D-D-L-D-S-V-N-G-Q-A-K
       66       70                      80

[3]   (W)-P-I-E-G-Y-D-F-N-E-S-K-A-V-R-N-N-Q-F-V-G-L-A-I-D
      120                     130                   140

[4]   (W)-V-S-Q-L-K-S-E-F-G-L
       156       160        166
```

Figure 1. Peptides isolated after cleavage of A. nidulans flavodoxin at Trp. Peptides 2 and 3 have not been sequenced to their C-termini. Numbering is established for peptides 1 and 2 since the sequence is known from the N-terminus; the other starting positions are assigned from the x-ray map at 2.0 Å and may be revised when the sequence is complete.

The chain conformation is being reexamined and the new sequences built into a 1.65 Å map using an Evans & Sutherland graphics system with the software developed by Jones (5). Phases for this map were calculated from a partial molecular model that included the backbone (excepting residues 27-29, 67-70, and 82-83), 92 side-chains, and 178 solvents. After partial refinement by group least squares (6), the R factor for this model was 0.335 (1.65 - 5.0 Å resolution). We are also working with 'deletion' maps in which residues to be rebuilt are omitted from the model as they are examined. The results of fitting residues 57-60, W-N-V-G, to a deletion map are shown in Figure 2. The backbone conformation of this region is essentially as described earlier (3).

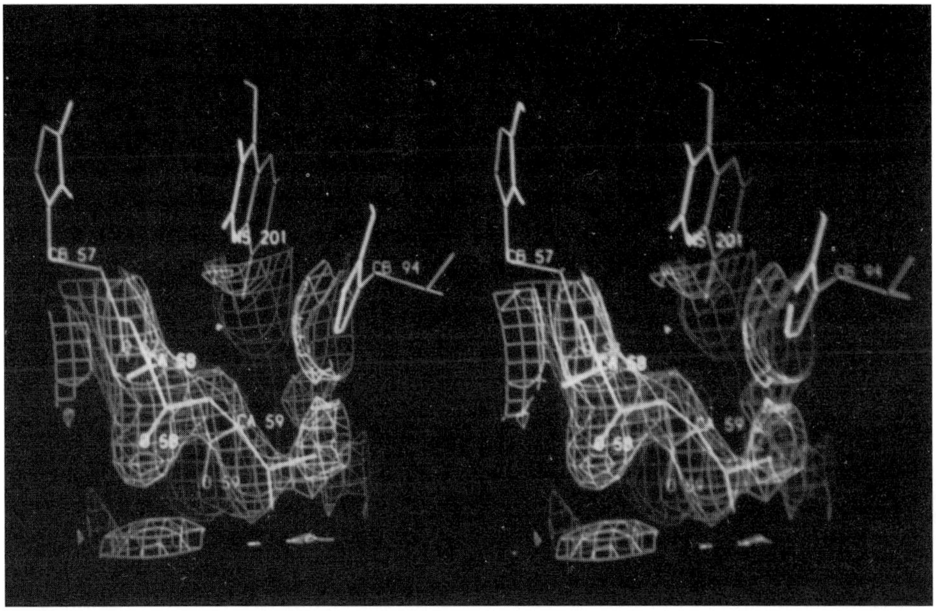

Figure 2. A stereoview of the electron density in a 1.65 Å map of oxidized algal flavodoxin. Atoms of residues 57-62 were omitted from the phase calculation. Directions of the carbonyl groups of residues 57-59 correspond to a III' conformation (or left-handed 3_{10} helix), and 58 O can be seen pointing away from the flavin N(5). Only the C_β of Asn 58 is shown in this view.

The Semiquinone Map at 3.0 Å Resolution

Data were collected from a crystal reduced with excess dithionite, and mounted in a glove box under nitrogen. Because crystals in the three oxidation states are isomorphous (7), we looked for local changes accompanying reduction by computing maps with coefficients $(|F_{sq}| - |F_{ox}|)\exp i\alpha_{ox}$ or $(|2F_{sq}| - |F_{ox}|)\exp i\alpha_{ox}$. Data between 10 and 5 Å were phased with MIR phases, and data from 5.0 to 3.0 Å with phases computed from the molecular model described above.

The major positive and negative features in the difference map were in the vicinity of residues 58-61, in the same region where conformation changes occur on reduction of flavodoxin from Clostridium MP. The semiquinone map can be fit with a reverse turn having parameters near those for Type II' (8) at $C_\alpha 58$, as shown in Figure 3. Higher resolution studies will be required to verify this interpretation, which implies an inversion of the peptide unit connecting residues 58 and 59, as seen by comparison of Figures 2 and 3.

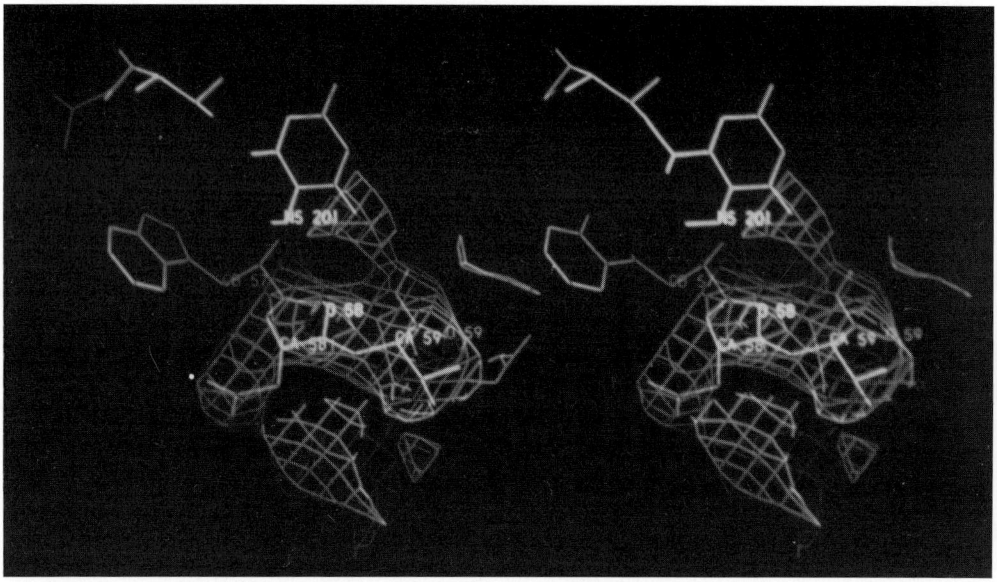

Figure 3. A stereoview of the electron density in a map at 3.0 Å, calculated with coefficients $(2|F_{sq}| - |F_{ox}|)$, showing the fit of a model in which O 58 points toward the isoalloxazine ring. The side chain of Val 59 has been rotated relative to its position in oxidized flavodoxin (Cf. Fig. 2).

Comparisons of *A. nidulans* and *Clostridial MP* Flavodoxins

The polypeptide folds of clostridial and algal flavodoxin can be aligned by superimposing the central β sheets (3,9), but to compare the local conformations in the chain reversal starting at residue 57 (56 in *C. MP* flavodoxin), we have calculated the best fit of the backbones using just the four residues comprising the turn. The similarity of the bends in algal and clostridial flavodoxins is apparent in Figure 4. From the torsion angles, the turn in oxidized algal flavodoxin is best classified as Type III', whereas the semiquinone model (at 58, the second position) resembles a Type II' turn (8). Oxidized flavodoxin from Clostridium MP approximates a Type III' conformation if we use coordinates from real space refinement (10); the bend in the semiquinone form is II' at the second position (G 57). The structurally related residues are aligned in Scheme [I].

[I]	Clostridium MP	M^{56}-G-D-E
	A. nidulans	W^{57}-N-V-G

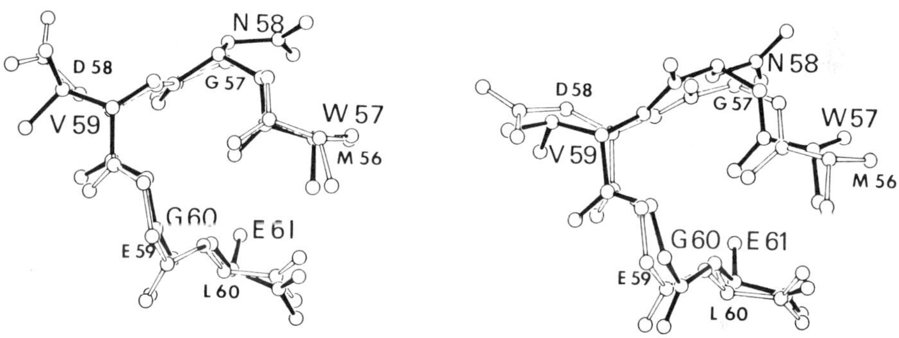

Figure 4. Comparison of the conformations of C. MP flavodoxin (56-60, open bonds) and A. nidulans flavodoxin (57-61) in the oxidized state (left) and the semiquinone state (right).

At the present stage of the structure analyses, we see that A. nidulans flavodoxin differs from clostridial flavodoxin in two ways that could affect the redox potentials for the oxidized/semiquinone equilibrium: the backbone O to NH(5) arrangement in the semiquinone state and the nature of the substituents at positions 2 and 3 of the reverse turn. Because the isoalloxazine ring is oriented differently in algal and clostridial flavodoxins (3), the distance from O 58 to N(5) is longer in the algal species. Numbers based on a map at 3.0 Å resolution are not very accurate, but the distance in the current model of algal semiquinone is 3.6 Å, in contrast to 2.7 Å for the NH(5)-O 57 distance in clostridial flavodoxin semiquinone. This FMN:protein interaction thus appears to be weaker in algal flavodoxin.

The side chains occurring in reverse bends have significant effects on the conformational energies (11,12). Gly, found in C. MP flavodoxin, is the favored residue at the second position of a Type II' bend since it avoids the close approach of a CH_β to the NH of the third residue of the bend; thus a Type II' bend with Asn, as in algal flavodoxin, would be expected to have a relatively high energy. That prediction is borne out by the tabulated occurrences of residues at this position in known protein structures (12). Further, Val occurs only rarely at the third site in reverse turns. We have built the Val in oxidized algal flavodoxin with C_γ atoms staggered relative to the main chain, and have turned the side chain in the semiquinone model by approximately 80°, as suggested by the difference maps. When the Val coordinates are established more accurately, we can include the contribution of both Val and Asn in more quantitative estimates of the relative stability of the oxidized and semiquinone conformations in A. nidulans flavodoxin. The qualitative impression from the present data is that conversion of the oxidized to the semiquinone conformation is less favorable in A. nidulans than in clostridial flavodoxin.

Acknowledgment: This research was supported by NIH Grant GM 16429.

References

1. Sykes, G.A., Rogers, L.J.:Biochem. J. 217, 845-850 (1984).

2. Entsch, B., Smillie, R.M.:Arch. Biochem. Biophys. 151, 378-386 (1972)

3. Smith, W.W., Pattridge, K.A., Ludwig, M.L., Petsko, G.A., Tsernoglou, D., Tanaka, N., Yasunobu, K.T.:J. Mol. Biol. 165, 737-755 (1983)

4. Fontana, A., Vita, C., Toniolo, C.:FEBS Lett. 32, 139-142 (1973).

5. Jones, T.A.:J. Appl. Cryst. 11, 268-272 (1978).

6. Hoard, L.H., Nordman, C.E.:Acta Cryst. A35, 1010-1015 (1979).

7. Smith, W.W., Crespi, H.L., Entsch, B., Ludwig, M.L., Nordman, C.E.:J. Mol. Biol. 94, 123-126 (1975)

8. Venkatachalam, C.M.:Biopolymers 6, 1425-1436 (1968).

9. Ludwig, M.L., Pattridge, K.A., Smith, W.W., Jensen, L.H., Watenpaugh, K.:Flavins and Flavoproteins (Massey, V., Williams, C.H., Jr., eds.) pp 19-27, Elsevier/North Holland, New York 1982.

10. Smith, W.W., Burnett, R.M., Darling, G.D., Ludwig, M.L.:J. Mol. Biol. 117, 195-225 (1977).

11. Smith, J.A., Pease, L.G.:CRC Critical Revs. in Biochemistry 8, 315-400 (1977).

12. Lewis, P.N., Momany, F.A., Scheraga, H.A.:Israel J. Chem. 11, 121-152 (1973).

13. Chou, P.Y., Fasman, G.D.:J. Mol. Biol. 115, 135-175 (1977).

PHOTOCHEMICAL FORMATION OF A STABLE RED DERIVATIVE OF FLAVODOXIN

Stephen G. Mayhew

Department of Biochemistry, University College Dublin,
Belfield, Dublin 4, Ireland

Vincent Massey

Department of Biological Chemistry, University of Michigan,
Ann Arbor, Michigan 48109, U.S.A.

Introduction

When flavodoxins are photo-irradiated in anaerobic solution in the presence of EDTA and a catalytic concentration of 5-deazaflavin, they are rapidly reduced first to the blue neutral semiquinone, and subsequently to the hydroquinone (1). Photo-irradiation in air-saturated solution causes a slow irreversible loss in the visible absorbance, and is attributable to destruction of the flavin. However, if the solution contains oxygen at a concentration lower than in air-saturated solution, the flavodoxin is rapidly converted to a stable red species. This red derivative is of interest because its absorption spectrum resembles those of orange-red proteins previously observed during the isolation of flavodoxins from two bacterial sources. This paper describes conditions which lead to the formation of the red derivative, and evidence that the derivative contains a new modified form of FMN.

Formation of red flavodoxin

The rate and extent of conversion of flavodoxin to the red derivative depends on pH and the oxygen concentration, and full formation requires a high concentration of a carboxylic

Flavins and Flavoproteins

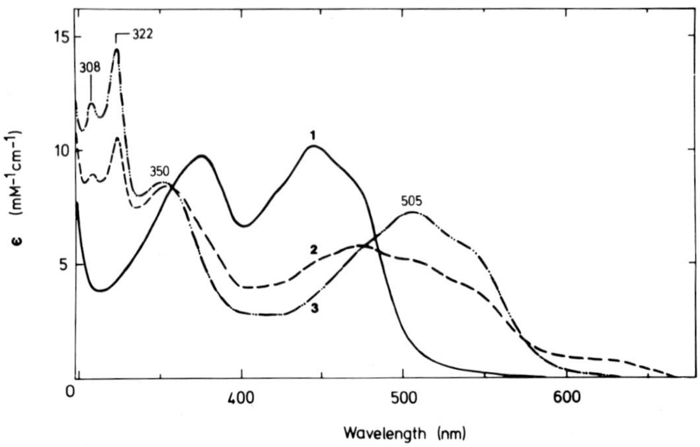

Fig. 1. Photochemical formation of red flavodoxin.
An anaerobic cuvette contained in 1 ml: 55 µM M.elsdenii flavodoxin; 1 µM 5-deazaflavin; 5 mM EDTA; 0.75 M sodium butyrate, pH 5. The cuvette was made 'partially anaerobic' by evacuating it for about 1 min at a vacuum pump, and it was then exposed at room temperature to light from a 150 W tungsten-halogen lamp in a slide projector. Curve 1; before light irradiation; curves 2 and 3, after 4.5 and 7 min light respectively.

acid or an alcohol. Fig. 1 shows that oxidized Megasphaera elsdenii flavodoxin is rapidly converted to the red species when it is irradiated in the presence of butyrate. Similar changes have been observed with flavodoxins from Clostridium tyrobutyricum, Clostridium MP, Clostridium kluyveri and Azotobacter vinelandii.

The effects of pH on the reaction have been examined with acetate and butyrate; in both cases conversion to red flavodoxin is greatly enhanced at acid pH. The effects of oxygen concentration have been examined qualitatively by varying the time during which air is removed from the reaction mixture before starting the reaction. After very short periods of air removal there is usually a lag before conversion to red flavodoxin begins; it seems likely that excess oxygen is reduced during the lag phase. If the period of air removal is long, only a

partial conversion occurs and the remaining flavodoxin is reduced to the semiquinone; admission of more air and further irradiation allows more of the red flavodoxin to be generated. Some semiquinone is always observed during the reaction.
The long wavelength absorbance of curve 2 in Fig. 1 shows that a transient formation of semiquinone occurred. When flavodoxin is first reduced to the semiquinone under anaerobic conditions and then partially aerated before further irradiation is carried out, much of the semiquinone is rapidly converted to oxidized flavin before the red derivative appears in solution. The semiquinone of flavodoxin reacts only slowly with oxygen in the dark, and it therefore appears that light irradiation under the conditions of Fig. 1 enhances this reactivity.

Red flavodoxin is not formed when Tris-HCl, pH 7.3, is used as the buffer and the photo-oxidizable substrate in the reaction. This buffer system was therefore used to test the effects of a range of carboxylic acids and alcohols in facilitating the reaction. Red flavodoxin was formed when the following compounds were included in the reaction mixture at concentrations in the range 0.1 - 0.5 M: acetate, propionate, butyrate, pentanoate, phenylacetate, EDTA, propanol, 1- and 2-butanol.
The red derivative was not formed with ethanol in the Tris HCl buffer, but its formation was enhanced by ethanol when the reaction was carried out at pH 5 in 10 mM acetate with EDTA (5 mM) as the photo-oxidizable substrate.

Lumiflavin-3-acetate can replace 5-deazaflavin in the reaction, showing that the strong reductant provided by 5-deazaflavin is not essential for the conversion to red flavodoxin.
The addition of superoxide dismutase and catalase to the photo-reaction did not prevent the formation of red flavodoxin, suggesting that reduced oxygen species such as superoxide and peroxide are not involved in the modification reaction.

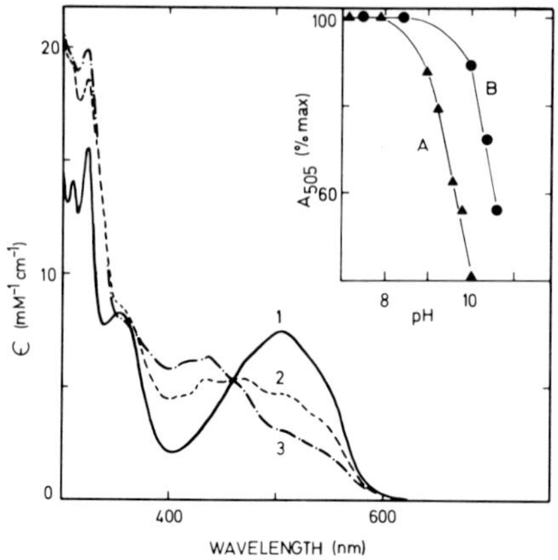

Fig. 2. Effect of pH on red flavodoxin
The pH of a solution of red M. elsdenii flavodoxin, made as in Fig. 1, or with sodium acetate replacing the butyrate, was adjusted stepwise by addition of solid sodium carbonate and solid citric acid. The spectra shown are for a titration with red flavodoxin made in acetate. Curve 1, pH 5.4; curve 2,pH 9.5 curve 3, pH 10. The inset shows a plot of absorbance at 505 nm (% max.) versus pH for titrations done with red flavodoxin made in acetate (curve A) and butyrate (curve B).

Properties of red flavodoxin

The absorption spectrum of the red derivative formed with M. elsdenii flavodoxin has absorption maxima at 308, 322, 350 and 505 nm with an absorption coefficient at 505 nm of 7.3 mM^{-1} cm^{-1} (Fig. 1). There is also a peak at 254 nm in addition to the peak at 272 nm which is found in native flavodoxin. The spectrum is independent of the carboxylic acids and alcohols that have been tested in the photochemical reaction. The spectrum changes reversibly at high pH (Fig. 2), but the pH ranges over which the effects occur for the derivative formed in acetate and butyrate are different (inset Fig. 2).

The derivative is stable in air for several weeks at 4°. However, an excess of ferricyanide causes a slow and partial

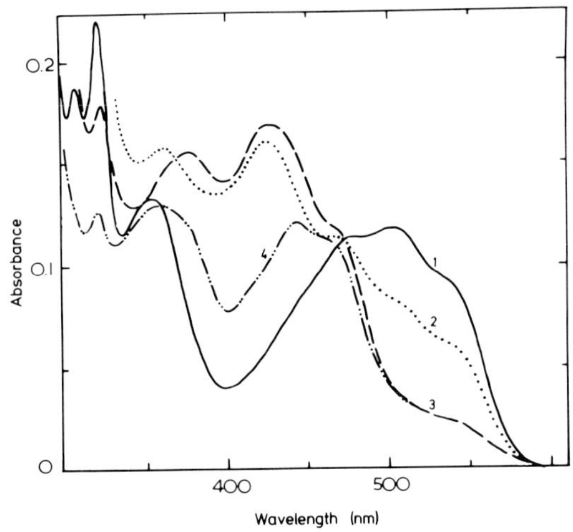

Fig. 3. Effect of ferricyanide on red flavodoxin.
Red M. elsdenii flavodoxin (18 μM), made as in Fig. 1, was treated at 20° in 80 mM Tris-HCl buffer, pH 8, with 0.1 mM $K_3Fe(CN)_6$. Curve 1, before adding ferricyanide; curves 2 & 3, 35 min and 14 hr after adding ferricyanide; curve 4, after subsequent dialysis versus 10 mM Tris-HCl buffer, pH 8.

conversion back to oxidized flavodoxin (Fig. 3), suggesting that the red derivative contains a reduced flavin. The red colour and the long wavelength absorbance of the derivative suggested that it might contain flavin semiquinone. However, the derivative did not give an ESR signal under conditions which allowed an ESR signal to be observed with the semiquinone of native flavodoxin (these measurements were carried out for us by Dr. R.C. Barklie, Trinity College, Dublin).

Treatment of the derivative with an excess of dithionite at pH 4.6 or pH 8 has no effect on the absorption spectrum. However, when the derivative formed in acetate at pH 4.6 was first treated with dithionite and then photo-irradiated, the absorption spectrum slowly changed to that of flavodoxin semiquinone.

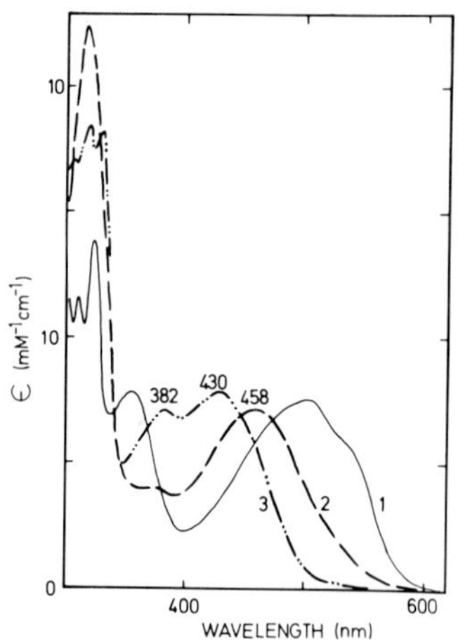

Fig. 4. Effect of pH on the acid extract of red flavodoxin. Red flavodoxin from M. elsdenii, made as in Fig. 1 but with acetate replacing butyrate, was treated with 5%(w/v) trichloroacetic acid at 0° and then centrifuged to remove apoflavodoxin. TCA was removed from the supernatant with ether, and the pH was then adjusted with solid $NaHCO_3$. Curve 1, untreated red flavodoxin; curve 2, TCA extract at pH 2.95; curve 3, TCA extract at pH 7.5

Properties of the protein-free chromophore

When red flavodoxin is treated with 5% trichloroacetic acid to precipitate the apo-enzyme, the chromophore remains in solution indicating that the modification does not involve a covalent interaction with the protein. Further, the chromophore is bound by M. elsdenii apoflavodoxin to give a complex whose absorption spectrum is similar to that of red flavodoxin; this observation indicates that the chromophore is a modified form of FMN.

The spectrum of the chromophore depends on pH (Fig. 4), and at pH 7.5 it has absorption maxima at 382 and 430 nm. Further spectral changes occur at higher pH. However, the spectrum changes slowly at 4° indicating that the chromophore is less stable in free solution than in the protein; in neutral solution, the absorbance around 370 nm increases, while the maximum at 430 nm declines.

Discussion

These results establish that red flavodoxin contains a modified form of FMN. The visible and near UV absorption spectra of the protein-bound and protein-free flavin do not resemble the spectrum of any flavin of known structure, and therefore the chemical nature of the modification is not yet known.

The spectrum of red flavodoxin generated photochemically is very similar to the spectrum of an orange-red protein isolated from the mother liquor of crystals of *C. kluyveri* flavodoxin (2). Small amounts of red flavodoxin also occur in preparations of flavodoxin from *Clostridium MP* (S.G. Mayhew, unpublished). The preparations from both organisms were carried out with minimal exposure of the proteins to light, and it therefore seems likely that in these cases the modification(s) occurred in the dark. It may be significant that both organisms produce large amounts of short-chain fatty acids, compounds which facilitate formation of red flavodoxin in the photochemical reaction.

References

1. Massey, V., Hemmerich, P.: Biochemistry 17, 9-17 (1978)
2. Schwan, T., Doctoral Thesis, University of Göttingen, 1975

DESULFOVIBRIO VULGARIS FLAVODOXIN. A ^{13}C AND ^{15}N NMR INVESTIGATION

Jacques Vervoort, Franz Müller, Jean LeGall, Adelbert Bacher and Helmut Sedlmaier

Department of Biochemistry, Agricultural University, De Dreijen 11, 6703 BC Wageningen, The Netherlands; Department of Biochemistry, University of Georgia, Athens, G.A. 30602, USA; Department of Organic Chemistry and Biochemistry, University of Munich, Lichtenbergstr. 4, D-8046 Garching, West-Germany

Introduction

Many techniques have been used to obtain information on the interaction between apoflavodoxin and FMN. Among these, NMR is one of the most versatile and powerful methods. Because of the small molecular mass of flavodoxins it is even possible to obtain structural information by 2-dimensional NMR (1). Here we report on a ^{13}C and ^{15}N NMR study on FMN bound to the apoflavodoxin from Desulfovibrio vulgaris.

Results and Discussion

The ^{13}C and ^{15}N chemical shifts of FMN free and bound to D.vulgaris flavodoxin are summarized in Table 1. Figure 1 shows the spectra of some ^{13}C enriched oxidized FMN molecules bound to D.vulgaris flavodoxin. The ^{13}C chemical shifts of D. vulgaris flavodoxin in the oxidized state show downfield shifts for almost all atoms of the xylene subnucleus and upfield shifts for C(4) and C(4a) as compared to free FMN. The difference in chemical shifts between free and bound FMN can be accounted for by the larger polarization of the flavin molecule in the flavodoxin. Electron density is withdrawn from the xylene subnucleus and reallocated on the C(2) carbonyl group and to a lesser extent on the C(4) carbonyl group. The ^{15}N chemical shifts show an upfield shift for N(1) and N(3), and a downfield shift for N(5) and N(10). The shifts indicate a hydrogen bond to N(1), somewhat stronger than in H_2O, a hydrogen bond to N(5), not as strong as in H_2O and a weak hydro-

Flavins and Flavoproteins

Table 1. ^{13}C and ^{15}N chemical shifts of various selectively enriched free FMN and bound to D.vulgaris flavodoxin. ^{13}C chemical shifts are relative to TMS. ^{15}N chemical shifts are relative to liquid NH_3.

C-Atom	OXIDIZED		REDUCED	
	D.vulgaris flavodoxin	FMN(4)	D.vulgaris flavodoxin	$FMNH^-$(4)
C(2)	159.7	159.8	157.5	157.9
C(4)	162.4	163.7	155.0	157.2
C(4a)	134.3	136.2	102.7	101.4
C(5a)	137.4	135.4	134.6	133.7
C(6)	132.5	132.8	114.5	116.5
C(7)	142.0	140.6	130.7	132.8
C(7α)	20.5	19.9	19.4	19.1
C(8)	154.0	151.9	126.7	130.8
C(8α)	23.3	22.0	20.3	19.8
C(9)	117.2	118.2	114.7	117.3
C(9a)	131.9	131.5	129.1	130.1
C(10a)	152.3	152.0	154.0	154.9
N(1)	188.0	190.8	186.6	182.6
N(3)	159.9	160.5	148.3	149.3
N(5)	341.1	334.7	62.1	57.7
N(10)*	165.6	164.6	98.4	97.2

*7-methyl-10-ribityl-isoalloxazine-5'-phosphate derivative.

gen bond to the N(3)H group. The main difference between the flavodoxins from M.elsdenii and Clostridium MP on the one hand and from D.vulgaris and A.vinelandii on the other hand is the lack of a hydrogen bond towards N(5) in the first group in the oxidized state. As compared to the chemical shift of N(10) in free FMN the corresponding chemical shift in D.vulgaris flavodoxin is shifted downfield by 1 ppm, indicating a small positive charge on N(10). This is reflected in particular in the downfield shift of the resonance due to C(7) and C(5a). In the reduced state the ^{13}C chemical shifts again indicate a strong hydrogen bond to the C(2) carbonyl group and a

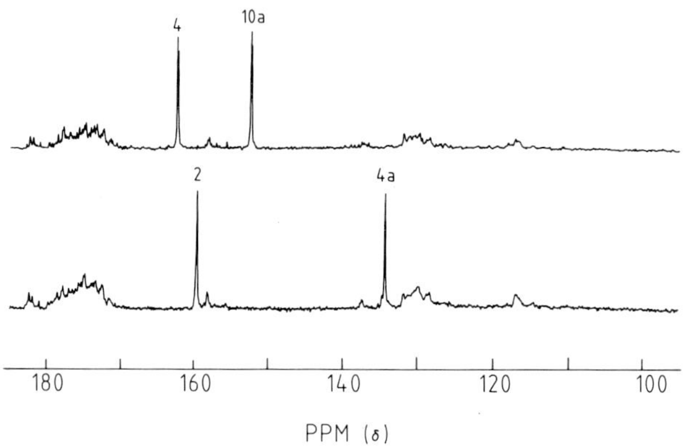

Figure 1. ^{13}C NMR spectra of D.vulgaris apo-flavodoxin reconstituted with [4,10a-^{13}C] FMN and [2,4a-$^{13}C_2$] FMN (oxidized state).

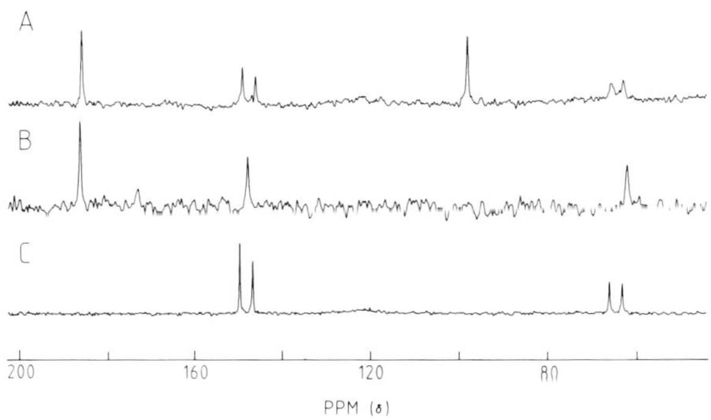

Figure 2. ^{15}N NMR spectra of D.vulgaris apoflavodoxin reconstituted with [1,3,5,10-$^{15}N_4$]FMN(A) and [1,3,5-$^{15}N_3$]FMN (B,C). All spectra were obtained in the reduced state. Spectrum B was taken under broad band ^1H decoupling and spectrum C was obtained applying the DEPT method.

weak one, if any at all, to the C(4) carbonyl group. Similar observations were made with the flavodoxin from M.elsdenii (2). Remarkable is the rather drastic upfield shift of the resonance due to C(8). The resonances due to C(6), C(7) and C(9) also shift upfield, about 2 ppm. This indicates that the electron density at these positions increases much more in flavodoxin than in free flavin upon two electron reduction. The ^{15}N chemical shifts indicate a decrease in electron density at the N(1) position, a weak hydrogen bond to the N(3)H group and a strong hydrogen bond to the N(5) atom. The ^{13}C and ^{15}N chemical shifts clearly show that the flavin is ionized in reduced D.vulgaris flavodoxin. Figure 2 shows the ^{15}N NMR spectra of D.vulgaris flavodoxin. Although ^{15}N NMR is a rather insensitive method, the existence of a clear coupling of the ^{15}N atom with the attached covalently bound hydrogen atoms enables one to enhance the ^{15}N signals by polarization transfer using the DEPT method (3). The degree of hybridization of the N(5) atom in reduced flavodoxin can be estimated from the $^{15}N(5)$-^{1}H coupling constant. This value is 86.2 ± 2Hz for D.vulgaris flavodoxin and 92.0 ± 2Hz for M.elsdenii flavodoxin (4), which indicates that N(5) in D.vulgaris flavodoxin is more sp^3 hybridized than in M.elsdenii flavodoxin. It is suggested that this observation is related to the fact that the electron exchange reaction between the semiquinone and the reduced D.vulgaris flavodoxin is much slower than the corresponding one in M.elsdenii flavodoxin (5).

Acknowledgements

We thank Miss Lyda Verstege for typing of the manuscript and Dr. C.A.H. Rasmussen for carefylly reading the manuscript. This study was supported in part by the Dutch Foundation for Chemical Research (SON) with financial aid from the Dutch Organization for the Advancement of Pure Research (ZWO) and by the Deutsche Forschungsgemeinschaft.

References

1. Moonen, C.T.W., Scheek, R.M., Boelens, R., Müller, F.: Eur.J.Biochem. 144, 323-330 (1984).
2. Van Schagen, C.G., Müller, F.: Eur.J.Biochem. 120, 33-39 (1981).
3. Doddrell, D.M., Pegg, D.T., Bendell, M.R.: J.Magn.Reson. 48, 323-327 (1982).
4. Franken, H.-D., Rüterjans, H., Müller, F.: Eur.J.Biochem. 138, 481-489 (1984).
5. Moonen, C.T.W., Müller, F.: Eur.J.Biochem. 140, 303-309 (1984).

^{13}C-NMR STUDY ON THE INTERACTION OF RIBOFLAVIN WITH RIBOFLAVIN BINDING PROTEIN

Retsu Miura, Hiromasa Tojo*, Shigeru Fujii*, Toshio Yamano*, and Yoshihiro Miyake

National Cardiovascular Center Research Institute, Fujishirodai, Suita, Osaka, Japan and *Osaka University Medical School, Kita-ku, Osaka, Japan

Introduction

As extension of our recent ^{19}F-NMR study on the interaction of 8-fluoro-8-demethylriboflavin (1) with egg white riboflavin binding protein (RBP), we report here the ^{13}C-NMR study on the interaction of RBP with riboflavin (Rf) selectively ^{13}C-enriched at the 2-, 4-, 4a-, and 10a-positions and with 3-([^{13}C]methyl)Rf. We also studied fluorometrically the binding between RBP and Rf or 3-methyl-Rf (MRf), and compared the results with those from the NMR study.

Results and Discussion

Figure 1 illustrates the ^{13}C-NMR spectra of [2-^{13}C]Rf free (a) and bound to RBP (b). Spectrum c is for the mixture of RBP and [2-^{13}C]Rf in slight molar excess over RBP. Figure 2 shows the titration profiles of 2-^{13}C signal of Rf in the free (open circles) and RBP-bound (solid circles) forms. The solid line for open circles was obtained with a pK value of 9.90 from the fitting to the Henderson-Hasselbalch equation. This pK corresponds to the N(3)-H ionization. The results in Fig. 2 indicate that the chemical shift of 2-^{13}C is sensitive to N(3)-ionization and that the pK of N(3)-H in Rf is shifted to the alkaline side when Rf is bound to RBP. We next measured

Flavins and Flavoproteins

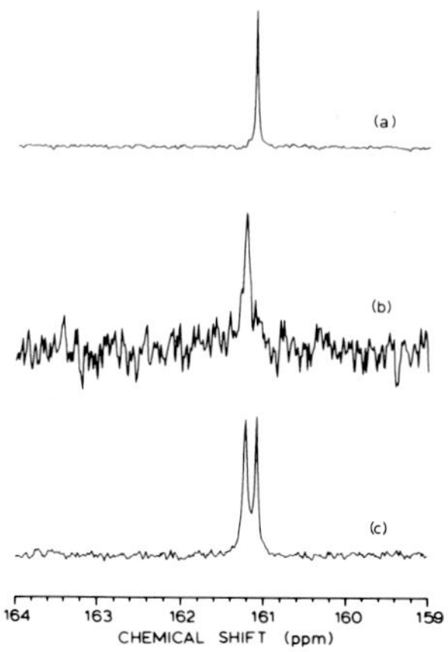

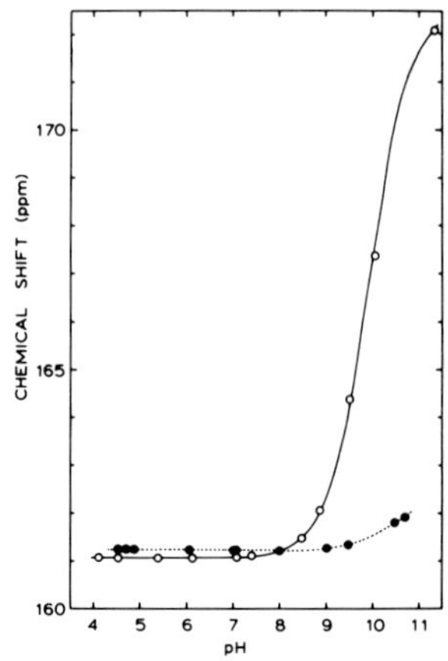

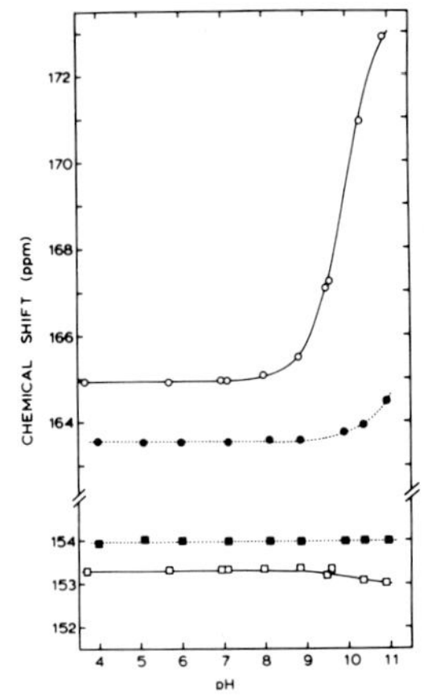

Fig. 1 (left). ^{13}C-NMR spectra, measured at 67.8 MHz, of [2-^{13}C]Rf in the free (a), RBP-bound (b) forms and the mixture thereof (c). Chemical shifts are relative to the methyl signal of internal or external 3-(trimethylsilyl)propionic acid-d_4.
Fig. 2 (middle). The pH-dependence of 2-^{13}C chemical shift of Rf free (open circles) and bound to RBP (solid circles). The solid line is the theoretical curve with a pK of 9.90.
Fig. 3 (right). The pH-dependence of 4- (circles) and 10a-^{13}C (squares) chemical shifts of Rf free (open symbols) and bound to RBP (solid symbols). The solid lines for open circles and open squares are theoretical curves with pK's of 9.95 and 9.90, respectively.

^{13}C-NMR spectra of [4,10a-$^{13}C_2$]Rf and [4a-^{13}C]Rf in the free and RBP-bound states. Figures 3 and 4 show the pH-dependence of the chemical shifts of 4- and 10a-^{13}C, and 4a-^{13}C, respectively, either in the free (open symbols) or RBP-bound (solid symbols) states. The solid lines for open circles (4-^{13}C, free form) and for open squares (10a-^{13}C, free form) in Fig. 3 are the theoretical curves fitted with pK's of 9.95 and 9.90, respectively. Similarly, the solid line in Fig. 4 is the theoretical curve for 4a-^{13}C (free form) with a pK of 10.00. These results demonstrate that these signals are also sensitive to N(3)-ionization and that pK for N(3)-H shifts to the alkaline side in the RBP-bound form. The characteristic profile is noted for 4a-^{13}C in the RBP-bound form in the acidic to neutral pH region. We investigate further the ^{13}C-NMR of 3-([^{13}C]methyl)Rf. Results are shown in Fig. 5. Unlike the

2-, 4-, 4a-, or 10a-^{13}C signal of Rf, the 3-[^{13}C]methyl signal of MRf shows no pH-dependent chemical shift change either in the free (open circles) or RBP-bound (solid circles) forms. It is suggested from these results that the alkaline shift of pK for N(3)-H of Rf bound to RBP is due to the hydrophobic environment around the N(3) region of Rf bound to RBP.

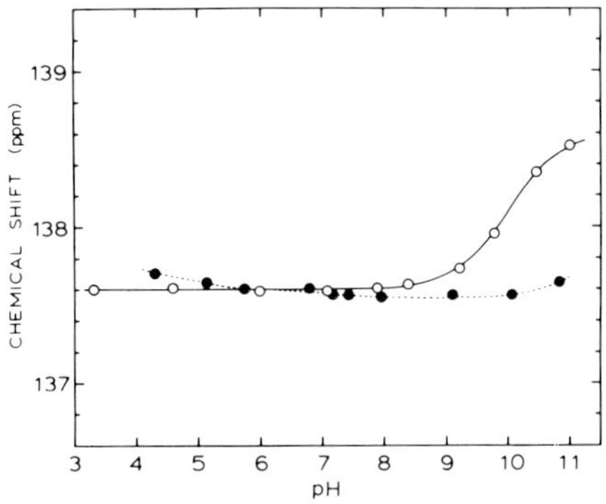

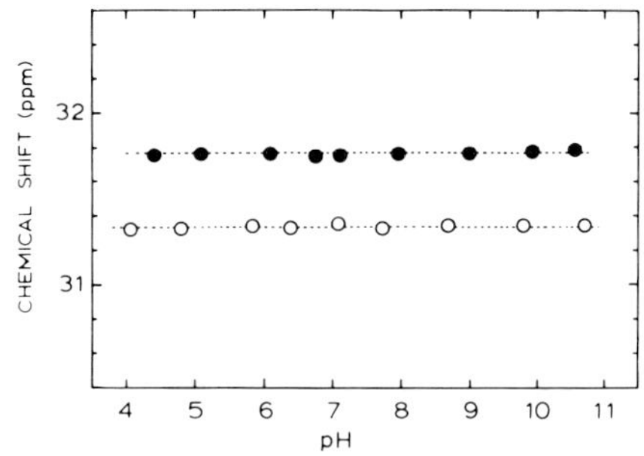

Fig. 4 (left). The pH-dependence of 4a-^{13}C chemical shift of Rf free (open circles) and bound to RBP (solid circles). The solid line is the theoretical curve with a pK of 10.00.
Fig. 5 (right). The effect of pH on the N(3)-^{13}C chemical shift of MRf free (open circles) and bound to RBP (solid circles).

For further analysis of the NMR results, we measured the association constant (Ka) for Rf-RBP or MRf-RBP complexes by fluorometric titration at various pH's following the procedure reported previously (2, 3). Figure 6 illustrates the results. The profile for Rf-RBP (Fig. 6A) is essentially identical to that reported by Becvar and Palmer (4) and exhibits participation of two ionizable groups in the binding. The non-participation of the group with a basic pK in the binding of MRf-RBP clearly demonstrates that N(3)-H of Rf and an acidic group in RBP play important roles in the binding of Rf to RBP. Thus the results in Fig. 6 (solid circles) were analyzed by non-linear least squares fitting according to Schemes I and II. The solid line in Fig. 6A is the theoretical curve with the best fit values of 4.89, 9.95, 2.31 x 10^7, 1.42 x 10^9, 9.01 x 10^6 and 12.15 for pK_1, pK_2, Ka_3, Ka_4, Ka_5 and pK_3 (Scheme I), respectively. The solid line in Fig. 6B is the theoretical curve with 5.04 and 1.01×10^{9} for pK_1 and Ka_2 (Scheme II), respectively.

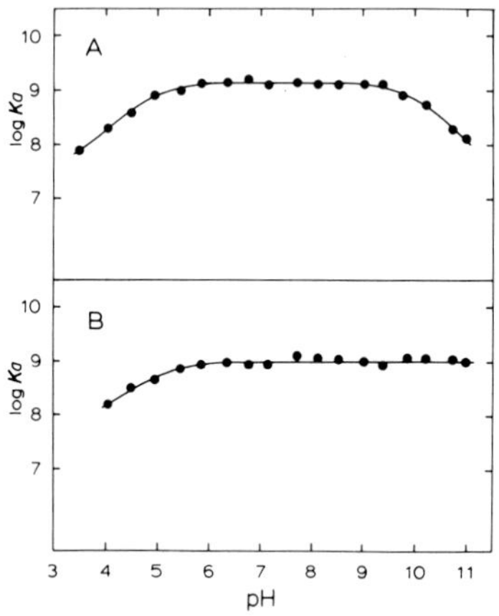

Fig. 6. The pH-dependence of the association constant (Ka) for Rf-RBP (A) and MRf-RBP (B). Solid lines in A and B are the theoretical curves with the values given in the text.

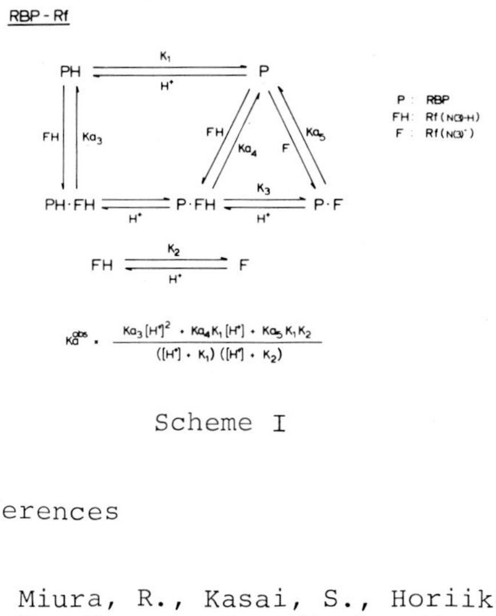

Scheme I

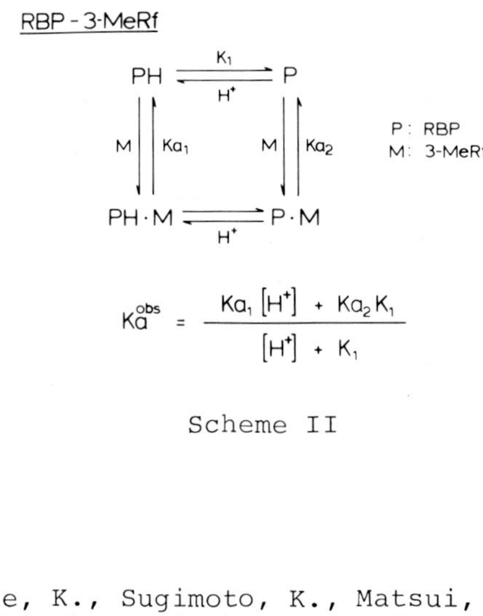

Scheme II

References

1. Miura, R., Kasai, S., Horiike, K., Sugimoto, K., Matsui, K., Yamano, T., Miyake, M.: Biochem. Biophys. Res. Commun. 110, 406-411 (1983).

2. Edmondson, D. E., Tollin, G.: Biochemistry 10, 124-132 (1971).

3. Nishikimi, M., Kyogoku, Y.: J. Biochem. 73, 1233-1242 (1973).

4. Becvar, J., Palmer, G.: J. Biol. Chem. 257, 5607-5617 (1982).

THE STRUCTURE OF GLYCOLATE OXIDASE FROM SPINACH

Ylva Lindqvist, Carl-Ivar Brändén

Departement of Chemistry and Molecular Biology
Swedish University of Agricultural Sciences
S- 750 07 Uppsala, Sweden

Summary

A preliminary high resolution structure determination of glycolate oxidase from spinach is reported. X-ray data were collected on films at the synchrotron radiation source in Daresbury. The structure was solved using two heavy atom derivatives and a solvent flattening procedure developed by Bi-Cheng Wang. The subunit structure is essentially a one-domain structure of the eight stranded α/β - barrel type first described for triose phosphate isomerase. In addition there are three extra large loop regions at both ends of the barrel. The coenzyme FMN and a substrate analogue, thioglycolate, are both bound to this barrel structure at the carboxyl ends of the parallell strands. The molecule is octameric with 422 symmetry having a hole in the middle of about 15 - 20 Å diameter.

Introduction

Glycolate oxidase is a peroxisomal enzyme that catalyzes the oxidation of α-hydroxyacids using FMN as cofactor. The main sources for studies on this enzyme have been mammalian livers and some plant species. The human liver enzyme is a tetramer built up of identical subunits of molecular weight 43.000 D (1). It is believed to be involved in the metabolic production

Flavins and Flavoproteins

of oxalate by the oxidation of glycolate through glyoxylate (2). A number of inhibitor studies of the liver enzyme has been made (1,3,4) since inhibition of this enzyme might be useful for treatment of hyperoxaluria, a genetic disease in which overproduction of oxalate results in large deposits of calcium oxalate.

Pure plant enzyme has been prepared from spinach (5,6),pea (7) and cucumber (8). The enzyme from cucumber has a subunit molecular weight of 43.000 D which aggregates to two predominant species in solution of molecular weights 180.000 and 700.000 D (8). Ultracentrifugation studies on the spinach enzyme (5) also led to the conclusion that there is an equilibrium between different states of aggregation;one of molecular weight 140.000 and another twice as large. In green plants glycolate oxidase is one of the key enzymes in photorespiration,where it oxidizes glycolate to glyoxylate. Due to this pathway for carbon metabolism, which is initially catalyzed by the oxygenase reaction of ribulose-1,5-bisphosphate carboxylase/oxygenase,Rubisco, net photosynthesis is drastically reduced (9).Detailed knowledge at the molecular level of the enzymatic processes involved in photorespiration is required both at the gene (10) and enzyme level in order to be able to work out procedures for an effective control of photorespiration _in vivo_ (11).

We have thus undertaken X-ray crystallographic studies to determine the three-dimensional structures of the two key enzymes in this pathway, Rubisco (12,13) and glycolate oxidase. The primary sequence of glycolate oxidase is currently being determined in collaboration with H. Jörnvall at Karolinska Institutet,Stockholm. Two crystal forms,A and B, suitable for X-ray investigation have been obtained as previously described (6). Form A crystals are easily formed, contain FMN and diffract to high resolution. These crystals which have been used in the studies reported here are tetragonal with unit cell dimensions of a = b = 148.1 Å and c =

134.9 Å. The spacegroup is I422 with one subunit in the asymmetric unit. Crystals containing a substrate analogue, thioglycolate, were obtained by cocrystallization (14).

Low resolution X-ray studies

We initially solved the structure of glycolate oxidase to a resolution of 5.5 Å (14). In addition we located the substrate binding site at this resolution. Intensity data to 5.5 Å resolution, comprising 2670 independent reflexions, were measured on a diffractometer at 4°C for the native and thioglycolate crystals. For the two heavy atom derivatives, $K_2Hg(CN)_4$ and $Pt(NH_3)_2Cl_2$, data collection included Friedel pairs in order to use the anomalous scattering effect of the heavy atoms in the phase determination. These data were processed and the heavy atom parameters were located and refined in the usual way. The electron density map showed very clear boundaries between protein and solvent.
From this map it was apparent that the glycolate oxidase molecules in our crystals are octamers with a diameter of approximately 100 Å and with 422 symmetry in the subunit arrangement. We could also deduce from this map that the subunit molecular weight is around 40.000 D. There are close contacts between the subunits related by the fourfold symmetry whereas there are fewer interactions between the two tetramers that form the octamer.
Usually a low resolution study does not yield any information about the fold of the polypeptide chain. However, we noticed a fairly large number of features in the map which could be interpreted as helical segments. We then searched for known domain structures that would fit this arrangement of helices. During recent years evidence has accumulated that long polypeptide chains generally fold into several discrete domains and that there seems to be a limited number of different folding patterns for such domains (15). Several of these

patterns have a characteristic arrangement of helices such as the four helices in the dinucleotide binding domains (16) or the eight helices of the triose phosphate isomerase barrel structure (17). We could show that eight of the helical segments in our low resolution map of glycolate oxidase were arranged in the same characteristic barrel shape and dimensions as those of the triose phosphate isomerase structure. Furthermore we showed that the substrate binding site was at one end of this barrel similar to the substrate binding site in triose phosphate isomerase. The eight helices and strands of this barrel structure comprise about two thirds of the glycolate oxidase subunit structure. We could not, however, from the low resolution map accomodate the remaining density in terms of any known domain structure nor could we locate the position of the FMN molecule.

High resolution X-ray studies

All high resolution X-ray data were collected with film methods. Native data to 2.2 Å resolution were obtained using synchrotron radiation in Daresbury. The data were collected as 1°-oscillations with the fourfold axis along the spindle. All films were measured on an Optronics drumscanner and processed using a program originally written by M. Rossmann (18) and modified by S.J.Remington. A total of 65 native film packs have been measured which comprises 100.000 reflexions 35.000 of which are independent. This corresponds to 75% of all reflexions to 2.2 Å resolution. The R- value for the merged dataset is 10% on intensities. A dataset to 2.7 Å resolution for the platinum derivative was also collected at the synchrotron in Daresbury. These crystals are very radiation sensitive and it would have been quite difficult to collect this dataset with a conventional X-ray source. 62% of the data giving a merged R-value of 11% was obtained.

Finally data for the mercury derivative were collected to 2.5 Å resolution on a rotating anode in our laboratory. This data-set comprises 75% of the total and the merged R-value is 9%. Reflexions to 5.5 Å resolution were extracted from these three film data sets and scaled to the corresponding diffractometer data. We obtained an R-value of 8% on F for all three data sets showing that there are no substantial systematic errors in the data which are due to the method of data collection. The heavy atom parameters were refined using a program originally written by M. Rossmann and modified by L.F.Ten Eyck and S.J.Remington. All the data including the low resolution diffractometer data were used in the refinement. The figure of merit is 0.40 and the Cullis R-values (19) are 0.54 for the mercury and 0.73 for the platinum derivative.

The electron density map obtained has the quality that can be expected from a two-derivative map without anomalous information. It was possible to build in the β - strands and some of the helices of the α/β-barrel but several helices were poorly defined. For density not belonging to the barrel it was not possible to trace the main chain. Since our platinum derivative did not contribute much to the phasing at higher resolution we had in actual fact essentially SIR phases. We therefore tried the ISIR-ISAS program package written by Bi-Cheng Wang (Acta Cryst A, in press). This program defines a protein-solvent boundary from the map, flattens the solvent and after backtransformation of the filtered map combines the phases using Sim weighting. This is done in several iterative cycles.

The result on our data was striking and gave a map where the helices of the barrel were all clearly defined. Furthermore the FMN molecule could be easily identified from the density corresponding to the flat ring and the heavy phosphate group. Pieces of polypeptide chain could be traced through most of the remaining density, which is mainly situated at the top and bottom of the barrel. A tentative model of the main chain atoms has been built on a Vector General graphics display using the FRODO program package (20,21). This model currently comprises 323 residues. In the absence of an aminoacid sequence there are still some uncertainties about connections in a few of the loop regions as well as the end regions. There is no doubt about the positions of the eight strands and helices of the barrel and the FMN molecule nor of the positions of the remaining helices. However, the order of these elements of secondary structure along the polypeptide chain given in Fig 1 should be regarded as tentative and could change. It is given here more for convenience of description than as a final result.

Description and discussion of the structure

Fig 1 gives a schematic diagram of our tentative model of the subunit structure. The main features of this model are the eight helices and strands of the barrel structure. The remining residues have been built into electron density which is mainly present at the top and bottom of the barrel. They have been assigned in our model to three regions; the first thirty residues at the N-terminus, an extra loop region of about thirty residues between strands six and seven and a loop region between strands two and three comprising about forty residues in addition to the corresponding barrel helix. The first two regions are essentially at that end of the barrel where the carboxyl ends of the strands are. The last region is at the other end of the barrel.
In order to obtain a quantitative measure of the similarities between the barrel structure of triose phosphate isomerase and that in our model we superimposed the $C\alpha$ -atoms of the two structures using the programme HOMO written by M.Rossmann. We used 40 atoms from the eight β -strands for the initial alignement. Since the α/β -barrel has pseudo 8-fold symmetry equivalences in structure can be produced by 45° rotations

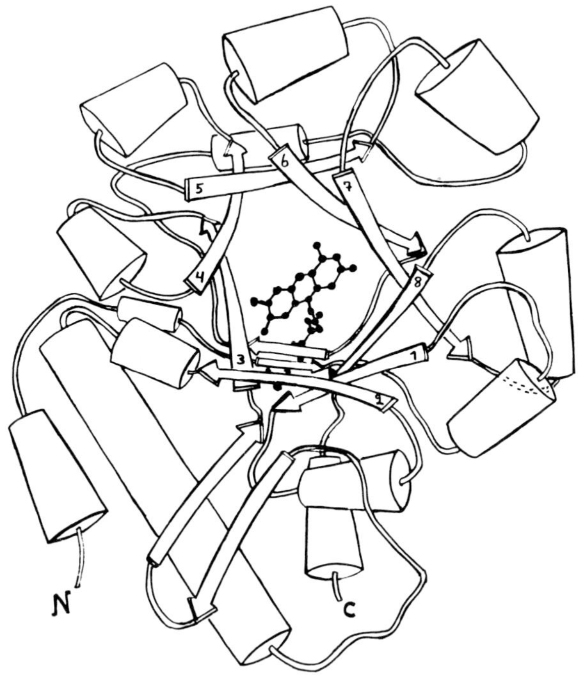

Figure 1. A schematic diagram of the subunit structure of glycolate oxidase from spinach. The structure is viewed down the axis of the barrel with FMN bound at the bottom of the barrel.

about the 8-fold axis. We chose one of these possible alignements which superimposes our long helix between strands 2 and 3 with the long helix F1 comprising residues 177 - 196 in triose phosphate isomerase. After refinement of the rotation parameters and maximizing the number of overlaps we obtained 140 consequtive equivalences of $C\alpha$ -atoms with a root mean square difference of 2.8 Å. This value compares well with that obtained for similar studies of structural similarities between on the one hand triose phosphate isomerase and on the other hand domain A of pyruvate kinase (160 equivalences with

an r.m.s. of 3.0 Å) or phosphogluconate aldolase (150 equivalences with an r.m.s. of 3.0 Å) (22,23).
An α/β-barrel structure has now been found in a number of different enzymes, triose phosphate isomerase (17), pyruvate kinase (22), phosphogluconate aldolase (23), glycolate oxidase, α-amylase (24) and xylose-isomerase (25). With the possible exception of α-amylase these enzymes all fall into a functional pattern which Rose first pointed out (26). An essential part of the mechanism of these enzymes is the abstraction of a proton next to a carbonyl group, immobilazation of the intermediate in a highly planar form and a similarity in the position of the functioning base for those cases where this is known. Muirhead (27) has discussed the requirements of the glycolate oxidase structure to fit into this pattern. We hope that this will be clarified from high resolution studies of substrate analogues in combination with the amino acid sequence.
The FMN molecule is bound to the barrel structure (Fig 2) at

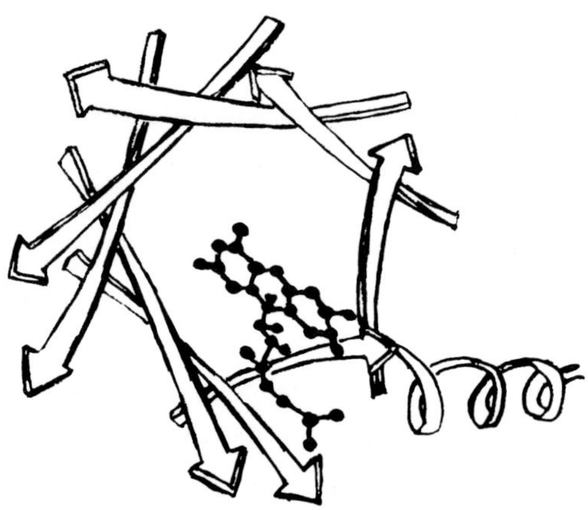

Figure 2. A schematic diagram of bound FMN in relation to the strands of the barrel viewed from the carboxyl ends of the strands.

the carboxyl ends of the strands. The barrel axis passes through the isoalloxazine ring with the ring plane almost parallell to this axis. The N5 edge of the ring is most exposed whereas the ribityl side chain is more buried in the barrel. The phosphate group is located outside the amino end of the short helix immediately after strand number two. Part of the isoalloxazine ring and the ribityl side chain are shielded from solution by the extra loop regions at this end of the barrel.

The broad peak in the low resolution difference Fourier map that corresponds to the position of bound thioglycolate is positioned very close to the N5 edge of the flavin ring. There is thus no doubt that this is the active site region of the subunit.

There are several surprising aspects of the subunit structure of glycolate oxidase. Firstly it is comprised essentially of one domain which binds both FMN and the substrate. This is in contrast to other enzyme classes such as NAD^+ dependent dehydrogenases where coenzyme and substrate are bound by separate domains (16). Secondly this domain structure belongs to a different structural class than the other flavin-coenzyme binding domains. In flavodoxin (28) , glutathione reductase (29) and p-hydroxybenzoatehydroxylase (30) the flavin coenzyme is bound to an α/β- domain structure which is built from a parallell pleated sheet arranged in a twisted plane with helices on both sides. The phosphate groups of the coenzyme bind in a crevice created by a change of the strand order (31). In contrast the α/β-domain structure that binds FMN in glycolate oxidase is arranged in a closed barrel form where no change of strand order occurs.

A view of the Cα -atoms of the octameric molecule projected down the 4-fold axis is shown in Fig 3. The close contacts between those subunits that are related by the fourfold axis arise from contacts between the extra loop regions. The large hole in the middle of the molecule is about 15 -

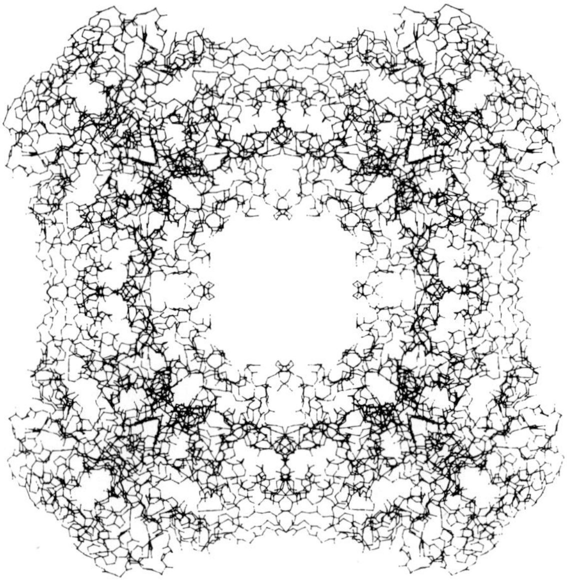

Figure 3. A diagram of the main chain atoms of the octameric glycolate oxidase molecule viewed along the 4-fold symmetry axis.

20 Å in diameter. This hole arises as a natural consequence of packing roughly spherical large objects around a 4-fold symmetry axis. To our knowledge this is the first structure determination of an octameric protein molecule with strict 422 symmetry.

Acknowledgements

We wish to thank the staff at the synchrotron radiation source in Daresbury and in particular John Helliwell for generous help in using the equipment. We also wish to thank M.Rossmann,J.Remington, A.Jones and Bi-Cheng Wang for use of their computer programs. It is a pleasure to acknowledge the technical assistance of Mrs I. Kurland. This work was supported by a grant from the Swedish Natural Science Research Council (grant no E-EB 2767).

References

1. Schwam,H.,Michelson,S., Randall,W.C., Sondey,J.M., Hirschman,R.: Biochemistry 18, 2828-2833 (1979).
2. Liao,L.L., Richardson,K.E.: Arch. Biochem. Biophys. 154, 68-75 (1973).
3. Schuman,M., Massey,V.: Biochim. Biophys. Acta 227, 521-537 (1971).
4. Randall,W.C., Streeter,K.B., Cresson,E.L., Schwam,H., Michelson,S.R., Anderson,P.S., Cragoe,E.J.Jr., Williams, H.W.R., Eichler,E., Rooney,C.S.: J. Med. Chem. 22, 608-614 (1979).
5. Frigerio,N.A., Harbury,H.A.: J. Biol. Chem. 231, 135-155 (1958).
6. Lindqvist,Y., Brändén,C.-I.: J. Biol. Chem. 254, 7403-7404 (1979).
7. Kerr,M.W., Groves,D.: Phytochemistry 14, 359-362 (1975)
8. Behrends,W., Rausch,U., Loeffler,H.-G., Kindl,H.: Planta 156, 566-571 (1982).
9. Miziorko,H.M., Lorimer,G.H.: Annu Revs. Biochem. 52, 507-536 (1983).
10. Somerville,C.R., Somerville,S.C.: Mol. Gen. Genet. 193, 214-219 (1984).
11. Somerville,C.R., Ogren,W.L.: Trends in Biochem. Sci. 7, 171-174 (1982).
12. Andersson,I., Brändén,C.-I.: J. Mol. Biol. 172, 363-366 (1984).
13. Schneider,G., Brändén,C.-I., Lorimer,G.H.: J. Mol. Biol. 175, 99-102 (1984).
14. Lindqvist,Y.,Brändén,C.-I.: J. Mol. Biol. 143, 201-211 (1980).
15. Richardson,J.S.: Adv. Protein Chem. 34, 167-339 (1981).
16. Rossmann,M.G., Liljas,A., Brändén,C.-I., Banaszak,L.J.: in The Enzymes (Boyer,P.D., ed.) 3rd edit. Vol 11, 61-102, Academic Press, New York and London (1975).
17. Phillips,D.C., Sternberg,M.J.E., Thornton,J.M., Wilson,I.A.

: J. Mol. Biol. 119, 329-351 (1978).

18. Rossmann,M.G.: J. Appl. Crystallogr. 12, 225-238 (1979).

19. Cullis,A.F., Muirhead,H., Perutz,M.F., Rossmann,M.G., North,A.C.T.: Proc. Roy. Soc. London A265, 15-31 (1961).

20. Jones,T.A.: J. Appl. Crystallogr. 11, 268-272 (1978).

21. Jones,T.A.: in Computational Crystallography (Sayre,D.,ed) pp 303-317, Oxford University Press (1982).

22. Stuart,D.I., Levine,M., Muirhead,H., Stammers,D.K.: J. Mol. Biol. 134, 109-142 (1979).

23. Lebioda,L., Hatada,M.H., Tulinsky,A., Mavridis,I.M.: J. Mol. Biol. 162, 445-458 (1982).

24. Matsuura,Y., Kusunoki,M., Harada,W., Tanaka,N., Iga,Y., Yasuoka,N., Toda,H., Narita,K., Kakudo,M.: J. Biochem. Japan 87, 1555-1558 (1980).

25. Carrell,H.L., Rubin,B.H.,Hurley,T.J., Glusker,J.P.: J. Biol. Chem. 259, 3230-3236 (1984).

26. Rose,I.A.: Phil. Trans. Roy. Soc. Lond. B293, 131-143 (1981).

27. Muirhead,H.: Trends in Biochem.Sci. 8, 326-330 (1983).

28. Burnett,R.M., Darling,G.D., Kendall,D.S., LeQuesne,M.E., Mayhew,S.G., Smith,W.W., Ludwig,M.L.: J. Biol. Chem. 249, 4383-4392 (1974).

29. Schulz,G.E., Schirmer,R.H., Saschenheimer,W., Pai,E.F.: Nature 273, 120-124 (1978).

30. Wierenga,R.K., DeJong,R.J., Kalk,K.H., Hol,W.G.J., Drenth, J.: J. Mol. Biol. 131, 55-73 (1979).

31. Brändén,C.-I.: Quart.Revs. Biophys. 13, 317-338 (1980).

ENDOR STUDIES OF FLAVOPROTEINS

Norbert Bretz, Harry Kurreck
Institut für Organische Chemie, Freie Universität Berlin
Takustr. 3, 1000 Berlin 33, West Germany

Electron nuclear double resonance (ENDOR) spectroscopy is a powerful tool in the investigation of paramagnetic organic and bioorganic molecules. Whereas application of EPR spetroscopy is often hampered by the limited resolution of this technique ENDOR spectra show a much better resolution even of very large molecules with low symmetry. In the field of flavins and flavoproteins Ehrenberg, Hyde et al. were the first to study paramagnetic derivatives of, e.g., NADPH dehydrogenase and flavodoxins by means of ENDOR in disordered solids. Recently we have shown that modern ENDOR equipment (especially signal to noise improvement and resolution enhancement) allows a more detailed study of the magnetic properties of flavoproteins and moreover detection of flavoprotein/substrate interactions. It could be demonstrated that the rigid matrix ENDOR spectrum of methanol oxidase is significantly affected if hydroxylamine is added acting as a kind of "pseudosubstrate".

In this paper we present new results of our ENDOR studies on D-amino acid oxidase and glucose oxidase. Both enzymes are known to occur in different radical states of the prosthetic group, i.e., neutral or anionic, depending on the pH of the solution. Thus, by changing the pH of the buffer (glucose oxidase) or by adding p-fluorobenzoate (D-amino acid oxidase) pronounced changes of the ENDOR spectra are observed. Whereas one radical species remains unaffected by the modified conditions, in addition a different radical is apparently generated.

Flavins and Flavoproteins

Since these enzymes contain more than one prosthetic group per molecule it may be assumed that the latter occur in different enviroments exhibiting different chemical and magnetic properties. Obviously only part of the flavins within the protein are accessible to the substrate molecules. The enzyme/substrate interactions clearly cause redistributions of the unpaired spin density within the isoalloxazine moiety which is reflected in the ENODR spectra.

The authors wish to thank Professor Dr. F. Müller, University of Wageningen, for collaboration.

References

1. Kurreck, H., Bock, M., Bretz, N., Elsner, M., Kraus, H., Lubitz, W., Müller, F., Geissler, J., Kroneck, P.M.H.: J. Am. Chem. Soc. 106, 737 (1984).

2. Kurreck, H., Kirste, B., Lubitz, W.: Angew. Chem. Int. Ed. Engl. 23, 173 (1984).

3. Massey, V., Palmer, G., Williams, C.H., Swoboda, B.E.P., Sands, R.H.: Flavins and Flavoproteins 1966, 133.

4. Yagi, K., Takai,A., Ohishi, N.: Biochim. Biophys. Acta 289, 37 (1972).

RESONANCE RAMAN STUDY ON THE COMPLEXES OF D-AMINO ACID OXIDASE

Yasuzo Nishina, Kiyoshi Shiga,* Retsu Miura,** Hiromasa Tojo,*** Yoshihiro Miyake,** Toshio Yamano,*** and Hiroshi Watari

National Institute for Physiological Sciences, Okazaki, Aichi 444, *Department of Physiology, Kumamoto University Medical School, Kumamoto 860, **Department of Biochemistry, National Cardiovascular Center Research Institute, Suita, Osaka 565, and ***Department of Biochemistry, Osaka University Medical School, Kita-ku, Osaka 530, Japan

Introduction

We have previously measured the resonance Raman (RR) spectra of the purple intermediate of D-amino acid oxidase (DAO) with D-alanine or D-proline, and concluded that the purple intermediate consists of the reduced enzyme and substrate intermediate, an imino acid (1-3). We have also postulated that the plane defined by C(10a)=C(4a)-C(4)=O of the reduced flavin and one containing $N=C-CO_2^-$ of the substrate intermediate are arranged in close contact to each other. We have further postulated that the imino acid should have a protonated imino group and the carboxylate of the imino acid in contact with the flavin moiety interacts with the enzyme through proton(s) (3). In the present study we observed the RR spectra of some complexes between DAO and ligands to clarify the details of their structures.

Results and Discussion

Resonance Raman spectra of the purple intermediate of D-amino acid oxidase with N-methyl-D-alanine are shown in Fig. 1. Judging from the assignment of the RR bands of the purple

Flavins and Flavoproteins

intermediate with D-alanine or D-proline (1), we can assign the bands at 1671 and 1338 cm^{-1} to those associated with C=N stretching and COO^- symmetric stretching, respectively, of an imino acid N-methyl-α-iminopropionate. The band at 1606 cm^{-1} is associated with C(10a)=C(4a)-C(4)=O moiety of reduced flavin (2). The downshift of the 1671 cm^{-1} band in D_2O indicates that the imino acid should have a protonated imino group, as indicated in the intermediate with D-proline. The

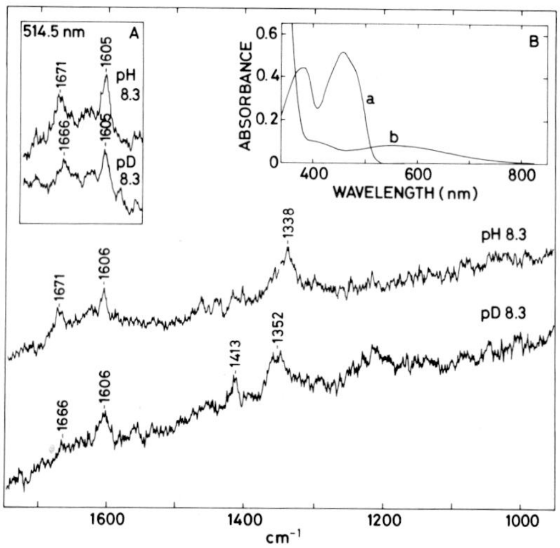

Fig. 1 Resonance Raman spectra of the purple intermediate of D-amino acid oxidase with N-methyl-D-alanine, excited at 632.8 nm. The inset (A) shows the RR spectra excited at 514.5 nm and the other inset (B) shows the visible absorption spectra in the oxidized state (a) and the purple intermediate (b).

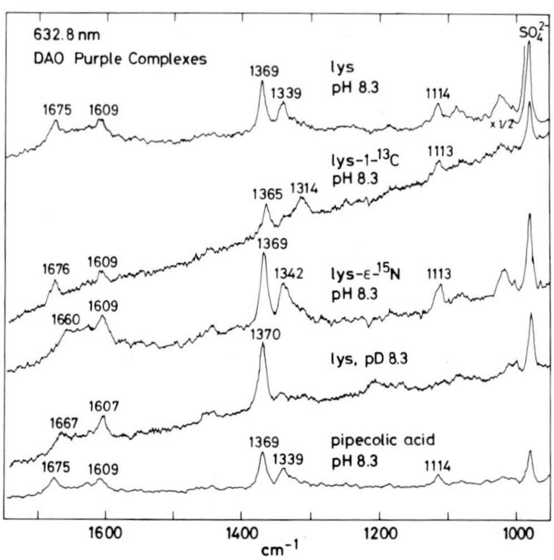

Fig. 2 Resonance Raman spectra excited at 632.8 nm of the purple complex of D-amino acid oxidase with D-lysine or pipecolate.

large downshift (29 cm^{-1}) of 1692 cm^{-1} band in D_2O for the purple intermediate with alanine was previously reported (3), and can be explained by the exchange of the two iminium hydrogen atoms by deuteriums. That the band at 1338 cm^{-1} shifts in D_2O suggests that the carboxylate group of the imino acid interacts with the enzyme through some proton(s) as indicated in the case of proline (1).

Figure 2 shows the RR spectra of the purple complex with D-lysine or pipecolate. The RR spectra of the complex with D-lysine was essentially similar to that of the complex with pipecolate. The 1675 cm^{-1} band shifts to 1660 cm^{-1} upon ε-^{15}N substitution, but does not shift upon α-^{15}N substitution of lysine (not shown). Therefore, the ligand is not α-imino-ε-

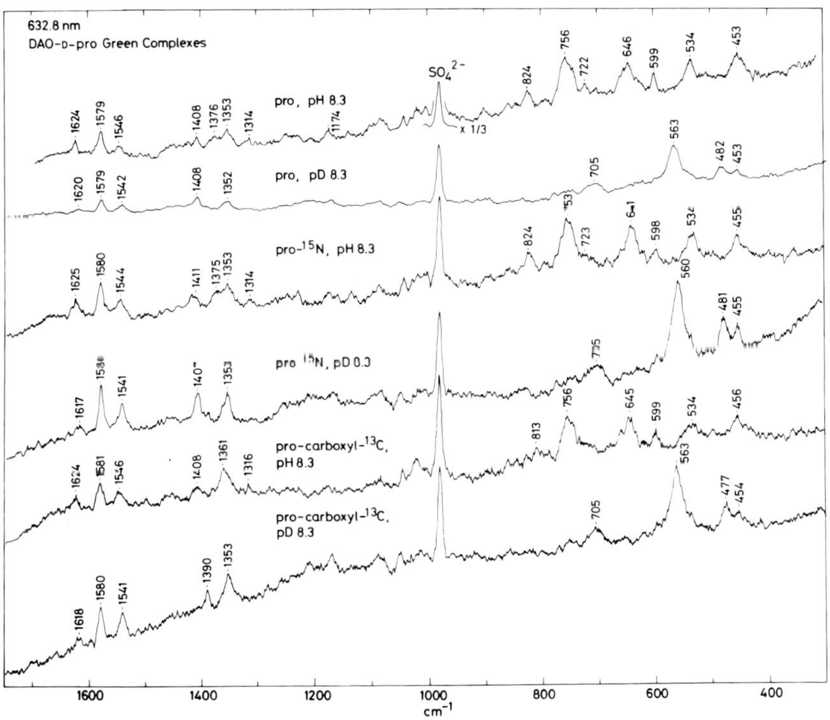

Fig. 3 Resonance Raman spectra excited at 632.8 nm of the green complexes of D-amino acid oxidase with a product of D-proline.

aminocaproate, but Δ^1-piperideine-2-carboxylate. The 1675 cm^{-1} band also shifts to 1667 cm^{-1} in D_2O. The 1369 and 1339 cm^{-1} bands shifted to 1365 and 1314 cm^{-1}, respectively, upon using lysine-1-^{13}C. Disappearance of the 1339 cm^{-1} band associated with COO^- group in D_2O also suggests that the carboxylate is not free, but interacts with the enzyme through proton(s) in the enzyme.

It is known that the oxidized DAO forms a green complex with the Δ^1-pyrrolidine-2-carboxylate derived from D-proline or Δ^1-piperideine-2-carboxylate derived from D-lysine (4,5). RR spectra of the green complexes of the enzyme with isotopically labeled substrate (proline (Fig. 3) and lysine) were measured. RR spectra were also obtained for the complexes of the enzyme reconstituted with isotopically labeled FAD's. From these data, we can assign the bands at 1579, 1546, and 1353 cm^{-1} are associated with oxidized flavin. The 1376 and 1388 cm^{-1} bands for proline and lysine, respectively, which are associated with COO^-, are affected in D_2O solution. The carboxylate of the ligand in the green complex also probably interacts with the enzyme through proton(s) of some amino acid residue.

References

1. Nishina, Y., Shiga, K., Watari, H., Miura, R., Tojo, H., Miyake, Y., Yamano, T.: Biochem. Biophys. Res. Comm. 106, 818-822 (1982).

2. Miura, R., Nishina, Y., Ohta, M., Tojo, H., Shiga, K., Watari, H., Yamano, T., Miyake, Y.: Biochem. Biophys. Res. Commun. 111, 588-594 (1983).

3. Nishina, Y., Shiga, K., Miura, R., Tojo, H., Ohta, M., Miyake, Y., Yamano, T., Watari, H.: J. Biochem. 94, 1979-1990 (1983).

4. Massey, V., Ganther, H.: Biochemistry, 4, 1161-1173 (1965).

5. Yagi, K., Okamura, K., Naoi, M., Takai, A., Kotaki, A.: J. Biochem. 66, 581-589 (1969).

THE ROLE OF ARGININES IN D-AMINO ACID OXIDASE

Mirella P. Simonetta, Stefania Galliani, Maria A. Vanoni, Severino Ronchi** and Bruno Curti

Department of General Physiology and Biochemistry
**Institute of Veterinary Physiology and Biochemistry
University of Milano - Milano, Italy

A variety of kinetic and structural studies has pointed out the involvement of arginine residues in the active site of D-amino acid oxidase (DAAO,EC 1.4.3.3) (1,2). Ferti *et al.* (3) showed that incubation of the enzyme with 1,2-cyclohexanedione (1,2-CHD) specifically modifies an active site arginine with a loss of activity, and that benzoate protects from inactivation. Evidence has been given (4) that this arginine is the postulated positively charged group in the active site of DAAO interacting with the FAD N(1)-C(2)=locus. The identification of the 1,2 CHD modified arginine in the primary structure of the protein was attempted: the results didn't allow the identification of this arginine due to the poor stability of the adduct and the low sensitivity of its detection method. Phenylglyoxal (PGO) was chosen as the alternative chemical modifier of arginine residues since it gives a less unstable derivative compared to 1,2-CHD adduct and it is readily available in radioactive form. This structural research requires previous kinetic analyses of the modification reaction.
Fig.1A shows the inactivation of apoprotein with various PGO concentrations under pseudo-first order conditions. The reaction is first order with respect to PGO concentration and the measured second order rate constant is 11.6 $M^{-1}min^{-1}$ (Fig.1B). Some variability in apoprotein reactivity was observed when different enzyme preparations were used. In some cases biphasic inactivation patterns were observed at low PGO concentrations.

Flavins and Flavoproteins

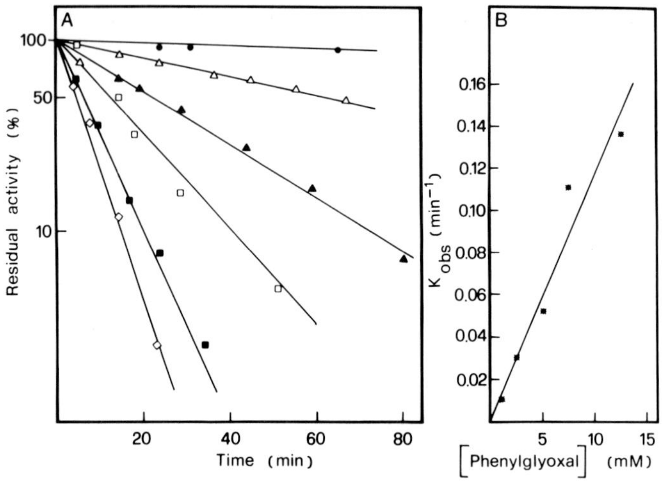

Fig.1. Inactivation of apoprotein by PGO. (A) 200 µg of enzyme were incubated at 25°C in the dark in 0.2 ml of a final volume mixture containing 150 mM potassium phosphate buffer pH 8.0 and PGO in different concentrations. The molar PGO:enzyme ratios were: (△) 40:1; (▲) 100:1; (□) 200:1; (■) 300:1; (◇) 500:1; (●) no PGO. (B) Plot of the apparent first order inactivation rate constant (K_{obs}) vs. PGO concentration.

Fig.2. Inactivation of holoenzyme by PGO. (A) Incubation conditions as for Fig.1. The molar PGO:enzyme ratios were: (○) 100:1; (△) 300:1; (▲) 400:1; (□) 450:1; (■) 500:1; (◇) 550:1; (●) no PGO. (B) Plot of the apparent first order inactivation rate constant (K_{obs}) vs. PGO concentration.

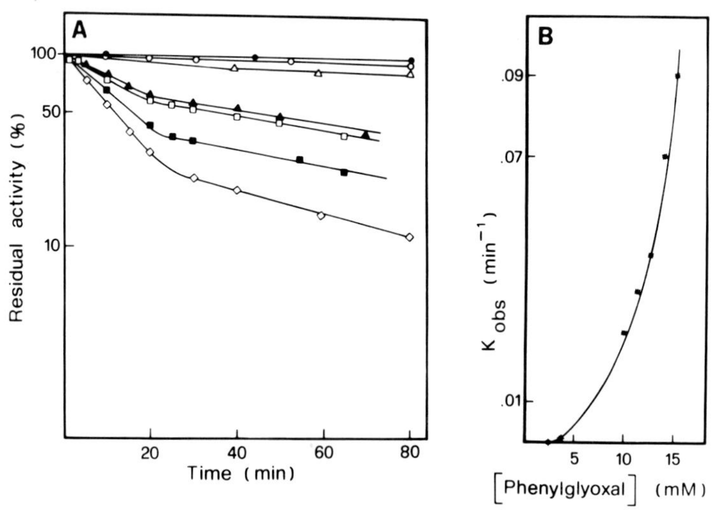

Reaction of holoenzyme with PGO follows a pseudo-first order kinetics and is biphasic under all conditions (Fig.2A). In comparison to apoprotein higher PGO concentrations were required to observe an appreciable inactivation rate. The plot of K_{obs} vs. PGO concentration is not linear and indicates that the inactivation is second order in PGO (Fig.2B). Benzoate protects from inhibition, while additional amounts of coenzyme are not effective. In the case of apoprotein the addition of FAD strongly protects against inactivation, giving a 10-fold decreased inactivation rate, and producing holoenzyme-like patterns; the presence of benzoate enhances to some extent this protection. This additional protective effect of benzoate might reflect a tighter enzyme-coenzyme binding, a conformational change or merely the fact that benzoate shields the arginine residue. Kinetic parameters of the modified apoprotein have been determined in comparison to the native enzyme: inactivation by PGO results in a proportional decrease in the turnover number with no indication of change in the K_m value. Thus the inactivation of apoenzyme by PGO is a 'all or none' process. Amino acid analysis and ^{14}C-PGO incorporation studies were performed at different times of apoenzyme inactivation.

Table I. Amino acid analysis and ^{14}C-PGO incorporation measurements on PGO inactivated D-amino acid oxidase. Incubation conditions as in Fig.1A. Molar PGO:enzyme ratio 100:1.

Reaction time	FAD	FAD+ benzoate	Residual activity	Amino acid analysis	Radioactivity
min			%	Δarg	mol PGO/ mol enzyme
0	-	-	100		
0.5	-	-	90	0.43	0.84
7.0	-	-	72	1.16	1.03
16.5	-	-	50	2.11	1.66
51.5	-	-	17	3.25	4.73
51.5	+	-	68	2.47	n.d.
51.5	-	+	76	1.41	2.65

Loss of activity is followed by the disappearance of arginine residues, with no other variation in amino acid content. The amino acid analysis data were analyzed according to Tsou (5); one out of a total of four arginines appears to be essential for catalytic activity. Addition of FAD and FAD+benzoate in the apoprotein-PGO incubation mixture decreases the loss of arginine of about 1 and 1.7 residues respectively: thus one arginine appears to be shielded by the coenzyme moiety. Radioactivity measurements well agree with amino acid analysis taking into account the incorporation of 1 mol PGO/mol Arg missing in the apoprotein and 2 mol PGO/mol Arg incorporated in the presence of FAD+benzoate. The different reaction stoichiometry in the apoenzyme and in the reconstituted holoenzyme can be viewed, in the holoenzyme, as a reflection of a different accessibility for a second molecule of PGO to condense with the 1:1 complex giving a 2:1 adduct. This aspect needs further investigations, since it was not available a high purity radioactive batch. Experiments designed to identify the modified essential residue within DAAO primary structure are being undertaken. The lability of the adduct under the usual conditions of chemical and enzymatic digestion implies the set up of new handling procedures. Preliminary results suggest that this essential arginine residue is present in the second half of the polypeptide chain.

Acknowledgments

This work was supported by grants from Consiglio Nazionale delle Ricerche and Ministero della Pubblica Istruzione.

References

1. Kotaki, A., Harada, M., Yagi, K.: J. Biochem. (Tokio) 64, 537 (1968).

2. Nishino, T., Massey, V., Williams, C. H.,Jr.: J. Biol. Chem. 255, 3610 (1980).

3. Ferti, C., Curti, B., Simonetta, M. P., Ronchi, S., Galliano, M., Minchiotti, L.: Eur. J. Biochem. 119, 553 (1981).

4. Fitzpatrick, P. F., Massey, V.: J. Biol. Chem. 258, 9700 (1983).

5. Tsou, C.-L.: Sci. Sin. 11, 1535 (1962)

6. Peters, R. G., Warren, C. J., Cromartie, T. H.: Biochemistry 20, 2564 (1981).

THE EFFECT OF THE METHYLATION OF HISTIDINE-217 IN PIG KIDNEY D-AMINO ACID OXIDASE ON LIGAND BINDING AND ON CATALYSIS.

Richard P. Swenson, Charles H. Williams, Jr. and Vincent Massey
Veterans Administration Medical Center and Dept. of Biological Chemistry, University of Michigan, Ann Arbor 48105, Michigan, U.S.A.

D-Amino acid oxidase (E.C.1.4.3.3) from pig kidney can be inactivated by covalent modification with the active site-directed reagent methyl-p-nitrobenzenesulfonate. Inactivation is concomitant with the methylation of the 1- and 3-nitrogen of histidine-217 in a mole ratio of 0.64:0.36 (1). Unlike several other covalently-modified derivatives of this enzyme, methylated enzyme is capable of binding small aromatic carboxylic acids. Titration with either benzoate or anthranilate causes spectral pertubations in the visible region similar to those observed for unmodified enzyme. However, Scatchard plots of data derived from the absorbance changes that occur at 495 nm or 560 nm during the titration of the methylated enzyme with benzoate or anthranilate, respectively, were nonlinear at all pH values tested. Representative Scatchard plots from the titration with benzoate at pH 5.3 and 8.4 are shown in Fig 1A. The data were adequately fit by nonlinear regression analyses to a model consisting of two independent binding populations. Within statistical error, the mole fraction of each population remained essentially constant over the pH range of 5.3 to 9.6. Average values for each mole fraction, as determined by this method, were 0.63±0.12 and 0.35±0.12 from the benzoate data, and 0.68±0.06 and 0.32±0.05 from the anthranilate data (pH 8.5 only). These values are quite similar to the mole fractions of the 1- and 3-methyl derivatives of histidine-217 observed during the characterization of the methylated enzyme (1), suggesting that the two methylated forms of histidine-217 are responsible for the nonhyperbolic binding results. As shown in Fig 1B, the apparent dissociation constants for the benzoate complex of the two binding populations remained essentially independent of pH between 5.3 and 8.4. This observation is in marked contrast to the situation for native enzyme in which complex stability is influenced by an ionizable group with an apparent pK_a of 6.3 (2). Because the methylation of histidine-217 is responsible for this difference, the results suggest

Flavins and Flavoproteins

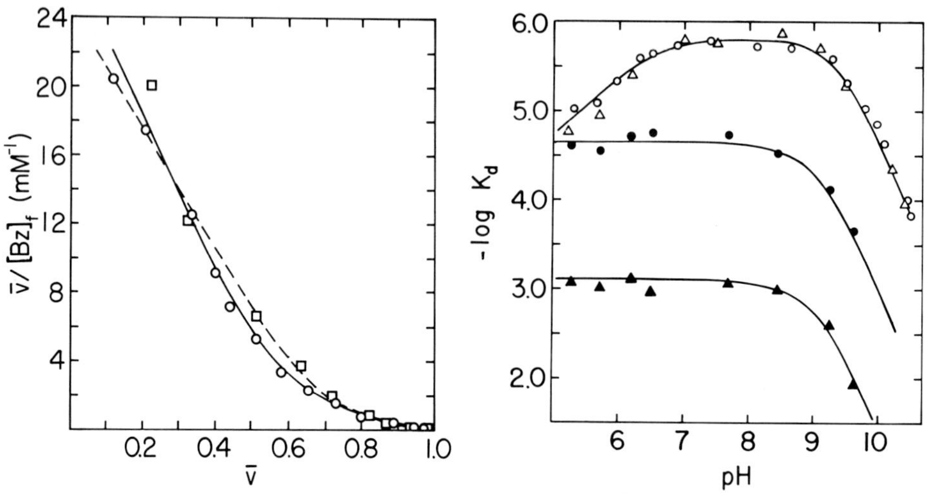

Figure 1. Panel A: Scatchard plots of data from the spectrophotometric titration of methylated D-amino acid oxidase with sodium benzoate at pH 8.4 (○) and 5.3 (□), 19°C. The lines are theoretical curves fit to the data by nonlinear, least square regression analyses using a model consisting of two independent binding populations. Panel B: The pH dependence of the apparent dissociation constant for the benzoate complex of native (○,△) and of the two apparent binding populations of methylated enzyme (n_1=0.63,●;n_2=0.35,▲) The data for native enzyme are replotted from Quay and Massey (2).

that it is the ionization state of this residue that affects complex stability in the acidic region.

D-Alanine was able to reduce the bound flavin in the methylated enzyme, but at a much slower rate than for native protein. The rate of reduction of the modified enzyme in the absence of oxygen, as determined by the appearance of the reduced enzyme-imino acid product charge transfer complex, was 20 s^{-1} (Fig 2), to be compared to 1170 s^{-1} determined for the native enzyme (3, 4). The decay of this complex was biphasic. Approximately 62% of the decrease in absorbance at 550 nm occured at a rate of 0.36 s^{-1}, the remainder at 0.015 s^{-1}. The latter rate is similar to that observed for native enzyme (3). A 2-fold deuterium isotope effect on the steady state turnover number for the modified enzyme was observed when the α-hydrogen of D-alanine was substituted, suggesting that C-H bond cleavage during flavin reduction is at least partially rate limiting in catalysis by the methylated enzyme. Product release (k_{off} = 10 s^{-1}) limits turnover in unmodified enzyme (3). No significant differences in

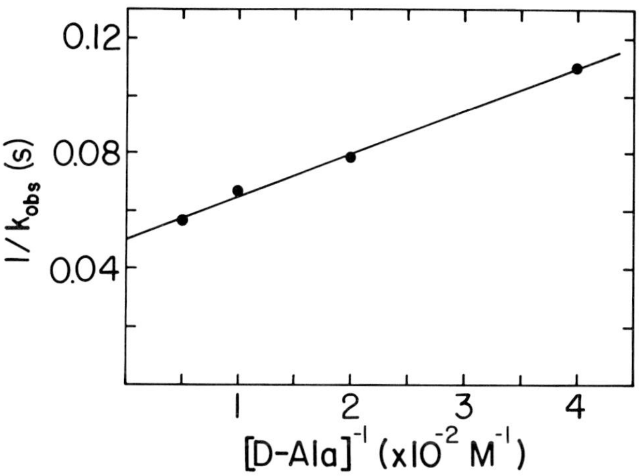

Figure 2. Double reciprocal plot of the reductive half-reaction data for methylated D-amino acid oxidase using D-alanine as substrate at pH 8.5, 25°C. Rates were determined from the exponential increase in the absorbance at 550 nm upon mixing anaerobic solutions of substrate and methylated enzyme in a stopped flow spectrophotometer.

the oxygen half reaction were indicated by the steady state turnover data for nondeuterated substrate.

The above results indicate that some aspect of the electron transfer reaction is impeded by the methylation of histidine-217. To test the hypothesis that methylation may interfere with the proton transfer step associated with the transient formation of the α-carbanion of the substrate during the reduction of the flavin, studies were conducted using the substrate analogue nitroethane. The preformed carbanion of this nitroalkane is stabilized in solution by charge delocalization and is capable of reducing native D-amino acid oxidase <u>via</u> an N^5-flavin adduct (5). This reaction can proceed without C-H bond cleavage or the initial proton transfer event. When nitroethane anion was mixed anaerobically with the methylated enzyme, reduction of the flavin was observed; however, the kinetics are somewhat more complex than for the native protein. In addition to a rapid (<3 ms) decrease in the absorbance at 450 nm associated with the binding of the reagent (5), <u>two</u> nitroethane anion concentration-dependent phases were observed at this wavelength. Within these two phases, approximately 40% of the absorbance decrease proceeded at nearly the same rate as determined for native protein (2.2±1.6 <u>vs</u> 6.2 s^{-1} (5)), whereas the remainder occurred at approximately one-tenth this rate. Although it cannot be stated with certainty, the biphasic reduction of methylated enzyme may be due to the heterogeneity of the methylation of histidine-217. If so, then these results suggest that a major fraction of the enzyme is capable of as nearly an efficient electron transfer from preformed carbanion as unmodified protein. Indeed, the steady state turnover of this compound by methylated enzyme was observed to occur at approximately the same rate (by extrapolation to infinite nitroethane anion concentration) as for unmodified protein (0.7 ± 0.5 vs $1.1 \pm 0.1\ s^{-1}$, in air-saturated buffer, pH 8.5, 25°C), despite the fact that flavin reduction is partially rate limiting in turnover (5). The nitroethane results for at least one species of modified protein seem to be in contrast to those obtained with D-alanine for which flavin reduction was hindered. Although indirect effects cannot be entirely excluded as the cause of these observations, it is, none-the-less, tempting to conclude that histidine-217 is involved in the catalytic steps leading to formation of reduced enzyme, perhaps functioning as the proton acceptor in the formation of the substrate carbanion during the electron transfer step.

Acknowledgement

This work has been supported by the Medical Research Service of the Veterans Administration and by Grants GM21444 (C.H.W.) and GM11106 (V.M.) from the National Institute of General Medical Sciences.

References

1. Swenson, R. P., Williams, C. H., Jr., and Massey, V.: J. Biol. Chem. 259, 5585-5590 (1984).

2. Quay, S., and Massey, V.: Biochemistry 16, 3348-3354 (1977).

3. Porter, D. J. T., Voet, J. G., and Bright, H. J.: J. Biol. Chem. 252, 4464-4473 (1977).

4. Fitzpatrick, P. F., and Massey, V.: J. Biol. Chem. 257, 12916-12923 (1982).

5. Porter, D. J. T., Voet, J. G., and Bright, H. J.: J. Biol. Chem. 248, 4400-4416 (1973).

STOICHIOMETRY OF THE SELF-ASSOCIATION OF D-AMINO ACID OXIDASE

Hiromasa Tojo, Kihachiro Horiike*, Kiyoshi Shiga**, Yasuzo Nishina**, Hiroshi Watari**, and Toshio Yamano

Department of Biochemistry, Osaka Unversity Medical School, Osaka 530, Japan, *Department of Biochemistry, Shiga University of Medical Science, Shiga, Japan, and **National Institute for Physiological Sciences, Okazaki, Japan

Introduction

Hog kidney D-amino acid oxidase undergoes a self-association in aqueous solutions (1, and references cited therein). The association system has hardly been quantitatively characterized (2,3). There has been also the discrepancy between the stoichiometries evaluated by gel filtration studies (monomer-dimer) and by light scattering studies (equilibrium involving the higher oligomers) (2-5). Therefore, we performed gel filtration with a short column to clarify the apparent discrepancy of the stoichiometry. Furthermore, we have analyzed the apparent weight-average molecular weight (M_w^a) vs. protein concentration (c) data from a light scattering study (3) in terms of the quantitative models including the effect of solution nonideality by non-linear least-squares technique.

Results and Discussion

Gel chromatographic evidence for the self-association past the dimer. The apparent Stokes radius (R_e) tended to increase continuously with increase in the peak concentration of the eluted holoenzyme and did not level off over the concentration range examined (Fig. 1). We observed the apparent R_e value of

Flavins and Flavoproteins

61.5 Å at the largest loading mass tested. On the other hand, decrease of the loading concentration or the sample volume applied resulted in decrease of the apparent R_e values to that of the monomeric unit. The R_e value extrapolated to zero enzyme concentration was estimated to be 23.5-25.5 Å in accordance with a previous work (25.1 Å) (6). Since the linear arrangement of the subunits has the largest frictional coefficient, we can evaluate approximately an upper limit of the R_e value for the each n-mer. To evaluate qualitatively the lower limit of the extent of polymerization, we therefore calculated the R_e values of the oligomers with the linear arrangement (based on the R_e value of the monomer of 25.5 Å) from the frictional coefficients calculated by the Kirkwood theory (7). Although the Kirkwood theory generally underestimates the frictional coefficients, the difference between the values calculated by the Kirkwood theory and by the more rigorous theory is at most 7% for proteins (7). The largest value among the apparent R_e observed would correspond to the calculated value (62.5 Å) for the hexamer with linear subunit arrangement. It is evident from these considerations that the higher polymers greater than the dimer participate in the self-associating system contrary to the previous gel chromatographic studies (e.g., 4).

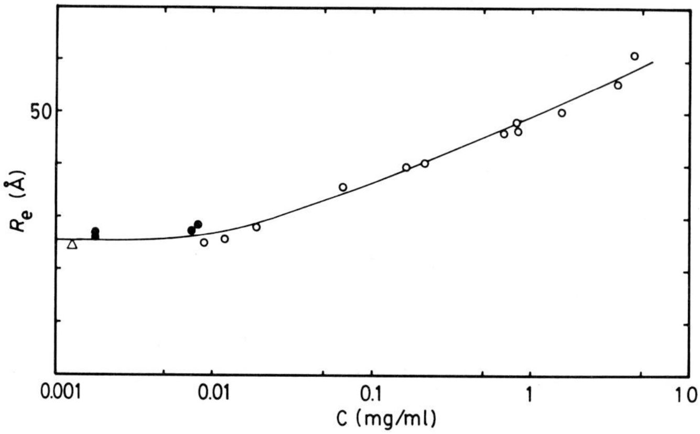

Fig. 1. Dependence of apparent Stokes radii on the peak concentration, c, of the D-amino acid oxidase eluted from a 75 mm-long column of TSK-GEL G3000 SW in 0.1 M sodium pyrophosphate, pH 8.3. ○, with FAD (31.2 μM); ●, without FAD; △, containing 0.3 M NaCl without FAD.

Analysis of light scattering data in terms of the models. The models tested included those for indefinite self-association of the specified oligomer (ID model) and the attenuated equilibrium constant model of indefinite self-association of the monomer (AK model). In the ID model a specified m-mer, M_m, as a basic unit associates sequentially with an equal molar association constant, $\underline{K}_m^{ID}$, for each step, and the formation of the basic unit from the monomer, M_1, is governed by the molar association constant ($\underline{K}_m$): $mM_1 \rightleftarrows M_m$, $2M_m \rightleftarrows M_{2m}$,---, $M_{(i-1)m} + M_m \rightleftarrows M_{im}$,---. In the AK model the entropy change for each step, $M_{i-1} + M_1 \rightleftarrows M_i$ (i=2,3,---), is assumed to decrease successively. This leads to the sequential attenuation of the isodesmic association constant, that is, $K_i = K_1^{ID}/i$ (8). The M_w^a vs. c data obtained from a light scattering study for the holoenzyme were directly fitted to the equation: $M_w^a = M_w/(1+2BcM_w)$, where M_w is the weight-average molecular weight, B is the colligative virial coefficient, and c is the enzyme concentration. M_w and c are the functions of the association constants and the monomer concentration, and are dependent on the models. The closed solution for each model is readily derived. The previously-reported value of the monomer molecular weight was used (9). The criteria for the goodness of fit as reported (10) were adopted. For ID models the increase of the degree of polymerization of a basic unit (m≥4) lead to a poorer fit. The AK models examined also did not fit the exerimental data. The minimal value of the residual sum of squares and of the variance were obtained for the model ID with m=1 and $K_2 \neq K_1^{ID}$ (model ID-1). Both ID models with m=2 and m=3 (model ID-2 and ID-3, respectively) fitted the data to a comparable extent to the case of model ID-1. The best-fit data for the three models are given in Table I. Any one of the models ID-1, ID-2, and ID-3 cannot be excluded on the basis of the criteria of the goodness of fit employed in this study.

The sedimentation boundary profiles observed were compared with those calculated by the method of Gilbert with the better models. Model ID-2 appeared to be favored because there is a

Table I. Best-fit parameters for the successful models.

Models	K_2x10^{-5}	K_3x10^{-10}	K_m^{ID}x10^{-4}	Bx10^8	SSx10^{-8}	RMSx10^{-4}
	(M^{-1})	(M^{-2})	(M^{-1})	($\frac{mol \cdot l}{g^2}$)		
ID-1 (m=1, $K_2 \neq K_1^{ID}$)	2.42 (0.63)	-	13.0 (0.14)	-3.94 (0.39)	25.1	0.964
ID-2 (m=2)	7.28 (2.66)	-	4.34 (0.27)	-3.05 (0.36)	28.9	1.11
ID-3 (m=3)	-	4.05 (1.13)	2.46 (0.18)	-3.86 (0.44)	27.9	1.07

SS: residual sum of squares; RMS: root mean square of variance.
The values in parentheses are the standard deviations of the parameters.

distinct shoulder in the trailing boundary in agreement with the observed pattern. However, since the method ignores the effect of diffusion, it should be borne in mind that there is a possibility that the characteristic boundary profiles predicted by the method are somewhat obscured by diffusional spreading of the boundary.

References

1. Tojo, H., Horiike, K., Shiga, K., Nishina, Y., Watari, H., Yamano, T.: FEBS Lett. 114, 4-6 (1980).
2. Henn, S.C., Ackers, G.K.: Biochemistry 8, 3829-3838 (1969).
3. Tojo, H., Horiike, K., Shiga, K., Nishina, Y., Miura, R., Watari, H., Yamano, T.: Flavins and Flavoproteins (Massey, V., Williams, C.H.,Jr., eds.), pp. 175-178, Elsevier North-Holland, New York 1982
4. Fonda, M.L., Anderson, B.M.: J. Biol. Chem. 243, 5635-5643 (1968).
5. Antonini, E., Brunori, M., Bruzzesi, M.R., Chiancone, E., Massey, V.: J. Biol. Chem. 241, 2358-2366 (1966).
6. Henn, S.C., Ackers, G.K.: J. Biol. Chem. 244, 465-470 (1969).
7. Teller, D.C., Swanson, E., de Haen, C.: Methods Enzymol. 61, 103-124 (1979).
8. Garland, F., Christian, S.D.: J. Phys. Chem. 79, 1247-1252 (1975).
9. Tojo, H., Horiike, K., Shiga, K., Nishina, Y., Miura, R., Watari, H., Yamano, T.: J. Biochem. 92, 1741-1752 (1982).
10. Mannervik, B.: Methods Enzymol. 87, 370-390 (1982).

^{31}P NMR AND CHEMICAL STUDIES ON THE PHOSPHORUS RESIDUES IN MILK XANTHINE OXIDASE

Dale E. Edmondson and Michael D. Davis
Department of Biochemistry, Emory University, Atlanta, Georgia 30322 U.S.A.

Franz Müller
Department of Biochemistry, Agricultural University
Wageningen, The Netherlands

Introduction

As an extension of previous studies (1-3) on the use of ^{31}P NMR in the study of flavoenzymes, we report here ^{31}P NMR experiments on xanthine oxidase. In addition to the molybdenum and two Fe_2-S_2 redox centers in the enzyme per catalytic center, the enzyme contains FAD, a monophosphorylated molybdenum cofactor (MoCo) (4,5), and a covalently-bound phosphoseryl residue (this work). Thus, the use of ^{31}P NMR spectroscopy provides a probe to study the environments of the respective phosphorylated residues which could lead to new insights into the structure and function of this complex metalloflavoenzyme. The data presented here shows that the phosphoseryl residue is proximal to the Mo center and suggests that it might function as the previously postulated (6) but structurally unknown nucleophile in catalysis.

Results and Discussion

Milk xanthine oxidase was purified from fresh, raw buttermilk as previously described (7). Enzyme preparations routinely showed a 50-60% functionality, exhibited A_{280}/A_{450} spectral

Flavins and Flavoproteins

ratios of 5.2-5.5, and the expected stoichiometries of Mo, FAD, and Fe/S residues. Analysis for phosphorus was done using Bartlett's procedure (8). ^{31}P NMR spectra were obtained at 81 MHz using an IBM/Bruker WP-200SY spectrometer using quadrature phase detection at ambient temperature in the absence of broad band proton decoupling. All chemical shift values are relative to an external standard of 85% H_3PO_4.

Chemical analysis showed the presence of three moles of dissociable phosphorus per mole of flavin or Mo which are due to the FAD residue (7) and to the MoCo (4,5). One mole of phosphorus remained with the protein after repeated washing of the precipitate with 5% tricholoroacetic acid. Analysis for the phosphoamino acid after acid hydrolysis of the phosphopeptide isolated from proteolytic digests by thin-layer electrophoresis demonstrated the presence of phosphoserine. pH-Dependent chemical shift data of the phosphopeptide and of the isolated MoCo demonstrated each to contain monophosphoesters (data not shown).

As suggested by the analytical data, four resonances are observed in the ^{31}P NMR spectrum of the native enzyme of 50-60% functionality (Fig. 1A). The upfield resonances occurring at +8.8 and +13.5 ppm are assigned to the pyrophosphate moiety of FAD by analogy to previous data with glucose oxidase (2) and from the observation that these resonances are missing in the spectrum of the deflavoenzyme (data not shown). The two remaining resonances at $\tilde{-}$ 3 ppm and at $\tilde{}$+ 1 ppm are due to the phosphorus groups on the MoCo and the phosphoseryl moieties. The relative intensity of the -3 ppm resonance is observed to increase (relative to the other resonances) on desulfuration of the Mo center (Fig 1A-C). The line width ($\tilde{}$10Hz) of the -3 ppm resonance suggests it to exhibit a greater mobility (about 20-fold) than expected if it were to assume a rotation coincident with the rotational correlation time of the enzyme.

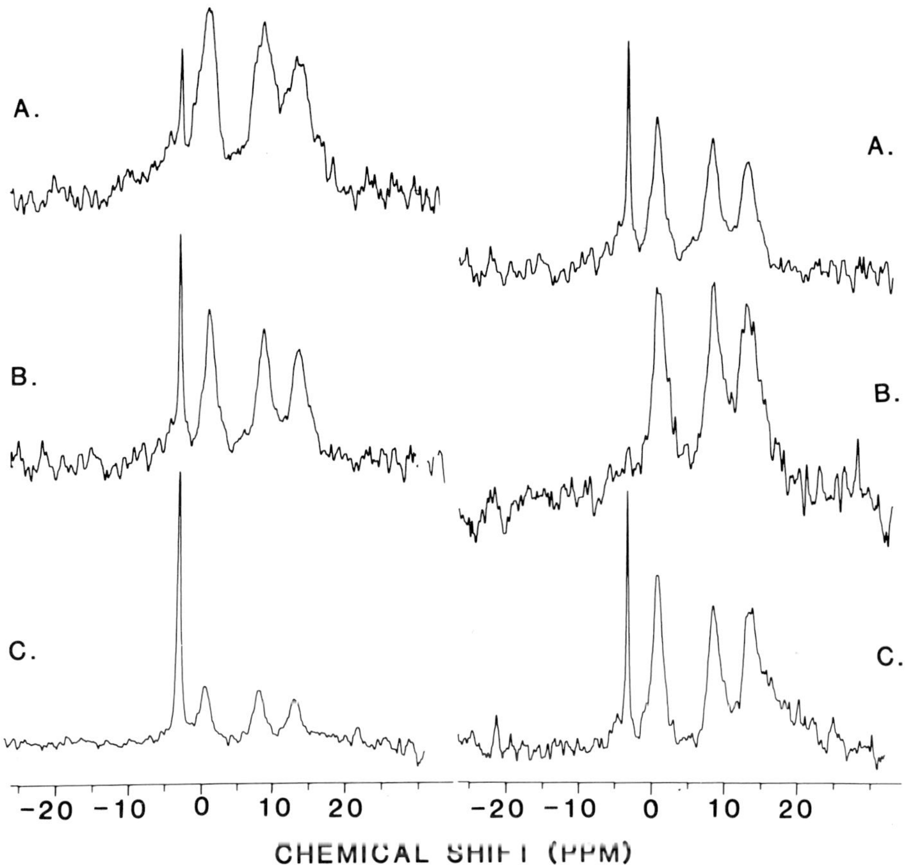

Figure Legends

Fig. 1. (Left) Effect of active site functionality (the cyanolyzable sulfur at the Mo center) on the ^{31}P NMR spectrum of xanthine oxidase. Enzyme samples were dissolved in 50 mM Tris-acetate, pH 8.0. An exponential broadening equivalent to 25Hz was used prior to Fourier transformation. A. 60% functional enzyme; 1.2 mM E.FAD; 96,600 acquisitions; B. 20% functional enzyme; 0.65 mM E.FAD; 67,400 acquisitions; C. desulfo xanthine oxidase; 0.8 mM E.FAD; 77,500 acquisitions.

Fig. 2. (Right) Effect of Mn(II) on the ^{31}P NMR spectrum of 20% functional xanthine oxidase (0.65 mM E.FAD) A. Before the addition of Mn(II); B. In the presence of 65 μM (Mn(II); C. After removal of Mn(II) by Chelex chromatography. The buffer and spectral acquisition parameters are the same as in Fig. 1.

The FAD resonances and the ~+1 ppm resonance exhibit linewidths expected for the molecular size of xanthine oxidase (150-175Hz). These data suggest the sharp resonance to be due to the phosphoseryl residue and the ~+1 ppm peak due to the MoCo since the Mo center remains firmly attached to the enzyme on desulfuration and, one could argue, that the MoCo also would.

Further evidence for this suggestion is apparent in experiments where Mn(II) is added to the enzyme. Any phosphorus residues exposed to the solvated Mn(II) ion would undergo extensive line-broadening on interaction with the paramagnetic ion. Previous work on Azotobacter flavodoxin (9) and on glucose oxidase (2) has shown the protein-bound phosphate resonances are susceptible to broadening in the presence of Mn(II) while the flavin coenzyme phosphates are not. The data in Fig. 2 show that only the ~- 3 ppm peak disappears on the addition of Mn(II) while the other resonances are unaffected. Furthermore, this effect is fully reversible on removal of the Mn(II) ion from solution by Chelex treatment (Fig. 2C). Similar results were obtained when the same experiment was performed with the desulfoenzyme. Thus, as observed with other flavoenzymes (2,9) the FAD pyrophosphate moiety is "buried" in the protein matrix as is the MoCo phosphate while the mobile phosphoseryl moiety is exposed to solvent. Efforts to selectively remove the phosphoseryl phosphate by alkaline phosphatase or phosphoprotein phosphatase treatment have, to date, been unsuccessful. This suggests the phosphoseryl moiety to be exposed to solvent yet be in an environment not accessible to the active sites of these phosphatases.

The mobile phosphoseryl resonance at -3 ppm appears to be due to only desulfo enzyme in the samples (Fig. 1). Presumably, it would be absent in the ^{31}P NMR spectrum of fully functional enzyme in which the cyanolyzable sulfur ligand to the Mo were present in all of the Mo centers. A careful examination of the spectrum in Fig. 1A shows that a broad resonance centered at +5 ppm is present in the 50-60% functional enzyme, but absent in that of the desulfo enzyme (Fig. 1C). Thus, the phosphoseryl resonance in the functional enzyme appears to be in a strongly immobilized form with a drastically altered environment relative to that of the desulfo form.

To further probe of the proximities of the various phospho groups to the Mo center, the desulfo enzyme was converted to the stable glycol inhibited Mo(V) form described by Lowe _et al._ (10). In this form, the Fe/S centers and the FAD center are oxidized while the Mo is in a stable MO(V) paramagnetic form with an electron spin relaxation time sufficiently long (~10^{-7}sec) to result in extensive line broadening of any resonances due to phosphorus nuclei within a 10Å radius. The spectrum in Fig. 3 (which should be compared to that of Fig. 1C) shows that the FAD resonances are unaffected by the presence of Mo(V) while the MoCo and the mobile phosphoseryl resonances are extensively broadened. The residual peaks observed are probably due to the incomplete formation of Mo(V) which is estimated to be 50% of the total Mo in the sample by double integration of the ESR signal. These data show the phosphoseryl and the MoCo moieties to be in close proximity to the Mo center while the FAD pyrophosphate moiety is more remote.

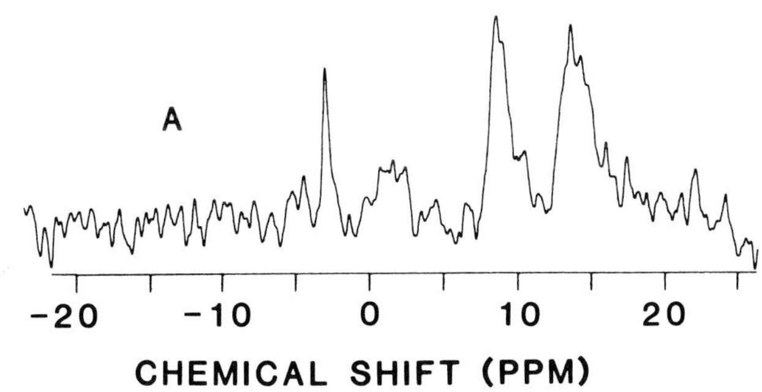

Fig. 3. ^{31}P NMR spectrum of desulfo xanthine oxidase (0.8 mM E.FAD) after treatment with dithionite and ethylene glycol to form the "inhibited' Mo(V) species (10) and removal of these reagents by gel filtration. The Mo(V) content was determined to be 50% of the total Mo by double integration of the ESR signal using a Cu-EDTA standard. The same buffer and spectra acquisition parameters as in Figs. 1 and 2 were used.

The proximity of the phosphoseryl residue to the Mo center of xanthine oxidase opens up a new line of investigation to probe whether it has any functional role in the catalytic mechanism of the enzyme. Phosphate is known to bind to the Mo center of sulfite oxidase in an inhibitory manner (10) so there is precedent to suspect the phosphoryl group on the phosphoseryl moiety to interact with the Mo center on xanthine oxidase. Indeed, it is tempting to speculate that the phosphoseryl residue could be the unknown nucleophilic group "X" suggested by Bray and colleagues (6) in their proposed mechanism for xanthine oxidase-catalyzed hydroxylation in which solvent oxygen is incorporated into the product. If the phosphoseryl residue were to function in this capacity, one might expect to observe solvent oxygen incorporation into the phosphate group during catalytic turnover. An NMR experiment was designed to test this possibility. Xanthine oxidase (2 moles E. FAD) 60% functionality) was dissolved in 40% enriched ^{17}O H_2O and allowed to undergo 10 turnovers with xanthine as substrate under aerobic conditions. A control sample with ^{16}O H_2O was treated identically. Any incorporation of ^{17}O (I=5/2) would be expected to extensively broaden the ^{31}P resonance resulting in a reduction of peak intensity in line with the number of ^{17}O nuclei incorporated. Any non-catalytic solvent oxygen exchange into the serine phosphate would be expected to be nil under the conditions of the experiment (12). The ^{31}P NMR spectrum was determined for each sample after removal of the products and solvent and were superimposable as judged from the manual overlay of the two spectra and from the computer subtraction of one spectrum from the other. This behavior would be in line with our observation (Fig. 1) that the "sharp" phosphoseryl resonance at -3 ppm in the 60% functional enzyme is due to desulfo or inactive enzyme while the phosphoseryl resonance in the functional enzyme is broad and a further broadening by any ^{17}O incorporation would not be observable. If any ^{17}O were incorporated into the phosphoseryl group of the functional enzyme, it would only be observable by comparing the relative intensitities of the sharp -3 ppm resonances after desulfuration of the Mo center by CN^- treatment. The spectra in Fig. 4 on the samples after CN^- treatment shows

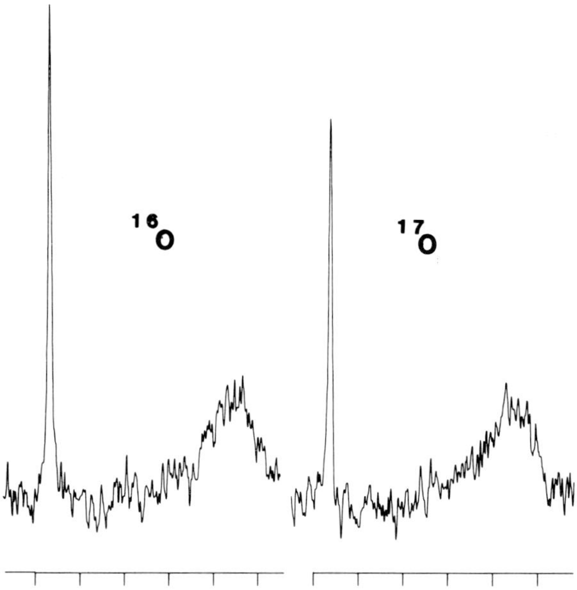

Fig. 4. A comparison of the ^{31}P NMR spectral intensities of xanthine oxidase (1 mM E.FAD) that had been incubated with 10-fold molar ratio of xanthine in ^{17}O and ^{16}O H_2O and subsequently converted to the desulfoenzyme by CN^- treatment. The sharp resonance is assigned to the phosphoseryl residue while the broad resonance shown is assigned to the MoCo. An exponential broadening equivalent to 1 Hz was used prior to Fourier transformation. Identical acquisition parameters and the number of acquisitions were used in obtaining each spectrum.

the intensity of the "mobile" phosphoseryl resonance in the sample incubated in ^{17}O H_2O to be reduced in intensity by ~25% relative to the ^{16}O control. This magnitude of decrease in intensity suggests the incorporation of a single ^{17}O nucleus into the functional phosphoseryl residue since the enzyme was of 60% functionality and the isotopic enrichment of the ^{17}O H_2O used was 40%.

The incorporation of more than one ^{17}O into the phosphoseryl group would result in a greater decrease in peak intensity.

Further work needs to be done to reconcile these findings with the effect of ^{17}O H_2O on the ESR spectral properties of Mo(V) by Bray and coworkers (13). Similarly, further work is required to elucidate the catalytic role of the phosphoseryl residue and its interaction with the Mo center. Of interest is the recent finding (14) that a group with an apparent pK_a=6.6 is observed in studies on the pH-dependence of the rate of reactivation of butyraldehyde-inhibited xanthine oxidase in which the aldehyde is associated with the stabilization of the Mo(V) inhibited form. This pK_a is within the range expected for a monophosphate ester such as phosphoserine.

The results presented here would appear to open up a new area of investigation regarding the role of the phosphoseryl residue in xanthine oxidase catalysis. Further efforts in this direction are currently underway in our laboratories.

This work was supported by grants from the NSF PCM 83-16263, the NIH (GM-29433), the Dutch National Science Foundation (ZWO/SON), and from NATO (049.82).

References

1. Edmondson, D.E., James, T.L.: Proc. Natl. Acad. Sci. USA 76, 3786-3789 (1979).
2. James, T.L., Edmondson, D.E., Husain, M.: Biochemistry 20, 617-621 (1981).
3. Moonen, C., Muller, F.: Biochemistry 21, 408-414 (1982).
4. Rajagopalan, K.V., Johnson, J.L, Hainline, B.E.: Fed. Proc. 41, 2608-2612 (1982).
5. Ishizuka, M., Ishio, K., Toraya, T., Fukui, S.: Biochem. Biophys. Res. Commun. 111, 537-543 (1983).
6. Bray, R.C., Gutteridge, S., Stotter, D.A., Tanner, S.J.: Biochem. J. 177, 357-360 (1979).

7. Massey, V., Brumby, P.E., Komai, H., Palmer, G.: J. Biol. Chem. 244, 1682-1691 (1969).

8. Bartlett, G.R.: J. Biol. Chem. 234, 466-468 (1959).

9. Edmondson, D.E., James, T.L.: Dev. in Biochm. 21, 111-118 (1982).

10. Lowe, D.J., Barber, M.J., Pawlik, R.T., Bray, R.C.: Biochem. J. 155, 81-85 (1976).

11. Gutteridge, S., Lamy, M.T., Bray, R.C.: Biochem. J. 191, 285-288 (1980).

12. Gerothanassis, I.P., Sheppard, N.: J. Mag. Res. 46, 423-439 (1982).

13. Bray, R.C., Gutteridge, S.: Biochemistry 21, 5992-5999 (1982).

14. Morpeth, F.F., Bray, R.C.: Biochemistry 23, 1332-1338 (1984).

SPECIFIC MODIFICATION OF NAD BINDING SITE OF CHICKEN LIVER XANTHINE DEHYDROGENASE WITH 5'-p-FLUOROSULFONYLBENZOYLADENOSINE.

Tomoko Nishino, Takeshi Nishino and Keizo Tsushima
Department of Biochemistry, Yokohama City University School of Medicine, Yokohama 232, Japan

Introduction

Conversion of dehydrogenase (D-form) to oxidase form (O-form) is known to occur in rat, but not in chicken liver xanthine dehydrogenase. To clarify the structural difference of active sites around FAD domain between rat and chicken enzymes, chemical modification by 5'-Fluorosulfonylbenzoyladenosine (5'-FSBA), a NAD analogue, has been carried out (1,2,3.). In this paper 5'-FSBA was found to be very specific modifier of NAD binding sites of both enzymes. Modified amino acid residue of chicken enzyme by 5'-FSBA was identified as tyrosine.

Materials and Methods

Xanthine dehydrogenase was purified from livers of chicken fed a high protein diet by the procedure of Nishino (4). 5'-FSBA, N-acetyltyrosine were obtained from Sigma, ^{14}C-FSBA from New England Nuclear and all other chemicals were of analytical grade. Cellulose TLC plate from Eastman. The enzyme was assayed by measuring changes of absorbance at 340 nm under standard conditions (4). Inactivation reaction was performed at pH 7.8, 25°C with various concentrations of FSBA. Enzyme concentration was 2.5×10^{-6}M. Experiments of ^{14}C-FSBA incorporation was performed at 0°C. After various incubation periods, residual enzyme activities were checked and reactions

Flavins and Flavoproteins

were terminated by addition of ethanol (final 50%) and precipitates were washed thoroughly and radioactivities were counted by sintillation counter.
Carboxybenzenesulfonyl-tyrosine (CBS-Tyr) was synthesized by the procedure of Colman et al. (2). FSBA inactivated enzyme was hydrolysed for 20 hrs with 6N HCl at 110°C. Hydrolysate and synthesized CBS-Tyr were subjected first to thin layer electrophoresis in the horizontal direction at pH 6.3 in pyridine- acetate buffer. After drying the plate it was subjected to chromatograpy in butanol: acetic acid: water/ 4:1:5 in vertical direction (2). Amino acids and CBS-Tyr on TLC plate were detected by using ninhydrin reaction.

Results and Discussions

Fig. 1 shows time course of inactivation of xanthine dehydrogenase with various concentrations of 5'-FSBA. The reaction displayed pseudo-first order kinetics. In inset of Fig. 1, the double reciplocal plots of inactivation velocity versus 5'-FSBA concentration was extrapolated to origin. The results indicate that enzyme and 5'-FSBA do not form a complex reversibly prior to the inactivation process. As shown in Fig. 2, 1 mol of ^{14}C-FSBA per mol of FAD was incorporated into chicken enzyme for complete inactivation. Inactivation of xanthine dehydrogenase was completely protected by NAD (data was not shown). Such a stoichiometry of ^{14}C-FSBA incorporation and a protection by NAD indicated that 5'FSBA reacted with amino acid residue of NAD binding site specifically. 5'-FSBA inactivated xanthine dehydrogenase of rat and chicken liver were apparently very slowly and slightly reactivated by DTT, because the inactivation of both enzymes was caused by the reaction with DTT. Fig. 3 showed the isolation of CBS- Tyr from hydrolysates of 5'-FSBA modified chicken enzyme. Identification of CBS-Try was made by comparison with authentic sample by the method of Colman et al.

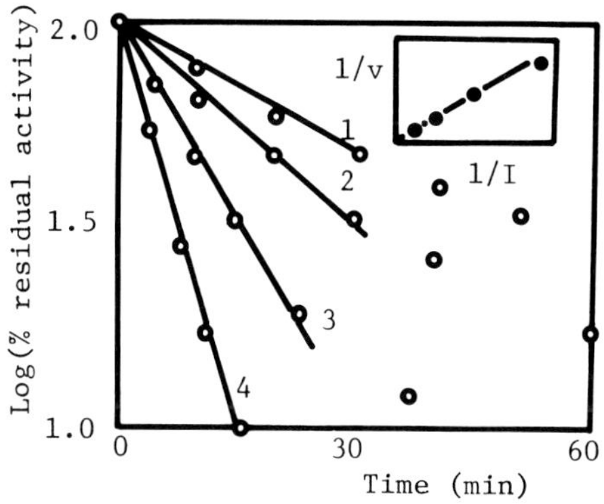

Fig. 1. Time course of inactivation of chicken xanthine dehydrogenase by 5'-FSBA.
5'-FSBA concentrations are, 1; 27.4μM, 2; 50.9μM, 3; 101.8μM, 4; 208.5μM, respectively.

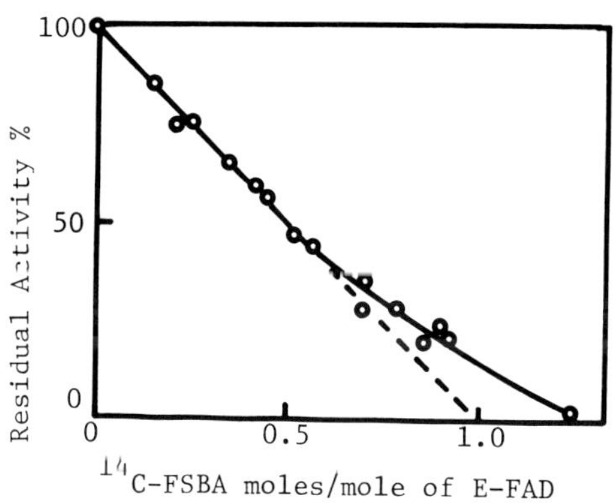

Fig. 2. Incorporation of ^{14}C-FSBA to chicken xanthine dehydrogenase. After reaction with ^{14}C-FSBA for various incubation periods at pH 7.8 at 0°C, residual activities were determined and quenched by ethanol, then radioactivities were counted.

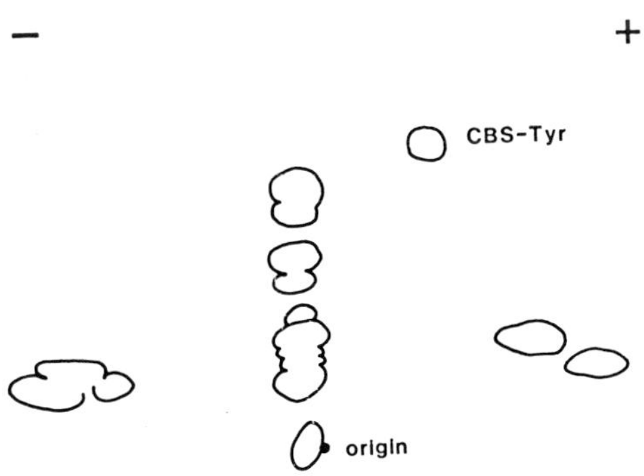

Fig. 3. Identification of modified amino acid. 5'-FSBA modified xanthine dehydrogenase was hydrolyzed and the sample was subjected first to thin layer electrophoresis in the horizontal direction and secondary to chromatography in vertical direction.

Inactivation of chicken and rat xanthine dehydrogenase by butanedione implicated active arginine residue in NAD binding site (unpublished data). Rat xanthine dehydrogenase inactivated by butanedione could be converted to O-form by SH-modifier, suggesting that NAD binding site is not concerned in D-O conversion. Furthermore, 5'-FSBA also inactivated rat liver xanthine dehydrogenase in a similar proportion to chicken enzyme. 5'-FSBA inactivated rat liver xanthine dehydrogenase could not be converted to O-form by SH-modifier treatment. In chicken enzyme, tyrosine was identified as an active amino acid residue at NAD binding site. These results also support that a different site from NAD binding site is concerned in O-D interconversion reaction.

Acknowledgement

This work was supported by a Grant-in-Aid for Scientific Research from the Japanese Ministry of Education, Science and Culture (No. 58480142)

References

1. Esch, S. F., Allison, S. W.: J. Biol. Chem. 253, 6100-6106 (1978)
2. Saradambal, V. K., Bednar, A. R., Colman, F. R.: J. Biol. Chem. 256, 11866-11872 (1981).
3. Colman, F. R.: Ann. Rev. Biochem. 52, 76-91 (1983).
4. Nishino T.: Biochem. Biophys. Acta 341, 93-98 (1974).

STUDIES BY ELECTRON-PARAMAGNETIC-RESONANCE SPECTROSCOPY OF THE ENVIRONMENT OF THE METAL IN THE MOLYBDENUM COFACTOR FROM XANTHINE OXIDASE

T.R. Hawkes and R.C. Bray
School of Chemistry and Molecular Sciences, University of Sussex, Brighton, BN1 9QJ, U.K.

The molybdenum cofactor (1) prepared (2) by denaturing xanthine oxidase by heat treatment or other methods, has been partially purified by anaerobic gel filtration in the presence of sodium dithionite, with little loss of activity. A range of products with different elution volumes was obtained. This behaviour is apparently related to association of the molybdenum cofactor with various residual peptides. E.p.r. signals from molybdenum(V) in the active cofactor, present either in crude preparations or in purified fractions, may be generated in dimethyl sulphoxide solution by controlled oxidation carried out on the molybdenum cofactor alone or in the presence of added thiols. The g-values of the spectra suggest that in the oxidized cofactor, molybdenum has one terminal oxygen ligand and four ligands from thiolate groups. It is proposed that two of these are from the organic part of the cofactor and two from cysteine residues in the protein or in residual peptides. An e.p.r. signal generated in high yield with little loss of cofactor activity in the presence of thiophenol has $g_{\parallel}$ 2.0258 and $g_{\perp}$ 1.9793. It is proposed that in this species two cysteine residues have been replaced by two thiophenol molecules. It is suggested that the thiophenol complex might be useful in further purification of the molybdenum cofactor.

Flavins and Flavoproteins

Full details of this work, which was supported by the Wellcome Trust and the Medical Research Council, will be published elsewhere (3).

References

1. Johnson, J.L., Hainline, B.E., Rajagopalan, K.V. and Arison, B.H.: J. Biol. Chem. 259, 5414-5422 (1984).

2. Hawkes, T.R. and Bray, R.C.: Biochem. J. 219, 418-493 (1984).

3. Hawkes, T.R. and Bray, R.C.: Biochem. J. (in press).

COUPLING BETWEEN Mo(V) AND REDUCED Fe/S 1 CENTRES IN ALDEHYDE OXIDASE AND XANTHINE OXIDASE

Graham N. George
School of Chemistry and Molecular Sciences, University of Sussex, Brighton, BN1 9QJ, U.K.

Introduction

The first spin-spin interaction discovered in a biological system was that between Mo(V) and the Fe/S 1 centres in xanthine oxidase (1). However, to date, only preliminary observations of an analogous interaction in the related molybdo/iron-sulphur/flavoprotein aldehyde oxidase have been reported (2).

In xanthine oxidase the fact that the two spins are oriented near the so called 'magic angle' (3) with respect to the Mo(V) $\underline{\underline{g}}$ tensor (4,5) (where the dipolar coupling is zero along the principal axes of $\underline{\underline{g}}$) has so far precluded any accurate estimation of distance from a simple measurement of the splitting at each of the three principal $\underline{\underline{g}}$ directions. However simulations of the full powder lineshape can distinguish distortions caused by dipolar splittings at orientations other than along the principal axes of $\underline{\underline{g}}$, thus providing an estimate of, or at least a lower limit for, the distance between the two spins. These distortions would be more evident at X band than at Q band microwave frequencies.

Materials & Methods

Samples of the desulpho forms of aldehyde oxidase and xanthine

Flavins and Flavoproteins

oxidase giving the Desulpho-Inhibited signal were made as in ref. 6. E.p.r. spectroscopy and computing was done as described in ref. 7. Simulations were calculated using a perturbation solution of the spin hamiltonian.

$$\mathcal{H} = \underline{H}\cdot\underline{\underline{g}}_1\cdot\underline{S}_1 + \underline{H}\cdot\underline{\underline{g}}_2\cdot\underline{S}_2 + \mathcal{H}_{dip} - J\ \underline{S}_1\cdot\underline{S}_2$$

where $\underline{\underline{g}}_1$ and $\underline{\underline{g}}_2$ are the $\underline{g}$-tensors for the two spins, $\mathcal{H}_{dip}$ is the dipolar interaction and J the exchange coupling. Best fit parameters were obtained by iterative minimisation of the error between experimental and calculated lineshapes.

Results & Discussion

The xanthine oxidase Mo(V) signal that has been best studied (5), with respect to its magnetic coupling to the Fe/S 1 site, is the signal called Desulpho Inhibited. This signal is unique in that it is stable to reducing or oxidising agents and thus can be studied under conditions where the Fe/S moiety is either oxidised or reduced. For this reason the analogous signal in aldehyde oxidase (2,6) was chosen for this study.

In samples where both Fe/S and molybdenum were reduced a splitting was observed of the Mo(V) signal and of the g_x-g_y region of the Fe/S 1 e.p.r. signal. However, when only one of the two sites was reduced no interaction was observed (Figs. 1 & 2). In addition, the temperature dependence of the split Mo(V) signal was fully consistent with an interaction only with Fe/S 1 (1). Similar results have been observed for xanthine oxidase (1,5), except that only a broadening of the Fe/S 1 signal is seen instead of a resolved splitting (5).

Simulations of both X and Q band data (using the parameters in Scheme 1) are shown in Figs. 1 & 2. Adequate simulations could be obtained without including noncolinearity between $\underline{\underline{g}}_1$

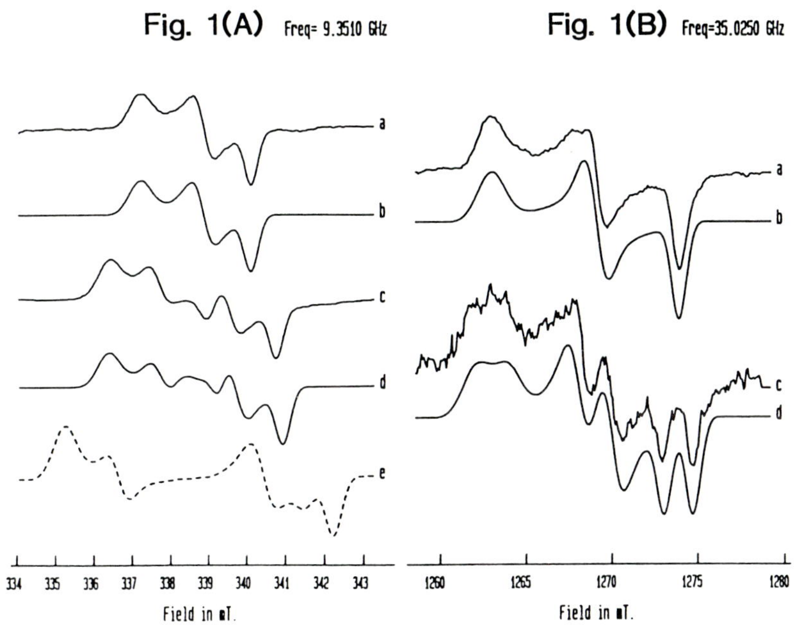

Fig. 1. Splitting of the Mo(V) e.p.r. signal of Aldehyde Oxidase by Fe/S 1.
(A) at X Band:
(a) shows the e.p.r. lineshape when no interaction is present (no Fe/S reduced), and (b) is a simulation of (a). (c) shows the splitting of the Mo(V) signal caused on reducing the Fe/S centres with dithionite, and (d) is the corresponding simulation (see Scheme 1 for parameters). (e) is an alternative simulation with all parameters the same as (d) except that r=8Å. Running conditions were temperature 25K, and power 10μW.
(B) at Q band:
(a)-(d) correspond to (a)-(d) in (A) except that running conditions were 30K and attenuation 50dB.

and $\underline{\underline{g}}_2$, although this is probably not the case in reality, and it was possible to calculate a lower limit for the distance between the sites of 15Å. Simulations were also performed of the corresponding signal from xanthine oxidase (not illustrated) and the parameters obtained are given in Scheme 1.

Previous studies of the interaction in xanthine oxidase, based

Fig. 2 Freq= 9.3310 GHz

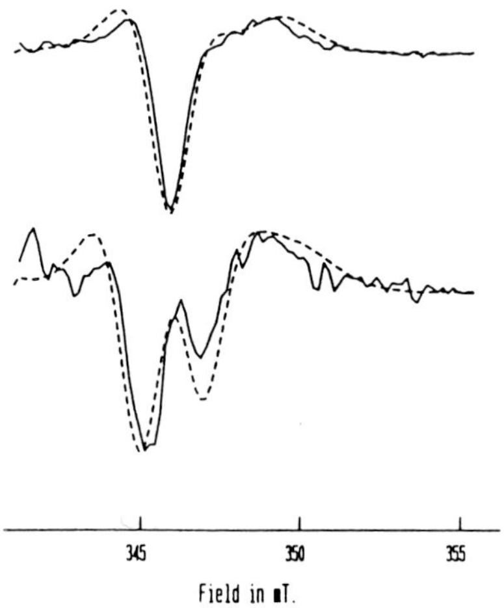

Fig. 2. Splitting of the Fe/S 1 signal by Mo(V).

The upper pair of traces (— experimental, -- simulation) show the g_x-g_y region of the second derivative Fe/S 1 spectrum when no interaction is present (no Mo(V) present). The lower pair of traces shows the splitting caused when Mo(V) is present.

on simulations of the Q-band data in ref. 1, and on the magnitude of the exchange interaction, have provided distance estimates of 8-15Å (4). Estimates of 8-15Å have also been made from the change in relaxation properties of the Mo(V) signal on reduction of Fe/S 1 (8), however it should be noted that the error in this type of calculation is large.

The numbers obtained in the present study agree with previous estimates, though our simulations indicate a higher lower limit for the spin-spin distance of about 15Å for both xanthine oxidase and aldehyde oxidase. The similarity of the nature of the coupling in aldehyde oxidase to that in xanthine oxidase is indicative of the close similarity of these two enzymes.

Scheme 1

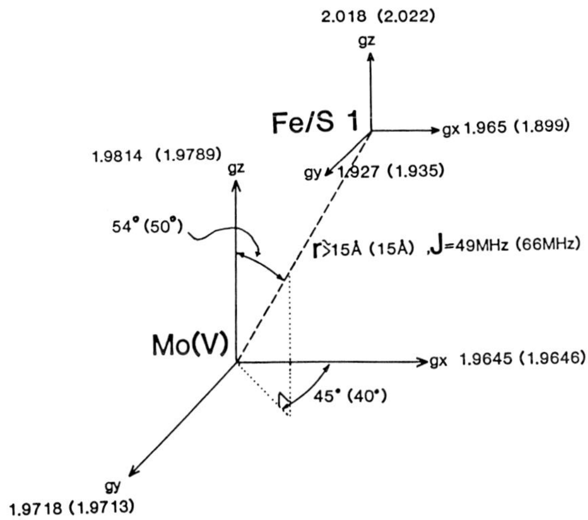

Scheme 1. Proposed spatial arrangement for Mo(V) and Fe/S 1 in aldehyde oxidase and xanthine oxidase.
The parameters obtained by fitting the data in figures 1 & 2 are indicated, those numbers in parentheses are the corresponding parameters obtained for xanthine oxidase.

Acknowledgements

The author thanks Dr. R.C. Bray for helpful discussions and for use of e.p.r. facilities (provided by the MRC). The work was supported by a grant to RCB from the SERC.

References

1. Lowe, D.J., Lyndel-Bell, R.M., Bray, R.C.: Biochem.J. 130, 239-249 (1972).
2. Barber, M.J., Coughlan, M.P., Rajagopalan, K.V., Siegel, L.M.: Biochemistry 21, 3561-3568 (1982).
3. Coffman, R.E., Beuttner, G.R.: J.Phys.Chem. 83, 2387-2392 (1979).

4. Coffman, R.E., Beuttner, G.R.: J.Phys.Chem. 83, 2392-2400 (1979).

5. Lowe, D.J., Bray, R.C.: Biochem.J. 169, 471-479 (1978).

6. Bray, R.C., George, G.N., Gutteridge, S., Norlander, L., Stell, J.P.G., Stubley, C.: Biochem.J. 203, 263-272 (1982).

7. Bray, R.C., George, G.N.: Biochem.Soc.Trans. in press.

8. Barber, M.J., Salerno, J.C., Siegel, L.M.: Biochemistry 21, 1648-1656 (1982).

PARTIAL AMINO ACID SEQUENCE OF L-LACTATE OXIDASE FROM M. SMEGMATIS

David A. Giegel, Vincent Massey and Charles H. Williams, Jr.

Veterans Administration Medical Center and the Department of Biological Chemistry, University of Michigan, Ann Arbor, Michigan

INTRODUCTION

L-Lactate oxidase (LO) is a member of the dehydrogenase/oxidase class of flavoproteins (1). It catalyzes the oxidative decarboxylation of L-lactate to acetate, carbon dioxide and water. The proposed first step in the catalytic mechanism is the abstraction of the alpha proton of the substrate with subsequent transfer of a pair of electrons to the flavin (2). There is some evidence (3,4) that a histidine residue may function as the base that removes the alpha proton of the substrate.

D-Amino acid oxidase (DAAO) is a member of the same class of flavoproteins. LO appears to be mechanistically very similar to DAAO (5). The complete primary structure of DAAO has been determined (6,7). Histidine 217 in DAAO has been implicated as being involved in the catalytic cycle of the enzyme (8). We are therefore interested in determining if there are any homologies between the protein sequences of DAAO and LO.

DIGESTION AND TRYPTIC MAPPING

L-Lactate oxidase was carboxymethylated with $[^{14}C]$-iodoacetic acid and suspended in 1.75 M urea + 1% ammonium bicarbonate. Trypsin was added (6% final concentration by weight) and the digestion monitored by removing aliquots every hour and injecting these samples onto a C-18 HPLC column. Peptides were eluted with a linear gradient from 0.1% trifluoroacetic acid to 50% acetonitrile in 0.1% trifluoroacetic acid. Figure 1 shows the tryptic peptide map obtained after a digestion time of eight hours. Preparative amounts (approximately 20 nanomoles) were run at low pH and individual peaks collected. An analytical HPLC column at pH 6.2 was used to check the purity of the low pH fractions. If a fraction was clean after the low pH isolation, amino acid analysis and sequencing were performed directly on the peptide. If two or more peptides were present at high pH, the peak was purified over a high pH column then desalted on a low pH column before sequencing and amino acid analysis.

Flavins and Flavoproteins

FIGURE I: Tryptic Map After 8 Hours of Digestion

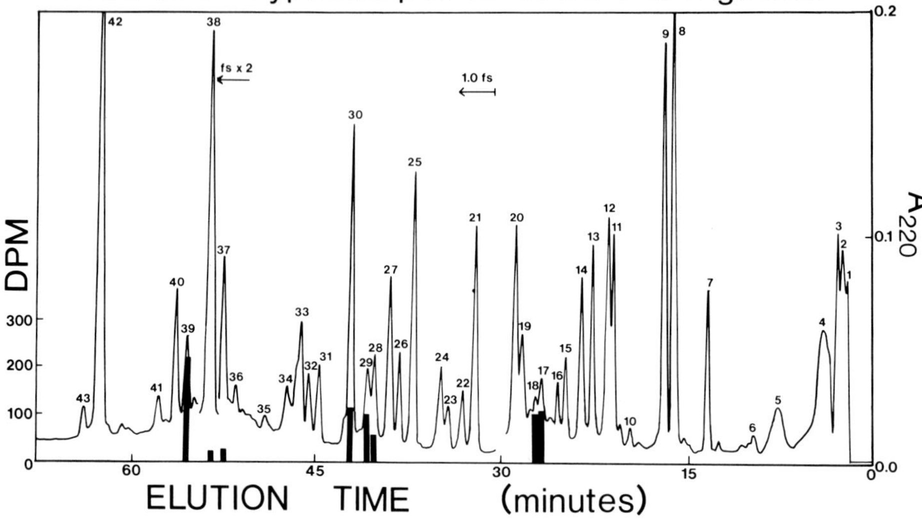

Conditions: Solvent A - 0.1% trifluoroacetic acid in water, Solvent B - 0.1% trifluoroacetic acid in acetonitrile. 0 to 50% linear gradient in B, total time - 70 minutes. Approximately 20 nanomoles of digest injected.

SEQUENCING

All sequencing was performed by the manual methods described by Tarr (9). PTH-amino acids were detected as described by Swenson, et al (10). Table I lists the peptides that have been fully or partially sequenced so far. The sequence of the amino terminus of the protein was determined by sequencing the intact, carboxymethylated protein.

Three pairs of peptides appear to be very closely related in sequence. Peptides 11 and 13 have identical sequences through the first six residues. Additionally, the kinetics of digestion showed that the heights of these two peaks were reciprocally related. As peak 11 increased with time, number 13 decreased (data not shown). It is possible that peptide 11 is a digestion product of 13.

Peptides 26 and 27-1 have identical sequences for 11 residues. The only difference is in the amino terminus of peptide 26, where a lysine is found that is not present in peptide 27-1. Finally, peptides 31 and 32 are identical for the first 11 residues. The only difference in the amino acid analyses of the two peptides is that 31 contains both an arginine and a lysine while 32 contains only an arginine.

TABLE I: SEQUENCES OF TRYPTIC PEPTIDES

PEAK NO.	SEQUENCE	PEAK NO.	SEQUENCE
7	T G A D V V K	23	D L A E S F I R
8	G I Q H P D A D R	24	A S G D T P V L F D - G
9	G I E H V A R	25	D L S V E L G K
11	A H S G V E A E G L L R	26	K H A G D T S A Y F Q L Y
12	A N V E A F K	27-1	H A G D T - A Y F Q L Y
13	A H S G V E	30-14	A L A M G A - A V G I
14	M L M A A T E R	31	T G V P Y I T S T L A V S
20	M P V I L K	32	T G V P Y I T S T L A
21	H A G	39-10	G L A L P M F V P I G
22	E L T I D A L R P T R	N-TERMINUS	S N W G D Y E N E I

TABLE II: HOMOLOGIES WITH D-AMINO ACID OXIDASE

		SIGMA VALUE	PERCENT IDENTITY
	50		
DAAO	G L W Q P Y T S E P		
LO, N-TERMINUS	S N W G D Y E N E I	4.0	30
	245		
DAAO	G N W N E I N N I		
LO, N-TERMINUS	S N W G D Y E N E I	4.0	40
	224		
DAAO	Y N S P Y I I P G L Q A V T		
LO, PEPTIDE 31	T G V P Y I T S T L A V S	3.8	43
	214		
DAAO	I I T H D L E R		
LO, PEPTIDE 8	G I Q H P D A D R	3.6	44

HOMOLOGIES WITH DAAO

Table II lists a few tentative homologies between DAAO and LO. The sequences were compared with the computer program that was used by Williams, et al (11). The major homology of interest is between peptide eight of LO and the His-217 peptide of DAAO. The sigma value for the comparison is low so the homology is only tentative. If allowance is made for the proline at position five of the LO peptide, a 44% identity is seen between the two stretches. The secondary structure prediction that was performed on the completed DAAO sequence localized His-217 near the end of a β-sheet region (12). If the same structure holds for LO, introduction of a proline after the histidine should not be expected to alter the majority of the β-sheet region. It will be interesting to determine if this histidine in LO is actually one of the ones that is modified by fluorodinitrobenzene (3,4). The tentative homology between the N-terminus of LO and the DAAO sequence beginning at residue 50 is also worthy of note, since the adenine nucleotide binding region of DAAO would appear to involve the first 31 residues of the DAAO sequence (12). As lactate oxidase contains FMN as cofactor, homology between the two proteins would be expected to begin later in the DAAO sequence.

REFERENCES

1. Hemmerich, P. and Massey, V.: Oxidases and Related Redox Systems (T.E. King, et al, eds) Pergamon Press, pp 379-405 New York, 1982.

2. Walsh, C., Lockridge, O., Massey, V. and Abeles, R.H.: J. Biol. Chem. 248, 7049-7054 (1973).

3. Murphy, C.J., Shepherd, M.G. and Sullivan, P.A.: Biochemistry 22, 1665-1669 (1983).

4. Soon, C.Y., Shepherd, M.G. and Sullivan, P.A.: Biochem. J. 165, 385-393 (1977).

5. Walsh, C.T., Schonbrunn, A. and Abeles, R.H.: J. Biol. Chem. 246, 6855-6866 (1971).

6. Swenson, R.P., et al: J. Biol. Chem. 257, 8817-8823 (1982).

7. Ronchi, S., et al: J. Biol. Chem. 257, 8824-8834 (1982).

8. Swenson, R.P., Williams, C.H., Jr. and Massey, V.: J. Biol. Chem. 259, 5585-5590 (1984).

9. Tarr, G.E.: Methods in Enzymology, Vol. 47 (C.H.W. Hirs and S.N. Timasheff, eds) Academic Press, pp 335-357, 1977.

10. Swenson, R.P., Williams, C.H., Jr. and Massey, V.: J. Biol. Chem. 257, 1937-1944 (1982).

11. Williams, C.H., Jr., Arscott, L.D. and Schulz, G.E.: Proc. Natl. Acad. Sci. USA 79, 2199-2201 (1982).

12. Swenson, R.P., et al: Flavins and Flavoproteins (V. Massey and C.H. Williams, Jr. eds) Elsevier North Holland, Inc., pp. 72-77, 1982.

CHEMICAL MODIFICATION OF SULFHYDRYLGROUPS IN P-HYDROXYBENZOATE HYDROXYLASE FROM PSEUDOMONAS FLUORESCENCS

Willem J.H. van Berkel, Franz Müller, Wicher J. Weijer, Peter A. Jekel and Jaap J. Beintema

Departments of Biochemistry, Agricultural University, De Dreijen 11, 6703 BC, Wageningen and State University of Groningen, Nijenborgh 16, 9747 AG Groningen, The Netherlands

Introduction

The three-dimensional structure of p-hydroxybenzoate hydroxylase from P.fluorescens is known at a resolution of 0.25 nm (1). The sequence has also been elucidated (2). These results form a good basis for a search of the amino acid residues involved in the catalysis of the enzyme. In a previous paper (3) we have shown that four out of nine histidine residues are alkylated by diethylpyrocarbonate. At least one of the four histidine residues is involved in the binding of NADPH. The enzyme contains five cysteine residues (4). One of these residues reacts selectively with N-ethylmaleimide. Mercurial compounds react with additional cysteine residues. The effects of the cysteine modification on the enzymic activity are described.

Results and Discussion

The reaction of the enzyme with N-[^{14}C]ethylmaleimide yields a preparation which has incorporated about one mol label/mol proteinbound FAD, confirming previous results (4). The reaction is rather fast, not allowing characterization of the kinetics of the reaction. The modified cysteine residue was identified as Cys-116 in the sequence (5). Chemical modification of Cys-116 does not affect the enzymic activity of the enzyme. It is interesting to notice that the preparation of heavy atom derivatives of the enzyme for crystallographic studies (1) using mercurial compounds, did not lead to the modification of Cys-116 but mainly of Cys-332 . This strongly indicates that Cys-116 had been oxidized by air in the single crystal used. This is

Flavins and Flavoproteins

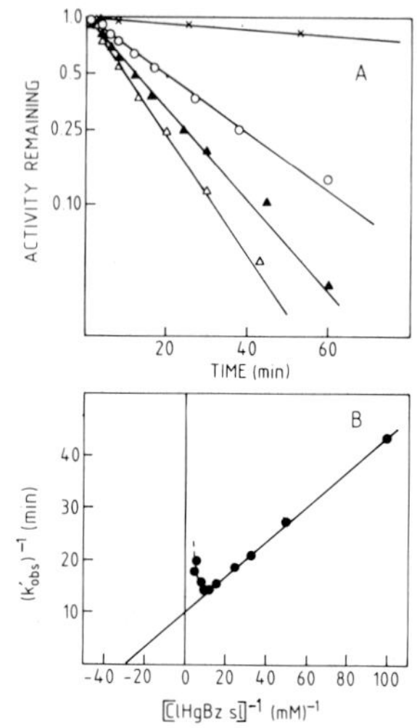

Figure 1. The time-dependent inactivation of p-hydroxybenzoate hydroxylase as a function of the concentration of ClHgBzsl.
A: 1.6 μM N-ethylmaleimide prelabeled enzyme in 80 mM Hepes buffer, pH 7.0, I = 0.1 M, was incubated with the following concentrations of the mercurial compound: 20 μM, o-o-o; 40 μM, ▲-▲-▲; 80 μM, Δ-Δ-Δ, 20 μM in the presence of 0.8 mM p-hydroxybenzoate, x-x-x.
B: Doubly reciprocal plot of the results in A.

in agreement with the observed chemical reactivity of Cys-116 towards molecular oxygen.

The mercurial compounds p-chloromercuribenzoate (ClHgBzOH), a spinlabel derivative thereof (ClHgBzsl) and $HgCl_2$ react with Cys-116 and other cysteine residues as well. In order to identify and to study the kinetics of the enzyme modification by these reagents Cys-116 was prelabeled with N-ethylmaleimide. The modification of the enzyme by the mercurial compounds leads to a complete loss of the enzymic catalysis (5).

ClHgBzsl reacts more selectively with the enzyme than ClHgBzOH, i.e. in the former case one cysteine residue is modified, whereas in the latter case 1-2 residues are modified. The kinetics of both reactions show some remarkable features. The rate of inactivation of the enzyme by the two mercurial compounds does not linearly increase with the concentration of the modifying reagents (figure 1). The results indicate that initially an enzyme-inhibitor complex is formed leading to an inactive enzyme where the inhibitor is covalently bound. Apparent dissociation constants (K_i) of 33 μM and 22 μM are calculated for ClHgBzsl and ClHgBzOH, respectively. The corresponding first-order rate constants at saturation are 0.10 min^{-1} and 0.204 min^{-1}, and the second-order rate constants are 50 $M^{-1}s^{-1}$ and 163 $M^{-1}s^{-1}$, respectively. It should be noted that the modification of the enzyme can be fully reversed by incubation in the presence of dithiothreitol.

The inactivation reaction is inhibited by the mercurial compounds themselves. Using a 2 μM enzyme solution the inhibition is observed at a concentration of mercurial reagents exceding 80 μM. This inhibition reaction resembles

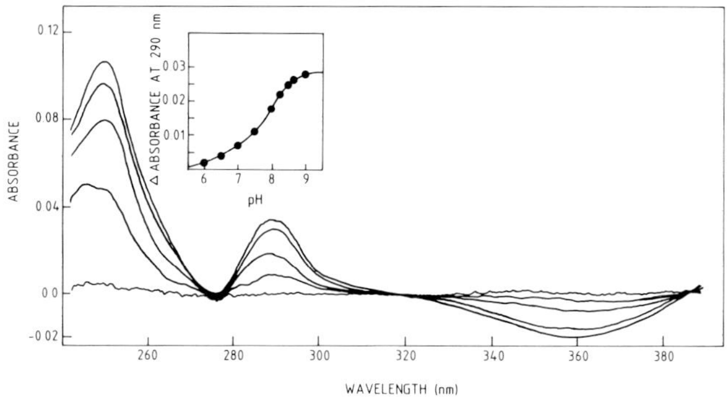

Figure 2. Ultraviolet difference spectra observed during the inactivation of the enzyme by ClHgBzOH. The final concentration of the enzyme was 10 μM in 80 mM Hepes buffer, pH 8.5, I = 0.1 M. The final concentration of ClHgBzOH was 40 μM. The spectra shown were recorded 15s, 130s, 520s and 1010s after mixing of the solutions. Inset: pH-dependent absorbance difference of the peak at 290 nm.

that observed in normal assays using a high concentration of substrate. It is suggested that the mercurial compounds also bind to the substrate binding site, which is devoid of cysteine residues (1). This idea is supported by the facts that inactivated enzyme shows very little affinity for the substrate and that the substrate protects the enzyme from fast inactivation by the mercurial compounds (figure 1).Cys-152 is the most probable candidate responsible for the loss of activity upon modification (5). The rate of the inactivation reaction is almost independent on the pH of the solution. Modification of the enzyme by ClHgBzOH in particular leads to additionl absorption bands in the difference absorption spectrum (figure 2). Besides the expected band at 250 nm, peaks at 290 nm and 360 nm are observed.The intensities of the two latter peaks are strongly dependent on pH, reaching an optimum at about pH 9 (figure 2, insert). The results suggest that the ionizing group possesses a pK value of ∿ 7.6. The molar difference absorption coefficient at pH 9 is 3200 $M^{-1}cm^{-1}$ at 290 nm and ∿ 1600 $M^{-1}cm^{-1}$ at 360 nm. The fact that the flavin absorption properties are affected by the ionization of the unknown group in the enzyme indicates that this group is located in the vicinity of the prosthetic group. According to the three-dimensional structure (1) Tyr-201, Tyr-222 and Tyr-385 are located in the active site of the enzyme. Therefore it is tentatively proposed that the

ionizing group is due to at least one of these tyrosine residues, in agreement with the observed $\Delta\varepsilon$ value.
According to the three-dimensional structure (1) none of the five cysteine residues seem to be in a position expected to be crucial for the enzymatic activity. The fact that modification of Cys-152 affects the activity strongly indicates that the loss of activity is a result of long-distance effects affecting the substrate-binding site. It is also concluded from the results that Cys-152 and the substrate-binding site are interdependent. This conclusion is supported by the fact that Cys-152 is not modified when crystals of the enzyme-substrate complex are soaked with mercurial compounds. Instead, as already mentioned, Cys-332 is labeled. From the results presented in this paper it can be concluded that binding of the substrate to the enzyme influences the accessibility of Cys-152 for the reagents, i.e. Cys-152 becomes more buried in the protein, as has also been observed for SH groups in phenol hydroxylase.

Acknowledgements

We thank Miss Lyda Verstege for typing the manuscript, Dr. C.A.H. Rasmussen for carefully reading the manuscript, Mr. M.M. Bouwmans for preparing the figures. This study was carried out under the auspices of the Dutch Foundation for Chemical Research (SON) with financial aid from the Dutch Organization for the Advancement of Pure Research (ZWO).

References

1. Wierenga, R.K., de Jong, R.J., Kalk, K.H., Hol, W.G.H., Drenth, J.: J.Mol.Biol. 131, 55-73 (1979)
2. Hofsteenge, J., Weijer, W.J., Jekel, P.A., Beintema, J.J.: Eur.J. Biochem. 133, 91-108 (1983).
3. Wijnands, R.A., Müller, F.: Biochemistry, 21, 6639-6646 (1982).
4. Müller, F., Voordouw, G., van Berkel, W.J.H., Steennis, P.J., Visser, S., van Rooijen, P.: Eur.J.Biochem. 101, 234-244 (1979).
5. Van Berkel, W.J.H., Weijer, W.J., Müller, F., Jekel, P.A., Beintema, J.J.: Eur.J.Biochem. 1984, in press.

CHEMICAL MODIFICATION OF PHENOL HYDROXYLASE BY p-NITROBENZENESULFONYL FLUORIDE

C. Torsten Sejlitz and Halina Y. Neujahr
Department of Biochemistry, The Royal Institute of Technology, S-100 44 Stockholm, Sweden

INTRODUCTION

Phenol hydroxylase (EC 1.14.13.7), isolated from the soil yeast *Trichosporon cutaneum*, is a dimeric enzyme (M_r=2x76000) containing two FAD per dimer. It catalyzes the reaction:

$$\text{phenol} + \text{NADPH} + O_2 + H^+ \longrightarrow \text{catechol} + \text{NADP}^+ + H_2O$$

Besides being a substrate, phenol acts as an inhibitor (1,4), and also as an effector, causing conformational changes, with concomitant changes in the affinity for NADPH (1), in the reactivity of essential residues (2,3) and in the FAD-spectrum of the enzyme (1,4).

In the present investigation, phenol hydroxylase was treated with the tyrosyl reagent p-nitrobenzenesulfonyl fluoride (pNBSF), which gives a stable O-sulfonyl-derivative with tyrosine (NBS-tyr)(5). The modified enzyme was analyzed for activity, for its capacity to bind phenol, and for modified cysteinyl and tyrosyl residues, as well as for changes in the FAD-spectrum and its perturbation upon titration with phenol.

RESULTS

Modification of cysteinyl residues

About two cysteinyl residues per enzyme dimer were blocked by pNBSF. By reaction rate comparisons, two of them were shown to be identical with the two most reactive towards Ellman's reagent and are, as previously shown (2), not essential for enzyme activity. A slight modification of essential cys residues was also observed and, accordingly (2), some FAD was lost. Hence, all activity and spectral data were normalized to the FAD-content of the samples.

Flavins and Flavoproteins

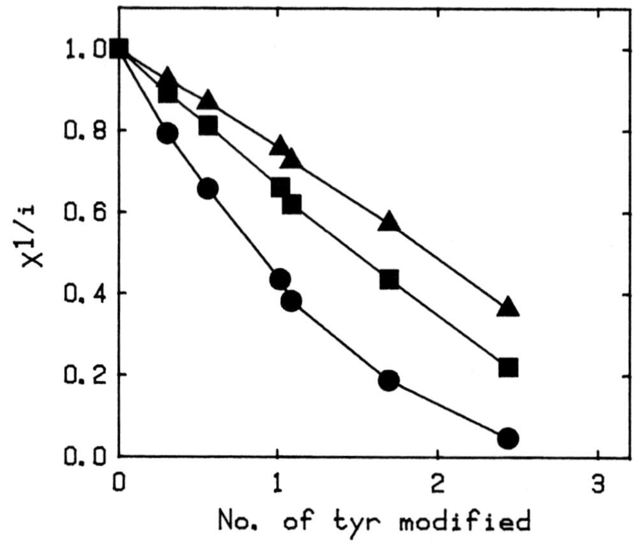

Fig 1. Tsou plot of activity versus the number of modified tyrosyl residues. Activity is plotted as $X^{1/i}$, where X denotes activity in fraction of control and i=1 (●), 2 (■) and 3 (▲).

Modification of tyrosyl residues

Up to 2.5 tyrosyl residues of the total 39 were modified, leading to an almost total loss of activity. The Tsou plot (6) indicates that three or four tyrosyl residues are involved, of which one is not essential (fig 1).

When studying the kinetics of inactivation, side reactions had to be considered. Experimental data were evaluated by numerical simulation, employing the following reaction model:

$$\text{enzyme}\cdot\text{phenol} \underset{K_d}{\rightleftharpoons} \text{enzyme} + \text{phenol}$$

$$\text{enzyme} + \text{pNBSF} \xrightarrow[k_E]{} \text{inactive enzyme}$$

$$\text{enzyme}\cdot\text{phenol} + \text{pNBSF} \xrightarrow[k_{EP}]{} \text{inactive enzyme}$$

$$\text{pNBSF} + H_2O \xrightarrow[k_W]{} \text{products}$$

$$\text{pNBSF} + \text{phenol} \xrightarrow[k_P]{} \text{products}$$

An enzyme·phenol complex with K_d around 5 µM was found to be protected towards pNBSF, with an inactivation rate of <70%.

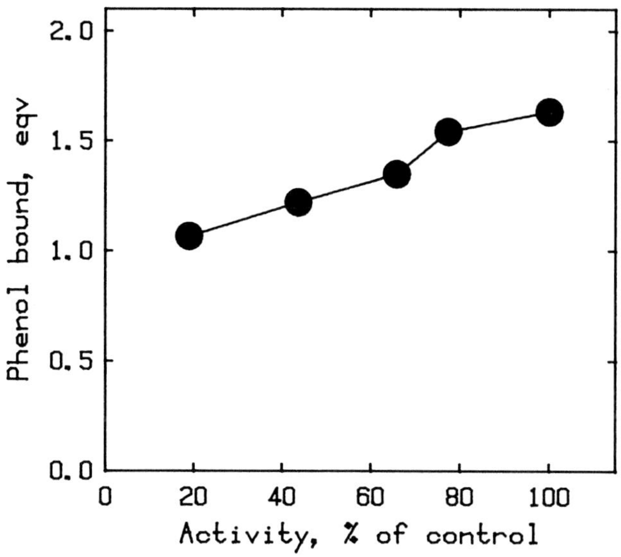

Fig 2. Binding of phenol to phenol hydroxylase. The number of equivalents of ^{14}C-phenol bound is plotted against fraction remaining activity after modification.

Binding of phenol to modified enzyme

The binding of phenol to the enzyme was studied by equilibrium dialysis with 5 μM ^{14}C-phenol. It was found, that while the native enzyme binds approximately two equivalents of phenol per dimer, the modified species, when extrapolating to zero activity, binds approximately one equivalent (Fig 2).

The normal perturbation of the FAD-spectrum of the enzyme, caused by the effector function of phenol (4), was not affected. Also, the binding constant of the effector complex was practically unaffected.

DISCUSSION

From the data above, it is proposed, that phenol hydroxylase has two binding sites for phenol. At least one of them involves one or more tyrosyl residues. Since 100% inactivation correlates with the blocking of one site, the two sites cannot be both identical and independent. Further, the binding of phenol to one of the sites has to be sufficient for effecting the conformation change. When trying to interpret these observations, one has to consider two possibilities; either the two sites are identical and interactive, or they are different and have different functions. If the two sites are identical, this implies, that the modification of one site protects the other without affecting the response of the enzyme to phenol as effector. If they are not, this may indicate that one site is responsible for the formation of the effector complex, and that the other is a substrate site, where the hydroxylation occurs. The third effect of phenol on phenol hydroxylase; the substrate inhibition, is not observed at the phenol concentrations used in this study.

METHODS

Phenol hydroxylase (1-10μM) was treated with .5-2mM pNBSF in 50 mM Tris·SO_4, pH8, at 10-22°C (5).

NBS-tyr was determined by HPLC after acid hydrolysis (5).

The number and properties of modified cysteinyl residues were determined by differential labelling with Ellman's reagent of denatured and native enzyme samples, respectively.

Spectra were taken at pH 7.6, 10°C after gel filtration.

REFERENCES

1. Neujahr,H.Y. and Kjellén,K.G. Biochemistry 253, 8835-8841 (1978)
2. Neujahr,H.Y. and Gaal,A. Eur.J.Biochem. 58, 351-357 (1975)
3. Neujahr,H.Y. and Kjellén,K.G. Biochemistry 19, 4967-4972 (1980)
4. Neujahr,H.Y. Biochemistry 22, 580-584 (1983)
5. Liao,T.-H., Ting,R.S., and Hiung,J.E. J.Biol.Chem. 257, 5637-5644 (1982)
6. Tsou,C.-L. Sci.Sin. (Engl. Ed.) 11, 1535-1558 (1962)

Abbreviations:

pNBSF:	4-Nitro-benzenesulfonyl fluoride
NBS-tyr:	O-(4-Nitro-benzenesulfonyl)-tyrosine
phenol hydroxylase:	Phenol,NADPH:oxygen oxidoreductase (2-hydroxylating) (EC 1.14.13.7)

STUDIES OF 2,5-DIKETOCAMPHANE MONOOXYGENASE FROM *PSEUDOMONAS PUTIDA* ATCC 17453

David G. Taylor and Peter W. Trudgill

Department of Biochemistry, University College of Wales,
Aberystwyth, Dyfed, SY23 3DD, United Kingdom

Abstract

Cleavage of both carbocyclic rings of the (+)-camphor molecule by *P. putida* ATCC 17453 requires the participation of three monooxygenases. Two of these, (+)-camphor 5-hydroxylase and 2-oxo-Δ^3-4,5,5-trimethylcyclopentenylacetyl-CoA monooxygenase have been thoroughly characterized. However, the precise composition of the third, 2,5-diketocamphane 1,2-monooxygenase, is still uncertain some twenty years after it was first reported. It has variously been described as consisting of two and three protein components, all of which may bind flavin, and there have been suggestions that it carries catalytically functional non-heme iron. All workers agree that the complex readily dissociates in dilute protein solutions and during purification. One component of the complex, an NADH oxidase has been purified in isolation and shown to be a single polypeptide chain (M_r 36000) and, although it is always obtained flavin free, it does bind FMN weakly (K_D 450 nM) and will carry electrons from NADH to excess free FMN.

Our initial aim was to purify the intact complex to homogeneity. In this we were not really successful for although we did obtain [by a combination of $(NH_4)_2SO_4$ fractionation, DEAE-cellulose chromatography and preparative FPLC] a pure preparation that displayed tightly coupled oxygenating activity, it contained little of the NADH oxidase (E1); we calculated it to be about

Flavins and Flavoproteins

1% of the total protein. As a consequence of this, with so little of the oxygenating component (E2) in the coupled form, the specific activity was low. However, this did allow us to fully characterize E2 and to study the reconstitution of the intact complex by addition of E1, purified in isolation by an alternative procedure.

Ultracentrifugal analysis of E2 gave M_r 78000 and dissociation of the flavin from this bright yellow protein, followed by HPLC and spectral analysis showed it to be FMN with one mole bound to each mole of E2; iron and copper were virtually absent. SDS-polyacrylamide gel electrophoresis gave a single subunit band M_r 36000 and the native protein thus consists of two equal size subunits. Electrophoresis of native E2 on gels prepared from 4.5% to 9% acrylamide gave two flavin-free protein bands, as previously reported. 'Ferguson plots' of these together with standard proteins showed them both to have M_r 78000. Excision of the two bands from slab gels, followed by extraction with buffer and assay in the presence of FMN, purified E1 and 2,5-diketocamphane showed them both to have oxygenase activity. Titration of purified E2 with E1 established that both forms of E2 combine with E1 in an equimolar ratio. Additional FMN is not required in these experiments, the NADH oxidase (E1) couples directly with the flavoprotein oxygenating component (E2), presumably at its flavin binding region. No evidence for the previously reported additional discrete third protein component was obtained.

RECENT PROGRESS IN BIOLUMINESCENCE: CLONING OF THE STRUCTURAL GENES ENCODING BACTERIAL LUCIFERASE, ANALYSIS OF THE ENCODED SEQUENCES, AND CRYSTALLIZATION OF THE ENZYME

Thomas O. Baldwin, Timothy C. Johnston, and Rosemarie Swanson

Department of Biochemistry and Biophysics, Texas A&M University, and the Texas Agricultural Experiment Station, College Station, TX., USA, 77843

SCOPE

The purpose of this paper is to summarize progress in bacterial bioluminescence in the past three years, a subject too large for the broad scope of this book. Of necessity the subject of this paper has been confined to structural studies on bacterial luciferase; however, references for some of the more exciting new developments in the broader field follow. The literature has been reviewed recently (1,2). E.A. Meighen has made dramatic progress in understanding the metabolism of the aldehyde substrate in the bioluminescent bacteria (3). In addition, much has been done in the area of clinical applications of bioluminescence and it now appears that luciferases will soon be used in clinical diagnostics (4). Kürfurst, Ghisla and Hastings (5) have obtained evidence of an intermediate which they report to be the emitter in the bioluminescence reaction; they suggest that it is a 4a-hydroxyflavin that decays more slowly than bioluminescence. S.-C. Tu has done a most interesting experiment with a photoactivatable aldehyde analog which suggests that the aldehyde binding site resides at the interface between the two subunits(6).

INTRODUCTION

Three years ago at the Flavin Symposium in Ann Arbor, we presented amino acid sequence data obtained from fragments of the α subunit of luciferase that resulted from limited action of chymotrypsin (7). One of those fragments contained the sequence Met-Asp-Cys-Trp-Tyr-Asp. This sequence

Flavins and Flavoproteins

allowed us to synthesize an oligonucleotide probe that would hybridize to the DNA in Vibrio harveyi that encodes the region of the α subunit from which the proteolytic fragment derives (see Fig. 1). From that initial cloning experiment (8) has come the first solid insight into the biochemistry and genetics of the regulation of expression of the bioluminescence system; the entire coding region of the luciferase subunits has been sequenced and from the DNA sequence, the sequence of amino acids has been deduced. The location and functions of the genes encoding the accessory enzymes involved in the synthesis of the aldehyde substrate have been largely determined (9,10), and luminescence systems from other organisms have been cloned, even from a symbiotic bacterium that has never been cultured in the laboratory (11). The past three years have indeed been exciting because of the speed with which we have been obtaining answers to questions of long standing. However, we are now only beginning our efforts to understand the molecular mechanism of the bacterial bioluminescence reaction in terms of the chemistry of the enzyme; this report is therefore a progress report, not the final word on luciferase structure and function.

ISOLATION OF THE LUCIFERASE GENES

Use of the Synthetic Oligonucleotide to Isolate the First Clone of Luciferase--Bacterial luciferase is very susceptible to inactivation by proteolytic enzymes (12). Virtually all proteases inactivate luciferase by hydrolyzing a few bonds in the α subunit that appear to be near the active center (13). Inactivation is due not to gross structural alterations associated with bond scission, but rather to subtle structural movements that are barely discernible (14). Even the denaturability of the inactive enzyme by urea is the same as native enzyme (15). S.K. Rausch has purified the large proteolytic fragments of the α subunit from sodium dodecyl sulfate gels, and determined the N-terminal sequence of each fragment; based on these data, he has proposed a model (Fig. 1) for the location of the protease sensitive bonds in the α subunit (7,16).

Based on the sequence of the N-terminal region of the δ_H fragment, Rausch

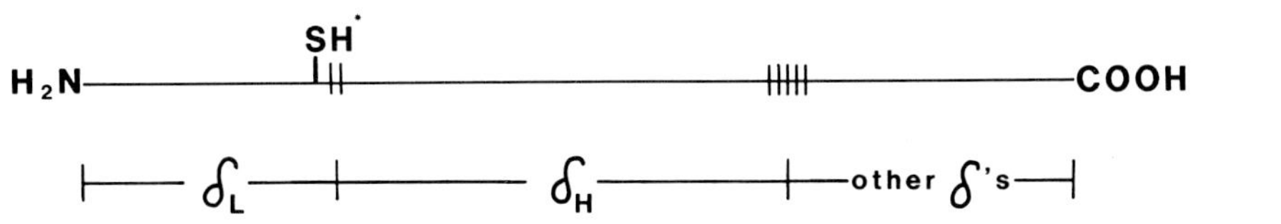

Fig. 1. Model of locations of the protease-labile regions of the α subunit.

proposed the following oligonucleotide as a good candidate for isolation of the luciferase gene (17):

PEPTIDE SEQUENCE:	Met	Asp	Cys	Trp	Tyr	Asp	
mRNA:	5'AUG	GA$^{C}_{T}$	UG$^{C}_{T}$	UGG	UA$^{C}_{T}$	GA	'3
OLIGONUCLEOTIDE:	3'TAC	CT$^{A}_{G}$	AC$^{G}_{A}$	ACC	AT$^{G}_{A}$	CT	'5

Fig. 2. Synthetic oligonucleotide used to clone luciferase.

This oligonucleotide was synthesized and used to screen a genomic clone bank in phage lambda containing EcoRI fragments of V. harveyi DNA. The isolated cloned fragment of DNA contained the sequence complementary to the synthetic oligonucleotide, and sequence analysis demonstrated that the fragment had all of the α subunit coding region and the coding region for the N-terminal 13 amino acids of the β subunit. The location of these coding regions in the isolated fragment is depicted below.

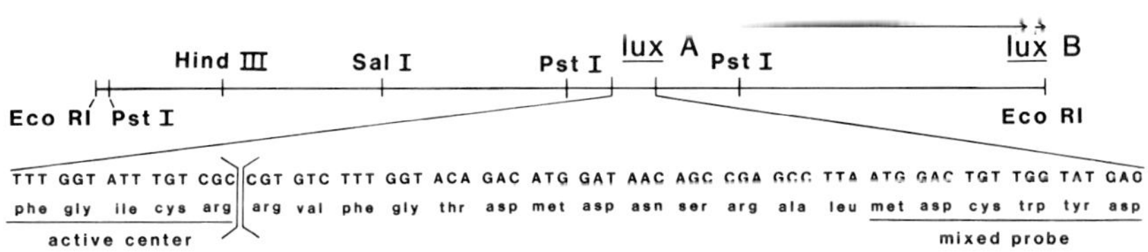

Fig. 3. Location of the first protease-labile region and the reactive thiol within the coding region for the α subunit (lux A).

There are several interesting points about the structure of the active center of luciferase that are immediately apparent from this limited sequence. First, the trypsin-sensitive bond (Arg-Val) is located only two residues from the chymotrypsin-sensitive bond (Phe-Gly). Second, the protease-sensitive region is located very close in the primary sequence to the reactive thiol (Ile-Cys-Arg) that is thought to reside in or near the active center (15,18). The position of the reactive thiol relative to the protease labile region was predicted on the basis of fluorescence polarization and EPR measurements on luciferase labeled at the reactive thiol and then treated with either trypsin or chymotrypsin (15,19).

Cloning of the Luminescence System from *V. harveyi* and the Expression of Bioluminescence in *E. coli*--The restriction map shown above has a single HindIII site that is upstream from the *lux* A gene. Since the synthetic

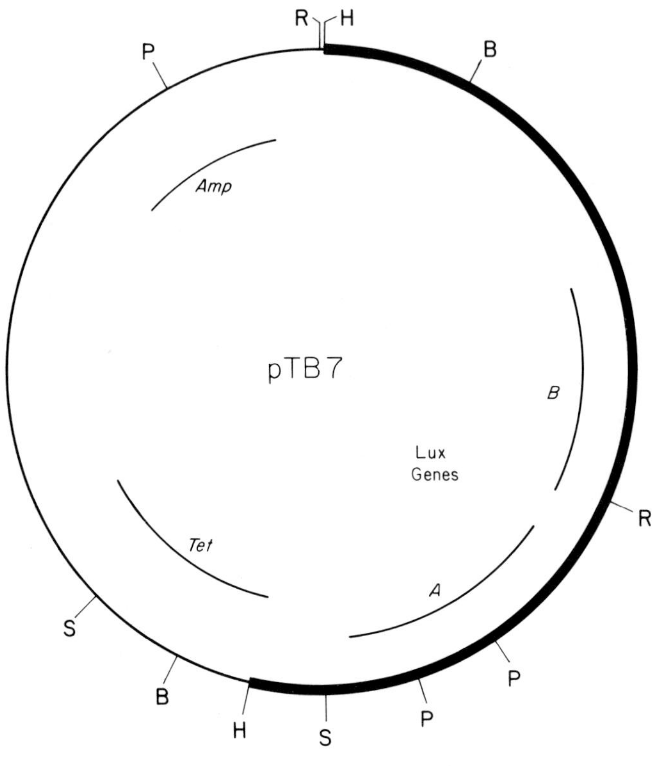

Fig. 4. Restriction map showing the locations of the *lux* A (A; luciferase α subunit) and *lux* B (B; luciferase β subunit) on the pBR322 derived plasmid pTB7. Transcription of *lux* A and *lux* B is counterclockwise. *E. coli* cells carrying pTB7 express resistance to ampicillin and, upon stimulation with a long-chain aldehyde, intense bioluminescence.

oligonucleotide hybridized to a 4 kb HindIII fragment by Southern blot analysis, it was clear that the entire α and β subunit coding regions had to be contained on that HindIII fragment. We cloned genomic HindIII fragments from *V. harveyi* into the HindIII site of plasmid pBR322 and found 12 colonies (of roughly 6000 colonies screened) that expressed bioluminescence when supplied with aldehyde vapor. One of these colonies harbored pTB7 (Fig. 4), which has been analyzed in detail (20). It was obvious from the requirement of aldehyde for bioluminescence that our 4 kb HindIII fragment did not express the structural genes for the aldehyde-biosynthetic apparatus; however, it was clear that the luciferase in *E. coli* was being supplied with $FMNH_2$. Specific deletion experiments demonstrated that the flavin reductase activity was not encoded on the recombinant plasmid; rather the flavin reductase activity was present in lysates of *E. coli* without any plasmid, demonstrating that *E. coli* carries an NADH:FMN oxidoreductase activity similar to that in *V. harveyi*.

Several recent publications have suggested that light emission does not occur directly from luciferase but that an energy transfer system might be involved in bacterial bioluminescence [see Ziegler & Baldwin (1981) for review]. Isolation of pTB7 has provided an excellent opportunity to test this hypothesis by comparison of the bioluminescence emission spectra of *V. harveyi* and of *E. coli* carrying pTB7. There was no discernible difference either *in vivo* or *in vitro*, indicating that in *V. harveyi*, light emission is from a luciferase-bound flavin chromophore, not from a secondary emitter protein.

Another question of interest that can potentially be answered with these experiments pertains to the stability *in vivo* of a single luciferase subunit in the absence of the other. To analyze the fate of the α subunit *in vivo* in the absence of β, we digested pTB7 with EcoRI, which deleted the β subunit gene but left the α subunit gene and the promoter intact. The content of the α subunit (or fragments) within *E. coli* carrying this plasmid was analyzed by Western blot analysis (21). No α subunit could be detected. Furthermore, ^{35}S incorporation studies and analysis by two-dimensional gel electrophoresis did not indicate accumulation of measurable α subunit. This observation suggests that the α subunit is unstable in the absence of β and is degraded to fragments that are not recognizable by antiluciferase antibody.

ANALYSIS OF THE PRIMARY STRUCTURE

The Amino Acid Sequence of the Alpha Subunit of Bacterial Luciferase--The complete amino acid sequence of the α subunit of bacterial luciferase has been determined from the nucleotide sequence of the DNA fragment depicted above (Fig. 3) and from sequences of numerous small peptides from the α subunit (17,22). The amino acid sequence together with the secondary structure predicted by the modified method of Chou and Fasman (23) is shown in Fig. 5.

The residues drawn in slanted areas are those that are predicted to be in alpha-helical structure, those drawn in the broad arrow-like structures are those regions predicted to be in beta structure, and the residues in the narrow parallel lines are those predicted to be neither alpha-helical nor beta structure. Amino acids that appear as black letters on a white background are those residues found most often in contact with the aqueous environment, while residues that appear as white letters on a black background are those that are found more often inside globular proteins than in contact with the aqueous solvent (24). The predicted secondary structure has 34% of the residues in alpha-helical configuration and 12% in the beta sheet configuration, in reasonable agreement with the measurements of 28% alpha-helix and 14% beta sheet reported by Holzman and Baldwin for the dimer (14).

The α subunit had a calculated molecular weight of 40,108, compared with 42,000 determined by electrophoretic mobility in the presence of sodium dodecyl sulfate, and comprised a total of 355 amino acid residues. The amino acid composition of the α subunit determined from the sequence compared very well with the chemically determined composition. The sequence gave values for the acidic residues and amides; there were 19 Asn,

Fig. 5. The complete amino acid sequence and predicted secondary structure of the α subunit of bacterial luciferase.

28 Asp, 24 Glu, and 13 Gln. The basic amino acids (20 Lys, 13 Arg, and 11 His) are too few to compensate, and even if all histidinyl residues were fully protonated at the isoelectric point, there would still be 8 excess acidic residues, which explains why the α subunit is one of the most acidic of all proteins in E. coli carrying pTB7; it has an isoelectric point of about 4.8 (25).

Approximately 2/3 of the α subunit was sequenced by protein chemical techniques. The amino acid sequence implied by the DNA sequence was, with few exceptions, in agreement with the chemically determined sequence. The regions of the α subunit thought on the basis of chemical modification and limited proteolysis studies to comprise the active center were found to reside in two discrete and relatively basic regions, one from around residues 100-115 and the second from around residues 280-295. The relative positions of the two protease labile regions as predicted by Rausch (7,16,17) are shown in Figure 1.

The sequence surrounding the first protease labile region and its position relative to the active center thiol are depicted in Figure 3, above. Chymotrypsin cleaved the bond between residues 117 & 118 (Phe-Gly) in the first protease labile region while trypsin cleaved the bond between residues 115 & 116 (Arg-Val) in the same region. There were many more bonds cleaved in the second protease labile region and the only ones identified with certainty were between residues 280 & 281 (Phe-Val) and between residues 282 & 283 (Leu-Lys) for cleavage with chymotrypsin and between residues 279 & 280 (Arg-Phe) for cleavage with trypsin.

Comparison of the α and β Subunit Structures--Based on N-terminal amino acid sequence comparisons, it has been predicted that the two subunits of luciferase have arisen by gene duplication and subsequent divergence (26). This interpretation was based on a statistical analysis of the liklihood of the sequence similarities between the two subunits being due to random chance, but the available data were only from the N-terminal regions of the two polypeptides. A large body of chemical modification data and mutant enzyme analyses has shown convincingly that the active center resides primarily if not exclusively on the α subunit. If the two subunits have arisen by gene duplication, then one would expect some remnants of structural similarity to remain, and perhaps to find a flavin binding site on the β subunit. This hypothesis has been advanced several times following the original assertion of homology between the two subunits. It was therefore with some interest that we awaited the comparisons of the sequence in the β subunit with the region in the α subunit surrounding the reactive thiol and the first protease labile region. The comparison is shown in Fig. 6, with the amino acid sequences aligned without introduction of any gaps, from the first amino acid (residue 1), through the reactive thiol region (residue 106 in α) and the first protease labile region, into the area between the two protease labile regions. Comparison of these two sequences shows 71 of 200 pairs, or 36%, to be identical. Another comparison using the hydrophobicity correlation method of Sweet and Eisenberg (27) gives a correlation coefficient of 0.66, with a variance of 0.06 when various long stretches of sequence are compared rather than the entire 200

10 20

α MET-LYS-PHE-GLY-ASN-PHE-LEU-LEU-THR-TYR-GLN-PRO-PRO-GLU-LEU-SER-GLN-THR-GLU-VAL-

β MET-LYS-PHE-GLY-LEU-PHE-PHE-LEU-ASN-PHE-MET-ASN-SER-LYS-CYS-SER-SER-ASP-GLN-VAL-

30 40

α MET-LYS-ARG-LEU-VAL-ASN-LEU-GLY-LYS-ALA-SER-GLU-GLY-CYS-GLY-PHE-ASP-THR-VAL-TRP-

β ILE-GLU-GLU-MET-LEU-ASP-THR-ALA-HIS-TYR-VAL-ASP-GLN-LEU-LYS-PHE-ASP-THR-LEU-ALA-

50 60

α LEU-LEU-GLU-HIS-HIS-PHE-THR-GLU-PHE-GLY-LEU-LEU-GLY-ASN-PRO-TYR-VAL-ALA-ALA-ALA-

β VAL-TYR-GLU-ASN-HIS-PHE-SER-ASN-ASN-GLY-VAL-VAL-GLY-ALA-PRO-LEU-THR-VAL-ALA-GLY-

70 80

α HIS-LEU-LEU-GLY-ALA-THR-GLU-THR-LEU-ASN-VAL-GLY-THR-ALA-ALA-ILE-VAL-LEU-PRO-THR-

β PHE-LEU-LEU-GLY-MET-THR-LYS-ASN-ALA-LYS-VAL-ALA-SER-LEU-ASN-HIS-VAL-ILE-THR-THR-

90 100

α ALA-HIS-PRO-VAL-ARG-GLN-ALA-GLU-ASP-VAL-ASN-LEU-LEU-ASP-GLN-MET-SER-LYS-GLY-ARG-

β HIS-HIS-PRO-VAL-ARG-VAL-ALA-GLU-GLU-ALA-CYS-LEU-LEU-ASP-GLN-MET-SER-GLU-GLY-ARG-

110 120

α PHE-ARG-PHE-GLY-ILE-CYS-ARG-GLY-LEU-TYR-ASP-LYS-ASP-PHE-ARG-VAL-PHE-GLY-THR-ASP-

β PHE-ALA-PHE-GLY-LEU-SER-ASP-CYS-GLU-LYS-SER-ALA-ASP-MET-ARG-PHE-PHE-ASN-ARG-PRO-

130 140

α MET-ASP-ASN-SER-ARG-ALA-LEU-MET-ASP-CYS-TRP-TYR-ASP-LEU-MET-LYS-GLU-GLY-PHE-ASN-

β THR-ASP-SER-GLN-PHE-GLN-LEU-PHE-SER-GLU-CYS-HIS-LYS-ILE-ILE-ASN-ASP-ALA-PHE-THR-

150 160

α GLU-GLY-TYR-ILE-ALA-ALA-ASP-ASN-GLU-HIS-ILE-LYS-PHE-PRO-LYS-ILE-GLN-LEU-ASN-PRO-

β THR-GLY-TYR-CYS-HIS-PRO-ASN-ASN-ASP-PHE-HIS-SER-PHE-PRO-LYS-ILE-SER-VAL-ASN-PRO-

170 180

α SER-ALA-TYR-THR-GLN-GLY-GLY-ALA-PRO-VAL-TYR-VAL-VAL-ALA-GLU-SER-ALA-SER-THR-THR-

β HIS-ALA-PHE-THR-GLU-GLY-GLY-PRO-ALA-GLN-PHE-VAL-ASN-ALA-THR-SER-LYS-GLU-VAL-VAL-

190 200

α GLU-TRP-ALA-ALA-GLU-ARG-GLY-LEU-PRO-MET-ILE-LEU-SER-TRP-ILE-ILE-ASN-THR-HIS-GLU-

β GLU-TRP-ALA-ALA-LYS-LEU-GLY-LEU-PRO-LEU-VAL-PHE-ARG-TRP-ASP-ASP-SER-ASN-ALA-GLN-

Fig. 6. Alignment of the N-terminal 200 residues of the α and β subunits of V. harveyi luciferase.

amino acid residue stretch shown above. Identity between two sequences gives a correlation of 1.0; hydrophobicity correlation coefficients are as low as about 0.55 for structurally similar but widely divergent sequences (27).

THREE DIMENSIONAL STRUCTURE

The structure and function of bacterial luciferase have received considerable attention by researchers in at least four laboratories during the past 10-15 years and there is therefore much known about the structure of the enzyme (1). Chemical modification studies on the wild-type and mutant forms of luciferase, and detailed mutant characterizations, though yielding valuable information regarding the amino acid residues in or near the active center, allow one only to infer the mechanism in rather vague molecular terms. Understanding stereochemical details of the bacterial bioluminescence reaction requires knowledge of the three-dimensional structure of the enzyme.

Crystallization Procedures--Crystals of luciferase have been obtained (28) at room temperature in the pH range 6.2 to 7.4 from phosphate, ammonium sulfate and/or citrate solutions, 0.5 to 5 mM in dithiothreitol, using microdialysis or vapor diffusion. The method that yielded large crystals with greatest reliability was the hanging-drop vapor diffusion technique, with starting protein concentrations of 50 to 100 mg/ml in 0.4 M phosphate, over approximately 2.0 M phosphate in the well. Crystals have been seen to form in as short a time as 10 hours, but the drops with largest crystals took from 50 to 100 hours to show crystals, and the crystals in some cases continued to grow for two months. The trials were set up as 5 μl drops over 0.6 ml wells with 0.45 ml of liquid to reduce air exposure. The optimum phosphate concentration in the well varied by ± 0.2 M from 2.0 M, depending on protein concentration and unknown factors. Crystal-producing concentrations of citrate were about 0.9 M, and of ammonium sulfate, 1.3 to 2.1 M (i.e., 30 to 50% saturation). Two additional factors which seem to favor large crystal formation are the small drop size and the presence of an unusually high concentration of dithiothreitol (ca. 5 mM).

X-ray Characterization--Although two of the cell constants of the crystals are of reasonable lengths (ca. 60 Å and 111 Å), the third cell constant is 302 Å, resulting in diffraction that is not resolvable with ordinary techniques. However, the use of an adjustable-slit collimator set to 0.1 mm x 0.2 mm, with the narrow dimension parallel to the long axis, clearly resolved the diffraction pattern. The crystals diffract on still photographs to about 2.5 Å and on oscillation pictures to about 3.0 Å. The diffraction is good for about 30 hours of beam time, after which the higher angle data begins to weaken noticeably. The physical appearance of the crystals also changes, albeit more slowly than the resolution loss.

The cell constants (a = 59.6 ± 0.4 Å, b = 112 ± 0.7 Å, and c = 302 ± 2 Å) and space group ($P2_12_12_1$) were determined from precession photographs of the three major zones, all of which show mm symmetry and axial extinctions consistent with the existance of three two-fold screw axes. The density of a crystal (cross-linked with 1% glutaraldehyde for 90 min and soaked in water for 2 h) in a water-saturated gradient of toluene, dichloroethane, and carbon tetrachloride was 1.114 ± 0.009 g/ml. This value implies (29) that the number of molecules in the asymmetric unit is 2, corresponding to a volume per dalton of 3.3 $Å^3$. The crystals are quite strong mechanically, despite their large volume per dalton, moderate range of diffraction and short beam life.

CONCLUSIONS AND PERSPECTIVES

We have discussed here recent advances in understanding the structure of bacterial luciferase. The rapid advances came as the result of the recent cloning of the luciferase genes and the application of the techniques of recombinant DNA to determination of the structure of the enzyme. The covalent structures of both subunits have been determined, initial advances have been made into an understanding of the mode of folding of the subunits in the living cell, and the enzyme from Vibrio harveyi has been crystallized and collection of diffraction data at 3.0 Å resolution initiated. The two subunits are strikingly homologous, and will probably have similarities in three dimensional structure as well as sequence. Since the two subunits are functionally quite distinct, it was not surprising, although gratifying, to find that the sequence similarities were minimal in those regions that were thought to possess residues contributing to the active center of the α subunit.

In other laboratories much progress has been made as well. Dr. Kenneth Nealson at Scripps Institution of Oceanography and his colleagues at The Agouron Institute in La Jolla are actively studying the control of the bioluminescence system using recombinant DNA technology. Dr. E.A. Meighen at Magill University in Montreal has made substantial progress in understanding the metabolism of the aldehyde substrate. Recently, he has cloned the luminescence systems from several species of luminous bacteria and is studying the aldehyde metabolism in E. coli by various plasmid construction techniques. Dr. J.W. Hastings at Harvard has also cloned the luciferase genes from V. harveyi.

The advances in recombinant DNA technology in recent years have led to much excitement about determination of primary structures of enzymes by DNA sequencing rather than protein sequencing techniques, and in fact an enormous amount of progress has been made. We worked for five years to determine the amino acid sequence of luciferase, and due to poor solubility of fragments of luciferase, were forced to determine the sequence from the DNA. We are now in a position to ask "What next?".

Our immediate goal is to understand the interactions between bacterial luciferase and the flavin and how those interactions stabilize the 4a-peroxydihydroflavin intermediate and lead ultimately to light emission (see ref. 1). To that end we are currently collecting and analyzing x-ray diffraction data (in collaboration with Drs. Rosemarie Swanson and Brian Matthews). Also, we are altering specific amino acid residues in bacterial luciferase by oligonucleotide-directed site-specific mutagenesis in an attempt to alter specific activity-related interactions. In these studies, it is readily apparent that recombinant DNA technology will allow us to ask very specific questions about enzyme structure and function and then to design an enzyme to the desired specifications. The limiting ingredient in these studies will likely very quickly become our sophistication in understanding enzyme structure well enough to be able to design the proper variants. Clearly, that sophistication will not come from site-directed mutagenesis, but from classical protein chemistry and mutant enzyme analysis. From the information gained from these studies, more precise questions can be formulated, and we hope, answered through the proper combination of protein chemistry and recombinant DNA technology.

We acknowledge with gratitude the contributions of our friends and colleagues Dr. Larry Weaver, Dr. Brian Matthews, Dr. Miriam Ziegler, and students past and present. Our studies are supported by the NSF, NIH, Robert A. Welch Foundation, and a Hatch Grant. This manuscript was skillfully prepared by Ms. Grace Wood.

References

1. Ziegler, M.M., Baldwin, T.O.: Current Topics in Bioenergetics, Vol. 12, (D. Rao Sanadi, ed.), pp. 65-113, Academic Press, New York (1981).

2. Hastings, J.W., Gupta, S.C., Potricus, C.J., Makemson, J.: in Advances in Microbial Biology (1984, in press).

3. Byers, D., Meighen, E.: J. Biol. Chem. 259, 7109-7114 (1984).

4. Proceedings of the Third International Symposium on Applications of Chemiluminescence and Bioluminescence (T. Whitehead and L. Kricka, eds.), W. de Gruyter & Co., Berlin (1984, in press).

5. Kurfurst, M., Ghisla, S., Hastings, J.W.: Proc. Natl. Acad. Sci. U.S.A. 81, 2990-2994 (1984).

6. Tu, S.-C., Henkin, J.: Biochemistry 22, 519-523 (1983).

7. Rausch, S.K., Dougherty, J.J., Jr., Baldwin, T.O.: in Flavins and Flavoproteins (V. Massey and C.H. Williams, eds.), pp. 375-378, Elsevier, Amsterdam (1982).

8. Cohn, D.H., Ogden, R.C., Abelson, J.N., Baldwin, T.O., Simon, M.I., Nealson, K.H., Mileham, A.J.: Proc. Natl. Acad. Sci. U.S.A. 80, 120 123 (1983).

9. Engerbrecht, J., Nealson, K.H., Silverman, M.: Cell 32, 773-781 (1983).

10. Rodriguez, A., Wall, L., Carey, L., Boylan, D., Byers, D., Meighen, E.: in Proceedings of the Third International Symposium on Analytical Applications of Bioluminescence and Chemiluminescence (T.P. Whitehead and L. Kricka, eds.), W. de Gruyter, Berlin (1984, in press).

11. Nealson, K.H., Cassin, R.C.: in Proceedings of the Third International Symposium on Analytical Applications of Bioluminescence and Chemiluminescence (T.P. Whitehead and L. Kricka, eds.), W. de Gruyter, Berlin (1984, in press).

12. Baldwin, T.O., Hastings, J.W., Riley, P.L.: J. Biol. Chem. 253, 5551-5554 (1978).

13. Holzman, T.F., Baldwin, T.O.: Proc. Natl. Acad. Sci. U.S.A. 77, 6363-6367 (1980).

14. Holzman, T.F., Riley, P.L., Baldwin, T.O.: Arch. Biochem. Biophys. 205, 554-563 (1980).

15. Baldwin, T.O., Dougherty, J.J., Jr., Merritt, M.V., Rausch, S.K., Ziegler, M.M.: in preparation (1984).

16. Rausch, S.K., Baldwin, T.O.: in preparation (1984).

17. Rausch, S.K.: Ph.D. Thesis, University of Illinois, Urbana (1983).

18. Nicoli, M. Ziegler: Ph.D. Thesis, Harvard University, Cambridge, MA. (1972).

19. Dougherty, J.J., Jr., Rausch, S.K., Baldwin, T.O.: in Flavins and Flavoproteins (V. Massey and C.H. Williams, eds.), pp. 379-382, Elsevier, Amsterdam (1982).

20. Baldwin, T.O., Berends, T., Bunch, T.A., Holzman, T.F., Rausch, S.K., Shamansky, L., Treat, M.L., Ziegler, M.M.: Biochemistry (1984, in press).

21. Towbin, H., Staehelin, T., Gordon, J.: Proc. Natl. Acad. Sci. U.S.A. 76, 4350-4354 (1979).

22. Cohn, D.H., Mileham, A.J., Simon, M.I., Nealson, K.H., Rausch, S.K., Baldwin, T.O.: J. Biol. Chem. (1984, submitted).

23. Garnier, J., Oglethorpe, D.J., Robson, B.: J. Mol. Biol. 120, 97-120 (1977).

24. Swanson, R.: Bull. Math. Biol. 46, 187-203 (1984).

25. Hunt, J., Johnston, T.C., Baldwin, T.O., unpublished observations.

26. Baldwin, T.O., Ziegler, M.M., Powers, D.A.: Proc. Natl. Acad. Sci. U.S.A. 76, 4887-4889 (1979).

27. Sweet, R.M., Eisenberg, D.: J. Mol. Biol. 171, 479-488 (1983).

28. Swanson, R., Weaver, L., Matthews, B., Baldwin, T.O.: in preparation (1984).

29. Matthews, B.W.: J. Mol. Biol. 82, 513-526 (1974).

PROBING THE BACTERIAL LUCIFERASE ALDEHYDE SITE BY AFFINITY AND PHOTOAFFINITY LABELING

Shiao-Chun Tu, Aaron Fried

Department of Biochemical and Biophysical Sciences, University of Houston-University Park, Houston, Texas 77004, U.S.A.

Introduction

Bacterial luciferase, an αβ heterodimer, catalyzes a monooxygenation reaction in which $FMNH_2$ and a long chain aldehyde are oxidized by O_2 to yield FMN, carboxylic acid, water, and light. For <u>Vibrio harveyi</u> luciferase, earlier hybridization studies using chemically (1) and mutationally (2) modified subunit variants indicate that α subunit (Mr 42,000) participates directly in catalysis whereas the function of the β subunit (Mr 37,000) is unclear. More recently, a functional role in the $FMNH_2$ binding has been indicated for the β subunit (3,4,5). The subunit location of and essential amino acid residue(s) present at the aldehyde site, however, remain unknown. In this report, we describe the results of photoaffinity and affinity labeling studies carried out recently by us to delineate the nature of luciferase aldehyde site (6,7).

Materials and Methods

Luciferase was purified from <u>V. harveyi</u> cells as described previously (8). Luciferase activities, in quanta per sec (q/s) were determined by two assay methods. The standard assay involves the injection of 1 ml of 50 µM $FMNH_2$ (photoreduced in the presence of 2 mM EDTA) into 1 ml sample solution containing luciferase, 0.2% bovine serum albumin (BSA), and 0.001% decanal under aerobic conditions (8). For the dithionite assay (9), 10 µl aliquot of an anaerobic dithionite solution (1.5%) was added to 1 ml sample solution containing luciferase and FMN to reduce the latter. To this, 1 ml of a decanal solution (0.01%) was quickly injected

Flavins and Flavoproteins

via a syringe under aerobic conditions to initiate the bioluminescence reaction. Unless stated otherwise, all enzyme experiments and assays were carried out at 23°C in 0.05 M phosphate buffer, pH 7.

For the photoaffinity labeling studies, 1-diazo-2-oxoundecane (DOU) was synthesized as a probe (6). Crystalline decanoic acid was first reacted with thionyl chloride to form the acid chloride. This product was then reacted with diazomethane to yield DOU. The 1-diazo-2-oxo[1-^{14}C]undecane ([^{14}C]DOU) was similarly synthesized using radioactive diazomethane which was obtained by reacting N-[^{14}C]methyl-N-nitroso-p-toluenesulfonamide with ethanolic NaOH. The purified [^{14}C]DOU had a specific activity of 0.35 mCi/mmol.

2-Bromo[1-^{14}C]1-decanal ([^{14}C]BrDA) was synthesized by the bromination of [1-^{14}C]decanal as an affinity labeling probe for luciferase (7). The [1-^{14}C]decanal precursor was obtained by first reacting 1-chlorononane with sodium [^{14}C]cyanide to form [1-^{14}C]decanenitrile, and then by sequentially converting the nitrile to carboxylic acid, alcohol, and the corresponding aldehyde. The final [^{14}C]BrDA product had a specific activity of 1.03 mCi/mmol. 2-Bromo-1-decanal (BrDA) was similarly synthesized using nonlabeled decanal as a starting material.

Results and Discussion

In luciferase assays carried out under dim light, DOU functioned as an inhibitor competitive against decanal (Fig. 1) with a K_i of 0.7 ± 0.1 μM. DOU thus appears to specifically and reversibly bind to the aldehyde site of luciferase. The binding of DOU to luciferase aldehyde site was also demonstrated in another experiment. The bacterial bioluminescence reaction involves 4a-hydroperoxyFMN as a key intermediate (10), designated II. Several types of non-aldehyde long chain aliphatic compounds have been shown to markedly stabilize intermediate II by binding, reversibly, to the aldehyde site (11,12,13). We have found that intermediate II can also be greatly stabilized by complexing with DOU; the half-life of II was lengthened from 9 s to 12 min at 23°C by the addition of DOU. Furthermore, the II-DOU complex exhibited full bioluminescence activity upon the addition of excess amounts of decanal, indicating that the bound DOU cab be displaced by decanal.

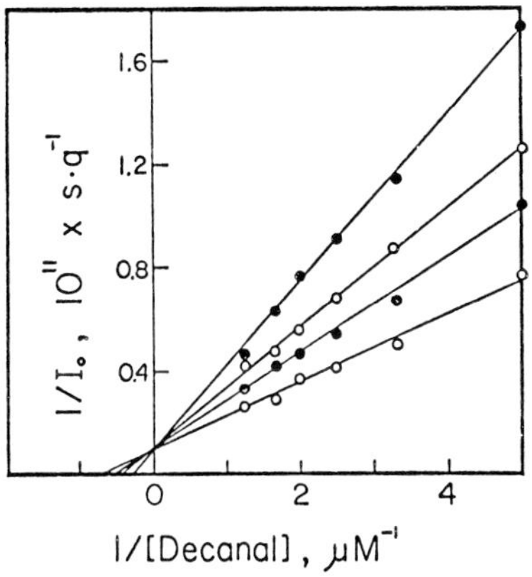

Fig. 1. DOU as a competitive inhibitor for decanal. Luciferase activities were determined by the standard assay without using BSA. Reaction solutions, 1 ml each, contained 10 µg luciferase and, from the bottom upward, 0, 0.35, 0.58, and 0.81 µM DOU.

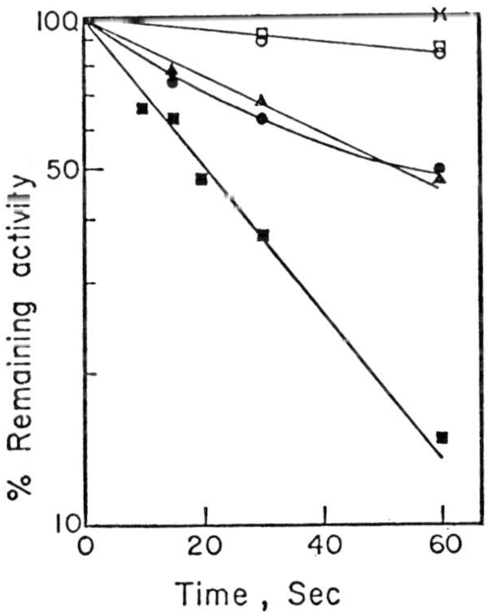

Fig. 2. Kinetics of luciferase photoinactivation. Samples contained 10 µM enzyme, 65 µM DOU, and in the absence (■) or presence of 0.5 mM decanal (●) or 13 mM 2-ME (▲) were each irradiated, and the remaining activities were determined after different times. An enzyme sample (○) or enzyme plus decanal (□) was each similarly irradiated as a control. Enzyme plus DOU kept in the dark showed no irreversible inactivation (×).

Photoaffinity labeling was carried out by the irradiation of samples containing luciferase and DOU, in the presence or absence of other reagents as indicated, using a lamp (RPR-2537 from Southern New England UV Co.) with a 254-nm emmision peak. An irreversible inactivation of luciferase occurred following apparent first-order kinetics when the enzyme was irradiated in the presence of DOU (Fig. 2). Such inactivation was retarded in the presence of decanal. The addition of 2-mercaptoethanol (2-ME) as a ketene scavenger also decreased the photoinactivation rate due, at least in part, to the significant absoption of UV light by 2-ME thus reducing the light flux reaching DOU. Only low levels of photoinactivation were detected when luciferase was irradiated either alone or in the presence of decanal.

Photolabeling of luciferase by DOU may proceed by both specific and nonspecific pathways as depicted in the Scheme 1. Upon photolysis (step

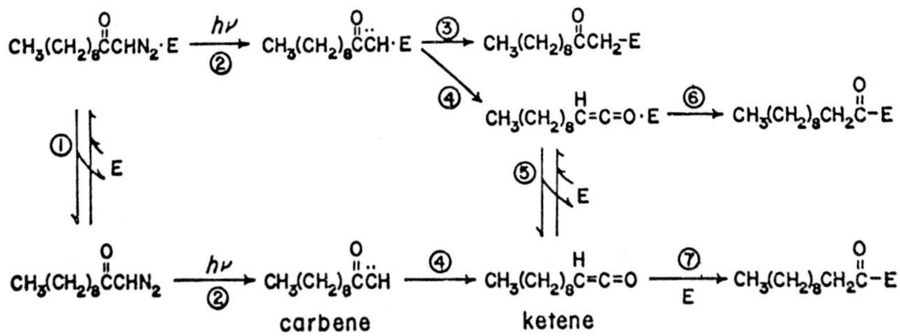

(1)

Scheme 1

2), both free and bound DOU are activated to generate the highly active carbene. The bound carbene may then specifically tag amino acid residue(s) at the active center (step 3). Since carbenes are very short-lived in water, nonspecific labelings by free carbene are probably insignificant. However, both free and bound carbene can undergo the Wolff rearrangement to form the electrophilic ketene (step 4). While the bound ketene may label the active center specifically (step 6), the free ketene is, due to its longer lifetime, capable of reacting nonspecifically with nucleophilic groups not located at the active center (step 7).

Irradiation of [^{14}C]DOU and luciferase resulted in covalent attachment of the probe to both α and β subunits (Fig. 3A). The labeling of α or β alone was insufficient to quantitatively account for the extent of luciferase inactivation. On the other hand, a molar ratio of 1.4 for the total probe incorporated into α and β over enzyme inactivated was observed, indicating the existence of nonspecific labeling. Decanal

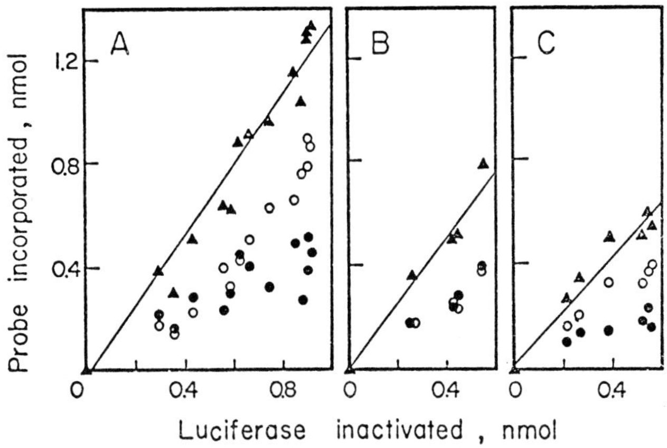

Fig. 3. Correlation of the amount of luciferase photoinactivation with that of probe incorporation into α (○) or β (●). (A), 10.7 μM enzyme + 65 μM [^{14}C]DOU; (B) = (A) + 0.5 mM decanal; (C) = (A) + 13 mM 2-ME. The total amount of probe incorporated into α and β are shown as (▲).

retarded both the rates of enzyme inactivation and photolabeling. However, the molar ratio of probe incorporated into α and β over enzyme inactivated remained at 1.3 (Fig. 3B), and the quantities of probe attached to α and β were each about the same as that observed with the aldehyde-free sample at corresponding degrees of inactivation (Figs. 3A,B). Therefore, the addition of decanal was not expected to and apparently did not prevent the nonspecific labeling. When 2-ME (a ketene scavenger) was added to quench the nonspecific labeling, the degrees of photolabeling of α were quite comparable to those shown in Figs. 3A,B but significantly smaller amounts of probe tagged the β subunit (Fig. 3C). Most importantly, the total amounts of probe incorporated into α and β in the 2-ME-added sample correlated stoichiometrically with the quantities of enzyme inactivated.

These results are interpreted to indicate that the aldehyde site is located at the subunit interface of luciferase. The photoactivated probe generated in situ at the active center may covalently tag either subunit via the specific pathways (steps 3,6), thus preventing further bindings of DOU or aldehyde. Consequently, the specific labeling of α and that of β are mutually exclusive, and each is a killing event regarding enzyme activity. In the absence of an effective scavenger, free ketene apparently leads to a lower but significant level (~40%) of nonspecific labeling of luciferase, primarily on the β subunit.

To further characterize the luciferase aldehyde site, BrDA was used as an affinity labeling reagent (7). When luciferase was reacted with an excess amount of BrDA, enzyme inactivation occurred following apparent first-order kinetics (Fig. 4). Such an inactivation was significantly

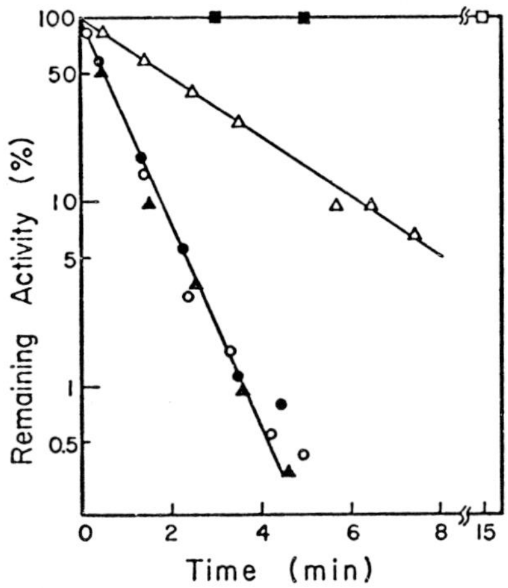

Fig. 4. Kinetics of luciferase inactivation by BrDA. One µM enzyme was treated with 25 µM BrDA and remaining activities were determined at different times (○). Other samples contained an additional 50 µM dithiothreitol (■), 70 mM 2-ME (□), or 0.2 mM of FMN (▲), butanal (●), or decanal (△).

retarded if decanal was present. However, the addition of butanal, which exhibited very poor binding and substrate activities, resulted in no detectable protection. The presence of 0.2 mM FMN, a bioluminescence reaction product derived from $FMNH_2$, also failed to protect the enzyme.

When luciferase was first mixed with dithiothreitol (DTT) or 2-ME and then reacted with BrDA, no inactivation was observed (Fig. 4).

In order to establish BrDA as an active site-directed labeling reagent, it is important to demonstrate that this probe forms a complex with luciferase before the step of covalent labeling as shown in Scheme 2

$$E + P \underset{k_2}{\overset{k_1}{\rightleftharpoons}} E{:}P \xrightarrow{k_3} E\text{-}P \qquad (2)$$

where E and P are free species of enzyme and probe, respectively, and E:P and E-P are noncovalent complex of E and P and the covalently labeled adduct, respectively. In the presence of excess amounts of probe, the rate of enzyme inactivation is expected to show a saturation phenomenon at increasing levels of probe and the observed apparent first-order rate constant, k_{obs}, can be expressed as

$$\frac{1}{k_{obs}} = \frac{1}{k_3} + \frac{K_d}{k_3} \cdot \frac{1}{[P]} \qquad (3)$$

where K_d is the dissociation constant for the noncovalent binding and [P] is the probe concentration. When rates of luciferase inactivation at pH 6 and 23°C were measured at various BrDA concentrations, the saturation phenomenon was indeed observed (Fig. 5 inset). Therefore, the double reciprocal plot of k_{obs} versus [P] (Fig. 5) allows the determination of both K_d (23 µM) and k_3 (1 min^{-1}). In another experiment, the displacement of a bound aldehyde by BrDA added secondarily was demonstrated. Luciferase was first mixed with octanal, and the bioluminescence reaction was initiated by the addition of $FMNH_2$. During the course of light emission, a solution of of BrDA was introduced, resulting in prompt inhibition of the bioluminescence reaction. When a buffer solution was similarly added as a control, no change in the light emission kinetics was detected. The above described results clearly indicate that BrDA specifically binds to the luciferase aldehyde site.

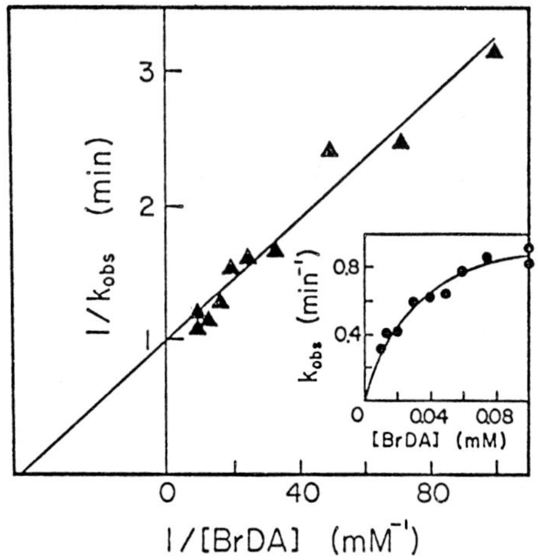

Fig. 5. Effect of BrDA concentration on the rate of luciferase inactivation. Enzyme samples, at 0.6 or 1.2 µM, were treated with BrDA at 23°C and pH 6.

The remarkable protection of luciferase activity by DTT or 2-ME suggests that a cysteinyl residue may be the target of BrDA labeling. Therefore, experiments were carried out to correlate the amount of probe incorporation with the loss of enzyme sulfhydryl groups. Fifty µM luciferase in 1 ml 0.05 M Pi, pH 8.1, was treated with 0.1 mM [^{14}C]BrDA. When ≥95% inactivation was reached, 10 µl of 2-ME was added to stop any further inactivation. Subsequently 0.4 g of urea was added to dissociate any noncovalently bound BrDA, and the sample was subject to Sephadex G-25 column chromatography to separate luciferase from free micromolecules (treatment 1). The fraction that contained the highest concentration of luciferase was used for determinations of the amount of probe incorporated by ^{14}C counting and the number of remaining sulfhydryl residues by reacting with 5,5'-dithiobis(2-nitrobenoic acid). One control run was carried out under identical conditions except that luciferase was first mixed with 2-ME and then reacted with BrDA (treatment 2). Luciferase was also similarly treated with 2-ME in the absence of BrDA (treatment 3) as another control. Results shown in Table 1 indicate that luciferase inactivation by BrDA was accompanied by both the incorporation of about 1.2

molecules of probe due to specific labeling and the loss of 0.8-1.1 residues of cysteinyl residues. Subsequent to treatment 1, the luciferase

Table1. Correlation of luciferase inactivation with [^{14}C]BrDA incorporation and cysteinyl group modification

Treatment	Remaining activity	[Probe]/[E]	[Cys]/[E]
	%		
(1) E + [^{14}C]BrDA ⟶ + 2-ME	≤5		
⟶ + urea ⟶ Sephadex G-25		1.3	12.6
(2) E + 2-ME ⟶ + [^{14}C]BrDA	100		
⟶ + urea ⟶ Sephadex G-25		0.1	13.4
(3) E + 2-ME	100		
⟶ + urea ⟶ Sephadex G-25			13.7

sample was also subject to SDS slab gel electrophoresis to separate the α from β subunit. The gel was first stained for protein bands and then the fluorogram was taken. Only the α subunit band showed a positive ^{14}C image, thus identifying the essential sulfhydryl group labeled by [^{14}C]BrDA being on α.

The existence of one particularly reactive essential cysteinyl residue on the α subunit of luciferase has been well established previously (14). However, the location with respect to specific substrate subsites and the potential functional role of this residue have not been established. The question that whether the same sulfhydryl group is the site of BrDA affinity labeling has been examined. This essential sulfhydryl residue was first modified by a reversible reagent sodium tetrathionate (TET) as described previously (15), leading to ~94% inactivation. Removal of unreacted TET by Sephadex G-25 column chromatography and subsequent reactivation of the treated enzyme under reductive conditions (15) led to ≥80% recovery of the initial activity. Initial inactivation by BrDA followed by reactivation under identical conditions led to no activity recovery. If TET and BrDA both modify the same cysteinyl residue, premasking of this group by TET should prevent the subsequent labeling of luciferase by BrDA. When luciferase was sequentially treated with TET, Sephadex G-25 chromatography, and BrDA, subsequent reactivation indeed restored ~90% of the initial activity. A similar experiment was also carried out using another reversible sulfhydryl modifying reagent 2-(2'-pyridyldithio)benzyldiazoacetate (2-PDBD) (16) in place of TET for the masking of the essential cysteinyl residue. Again, reductive reactivation of luciferase which had been sequentially treated with 2-PDBD and BrDA led to ~90% of activity recovery (7). Therefore, it appears that the essential sulfhydryl group previously identified by chemical modification was the same residue accessible to BrDA affinity labeling. Alternatively, the premasking of one essential group by TET or 2-PDBD may result in an enzyme conformational change which renders a second essential sulfhydryl residue inaccessible to BrDA. Although this possibility can not be ruled out, it is somewhat unlikely that both the inorganic salt TET and the organic mixed disulfide 2-PDBD cause a similar conformational change in luciferase to prevent the subsequent modification by BrDA.

In the bioluminescence reaction, the 4a-hydroperoxyFMN intermediate transfers one oxygen atom to the aldehyde to form the corresponding carboxylic acid. The flavin subsite and the aldehyde subsite must be within a very close vicinity. In the previous chemical modification study, FMN (at 26 μM) effectively protected luciferase against inactivation by N-ethylmaleimide (14, 17). However, no protective effect was observed at 200 μM FMN against BrDA inactivation (Fig. 4). It appears that the bound FMN at least partially blocks the binding of the bulkier N-ethylmaleimide whereas the binding of BrDA is specific for the aldehyde subsite.

Based on the above described results, Scheme 4 is postulated to illustrate the nature of affinity labeling of luciferase aldehyde site by BrDA. Species I, II, and III refer to the native enzyme, the luciferase:BrDA noncovalent complex, and luciferase with the essential sufhydryl residue labeled by BrDA, respectively. Taking the results of photoaffinity labeling study into consideration, both the essential cysteinyl group and the bound BrDA are located near the subunit interface. The pK_a of this sulfhydryl group has been previously determined to be 9.4 which does not show any apparent correlation with several cataly-

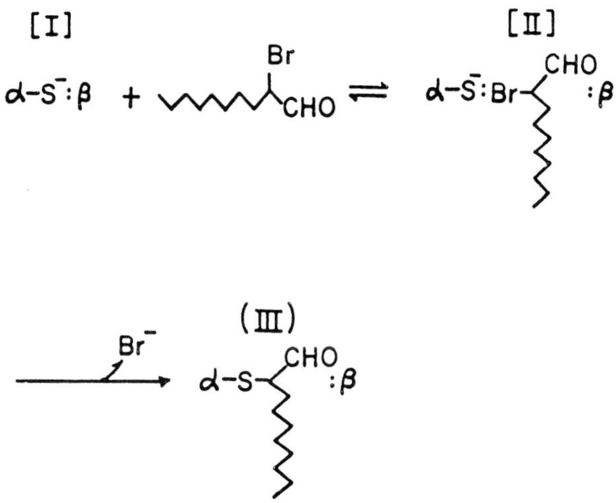

(4)

tic parameters (14). This residue has now been demonstrated to be at the aldehyde site and, more specifically, near the aldehyde functional group. However, whether this residue performs any catalytic function in the bioluminescence reaction remains to be delineated.

Acknowledgments

This work was supported by grants GM25953 and K04 ES00088 from N.I.H. and, in part, by a Robert A. Welch grant E-738. We are greatful to Dr. Jack Henkin for the synthesis of DOU and [^{14}C] DOU.

References

1. Meighen, E. A., Nicoli, M. Z., Hastings, J. W.:Biochemistry 10, 4069-4073 (1971).

2. Cline, T. W., Hastings, J. W.:Biochemistry 11, 3359-3370 (1972).

3. Meighen, E. A., Bartlet, I.:J. Biol. Chem. 255, 11181-11187 (1980).

4. Welches, W. R., Baldwin, T. O.:Biochemistry 20, 512-517 (1981).

5. Watanabe, H., Hastings, J. W., Tu, S.-C.:Arch. Biochem. Biophys. 215, 405-413 (1982).

6. Tu, S.-C., Henkin, J.:Biochemistry 22, 519-523 (1983).

7. Fried, A., Tu, S.-C.:J. Biol. Chem. in press (1984).

8. Hastings, J. W., Baldwin, T. O., Nicoli, M. Z.:Methods Enzymol. 57, 135-152 (1978).

9. Tu, S.-C., Hastings, J. W.:Biochemistry 14, 4310-4316 (1975).

10. Hastings, J. W., Balny, C., LePeuch, C., Douzou, P.: Proc. Natl. Acad. Sci. U.S.A. 70, 3468-3472 (1973).

11. Hastings, J. W., Gibson, Q. H., Friedland, J., Spudich, J.:in Bioluminescence in Progress (Johnson, F. H., Haneda, Y., eds) pp. 151-186, Princeton University Press, Princeton, NJ (1966).

12. Baumstark, A. L., Cline, T. W., Hastings, J. W.:Arch. Biochem. Biophys. 193, 449-455 (1979).

13. Tu, S.-C.:Biochemistry 18, 5940-5945 (1979).

14. Nicoli, M. Z., Meighen, E. A., Hastings, J. W.:J. Biol. Chem. 249, 2385-2392 (1974).

15. Cousineau, J., Meighen, E.: Can, J. Biochem. 55, 433-438 (1977).

16. Henkin, J.:J. Biol. Chem. 252, 4293-4297 (1977).

17. Nicoli, M. Z., Hastings, J. W.:J. Biol. Chem. 249, 2393-2396 (1974).

THE CATALYTIC TURNOVER OF BACTERIAL LUCIFERASE PRODUCES A QUASI-STABLE SPECIES OF ALTERED CONFORMATION

Nabil K. AbouKhair, Miriam M. Ziegler, and Thomas O. Baldwin

Department of Biochemistry and Biophysics, Texas A&M University, and the Texas Agricultural Experiment Station, College Station, TX 77843.

Introduction

Previous studies have shown that modification of a reactive thiol on bacterial luciferase results in inactivation of the enzyme, and that aerobic reaction of luciferase with $FMNH_2$ results in longer-lived protection of the enzyme than can be accounted for by the lifetime of the 4a-peroxydihydroflavin intermediate or by the simple binding of oxidized flavin to the enzyme (1). Two hypotheses were offered as explanations for this phenomenon: after a catalytic cycle, either the complex of the enzyme with the product FMN (E-FMN) dissociated more slowly than would be predicted from the K_d for oxidized flavin, or the enzyme had undergone a conformational change which gave it the "protected" character. The present experiments were undertaken to distinguish between the two hypotheses.

Results and Discussion

A mixture of luciferase and one substrate, $FMNH_2$ (produced by reduction of FMN by dithionite), was chromatographed at $0^{o}C$ on a Sephadex G-25 column equilibrated with the other two substrates for the bioluminescent reaction, O_2 and decanal, essentially as described by Kurfurst _et al._, who were studying the enzyme-hydroxyflavin intermediate postulated to be the emitter in the luminescent reaction (2). The chromatographic conditions were such that if the E-FMN complex were dissociating much more slowly than the decay of light emission, then the eluted protein fraction should contain about equivalent concentrations of enzyme and FMN; however, spectrophotometric

Flavins and Flavoproteins

and spectrofluorometric analyses of the eluted enzyme fraction showed that the enzyme:flavin ratio was always greater than 2:1. Furthermore, the rate of dissociation of the E-FMN complex was determined to be similar to that of the enzyme-hydroxyflavin intermediate (data not shown), suggesting that the FMN did not remain associated with the "spent" enzyme to an extent which could account for the observed long-lived protection from alkylation.

When the enzyme was permitted to react with $FMNH_2$ and O_2 (no aldehyde) for 30 sec at $0^{\circ}C$, the previously demonstrated (1) monophasic inactivation by N-ethylmaleimide was observed (Fig. 1), with a much longer lifetime than that of the peroxyflavin intermediate, indicating the existence of a protected "spent" form of the enzyme following nonluminescent turnover. When the enzyme-$FMNH_2$-O_2 reaction mixture was incubated at $0^{\circ}C$ for 20 min prior to addition of N-ethylmaleimide, the inactivation was biphasic, the rapid

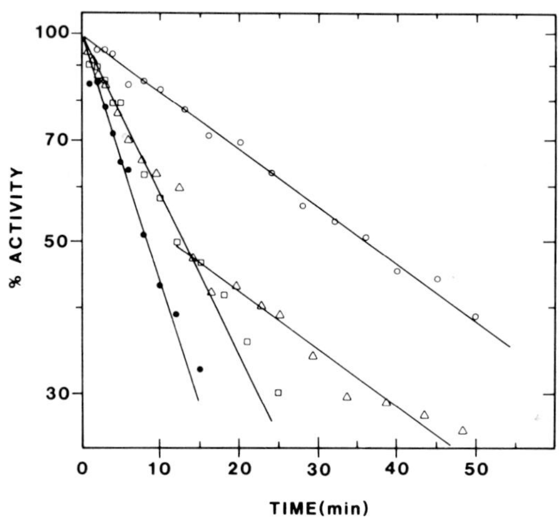

Figure 1. Protection of luciferase from inactivation by N-ethylmaleimide (NEM) at $0^{\circ}C$. $FMNH_2$ (0.5 ml) was rapidly injected into an aerobic solution of luciferase (0.49 ml). NEM (10 µl) was added at time = 0, either 30 sec (○) or 20 min (△) after flavin addition. Protection by FMN + H_2O_2 (□) was obtained by injection of $FMNH_2$ into the NEM, followed after 30 sec (time = 0) by addition of luciferase. Controls (no flavin, ●) had buffer injected instead of flavin, followed by addition of NEM at time = 0. Final concentrations in the reaction mixtures were as follows: luciferase, 1.7 µM; NEM, 1.0 mM; flavin, 24 µM. First order rate constants were 0.083 min^{-1} (no flavin); 0.053 min^{-1} (FMN + H_2O_2); 0.019 min^{-1} ($FMNH_2$ + O_2 + 30 sec); 0.053 min^{-1} and 0.019 min^{-1} (biphasic inactivation, $FMNH_2$ + O_2 + 20 min).

rate being similar to that seen with protection by oxidized FMN and the slow rate similar to the inactivation rate of the "spent" enzyme (Fig. 1).

Luciferase is rapidly inactivated by limited cleavage by a wide variety of proteases (3). When "spent" luciferase obtained either from nonluminescent (no aldehyde) turnover of the enzyme (data not shown) or from Sephadex G-25 chromatography of a luminescent reaction mixture was treated with chymotrypsin, long-lived protection similar to the protection from N-ethylmaleimide was observed (Fig. 2), suggesting that the "spent" enzymes from nonluminescent and luminescent turnover had similar characteristics. A 2-fold

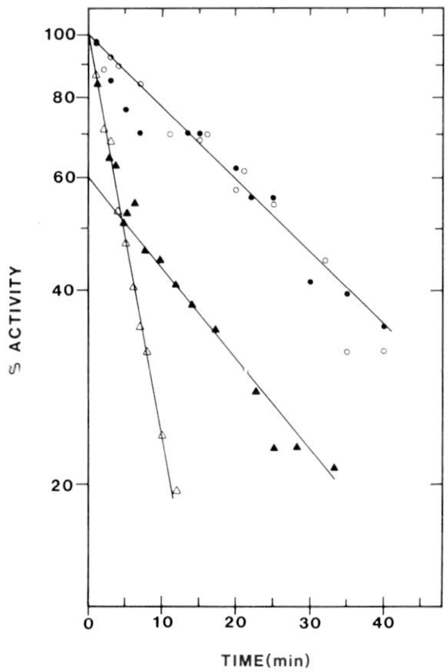

Figure 2. Inactivation of native and "spent" luciferase by chymotrypsin at 0°C. A 260 µl mixture of luciferase (0.1 mM) and FMN (0.55 mM) reduced by $Na_2S_2O_4$ was chromatographed at 1°C on a 1 x 12 cm column of Sephadex G-25 preequilibrated with 0.01% (v/v) decanal, 0.35 M phosphate, pH 7.0. The eluted protein fraction ("spent" luciferase) was collected and either mixed immediately (time = 0) with chymotrypsin (final concentration 0.1 mg/ml, ●; 0.2 mg/ml, ○) or permitted to stand at 0°C for 20 min prior to addition (at time = 0) of chymotrypsin (final concentration 0.2 mg/ml, ▲). Final concentration of "spent" luciferase was 8.5 µM. Native enzyme (10 µM) in the column elution buffer was also inactivated by chymotrypsin (final concentration 0.2 mg/ml, △). First order rate constants were 0.146 min^{-1} (native enzyme); 0.026 min^{-1} ("spent" enzyme immediately after elution from column, for both concentrations of protease); 0.146 min^{-1} and 0.032 min^{-1} ("spent" enzyme after 20 min at 0°C, biphasic inactivation).

decrease in protease concentration did not further decrease the rate of inactivation of the "spent" enzyme (Fig. 1), indicating that the rate-limiting step was not the proteolytic cleavage of the enzyme, but some slower step, presumably the relaxation of the "spent" enzyme to a protease-sensitive (native) conformation, with a $t_{1/2}$ of about 25 min at $0^{O}C$. As with N-ethylmaleimide inactivation of the enzyme following nonluminescent turnover, incubation for 20 min at $0^{O}C$ of the eluted "spent" enzyme following luminescent turnover, prior to chymotrypsin addition, gave a biphasic inactivation rate, the rapid phase presumably reflecting the amount of enzyme which had relaxed to the native state during the 20 min incubation.

We propose that the catalytic turnover of luciferase by either a luminescent or nonluminescent pathway produces a quasi-stable enzyme species of altered conformation (E'):

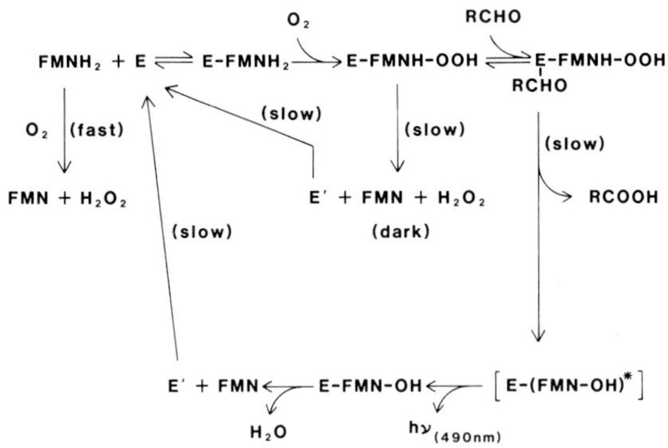

(This work was supported by grants from NSF, NIH and the Welch Foundation.)

References

1. Nicoli, M.Z., Meighen, E.A., Hastings, J.W.: J. Biol. Chem. 249, 2385-2392 (1974).

2. Kurfurst, M., Ghisla, S., Hastings, J.W.: Proc. Natl. Acad. Sci. U.S.A. 81, 2990-2994 (1984).

3. Holzman, T.F., Baldwin, T.O.: Proc. Natl. Acad. Sci. U.S.A. 77, 6363-6367 (1980).

ELECTRON MICROSCOPY AND X-RAY DIFFRACTION STUDIES ON HEAVY RIBOFLAVIN SYNTHASE FROM BACILLUS SUBTILIS

Rudolf Ladenstein[1], Birgit Meyer[1], Robert Huber[1], Harald Labischinski[2], Klaus Bartels[3], Hans-Dieter Bartunik[3], Luis Bachmann[4], Heide C. Ludwig[5] and Adelbert Bacher[5].

[1]Max-Planck-Institut für Biochemie, D-8033 Martinsried
[2]Robert Koch-Institut, D-1000 Berlin
[3]European Molecular Biology Laboratory, Outstation DESY, D-2000 Hamburg
[4]Lehrstuhl für Technische Chemie, Technische Universität München, D-8046 Garching
[5]Lehrstuhl für Organische Chemie und Biochemie, Technische Universität München, D-8046 Garching

Introduction

Heavy Riboflavin synthase from Bacillus subtilis has a molecular weight of about 1 000 000 and consists of 3 α-subunits and approximately 60 ß-subunits (1). Electron microscopic studies show spherical particles with a diameter of roughly 150 Å. Immunochemical evidence indicates that the α-subunits are not exposed to the molecular surface. This suggests that the α-subunits form the internal part of the enzyme, whereas the ß-subunits form a capsid-like protein shell with icosahedral symmetry which envelops the α-subunits. In this paper the dimensions and molecular parameters of heavy Riboflavin synthase will be reported and the hypothesis (2) proven, that the ß-subunits of the native heavy enzyme form a capsid structure which obeys icosahedral 532 symmetry. Furthermore, the crystal packing as well as the particle orientation relative to the crystal cell axes are shown.
The results are based on small angle X-ray scattering data, electron microscopy of freeze-fractured three-dimensional crystals and X-ray diffraction data to 3.5 Å resolution analyzed by Patterson correlation methods.

Flavins and Flavoproteins

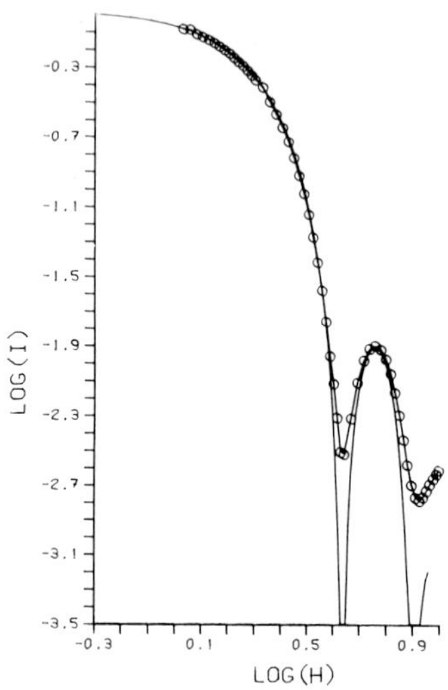

Table 1 Molecular Parameters of heavy Riboflavin synthase

R_G radius of gyration	65.9 Å
R_a outer radius	84.0 Å
R_i/R_a	0.3

Figure 1. Experimental scattering curve of heavy Riboflavin synthase in 0.1 M phosphate buffer, pH=7.0 c=4.0 mg/ml, overlaid with the calculated curve for a hollow sphere

Results

Small angle X-ray data were evaluated with the assumption that the overall structure of the oligomer can be described by a hollow sphere model, defined by the ratio of inner and outer radius, R_i/R_a. The model fitted the experimental scattering curves (Fig. 1) in a very satisfactory way and allowed the determination of several molecular parameters shown in Table 1. The shape of the scattering curves at large values of the scattering angle suggests an elongated form of the ß-subunits.

A careful inspection of electron micrographs, showing the ab and ac planes of the hexagonal crystals showed, that the individual molecules, appearing approximately spherical, are packed in hexagonal layers with particle positions (0,0,1/4) and (0,0,3/4) or (0,0,0) and (0,0,1/2) on the sixfold screw axes of the cell. The lattice constants found from the micrographs (a=b=156 Å, c=274 Å) are in good agreement with the values derived from X-ray diffraction data (3). In view of their small

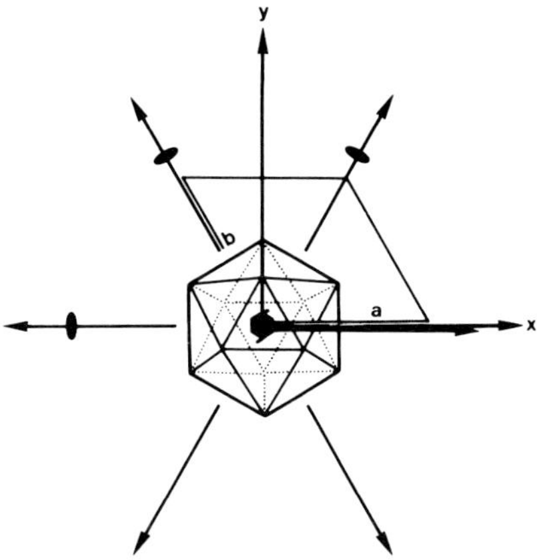

Figure 2. Orientation of the icosahedral particles relative to the crystal cell axes. Three local twofolds lying in the ab plane are indicated.

size (M_{β}=16000 dalton) it is not surprising, that single β-subunits cannot be recognized at the present resolution.

Rotation function calculations in Patterson space ($F^2 > 1\sigma$, integration radius R=10-50 Å) show a set of peaks for twofold, threefold and fivefold local rotation axes, accurately consistent with icosahedral symmetry and with the particle orientation described by Figure 2. The crystal packing can be described in short terms as follows: Enzyme particles with icosahedral symmetry (point group 532) are located at points 32 of the hexagonal cell, corresponding to positions (0,0,0) and

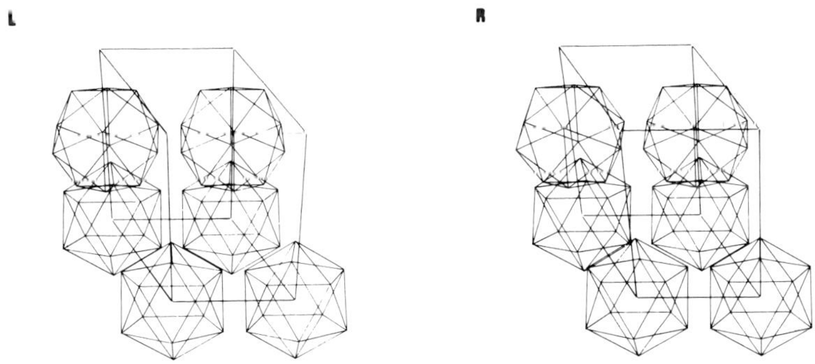

Figure 3. Stereo drawing of the crystal packing.
Four particles at z=0, and two particles at z=1/2 are shown.

(0,0,1/2) on the sixfold screw axes. Two orientations of the two particles in a unit cell are possible, which differ by the crystallographic rotation/translation operation described by the symbol 6_3 (Fig. 3)

From the data reported in this paper it may be concluded that the 60 ß-subunits form an icosahedral capsid with triangulation number T=1. At the present stage it is not possible to describe the location of the α subunits in closer detail. It is conceivable that they may not obey the crystallographic symmetry (3).

References

1. Bacher, A., Baur, R., Eggers, U., Harders, H., Otto, M.K. and Schnepple, H.: J. Biol. Chem. 255, 632-637 (1980).

2. Bacher, A., Mailänder, B., Otto, M. K., Schnepple, H. and Ben-Shaul, Y. Flavins and Flavoproteins (Yagi, K. and Yamano, T., eds) pp. 579-586, Japan Scientific Societies Press (1980).

3. Ladenstein, R., Ludwig, H. C. and Bacher, A.: J. Biol. Chem. 258, 11981-11983 (1983).

HEAVY RIBOFLAVIN SYNTHASE FROM *BACILLUS SUBTILIS*.
Primary structure and reaggregation of the β-subunits.

Heide C. Ludwig*, Friedrich Lottspeich*, Agnes Henschen* Rudolf Ladenstein*, and Adelbert Bacher#

*Max-Planck-Institut für Biochemie, D-8033 Martinsried, West Germany, and #Lehrstuhl für Organische Chemie und Biochemie der Technischen Universität München, D-8046 Garching, West Germany

Heavy riboflavin synthase of *Bacillus subtilis* is a 1,000,000 Dalton protein and consists of about 3 α-subunits and 60 β-subunits (1). The β-subunits catalyze the formation of the lumazine **2** by condensation of the pyrimidine **1** with a 4-carbon unit of hitherto unknown structure (2-4). The α-subunits catalyze the dismutation of the lumazine yielding riboflavin (**3**) and 5-amino-6-ribitylamino 2,4(1H,3H)-pyrimidinedione.

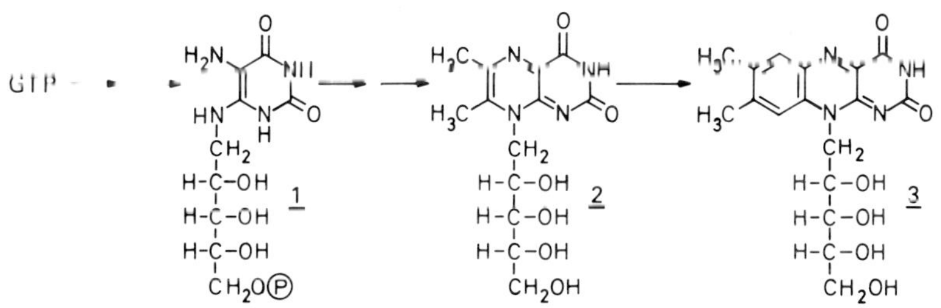

Results

Heavy riboflavin synthase dissociates at pH values above 7.5. Thus, α and β subunits can be easily separated by sucrose gradient centrifugation in appropriate buffers.

Flavins and Flavoproteins

The complete amino acid sequence of the β-subunits was determined by automated liquid phase Edman degradation (Fig. 1). The sequence was derived from peptides obtained by cleavage of the S-methylated and of the untreated β-subunits with cyanogen bromide, trypsin (with and without prior blocking of lysine residues) and Staphylococcus aureus V8 proteinase. The secondary structure was predicted by two methods with good agreement (5-7). The subunit molecular weight is 16,215.

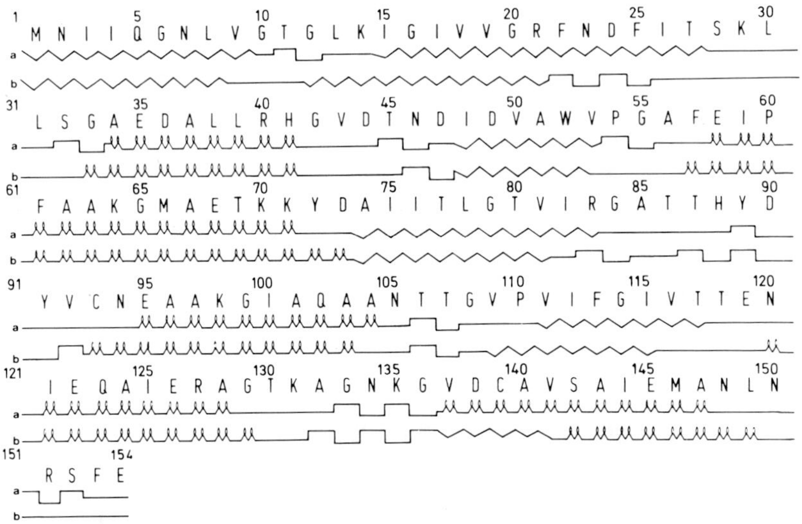

Fig. 1

Amino acid sequence and predicted secondary structures of heavy riboflavin synthase β-subunits. Loops, α-helix; zigzags, β-sheet; steps, β-turn structure; straight lines, random coil.

Isolated β-subunits form high molecular weight aggregates. Electron micrographs show large, apparently hollow, spherical particles with diameters of 22-33 nm. The sedimentation pattern in the analytical ultracentrifuge is complex. A broad boundary with a sedimentation velocity of about 70 S and a relatively sharp boundary with a sedimentation velocity of 48 S were observed. Gel electrophoresis under non-denaturing conditions shows multiple bands. The data suggest the presence of multiple molecular forms in equilibrium. Meniscus depletion equilibrium sedimentation shows a weight average molecular weight of 5,600,000.

Reaggregation of isolated β-subunits can be induced by the substrate analogue, 5-nitroso-6-ribitylamino-2,4(1H,3H)-pyrimidinedione (**4**) and yields smaller molecules sedimenting at a velocity of about 26 S (subsequently designated as 26 S aggregate). This 26 S aggregate was obtained in pure form by renaturation of β-subunits from 6.4 M urea in the presence of the ligand **4**. These molecules are stable in the presence of the ligand. However, removal of the ligand is accompanied by the reappearance of the larger molecules described above. The 26 S β-subunit aggregates and heavy riboflavin synthase migrate at similar velocities in polyacrylamide gel electrophoresis. The sedimentation velocity of the 26 S aggregate is about 7 % smaller as compared to the native heavy riboflavin synthase. The molecular weight of the 26 S aggregate was 960,000. On the basis of the subunit molecular weight of 16,215, it follows that the aggregate consists of 60±1 β-subunits. Electron micrographs of the 26 S aggregate show apparently spherical molecules, which are closely similar to native heavy riboflavin synthase. Some of the micrographs show tentative evidence of a central cavity. The spherical shape of the molecule has also been confirmed by small angle X-ray scattering experiments (8).

On the basis of stoichiometry considerations and of immunochemical evidence, we have suggested earlier that native heavy riboflavin synthase consists of an icosahedral shell of 60 β-subunits with a central core of 3 α-subunits. The icosahedral 532 symmetry of the native enzyme has been confirmed by X-ray diffraction analysis (10). The present data suggest that the 60 β-subunits of the 26 S aggregate obey the same icosahedral symmetry as the native heavy riboflavin synthase. Thus, the 26 S aggregate represents a useful model for the β-subunit capsid of the native enzyme.

This work was supported by the Deutsche Forschungsgemeinschaft and by the Fonds der Chemischen Industrie.

References

1. Bacher, A., Baur, R., Eggers, U., Harders, H., Otto, M. K., Schnepple H.: J. Biol. Chem. **255**, 632-637 (1980)
2. Bacher, A., Mailänder, B.: J. Bacteriol. **134**, 476-482 (1978)
3. Bacher, A., Ludwig, H.: Eur. J. Biochem. **127**, 539-545 (1982)
4. Neuberger, G., Bacher, A., manuscript in preparation
5. Chou, P. Y., Faßman, G. D.: Adv. Enzymol. **47**, 45-148 (1978)
6. Garnier, D., Osguthorpe, D. J., Robson, B., J. Mol. Biol. **120**, 97-120 (1978)
7. Nishikawa, K.: Biochim. Biophys. Acta **748**, 285-299 (1983)
8. Ladenstein, R., Meyer, B.: unpublished data
9. Bacher, A., Mailänder, B., Otto, M. K., Schnepple, H., and Ben-Shaul, Y. in Flavins and Flavoproteins, ed. Yagi, K., and Yamano, T., Japan Scientific Societies Press, 579-586 (1980)
10. Ladenstein, R., Meyer, B., Huber, R., Labischinski, Bartels, K., Bartunik, H. D., Bachmann, L., Ludwig, H., Bacher, A.: this volume

ENZYME REACTION MECHANISMS

MECHANISM OF α,ß-DEHYDROGENATION OF FATTY ACID CoA DERIVATIVES BY FLAVIN ENZYMES

Sandro Ghisla

Faculty of Biology, University of Konstanz
D-7750 Konstanz, FRG

Introduction

From the mechanistic point of view, the dehydrogenation of fatty acids (as CoA esters), and of a few other biologically relevant molecules, such as succinate, is unique in that it involves the rupture of two kinetically stable C-H bonds. This is contrary to the flavin dependent dehydrogenation of most other substrates, as has been discussed elsewhere (1,2). Generally a single C-H bond, which is flanked by a (kinetically) labile -X-H (or X^-) group (X = O or NR), is broken during catalysis. The stereochemistry of dehydrogenation is pro-2R, pro-3R, trans (3,4), and its minimal formulation is as follows:

$$\text{-}\overset{H}{\underset{H}{C}}\text{-}\overset{H}{\underset{H}{C}}\text{-COX} + Fl_{ox} \rightleftharpoons \text{-}\underset{H}{C}\text{=}\underset{H}{C}\text{-COX} + Fl_{red}H + H^{+} \qquad (X = S\text{-CoA})$$

With respect to its molecular mechanism the following questions are of particular importance:

a) What is the mode of rupture of the two C-H bonds? (hydride transfer, radical abstraction, carbanion mechanism or a combination of the same).
b) Is the mechanism sequential or synchronous/concerted? (do intermediates occur?). What is the breaking sequence of the C-H bonds?
c) Which of the two hydrogens (if any) is transferred to the flavin?
d) By which mechanism are the redox equivalents transferred from the substrate to the flavin acceptor (hydride or radical transfer, or covalent catalysis)?

Flavins and Flavoproteins

Dehydrogenation is Initiated by Abstraction of the α-Hydrogen as a Proton

The α C-H bond of fatty acid CoA derivatives is weakly acidic due to the presence of the thioester group; its pK might be close to that of ketones (5). The similarity of thioesters to ketones is evidenced by the finding that the latter function can be substituted to the former in a thioester analog, which retains full substrate activity for general acyl-CoA dehydrogenase (6). Deprotonation at position αC-H will be assisted by an enzyme active center base, probably a carboxylate (7), and leads to a carbanion, which is stabilized by mesomerisation to the enolate (Scheme 1). It is important to point out, however, that the pK of the α C-H group is likely to be > 16 (5). Even if the protein might affect it, the difference between the microscopic pK's of the protein base and that of the α C-H will still be large. Consequently, the equilibrium of Scheme 1 will lie far to the left, and the carbanion should be considered a transient species rather than a true intermediate.

Scheme 1:

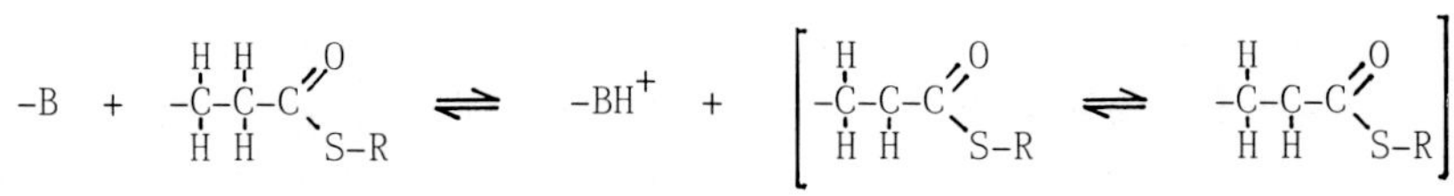

This consideration already suggests part of the answer to question a) raised in the Introduction: Deprotonation of the α C-H bond is a thermodynamically feasible event. Contrary to this, the reactivity of the ß-C-H bond is essentially that of an alkane ($pK \gg 20$). It might, however, be affected by chemical modification at the α carbon.

In the past evidence has accumulated in favour of a carbanion initiated mechanism, and this is now accepted by general consent. This evidence has been reviewed elsewhere and will only be summarized here:

Fendrich, and Abeles have shown (7), that vinylacetyl-CoA is tautomerised to cronotyl-CoA (i, Scheme 2), and that fluoride is eliminated from 3-fluoro-thioesters (ii, Scheme 2) in the presence of butyryl-CoA dehydrogenase (7). The allenic 3,4-Pentadienoyl-CoA is isomerised to the 2,4-diene by general acyl-CoA dehydrogenase (iii, Scheme 2)(8,9). 3-Ynoyl-thioesters inactivate acyl-CoA dehydrogenases, probably via isomerisation to the 2,3-allenyl-derivatives (iv, Scheme 2) (5,10).

Scheme 2:

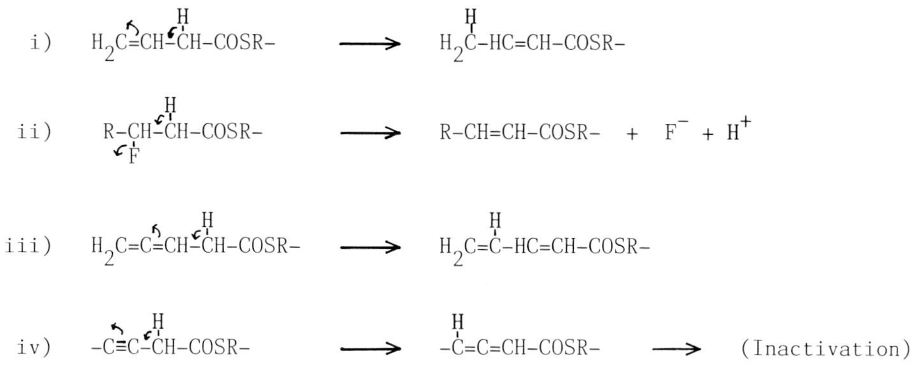

Possible Modes of Transfer of Reducing Equivalents between Substrate and Flavin

As discussed above catalysis is likely to be initiated by abstraction of the α-proton (Scheme 1). For the transfer of reducing equivalents from the transient carbanion to the oxidized flavin (and for the reversal of this reaction) several routes are conceivable, as shown in Scheme 3, and as pointed out elsewhere and by others (2,11,12,13).

The simplest alternative consists in the expulsion of a hydride from the carbanion transient, and its addition to the flavin position N(5) (Scheme 3, A). While this mechanism indeed appears most reasonable, the contention should be kept in mind, that oxidised flavin is a poor hydride acceptor (12,14). This hydride mechanism might be verified by demonstrating incorporation of labelled ß-hydrogen from substrate, as will be discussed below.

Oxidation of the transient carbanion by transfer of one electron to the flavin to form a radical pair could proceed by two routes, as shown in Scheme 3, (B,a), and (B,b). This is similar to the very first mechanistic proposal for the oxidation of activated fatty acids put forward by Cornforth in 1959 (15). It also is a reasonable mechanistic alternative, since radical oxidation of carbanions is a feasible reaction (11).

Depending on whether such a reaction would proceed via a neutral (Scheme 3, B,a), or a radical anion (B,b), incorporation of the α-or of the ß-hydrogen into the flavin might result. Collapse of the radical pair, on the other hand, would lead to covalent intermediates (Scheme 3, C). A (negative) test of such a radical transfer mechanism might involve the use of 5-deaza-5-carba-flavin analogs, for which a radical transition state would be very unfavourable energetically (16). Formation of flavin radicals upon addition of substrate to acyl-CoA dehydrogenase has been recently claimed (17).

Scheme 3:

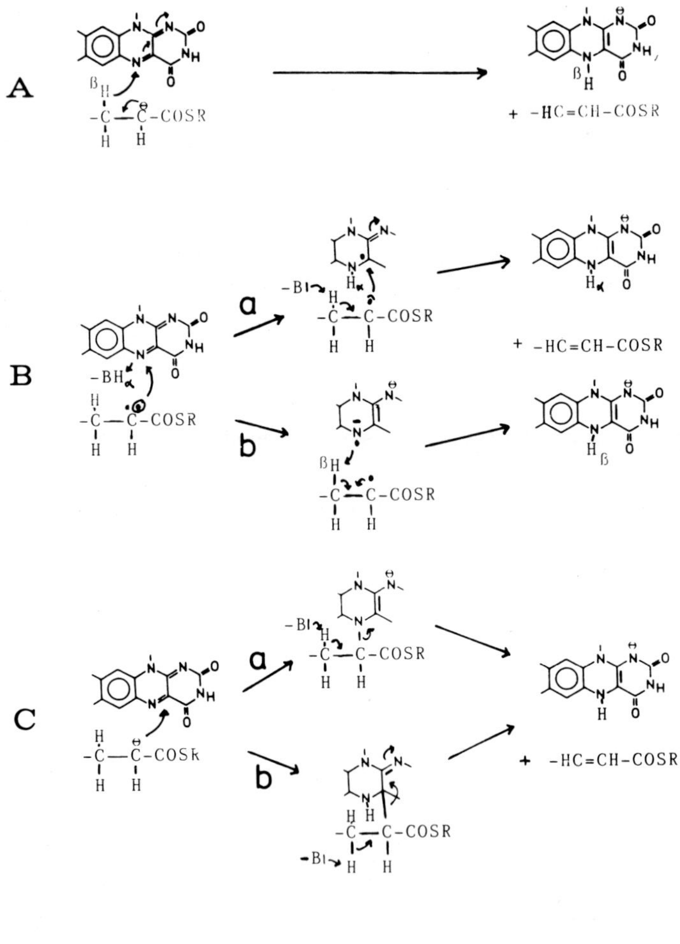

In general, transfer of reducing equivalents via covalent intermediates is an attractive mechanism which has been postulated to function in the the dehydrogenation of α-hydroxy- and α-amino acids (1,18,19). Covalent catalysis, as shown on Scheme 3, (C,a,b) has been explored by Venkataram and Bruice for the reaction of reduced flavins with maleimides, as models for the α,ß-dehydrogenation reaction (13). In this model system it is clear, that flavin C(4a)-adducts are intermediates, while N(5)-adducts do not seem to occur (13). While this route might be viable for the oxidation of succinate, in which the ß-hydrogen also is relatively acidic, in the case of fatty acid derivatives the low acidity of the ß-C-H group in adducts such as those depicted in Scheme 3 (C), might be a difficult problem to overcome. The intrinsic acidity of this group is not likely to be strongly affected by formation of the bond to the flavin N(5), since this nitrogen does not carry a positive charge. The pK of ß-C-H is likely to be comparable to that of the same function of (N-unprotonated) amino acids. An experimental test of covalent mechanisms also might involve the use of 5-deazaflavins. This is a reasonable expectation since substrate analogs, such as 3,4-pentadienoyl-CoA, have been found (8,9) to react with normal, as well as with 5-deaza-FAD reconstituted general acyl-CoA dehydrogenase. In both cases a covalent, N(5) or C(5) adduct was formed, which, in the second case was stable and could be isolated (8,9). Quite in general, intermediates such as those in Scheme 3, middle vertical row, should be accessible to detection by spectroscopic methods if they were to occur in appreciable concentrations during catalysis.

Spectral Course of the Reaction of General Acyl-CoA Dehydrogenase (GAD) with the Substrate Butyryl-CoA

The method of choice for the spectroscopic study of reaction intermediates occuring in flavin catalysis is repetitive scanning, since in most cases sufficient spectral changes can be expected. With general acyl-CoA dehydrogenase the oxidized (20), semiquinone (21), and reduced species (20) are well characterized (see also Fig. 1). Under standard conditions (pH 7.6, 25°, octanoyl-CoA as substrate) reduction is, however, much too fast to be followed with conventional spectrophotometers. We succeeded in slowing it down sufficiently by running it at -2° to 0°, pH 6.1, and by using fully deuterated butyryl-CoA as a substrate (22).

Fig. 1 shows such an experiment. Immediately upon addition of substrate the spectrum of the oxidized enzyme is perturbed in a fashion typical for its complexation with CoA derivatives (6,20). Thereafter the reaction proceeds strictly isosbestic in the region 310-500 nm. As will be detailed later, this reaction is typically triphasic (22,23). In the range 510-700 nm a weak charge transfer absorption, which is characteristic of the reduced enzyme/product complex develops first, and subsequently decays biphasically (23). The final spectrum is identical to that reported by Thorpe et al. for the reaction with (protio) butyryl-CoA (20). The rates of reduction at pH 7.6 and at pH 9.0 are faster, as observed also for the pig liver enzyme (24). The spectral course of the reaction, however, was qualitatively the same at all pH values studied. It is important to note that in no case could evidence for intermediates be obtained. Also, no indications were found for the occurrence of intermediates of the type proposed by Reinsch et al. (24) to occur in the reaction of pig liver enzyme with the same substrate.

Studies with General Acyl-CoA Dehydrogenase
and Selectively Deuterated Butyryl-CoA

Deuterated substrates have been used at various occasions for the study of α,ß-dehydrogenation. With the bacterial enzyme butyryl-CoA dehydrogenase, small isotope effects ($k_H/k_H \sim 1.2-2$) were found in turnover and single turnover experiments (25). For the medium chain length specific dehydrogenase Murfin (26) reports values of 3.5, 5.8, and 17 for α, ß- and α +ß deuterated substrates. Reinsch et al (24) report values of 30-50 for perdeutero butyryl-CoA. In view of these inconsistencies, we have investigated the reaction of general acyl-CoA dehydrogenase from pig kidney. This enzyme is different as compared to other acyl-CoA dehydrogenases investigated earlier (27,28), in that it does not show the typical long-wavelength absorbance in its native state (20). This spectral effect is due to tight binding of a CoA-persulfide to the oxidized enzyme (29). Thus kinetic complications due to interaction with this ligand should not arise with pig kidney GAD.

1) Turnover studies.

The results of turnover studies carried out by Pohl (21) under standard conditions (20) using selectively deuterated substrates are summarized in

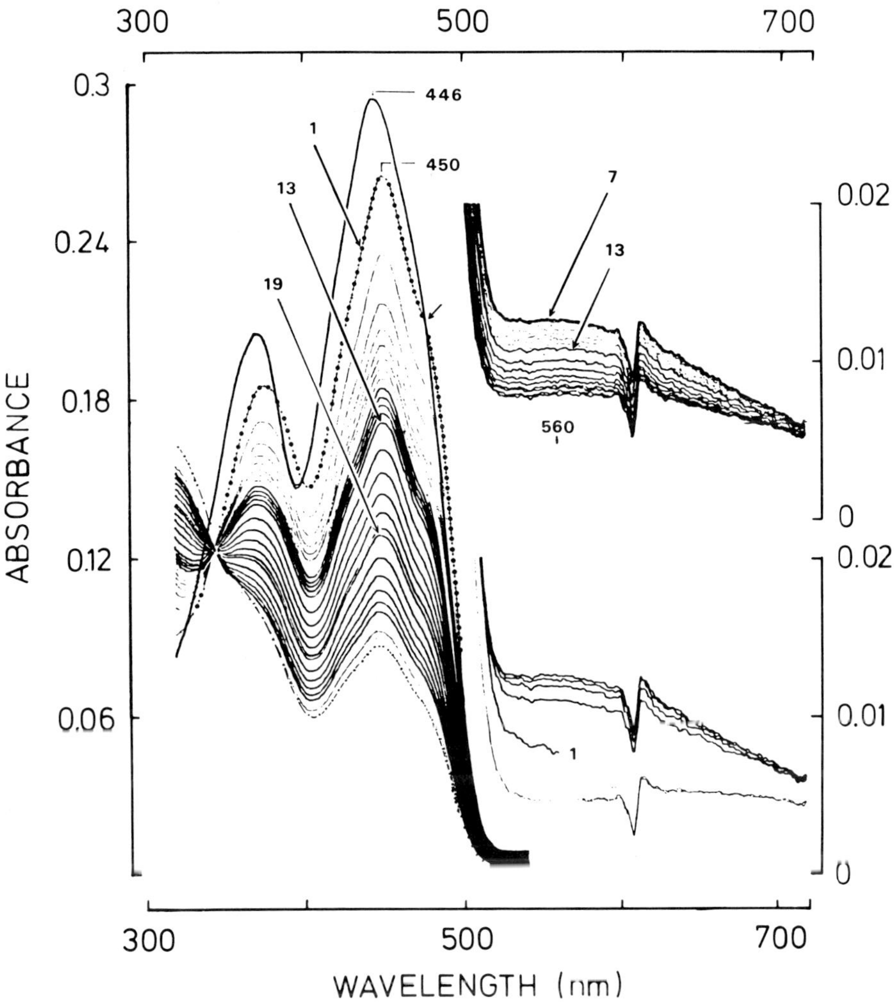

Figure 1: Spectral course of the anaerobic reaction of General Acyl-CoA Dehydrogenase with Butyryl-CoA

The enzyme (20 µM in 50 mM phosphate buffer pH 6.1) was reacted at 0° with perdeutero butyryl-CoA (200 µM). The figure is composed from two identical experiments in which the spectral ranges 720-520 (right hand side), and 540-310 nm (left hand side) were scanned. The curves shown were recorded at progressively increasing intervals (Beginning each 40 s; end each 10 min). Examples: Curves 1: 11.3 sec at 446 nm and ~ 8 s at 560 nm; curves 2: 47 s and 26 sec (446 resp. 560 nm); curves 7: 3.9 min and 5.5 min (446 resp. 560 nm); curves 13: 8.8 and 18.2 min (446 resp. 560 nm) and curves 19: 24.5 and 54 min (446 resp. 560 nm). For detailed conditions we refer to (23). Note that the signal at ~ 600 nm is due to an instrumental artefact.

Table 1. They indicate that substantial isotope effects are connected with the cleavage of the α- and of the ß-C-H bonds. In their magnitude they are comparable to those found by Murfin for the "medium chain" dehydrogenase (26).

Table 1. Kinetic constants for the reaction of General Acyl-CoA Dehydrogenase with selectively deuterated Butyryl-CoA Substrates

Substrate	V_{max} (min^{-1})	$V_{max}(H)$ / $V_{max}(D)$	K_m (μM)	*k_{obs} (s^{-1})	$k_{obs}(H)$ / $k_{obs}(D)$
CH_3-CH_2-CH_2-COSCoA	78	-	50	10.0	-
CD_3-CH_2-CH_2-COSCoA	78	-	50	nd	-
CH_3-CH_2-CD_2-COSCoA	33	2.4	50	5.5	1.8
CD_3-CD_2-CH_2-COSCoA	22	3.5	50	0.7	14.3
CD_3-CD_2-CD_2-COSCoA	9	8.7	33	0.35	28.6
CH_3-C(R)D,H-C(R)D,H-COSCoA	nd	nd	nd	0.35	28.6

*The rate k_{obs} is that of the first rapid reduction phase, compare with Fig. 2.

2) Single turnover (stopped flow) studies

When oxidized enzyme is reduced anaerobically with the butyryl-CoA derivatives listed in Table 1, a triphasic decrease of the absorption of the oxidized flavin is observed (Fig., 1 and 2). The third phase of this process has a $t_{1/2}$ of 2-3 min, it is independent from deuteration and, at first approximation, also from the concentration of the substrate. Similar to this is the slowest phase of absorbance decay at 560 nm. This slow phase cannot be involved in catalysis since it is slower than V_{max} by several orders of magnitude; it might be related to slow dissociation of crotonyl-CoA from the reduced enzyme complex and its replacement with butyryl-CoA, as proposed by Thorpe et al (20), and as shown in Scheme 5. The kinetic traces were analyzed as shown in Fig. 2.B, the rate constants

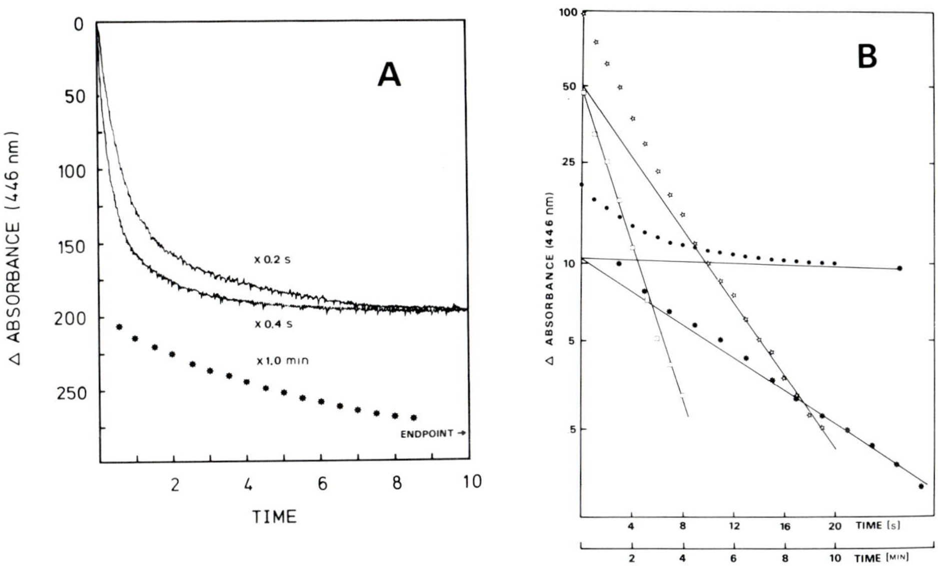

Fig. 2A: Kinetic course of the reaction of General Acyl-CoA Dehydrogenase with Butyryl-CoA.

The enzyme ($3x10^{-5}$ M in 50 mM phosphate buffer, pH 7.6) was reacted with butyryl-CoA ($7.5x10^{-4}$ M) at 25°anaerobically in the stopped-flow spectrophotometer.

Fig. 2B: Semilogaritmic analysis of spectral changes occurring during the reaction of general acyl-CoA dehydrogenase with butyryl-CoA.

The conditions were as in Fig. 2A, however perdeutero butyryl-CoA was used. Curve (✱) shows the slowest phase of the reaction ($t_{1/2}$ = 4.6 min, lower time scale). Curve (●): rapid phases plotted on the faster time range. Curve (✫) represents the plot of the previous curve (●) after subtraction of the slowest phase (✱) ($t_{1/2}$ = 3.3 s). Curve (□): Plot of the preceding curve (✫) after substraction of the slow phase of curve (●) ($t_{1/2}$ = 1.9 s).

obtained for the first rapid phase are listed in Table 1. Similar isotope effects were observed also on the second, slower phase. A main feature of kinetic traces, such as those shown in Fig. 2, is the dependence of the "extent of reduction" (i.e. the Δ OD of phase 1 + phase 2) from the concentration of the substrate. This dependence saturates at concentration $5x10^{-4}$ M (not shown). Importantly, the end point of phases 1 + 2 was constant when the same concentration of protio and deuterated butyryl-CoA substrates was employed. Such a saturation behaviour is compatible with the fully and rapidly reversible equilibria of Scheme 4.

Scheme 4: $$A \underset{k_2}{\overset{k_1}{\rightleftharpoons}} B \underset{k_4}{\overset{k_3}{\rightleftharpoons}} C$$

In such a system it would be required that the ratio k_3/k_4 is close to 1 for butyryl-CoA as a substrate. Furthermore, this indicates that step k_4 must have an isotope effect similar to that on k_3. This interpretation is in line with the proposals of Thorpe et al. (20), and in particular with the observations, that the extent of reduction depends on the chain length of substrate (20), and on the redox potential of the flavin bound to the enzyme (30). Also, simulations of the kinetic traces using the kinetic scheme shown above (Scheme 4), requires step k_4 to have a similar isotope effect as k_3, in agreement with a small equilibrium isotope effect. The finding of such an isotope effect on k_4 suggests that no (rapid) hydrogen exchange can occur at the level of C, and that the base(s) functioning in H-abstraction are monofunctional. A general discussion of the kinetic behaviour of this enzyme will be published elsewhere (23). The data presented in Table 1 were selected in order to illustrate the main points emerging from the investigations. First, the breaking of both the α- and the ß-C-H bond take place concomitantly with flavin reduction. The observation that α and ß isotope effects are multiplicative is compatible with a concerted mechanism as proposed by Murfin (26). Secondary isotope effects, if present, could not be detected with our system. The difference in isotope effect between rates of turnover, and rate of enzyme reduction indicates that the rate of reoxidation (k_5, Scheme 5) is partially rate limiting. The rate of k_5 can be estimated using the appropriate turnover equation as 3.8-5 s^{-1}, and is largely independent from the butyryl-CoA used. It thus does not show a relevant isotope effect (~1.3). Consequently this rate k_5 does not depend on the rupture of an X-H bond the -H of which resulting from the substrate. K_m values calculated for the various deuterated substrates using the equation $K_m = (k_2k_4 + k_2k_5 + k_3k_5)/k_1(k_3 + k_4 + k_5)$ and kinetic rates used for the simulation of stopped-flow results are 76, 86, 140, and 97 μM for protio, $\alpha{}^2H_2$, $\beta{}^2H_2$, and perdeutero butyryl-CoA respectively. They compare with corresponding K_m values of 50, 50, 50, and 33 μM estimated graphically from double reciprocal plots of turnover studies (22).

We consider this a reasonable agreement, in particular in view of the difficulties involved in the evaluation of multiphasic kinetic traces (cf. Fig. 2). The kinetic data can thus be described by the sequences of Scheme 5, as also proposed by others (20).

Scheme 5:

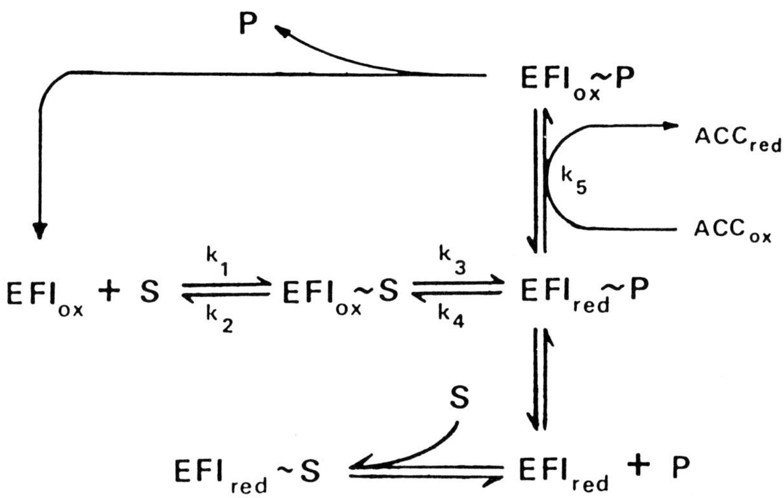

No attempt is made here to explain the biphasic kinetic behaviour observed in pre steady state experiments (Fig. 2). This will be attempted in a forthcoming publication (23).

Exchange of Protons of Butyryl-CoA Catalyzed by Butyryl-CoA Dehydrogenase

The exchange of the α- and of the ß-H of butyryl-CoA has been demonstrated by Gomes et al. to occur in the presence of bacterial butyrl-CoA dehydrogenase (31). However no further details were reported on this reaction. We have studied the incorporation of solvent ^{3}H under anaerobic conditions in the presence of the same enzyme and have found that two equivalents of label are incorporated in each substrate molecule (32). The time dependence of incorporation is monophasic over 5-6 half lives, indicating that the two hydrogens are exchanged at the same rate. This rate of exchange is ~300 substrate molecules exchanged/min/molecule of enzyme and thus at least as fast as the rate of turnover (~200 moles substrate/min/mole enzyme (27)). ^{1}NMR analysis of the substrate after anaerobic incubation in ^2H_2O shows that one hydrogen has been exchanged at each of the positions α and ß (32).

Exchange of Protons of Butyryl-CoA Catalyzed by General Acyl-CoA Dehydrogenase

While the exchange catalyzed by the bacterial dehydrogenase is fast, in the presence of pig kidney dehydrogenase only a slow exchange of the α-H could be demonstrated by direct ^{1}HNMR analysis (32). In the presence of an 500 fold excess substrate the half time of exchange was approx. 45 min (32). The results of these experiments are summarized in Table 2, where they are also compared to exchange experiments carried out with acyl-CoA dehydrogenases from other sources.

Table 2:

Exchange of α- and ß-Hydrogens Catalyzed by Acyl-CoA Dehydrogenases

Enzyme (Source)	Exchange (position)	Rel. rate of exchange	References
Bu-CoA DH (pig liver)	α and ß	α » ß	Biellmann & Hirth (3)
Bu-CoA DH (Clostr. kluyvery)	none	---	Bühler (33)
Enoate reductase (Clostr. Lal)	none	---	Bühler (33)
Bu-CoA DH (M. elsdenii)	α and ß	n.d.	Gomes et al. (31)
Bu-CoA DH "	α and ß	α = ß	Ghisla et al. (32)
GAD (ox liver)	α and ß	α > ß	Murfin (26)
" (rat liver)	α		Ikeda & Tanaka (34)
" (pig kidney)	α		Ghisla et al. (32)
" (5-deaza-FAD enzyme)	none	---	Ghisla and Thorpe, unpublished

Evidence for the Transfer of the ß-Hydrogen from Butyryl-CoA to the Flavin Position 5

The analysis of transfer of a labelled substrate hydrogen to the flavin N(5) cannot be carried out with normal FAD. In 1,5-dihydroflavin N(5)-H is likely to exchange fast with solvent. Furthermore, analysis of dihydroflavins is difficult due to their sensitivity to O_2. Oxidation would also remove the N(5) hydrogen. Therefore 5-deaza-FAD was used for this experiment; with this analog rapid exchange with the solvent does not occur, and reoxidation is very slow. On the other hand, due to their low redox potential 5-d-FAD bound to apo-GAD is not reduced by substrate (32). The reaction was thus conducted in the reverse direction, i.e. starting from reduced, C(5)-labelled 5-d-FAD-GAD and crotonyl-CoA:

Scheme 6:

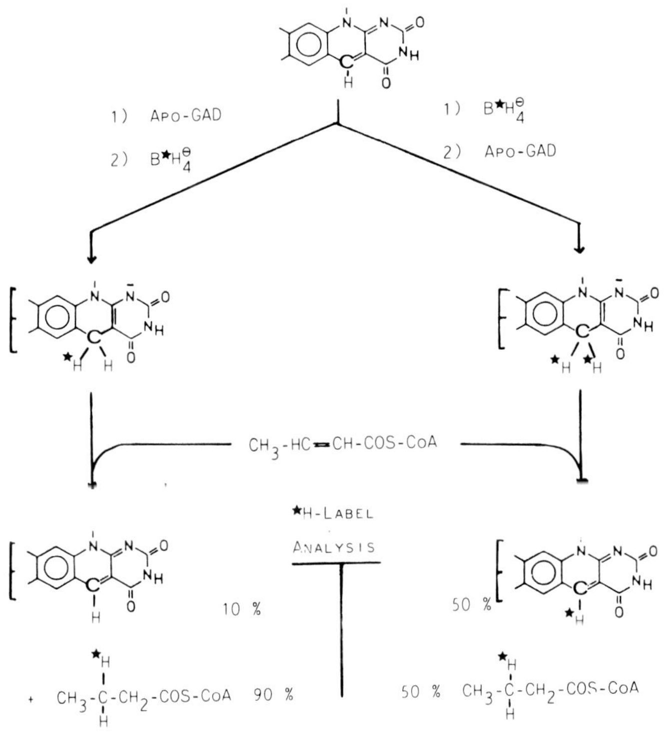

Labelled reduced 5-d-FAD enzyme was obtained either by direct reduction of oxidized 5-d-FAD-GAD with NaB^3H_4, or by reduction of free oxidized 5-d-FAD with the same reagent and subsequent reconstituion of holoenzyme with apo-GAD (Scheme 6). In the first case 0.9 equivalents were incorporated into butyryl-CoA. This shows that BH_4^- reduction of enzyme bound flavin is selective for the flavin side (Re or Si), which also interacts with substrate. In the experiment using prereduced 5-d-FAD 50 % of the label present in reduced 5-d-FAD-GAD was incorporated into butyryl-CoA, and the residual 50% remained with the (oxidized) 5-d-FAD-enzyme (32). This is as as expected for random labelling of C(5)-H_2 in reduced 5d-FAD. The butyric acid position into which the *H label was incorporated was determined by incubation of the product with strong alkali at 110° (26). Under these conditions only the α-H should exchange. Since only little label was released, it must have been incorporated into the ß-position (Scheme 6). In a companion control experiment, which was carried out using ($2R^3$, $3R^3$) butyryl-CoA, 50 % of the label was released (32). The label introduced into butyryl-CoA in this experiment (and originating from 5-d-FAD) was removed completely when the butyryl-CoA was incubated in the presence of electron acceptors with either GAD or butyryl-CoA from M. elsdenii. This proves that these enzymes have the same stereochemistry for the substrate ß-position.

Reduced 5-d-FAD-General Acyl-CoA Dehydrogenase and Reduced Native Enzyme React at Similar Rates with Crotonyl-CoA

Reoxidation of reduced 5-d-FAD-GAD is essentially complete with one equivalent crotonyl-CoA (32). This is contrary to the reaction of reduced native butyryl-CoA dehydrogenase, where an internal equilibrium is attained consisting of ~ 80 % oxidized and 20 % reduced enzyme (25). The rate of reoxidation of native reduced GAD by Crotonyl-CoA is 2.6 sec^{-1}, a value which is in good agreement with the rate of 3.7 s^{-1} obtained by computer simulation of the reduction of oxidized enzyme with butyryl-CoA (23). The rate of the reoxidation of reduced 5-d-FAD-GAD by crotonyl-CoA is 11.1 s^{-1} (32) and that of the deutero 5-d-FAD enzyme 2.2 s^{-1}, yielding an isotope effect of 5.

Mechanistic Conclusions

The results discussed above clearly establish, that the ß-hydrogen of substrate is transferred to the flavin position 5. The question whether this transfer is direct, or proceeds via e.g. covalent intermediates, as proposed by Venkataram and Bruice (13) cannot, as in many analog cases, be answered directly. Intermediates could occur transiently, and thus escape detection. In our opinion, however, there are two arguments that speak for a direct hydride transfer: First, as already pointed out above, the ß-hydrogen in a covalent adduct should not be particularly acidic, suggesting that covalent adduct formation might not be a convenient way for lowering the activation energy of catalysis. Secondly, if the ß-hydrogen is transferred to the position 5 of the flavin via a covalent adduct involving position C(4a), then this hydrogen must be abstracted as a proton, and the flavin position 5 should act as a base. This might be conceivable in the case of normal flavin, however not with 5-d-flavin. This brings up the next point, i.e. whether the reaction observed with normal and with 5-deaza-flavins occur by the same mechanism. This long disputed point (14) has not yet been answered satisfactorily. However, in the present case, a very important argument speaks in favor of the same mechanism. Both sets of reactions proceed at closely similar rates. If they were to proceed by basically different pathways, it is most likely, that different orientations of substrate, active center bases, and flavin would be required, and this would most probably also result in different reaction rates. With respect to the mechanism of exchange of substrate α-, and ß-hydrogens, it is most probable that this occurs at the level of the complex of reduced flavin and enoyl-CoA (Scheme 7).

Scheme 7:

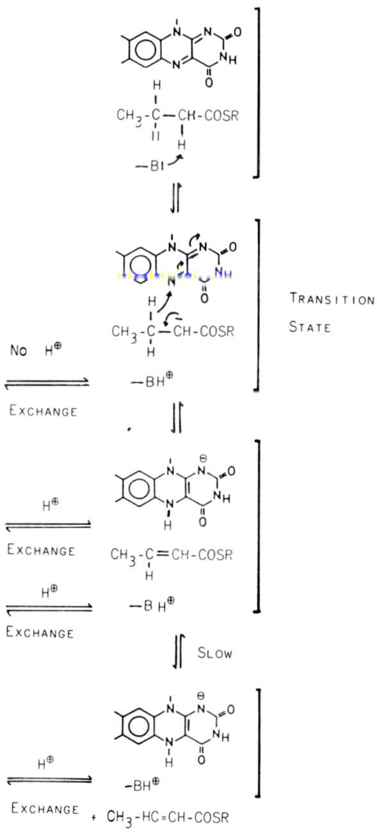

Exchange of the α-hydrogen at the level of the carbanion-oxidized flavin complex is not supported by the kinetic results, and is also not reasonable, since such a complex should be a true transient state, contrary to the proposal of others (24). Exchange at the level of free reduced enzyme also is improbable, at least in the case of pig kidney GAD, since dissociation of product has been proposed to be slow (20), and exchange is not influenced by the presence of oxygen. We (32) thus propose that exchange occurs at the level of the reduced enzyme enoyl-CoA complex, and that the different patterns of exchange observed with butyryl-CoA dehydrogenase and GAD result from a different degree of exposure to the solvent of the α-proton, which has been abstracted by a base, and of the ß-proton, which should be bound to the flavin N(5) position (Scheme 8).
The mechanism of α,ß-dehydrogenation is thus proposed to proceed in a concerted way, as shown in Scheme 8, i.e. in a complex in which the substrate is placed between the flavin plane and a base functioning in the abstraction of the α-proton.

Scheme 8:

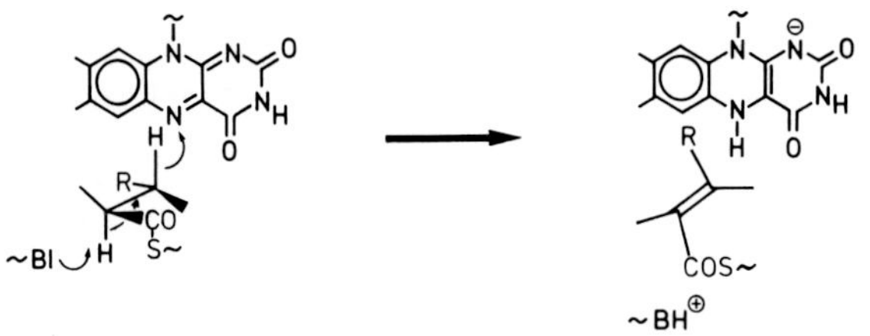

References

1. Ghisla, S.: Flavins and Flavoproteins (Massey, V., Williams, C.H., eds) Elsevier, Amsterdam, pp. 133-142 (1982).
2. Massey, V., Ghisla, S.: Biological Oxidations, 34. Mosbach Colloquium (Sund, H., Ullrich, V., eds) Springer, Berlin, pp. 114-139 (1983).
3. Biellmann, J.F., Hirth, C.G.: FEBS Letters 9, 55-56, and 335-336 (1970).
4. Bucklers, L., Umani-Ronchi, A., Retey, J., Arigoni, D.: Experientia 26, 931-933 (1970).
5. Jones, J.R.: The ionization of Carbon Acids, Academic Press, London, (1973).
6. Thorpe, C., Ciardelli, T.L., Stewart, C.T., Wieland, T.: Eur. J. Biochem. 118, 279-282 (1981).
7. Fendrich, G., Abeles, R.H.: Biochemistry 21. 6685-6695 (1982).

8. Wenz, A., Thorpe, C., Ghisla, S.: Flavins and Flavoproteins (Massey, V., Williams, C.H. eds) Elsevier, Amsterdam, pp. 605-608

9. Wenz, A., Thorpe, C., Ghisla, S.: submitted to Eur. J. Biochem. (1984).

10. Frerman, R.E., Miziorko, H.M., Beckmann, J.D.: J. Biol. Chem. 255, 5688-5696 (1980).

11. Bruice, T.C.: Advances in Chemistry Series, Biomimetic Chemistry 191, 89-118 (1980).

12. Farng, O.L., Bruice, T.C.: J. Chem. Soc., Chem. Commun. 185-186 (1984).

13. Venkataram, U.V., Bruice, T.C.: submitted

14. Hemmerich, P., Massey, V., Fenner, H.: FEBS Letters, 84, 5-21 (1977).

15. Cornforth, J.W.: J. Lipid Research 1, 1-28 (1959).

16. Blankenhorn, G.: Eur. J. Biochem. 80, 67-80 (1976).

17. McKean, M.,C., Sealy, R.C., Frerman, F.E.: Flavins and Flavoproteins (Massey, V. Williams, C.H., eds), Elsevier, Amsterdam pp 614-617 (1982).

18. Walsh, C.: Acc. Chem. Res. 13, 148-155, (1980).

19. Porter, D.H.T., Voet, J.B., Bright, H.J.: J. Biol. Chem. 248, 4400-4416 (1973).

20. Thorpe, C., Matthews, R., Williams, C.H.: Biochemistry 18, 331-337.

21. Mizzer, J.P., Thorpe, C.: Biochemistry 20, 4965-4970 (1981).

22. Pohl, B.: Ph. D. Thesis, Univ. Konstanz, (1984)

23. Pohl, B., Ghisla, S., Raichle, Th. in preparation.

24. Reinsch, J., Katz, A., Wean, J., Aprahamian, G., McFarland, J.R.: J. Biol. Chem. 255, 9093-9097 (1980).

25. Raichle, Th.: Ph. D. Thesis, Univ. Konstanz, (1981).

26. Murfin, W.W., Ph. D. Thesis, Washington University, (1974).

27. Engel, P., Massey, V.: Biochem. J. 125, 879-887 (1971).

28. Beinert, H.: The Enzymes, vol. 7, 2nd ed. 447-476 (1963).

29. Williamson, G., Engel, P.C., Mizzer, J.P., Thorpe, C., Massey, V.: JBC 257, 4314-4320.

30. Thorpe, C., Massey,: Biochemistry 22, 2972-2978 (1983).

31. Gomes, B., Fendrich, G., Abeles, R.H.: Biochemistry 20, 1481-1490. (1981).

32. Ghisla, S., Thorpe, C., Massey, V.: Biochemistry (1984) in press.

33. Bühler M.: Ph. D. Thesis, Technische Universität München, pp. 186-188 (1981).

34. Ikeda Y., Tanaka, K., (1984) in press.

BUTYRYL-CoA DEHYDROGENASE: ASPECTS OF ACCEPTOR AND SUBSTRATE SPECIFICITY

Paul C. Engel, Gary Williamson* and Lee Shaw

Department of Biochemistry, University of Sheffield
Sheffield England

*present address:-

Department of Biochemistry, Emory University, Atlanta, Georgia, U.S.A.

Introduction

The early studies of mammalian mitochondrial acyl-CoA-dehydrogenases by Green, Beinert and coworkers (1,2) revealed various striking properties. Among these is the extraordinary tightness of substrate binding (3). This seems likely to be, in part, an adaptation to the fact that a limited total pool of coenzyme A has to be shared among many CoA and acyl-CoA-dependent enzymes; there are, for example, 32 intermediates in the β-oxidation of stearoyl-CoA to acetyl CoA. Another striking property was the specificity for the acceptor, ETF, another flavoprotein (2). ETF receives its reducing equivalents from several dehydrogenases, but donates them only to the respiratory chain via ETF dehydrogenase (4). Equally, each acyl-CoA dehydrogenase will only pass on reducing equivalents to ETF. The fidelity of these donor-acceptor specificities ensures that reducing equivalents removed from acyl-CoA substrates are funnelled unerringly into the respiratory chain. This means that, even though the acyl-CoA dehydrogenases and ETF are not membrane-bound, mitochondrial β-oxidation should be subject to strict respiratory control in a way that does not apply to NAD^+-linked oxidation, since

Flavins and Flavoproteins

there are alternative possibilities for the re-oxidation of NADH.

Implicit in the specificity for ETF is the lack of reaction with oxygen. The mitochondrial acyl-CoA dehydrogenases are autoxidisable if they are reduced with dithionite, but when reduced by acyl-CoA substrates they react with O_2 extremely slowly (1). This is a second important reason for tight binding of substrates.

Sometimes fatty acid oxidation is required more for the provision of C-2 units than for ATP supply. There is accordingly a separate β-oxidation system in peroxisomes/glyoxysomes/microbodies which involves an acyl-CoA oxidase (5-7) instead of a dehydrogenase, thus escaping the rigid respiratory control alluded to above.

We have purified the short-chain acyl-CoA dehydrogenase from two sources, ox-liver mitochondria (8) and the anaerobic bacterium Megasphaera elsdenii (9). This allows an interesting comparison of oxygen reactivity in the two enzymes, one of which never encounters oxygen normally, whilst the other has to avoid oxidation by O_2 *in vivo* in order to fulfil its physiological role.

Results and Discussion

It was shown some years ago (10) that, in aerobic buffer, butyryl-CoA dehydrogenase from M.elsdenii formed an oxidised-enzyme-oxidised-product complex within the time of manual mixing with butyryl-CoA. We have recently measured the rate of the oxidase reaction implicit in that observation. The results, presented elsewhere in this Symposium (11), show that, with butyryl-CoA, the V_{max} value for this oxidase reaction is about 10% of that for the standard dye-coupled reaction, and the K_m values are essentially identical. No evidence for superoxide production was obtained and the stoicheiometry is 1 mol H_2O_2 produced per mol substrate oxidised.

When butyryl thioesters of truncated CoA analogues were used, higher K_m's were seen in both the dye-coupled assay and the oxidase assay, but in the latter, in marked contrast to the former, the maximum rate increased! Again, hexanoyl-CoA, a much poorer substrate than butyryl-CoA in the dye assay, gave an equally high V_{max} in the oxidase assay (11). In general, it appears that even in the case of this anaerobic enzyme, oxygen reactivity tends to be suppressed by tight substrate binding.

In the case of the mammalian enzyme, where the suppression of the oxidase reaction is virtually total, it was of interest to see whether the same generalisation applied. Although the oxidase rates are much lower than for the bacterial enzyme, once again they increase with the progression from butyryl-CoA down to S-butyryl-N-propionylcysteamine (11). For this admittedly poor substrate, V_{max} values in the oxidase assay and the dye-coupled assay were in fact identical.

In keeping with the catalytic properties, butyryl-CoA and butyryl-pantetheine give very different spectral changes when they are added aerobically to a concentrated solution of the ox-liver dehydrogenase. Reduction of butyryl-CoA causes the emergence of a pronounced long-wavelength band (Fig.1), and, as shown already in the 1950's (1), reoxidation of this substrate-reduced form of the enzyme takes many hours at room temperature. On the other hand, the spectrum seen after addition of a 2-fold molar excess of butyryl-pantetheine (Fig.2) is characteristic of a complex with a crotonyl thioester (3,10), indicating that turnover has been completed. When the butyryl-pantetheine is added anaerobically in 5-fold molar excess, the extent of reduction is not very great, as judged either by bleaching of the 450nm band or formation of long-wavelength absorption.

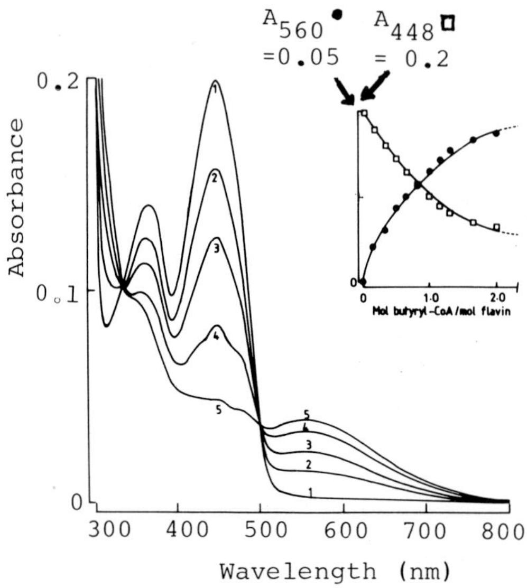

Fig. 1 Ox liver butyryl-CoA dehydrogenase: titration with butyryl-CoA

Enzyme (13.4 nmol bound FAD) was titrated aerobically with butyryl-CoA at pH 7.6 (50mM K P_i). Selected curves 1-5 are for 0, 0.33, 0.82, 1.3 and 3.6 mol substrate per mol FAD. Inset of A_{560} and A_{448} corrected for dilution includes more titration points.

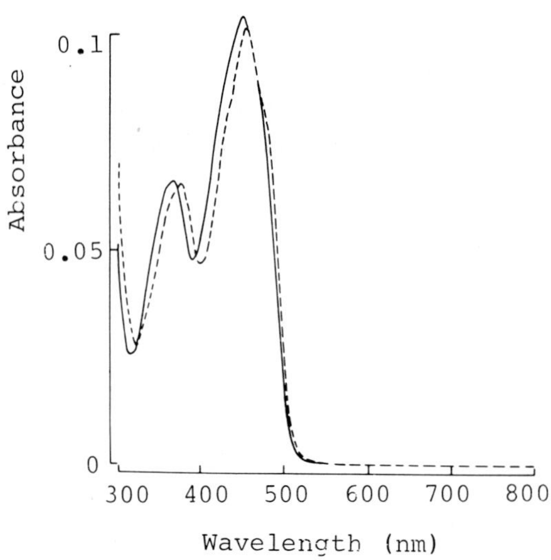

Fig. 2 Ox liver butyryl-CoA dehydrogenase: addition of butyryl-pantetheine

15 nmol butyryl-pantetheine was added aerobically to enzyme (7 nmol bound FAD). Conditions as Fig. 1. Before (——) and after (----).

Binding studies offer another approach to establishing the relative importance of the various portions of the thioester substrate molecule in its interaction with the enzyme. A detailed study of the M.elsdenii enzyme has been carried out with acetoacetyl-CoA, a competitive inhibitor, and a range of acetoacetyl thioesters of CoA fragments (12). For such studies, however, the enzyme preparation has to be homogeneously free of CoA-containing ligands, and, as isolated from the bacteria, this enzyme contains a mixture of such ligands (10), CoA, CoA persulphide (13,14), and occasionally acetoacetyl-CoA (15). We now have a procedure for quantitative stripping of such ligands from the enzyme (16). This involves reduction with dithionite, which converts CoA persulphide to CoA, followed by passage down a thiopropyl-Sepharose column in the pyridyl-disulphide-protected form. The latter step leaves CoA behind, disulphide-linked to the column material. The efficacy of this method in removing the last traces of CoA is demonstrated in Fig.3. This shows the re-conversion of the stripped, yellow enzyme to the green persulphide complex. The straight line relating increase in A_{710} to the incorporation of CoA passes accurately through the origin. This may be contrasted with a similar plot for unstripped enzyme, shown in the proceedings of the Ann Arbor symposium (17), in which the presence of residual CoA resulted in a sizeable positive intercept on the ordinate axis.

The CoA-free form of the M.elsdenii enzyme was used for spectrophotometric titrations with acetoacetyl thioesters. The change in the absorption spectrum was qualitatively remarkably similar for the series of thioesters that were tested:the main visible peak was shifted to 430nm and a new peak appeared at 580nm, as previously described for acetoacetyl-CoA (10). The amplitude of the new peak at saturation was similar for the various titrants - e.g. A_{580}: A_{430} = 0.36 for acetoacetyl-CoA and 0.28 for S-acetoacetyl N-acetyl cysteamine. The change in A_{580} were therefore used as a means of following the extent of saturation with each of the thioesters.

The titration, first of all, with acetoacetyl-CoA gives a clear end-point at 1 mol titrant per mol enzyme-bound FAD (Fig.4A). Binding is so tight that only when the extent of saturation exceeds 90% is it possible to estimate the concentration of the free enzyme. The last few points give an apparently linear Scatchard plot (Fig.4B) and an estimate of 0.12μM for the dissociation constant. This is close to an approximate estimate of 0.16μM for the K_i determined in rate measurements (12). Similar titrations with unstripped yellow enzyme (P.C.Engel, unpublished) do not give a clear end-point.

Titrations with acetoacetyl thioesters of CoA fragments show weaker binding, as might be expected. What was unexpected, however, was that the binding was in all cases complex. With S-acetoacetyl pantetheine, for example, extrapolation of the tight-binding portion of the plot indicated that saturation was again achieved with 1 mol titrant per mol FAD, but the Scatchard plot indicated at least two apparent dissociation constants, 1.7μM at low titrant concentration and 4.25μM closer to saturation. With the even smaller molecule of S-acetoacetyl N-acetyl cysteamine, there was, as in the other cases, a good pair of isosbestic points, at 427 and 502nm, but a markedly curved Scatchard plot was obtained (Fig.5). Tangents to different portions of this curve give apparent dissociation constants ranging from 0.19mM to 3.7mM.

These titrations with the thioesters of truncated CoA analogues give the appearance of negative cooperativity (18,19). There could be a non-allosteric explanation, however, It is possible that the range of binding constants observed may reflect the slightly different environments of four subunits within a tetramer. Why would this be seen only with the weak-binding ligands and not with acetoacetyl-CoA? In fact we cannot assert with any confidence that the binding of aceto-

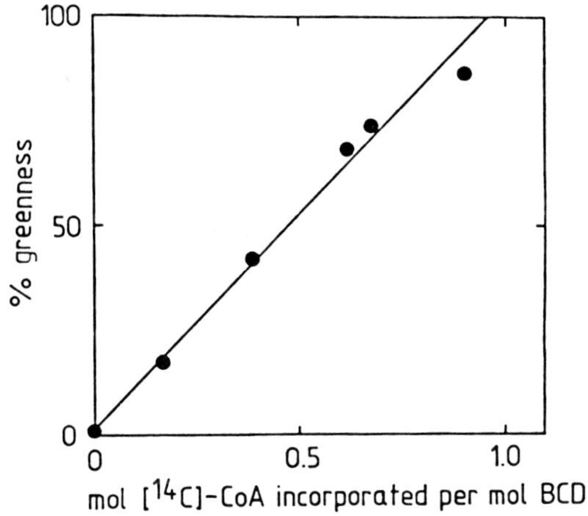

Fig.3 M.elsdenii butyryl CoA dehydrogenase:correlation of re-greening and ^{14}C-CoA uptake.

Enzyme was first de-greened and stripped of CoA. 6 samples (81μM) were incubated at 25°C with 2mM Na_2S and ^{14}C-CoASH (0-1.5 fold molar excess). After gel filtration the absorption spectrum and radioactivity were measured. 100% green enzyme has A_{710} : A_{430} = 0.54.

acetyl-CoA is simple, since our experiment only allows us to estimate the strength of binding of the final 10% of the full complement of thioester.

One of the original intentions was to assess the contribution of each part of the CoA molecule to the overall strength of binding. The complexities just mentioned mean that such deductions have to be treated with caution. Nevertheless free energies of binding were calculated for each thioester from the dissociation constant for the weakest binding detected in the Scatchard plot. These calculations suggest, perhaps surprisingly, that the adenosine phosphate moiety contributes only about a quarter of the total binding energy (Table 1). A fuller interpretation of these observations will have to await the determination of the enzyme's three-dimensional structure.

Fig.4A Titration of CoA-free M.elsdenii butyryl-CoA dehydrogenase (26.4 μM) with acetoacetyl CoA

0.1M KP_i pH 7. Dashed line, before addition; solid line, after adding 3.3-fold molar excess of acetoacetyl CoA.

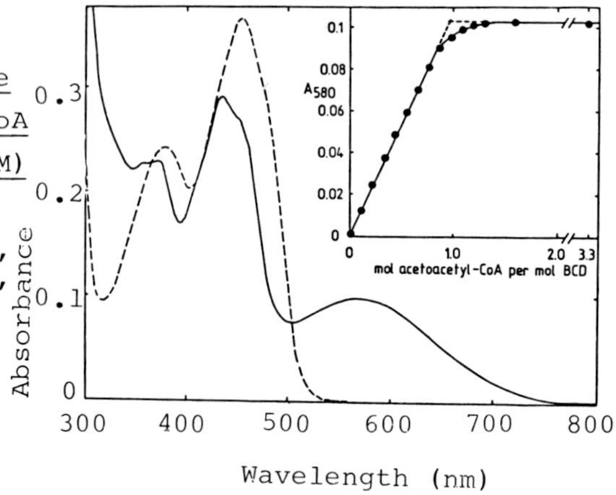

Fig.4B Scatchard plot of data from Fig.4A

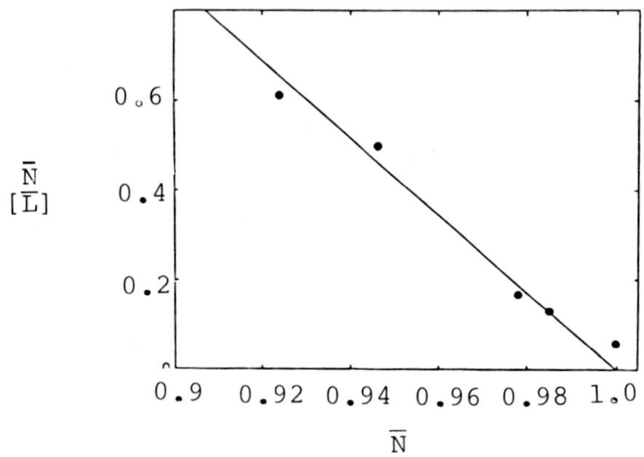

Fig.5 Scatchard plot for S-acetoacetyl-N-acetyl cysteamine

Conditions as Fig.4. Enzyme concentration 58.3μM, titrant 0-18.4mM (saturating).

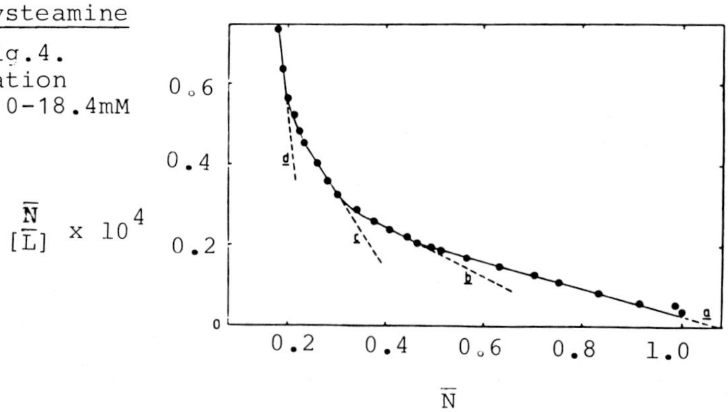

Table 1 Tightness of binding of various acetoacetyl thioesters to butyryl-CoA dehydrogenase from Megasphaera elsdenii

Thiol portion of thioester	Apparent K_d (μM)	Apparent ΔG^o (kJ mol^{-1})
Coenzyme A	0.12	-39.5
Pantetheine	4.25	-30.7
N-acetoacetyl cysteamine	116	-22.5
N-acetylcysteamine	3660	-13.9

Acknowledgement

We wish to thank the Science and Engineering Research Council for their support through Project Grants to P.C.E. and training awards to G.W. and L.S.

References

1. Beinert, H. (1963) The Enzymes (2nd Edn.) 7, 447-466
2. Beinert, H. (1963) The Enzymes (2nd Edn.) 7, 467-476
3. Steyn-Parvé & Beinert, H. (1958) J.Biol.Chem. 223, 843-852
4. Ruzicka, F.I. & Beinert, H. (1977) J.Biol.Chem. 252, 8440-8445.
5. Lazarow, P.B. & DeDuve, C. (1976) Proc.Nat.Acad.Sci. USA 73, 2043-2046
6. Osumi, T. & Hashimota, T. (1978) Biochem.Biophys.Res. Commun. 83, 479-485
7. Inestrosa, N., Bronfman, M. & Leighton, F. (1979) Biochem. J. 182, 779-788
8. Shaw, L. & Engel, P.C. (1984) Biochem.J. 218, 511-520
9. Engel, P.C. (1981) Methods Enzymol. 71, 359-366
10. Engel, P.C. & Massey, V. (1971) Biochem. J. 125, 889-902
11. Ellison, P.A., Shaw, L., Williamson, G. & Engel, P.C. (1984) this Symposium pp 413-416

12. Williamson, G. (1983) Ph.D. Thesis, University of Sheffield

13. Williamson, G., Engel, P.C., Mizzer, J.P., Thorpe, C. & Massey, V. (1982) J.Biol. Chem. 257, 4314-4320

14. Williamson, G., Engel, P.C., Nishina, Y. & Shiga, K. (1982) FEBS Lett. 138, 29-32

15. Engel, P.C. (1980) 'Flavins and Flavoproteins' (ed. K. Yagi & T.Yamano) 423-430 Japan Scientific Societies Press.

16. Williamson, G. & Engel, P.C. (1982) Biochim.Biophys.Acta 706, 245-248

17. Williamson, G., Engel, P.C., Mizzer, J.P., Thorpe, C. & Massey, V. (1982) in 'Flavins and Flavoproteins' (ed. V. Massey & C.H. Williams, Jr.) 593-598

18. Dalziel, K. & Engel, P.C. (1968) FEBS Lett. 1, 349-352

19. Conway, A. & Koshland, D.E. Jr. (1968) Biochemistry 7, 4011-4022

OXYGEN REACTIVITY OF BUTYRYL-CoA DEHYDROGENASE FROM MEGASPHAERA ELSDENII AND FROM OX-LIVER MITOCHONDRIA

Patricia Ellison, Lee Shaw, Gary Williamson and Paul C. Engel

Department of Biochemistry, University of Sheffield
Sheffield, England

Introduction

Butyryl-CoA dehydrogenase (BCD) catalyses the oxidation of saturated acyl thioesters to 2,3-unsaturated products. M.elsdenii BCD (1) and the analogous mammalian short-chain acyl-CoA dehydrogenase (SCAD) (2) display many similarities (3,4,5,6,7) but differ in the stability of their substrate-reduced forms to reoxidation by O_2 (8) whereas ox-liver SCAD is only reoxidised over a period of hours. The oxidase activity of CoA-free M.elsdenii BCD (10,11) and ox liver SCAD (3) was investigated.

Results

Catalytic reaction with various butyryl thioesters.
Dehydrogenase activity of BCD and SCAD was determined using a dye-coupled assay with PES and DCPIP (12,13). Oxidase activity was measured by following the increase in A_{263} in air-saturated buffers due to the formation of a 2,3-enoyl thioester product (table 1).
For all substrates, oxidase activity of ox-liver SCAD is very much lower than that of M.elsdenii BCD. The oxidase activity of both enzymes increases with truncated CoA analogue substrates, the best oxidase substrate being S-butyryl N-propionyl cysteamine.

Flavins and Flavoproteins

Catalytic reaction with various acyl-CoA substrates. Dehydrogenase and oxidase activity of <u>M.elsdenii</u> BCD with various acyl-CoA substrates was determined (table 1). For C_4 and C_5 acyl CoA, V_{max} in the oxidase assay was 5-10% of the value for the dye-coupled assay, with similar K_m values.

Table 1 Kinetic parameters for oxidase and dehydrogenase activity of (A) Ox-liver SCAD and (B) <u>M.elsdenii</u> BCD with various substrates.

(A) Ox-liver SCAD

	dehydrogenase		oxidase	
	K_m (μM)	V_{max} (min^{-1})	K_m (μM)	V_{max} (min^{-1})
S-butyryl-CoA	2.7	150	-	0
S-butyryl pantetheine	38	10.8	23	0.28
S-butyryl N-propionyl cysteamine	149	2.1	143	2.0

(B) <u>M.elsdenii</u> BCD

	dehydrogenase		oxidase	
	K_m (μM)	V_{max} (min^{-1})	K_m (μM)	V_{max} (min^{-1})
S-butyryl-CoA	16	331	14	30
S-butyryl pantetheine	243	169	27	56
S-butyryl N-propionyl cysteamine	948	27	650	71
Propionyl-CoA (C3)	130	1.6	-	
Pentanoyl-CoA (C5)	15	149	36	7.1
Hexanoyl-CoA (6)	53	14	133	29

Reaction with oxygen.

Oxygen consumption by M.elsdenii BCD was measured with an O_2 electrode, 2ml samples of BCD (30-80μM) were separately titrated with S-butyryl-CoA and S-butyryl pantetheine (table 2) The stoichiometry of O_2 consumption in the presence and absence of 20μg of catalase suggests that H_2O_2 is the product of the oxygen reaction.

$$\text{Butyryl-thioester } O_2 \xrightarrow{\text{BCD}} \text{crotonyl-thioester} + H_2O_2 \quad (1)$$

$$H_2O_2 \xrightarrow{\text{catalase}} H_2O + 1/2\ O_2 \quad (2)$$

The previous results suggest that in the case of butyryl-CoA this reaction is largely suppressed as a result of the binding of the adenosine moiety of the substrate molecule.

Table 2 O_2 consumption of M.elsdenii BCD oxidase reaction (mols O_2/mol substrate oxidised)

	+ catalase	- catalase
S-butyryl-CoA	0.56	1.0
S-butyryl pantetheine	0.45	0.9

Acknowledgement. We are indebted to the Science and Engineering Research Council who supported this work through a Project Grant to P.C.E. and training awards to G.W., L.S. and P.A.E.

References

1. Baldwin, R.L. & Milligan, L.P.:Biochim.Biophys.Acta. 92, 431-432 (1964).

2. Beinert, H.: in the Enzymes, 2nd edn. (Boyer, P.D. Lardy H. and Myrback, K., eds) vol.7 pp. 447-473 Academic Press N.Y. (1963).

3. Shaw, L. & Engel, P.C. : Biochem.J. 218, 511-520 (1984).

4. Williamson, G. : Ph.D thesis, University of Sheffield.

5. Steyn-Parvé, E.P. and Beinert, H. : J.Biol.Chem. 223, 853-861 (1958).

6. Engel, P.C. and Massey, V. : Biochem.J. 125, 889-902 (1971).

7. Williamson, G., Engel, P.C., Mizzer, J.F., Thorpe, C. & Massey, V. : J.Biol.Chem. 257, 4314-4320 (1982).

8. Engel, P.C. and Massey, V. Biochem.J. 125, 879-887 (1971).

9. Engel, P.C. : Biochem.Soc.Trans. 6, 76-80 (1978).

10. Engel, P.C. : Methods Enzymol. 71, 359-366 (1981).

11. Williamson, G. & Engel, P.C. : Biochim.Biophys.Acta. 706, 245-248 (1982).

12. Williamson, G. & Engel, P.C. : Biochem.J. 218, 521-529 (1984).

SUICIDE INACTIVATION OF SHORT-CHAIN ACYL-CoA DEHYDROGENASE BY PROPIONYL-CoA

Formation of a substrate-flavin adduct

Lee Shaw and Paul C. Engel

Biochemistry Department
University of Sheffield
Sheffield S10 2TN, U.K.

The inactivation reaction

Ox liver short-chain acyl-CoA dehydrogenase (SCAD) catalyses the α,β-dehydrogenation of saturated C_4 - C_8 acyl-CoA substrates during mitochondrial β-oxidation (1,2). Although this enzyme exhibits almost maximal activity towards butyryl-CoA ($V = 150\ min.^{-1}$), its V with propionyl-CoA is much lower ($1.8\ min.^{-1}$). This might simply result from a decreasing energy of interaction of propionyl-CoA with SCAD. However, the results presented in this paper demonstrate that the situation is more complicated.

Time-courses for propionyl-CoA oxidation were markedly curved, suggesting the possibility of enzyme inactivation. Incubation of SCAD and propionyl-CoA in the absence of electron acceptors caused a slow decline in activity to 22% of the initial value. No further inactivation could be obtained by addition of more propionyl-CoA. Using this final activity as an equilibrium value, the decline in catalytic activity was shown to exhibit pseudo-first-order kinetics (3). Incorporation of tightly bound [^{3}H]CoA-propionyl-CoA paralleled SCAD inactivation.

Flavins and Flavoproteins

Spectral changes caused by propionyl-CoA

The aerobic addition of an excess of propionyl-CoA to SCAD caused a slow bleaching of the flavin A_{448} to 30% its initial value (Figure 1). Using the final A_{448} as an equilibrium value, the decline in flavin absorbance also exhibited pseudo-first-order kinetics. Since the apparent first-order rate constants (k_{app}) for flavin bleaching and inactivation were almost identical, these are related events. An analysis of the kinetics for bleaching at several propionyl-CoA concentrations suggests that a Michaelis complex forms in a saturable initial encounter with propionyl-CoA. The extrapolated k_{app} for saturating propionyl-CoA was 8.9×10^{-3} sec^{-1}. In the absence of electron acceptors, the inactive reduced complex is formed. However, in their presence the Michaelis complex partitions between catalysis and inactivation in the ratio 4.3:1.

Addition of acetoacetyl-CoA, which has a very high affinity for free SCAD (1) (K_D = 0.15μM), caused a slow reversion of propionyl-CoA-modified enzyme to the oxidised form complexed with acetoacetyl-CoA (1). Inactivation by propionyl-CoA appears to be reversible, and the consequent equilibrium between modified and unmodified forms explains the incompleteness of inactivation and flavin reduction.

Nature of the flavin released from propionyl-CoA-treated enzyme

The visible absorption spectrum of the flavin released aerobically with trichloroacetic acid was reminiscent of a neutral flavosemiquinone (Figure 2). This species indeed gave an epr signal (see chapter by George et al., : this symposium). By analogy with several model flavins (5,6) and also the case

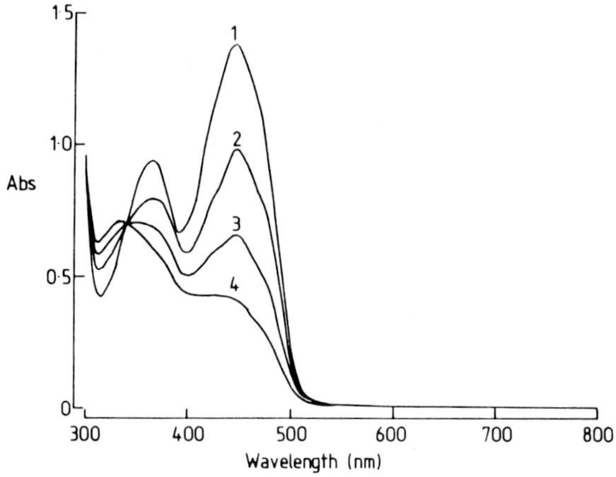

Figure 1 : Serial spectra of 93μM-SCAD & 466μM-propionyl-CoA after 0 (1), 95 (2), 335 (3) and 815 seconds (4).

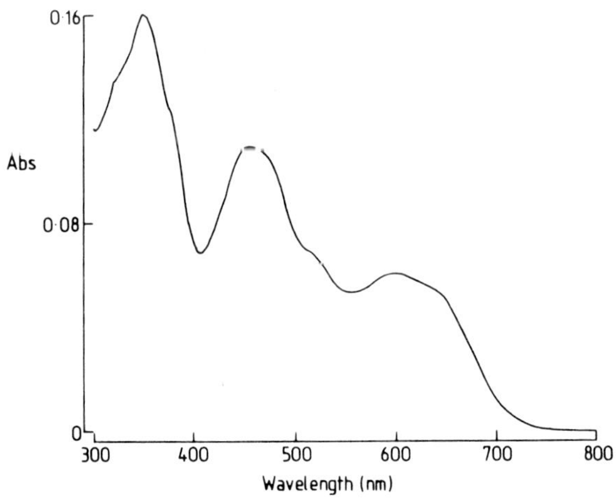

Figure 2 : Absorption spectrum of flavin extracted with trichloroacetic acid from 45μM-SCAD modified with propionyl-CoA. Protein precipitate removed by centrifugation and the supernatant was neutralised.

of lactate oxidase and glycollate (7), it seems probable that a reduced N-5 alkyl flavin adduct is formed with propionyl-CoA. This is further supported by the appearance of an absorbance maximum at 335 nm (Figure 1). Release of the adduct allows the attack of oxygen to form a stable N-5-alkyl flavosemiquinone. When [^{3}H]CoA-propionyl-CoA was used, HPLC purification demonstrated that CoA was retained in the released adduct. Acrylyl-CoA, the product of propionyl-CoA oxidation does not covalently modify reduced or oxidised enzyme, and is thus unlikely to be involved in adduct formation.

Summary

These results suggest that propionyl-CoA is a pseudosuicide substrate for SCAD and that inhibition occurs by the formation of an N-5 alkyl flavin adduct.

References

1. Shaw, L., Engel, P.C. : Biochem.J. 218, 511-520 (1984).
2. Green, D.E., Mii, S., Mahler, H.R., Bock, R.M. : J.Biol Chem. 206, 1-12 (1954).
3. Frost, A.A., Pearson, R.G. : Kinetics and Mechanism, 2nd edn., p186 London, Macmillan (1961)
4. Benecky, M., Li, T.Y., Schmidt, J., Frerman, F., Watters, K.L., McFarland, J.T. : Biochemistry 18, 3471-3476 (1979)
5. Hemmerich, P., Massey, V., Weber, G. : Nature 213, 728-730 (1967
6. Walker, W., Hemmerich, P., Massey, V. : Eur.J.Biochem. 13, 258-266 (1970)
7. Massey, V., Ghisla, S., Kieschke, K. : J. Biol. Chem. 255, 2796-2806 (1980)

STRUCTURE OF THE FLAVIN N-5 ADDUCT FREE RADICAL OBTAINED FOLLOWING INHIBITION OF SHORT-CHAIN ACYL-CoA DEHYDROGENASE BY PROPIONYL-CoA

Graham N. George*, Lee Shaw, Robert C. Bray* and Paul C. Engel

School of Chemistry and Molecular Sciences, University of Sussex, Brighton, BN1 9QJ, U.K.* and

Department of Biochemistry, University of Sheffield, Sheffield S10 2TN, U.K.

Summary

Shaw and Engel (this volume) have shown that ox liver short-chain acyl-CoA dehydrogenase is progressively inactivated by the substrate, propionyl-CoA. During this process, a species accumulates which, on denaturation in the presence of air, yields a stable flavin radical, believed to be an N-5 adduct. We have carried out a quantitative analysis of the e.p.r. spectrum, with the help of isotopic substitution experiments in $^{1}H_2O$ and $^{2}H_2O$, employing propionyl-CoA substituted in the 2- and 3 positions of the propionyl residue with ^{13}C or ^{2}H. We conclude that the adduct radical has the structure of an acrylyl-CoA adduct, having the C-2' atom linked to N-5 of the flavin. The formation of this species is briefly discussed in relation to the mechanism of the catalytic reaction.

Introduction

The mechanism of action of fatty acyl-CoA dehydrogenase has been extensively studied (e.g. 1-4). Shaw and Engel (5) have shown that in the case of ox liver short-chain acyl-CoA dehydrogenase, the enzyme is progressively inactivated during

Flavins and Flavoproteins

treatment with, and slow turnover of, the 'poor' substrate, propionyl-CoA. During this process a species accumulates which, on denaturation in the presence of air, yields a stable free radical which these workers suggested was a flavin N-5 adduct. In the present investigations we have carried out a detailed analysis of the form of the e.p.r. signal from this flavin radical species, enabling us to deduce its structure. Relevance of this species to the mechanism of action of the enzyme is discussed.

Materials and Methods

Chemicals. Isotopically enriched propionic acids were used to the following enrichments: 90% for [2-^{13}C] and [3-^{13}C], 98.9% for [2-$^{2}H_2$] and 99.9% for [3-$^{2}H_3$]. Normal propionyl-CoA was prepared (6) from propionic anhydride and the isotopically enriched derivatives were synthesised by the mixed anhydride method (7).

Formation of the flavin adduct free radical. Short chain acyl-CoA dehydrogenase was prepared as described by Shaw and Engel (8). The enzyme (150-200 μM) was treated at 20-25°C with normal or isotopically substituted propionyl-CoA (five-fold molar excess) in 200 mM potassium phosphate, pH 7.6 for 20 min. Trichloroacetic acid was added to a final concentration of 5%. After centrifugation to remove the protein precipitate, the supernatant solution was extracted with ether to remove trichloroacetic acid. For experiments in $^{2}H_2O$ the enzyme was dialysed against 200 mM potassium phosphate buffer in $^{2}H_2O$ and propionyl-CoA was dissolved in $^{2}H_2O$. Denaturation was performed by using 5% trichloroacetic acid in $^{2}H_2O$.

E.p.r. spectra. Spectra were recorded (in the normal 1st derivative form), on a Varian E9 e.p.r. spectrometer linked to a Digital Equipment Co. PDP11/10 computer, for samples of the adduct radical solution contained in silica e.p.r. tubes (1 mm internal diameter) or in a small silica flat cell. Recording conditions were generally as follows: microwave power 5 mW, modulation amplitude 0.05 mT, temperature 20°C. For some samples, time-averaging was used to improve signal-to-noise ratios. Manipulation of spectra was performed on the University of Sussex Digital Equipment Co., VAX11/780 computer, after transfer from the PDP via a link. Simulations were calculated correct to first order perturbation theory by using software which minimised iteratively the sum of squares error between smoothed experimental and calculated spectra. For all the simulations illustrated in Figs. 1-3, Lorenzian line shape, with a line width of 0.15 mT was assumed.

Results and Discussion

E.p.r. spectrum of the free radical from reaction of propionyl-CoA with the enzyme. We prepared the free radical species, described by Shaw and Engel (5) by treating ox liver short chain fatty acyl-CoA dehydrogenase with the 'poor' substrate, propionyl-CoA. We then liberated the flavin from the protein, as described in the Materials and Methods section, and studied the e.p.r. spectrum of the product. Two forms of the flavin adduct free radical were observed, each having a characteristic absorption and e.p.r. spectrum. In acid solution in the presence of trichloracetic acid the radical was red. After removal of the trichloracetic acid by extraction with ether, the pH of the solution reverted to a value of 6 to 7 and the solution became green. Under these conditions the e.p.r. spectrum of Fig. 1(a) was observed. As is described below, we employed isotopic substitution experiments to assist in the analysis of the hyperfine couplings in this form of the free radical.

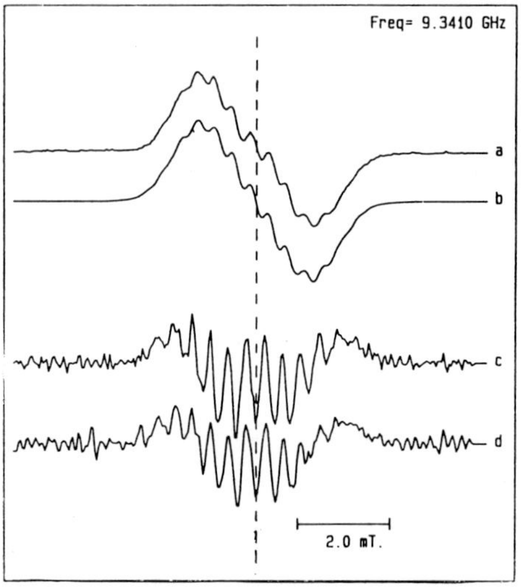

Fig. 1. E.p.r. spectrum of the adduct radical in $^{1}H_2O$ (a and c) and in $^{2}H_2O$ (d).
(a) is the 1st derivative spectrum, as recorded, of the radical in $^{1}H_2O$ and (b) is the corresponding computer simulation (see Scheme 2 for parameters). (c) is the 2nd derivative spectrum corresponding to (a) and (d) is the corresponding spectrum of a sample in $^{2}H_2O$. The vertical line corresponds to g 2.0037.

Hyperfine couplings in the spectrum of the radical are not well resolved. This is not surprising. Müller et al. (9) noted that radicals from the natural flavin coenzymes give broader lines than does that from lumiflavin. The presence of the propionyl-CoA residue linked (5) to FAD makes the molecule with which we are dealing quite a large and complex one. Hyperfine coupling patterns in the radical are expected to be complicated, with numerous couplings to ^{1}H and to ^{14}N atoms involved. To assist in analysis of the spectrum, we employed second derivative spectra (first derivative of the spectra as recorded). The considerable enhancement of spectral features resulting from taking the second derivative is apparent (compare Figs. 1(c) and (a)).

The e.p.r. spectrum of our radical was stable for extended

periods at 20-25°C, without exclusion of oxygen and these conditions were therefore used routinely. When the temperature was raised to 80°C, no decrease in linewidth was observed, thus eliminating dynamic effects as a source of line broadening.

Hyperfine coupling to nuclei of the flavin ring system and to protons of the propionyl residue. We first tested the effect of developing the radical in $^{2}H_2O$. As shown in Fig. 1(d), this did not modify the spectrum in any way (cf. Fig. 1(c)). Since N-5 flavin radical adducts lack the strongly coupled exchangable proton in the 5-position (10) this finding supports strongly the view of Shaw and Engel (5) that we are indeed dealing with such a species.

Strong coupling (9,10) of the N-5 atom of the flavin nucleus means that in N-5 adducts, there will be appreciable spin density on atoms in the side chain, in addition of course to that on atoms of the ring system. That coupling to atoms in the side chain is present in our species is shown by the data in Fig. 2(d). Thus, employing propionyl-CoA substituted in the C-3 position by deuterium,resulted in marked changes in the spectrum in comparison with that from unlabelled propionyl-CoA (Fig. 2(a)).

Our primary aim in the present work was to analyse hyperfine coupling of atoms in the N-5 side chain. Since effects of these atoms on the spectrum are compounded with those of atoms in the flavin nucleus, it is necessary clearly to distinguish couplings arising from the two groups of atoms. Couplings of corresponding atoms in radicals of different flavin derivatives differ significantly from one another (9,10) and it therefore seemed unwise to rely entirely on literature values for these. Accordingly, we adopted the following procedure.

We started by analysing the spectrum of Fig. 2(d). We fitted this spectrum, by using an iterative procedure (see the

Materials and Methods section), starting from couplings similar to those given by Dunham et al. (11). Having obtained a reasonable fit, we turned our attention to the spectrum of Fig. 2(a), obtained with normal unenriched propionyl-CoA. Starting from the parameters obtained for the ^{2}H-substituted form, we then added proton couplings, as appropriate (at the same time decreasing the linewidth slightly to compensate for unresolved splittings from ^{2}H, which must be present in Fig. 2(d)). We continued varying the number and assumed couplings for additional protons until we arrived at a reasonable simulation. We found that a satisfactory fit could be obtained if, and only if, two strongly coupled protons, corresponding to those of the propionyl C-3 residue, were present. Fits could not be obtained if one or three protons were assumed (see below). Having obtained this preliminary fit, we then returned to Fig. 2(d), retaining the new line-width used in fitting Fig. 2(a) but including appropriate small splittings from two ^{2}H atoms, to correspond to those on the C-3 position of the propionyl residue. We then repeated the iterative procedure to obtain a new fit. The entire procedure was carried through several cycles of fitting until the final simulated spectra illustrated in Fig. 2(b) and (e) were obtained.

Up to this point, we made rather arbitrary and unchecked assumptions about possible coupling from protons of the C-2 position of the propionyl residue. We next turned our attention to these, by employing propionyl-CoA substituted with ^{2}H in the C-2 position. In contrast to substitution in the C-3 position, this had no detectable effect on the spectrum (Fig. 2c).

Values of the splittings from atoms of the propionyl residue are considered below. We concentrated our attention on analysis of the splittings from these side chain atoms, and precise analysis of couplings from atoms of the flavin nucleus

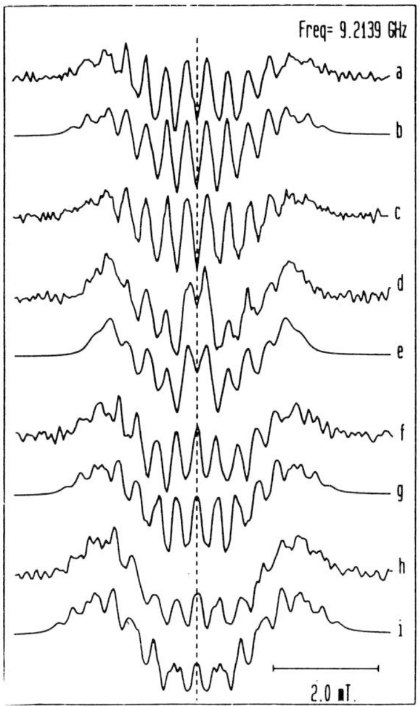

Fig. 2. Effect of isotopic substitutions on the 2nd derivative e.p.r. spectrum of the adduct radical.

Samples were prepared by using (a) unenriched propionyl-CoA (as in Fig. 1c), (c) $[2\text{-}^2H_2]$-propionyl-CoA, (d) $[3\text{-}^2H_3]$-propionyl-CoA, (f) $[3\text{-}^{13}C]$-propionyl-CoA, (h) $[2\text{-}^{13}C]$-propionyl-CoA. Computer simulations (using the parameters of Scheme 1), shown in (b), (e), (g) and (i), correspond to (a), (d), (f) and (h), respectively. The vertical line corresponds to g 2.0037.

is probably not critical to this. Nevertheless final values (Scheme 1) for the couplings from atoms in the riboflavin ring system are fully consistent with literature values (9-11). Though neutral forms of the radical have been relatively neglected, splittings in their spectra appear to differ from those in the cationic form mainly in the smaller coupling for the former from the nitrogen in the 10-position. In this context, the fact that in error we started from a coupling of 0.50 mT for N-10, appropriate (9) to the cationic radical, but ended up at the end of our fitting procedure with a value of 0.396 mT, appropriate to the neutral radical (10), seems highly significant. Since the neutral radical is expected under our conditions, this gives us considerable confidence in the overall reliability of our procedures for extracting hyperfine couplings from the spectra, despite the large line-widths.

Effect of ^2H_2O on the e.p.r. spectrum. As already noted (Fig. 1(c) and (d), the spectrum was quite unchanged if the adduct was formed in ^2H_2O and the spectrum was recorded in this medium. Similarly, the spectrum from 2-^2H-propionyl-CoA and the 3-^2H derivative were unchanged if the adduct was formed in ^2H_2O (spectra not shown). This is particularly significant in relation to the structure of the adduct (see below), since it is known (e.g. 2,12) that acyl-CoA dehydrogenases cause exchange with the solvent of protons in the propionyl C-2 and C-3 positions.

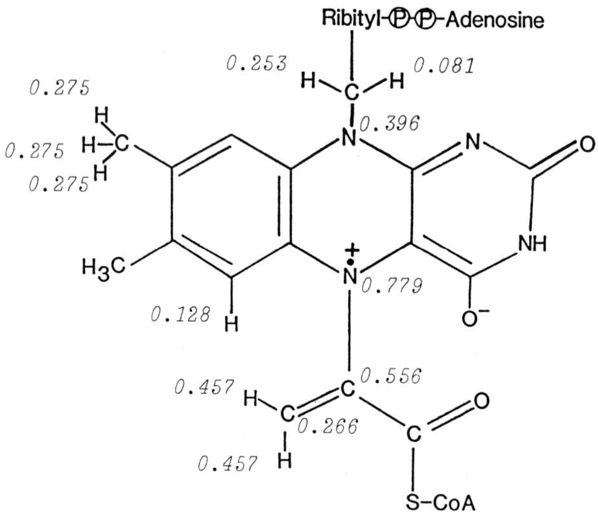

Scheme 1.

Structure deduced for the adduct radical. (numbers indicate hyperfine splittings in mT from ^{1}H, ^{13}C and ^{14}N atoms)

Coupling to ^{13}C nuclei of the propionyl residue. To obtain further information on the structure of the adduct radical, we obtained spectra from propionyl-CoA substituted in the 3- or the 2-position with ^{13}C (Fig. 2(f) and (h)). Couplings to ^{13}C were estimated by iterative fitting, starting from the parameters deduced as described above. Couplings are indicated in Scheme 1 and simulations are given in Fig. 2(g) and (i), respectively.

Structure of the signal-giving species. Four alternative structures might be suggested for the N-5 adduct free radical, as illustrated in Scheme 2(a-d). Only one of these, Scheme 2(a) (see also Scheme 1), is compatible with our results. This structure is in keeping with stronger coupling of ^{13}C in the C-2 position than that in the C-3 position, with two coupled protons in the C-3 position and with no protons in the C-2 position. The structures of Scheme 2(b), (c) and (d) are not consistent with the data, since wrong numbers of coupled protons would be expected in one or both positions.

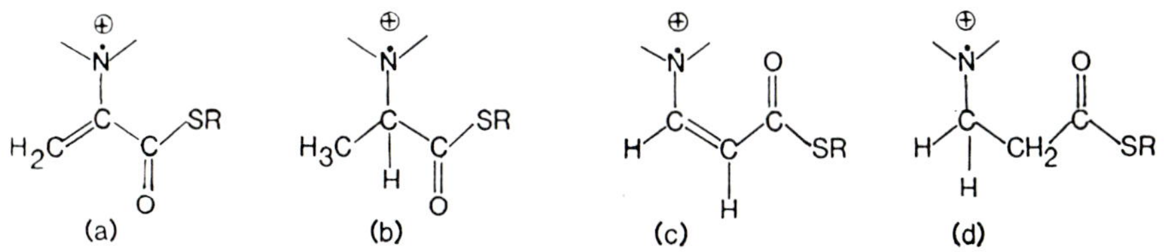

Scheme 2. Possible structures considered for the adduct radical.

We note that a single proton on the C-2' position, linked to nitrogen in the propionyl N-5 adduct (as in Scheme 2(b)), might have a vanishingly small coupling (10). Therefore, distinguishing among the different structures of Scheme 2 depends heavily on the correctness of our conclusion, above, that there are two, and only two, coupled protons in the C-3' position of the propionyl residue.

Additional data on this point will therefore be summarized. There is no doubt (compare Figs. 2a and d) that strongly coupled protons in the 3'-position are involved, and our conclusion (Scheme 1) is that there are two of them, each giving a splitting of 0.457 mT. There clearly cannot be three protons with such a splitting, since the simulation (not illustrated) is then impossibly bad. 0.457 mT corresponds very roughly to the spacing of the lines in the spectrum (Fig. 2a), and addition of coupling by an extra proton therefore causes 'upward' and 'downward' features to be completely

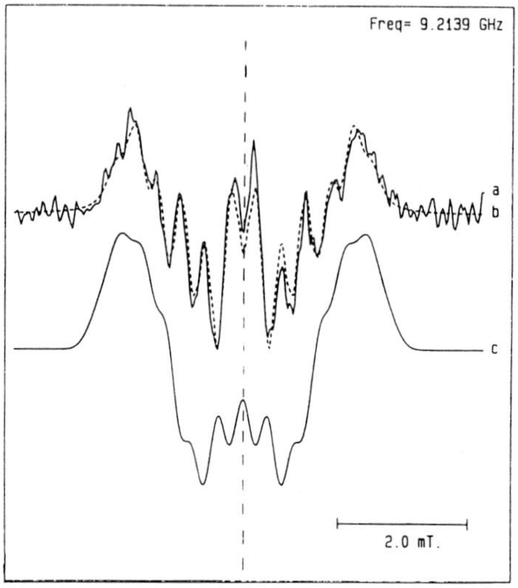

Fig. 3. Alternative simulation of the 2nd derivative e.p.r. spectrum from the radical prepared with $[3-{}^2H_3]$-propionyl-CoA.

(a) (full line) and (b) (dashed line) show respectively the experimental spectrum and the simulation of Figs. 2(d) and (e). (c) is a possible alternative, though markedly inferior simulation, based on three coupled protons $[A({}^2H)$ 0.074 mT] in the 3' position and one $[A({}^1H)$ 0.457 mT] in the 2' position (cf. Scheme 2b). The vertical line corresponds to g 2.0037.

out of step with one another in the simulated and experimental spectra. Conversely, with four protons, each again giving a splitting of 0.457 mT, the simulation (not illustrated) differs mostly in the extreme outer regions from that of Fig. 2(b), and is therefore not unequivocally incompatible with the experimental spectrum (Fig. 2a). Four equivalent protons in the propionyl residue is compatible with Scheme 2(b), with the assumption that the single C-2' proton has the same coupling as do the three C-3' protons. However, the deuterium-substitution results are not compatible with such a structure. The simulation of the spectrum of the C-3' ^{2}H-compound, based on three ^{2}H atoms in this position and retaining one ^{1}H in the C-2' position (Fig. 3c), is markedly inferior to the simulation already given (Fig. 3b). Still more convincingly, the fact that as already noted, deuterium-substitution in the C-2' position has no effect on the spectrum (Fig. 2c) is obviously quite incompatible with a proton with a coupling of 0.457 mT in this position. Thus, our data do indicate unequivocally that the structure of the adduct is the one given in Schemes 2(a) and 1.

Mechanism of the formation of the adduct. Flavin N-5 adduct radicals prepared chemically (e.g. 10) are, like our adduct, relatively stable in air. Flavin N-5 adducts have been assigned (13,14) important roles as intermediates in the turnover of substrates, especially glycolate, by the enzyme L-lactate oxidase, and by implication, in the mechanisms of other flavoprotein dehydrogenases also. In this context it is particularly significant that our adduct, prior to oxidation by air to the free radical, and in contrast to the adducts of these workers, is a species derived formally by reaction of oxidised flavin with, not the substrate, propionyl-CoA, but the product of normal turnover, acrylyl-CoA.

Kinetic data (5) indicate that the bound species is not an intermediate in normal turnover of the substrate. Though in due course further oxidation-reduction, kinetic, e.p.r. and if possible n.m.r. studies should lead to unambiguous interpretation of the inhibitory process, and thus perhaps of the enzymic mechanism, a number of interpretations of our results are at present possible. An acrylyl adduct of reduced flavin, if formed in the intact enzyme by a product molecule attempting to enter the catalytic cycle a second time, would presumably readily be converted, following denaturation, to the observed radical by 1-electron oxidation. Such a mechanism would tend to favour an N-5 adduct as an intermediate in the normal catalytic reaction. However, so far it has not proved possible (5) to obtain the adduct by treatment of resting or reduced short-chain acyl-CoA dehydrogenase with acrylyl-CoA. Though alternative routes to a bound acrylyl adduct of reduced flavin can be envisaged, the mechanism of removal of electrons from propionyl-CoA, and indeed their ultimate fate in the absence of added electron acceptors, is unclear. A total of four electrons would have to be removed from the substrate molecule, per flavin molecule inactivated, to yield such an enzyme-bound complex.

Alternatively, an enzyme-bound propionyl adduct, formed in a slow side reaction to the catalytic pathway, might be converted to the radical, following its liberation by denaturation, by a hypothetical 3-electron oxidation process. Such a scheme would favour an enzymic mechanism not involving a covalent intermediate in normal turnover. Work in hand should distinguish these possibilities.

Acknowledgements

We thank Dr. M. Blackburn for discussions and Mr. J. Blackburn for synthesizing [3-$^2H_3$]-propionic acid. Facilities for e.p.r. work were provided by the MRC; G.N.G. is supported by the SERC and L.S. received a studentship from the SERC.

References

1. Beinert, H.: The Enzymes 2nd Ed. 7, 447-466 (1963).
2. Gomes, B., Fendrich, G., Abeles, R.H.: Biochemistry, 20, 1481-1490 (1981).
3. Wenz, A., Thorpe, C., Ghisla, S.: J.Biol.Chem. 256, 9809-9812 (1981).
4. Reinsch, J., Rojas, C., McFarland, J.T.: Arch.Biochem. Biophys. 227, 21-30 (1983).
5. Shaw, L., Engel, P.C.: This volume (1984).
6. Simon, E.J., Shemin, D.: J.Amer.Chem.Soc. 75, 2520 (1953).
7. Stadtman, E.R. in Methods in Enzymol, ed. Colowick, S.P, Kaplan, N.O., 3, 931, Academic Press, New York (1957).
8. Shaw, L., Engel, P.C.: Biochem.J. 218, 511-520 (1984).
9. Müller, F., Grande, H.J., Harding, L.J., Dunham, W.R., Visser, A.J.W.G., Reinders, J.H., Hemmerich, P., Ehrenberg, A.: Eur.J.Biochem., 116, 17-25 (1981).
10. Müller, F., Hemmerich, P., Ehrenberg, A., Palmer, G., Massey, V.: Eur.J.Biochem., 14, 185-196 (1970).

11. Dunham, W.R., Harding, L.J., Grande, H.J., Müller, F.: in Flavins and Flavoproteins, ed. Massey, V., Williams, C.H., pp.568-572 Elsevier, New York (1982).

12. Rhead, W.J., Hall, C.L., Tanaka, K.: J.Biol.Chem., 256, 1616-1624 (1981).

13. Massey, V., Ghisla, S., Kieschke, K.: J.Biol.Chem., 255, 2796-2806 (1980).

14. Ghisla, S., Massey, V.: J.Biol.Chem., 255, 5688-5696 (1980).

PURIFICATION AND PROPERTIES OF FIVE DISTINCT ACYL-COA DEHYDROGENASES FROM RAT LIVER MITOCHONDRIA.

Yasuyuki Ikeda, Kazuko O. Ikeda and Kay Tanaka. Department of Human Genetics, Yale University School of Medicine, New Haven, CT 06510 USA.

Purification and physical properties.

Short chain (SCAD), medium chain (MCAD), and long chain (LCAD) and 2-methyl-branched chain acyl-CoA (2-meBCAD) and isovaleryl-CoA (IVD) dehydrogenases were purified to homogeneity from rat liver mitochondria (Fig. 1) by sequential chromatography on DEAE-Sephadex A-50, hydroxyapatite, Matrex Gel Blue A, Agarose-Hexane-CoA, and Bio-Gel A-0.5m (1-4). Physical and catalytic properties of these five enzymes are summarized in Table 1 (2-5). The native molecular weights of these five enzymes were 160,000, 180,000, 180,000, 170,000 and 175,000, respectively, as determined by gel filtration. On SDS-PAGE, both with and without 2-mercaptoethanol, the five enzymes each exhibited a single protein band with molecular weights of 41,000, 45,000, 45,000, 41,500 and 43,000, respectively, indicating that each is composed of four subunits of equal size (Fig. 1). The five acyl-CoA dehydrogenases were immunologically distinct from each other. The FAD content was calculated to be 1 mol/mol of subunit. While FAD binding by SCAD was very tight, that by the other four dehydrogenases was less tight. MCAD and LCAD were also purified to homogeneity as FAD-free apoenzymes in high yields. The apoenzymes were converted to the fully active holoenzymes by incubation with FAD (4).

Catalytic properties and the role of each dehydrogenases in the dehydrogenation of various acyl-CoA's.

The enzyme activity was determined using the dye-reduction assay. We also directly determined the rate of 2-enoyl-CoA product formation using gas chromatography. The activities determined by these two methods agreed well (5). IVD dehydrogenated isovaleryl- and n-valeryl-CoA's but it did not act essentially on any other acyl-CoA's with a straight or branched acyl chain. 2-meBCAD dehydrogenated S-2-methylbutyryl- and

Flavins and Flavoproteins

TABLE I

Properties of short chain acyl-CoA, medium chain acyl-CoA, long chain acyl-CoA, isovaleryl-CoA and 2-methyl-branched chain acyl-CoA dehydrogenases purified from rat liver mitochondria.

Property	FATTY ACID OXIDATION			LEUCINE OXIDATION	ISOLEUCINE AND VALINE OXIDATION
	Short Chain Acyl-CoA Dehydrogenase (SCAD)	Medium Chain Acyl-CoA Dehydrogenase (MCAD)	Long Chain Acyl-CoA Dehydrogenase (LCAD)	Isovaleryl-CoA Dehydrogenase (IVD)	2-Methyl-Branched Chain Acyl-CoA Dehydrogenase (2meBCAD)
Native	160,000	180,000	180,000	175,000	170,000
Subunit Mr.	41,000	45,000	45,000	43,000	41,500
Isoelectric point	4.8 ± 0.2	4.9 ± 0.2	4.9 ± 0.2	5.6 ± 0.2	5.5 ± 0.2
$A^{1\%}_{280nm}$	12.8	12.6	22.1	12.3	12.5
Spectral maxima (nm)	271:365:450	271:370:445	275:365:443	275:325:345:443	275:340:435
Ratios of the absorbances at the above maxima	6.3:0.75:1.0	6.3:0.75:1.0	10.1:0.75:1.0	12.5:1.5:0.9:1.0	10.3:1.3:1.0
FAD content/subunit	1	1	1	1	1
Fluorescence:					
Excitation	370 and 465nm	370 and 465nm	380 and 465nm	370 and 465nm	380 and 465nm
Emission	530nm	530nm	530nm	530nm	530nm
Intensity relative to equimolar FAD	2%	28%	29%	24%	28%
K_m (uM) and V_{max} (S^{-1})					
C_4CoA	10.7/28.3	135/3.8	/0	/0.3	/0.3
C_6CoA	285/8.5	9.4/31.2	/0	/0	/0
C_8CoA	/0	4.0/27.2	123/2.4	/0	/0
$C_{10}CoA$	/0	5.4/25.2	24.3/4.5	/0	/0
$C_{12}CoA$	/0	5.7/11.7	9.0/5.4	/0	/0
$C_{14}CoA$	/0	/0	7.4/6.0	/0	/0
$C_{16}CoA$	/0	/0	2.5/5.4	/0	/0
iC_5CoA	/0	/0	/0	33/7.8	/0
iC_4CoA	/0	/0	/0	/0	89/5.7
$2\text{-}meC_4CoA$	/0	/0	/0	/0	20/6.2
Immunological Properties:					
Antibody to SCAD	Positive reaction	No reaction	No reaction	No reaction	No reaction
Antibody to MCAD	No reaction	Positive reaction	No reaction	No reaction	No reaction
Antibody to LCAD	No reaction	No reaction	Positive reaction	No reaction	No reaction
Antibody to IVDH	No reaction	No reaction	No reaction	Positive reaction	No reaction

isobutyryl-CoA's with equally high activity but did not act on any other acyl-CoA's. The branched chain acyl-CoA's were not dehydrogenated by SCAD. Our short chain, medium chain, and long chain acyl-CoA dehydrogenases exhibited substrate specificities significantly more specific than those of the previous preparations isolated from other sources (Table 1). The physiological roles of the three straight chain acyl-CoA dehydrogenases in the dehydrogenation of various acyl CoA's were assessed by measuring the activities in mitochondrial sonicates precipitated by individual monospecific antibodies directed against each enzyme, and also by their $V_{max}^{app}/K_{m}^{app}$ values for various acyl-CoA (Table 1) substrates. Most of the acyl-CoA's, with an exception of lauroyl-CoA, appeared to be

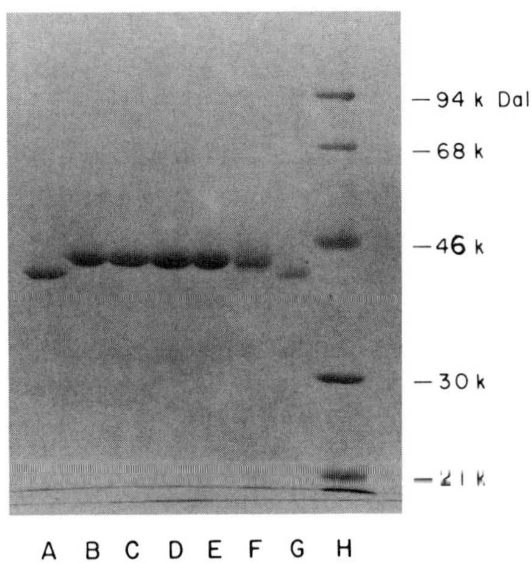

Fig. 1. Slab SDS-polyacrylamide gel electrophoresis of the purified SCAD, MCAD and LCAD. Symbols are as follows: A, holo-SCAD; B, Apo-MCAD; C, holo-MCAD; D, Apo-LCAD; E, holo-LCAD; F, IVD; G, 2-meBCAD; H, authentic standard proteins.

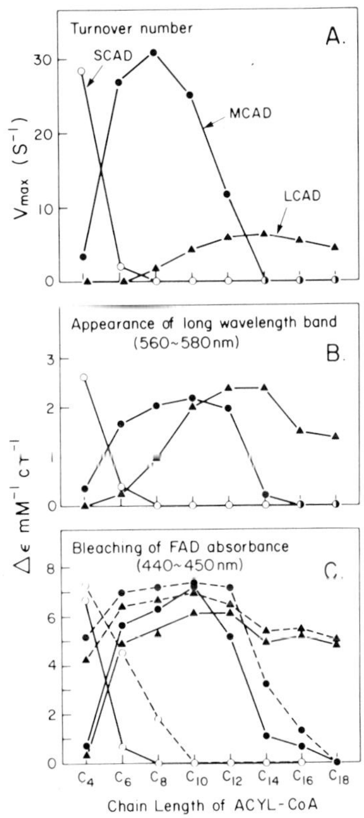

Fig. 2. The steady state turnover numbers (A), the extinction of long wavelength absorbance (580 nm) band (B) and the bleaching of the enzyme-bound FAD (C) of SCAD, MCAD and LCAD for various acyl-CoA's. Extinction coefficients for the bleaching of FAD which were induced by titration with stoichiometric amount (solid line) or 10 times molar excess (dotted line) of the acyl-CoA's indicated. o———o, o- - - -o, SCAD; ●———●, ●- - - -●, MCAD; ▲———▲, ▲- - - -▲, LCAD.

dehydrogenated mainly by a single dehydrogenase under physiological conditions. Butyryl-CoA was almost exclusively dehydrogenated by SCAD while acyl-CoA's with C_6-C_{10} were mainly dehydrogenated by MCAD. Acyl-CoA's with C_{14}-C_{22} were exclusively dehydrogeneated by LCAD. C_{24}-Acyl-CoA's were not dehydrogenated by LCAD. Lauroyl-CoA was jointly dehydrogenated by MCAD and LCAD. In each of these complete reactions, ETF is first reduced to red anionic semiquinone and then to hydroquinone. The pKa's and optimum pH's for the complete reaction were 7.0 and 8.0, respectively, for each enzyme.

Spectrophotometric titration with various substrates.

The three straight chain acyl-CoA dehydrogenases were titrated with various acyl-CoA substrates and the results were compared to the turnover numbers of the same enzyme/substrate combination which were determined by the dye-reduction assay(s). The long wavelength absorbance (centered around 580 nm) appeared only in the enzyme/substrate combinations in which enoyl-CoA was produced at a significant rate in the complete reaction. There was a good correlation between the magnitudes of the long wavelength absorbance and the turnover numbers (Fig. 2 A and B). In contrast, the bleaching of the flavin chromophore occurred not only in the titration with preferred substrates, but also in that with unfavorable substrates from which no 2-enoyl-CoA was produced (Fig. 2C). Neither the appearance of the long wavelength absorption nor the bleaching of flavin chromophore was observed in the interaction of the enzymes with 2-enoyl-CoA's.

References

1. Ikeda, Y., Dabrowski, C., Tanaka, K.: J. Biol. Chem. 258, 1066-1076 (1983).
2. Ikeda, Y., Tanaka, K.: J. Biol. Chem. 258, 1077-1085 (1983).
3. Ikeda, Y., Tanaka, K.: J. Biol. Chem. 258, 9477-9487 (1983).
4. Ikeda, Y., Ikeda, K.O., Tanaka, K.: J. Biol. Chem. (submitted).
5. Ikeda, Y., Ikeda, K.O., Tanaka, K.: J. Biol. Chem. (submitted).

MECHANISM OF ACTION OF SHORT CHAIN, MEDIUM CHAIN, AND LONG CHAIN ACYL-COA DEHYDROGENASES ISOLATED FROM RAT LIVER.

Yasuyuki Ikeda, David Hine, Kazuko O. Ikeda and Kay Tanaka. Department of Human Genetics, Yale University School of Medicine, New Haven, CT 06510.

Demonstration of Acyl-CoA Dehydrogenase Catalyzed C-2 Monodeuteration of Acyl-CoA Substrate in D_2O.

The mechanisms of the initial interactions of three rat liver acyl-CoA dehydrogenases (1-3) (short chain (SCAD), medium chain (MCAD), and long chain acyl-CoA dehydrogenases (LCAD)) and their fatty acyl-CoA substrates were studied using enzyme catalyzed deuterium exchange. The purified SCAD, MCAD, or LCAD (2) was incubated with an appropriate substrate in D_2O in the absence of electron acceptor. The reaction products were identified and quantitated using mass spectroscopy (Fig. 1) and ^{1}H-NMR (Fig. 2). A rapid monodeuteration of the substrate occurred to replace one of the prochiral C-2 hydrogens, while no C-3 hydrogens were exchanged with deuterium. The C-2 monodeuteration proceeded to the extent of 80% of the total amount of substrate added at 90 min (Fig. 3) and almost to completion at 120 min. The pKa's and optimum pD's for the C-2 proton/deuteron exchange reactions were 6.0 and 7.5, respectively, for each of the three acyl-CoA dehydrogenases. The apparent turnover numbers were 3.0, 3.3 and 0.5 S^{-1} for SCAD/valeryl-, MCAD/octanoyl- and LCAD/palmitoyl-CoA reactions, respectively. These results provide the first direct evidence for carbanion formation via abstraction of C-2 hydrogen by a base in the enzyme, as the first step of the catalytic pathway of acyl-CoA dehydrogenation.

The Physico-chemical Nature of the Acyl-CoA Dehydrogenase/substrate Complex.

When a large amount of each acyl-CoA dehydrogenase (23-38 nmol of enzyme-bound FAD) was reacted with a moderate excess of a suitable acyl-CoA (70-80 nmol) in D_2O in the absence of an electron acceptor, maximum bleaching of the FAD absorbance and the appearance of the long wavelength absorbance,

Flavins and Flavoproteins

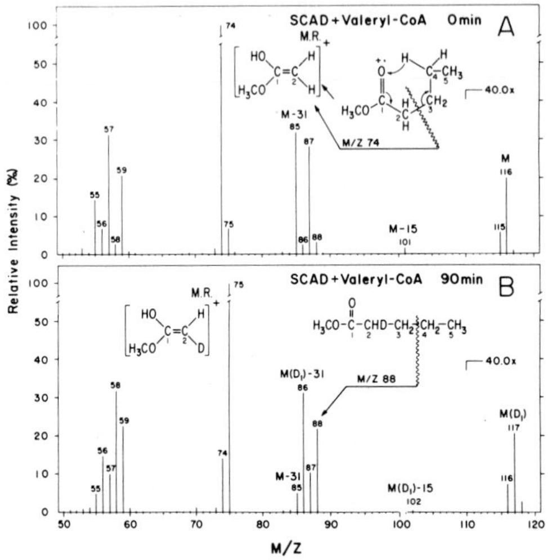

Fig. 1. Mass spectral identification of 2-monodeuterated valeryl-CoA produced in the proton/deuteron exchange reaction catalyzed by short chain acyl-CoA dehydrogenase in the absence of an electron acceptor. The pure short chain acyl-CoA dehydrogenase (0.22 nmol of enzyme-bound FAD) was incubated with valeryl-CoA (300 nmol) in a final volume of 1 ml of D_2O containing 100 mM potassium phosphate buffer pD 7.5, at 37°C. No electron acceptor was added. The methylated samples were analyzed by GC/MS/COM with electron impact ionization. A: Unlabelled substrate at 0 min. B: Product from the exchange reaction at 90 min.

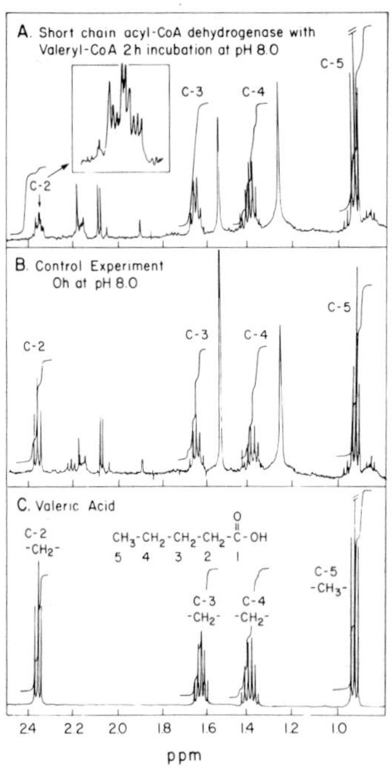

Fig. 2. Identification by ^{1}H-NMR (500 MHz) of 2-monodeuterated valeryl-CoA produced in the proton/deutron exchange reaction catalyzed by short chain acyl-CoA dehydrogenase in the absence of an electron acceptor. The pure short chain acyl-CoA dehydrogeanse (1.5 nmol of enzyme-bound FAD) was reacted with valeryl-CoA (500 nmol) in 1 ml of D_2O containing 100 mM potassium phosphate buffer, pH 8.0 at 37°C. After alkaline hydrolysis and acidification, the reaction mixture was extracted with $CDCl_3$. The spectrum was taken at 315°K. A: The product from the exchange reaction at 120 min. B: Unlabelled substrate at 0 min. C: Authentic unlabelled valeric acid standard. Chemical shifts are referenced to the tetramethylsilane signal in CCl_4 at 0 ppm.

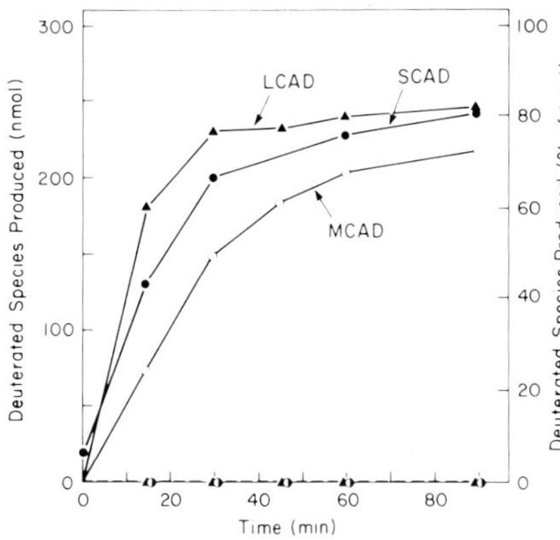

Fig. 3. Time course of the accumulation of 2-monodeuterated product of the proton/deuteron exchange catalyzed by short chain, medium chain, and long chain acyl-CoA dehydrogenases. The pure short chain- (SCAD) (0.22 nmol enzyme-bound FAD), medium chain- (MCAD) (0.16 nmol enzyme-bound FAD), and long chain acyl-CoA dehydrogenases (LCAD) (1.5 nmol enzyme-bound FAD) were reacted with valeryl-CoA (300 nmol), octanoyl-CoA (300 nmol), and palmitoyl-CoA (300 nmol), respectively, in 1 ml of D_2O containing 100 mM potassium phosphate buffer, pD 7.5, in the absence of an electron acceptor. At the time indicated, KOH was added to stop the reaction and to hydrolyze CoA thioester. After acidification, the fatty acid moiety of the reaction product was extracted and methylated as described in Fig. 1. The amount of the deuterated compound was determined by monitoring the abundance of the respective molecular ions. Symbols are: C-2 Monodeuterated species produced by SCAD (●———●), MCAD (o———o), and LCAD (▲———▲). Dideuterated species catalyzed by SCAD (●- - - -●), MCAD (o- - - -o), and LCAD (▲- - - -▲).

attributed to a charge transfer complex, were observed. However, the dehydrogenation products, 2-enoyl-CoA's were produced either not at all or in an amount which represented only a minor fraction of the amount of the enzyme added, while the substrates in the enzyme/substrate complexes rapidly turned over as indicated by the extensive C-2 monodeuteration. Unlike previous hypotheses, these results indicate that the hydride ion transfer from C-3 of the substrate to the enzyme-FAD is not yet complete in the charge transfer complex. We propose that acyl-CoA dehydrogenases and acyl-CoA's form donor-acceptor resonance hybrids (charge transfer complexes) via the intervening hydride ion. The transfer of the hydride

Scheme I : Proposed Donor-Acceptor Resonance Hybrid as Charge Transfer Complex

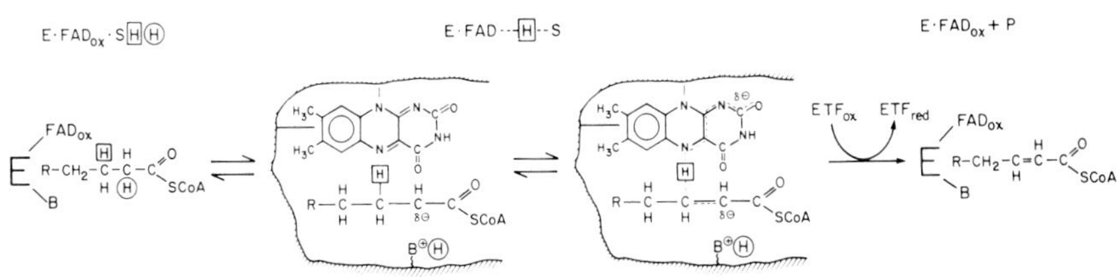

ion to alloxazine N-5 and the release of products are completed only when electrons are transferred from enzyme-FAD to ETF or another suitable electron acceptor as illustrated in Scheme I (4).

Demonstration of a Catalytically Essential Cysteine Residue in Three Acyl-CoA Dehydrogenases.

We observed that the apoenzymes of MCAD and LCAD were severely inactivated by NEM, pCMB and IAA. The holoenzymes were not susceptible to the action of these SH reagents. Preincubation of the apoenzymes with FAD completely prevented the inactivation. However, holo-SCAD was strongly inactivated by pCMB. Holo-MCAD and LCAD were inactivated by MMI and $HgCl_2$, respectively, without perturbing the FAD chromophore. The enzymes modified by these SH reagents were incapable of C-2 proton/deuteron exchange. The observed differences of the apo and holoenzymes in their susceptibility to the SH reagents, together with the results from the titration of the apo-MCAD and apo-LCAD with [^{14}C]NEM and that of the holo-SCAD with [^{14}C]pCMB, indicated that all three acyl-CoA dehydrogenases contain an essential cysteine residue in the vicinity of FAD-binding site in each subunit. This cysteine residue does not participate in the FAD-binding site per se, however. In addition, substrate strongly protected the enzymes from the inactivation by SH reagents, also indicating that the essential cysteine residue is located within the active center of the enzymes. Based on the above data, the cysteine residue may be considered to be a good candidate for the base with abstracts C-2 proton (5).

References

1. Ikeda, Y., Dabrowski, C., Tanaka, K.: J. Biol. Chem. 258,1066 (1983).
2. Ikeda, Y., Ikeda, K.O., Tanaka, K.: J. Biol. Chem. (submitted).
3. Ikeda, Y., Ikeda, K.O., Tanaka, K.: J. Biol. Chem. (submitted).
4. Ikeda, Y., Hine, D., Ikeda, K.O., Tanaka, K.: J. Biol. Chem. (submitted).
5. Ikeda, K.O., Ikeda, Y., Tanaka, K.: J. Biol. Chem. (submitted).

INACTIVATION OF PIG KIDNEY GENERAL ACYL-CoA DEHYDROGENASE BY 2-ALKYNOYL-CoA DERIVATIVES

Kurt Freund, John P. Mizzer, and Colin Thorpe
Chemistry Department, University of Delaware, Newark, Delaware 19716 U.S.A.

Introduction

General acyl-CoA dehydrogenase is a mitochondrial flavoprotein which catalyzes the oxidation of saturated acyl-CoA thioesters of medium chain length to their corresponding trans-2-enoyl derivatives. This reaction is probably initiated by abstraction of the pro-R proton at C-2 with transfer of the pro-R hydrogen at C-3 to the N-5 position of the flavin (1). The interaction of several suicide substrates with the dehydrogenase is consistent with an initial α-proton abstraction. Thus, for example, 3-alkynoyl thioester derivatives are believed to undergo enzyme catalyzed isomerization to the corresponding allene prior to irreversible attachment to the protein (2,3). Abeles and coworkers have shown that a glutamate residue is modified upon inactivation, and they suggest that this residue may be the active site base (3). In light of this work, we have examined the interaction of 2-alkynoyl-CoA thioesters:

$$R\text{-}CH_2\text{-}C{\equiv}C\text{-}CO\text{-}SCoA$$

with the general acyl-CoA dehydrogenase. These analogues are also potent inactivators of the enzyme, and their interaction with the dehydrogenase presents several interesting and unanticipated features.

Flavins and Flavoproteins

Results and Discussion

Figure 1 shows the spectral changes seen when 2-octynoyl-CoA (I) is mixed with general acyl-CoA dehydrogenase at 25°C. A charge transfer band appears centred at about 800 nm and then decays more slowly (see inset) yielding a somewhat resolved and red shifted oxidized flavin spectrum. At 4°C, a further species (E·I) is observed immediately after mixing with a spectrum rather similar to that of the final species shown in Figure 1 (curve 4). These data and results to be outlined later are consistent with the following minimal scheme:

$$E + I \underset{k_{-1}}{\overset{k_1}{\rightleftharpoons}} E\cdot I \xrightarrow{k_2} E\text{-}I^{*} \xrightarrow{k_3} E\text{-}I^{\ddagger} \qquad \text{Scheme 1}$$

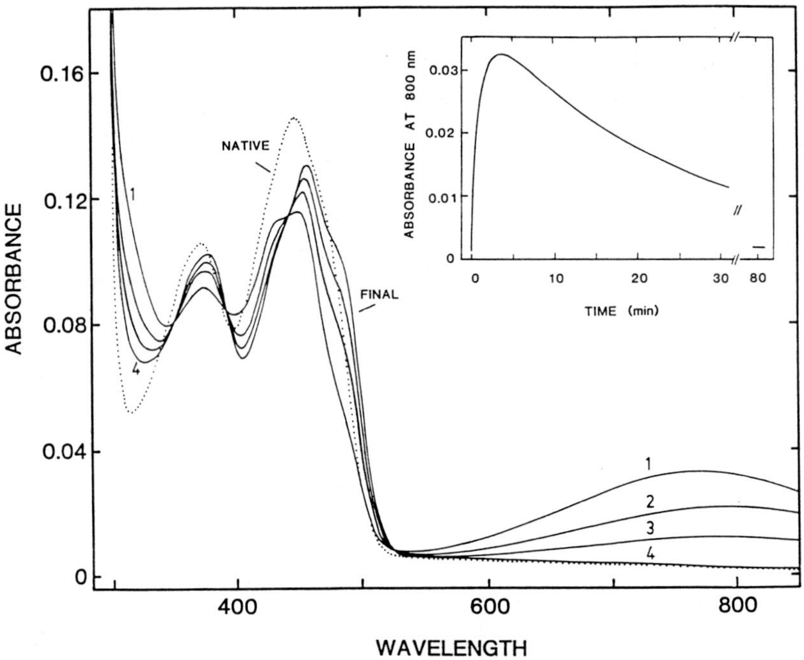

Figure 1. Spectral changes on the addition of 2-octynoyl-CoA to pig kidney general acyl-CoA dehydrogenase. The native enzyme in 50 mM phosphate buffer, pH 7.6 (····) was mixed with 2.9 equivalents (28 μM) 2-octynoyl-CoA: curves 1-4; 1, 13, 29, and 75 minutes after mixing respectively.

where $E-I^*$ and $E-I^{\ddagger}$ represent the 800 nm species and final form respectively. The stoichiometry of the interaction of 2-octynoyl-CoA with the dehydrogenase was determined spectrophotometrically as 1.1 per enzyme FAD by choosing an isosbestic point between $E-I^*$ and $E-I^{\ddagger}$ forms (at 442 nm) and recording the final absorbance after each addition of inhibitor.

The modified enzyme (curve 4) is inactive and does not regain activity nor lose FAD upon prolonged dialysis in phosphate buffer, pH 7.6. Although the native dehydrogenase is reduced rapidly and extensively by 1 equivalent of octanoyl-CoA (4), the spectrum of the treated enzyme is unchanged on prolonged incubation with an excess of this substrate. However, the enzyme derivative can be photoreduced readily to the dihydroflavin level and then rapidly reoxidized on the addition of the physiological electron acceptor, electron transferring flavoprotein. Thus the interflavin oxidative half-reaction appears largely unaffected upon modification of the dehydrogenase with 2-octynoyl-CoA.

Assays run during incubation of the enzyme with 2-octynoyl-CoA show that inactivation appears to be associated with the appearance of $E-I^*$, not in its slower decay to $E-I^{\ddagger}$ (Figure 1, inset). Further, no recovery of enzyme activity is seen on dilution of the enzyme to nanomolar levels in these assays. Evidence that inactivation results from covalent modification of the protein comes from experiments in which samples of octynoyl-CoA treated dehydrogenase were quenched with the acidic charcoal suspension used to prepare apoprotein (5). Apoprotein derived from the fully modified enzyme rebinds FAD with recovery of the spectrum of $E-I^{\ddagger}$, but no activity is regained. The rate of this covalent modification is clearly considerably faster than k_3 (Scheme 1) and thus may also be associated with the formation of $E-I^*$.

Surprisingly, the modified enzyme does not exhibit the increased absorbance at 260 nm expected for the incorporation of a CoA moiety. Preliminary experiments suggest that CoASH is eliminated in the slower phase of the reaction, i.e. with decay of the long wavelength absorbance. Possibly the charge transfer donor is the immediate product of attack of an anionic nucleophile on the acetylene, yielding:

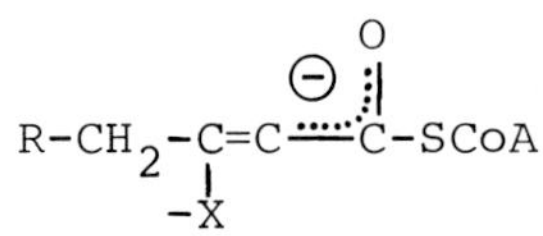

This could conceivably eliminate CoASH with loss of long wavelength absorbance. Identification of the target nucleophile(s) and isolation of labelled peptides must await the use of 1-[^{14}C]-2-octynoyl-CoA. Since acyl-CoA oxidase from *Candida tropicalis* is also inactivated irreversibly by 2-octynoyl-CoA, it will be interesting to compare the sequence of labelled peptides from both oxidase and dehydrogenase.

Acknowledgment

This work was supported in part by NIH grant GM 26643.

References

1. Ghisla, S., Thorpe, C., Massey, V.: Biochemistry, in press.
2. Frerman, F. E., Miziorko, H. M., Beckman, J.: J. Biol. Chem. 255, 11192-11198 (1980).
3. Fendrich, G., Abeles, R. H.: Biochemistry 21, 6685-6695 (1982).
4. Thorpe, C., Matthews, R. G., Williams, C. H.: Biochemistry 18, 331-337 (1979).
5. Mayer, E. J., Thorpe, C.: Anal. Biochem. 116, 227-229 (1981).

ON THE INACTIVATION OF GENERAL ACYL-CoA-DEHYDROGENASE FROM PIG KIDNEY BY METHYLENECYCLOPROPYL-ACETYL-CoA, A METABOLITE OF HYPOGLYCIN

Studies with flavin modified enzymes

Hans-Dieter Zeller and Sandro Ghisla

Fakultät für Biologie der Universität Konstanz
D-7750 Konstanz, FRG

Introduction

General acyl-CoA dehydrogenase from pig kidney (GAD), a typical fatty acyl-CoA dehydrogenase, has previously been shown to be irreversibly inactivated by methylenecyclopropyl-acetyl-CoA (MCPA-CoA), a metabolite derived from the poisonous amino acid hypoglycin (1). Inactivation results from addition of the inactivator to the flavin coenzyme. Elucidation of the structure of the flavin adduct(s) and of the detailed mechanism of inactivation were hampered by the instability of the adduct itself. The available evidence indicates that adduct formation involves position N(5) of the isoalloxazine moiety, and probably also another flavin function such as C(6), C(4a), or C(4)=O. The goal of our work is the elucidation of the structure of the flavin-MCPA adduct. We hope that this might aid in the understanding of the inactivation mechanism of this class of suicide inhibitors.

Results and Discussion

5-Carba-5-deaza-FAD-General Acyl-CoA Dehydrogenase

Inactivation with this analog was attempted since adducts formed with it are (more) stable as compared to those formed with native enzyme (1). This modified GAD does <u>not</u> react with MCPA-CoA, it also does not react with normal substrates (2), as opposed to the very rapid reaction of 3,4-pentadienoyl-CoA, a similar suicide inactivator (3). The lack of reactivity is attributed to the very low redox potential (-310 mV) of the 5-d-FAD (4).

Iso-FAD-General Acyl-CoA Dehydrogenase

This modified enzyme shows an enzymatic activity 12% that of native GAD. It is completely and irreversibly inactivated in a way similar to that observed with native enzyme (1). This allows the exclusion of position C(6) as a point of adduct formation. The time dependence of the inactivation is very complex and does not lead to complete disappearance of the oxidized iso-FAD absorbance, suggesting inactivation due to processes other than flavin modification. However, the apoprotein prepared from inactivated iso-FAD-GAD could be reconstituted with normal FAD to yield the same complex as native enzyme, as judged from its spectral properties (absorbance ratio 260/450 nm) and its specific catalytic activity. This indicates that either protein modification does not occur, or that it is reversible.

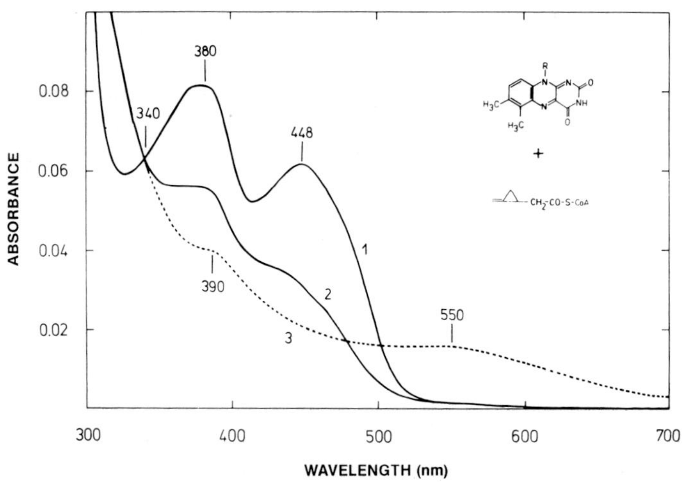

Fig. 1. Reaction of iso-FAD-General Acyl-CoA Dehydrogenase with MCPA-CoA

The enzyme, 7 μM in 0.1 M phosphate buffer pH 7.6 (curve 1) was incubated 2 h at 25° with 28.0 μM MCPA-CoA (curve 2). The reaction with 28 μM octanoyl-CoA (curve 3) is shown for comparison.

4-Thio-FAD-General Acyl-CoA Dehydrogenase

Substitution of normal FAD with 4-thio-FAD lowers the activity of GAD to 10 % the original value. 4-thio-FAD is bound only weakly by the protein. The reaction of this modified GAD with MCPA-CoA differs significantly in its course, from that observed with normal enzyme (1). First, an intermediate is formed with a half time of ~ 4 min. This species has an absorption spectrum (Fig. 2) similar to that observed upon reaction of 4-thio-FAD-GAD with octanoyl-CoA, i.e. to that of the reduced enzyme enoyl-CoA complex. In a second, slower phase, ($t_{1/2}$ ~ 110 min) a stable product is formed, the spectral properties of which are closely similar to those of an 4a-adduct formed with 4-thio-FMN lactate oxidase (5). HPLC analysis of the product(s) released upon denaturation of MCPA-CoA inactivated 4-thio-FAD-GAD shows three main peaks with retention times higher than those of 4-thio-FAD. The isolation and identification of these products is in progress.

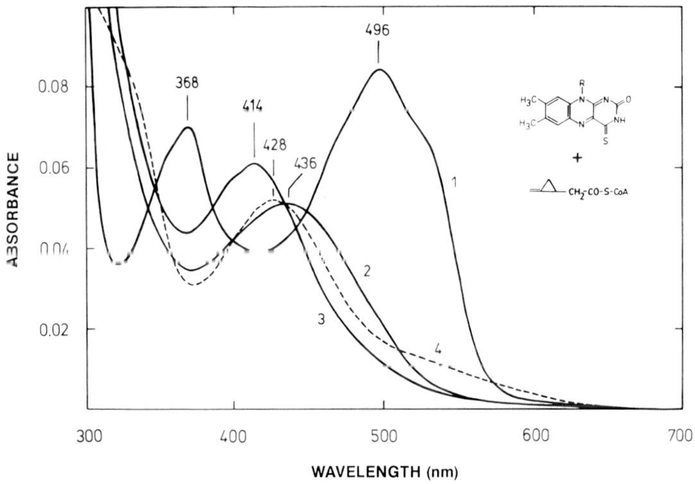

Fig. 2. Reaction of 4-thio-FAD-General Acyl-CoA Dehydrogenase with MCPA-CoA

The enzyme, 5.4 μM in 0.1 M phosphate buffer pH 7.6 (curve 1), was incubated at 25° with 22 μM MCPA-CoA. A first intermediate is formed within 30 min (curve 2). In a second, slower phase, a second species is formed, and its spectrum was recorded after 4 hr (curve 3). The product formed with 22 μM octanoyl-CoA (curve 4) is shown for comparison.

Conclusions

The kinetics of the reaction indicate that inactivation is a biphasic process. The reactivity observed with iso-FAD-GAD excludes position C(6) of the flavin as an obligatory point of adduct formation during inactivation with MCPA-CoA. When comparing with reduced 4-thioflavin models, the primary adduct most probably has an 1,5-dihydroflavin structure, the point of substitution being N(5). The primary adduct rearranges to a secondary 4a,5-dihydroflavin derivative:

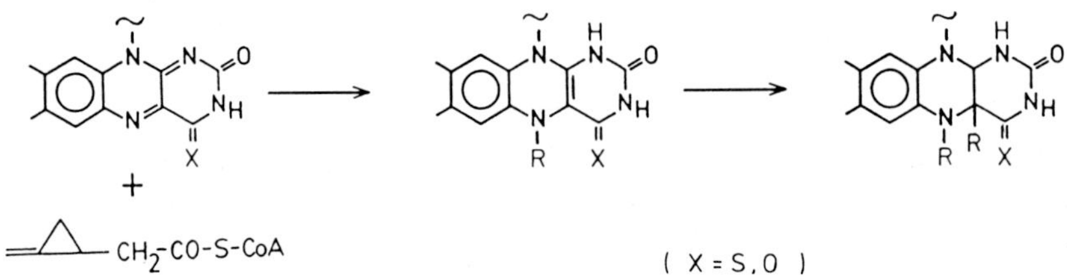

The secondary adduct is unstable upon protein denaturation; it decays to products the structure of which is presently under investigation. The fact that 5-d-FAD-GAD does not react suggests that α-proton abstraction is not the only event required for initiating inactivation. This could imply that MCPA-CoA dehydrogenation is involved in inactivation.

References

1. Wenz, A., Thorpe, C., Ghisla, S.: J. Biol. Chem. 256, 9809-9812 (1981).
2. Thorpe, C., Massey, V.: Biochemistry 22, 2972-2978 (1983).
3. Wenz, A., Ghisla, S., Thorpe, C.: "Flavins and Flavoproteins", (Massey, V., and Williams, C.H. eds), Elsevier, Amsterdam, pp 605-608 (1982).
4. Hemmerich, P., Massey, V., Fenner, H.: FEBS Letters 84, 5-21 (1977).
5. Massey, V., Claiborne, A., Biemann, M., Ghisla, S.: J. Biol. Chem., in press.

STRUCTURE-FUNCTION CORRELATION IN β-OXIDATION ENZYMES

Camilo Rojas, William Gustafson, Jack Schmidt, Dan Domanski, Benjamin A. Feinberg, and James T. McFarland

Department of Chemistry, University of Wisconsin-Milwaukee, Milwaukee, Wisconsin, USA, 53201

Our research group has been using a new chromophore producing pseudo substrate (β-2-furylpropionylCoA, FPCoA) to investigate the function of two enzymes, general fatty acylCoA dehydrogenase from porcine liver mitochondria and fatty acylCoA oxidase from Candida tropicalis, involved in β-oxidation. The purpose of these studies is to compare and contrast the functional and structural behaviour of these two similar but distinct enzyme systems in order to reach greater insight into the mechanism of each. The reactions of the two enzymes with FPCoA are shown below; the dehydrogenase in reaction 1 transfers electrons to ETF an FAD containing flavoprotein. This protein serves as a substrate for the electron transport chain while in reaction 2, oxidase transfers electrons to O_2.

$$R\text{-}CH_2\text{-}CH_2\text{-}\overset{O}{\overset{\|}{C}}\text{-}S\text{-}CoA + 2ETF \overset{DH}{\rightleftharpoons} 2ETF^- + R\text{-}CH{=}CH\text{-}\overset{O}{\overset{\|}{C}}\text{-}S\text{-}CoA \quad (1)$$

$$R\text{-}CH_2\text{-}CH_2\text{-}\overset{O}{\overset{\|}{C}}\text{-}S\text{-}CoA + O_2 \overset{OX}{\rightleftharpoons} H_2O_2 + R\text{-}CH{=}CH\text{-}\overset{O}{\overset{\|}{C}}\text{-}S\text{-}CoA \quad (2)$$

One of our first observations on the dehydrogenase enzyme was that the stoichiometric production of β-2-furylacryloylCoA, FACoA, during the reductive half reaction was followed by a catalytic reaction requiring oxygen and producing peroxide (1), i.e, by an oxidase reaction (1). We now have carried out isotope experiments which confirm our original suggestion that product dissociation from the enoylCoA reduced flavin charge transfer complex is rate limiting for the oxidase reaction. The current data show that although there is a primary isotope effect of 7.3 upon the stoichiometric production of FACoA, neither the breakup of the product charge transfer complex nor the catalytic oxidase reaction show a deuterium isotope effect upon reaction with tetradeutero FPCoA. The interesting question raised by the observation of oxidase activity in the dehydrogenase is the chemical basis of the functional difference between dehydrogenase and oxidase. Figure 1 shows the pH dependencies of the steady state kinetic reactions of the two enzymes. The use of the chromophoric pseudo substrate permits the direct comparison of the pH profiles since there is no need to couple the reaction to a dye because double bond production can be observed directly. Table 1 shows the primary isotope effect for both enzymes across the pH range assuring that the chemical step for both reactions can be compared directly. The most striking comparison in Figure 1

Flavins and Flavoproteins

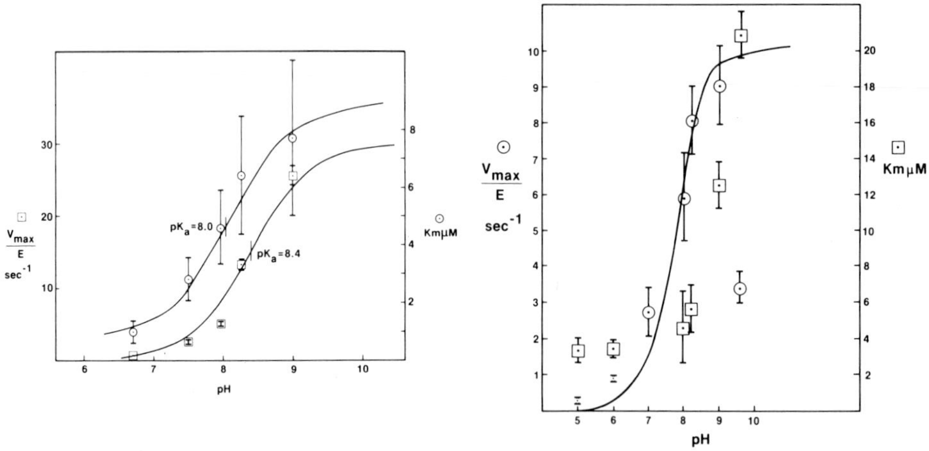

FIGURE 1. pH Dependencies of the Steady State Parameters for Dehydrogenase (left) and Oxidase (right).

is the similarity in the pH profiles for both kinetic parameters with these two enzymes. The data are consistent with a mechanism involving base catalyzed removal of the C-2 proton of substrate followed by hydride transfer to the electron acceptor.

TABLE I.

	pH 6.7		pH 8.5	
	FPCoA(H)	FPCoA(D)	FPCoA(H)	FPCoA(D)
Fatty AcylCoA Dehydrogenase	K_m = 1 M	K_m = 1 M	K_m = 7.0 M	K_m = 6.7 M
	V/E=.93 sec^{-1}	V/E=.25 sec^{-1}	V/E=20.5 sec^{-1}	V/E=3.3 sec^{-1}
	V(H)/V(D) = 3.7		V(H)/V(D) = 6.2	
Fatty AcylCoA Oxidase	K_m = 3.4 M	K_m = 3.3 M	k_m = 23.5 M	k_m = 19.6 M
	V/E=.97 sec^{-1}	V/E=.33 sec^{-1}	V/E=10.7 sec^{-1}	V/E=4.0 sec^{-1}
	V(H)/V(D) = 2.9		V(H)/V(D) = 2.7	

Figure 2 shows the electronic spectra of the product charge transfer complex formed upon the reductive half reaction between FPCoA and dehydrogenase and oxidase. The product charge transfer complex of dehydrogenase is not stable (1) and its spectrum has been determined using a stopped-flow spectrometer. The oxidase forms a kinetically stable complex with an electronic spectral transition at higher energy than that of the dehydrogenase. A previous investigation has shown that other substrates of the

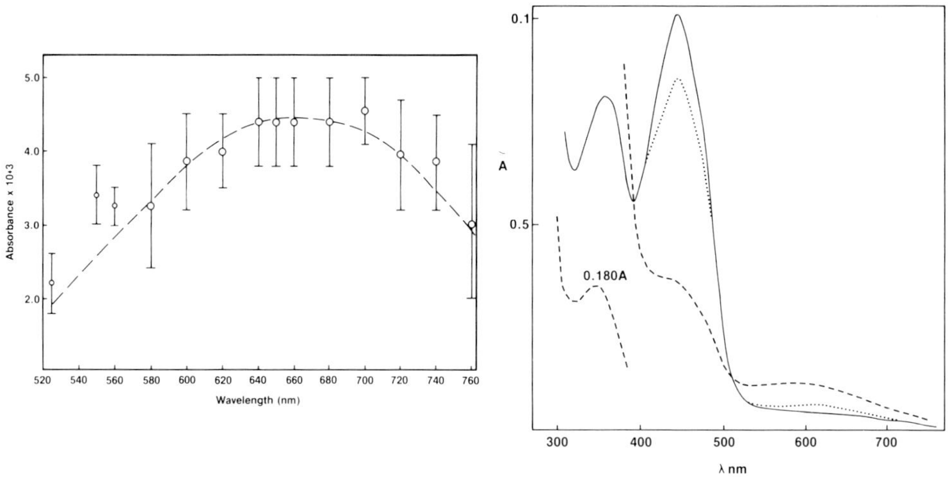

FIGURE 2. Charge Transfer Complexes of Dehydrogenase (left) and Oxidase (right).

oxidase do not form substantial quantities of the product charge transfer complex (2), but we believe this is the result of insufficiently anaerobic conditions. In fact, the production of FACoA permits the demonstration of substrate turnover in a particularly graphic manner demonstrating that present techniques for exclusion of O_2 are not sufficient to prohibit O_2 reaction with the oxidase. We also have compared the chemical reactivity of the two charge transfer complexes; the oxidase charge transfer complex reacts with O_2 but not with ETF while that for the dehydrogenase reacts with ETF but only reacts with O_2 after dissociation. Table II summarizes the chemical reactivity and the spectral and kinetic properties of each charge transfer complex.

TABLE II.

Product Charge Transfer Complex	
Dehydrogenase	Oxidase
Kinetically Unstable	Kinetically Stable
Electronic Transition: 650 nm	Electronic Transition: 590 nm
Reacts With ETF	Does Not React With ETF
Does Not React With O_2	Reacts With O_2
Does Not React With EnoylCoA	Reacts With EnoylCoA

The data summarized above suggest that the specificity of reaction in the two β-oxidation enzymes lies in the differences in the chemical reactivity of the product charge transfer complexes. The fact that the rate of reaction of the product charge transfer complex of dehydrogenase and ETF is comparable to the rate of the overall reaction for transfer of electrons from FPCoA to ETF supports the argument that the reaction takes place through the charge transfer complex. The same argument can be made for oxidase since the reaction of O_2 with charge transfer complex results in the reoxidation of flavin and the disappearance of the charge transfer spectral band.

We also have determined the formal reduction potentials for both enzymes using spectroelectrochemical techniques. The dehydrogenase has an $E^{o'}$ = -0.162 V while the oxidase is not as good an oxidizing agent with an $E^{o'}$ = -0.185 V. This difference in reduction potentials is consistent with our observation that the equilibrium constant for the enzyme reaction of FPCoA is very large for dehydrogenase, but not for oxidase. The functional data above can be correlated with structural data collected by resonance Raman, RR, spectroscopy (3). RR data show little or no hydrogen bonding between FAD and oxidase protein while the hydrogen bonding between FAD and dehydrogenase is more extensive and slightly less than between FAD and water in aqueous solution. In keeping with this observation and the functional data, hydrogen bonding between protein and flavin would be expected to result in electron delocalization from the FAD (4) resulting in an increase in formal electrode potential for the dehydrogenase. Likewise the dipoles that result from the formation of hydrogen bonds between dehydrogenase protein and FAD would be expected to stabilize a charged intermediate (5), such as the charge transfer product complex, resulting in the observed lowering of the electronic transition in the dehydrogenase.

References

1. McFarland, J.T., Lee, M.-Y., Reinsch, J., and Raven, W.: Biochem. 21, 1224-1229 (1982).

2. Jiang, Z.-Y., and Thorpe, C.: Biochem. 22, 3752-3758 (1983).

3. Schmidt, J., Coudron, P., Thompson, A.W., Watters, K.L., and McFarland, J.T.: Biochem. 22, 76-84 (1983).

4. Watanabe, Y., Nishimoto, K., Kashiwagi, H., and Yagi, K.: Flavins and Flavoproteins (Massey, W., and Williams, C., Ed.), 541-545, Elsevier/North-Holland, Amsterdam (1982).

5. Warshel, A., Proc. Natl. Acad. Sci. (USA) 75, 5250-5254 (1976).

PURIFICATION AND CHARACTERIZATION OF GLUTARYL-CoA DEHYDROGENASE, ELECTRON TRANSFER FLAVOPROTEIN AND ETF-Q OXIDOREDUCTASE FROM *PARACOCCUS DENITRIFICANS*

Mazhar Husain, Daniel J. Steenkamp

Molecular Biology Division, Veterans Administration Medical Center, San Francisco, CA 94121 and Department of Biochemistry and Biophysics, University of California, San Francisco, CA 94143

Introduction

The many similarities between the respiratory chain of *Paracoccus denitrificans* and that of the mitochondrial inner membrane (1) have stimulated much interest. There is, however, a paucity of information on the molecular characteristics of the respiratory proteins of *P. denitrificans* which are not associated with Complex IV. An indication that an ETF-dependent enzyme system may be present in *P. denitrificans* was obtained by Albract, *et al.* (2), who observed EPR signals characteristic of the $[Fe_4S_4]^{1+}$-cluster of ETF-Q oxidoreductase in membrane preparations from this organism. We report here the isolation of glutaryl-CoA dehydrogenase (GluCoADH), ETF and ETF-Q oxidoreductase from *P. denitrificans* and present studies on their cross reaction with components of the mammalian system.

Experimental Procedures

P. denitrificans cells grown on 0.2% glutarate were suspended in 10% ethylene glycol containing 0.1 mM phenylmethylsulfonylfluoride (PMSF) and broken in a French Press. The supernatant from high speed centrifugation was chromatographed on DEAE-cellulose DE-53. ETF and GluCoADH were successively eluted with a linear KCl gradient in 25 mM potassium phosphate, pH 6.6 containing 10% ethylene glycol and 0.1 mM PMSF. Further purification of ETF was affected by acid precipitation and chromatography on hydroxyapatite and Sephadex G-100. GluCoADH was further purified by chromatography on DEAE-Sepharose and hydroxyapatite. ETF-Q oxidoreductase was purified

Flavins and Flavoproteins

from *P. denitrificans* membranes by modification of the procedure of Ruzicka and Beinert (3). GluCoADH was assayed using PMS/DCIP-linked assay at a saturating concentration of glutaryl-CoA. ETF was detected by observing the decrease in absorbance in the visible region upon the addition of GluCoADH and glutaryl-CoA. ETF-Q oxidoreductase was assayed by observing the reduction of CoQ_1 at 275 nm with GluCoAD and ETF as electron donors.

Results

Fig. 1 shows that *Paracoccus* ETF is very similar in molecular size to the ETF from pig liver. The subunit molecular weight of GluCoADH was about 43,000, whereas the native protein had a molecular weight of about 175,000 as determined by molecular exclusion chromatography. The enzyme contained 4 FAD residues per mole. GluCoADH is therefore a tetramer of four identical subunits, each containing one FAD. The oxidation of glutaryl-CoA by GluCoADH proceeded with decarboxylation to produce a stoichiometric amount of crotonyl-CoA. The enzyme had a pH optimum in the range of 8.0-8.5 and was maximally stable at pH 5.0. At 30°C in 100 mM potassium phosphate, pH 7.5, the catalytic center activity was about 960 min^{-1} with either pig liver ETF or PMS as electron acceptor. The Km,app for glutaryl-CoA and pig liver ETF was about 1.2 μM and 2.5 μM, respectively.

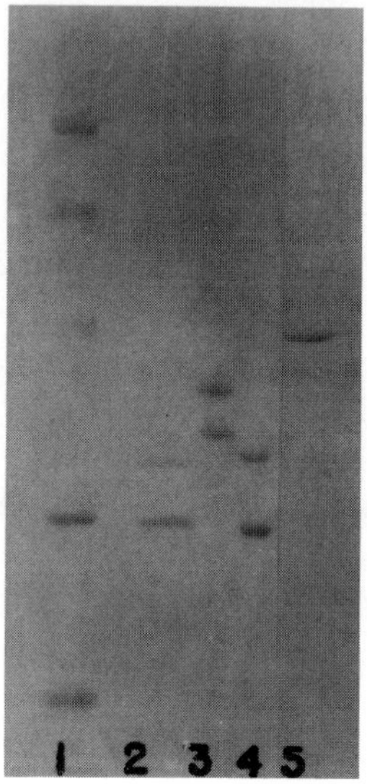

Fig. 1 SDS-PAGE on 10% polyacrylamide slab gel. 1, Marker proteins phosphorylase B, BSA, ovalbumin, carbonic anhydrase, soybean trypsin inhibitor and lysozyme; 2, pig liver ETF; 3, W_3A_1 ETF; 4, Paracoccus ETF; 5, Paracoccus GluCoADH.

Bacterium W_3A_1 ETF was not reduced. The formation of complexes of acetoacetyl-CoA with the oxidized form of the enzyme and of crotonyl-CoA with the fully reduced form were spectrophotometrically detected (Fig. 2). Similar to the ETF from pig liver or bacterium W_3A_1, *Paracoccus* ETF is comprised of two non-identical subunits (M_r 28000 and 32000) and contains one FAD per dimer. The amino acid composition of the three proteins showed marked similarities. The isoelectric point of *Paracoccus* ETF was 4.45 as determined by isoelectric focusing and reflects the changed proportion of acidic and basic amino acid residues relative to pig liver ETF, which has

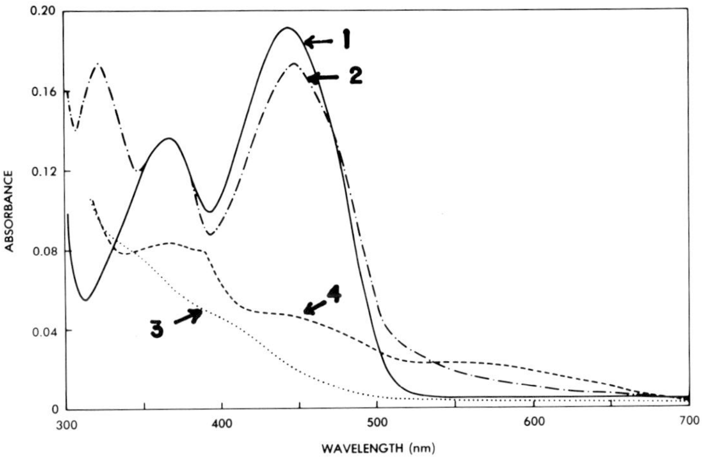

Fig. 2 The effect of acetoacety-CoA and crotony-CoA on the absorption spectra of GluCoADH. 1, Enzyme (14 nmoles) in 1 ml of 50 mM potassium phosphate, pH 7.0; 2, Same as in 1 but with 35 μM acetoacetyl-CoA; 3, Same as in 1 but after titration with dithionite; 4, Same as in 3 but with 0.15 mM crotonyl-CoA.

an isoelectric point of 6.8 (4). In common with other known electron transfer flavoproteins, *Paracoccus* ETF readily formed the anionic semiquinone, but was 13-fold less fluorescent than pig liver ETF. *Paracoccus* ETF accepted electrons from pig liver butyryl-CoA and octanoyl-CoA dehydrogenases (Fig. 3) and also from isovaleryl-CoA dehydrogenase (not shown), but not from dimethylglycine dehydrogenase. *Paracoccus* ETF was also inactive as an electron donor to pig liver ETF-Q oxidoreductase.

Partially purified *P. denitrificans* ETF-Q oxidoreductase catalyzed the reduction of CoQ_1 with GluCoADH and either pig liver or *Paracoccus* ETF as the electron donors (Fig. 3) and had an EPR spectrum similar to the mitochondrial enzyme (courtesy of Dr. Helmut Beinert). The isolation and initial characterization of GluCoADH, ETF and ETF-Q oxidoreductase from *P. denitrificans* extends the homology between the respiratory apparatus of this procaryote and that of mitochondria to the flavoprotein dehydrogenases and ETF associated with the electron transport chain.

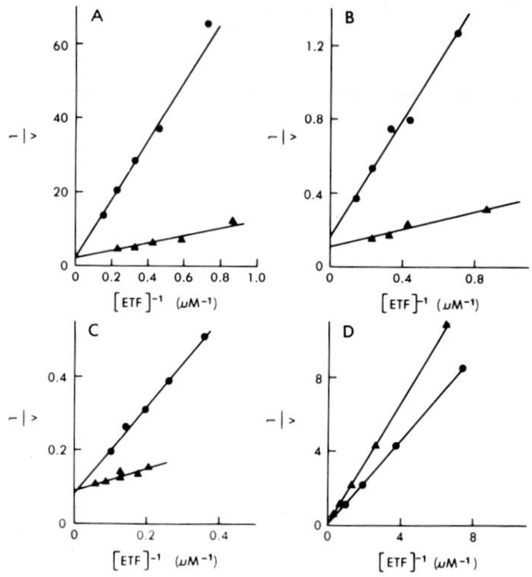

Fig. 3 Comparison of the activity of ETF from pig liver or Paracoccus *as electron acceptors from A, pig liver butyryl-CoA dehydrogenase; B, pig liver octanoyl-CoA dehydrogenase; C,* P. denitrificans *glutaryl-CoA dehydrogenase and, D, as electron carriers between* P. denitrificans *glutaryl-CoA dehydrogenase and ETF-Q oxidoreductase. The buffer used in assays in A, B, and C was 50 mM potassium phosphate, pH 7.0, containing 20 mM glucose and was made anaerobic by addition of one unit of glucose oxidase and 200 units of catalase. Substrate concentrations were 100 μM butyryl-CoA in A, 50 μM octanoyl-CoA in B and 30 μM glutaryl-CoA in C. The assay mixture in D contained 20 mM potassium phosphate, pH 7.0, 0.1 μM CoQ_1, 0.17 μM glutaryl-CoA dehydrogenase FAD and 30 μM glutaryl-CoA.*

References

1. John, P., Whatley, F.R.: Biochim. Biophys. Acta 463, 129-153 (1977).
2. Albracht, S.P.J., Van Verseveld, H.W., Hagen, W.R., Kalkman, M.L.: Biochim. Biophys. Acta 593, 173-186 (1980).
3. Ruzicka, F.J., Beinert, H.: J. Biol. Chem. 252, 8440-8445 (1977).
4. Husain, M., Steenkamp, D.J.: Biochem. J. 209, 541-545 (1983).

Acknowledgements

This work was supported by the Veterans Administration and by grants from the National Institutes of Health (HL-16251) and NSF (PCM 83-03225).

REACTIONS OF ETF AND ETF-Q OXIDOREDUCTASE

D.J. Steenkamp, R.R. Ramsay and M. Husain

Department of Biochemistry and Biophysics, University of California, San Francisco and Molecular Biology Division, Veterans Administration, San Francisco, California 94121

Introduction

ETF-Q oxidoreductase catalyzes the reoxidation of reduced ETF with Q_1 as the electron acceptor (1). It is essential for β-oxidation of fatty acids, for catabolism of some amino acids and for the operation of the mitochondrial one carbon cycle.

The available evidence suggests that the anionic semiquinone forms of ETF and Q are both physiologically important. ETF is rapidly reduced to a stable anionic semiquinone by ETF-dependent dehydrogenases and more slowly to the reduced form (2-4). It was of interest to determine whether the semiquinone and dihydroquinone forms of ETF are reoxidized at the same rate by ETF-Q oxidoreductase (ETFdh) with Q_1 as the electron acceptor.

Results

Glutaryl-CoA dehydrogenase (GDH) from *P. denitrificans* resembles in many properties the mitochondrial acyl-CoA dehydrogenases (5). It catalyzes reduction of ETF in two one-electron steps. For ETF_{ox} to $ETF^{\cdot -}$, V_{app} was 6.2 $\mu mol.min^{-1}mg^{-1}$, and for $ETF^{\cdot -}$ to $ETFH_2$, V_{app} was 1.3 $\mu mol.min^{-1}mg^{-1}$. As in all cases so far examined (2-4), reduction of ETF_{ox} is appreciably faster than that of $ETF^{\cdot -}$. GDH (with glutaryl-CoA) was used as the electron donor in the steady-state assay of ETFdh. When glutaryl-CoA is added to an assay mixture containing ETF, Q_1 and low levels of GDH a significant burst in the reduction of Q_1 is observed (Fig. 1). This reaction tapers off to

Flavins and Flavoproteins

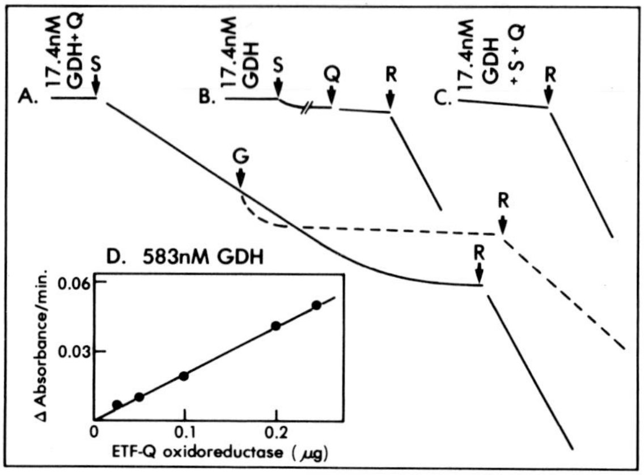

Fig. 1. Effect of concentration of Glutaryl-CoA dehydrogenase on the ETF-Q oxidoreductase assay. ETF and ETFdh from pig liver and GDH *from* P. denitrificans *were isolated as described (6,1,5). ETFdh was assayed at 25°C by following the decrease in absorbance at 275 nm due to the reduction of Q_1 ($E = 12{,}200\ M^{-1}cm^{-1}$). Assay mixtures (0.8 ml) contained 25 mM HEPES, pH 7.8, 10% ethylene glycol, 30 μM glucose, glucose oxidase (1 unit), catalase (24 units), 30 μM glutaryl-CoA, 0.9 μM ETF, 144 μM Q_1,* GDH *and ETFdh. Additions were S, 30 μM glutaryl-CoA; G, 122 nM* GDH; *Q, after 8 min preincubation, 144 μM Q_1; R, 0.025 μg ETFdh. The full scale represents 0.2 absorbance units at 275 nm. Inset D values were obtained at 0.58 μM* GDH.

a low zero order rate. The burst in the reduction of Q_1 is inhibited by excess GDH and is not seen when the reaction is started by adding Q_1. At higher levels of GDH the reaction rate is linearly dependent on ETFdh. Under these conditions double reciprocal plots, in which ETF or Q were varied at fixed concentrations of one another, indicated $K_{ETF} = 0.23$ μM, $K_Q = 6$ μM ($\alpha = 6.3$) and $V_{max} = 46$ μmol.min^{-1}mg^{-1}.

The reactions of $ETF^{\cdot -}$ and $ETFH_2$ with Q_1 and with ETFdh were investigated. $ETF^{\cdot -}$, generated by dithionite titration of ETF, was rapidly oxidized by Q_1. The reaction was pseudo first order at high levels of Q_1 and at pH 7.8 a rate constant of 1300 $M^{-1}sec^{-1}$ was determined. The reoxidation of $ETFH_2$ by Q_1 was very slow. Thus, reduction of Q_1 independent of ETFdh is observed in the assay (Fig. 1) only when $ETF^{\cdot -}$ is available, but not under conditions where $ETFH_2$ accumulates.

When ETFdh was added to $ETF^{\cdot -}$ rapid disproportionation of $ETF^{\cdot -}$ to a mixture of $ETF^{\cdot -}$, ETF and $ETFH_2$ was observed (Fig. 2). Molar extinction co-

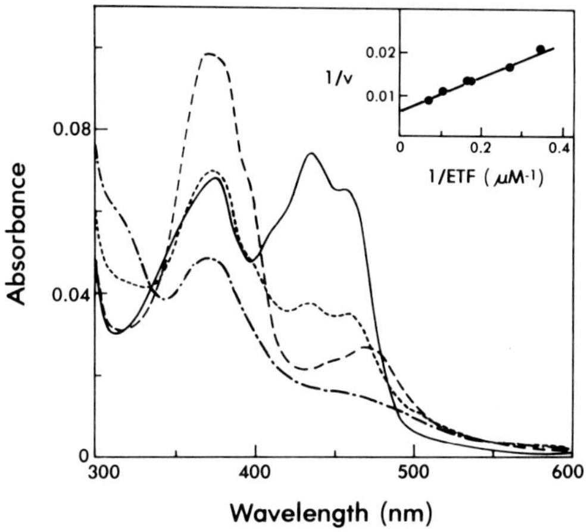

Fig. 2. Disproportionation of ETF$^{\cdot-}$ by ETF-Q oxidoreductase. Mixtures (0.8 ml) contained 25 mM HEPES (pH 7.8), 10% ethylene glycol, 30 µM glucose, glucose oxidase (1 unit), catalase (24 units) and 13.9 µM ETF reduced to ETF$^{\cdot-}$ by titration with dithionite. Disproportionation was initiated by adding 98 ng ETFdh. The Molar extinction coefficient was calculated to be 7300 at 372 nm. The spectra are: ——— ETF$_{ox}$; — — — ETF$^{\cdot-}$; -·-·- ETFH$_2$; ------ the equilibrium mixture after disproportionation. Inset. Double reciprocal plot for the rate of disproportionation (µmol.min^{-1} mg^{-1}).

efficients for ETF, ETF$^{\cdot-}$ and ETFH$_2$ were estimated at several wavelengths from spectra such as shown in Fig. 2. The equilibrium constant was about 1 at pH 7.8. In the disproportionation reaction the Michaelis constant for ETF was 8 µM and V was 200 µmoles.min^{-1}mg^{-1}. This unusual reaction is not a general property of oxidoreductases which interact with ETF. When GDH was added to 9.5 µM ETF$^{\cdot-}$ disproportionation proceeded at a rate of only 0.032 µmol.min^{-1}mg^{-1}.

ETFdh catalyzes the reoxidation of ETFH$_2$ with Q$_1$ as the acceptor. Determination of V for this reaction is both laborious and complicated by a non-linear dependence of the intercepts of the primary kinetic plots on Q$_1$. The V$_{app}$ of 50 µmol.min^{-1}mg^{-1}, estimated at the highest concentration used (146 µM), suggests that the reoxidation of ETFH$_2$ could be fast enough to account for the rate of the overall reaction. To establish whether disproportionation of ETF$^{\cdot-}$ by ETFdh occurs in the presence of Q the reaction was investigated by the stopped flow method. The spectral changes observed at different wavelengths during the rapid reaction were distinctly biphasic and the relative magnitudes were as anticipated from

Fig. 3 for the disproportionation of $ETF^{\cdot -}$.

Discussion

The data presented indicate that $ETF^{\cdot -}$ and $ETFH_2$ are formed at different rates by the ETF-dependent acyl-CoA dehydrogenases. $ETF^{\cdot -}$, but not $ETFH_2$, is rapidly reoxidized by Q. ETFdh catalyzes the disproportionation of $ETF^{\cdot -}$ at a rate which far exceeds that required for turnover. In the presence of Q, ETFdh also catalyzes the reoxidation of $ETFH_2$ but at a slower rate, close to that required for turnover. It is not yet clear whether disproportionation of $ETF^{\cdot -}$ followed by reoxidation of $ETFH_2$ in the presence of Q is an obligatory pathway in the catalytic mechanism of ETF-Q oxidoreductase, but, in view of the relative reaction rates, this unusual pathway is likely to contribute to its turnover.

References

1. Ruzicka, F.J. and Beinert, H.: J. Biol. Chem. 252, 8440-8445 (1977).
2. Beckmann, J.D., Frerman, F.E. and McKean, M.C.: Biochem. Biophys. Res. Commun. 102, 1290-1292 (1981).
3. Hall, C.L. and Lambeth, J.D.: J. Biol. Chem. 255, 3591-3595 (1980).
4. Steenkamp, D.J. and Husain, M.: Biochem. J. 203, 707-715 (1982).
5. Husain, M. and Steenkamp, D.J.: this volume, pp.
6. Husain, M. and Steenkamp, D.J.: Biochem. J. 209, 541-545 (1983).

Acknowledgements

This work was supported by the Veterans Administration and by grants from National Institutes of Health (HL-16251) and the National Science Foundation (PCM 81-10585 and PCM 83-03225).

CORRELATION BETWEEN REDOX STATE OF ETF AND DEHYDROGENATION OF OCTANOYL COA

Carole L. Hall
School of Chemistry, Georgia Tech
Atlanta, GA, 30332.

Introduction

Experiments to try to correlate dehydrogenation of octanoyl CoA with redox state of ETF were undertaken, combining the techniques of stopped flow spectrophotometry, rapid chemical quench, and HPLC analysis of the quenched reaction mixture to measure saturated and α,β unsaturated acyl CoA.

Results and Discussion

Experiments were performed using an Update Instruments stopped flow spectrophotometer with programmable ram. Data collection and analysis was dono with an OLIS 3820 Data System. HPLC analysis of quenched reaction mixtures was done using a modification of the method of Corkey, et al (1). Absorbance changes with time at 380 nm were measured upon rapidly mixing G-AD(1.25 μM[F]), previously treated with octanoyl CoA (10 μM), with ETF(10 μM) (2). Each trace in Fig. 1 consists of the mean of 3-4 "shots". The same experiment was carried out with 20 μM FeCN in the ETF syringe. Fig. 1 shows the records obtained in the absence (trace 1) and presence (trace 2) of FeCN on two different time scales. The patterns of absorbance change with time in the absence and presence of FeCN were the same as when G-AD was more concentrated (3).

Flavins and Flavoproteins

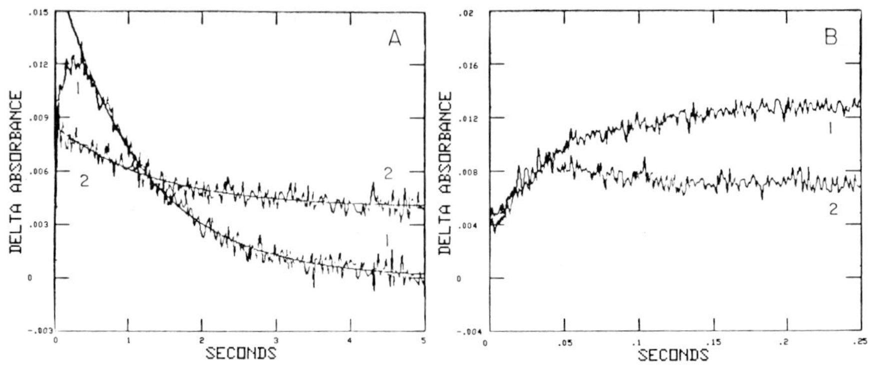

Figure 1. Delta Absorbance vs. Time at 380 nm in the Absence and Presence of FeCN.
A. The record obtained in the presence of FeCN (trace 2) was superimposed on the record obtained in the absence of FeCN (trace 1). The lines were drawn by the computer fitting program to determine the rate constant for the second phase. See the text.
B. Experimental conditions were as described in Fig. 1A except collection times were 0.25 sec.

Chemical quench experiments were performed using the same instrument as for the stopped flow absorbance studies. Absorbance changes with time were first measured in the stopped flow spectrophotometer as described above and then the instrument was reset so that enzymes were mixed using a "push-push" technique in which the mix of substrate-reduced G-AD and ETF were aged for stated time periods prior to mixing with 1 N HCl to stop the reaction. The length of the aging hose was such that at a ram velocity of 316 cm-1 the enzymes were mixed for 55 millisec before reaching the quenching mixing chamber when a straight push was used. Delays for the second push varied from 0.1 sec to 5 sec. Product formation was measured by HPLC analysis of the quenched reaction mixture. The amount of octenoyl CoA present was estimated from the integrated peak areas calculated for octenoyl CoA and octanoyl CoA. The peak area for octenoyl CoA was corrected for its increased extinction as compared to octanoyl CoA.

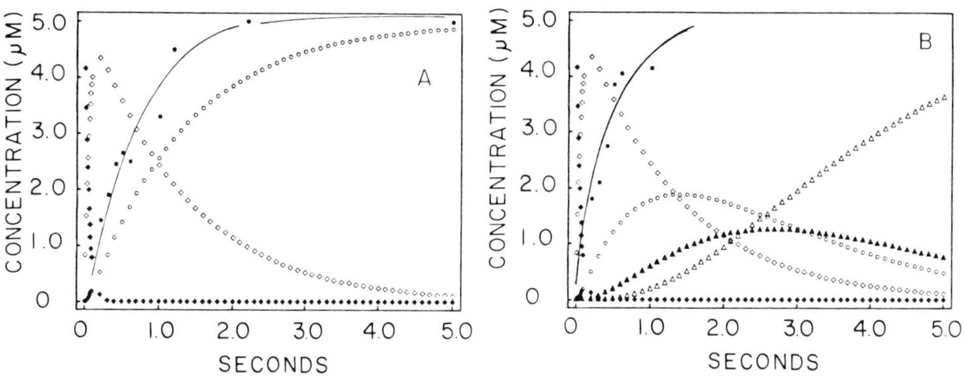

Figure 2A. Correlation between Octenoyl CoA Appearance and Calculated Concentrations for Starting Material, One Intermediate and Product vs. Time. See the text.
Figure 2B. Correlation between Octenoyl CoA Appearance and Calculated Concentrations vs. Time for a Sequential, Four Step Reaction. See the text.

Concentrations of starting material (ETF-ox, filled diamonds), one intermediate (ETF-SQ, open diamonds) and product (ETF-H_2, open circles) were calculated according to Eqns. 4.29 of Ref. 4 (using k_1 = 18 and k_2 = 0.74 s-1 obtained from the records shown as trace 1 in Fig. 1A & B) and plotted in Fig. 2A. Concentrations of octenoyl CoA calculated as described above were plotted as filled circles. The line was drawn to fit the points. It is apparent that in the absence of FeCN octenoyl CoA does not appear on the same time scale as ETF-SQ, but more nearly approaches that of ETF-H_2. Fig. 2B shows concentrations of starting material (ETF-ox, filled diamonds), intermediate I (ETF-SQ, open diamonds), intermediate II (ETF-H_2, open circles), intermediate III (ETF-SQ, filled triangles), and product (ETF-ox, open triangles) calculated according to Eqns. II.12 of Ref. 5 for a 4 step sequential reaction using the same rate constants for k_1 and k_2 as in Fig. 2A, and assuming k_3 and k_4 as 0.7 and 0.8 s-1 respectively. Concentrations of octenoyl CoA were obtained from the quenched reaction mixtures as described above and corrected to the concentration in the aging hose and were plotted as filled circles. Again, appearance of enoyl CoA is not correlated with appearance of SQ, but seems rather to parallel appearance of ETF-H_2 even in the presence of FeCN. The integrated areas for the sum of octanoyl plus octenoyl CoA were in good agreement with the integrated area for the octanoyl CoA originally added.

The results of other studies done at higher concentrations of G-AD and of computer modeling indicate that reduction of ETF to the fully reduced form occurs both in the absence and presence of FeCN (3). Results presented here show that appearance of the α, β double bond does not occur concomitantly with reduction of ETF to SQ in the absence or presence of FeCN but appears to correspond to full reduction of ETF. These results suggest fully reduced ETF is an obligate intermediate in the dehydrogenation of octanoyl CoA.

References

1. Corkey, B. E., Brandt, M., Williams, R. J. and Williamson, J. R.: Anal. Biochem. 118, 30-41 (1981).
2. Hall, C. L. and Lambeth, J. D.: J. Biol. Chem., 255, 3591-3595 (1980).
3. Hall, C. L.: In preparation.
4. Fersht, A.: Enzyme Structure and Mechanism, pp. 103-133, W. H. Freeman and Co., San Francisco (1977).
5. Rodiguin, N. M. and Rodiguina, E. N.: Consecutive Chemical Reactions, pp. 6-23, D. Van Nostrand Company, Inc., New York (1964).

Supported by USPHS grant GM 25494.

SOME OBSERVATIONS ON AN ACROYL-CoA REDUCTASE FROM CLOSTRIDIUM KLUYVERI AND AN NADH-DEPENDENT FUMARATE REDUCTASE FROM ENTEROBACTER AGGLOMERANS

Helmut Sedlmaier, Mathias Bühler, Richard Feicht, Johann Bader, Helmut Simon
Organisch Chemisches Institut, Techn. Univ. München, D 8046 Garching

Introduction

We are interested, for preparative purposes, in reductases which can be supplied with electrons by regenerable donors other than NAD(P)H. Examples are enoate reductase (EC 1.3.1.31) from Clostridium spec. La 1 or Clostridium kluyveri (1-3) or a 2-oxo-carboxylate reductase from Proteus species (3,4). Whereas enoate reductase accepts electrons from NADH as well as from reduced methylviologen ($MV^{+\cdot}$) the 2-oxocarboxylate reductase is readily reduced by $MV^{+\cdot}$ or benzylviologen ($BV^{+\cdot}$) but does not accept electrons from NAD(P)H. In Clostridium kluyveri we detected a flavoenzyme which seems to be an acroyl-CoA reductase. It cannot be reduced by NAD(P)H but by $MV^{+\cdot}$ or by reduced ferredoxin.

An NADH-dependent fumarate reductase containing FMN and FAD is present in an Enterobacter agglomerans. The enzyme also catalyzes the reduction of fumarate by $MV^{+\cdot}$.

Results

a) Acroyl-CoA reductase

Crude extracts of C. kluyveri (DSM 555) (grown according to l.c. (5)), separated on a DEAE-Sepharose CL-6B column by a linear gradient of KCl reveal two main activity areas catalyzing the reaction

$$C_2H_5COCH{=}CH_2 + 2H^+ + 2MV^{+\cdot} \longrightarrow C_2H_5COC_2H_5 + 2MV^{++}$$

The protein mixture eluting between 50-150 mM KCl was fractionated on a Sephacryl S-200 column. The area corresponding to a molecular weight of

Flavins and Flavoproteins

28 000 was further purified on Phenylsepharose CL-4B and Sephadex A-25. The so obtained acroyl-CoA reductase from C. kluyveri is different from the butyryl-CoA dehydrogenase from the same organism. On the DEAE-Sepharose CL-6B column the main activity of the butyryl-CoA dehydrogenase was eluted at a KCl concentration about 3 times higher and a volume about 3 times more than that for the acroyl-CoA reductase. By the different purification steps at least 5 different activities reducing ethyl vinyl ketone can be differentiated. Therefore the usual preparation of an enrichment scheme is not possible. Some properties of the purest fractions of the acroyl-CoA reductase (purity $>$ 90 % as jugded by gel electrophoresis) are summarized in Table 1. Acroyl-CoA is by far the best substrate as shown by the ratio V_{max}/K_m. The physiological electron donor seems to be reduced ferredoxin. A system containing 0.037 mM ferredoxin from C. spec. La 1 and hydrogenase from C. kluyveri, the reductase and acroyl-CoA reduced the latter at the same rate as the system in which ferredoxin was substituted by 2.5 mM methylviologen. The reduction of ethyl vinyl ketone in ^2H_2O buffer by $MV^{+\cdot}$ and the enzyme led to (2R)[1,2-^{2}H]-3-pentanone. The product showed the same optical rotation as (2R)[2-^{2}H]-3-pentanone for which the absolute configuration has been determined (6).

Table 1 Some properties and kinetic data of acroyl-CoA reductase (electron-donor = $MV^{+\cdot}$)

Molecular weight	28 400[a)]	1.7[b,c)] and 1.4[b)] mol FMN[d)]
Subunits	14 200	Less than 0.2 Fe[b)]

Substrate	Relative V_{max}(%)	Apparent K_m(mM)	$\frac{V_{max}}{K_m}$
Ethyl vinyl ketone	100	0.77	130
Acroyl-CoA	102	$<$0.001	$>$102000
Acroyl-N-acetylcysteamine	68	0.035	1950
Crotonyl-CoA	56	0.008	7000
$MV^{+\cdot}$	-	0.39	-

a) Average value obtained on two different gel columns; b) on the basis of 28 400 molecular weight; c) By spectrophotometric and fluorometric methods; d) Identification by TLC and HPLC (9).

This result shows for the α-carbon the same sterical course for the hydrogen addition as that determined for the reduction of crotonate in ^2H_2O buffer by whole cells of C. kluyveri (7) or that observed for mammalian butyryl-CoA dehydrogenase (8).
The reductase is astonishingly stable in the presence of air but rather unstable in the presence of $MV^{+\cdot}$. The reductase still has 75 % of its activity if stored under air in the presence of 30 % ethylene glycol and 50 mM phosphate buffer pH 7.0 at 10°C for 19 days. Under test conditions at pH 6.4 in the absence of oxygen 0.5 mM $MV^{+\cdot}$ inactivates the enzyme in 18 min up to 50 %. So only initial rates are given.

b) NADH-Dependent fumarate reductase

To the best of our knowledge so far an NADH-dependent fumarate reductase has not been characterized. Enterobacter agglomerans (NCTC 9381) contains such a reductase with a specific activity of about 0.8-1.0 U/mg protein in the crude extract when grown in the presence of 100 mM fumarate and a limited oxygen supply (80 ml air/min x 1). The enzyme could be enriched 450 times to a specific activity of 380 U/mg by chromatography on DEAE-Sepharose CL-6B, Sephacryl S 300, Procion Yellow and Procion Green (10) fixed to Sepharose 6B. From the latter two columns the enzyme was eluted with 2.0 and 4.0 mM NAD, respectively. Judged by electrophoresis the enzyme was about 90 % pure. The reductase catalyzes the following reactions:

$$NADH + fumarate + H^+ \longrightarrow NAD^+ + succinate$$

$$2\ MV^{+\cdot} + fumarate + 2\ H^+ \longrightarrow 2\ MV^{++} + succinate$$

$$NADH + APAD^+ \longrightarrow NAD^+ + APADH$$

$MV^{+\cdot}$ shows substrate inhibition. The reduction rate of fumarate by NADH or $MV^{+\cdot}$ is rather similiar if the concentration of $MV^{+\cdot}$ is in the range of 0.6-1.5 mM. The rate maximum can be observed with a concentration of $MV^{+\cdot}$ 0.2 mM. Above 1.5 mM fumarate exhibits substrate inhibition. NADPH does not function as an electron donor. 2-Methylfumarate or fumarate monomethyl ester also are not substrates.

The pro R hydrogen of NADH is mainly transferred to water as shown by the use of (4 S)- and (4 R)[4-^{3}H]NADH for the reduction of fumarate. By gel chromatography the molecular weight was estimated to be $3.0 \pm 0.2 \times 10^5$. Surprisingly 3.0 mol FAD and 1.9 mol FMN have been found per mol of the enzyme. By atomic absorption spectroscopy 24 Fe have been determined. SDS electrophoresis shows a complex pattern of subunits.

Acknowledgements. We thank Prof. A. Kröger for the hint that E. agglomerans contains an NADH-dependent fumarate reductase, Prof. W. Mannheim for several strains of Enterobacter and Deutsche Forschungsgemeinschaft (SFB 145) for financial support.

References

1. Tischer, W., Bader, J., Simon, H.: Eur. J. Biochem. 97, 103-112 (1979).
2. Bühler, M., Simon, H.: Hoppe Seyler's Z. Physiol. Chem. 363, 609-625 (1982).
3. Simon, H., Günther, H., Bader, J., Neumann, S.: N. Y. Acad. Sci, accepted.
4. Neumann, S., Simon, H.: FEBS Letters 167, 29-32 (1984).
5. Bader, J., Simon, H.: Arch. Microbiol. 127, 279-287 (1980).
6. Benner, S.A., Rozzell, J.D.,Jr.: J. Amer. Chem. Soc. 103, 993-994 (1981).
7. LaRoche, H.J., Kellner, M., Günther H., Simon H.: Hoppe Seyler's Z.Physiol. Chem. 352, 399-402 (1971).
8. Biellmann, J. F., Hirth, C. G.: FEBS Letters 9, 335-336 (1970).
9. Nielsen, P., Rauschenbach, P., Bacher, A.: Analyt. Biochem. 130, 359-368 (1983).
10. Clonis, Y.D., Lowe, C.R.: Biochim. Biophys. Acta 659, 86-98 (1981).

CHARACTERIZATION OF THE MODE OF ELECTRON TRANSPORT OF NADPH-ADRENODOXIN REDUCTASE

Toshio Yamano, Yasuki Nonaka and Shigeru Fujii

Department of Biochemistry, Osaka University
Medical School, Kita-ku, Osaka 530, Japan

Introduction

NADPH-adrenodoxin reductase (AdR) is an FAD-enzyme which plays an important role in electron transport from NADPH to adrenodoxin, namely, acting both in cytochrome $P\text{-}450_{scc}$- and in $P\text{-}450_{11\beta}$-linked hydroxylation systems. It is abundant in adrenocortical mitochondria and was first crystallized in our laboratory. Adrenodoxin affinity chromatography as well as 2',5'-ADP affinity chromatography were used successfully for its purification (1,2). Crystallized adrenodoxin reductase showed two protein bands in the isoelectric focusing. However, the analysis of the amino acid composition and the amino acid sequence of 37 residues at N-terminal region revealed that these reductases are almost or entirely identical with each other. Ferredoxin-related flavoproteins, ferredoxin-$NADP^+$ reductases, in plants or bacteria had been purified earlier (3). Crystallographical study on the latter has been started already (4). Molecular weight of adrenodoxin reductase of bovine adrenocortex is about 50,000 and higher by 17,000 than ferredoxin-$NADP^+$ reductase from Spirulina (2,5). The complete amino acid sequence of the adrenal reductase has not been determined yet, whereas that of the ferredoxin-$NADP^+$ reductase from Spirulina has been determined very recently (5).

Flavins and Flavoproteins

The elegant work performed by Lambeth and Kamin and their research groups, have disclosed the participation of reduced adrenodoxin reductase-$NADP^+$ charge transfer complex and introduced shuttle mechanism of electron transport of adrenodoxin between the reductase and adrenodoxin (6). We have established the occurrence of the semiquinone state of adrenodoxin reductase on the basis of EPR, resonance Raman spectroscopy as well as absorption spectroscopy (7,8).

$NADP^+$ binding to the reductase plays a crucial role in the function of this flavoprotein, not only as the electron donor but also in the electron transport mechanism. Sugiyama et al. (8) pointed out that the oxidation of reduced pyridine nucleotides is much higher with NADH than with NADPH in the presence of the reductase. Thus, $NADP^+$ prevents electron leakage to oxygen.

The dissociation constant for the complex of the reductase and $NADP^+$ is very small for that between the two-electron reduced state of the enzyme and $NADP^+$. The first step of the oxidation-reduction cycle of the reductase is that in which one mole of NADPH reduces the reductase and the produced $NADP^+$ is directly bound to the enzyme. Oxidation-reduction potential for the couple of the two-electron reduced state of the adrenodoxin reductase-$NADP^+$ complex and the oxidized reductase-$NADP^+$ complex is much higher than that for the reduced adrenodoxin/adrenodoxin couple (9). Therefore, the subject of the electron transfer among NADPH, the reductase and adrenodoxin may be understandable as restricted in the scope of the electron transfer among NADPH, the two electron-reduced FAD-$NADP^+$ complex of the reductase and oxidized adrenodoxin.

One of the objectives of this presentation is to propose a concept that $NADP^+$ may be classified as a co-prosthetic group in the two-electron reduced state of the reductase just as FAD bound to the reductase and the flavin semiquinone of the reductase may participate in the electron transfer.

Results and Discussion

1. NMR study of $NADP^+$ binding to the reductase.

^{31}P NMR spectra of adrenodoxin reductase showed a pair of doublets as shown in Fig. 1(d). This result shows that adrenodoxin reductase contains only one coenzyme, FAD, as a group giving ^{31}P NMR signals. Both of the doublets were shifted to upfield in the reductase compared to those of free FAD. Their chemical shifts in the bound form were -9.8 and -14.2 ppm. It may be a common feature for the coenzyme FAD in flavoproteins that one of the two ^{31}P signals of the pyrophosphate group undergoes a large upfield shift when FAD is bound to the apoprotein whereas the other shifts to upfield only slightly.

When $NADP^+$ was added to AdR, unfortunately the signals of $NADP^+$ pyrophosphate group partially overlapped with the downfield signal of AdR as shown in Fig. 1(c). However the upfield signal of AdR was observed to be scarecely shifted. This result shows that on formation of the complex of AdR with $NADP^+$, the environment of the pyrophosphate group was not greatly changed.

The pyrophosphate group of bound $NADP^+$ also gave resonances shifted from those of free $NADP^+$; one resonance was shifted to upfield and was observed at -12.8 ppm. Another resonance was overlapped with one of the resonances of AdR and was not resolved. However, it was scarecely shifted from those of free $NADP^+$. This feature of bound $NADP^+$ was similar to that of bound $NADP^+$ or bound NADPH in dihydrofolate reductase (10).

When $NADP^+$ was bound to AdR, the resonance of the 2'-phosphate group shifted to the lower field by 1.37 ppm from that of free $NADP^+$. Its pH-profile was remarkably different from that of free $NADP^+$ and the chemical shift value was constant in the observed pH region between pH 5.90 and pH 7.89. The resonance of 2'-phosphate of free $NADP^+$ was

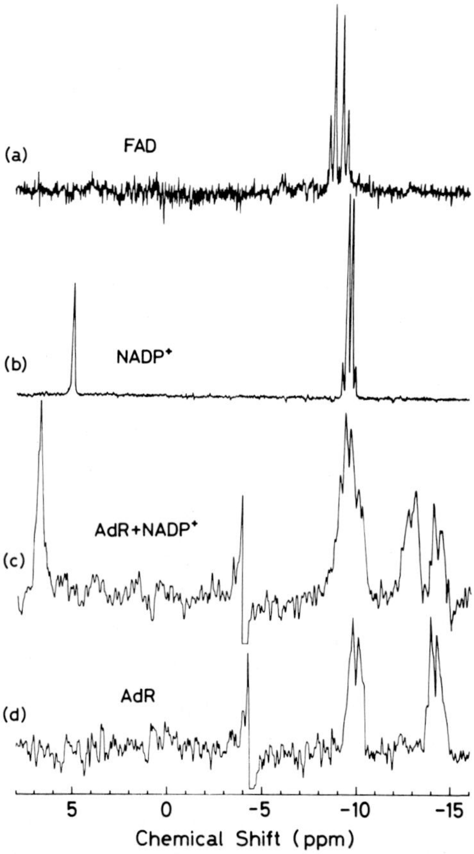

Fig. 1. ^{31}P NMR spectra measured at 80.98 MHz at 23 °C: (a) 0.5 mM FAD at pH 6.92 (100 mM KCl solution); (b) 2 mM $NADP^+$ at pH 6:92 (100 mM KCl solution); (c) 0.31 mM AdR and 0.31 mM $NADP^+$ at pH 6.93 (50 mM PIPES , 50 mM KCl solution); (d) 0.32 mM AdR at pH 6.93 (50 mM PIPES, 50 mM KCl solution).

observed as a singlet signal at a low field in a proton decoupled ^{31}P NMR spectrum as shown in Fig. 1(b). This signal shifted to downfield as pH was raised from 4 to 8. Its pH-profile shows that the 2'-phosphate group of free $NADP^+$ has a pK_a value of 6.05 at 23 °C between the monoanionic and the dianioic states. The resonance position for the 2'-phosphate of $NADP^+$ bound to the reductase suggests that the 2'-phosphate group of $NADP^+$ bound to AdR takes the dianionic state. Furthermore its pH-profile indicates that the dianionic state is strongly stabilized and the pK_a value of the 2'-phosphate of the bound $NADP^+$ is shifted to the acidic side from that of free $NADP^+$ by more than one pH unit. These results strongly suggest that there are some positively charged residues, such as arginyl, lysyl or histidyl residues, around the 2'-phosphate group of the bound $NADP^+$. This presumption is consistent with our previous report that arginyl and histidyl residues are essential for the enzymic activity and $NADP^+$ binding. Moreover, we found that the dissociation constant between AdR and $NADP^+$ became larger with raising the ionic strength with KCl. This result also supports the assumption that some ionic interaction takes part in the interaction between AdR and $NADP^+$.

The downfield shift of the 2'-phosphate of bound $NADP^+$ was also found in other two NADPH($NADP^+$) enzymes, dihydrofolate reductase and isocitrate dehydrogenase (10,11). The shift may be a common feature in $NADP^+$ or NADPH specfic enzymes. It has been revealed in the ^{31}P NMR studies of dihydrofolate reductases from various sources that the downfield shift of the 2'-phosphate resonance of bound $NADP^+$ or bound NADPH is mainly due to a histidyl residue interacting with the 2'-phosphate group. This result also suggests that there is a histidyl residue near the 2'-phosphate of bound $NADP^+$ in AdR. An excess amount of $NADP^+$ over AdR gave resonances of free $NADP^+$ in addition to those of bound $NADP^+$. Thus in the presence of excess $NADP^+$, bound $NADP^+$ and free $NADP^+$ were separately observed at the same time. This result shows that the exchange rate between bound $NADP^+$ and free $NADP^+$ was slow compared to the NMR time scale.

2. Characteristics of the two-electron reduced state of FAD-$NADP^+$ complex of AdR.

When adrenodoxin reductase was reduced by NADPH, the absorption spectrum was very much different from that of AdR reduced by dithionite.

As mentioned already, two-electron reduced FAD-$NADP^+$ complex of the reductase was first described by Lambeth et al. as a charge-transfer complex between AdR-$FADH_2$ and $NADP^+$ (12). The absorbance of the complex at 450 nm was about half of the oxidized FAD of the reductase and the broad band beyond 500 nm was observed. Those flavoproteins, such as glutathione reductase and lipoamide dehydrogenase, in which two sulfhydryl groups are directly involved in the redox cycle, show this sort of spectral property. Although ferredoxin-$NADP^+$ reductase or adrenodoxin reductase, which carries one electron to or from nonheme-iron protein, possesses a cysteinyl group (13,14) essential for the activity, evidence for the involvement of cysteinyl groups in the oxidation and reduction has not been demonstrated yet. The nature of this two-electron reduced species still remains obscure. The intensity of circular dichroism from 450 to 500 nm was about half of that of the oxidized reductase. Fluorescence intensity of the bound flavin of the reductase was reduced to about half of that of the oxidized form. The resonance Raman spectrum of this charge transfer complex exhibited a band at 1627 cm^{-1}, which conveys the character of the oxidized flavin.

3. Characteristics of the FAD semiquinone state of AdR.

As previously reported, the semiquinone state of AdR is very stable when $NADP^+$ is present in addition to the several-fold excess of NADPH. The results from the NMR study suggest that the flavin semiquinones of the reductase are kept shielded out of the access of the solvent water.

We measured ^1H NMR spectra, employing a 5-mm standard probe. In the presence of the NADPH genarating system with glucose-6-phosphate dehydrogenase, the change of ^1H NMR signals of adrenodoxin reductase was scarecely observed until equimolar NADPH with AdR was generated. As an excess of NADPH was generated, the signals of AdR decreased. This decrease was due to the line broadening by the formation of semiquinone of AdR. The formation of its semiquinone was confirmed by a direct observation of UV-visible absorption of the sample in the NMR cell. The visible spectrum of the sample with the ^{1}H NMR spectrum indicated that the blue semiquinone form of AdR was dominant. Thus, the semiquinone radical is located at a specific region of AdR and signals derived from the residues around the radical should disappear by line broadening. If the solvent, or the water molecules, easily comes in contact with the coenzyme flavin molecule, the signal of water (HDO) should become broader as the concentration of the semiquinone increases. With the formation of semiquinone, however, the observed signal of water did not become broader. This result shows that the bound FAD of AdR is buried in the protein which prevents access of solvent water to FAD in spite of the rapid electron transfer via AdR.

4. Titration of AdR with NADPH under anaerobic conditions.

Adrenodoxin reductase was titrated with NADPH under anaerobic conditions. The peaks at 450 and 376 nm decreased greatly with increased concentrations of NADPH. Isosbestic point was observed at 508 nm and the increment of the very flat absorbance was observed in the longer wavelength region from 508 to 900 nm. The spectral change at the ratio of NADPH/AdR less than one, was due to the charge transfer complex formation, as reported by Lambeth and Kamin (12). When excess NADPH was added, the decrease of the absorbance at 450 nm and the increase in the longer wavelength region was observed which were due to the further reduction of $AdRH_2 \cdot NADP^+$. The increment of the absorbance at 580 nm was characteristic of the blue, i.e., neutral, semiquione. When an extremely high concentration of NADPH (NADPH/AdR=10.8) was used, both absorbances at 580 and 780 nm decreased. This suggests that both the charge transfer complex and semiquinone disappeared by further reduction.

To elucidate the mechanism of semiquinone formation, we

performed the titration of adrenodoxin reductase with NADPH in the presence of various concentrations of $NADP^+$. The production of the charge transfer complex was monitored by the absorbance at 780 nm (Fig. 2 upper), since oxidized AdR or semiquinone has no absorbance at this wavelength, and since $AdRH_2$-NADPH complex shows little absorbance at 780 nm. At the first stage of reduction of AdR in which the charge transfer complex was formed with less than equimolar NADPH, we found that $NADP^+$ of various concentrations (0 to 4 mM) could not inhibit this process. So the dissociation constant of this process, represented by the equation,

$$AdR + NADPH \longleftrightarrow AdRH_2\text{-}NADP^+ \quad \cdots\cdots\cdots\cdots\cdots\cdots (1),$$

ought to be very small. When stoichiometric amount of NADPH was added to the reductase, almost all AdR was reduced to the charge transfer complex, two-electron containing species. When excess NADPH was added to the reductase, the absorbance at 780 nm decreased, indicating that the charge transfer complex decreased. This observation clearly showed that the reductase changed to a further reduced species. But when $NADP^+$ was added to the reaction mixture, the disappearance of the charge transfer complex was inhibited. These processes can be interpreted as follows. The dissociation process is involved before the further reduction of $AdRH_2$-$NADP^+$, as in equation 2;

$$AdRH_2\text{-}NADP^+ \longleftrightarrow AdRH_2 + NADP^+ \quad \cdots\cdots\cdots\cdots\cdots\cdots (2).$$

And further reduction occurs when excess NADPH is added

$$AdRH_2 + NADPH \longleftrightarrow AdRH_2\text{-}NADPH \quad \cdots\cdots\cdots\cdots\cdots\cdots (3).$$

This process could explain the decrease of the charge transfer complex at extremely high concentrations of NADPH. The absorbance around 580 nm, observed when excess NADPH was added to the reductase, demonstrate the formation of semiquinone state (Fig. 2 lower), which was further confirmed

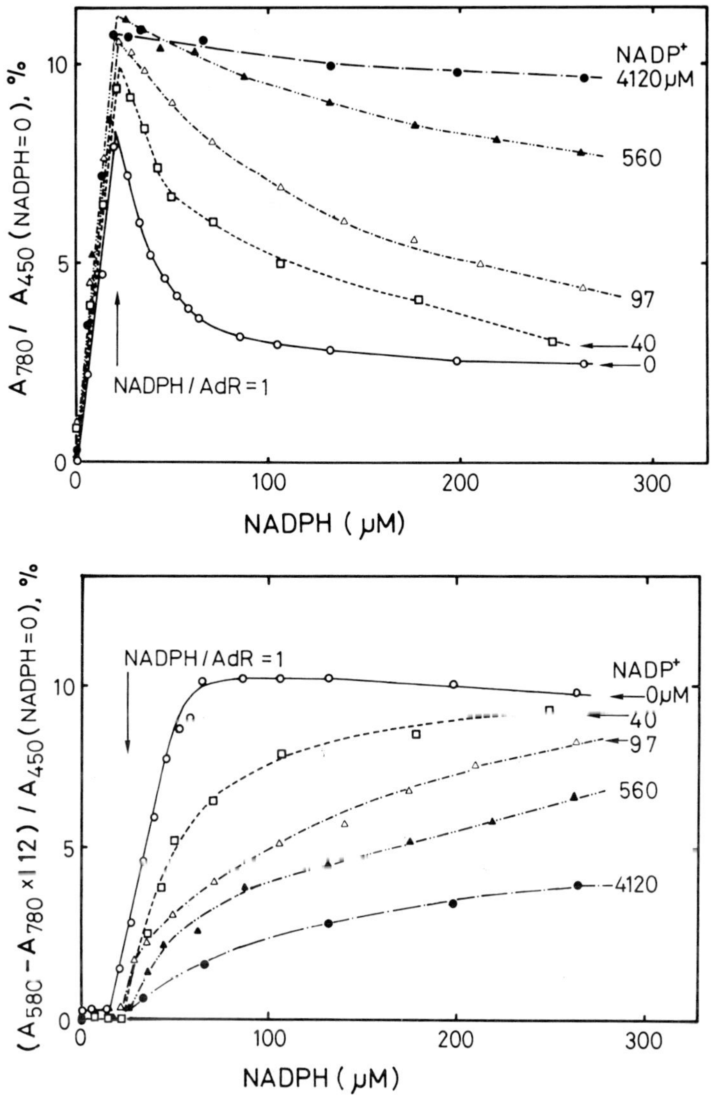

Fig. 2. Anaerobic titrations of AdR with NADPH in the presence of various concetrations of $NADP^+$: <u>upper</u>, CT complex formation; <u>lower</u>, semiquinone formation. AdR; 26 μM in 50 mM PIPES, pH 7.4 at 25 °C.

by EPR experiments. To estimate the formation of semiquinone, the absorbance of the charge transfer complex at 580 nm must be subtracted from the observed absorbace at 580 nm.

Hence,

aborbance of semiquinone at 580 nm =
observed absorbance at 580 nm - 1.12 x absorbance at 780 nm.

The plots of the absorbance of semiquinone against added NADPH under anaerobic conditions clearly showed that the semiquinone was formed considerably only with addition of excess NADPH and that addition of $NADP^+$ inhibited this process. So semiquinone formation must involve the two-electron containing species, $AdRH_2$-$NADP^+$, and more reduced species, that is, $AdRH_2$-NADPH, as expressed in equation 4.

$$AdRH_2\text{-}NADP^+ \;+\; AdRH_2\text{-}NADPH \;\longleftrightarrow\; 2\;AdRH\cdot\text{-}NADPH \quad \cdots\cdots(4)$$

The semiquinone was assumed to contain 3 electrons per AdR. For confirmation of the above hypothesis that the semiquinone contains more than two electrons, dithionite-reduced AdR ($AdRH_2$) was titrated with $NADP^+$ under anaerobic conditions. After the stoichiometric amount of dithionite was added to AdR, $NADP^+$ was added to the reaction mixture anaerobically. The absorbances at 450 and 780 nm increased, with the increase of $NADP^+$, and the spectrum showed that the charge transfer complex between reduced flavin and $NADP^+$ was formed during these processes. The increase of the absorbance at 450 nm level off essentially with the stoichiometric amount $NADP^+$ and the ratio of the 450 nm-absorbance to that before dithionite addtion reached the level of 50%, while the ratio of the 780 nm-absorption to the 450 nm absorption before dithionite addition reached the level of 10%. These levels were almost identical to those observed in the titration of AdR with NADPH in the presence of high concentrations of $NADP^+$. So this process was the charge transfer formation from $AdRH_2$ (eq.3). But no formation of semiquinone was observed. So the mechanism of semiquinone formation was not the reaction between $AdRH_2$, $AdRH_2$-$NADP^+$ and AdR-NADPH. And these plots allowed estimation of the dissociation constant between $AdRH_2$ and $NADP^+$, that is , about 0.5 µM. AdR was titrated again with $NADP^+$, but after dithionite in slight excess (1.1 mol/mol AdR) and excess NADPH (2.9 mol/mol AdR) were added. The absorbances at 450 and 780 nm increased with increase of $NADP^+$, and this showed the formation of charge transfer complex. And the semiquinone formation was observed concomitantly with the formation of the charge transfer complex. So the semiquinone was formed in the oxidation reduction process between the two-electron containing charge transfer complex and the four-electron containing complex, $AdRH_2$-NADPH.

THE MECHANISM OF SEMIQUINONE FORMATION

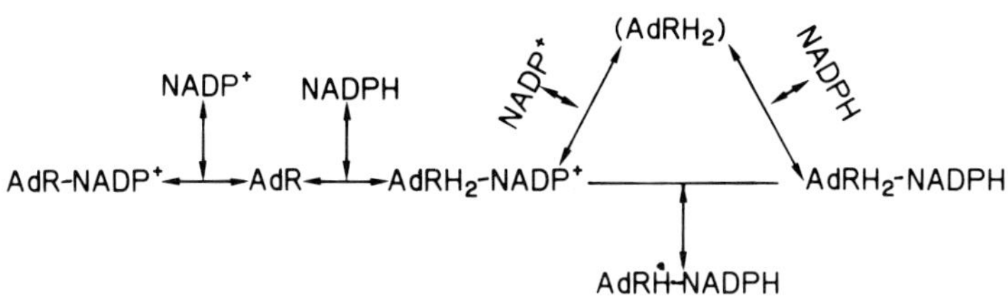

5. Electron transport from the reductase to adrenodoxin.

Fig. 3 illustrates the electron paramagnetic resonance (EPR) experiments. EPR spectrum of the mixture of reduced adrenodoxin and the reductase exhibited an equilibrium in which the reductase was mainly in the semiquinone form. When adrenodoxin was further added to this equilibrium to make the final concentration of adrenodoxin 1.5 times that of the reductase, the flavin semiquinone almost disappeared and a new equilibrium was attained instantaneously among the reduced adrenodoxin, trace of the flavin semiquinone and the increased level of the charge transfer complex of the reductase. It is not decisive whether the electrons to adrenodoxin were provided by the flavin semiquinone or the fully reduced flavin. In the latter case, a new equlibrium should be attained among the decreased NADPH, increased $NADP^+$ and

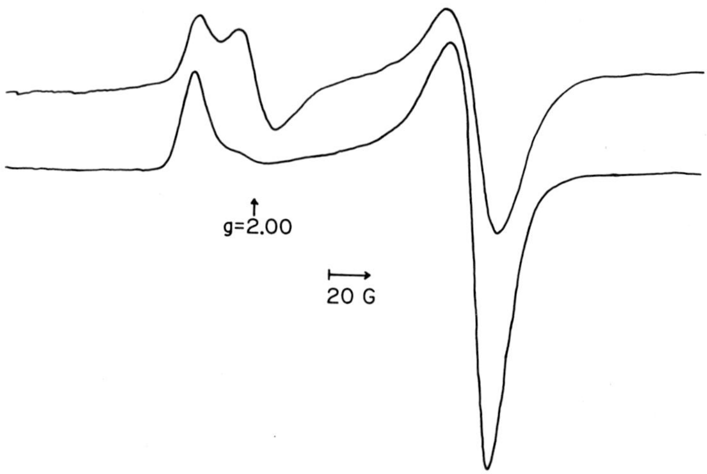

Fig. 3. EPR signal change of AdR semiquinone in the reaction with adrenodoxin: upper curve, AdR and adrenodoxin at the ratio of 2:1 were reduced with NADPH. AdR, 71 µM; lower curve, after anaerobic addition of adrenodoxin 71 µM, at -15 °C. EPR signals were observed at liquid nitrogen temperature, microwave power 50 mW, modulation 8 gauss.

the semiquinone species of the reductase: in the former case the semiquinone concentration should be decreased, which is consistent with the experimental result.

6. NADPH formation by the reverse reaction of the reductase with the reduced adrenodoxin, the reductase and $NADP^+$.

We have detected NADPH formation by photo irradiation in the reaction mixture containing $NADP^+$ (16), AdR and EDTA. The absorption at 340 nm increased by the prolonged irradiation. An aliquot from the solution was analyzed by HPLC and an elution peak was detected both by fluorescence and by absorbance at a retention time identical with that of the authentic NADPH. NADPH formation was also demonstrated in the reversed reaction, in which reduced adrenodoxin was used as the electron donor. The reduced adrenodoxin was prepared by the reduction of dithionite and the excess dithionite was decomposed by addition of formaldehyde. The reaction was performed by the addition of $NADP^+$ and AdR under anaerobic conditions. The increase in absorbance at 455 nm was observed by the addition of $NADP^+$. Without AdR, NADPH formation was hardly observable (For details, see another paper of Nonaka et al. in this volume). These facts may be interpreted by the presumption that one electron is provided to the three-electron reduced state of the reductase and that the four-electron reduced state is produced as the result. Then, probably shift of the equlibrium takes place to release the bound NADPH replaced by $NADP^+$.

In summary, there are three important oxidation-reduction states of adrenodoxin reductase in the presence of NADP(H) nucleotides: one is the two-electron reduced state of FAD-$NADP^+$ complex, and the other two are FADH·-NADPH and $FADH_2$-NADPH states of the reductase. These three states exist in the equilibrium. The electron transfer to adrenodoxin in the forward reaction and the formation of NADPH in the reverse reaction are demonstratable through one electron transfer and subsequent attainment of a new equilibrium.

References

1. Sugiyama, T., Yamano, T.: FEBS Letters, 52, 145-148 (1975).
2. Hiwatashi, A., Ichikawa, Y., Maruya, N., Yamano, T., Aki, K.: Biochemistry, 15, 3082-3090 (1976).
3. Shin, M., Tagawa, K., Arnon, D. I.: Biochem. Z., 338, 84-96 (1963).
4. Sheriff, S., Herriott, J. R.: J. Mol. Biol., 145, 441-451 (1981).
5. Yao, Y., Tamura, T., Wada, K., Matsubara, H., Kodo, K.: J. Biochem., 95, 1513-1516 (1984).
6. Lambeth, J. D., Seybert, D. W., Lancaster,Jr, J. R., Salerno, J., C., Kamin, H.: Molec. Cellular Biochem., 45, 13-31 (1982).
7. Kitagawa, T., Sakamoto, H., Sugiyama, T., Yamano, T.: J. Biol. Chem., 257, 12075-12080 (1982).

8. Sugiyama, T., Miura, R., Yamano, T.: J. Biochem., 86, 213-223 (1979).

9. Huang, Y.-Y., Kimura, T.: Anal. Biochem., 133, 385-393 (1983)

10. Hyde, E. I., Birdsall, B., Roberts, G.C.K., Feeney, J., Burgen, A.S.V.: Biochemistry, 19, 3746-3754 (1980).

11. Mas, M.T., Colman, R.F.: Biochemistry, 23, 1675-1683 (1984).

12. Lambeth, J. D., Kamin, H.: J. Biol. Chem., 251, 4299-4306 (1976).

13. Valle, M., Carrillo, N., Vallejos, R.H.: Biochim. Biophys. Acta, 681, 412-418 (1982).

14. Hiwatashi, A., Ichikawa, Y., Yamano, T., Maruya, N.: Biochemistry, 15, 3091-3096 (1976).

15. Nonaka, Y., Sugiyama, T., Yamano, T.: J. Biochem. 92, 1693-1701 (1982).

16. Sakamoto, H., Ohta, M., Miura, R., Sugiyama, T., Yamano, T., Miyake, Y.: J. Biochem., 92, 1941-1950 (1982).

STUDIES ON FORWARD AND REVERSE REACTIONS OF ADRENODOXIN REDUCTASE BY ELECTRONIC AND NMR SPECTROSCOPY

Y. Nonaka, S. Fujii, T. Yamano

Department of Biochemistry,
Osaka University Medical school

Introduction

Adrenodoxin reductase (EC 1.18.1.2) is an FAD containing flavoenzyme which is a component of steroid hydroxylating system in adrenocortical mitochondria(1,2). Adrenodoxin reductase (AdR) accepts two electrons from NADPH and transfers electrons to an iron-sulfur protein, adrenodoxin, which is linked to cytochrome $P\text{-}450_{scc}$ or cytochrome $P\text{-}450_{11\beta}$(3-5). Furthermore, adrenodoxin (AdX) can accept a single electron from AdR, and AdX forms a tight complex with AdR at a molar ratio of 1:1(6,7). To understand these electron transfer processes, it is crucial to know the mechanism of reduction of AdR and AdX. For this purpose these reaction mechanisms, we examined the reverse reaction of the electron transport system. The reverse reaction from reduced AdR to $NADP^+$ has already been reported from our laboratory (8). In this work, we demonstrate the reverse reaction from reduced AdX to $NADP^+$ via AdR.

Results and Discussion

To demonstrate the reverse reaction, that is, the NADPH formation from reduced AdX to $NADP^+$ via AdR, we attempted to modulate the NADPH level, by using an NADPH generating system

Flavins and Flavoproteins

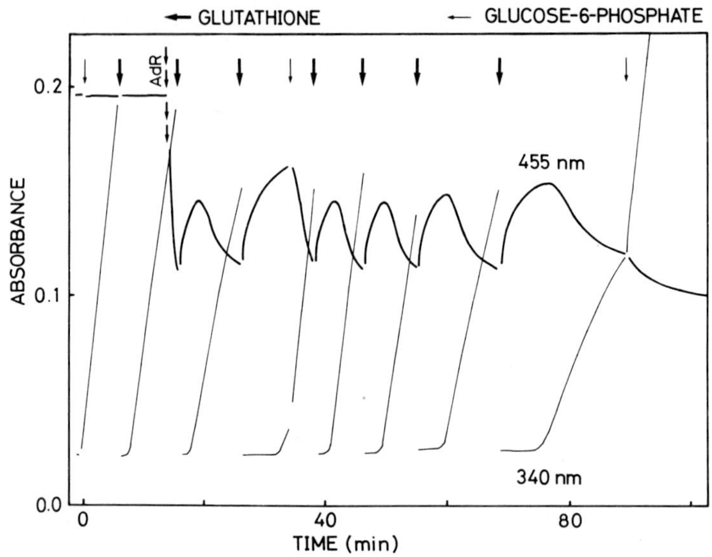

Fig. 1 Correlation of reduction levels of adrenodoxin and NADPH .

(glucose 6-phosphate dehydrogenase) and an $NADP^+$ generating system (glutathione reductase). The anaerobic conditions were attained by glucose-glucose oxidase system and bubbling and blowing of deoxygenated N_2 gas into the silicone rubber sealed cell. Fig. 1 shows the correlation of reduction level of AdX (21 µM) and the NADPH levels. The NADPH levels in the reaction mixture were monitored by the 340 nm aborbance (the thin line). On addition of G 6-P (the thin arrow) caused the increase in absorbance at 340 nm (increase of NADPH), whereas an addtion of oxidized glutathione (the bold arrow) decreased it very rappidly to almost zero level (the exhaustion of NADPH). The oxidation-reduction levels of AdX monitored by the absorbance at 455 nm (the bold line) did not change until AdR was added to the reaction mixture, even when G 6-P or glutathione was added. But in the presence of catalytic amount of AdR (0.24 µM), AdX was reduced or reoxidized repeatedly. Its oxidation-reduction levels corresponded to the changes of the NADPH levels; addition of oxidized glutathione

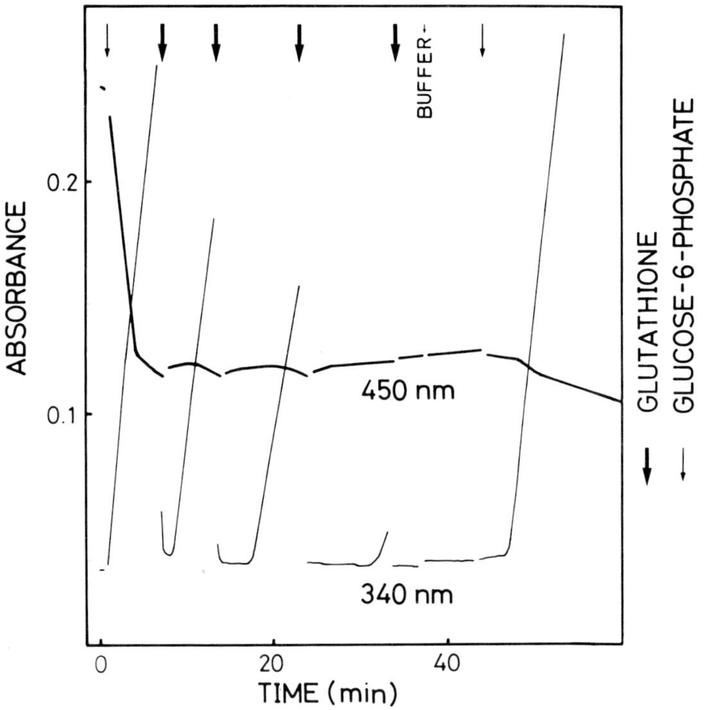

Fig. 2 Correlation of reduction levels of adrenodoxin reductase and NADPH..

decreased the NADPH level and increased the absorbance at 455 nm, and an addition of G 6-P caused the increase in the NADPH level and the decrease in the absorbance at 455 nm. The result showed that AdX was reoxidized via AdR when NADPH was exhausted by the $NADP^+$ generating system. AdR was essential for reduction or reoxidation of AdX by NADP(H). On the other hand, when AdX was absent in the system (Fig. 2), AdR (26 µM) was reduced from oxidized form to the two-electron containing form or to the more reduced form, when NADPH level was elevated, but reduced AdR was only very slowly reoxidized when the NADPH level was kept low. The spectrum of two-electron-containing AdR was observed with a low NADPH level. So it is unlikely that the reverse electron flow from reduced AdX to $NADP^+$ was via the enzymic cycle of oxidized AdR and tow-electron-containing form.

We therefore postulated the other enzymic cycle, that is, from semiquinone or $AdRH_2$-NADPH complex to the two-electron-containing form.

To investigate these enzyme species in terms of the molecular structure, we measured ^{31}P and ^{1}H NMR spectra of AdR under various conditions for better knowledge of the enzyme species involved. Analysis of the ^{31}P chemical shifts suggests that the conformation of the pyrophosphate group in FAD bound to the reductase is different from that of free FAD. The spectra of the complex of AdR with $NADP^+$ showed that the resonance of 2'-phosphate in $NADP^+$ bound to AdR was shifted to the lower field by 1.4 ppm compared with that in the dianionic state of free $NADP^+$. This indicates that the exchange between free and bound $NADP^+$ was slow in the time scale of NMR. And only one binding site of $NADP^+$ was found in the reductase.

REFERENCES

1. Suhara, K., Ikeda, Y., Takemori, S., Katagiri, M.: FEBS Lett. 28, 45-47 (1972)
2. Chu, J.-W., Kimura, T.: J. Biol. Chem. 248, 2089-2094 (1973)
3. Nakamura, Y., Otsuka, H., Tamaoki, B.: Biochim. Biophys. Acta 122, 34-42 (1966)
4. Suzuki, K., Kimura, T.: Biochem. Biophys. Res. Commun. 19, 340-345 (1965)
5. Omura,T., Sanders, E., Estabrook, R.W., Cooper, D.Y., Rosenthal, O.: Arch. Biochem. Biophys. 117, 660-673 (1966)
6. Sakamoto, H., Ichikawa, Y., Yamano, T. Takagi, T.:J. Biochem. 90, 1445-1452 (1981)
7. Chu. J.-W., Kimura, T.: J. Biol. Chem. 248, 5183-5187, (1973)
8. Sakamoto, H., Ohta, M., Miura, R., Sugiyama, T., Yamano, T., Miyake, Y.: J. Biochem. 92, 1941-1950 (1982)

TRANSIENT KINETICS OF FERREDOXIN-NADP$^+$ REDUCTASE REACTION

Shinya Yoshikawa, Norihiro Ohnishi, Aiko Morigiwa, Katsuhito Takeshima, Midori Matsumoto, Kyoko Nishiyama, Hiroshi Matsubara* and Keiun Kodo**

Department of Biology, Konan University, Higashinada-Ku, Kobe 658, *Department of Biology, Faculty of Science, Osaka University, Toyonaka, Osaka 560 and **Spirulina Research Lab., Osaka 532

Introduction

Ferredoxin-NADP$^+$ reductase is the terminal enzyme for the photosynthetic NADP$^+$ reduction. The enzyme (FNR) catalyzes the following reaction,

$$2\ Fd_{red} + NADP^+ + 2H^+ \longrightarrow 2\ Fd_{ox} + NADPH + H^+$$

where Fd_{red} and Fd_{ox} refer to the reduced and oxidized ferredoxin, respectively. Because of the strong autoxidizability of Fd_{red}, no detailed kinetic study on the electron transfer reaction from Fd_{red} to NADP$^+$ via FNR has been reported except for our steady state kinetic study(1). Therefore, we studied the electron transfer reaction with the stopped flow method under anaerobic conditions.

Materials and Methods

Ferredoxin and FNR were isolated from a blue green alga, Spirulina platensis (2, 3). Anaerobiosis was maintained with slight excess amount of dithionite and flushing N_2 to the reaction mixture placed in the reservoir of the stopped flow apparatus. The reaction medium contained 50 mM Tris-HCl buffer pH 7.4 and 25 mM $MgCl_2$ at 4°. The dead time of the stopped flow apparatus was 2.8 msec.

Flavins and Flavoproteins

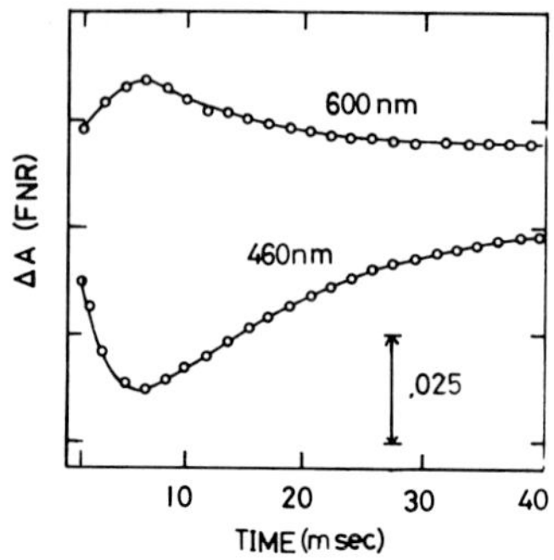

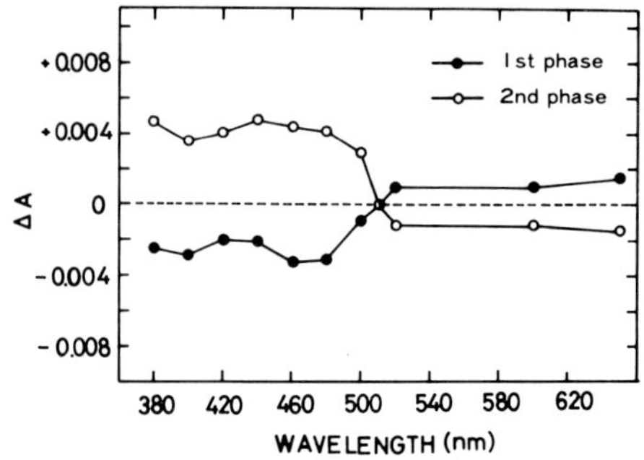

Fig. 1. The absorbance change due to FNR in the FNR reaction. Each point was obtained from the smoothed curve best fitted to the points calculated as indicated in the text. The experimental data (the absorbance change with time) were collected and processed with a microcomputer system.

Fig. 2.(right) The spectral change of FNR during the enzymic turnover. The absorbance changes for the first and the second phases are plotted against the wavelencth.

Results and Discussion

1) Reaction System I: Fd —— FNR —— $NADP^+$. Fd_{red} (8 μM) was reacted with FNR (4 μM) and $NADP^+$ (40 μM). The time course of the absorbance change depends on the wavelength. For example, a rapid increase ($t_{1/2}$ = 10 msec) was followed by a slower increase at 460 nm and decrease at 600 nm. The slower phase at 510 nm was not so prominent as at other wavelength. The change at 510 nm can be regarded as only due to the oxidation of Fd_{red}, since the difference between the molecular extinction coefficients of the oxidized, reduced and semiquinoide forms of FNR at 510 nm are much smaller than that between Fd_{red} and Fd_{ox}. However, the absorbance change due to FNR overlaps to that due to Fd at other than 510 nm. Contribution of Fd-absorbance change at other than 510 nm can be calculated by the absorbance change at 510 nm and the redox difference spectrum of Fd. Thus, the absorbance change of FNR with time is evaluated by subtracting the absorbance change due to Fd-oxidation from the experimental curve. Biphasic absorbance change with time was observed at

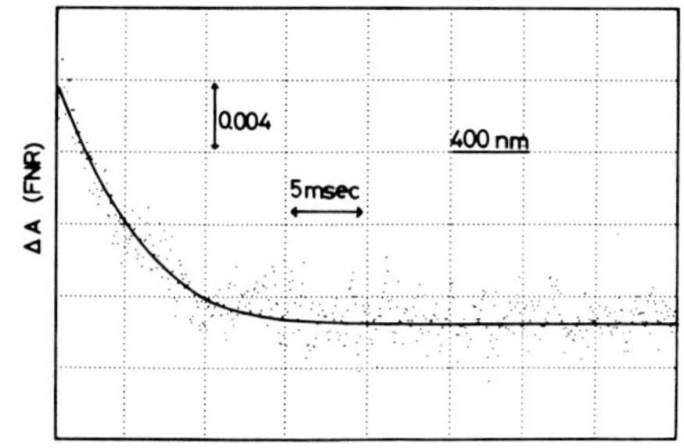

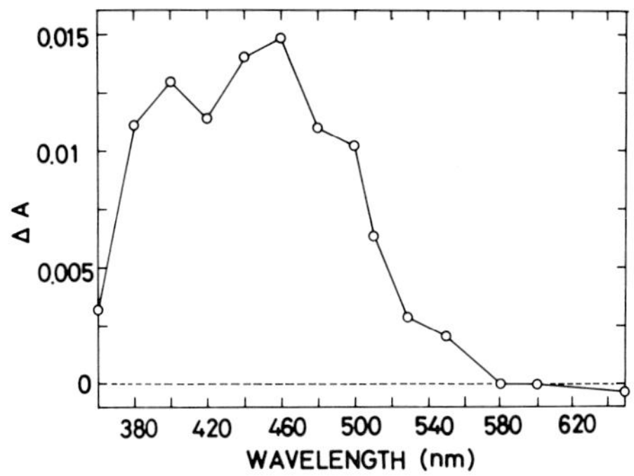

Fig. 3. (left) The absorbance change due to FNR in the absence of $NADP^+$. Each point was obtained by subtracting the contribution of absorbance change due to Fd-oxidation. Collection and processing of the experimental data were done with the microcomputor system. The solid line is the best fitted curve.

Fig. 4. (right) The spectral change of FNR during the reaction between Fd_{red} and FNR_{ox}.

each wavelength as shown in Fig. 1. The half life times for the first and second phases are 2.5 msec and 10 msec, respectively. The absorbance change during each phase is plotted against the wavelength, as shown in Fig. 2,regardless of the wavelength. The spectral change in the first phase, especially, the absorbance increase in the longer wavelength region resembles the difference spectrum of the oxidized form minus the semiquinone. The spectral change in the second phase is almost equal to that in the first phase in magnitude though opposite in sign. This result shows that FNR is reduced to semiquinoide form in the first phase and reoxidized to the oxidized form in the second phase. No spectral evidence was obtained for formation of fully reduced FNR during the enzymic turnover.

2)Reaction System II:Fd —— FNR. The reaction of Fd_{red}(8 μM) with the oxidized FNR(4 μM) was examined in the absence of $NADP^+$. The absorbance change due to FNR at each wavelength was evaluated by assuming the change at 600 nm is only due to Fd-oxidation. The time course is monophasic with $t_{1/2}$ = 5 msec at any wavelength examined, as typically shown in Fig. 3. The absorbance change during the reaction is plotted against the wavelength(Fig. 4). This spectrum resembles the difference spectrum between the oxidized and reduced forms of FNR. Therefore, FNR is reduced by Fd_{red} directly to the fully reduced state without any intermediate species.

3) Reaction Mechanism. We propose the following mechanism for FNR catalysis, based on the above results;

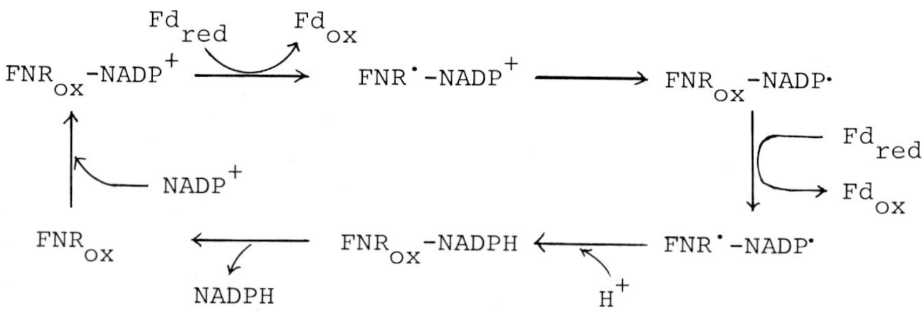

where FNR˙ and NADP˙ denote semiquinoide FNR and $NADP^+$ radical, respectively. On the other hand, in the absence of $NADP^+$, FNR is directly reduced to fully reduced form without forming semiquinone. These mechanisms indicate that $NADP^+$ is not only an electron acceptor for the FNR reaction but also an effector to modify the reaction path between Fd_{red} and FNR_{ox} to form semiquinone.

References

1. Masaki, R., Yoshikawa, S., Matsubara, H.: Biochim. Biophys. Acta 700,101-109 (1982).
2. Wada, K., Hase, H., Tokunaga, H., Matsubara, H. : FEBS Lett. 55, 102-104 (1975).
3. Wada, K., Tamura, T., Matsubara, H., Kodo, K.: J. Biochem. 94, 387-393 (1983)

SOME NEW IDEAS ABOUT THE POSSIBLE REGULATION OF REDOX POTENTIALS IN FLAVO-PROTEIN, WITH SPECIAL REFERENCE TO FLAVODOXINS

Chrit T.W. Moonen, Jacques Vervoort and Franz Müller
Department of Biochemistry, Agricultural University, De Dreijen 11,
6703 BC Wageningen, The Netherlands

Introduction

Flavin participates as a prosthetic group in a large variety of biological redox reactions. The redox potential of flavoproteins varies at least over a span of about ±250 mV with respect to free flavin. Understanding the regulation of the redox potential in flavoproteins is therefore desirable and will ultimately lead to a better insight into the action of flavin-dependent catalysis. The degree of planarity of free (1) and protein-bound reduced flavin is high, not contributing much to the energy needed to influence the redox potential, except for a few extreme cases. The reduced prosthetic group in flavodoxins and many other flavoproteins is ionized (2), the phosphate group of FMN in flavodoxins is in the dianionic state and highly immobilized (3), as is also the isoalloxazine moiety of FMN (4). These facts led us to consider the possible effects of these charges on the redox potential in flavodoxins.

Results and discussion

Our hypothesis will be worked out for the flavodoxins from Clostridium MP and M.elsdenii. Only the redox transition between the semiquinone and the two-electron reduced states is considered, since this transition is the biologically relevant one (5). The electrostatic interaction energy (W(r)) between the negative charge on N(1) of protein-bound reduced FMN and the two negative charges on the phosphate group of FMN can be calculated according to equation 1 (6):

$$W(r) = -qV(r) = -\frac{q^2 Z_1 Z_2}{4\pi\varepsilon_0\varepsilon_r} \qquad (1)$$

Flavins and Flavoproteins

where r is the distance between the charges, q the elementary charge and ε the dielectric permitivity. The distance r between the charges in Clostridium MP is about 0.8 nm (7) and ε is estimated to be about 20. The interaction energy is independent on the ionic strength. The calculation yields an interaction energy of -180 mV. The monopole-monopole interaction between N(1) and the negative charges of the protein is calculated by equation 2:

$$W = -qV = -\frac{Qe^2}{4\pi\varepsilon\varepsilon_0 R(1 + \kappa R)} \qquad (2)$$

where Q is the charge and R the radius of the protein. $\kappa = 0.33\sqrt{I}$, $Å^{-1}$ (at 294 K) in which I represents the ionic strength. Clostridium MP carries 16 negative charges at pH 7, excluding the two negative charges of the phosphate group of FMN and R = 1.8 nm (7). The dielectric permitivity is 80, since most or all of the charges are exposed to solvent water. The interaction energy is calculated as -160 mV for I = 0 and -57 mV for I = 100 mM. The third interaction is the monopole-dipole interaction (Eq. 3). It has been calculated that the dipole moment ($|\vec{P}_{tot}|$ is 459 Debye for Clostridium MP (L.W. van Leeuwen, personal communication). The negative pole points towards the flavin.

$$W(r,\Theta) = -qV(r,\Theta) = -\frac{3}{\varepsilon_1 + 2\varepsilon_2}\,\frac{Pr\cos\Theta}{4\pi\varepsilon_0 R^3} \qquad (3)$$

where P is the dipole moment, r(1.6 nm) the distance between the dipole moment of the protein and the center of the prosthetic group, Θ (150^0) the angle between the average dipole moment and the flavin, R (1.8 nm) the radius of the protein, ε_1 the electric permitivity of the interior of the protein and ε_2 (=80) that of the surrounding of the protein. The interaction energy between the monopole and the dipole is by approximation independent on the ionic strength. The calculated values thus depend on the estimated value of ε_1. If we assume ε_1 to be 5, we obtain -59 mV as contribution to the redox potential, ε_1 = 20 gives -54 mV. The calculated values consist of the corresponding redox potential of free FMN (-124 mV) and the three calculated contributions of the protein. The results are summarized in Table 1. The results show that the experimental values can

Table 1. ^{31}P chemical shifts and calculated distance between the isoalloxazine ring and the phosphate group of FMN in various flavodoxins and a comparison between calculated and experimentally (5) determined redox potentials (pH ∿8).

Flavodoxin	$^{31}P(\delta)$	Distance	Redox Potential (E_1)		
			Published	Calculated for I = 0	I = 100 mM
Clostridium MP	5.8 ppm	0.8 nm	-408 mV	-498 mV	-415 mV
M.elsdenii	5.3 ppm	0.78 nm	-375 mV	-498 mV	-395 mV
D.vulgaris	5.5 ppm	0.69 nm	-440 mV	-571 mV	-450 mV
A.vinelandii	6.5 ppm	0.69 nm	-464 mV	-571 mV	-450 mV

be reasonably explained by our theory, although it should be considered that the theoretical values are rough approximations since several assumptions had to be made in the calculations. The distance between the negative charge at N(1) and those of the phosphate group used in equation (1) is not so crucial for the calculated values. A deviation of ±0.1 nm effects the theoretical values by only ±20 mV. More important is the assumption of the electric permitivity in the protein for which no experimental values are available. It is reasonable to suggest that in e.g. M.elsdenii ε is higher than in A.vinelandii. The calculated values also nicely demonstrate the ionic strength dependence of the redox potentials. The finding of Edmondson and Tollin (8) that A.vinelandii flavodoxin is more easily reduced in the presence than in the absence of 3 M ammonium sulfate can be taken as a support of our theoretical approach. The theory however requires further refinement and experimental support. Our theory could also offer an explanation for the fact that all flavodoxins from exclusively the neutral semiquinone upon one-electron reduction. If the microenvironment of the protein would allow the formation of the anionic semiquinone then the semiquinone state would be strongly destabilized by the charge-charge interactions and the protein could probably no longer shuttle between the semiquinone and the hydroquinone state.

We believe that our theory can be applied more generally to other flavoproteins as well. The negative charges of the pyrophosphate group of FAD in glutathione reductase and p-hydroxybenzoate hydroxylase are stabilized by an arginine residue and a positive pole of an α-helix (9). Both prosthetic groups are ionized in the fully reduced enzyme. The redox potentials of both enzymes are more positive in the transition oxidized two-electron reduced state than that of the pyridine nucleotide couple. In p-hydroxybenzoate hydroxylase the negative charge on the reduced prosthetic group is stabilized by the positive pole of an α-helix. The redox potential of the couple two-electron/four-electron reduced states in lipoamide dehydrogenase and glutathione reductase is more negative than the couple oxidized/two-electron reduced states. This could be a consequence of the fact that the thiolate group destabilizes the four-electron reduced state. Similar regulatory mechanisms may be operating in more complex flavoproteins such as NADPH-cytochrome P-450 reductase.

Acknowledgements

We thank Mrs. J.C. Toppenberg-Fang for typing the manuscript and Dr. C.A.H. Rasmussen for carefully reading the manuscript. This study was carried out under the auspices of the Dutch Foundation for Chemical Research (SON) with financial aid from the Dutch Organization for the Advancement of Pure Research (ZWO).

References

1. Moonen, C.T.W., Vervoort, J., Müller, F.: Biochemistry 23 (1984), in press.
2. Van Schagen, C.G., Müller, F.: Eur. J. Biochem. 120, 33-39 (1981).
3. Moonen, C.T.W., Müller, F.: Biochemistry 21, 408-414 (1982).
4. Moonen, C.T.W., Müller, F.: Eur. J. Biochem. 133, 463-470 (1983).
5. Mayhew, S.G., Ludwig, M.: The Enzymes, 3rd Ed. Vol. 12, 57-118 (1975).
6. Van Leeuwen, J.W.: Biochim. Biophys. Acta 743, 408-421 (1983).
7. Smith, W.W., Burnett, R.M., Darling, G.D., Ludwig, M.L.: J. Mol. Biol. 177, 195-225 (1977).
8. Edmondson, D.E., Tollin, G.: Biochemistry 10, 133-145 (1971).
9. Wierenga, R.K., Drenth, J., Schulz, G.E.: J. Mol. Biol. 167, 725-739 (1983).

METHYLENETETRAHYDROFOLATE REDUCTASE: AN IMPERFECT ENZYME?

Maria A. Vanoni and Rowena G. Matthews

Biophysics Research Division and Department of Biological Chemistry, The University of Michigan, Ann Arbor, Michigan

Methylenetetrahydrofolate reductase is a flavoprotein which has been purified from pig liver (1). The enzyme catalyzes the NADPH-linked reduction of methylenetetrahydrofolate (CH_2-H_4folate), a reaction which commits tetrahydrofolate-bound one carbon units to use in the regeneration of the methyl group of adenosylmethionine (AdoMet). Reduction of the enzyme by NADPH is irreversible, proceeding with a free energy decrease of 9.0 kcal per mol, while reoxidation of the reduced flavoprotein by CH_2-H_4folate is freely reversible, being associated with a free energy decrease of 0.4 kcal. In addition to the physiological reaction, the enzyme also catalyzes NADPH-menadione and methyltetrahydrofolate-menadione oxidoreductase reactions. Rapid reaction studies of the reductive and oxidative half reactions associated with NADPH-CH_2-H_4folate oxidoreductase activity have been performed (2). Reduction of the enzyme by saturating concentrations of NADPH proceeds with a rate constant of 160 s^{-1} while reoxidation of photoreduced enzyme by saturating CH_2-H_4folate proceeds with a rate constant of 50 s^{-1}. The predicted rate of turnover if the enzyme utilizes a Ping Pong Bi Bi mechanism is 38 s^{-1}, in good agreement with the observed steady-state turnover.

Knowles and Albery (3) introduced the concept of perfection in enzyme catalysis. They have postulated that evolutionary pressures lead to improvements in the catalytic efficiency of

Flavins and Flavoproteins

enzymes until steps involving chemical catalysis are no longer rate limiting and the rate of approach of the less thermodynamically stable substrate to the enzyme active site limits the rate of catalysis. However, they specifically suggested that enzymes which are subject to metabolic regulation might be "imperfect" because constraints related to the regulatory process might take precedence over catalytic perfection. Methylenetetrahydrofolate reductase presents an opportunity to examine the catalytic perfection of a regulatory enzyme. This enzyme acts at a branch point in folate metabolism and its activity is allosterically regulated by AdoMet (4). Moreover, the overall physiological reaction can be dissected into its two half reactions. One may postulate that evolutionary forces might have pushed the two half reactions towards satisfaction of different needs: metabolic regulation and catalytic efficiency. This is particularly true because the irreversible half reaction, reduction of enzyme by NADPH, is only partially rate limiting in the overall catalysis of the physiological reaction and thus might be an appropriate target for metabolic control.

Analysis of the catalytic efficiency of NADPH oxidation can be accomplished utilizing the NADPH-menadione oxidoreductase reaction, since we have shown that reoxidation of enzyme by menadione occurs at a rate of 3230 s^{-1} at the menadione concentration used routinely (125 μM). Thus studies of the kinetic isotope effects associated with oxidation of NADPH provide information about the extent to which the hydrogen transfer is rate limiting in NADPH oxidation. Methylenetetrahydrofolate reductase stereospecifically removes the pro-S hydrogen from the 4 position of NADPH and this hydrogen is ultimately equilibrated with solvent, presumably because the reduced flavin exchanges rapidly with solvent. If methylenetetrahydrofolate reductase showed optimal catalysis of the oxidation of NADPH, such that approach of NADPH to the enzyme was rate limiting in steady-state turnover at physiological concentrations of NADPH (200 μM) one would expect that enzyme catalysis would be first order in NADPH in this concentration range and that the V/K isotope effect would be suppressed by a large forward commitment to catalysis (5). Competition experiments using [4S-^{3}H]NADPH showed a primary kinetic isotope effect on V/K of 10.8 ± 0.4. When [4S-^{2}H]NADPH was the substrate, the primary kinetic isotope effect was 4.78 ± 0.15 on V and 4.54 ± 0.59 on V/K. The primary kinetic isotope effects observed with tritiated and deuterated NADPH are of the magnitudes expected for fully expressed intrinsic isotope effects and, within the limits of experimental error, are in the ratio predicted by the Swain-Schaad relationship. Thus, they support the conclusion that NADPH oxidation is the rate limiting step in the reductive half reaction. Furthermore, at physiological NADPH concentration, both steady-state NADPH-menadione oxidoreductase turnover and pre-steady state reduction of the enzyme by NADPH are saturated in NADPH.

Methylenetetrahydrofolate reductase does not show maximum catalytic efficiency with respect to oxidation of NADPH. In particular, the overall rate of reduction of the enzyme is at least 29-fold slower than the minimum estimated value for the rate of approach of a physiological concentration of NADPH to the enzyme. However, the catalytic consequences of suboptimal efficiency are minimized by the fact that this half reaction is only partially rate limiting in overall catalysis so that the overall rate of turnover in the NADPH-CH_2-H_4folate oxido-reductase reaction is only 1.3-fold slower than would be expected if the efficiency of the reductive half reaction had been optimized. The catalytic efficiency of the oxidative half reaction has not yet been examined, nor is the effect of AdoMet on this half reaction known. However, we have shown that AdoMet inhibition leads to a 50,000-fold decrease in the rate of reduction of the enzyme by NADPH with no indication of an altered K_d for NADPH (6). Thus oxidation of NADPH appears to represent one site of allosteric regulation of the activity of this enzyme and the relative inefficiency of this catalytic step may reflect constraints related to the regulatory process.

This work has been supported in part by NIH grant GM 24908.

References

1. Daubner, S.C., and Matthews, R.G.: J.Biol.Chem. 257, 140-145 (1982).

2. Vanoni, M.A., Ballou, D.P., and Matthews, R.G.: J.Biol. Chem. 258, 11510-11514 (1983).

3. Knowles, J., and Albery, W.J.: Acc.Chem.Res. 10, 105-111 (1977).

4. Kutzbach, C., and Stokstad, E.L.R.: Biochim.Biophys. Acta 250, 450-477 (1971).

5. Northrup, D.B.: in Isotope Effects on Enzyme-Catalyzed Reactions (Cleland, W.W., O'Leary, M.H., and Northrop, D.B., eds.) pp. 122-152, University Park Press, Baltimore (1977).

6. Vanoni, M.A., Daubner, S.C., Ballou, D.P., and Matthews, R.G.: in Chemistry and Biology of Pteridines (Blair, J.A., ed.) pp. 235-239, Walter de Gruyter, Berlin (1983).

PROBING THE CATALYTIC MECHANISM OF GLUTATHIONE REDUCTASE WITH 2,4,6-TRINITROBENZENESULFONATE

Inger Carlberg and Bengt Mannervik

Department of Biochemistry, Arrhenius Laboratory, University of Stockholm, S-106 91 Stockholm, Sweden

Introduction

The natural electron-accepting substrate for glutathione reductase is glutathione disulfide (GSSG):

$$GSSG + NAD(P)H + H^{+} \rightarrow 2\ GSH + NAD(P)^{+}$$

Most other disulfides, except the mixed disulfide of coenzyme A and glutathione, are comparatively poor substrates for the enzyme (1, 2). However, non-physiological compounds, such as thionicotinamide adenine dinucleotide phosphate (thio-NADP^{+}), $Fe(CN)_6^{3-}$, 2,6-dichlorophenol-indophenol (DCIP), and 2,4,6-trinitrobenzenesulfonate (TNBS), may also serve as electron acceptors of glutathione reductase. The reactions in most cases are slow in comparison with the GSSG reduction (3). However, using TNBS as a mediator, rapid electron transfer to dioxygen or to cytochrome c has been demonstrated (see ref. 4). Thus, TNBS may be used as a tool in studies of the catalytic mechanism of glutathione reductase.

Results and discussion

TNBS, in concentrations between 0.03 and 0.25 mM, was reduced anaerobically with the same rate as aerobically, demonstrating that $O_2^{\cdot-}$ or H_2O_2 are not obligatory intermediates in TNBS reduction. On the contrary, reduction of dioxygen probably is mediated by electron transfer via TNBS (5). A limited excess of NADPH effects 4-electron reduction of TNBS in

Flavins and Flavoproteins

the absence of other acceptors, such as O_2 or cytochrome c. The product is presumably a monohydroxylamine derivative of TNBS.

At high TNBS concentrations, the rate of the TNBS-dependent NADPH oxidation approaches the rate obtained with GSSG; 3 mM TNBS giving approximately 50% of the corresponding rate of GSSG reduction. The most conspicuous finding in exploring the TNBS-dependent activity of glutathione reductase was a strong stimulatory effect of $NADP^+$ on the reaction rate (4). Table 1 shows that also the reductions of $Fe(CN)_6^{3-}$ and DCIP are stimulated 4 to 7-fold by $NADP^+$. The activating effect of $NADP^+$ is reminiscent of a regulatory role of NAD^+ proposed for lipoamide dehydrogenase (6). Stimulation of the physiological reaction with GSSG or of the transhydrogenase reaction with thio-$NADP^+$ was not observed (Table 1), but the possibility that product inhibition of $NADP^+$ may mask a smaller activiting effect has not been excluded. Evidently, $NADP^+$ in high concentrations will compete with NADPH for binding to the N-site of the protein (2). The stimulatory effect of $NADP^+$ on the TNBS-dependent activity increased monotonously with pH in the range of pH 4-9, except for $NADP^+$ concentrations $\geq$ 1 mM for which maximal activation was obtained at pH 8. Thus, the competition of $NADP^+$ with NADPH appears stronger at the higher pH values.

NADH, like NADPH, can support the glutathione reductase-catalyzed reduction of TNBS. Also this reaction was stimulated by $NADP^+$; at 0.1 mM NADH (and 0.5 mM TNBS, pH 7) addition of 0.025 mM $NADP^+$ enhanced the initial velocity 6-fold. NAD^+ had a smaller activating effect on the NADH-supported activity; 0.2 mM NAD^+ was required to reach a 3.5-fold stimulation. The same concentration of NAD^+ gave only 50% rate enhancement of the NADPH-dependent TNBS reduction, probably owing to the comparatively low affinity of NAD^+ for the enzyme.

Modification of glutathione reductase from human erythrocytes with ethoxyformic anhydride (7) to give 10% residual activity with GSSG as substrate resulted in 64% residual activity with TNBS. This modification of an essential histidine residue is believed to affect His 467' (8) in the G-site of the enzyme (2). An inactivation of 2-electron-reduced gluta-

Table 1. Effect of $NADP^+$ on different activities of glutathione reductase from human erythrocytes

Substrate pair	Activity (nmol/min)	
	Control	+ 0.2 mM $NADP^+$
0.1 mM NADPH/1 mM GSSG	212	196
0.01 mM NADPH/0.1 mM thio-$NADP^+$	1	0.5
0.1 mM NADPH/0.4 mM $Fe(CN)_6^{3-}$	1	4
0.1 mM NADPH/0.05 mM DCIP	2	15
0.1 mM NADPH/0.5 mM TNBS	11	48

thione reductase with iodoacetate giving 10% residual GSSG activity resulted in 25% residual TNBS activity. The modification affects the distal cysteine residue, Cys 58 (8), of the redox-active disulfide (9). These findings indicate that neither Cys 58 nor His 467' are required for electron transfer to TNBS. These amino acid residues have essential roles in the reduction of GSSG (2, 8, 9). The most likely mechanism appears to be direct one-electron transfer from the isoalloxazine ring of FAD to TNBS. The activating effect of $NADP^+$ is believed to originate from facilitation of electron transfer between the redox-active disulfide/dithiol and the isoalloxazine ring, as earlier suggested for the effect of NAD^+ on lipoamide dehydrogenase (6).

Acknowledgement

This work was supported by grants from the Swedish Natural Science Research Council.

References

1. Mannervik, B.: in Glutathione (Flohé, L., Benöhr, H.C., Sies, H., Waller, H.D., Wendel, A., eds.), pp. 120-131, Georg Thieme Publishers, Stuttgart 1974.

2. Mannervik, B., Boggaram, V., Carlberg, I., Larson, K.: in Flavins and Flavoproteins (Yagi, K., Yamano, T., eds.), pp. 173-187, Japan Scientific Societies Press, Tokyo 1980.
3. Boggaram, V., Larson, K., Mannervik, B.: Biochim. Biophys. Acta 527, 337-347 (1978).
4. Carlberg, I., Mannervik, B.: in Flavins and Flavoproteins (Massey, V., Williams, C.H., eds.) pp. 53-56, Elsevier/North Holland, New York 1982.
5. Mannervik, B., Boggaram, V., Carlberg, I.: in Flavins and Flavoproteins (Massey, V., Williams, C.H., eds.), pp. 145-148, Elsevier/North Holland, New York 1982.
6. Matthews, R.G., Ballou, D.P., Williams, C.H., Jr.: J. Biol. Chem. 254, 4974-4981 (1979).
7. Boggaram, V., Mannervik, B.: Biochem. Biophys. Res. Commun. 83, 558-564 (1978).
8. Pai, E.F., Schulz, G.E.: J. Biol. Chem. 258, 1752-1757 (1983).
9. Arscott, L.D., Thorpe, C., Williams, C.H., Jr.: Biochemistry 20, 1513-1520 (1981).

MULTIFUNCTIONALITY OF YEAST GLUTATHIONE REDUCTASE

C.S. Tsai, J.R.P. Godin and Y.H. Tsai

Department of Chemistry and Institute of Biochemistry
Carleton University, Ottawa, Canada K1S 5B6

Glutathione reductase (EC 1.6.4.2) promotes various NADPH-linked one-electron transfer and two-electron transfer reactions. These reactions are, likewise, facilitated by the free reduced FAD (1). Studies with heart lipoamide dehydrogenase reveal the modulation of the flavin reactivities by the apoenzyme structure (2). Since the complete amino acid sequence (3) and the three-dimensional structure (4) of glutathione reductase from erythrocyte are known, studies of glutathione reductase may further delineate the role of protein structures in regulating multi-functional activities of flavoenzymes.

The purified glutathione reductase from yeast exists in a single molecular form with an isoelectric pH of 6.08 (Figure 1). In addition to

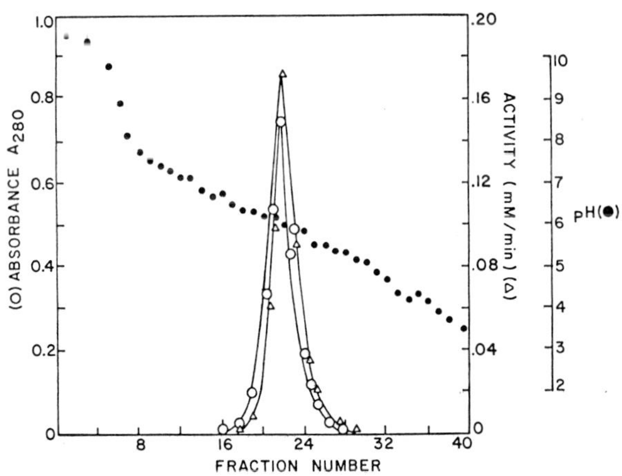

Fig. 1. Isoelectric focusing of yeast glutathione reductase

Flavins and Flavoproteins

its physiological reductase (RDase) activity, the yeast enzyme also displays nicotinamide nucleotide transhydrogenase (THase), inorganic electron transferase (ETase) and quinone diaphorase (DPase) activities (Table 1).

Table 1

Kinetic parameters (asymptotic) for reactions catalyzed by yeast glutathione reductase at pH 7.0

Multifunctional activities	V (μM min^{-1})	K_a (μM)	K_b (μM)	K_{ia} (μM)	V/E_t (min)
RDase	148	174	198	49.8	20.4×10^3
THase	1.89	6.83	49.4	0	152
ETase	47.3	154	445	0.68	4.5×10^3
DPase	9.50	106	297	44.4	909

Chemical modifications were carried out to correlate the multifunctional activities of glutathione reductase with critical amino acid residues (Table 2). Reductive carboxymethylation which monomerized the enzyme preferentially inactivated the RDase activity. The active site of glutathione reductase consists of flavin coenzyme, the redox active disulfide, Cys 58 and Cys 63 (5,6) and the catalytically competent His 467' (7). These residues are obligatory for the RDase activity as demonstrated by the elimination of the RDase activity by arsenite chelation which blocked the active-site disulfide, and sensitized photo-oxidation of histidine residues. X-Ray crystallographic study of erythrocyte glutathione reductase (8) reveals the localization of Tyr 197 in the binding pocket of NADP(H) and pairs of salt-bridges, Lys 66-Glu 201 and Arg 291-Asp 331, surrounding the active-site flavin. Treatment of the yeast enzyme with tetranitromethane which nitrated 1.2 ± 0.2 tyrosine residue per FAD abolished all multifunctional activities. Amidination with imidoesters which derivatized lysine residues, did not affect any of the multifunctional activities. However, amidation of carboxylic groups with ethylene diamine which reversed the charge

Table 2

Variation in Multifunctional Activities by Chemical Modifications

Enzyme derivatives	Effect	% Activity of Control RDase	THase	ETase	DPase
E_{rcM}	Monomer	0	0.15	24	5.6×10^3
$E_{h\nu}$	His	2.0	11	74	119
E_{As}	-S-S-	0	41	73	372
$E_{NH_3^+}$	COOH	19	200	724	112
E_{CH_3}	COOH	41	155	344	110
$E_{SO_3^-}$	COOH	38	133	198	108
E_{TNM}	Tyr	4.0	16	9	6
E_{ma}	Lys	100	104	106	91

Chemical modifications; reductive carboxymethylation (11) with iodoacetate (E_{rcM}), photooxidation (2) with rose bengal ($E_{h\nu}$), chelation (13) with arsenite (E_{As}), sensitized amidation (12) with ethylene diamine (E_{NH_3}+), propylamine (E_{CH_3}) and taurine (E_{SO_3}-), nitration (14) with tetranitromethane (E_{TNM}) and amidination (15) with methyl acetimidate (E_{ma}) were carried out as described in the literatures.

character, specifically enhanced the ETase activity.

An extensive sequence homology of the active site (6,9), albeit a detailed structural difference (Table 3), exists between lipoamide dehydrogenase and glutathione reductase (10). Their apoenzymes display similar structural perturbation effects on the multifunctional activities resulted from the reduced flavin. The nonphysiological ETase and DPase activities are suppressed by carboxylic groups and dimeric structure respectively. While the physiological RDase activity which requires the dimeric structure and the redox active disulfide is aided by the active-site histidine. An intermediate effect on the THase is generally observed.

Table 3

Empirical structural parameters for the redox-active disulfide containing octadecapeptides of heart lipoamide dehydrogenase and yeast glutathione reductase

Empirical parameter (Ave. per residue)	Lipoamide dehydrogenase	Glutathione reductase
P_{α}	0.92	0.94
P_{β}	1.02	1.13
Mol. vol. (nm) x 10^{-3}	62.4	65.2
Acc. surface area (nm^2)	0.376	0.375
Hydropathy index	0.61	1.04
Free energy (Kcal mol^{-1})	1.17	1.31

References

1. Dixon, M.: Biochim. Biophys. Acta, 226, 259-268 (1971).
2. Tsai, C.S., et. al.: Flavins and Flavoproteins (Editedy by Massey, V. and Williams, C.H.) Elsevier, New York, pp 149-152 (1982).
3. Krauth-Siegel, R.L.,et. al.: Eur. J. Biochem., 121, 259-267 (1982).
4. Thieme, R., et. al.: J. Mol. Biol., 152, 763-782 (1981).
5. Massey, V., Williams, C.H. Jr.: J. Biol. Chem., 240, 4470-4480 (1965).
6. Jones, E.T., Williams, C.H. Jr.: J. Biol. Chem., 250, 3779-3784 (1975).
7. Untacht-Grau, R., Schulz, G.E., Schirmer, R.H.: FEBS Lett., 105, 244-248 (1979).
8. Pai, E.F., Schulz, G.E.: J. Biol. Chem., 258, 1752-1757 (1983).
9. Williams, C.H. Jr., Arscott, L.D., Schulz, G.E.: Proc. Natl. Acad. Sci., U.S., 79, 2199-2201 (1982).
10. Rice, D.W., Schulz, G.E., Guest, J.R.: J. Mol. Biol., 174, 483-496 (1984).
11. Tsai, C.S., Templeton, D.M., Wand, A.J.: Arch. Biochem. Biophys., 206, 77-86 (1981).
12. Tsai, C.S., et. al.: Arch. Biochem. Biophys., 225, 554-561 (1983).
13. Massey, V., Palmer, G.: J. Biol. Chem., 237, 2347-2358 (1962).
14. Riordan, J.F., Vallee, B.L.: Meth. Enzymol., 25, 515-521 (1972).
15. Tsai, C.S., Redman, J., Templeton, D.M.: Arch. Biochem. Biophys., 209, 291-297 (1981).

THE REACTION BETWEEN NADPH AND MERCURIC REDUCTASE

Lena Sahlman and Sven Lindskog
Avdelningen för biokemi, Umeå Universitet
S-90187 Umeå, Sweden

Anne-Marie Lambeir and H. Brian Dunford
Department of Chemistry, University of Alberta
Edmonton, Alberta, Canada T6G 2G2

Introduction

Mercury resistance in bacteria (1) involves the plasmid-encoded enzyme mercuric reductase catalyzing the reaction:

$Hg^{2+} + NADPH \rightarrow Hg^{0} + NADP^{+} + H^{+}$.

Mercuric reductase, encoded on the transposon Tn *501* from *Pseudomonas aeruginosa*, is a dimeric enzyme with one molecule of FAD and a redox-active disulfide group per subunit (2). The enzyme requires a thiol compound for catalytic turnover (2-4). The nucleotide sequence of the gene *mer*A coding for mercuric reductase has been determined (5). The proposed amino acid sequence shows extensive homologies with the sequences of the related enzymes glutathione reductase and lipoamide dehydrogenase. These three enzymes also have similar spectral properties (6, 7).

Results

We have studied the reductive half reaction between mercuric reductase and NADPH at 5°C, pH 7.3, by rapid-scan and fixed-wavelength stopped-flow spectrophotometry with the aim to characterize reaction intermediates. Results of a rapid-scan experiment with a 10-fold molar excess of NADPH over FAD are shown in Fig. 1. During the initial 25 ms a well defined isosbestic point near 505 nm was observed, while the final phase of the

Flavins and Flavoproteins

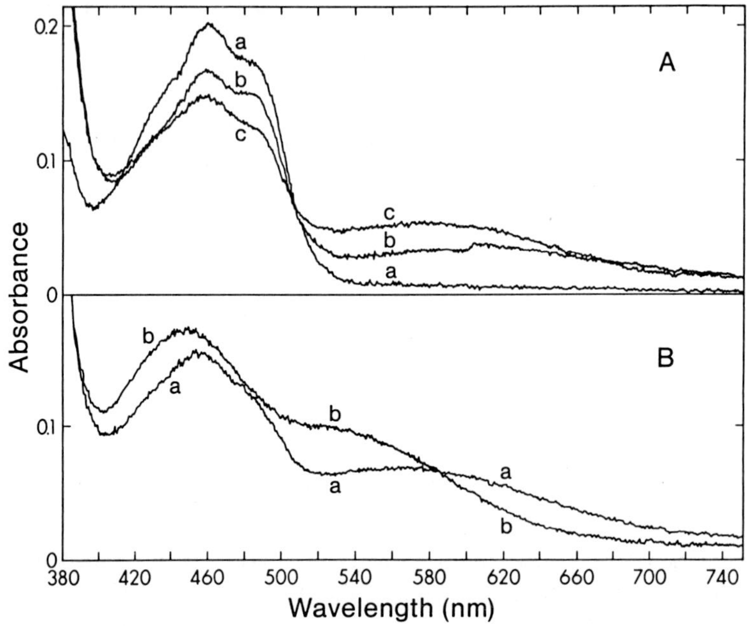

Fig. 1. Rapid-scan measurement of the reaction between mercuric reductase and a 10-fold molar excess of NADPH. Spectra are composed or several overlapping 96-nm regions obtained with 2-ms scanning time at 5.2^{o}C. Initial concentrations after mixing were 18 μM enzyme-bound FAD and 180 μM NADPH in 50 mM potassium phosphate buffer, pH 7.3. A; spectrum a, oxidized enzyme in absence of NADPH; spectrum b, at stop of flow in presence of NADPH; spectrum c, 24 ms after stop of flow. B; spectrum a, 44 ms after stop of flow; spectrum b, 286 ms after stop of flow.

reaction displayed an isosbestic point near 580 nm. The spectral changes can be described by a simple kinetic model involving three steps:

$$A \xrightarrow{k_1} B \xrightarrow{k_2} C \xrightarrow{k_3} D$$

where A, B, C, and D represent distinct spectral species.

The first step is virtually complete within the dead time of the apparatus. It is associated with a 16 % bleaching of flavin absorbance and the appearance of a very broad charge-transfer band (ε_{max} = 1.9 mM^{-1} cm^{-1}) centered near 600 nm. The second step involves a further, small bleaching of flavin absorbance and intensification of the charge-transfer band (ε_{max} = 3.3 $mM^{-1}cm^{-1}$) which becomes centered near 580 nm. The third step involves an increase of flavin absorbance and a blue-shift of the main

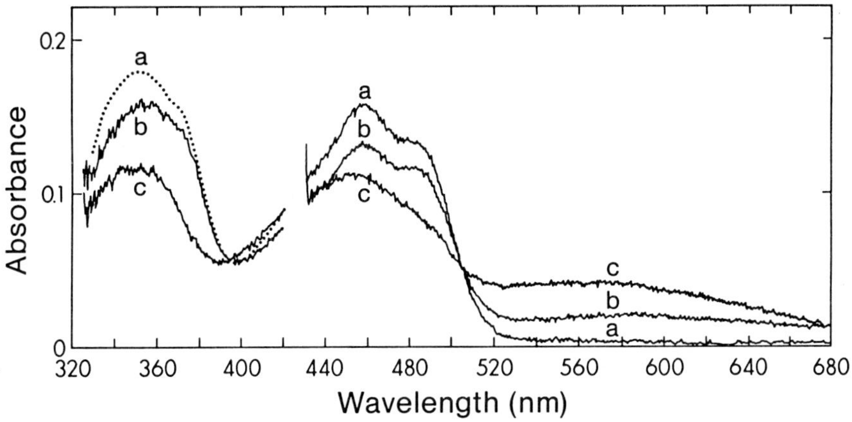

Fig. 2. Rapid-scan measurement of the reaction between mercuric reductase and an approximately equimolar amount of NADPH. Spectra were obtained at 5.2 ^{o}C as in Fig. 1. Initial concentrations after mixing were 13.8 μM enzyme-bound FAD and 14.4 μM NADPH in 50 mM potassium phosphate buffer, pH 7.3. Spectrum a is the sum of the spectra of oxidized enzyme and NADPH measured separately. Spectrum b was obtained at stop of flow and spectrum c 286 ms after stop of flow.

absorption maximum. Concurrently, the charge-transfer band increases in intensity and becomes centered near 530 nm (ε_{max} = 5.0 $mM^{-1}cm^{-1}$). As illustrated in Fig. 2, the first two steps require only one mol NADPH per mol FAD, whereas the third step requires a second equivalent of NADPH (8).

Steps 2 and 3 were studied separately in stopped-flow experiments at 577 nm and 507 nm, respectively, near the observed isosbestic points. Both steps were found to display well-behaved first-order kinetics. Both k_2 and k_3 were independent of the NADPH concentration in the investigated range (30 - 385 μM). Experimental values and standard errors at 5.0 ^{o}C, pH 7.3, were k_2 = 43 ± 4 s^{-1} with $\Delta H^{\neq}$ = 61.7 kJ/mol and $\Delta S^{\neq}$ = +9 ± 4 J/mol·K; k_3 = 7.8 ± 0.5 s^{-1} with $\Delta H^{\neq}$ = 69.9 kJ/mol and $\Delta S^{\neq}$ = + 23 ± 1 J/mol·K. Cysteine (1 mM) had only a marginal effect on k_2 and k_3. Preliminary experiments with specifically deuterated (4S) - [D]NADPH gave kinetic isotope effects (k_H/k_D) of 1.4 ± 0.2 in k_2 and 1.1 ± 0.1 in k_3. Thus, neither step 2 nor step 3 appear to be rate limited by hydride or proton transfer.

Treatment of the oxidized enzyme (E) with dithioerythritol (DTE) resulted in the formation of two-electron-reduced enzyme (EH_2). Addition of one molar equivalent of $NADP^+$ to DTE-generated EH_2 gave rise to a spectral species similar to the kinetic intermediate C, while addition of one molar equivalent of NADPH gave rise to a spectral species similar to D. On the basis of these results and comparisons with spectral forms reported in the literature on glutathione reductase and lipoamide dehydrogenase (9-11), we propose that the spectral intermediate B largely corresponds to an E-NADPH complex, C to an EH_2-$NADP^+$ complex and D to an EH_2-NADPH complex. It appears that intermediate species with reduced FAD or with a thiolate-flavin 4a-adduct do not accumulate during the reductive half reaction.

References

1. Summers, A.O., Silver, S.: Annu. Rev. Microbiol. *32*, 637-672 (1978).
2. Fox, B., Walsh, C.T.: J. Biol. Chem. *257*, 2498-2503 (1982).
3. Schottel, J.L.: J. Biol. Chem. *253*, 4341-4349 (1978).
4. Rinderle, S.J., Booth, J.E., Williams, J.W.: Biochemistry *22*, 869-876 (1983).
5. Brown, N.L., Ford, S.J., Pridmore, R.D., Fritzinger, D.C.: Biochemistry *22*, 4089-4095 (1983).
6. Williams, C.H.: The Enzymes, 3rd ed. *13*, 89-172 (1976).
7. Fox, B., Walsh, C.T.: Biochemistry *22*, 4082-4088 (1983).
8. Sahlman, L., Lindskog, S.: Biochem. Biophys. Res. Commun.: *117*, 231-237 (1983).
9. Matthews, R.G., Ballou, D.P., Williams, C.H.: J. Biol. Chem. *254*, 4974-4981 (1979).
10. Huber, P.W., Brandt, K.G.: Biochemistry *19*, 4568-4575 (1980).
11. Veeger, C., Massey, V.: Biochim. Biophys. Acta *67*, 679-681 (1963).

DEHYDROHALOGENATION AND INTERMOLECULAR HYDROGEN TRANSFER REACTIONS CATALYZED BY SOME LACTATE-OXIDIZING ENZYMES

Florence Lederer*

Centre de Génétique Moléculaire, CNRS
Gif-sur-Yvette, France

Introduction

At the preceding symposium, Ghisla reviewed the state of knowledge concerning the mechanism of flavoprotein-catalyzed dehydrogenations of the type $H\overset{|}{\underset{|}{C}}(ZH)-CO_2^- \longrightarrow Z{=}\overset{|}{C}-CO_2^-$ (1). Several mechanisms can be formally considered: one involving abstraction of $H^{\cdot}$ (radical mechanism), another one implicating hydride ion transfer, and finally the so-called carbanion mechanism, which proceeds by initial proton abstraction from the α-carbon. The carbanion could be oxidized by transfer of its electrons one by one to the flavin, or as a pair via a covalent adduct formed at flavin N(5).

The first piece of evidence in favour of the carbanion mechanism was provided by the study of the elimination reaction undergone by β-chlorosubstrates (2-6). It was proposed that a common intermediate, a carbanion, was partitioning between the normal oxidative reaction and the dehydrohalogenation reaction (2,6). For some substrates the normal oxidative reaction even seemed to be suppressed (5).

As stressed by Ghisla, however, the results were compatible with a carbanion mechanism but did not prove it (1). Bright and Porter (7) argued that the kinetic results obtained with D-amino acid oxidase were equally compatible with elimination

*Present address: INSERM U25, Hôpital Necker, Paris

Flavins and Flavoproteins

occurring from an E_{red}.P complex. In that case, no deduction was possible concerning the normal redox mechanism. Furthermore, rapid reaction studies provided results that were difficult to interpret in the frame of the non oxidative elimination hypothesis (8). Finally, flavocytochrome b_2 (9) and L-hydroxyacid oxidase from rat kidney (10) could not be shown to catalyze the non oxidative dehydrohalogenation of β-chlorolactate, even though they were expected to present the same mechanism as lactate oxidase. In both cases it was suggested that the putative carbanion was oxidized too fast for Cl^- elimination to be able to compete. Since, however, the first order rate constants for flavin reduction were found to be 100 s^{-1} and 233 s^{-1} for flavocytochrome b_2 and lactate oxidase respectively (11,12), with a similar deuterium isotope effect (8,11), the suggestion did not appear entirely satisfactory.

The study of the flavocytochrome b_2-catalyzed reverse reaction (13) has provided a tool to reinvestigate the matter. A quantitative analysis of the enzyme reaction with halogenolactates and halogenopyruvates (14), as well as preliminary qualitative experiments with lactate oxidase and hydroxyacid oxidase enable us to propose a rationalization for many of the discrepancies mentioned above. Furthermore an extension of the study has brought to light the fact that the hydrogen picked up from the substrate α-carbon is exchanged only slowly with solvent in flavocytochrome b_2 and lactate oxidase.

The Elimination Reaction.

Flavocytochrome b_2. When this enzyme is incubated with halogenopyruvates in the presence of lactate, it catalyzes their reduction to halogenolactates in a reaction which by-passes the heme (13). At higher enzyme and substrate concentrations, separation of a reaction mixture initially containing lactate and {2-^{14}C}bromopyruvate led to the detection of {2-^{14}C}-

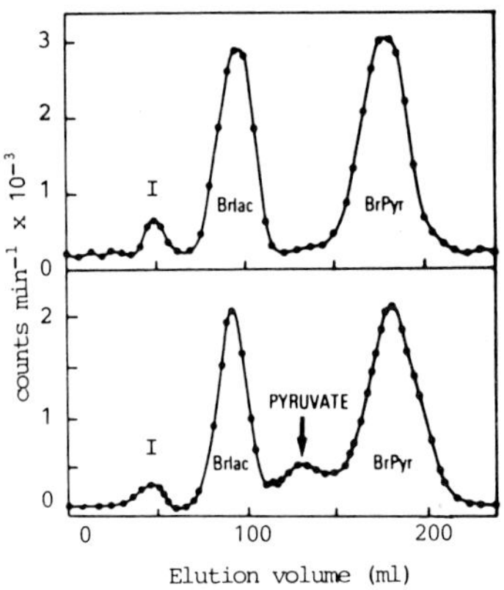

Figure 1. Reaction between reduced flavocytochrome b_2 and {2-^{14}C} bromopyruvate: separation of reaction products.
A : 13.7 mM bromopyruvate (51.3 x 10^3 counts min^{-1} μmol^{-1}), 24 mM L-lactate, 14.5 μM enzyme were incubated in 1 ml standard buffer (0.1 M phosphate buffer, 1 mM EDTA pH 7) for 15 min at 30°C under argon.
B : 10.6 mM bromopyruvate (84 x 10^3 counts min^{-1} μmol^{-1}), 24 mM L-lactate, 11 μM enzyme were incubated in 0.8 ml standard buffer for 90 min at 30°C under argon.
A and B : After acidification the reaction mixtures were separated on a column (16 x 0.9 cm) of AG1 X10 resin eluted at 60 ml/h with a 400 ml gradient of 0 to 2N HCO_2H (13). The effluent was analysed starting at the first acidic fraction. Peak I corresponds to a contaminant present in the initial {2-^{14}C}bromopyruvate solution (control not shown).

pyruvate (figure 1). That the reaction was due to a reductive elimination process was attested by an increase in total ketoacid concentration, whereas the transhydrogenation reaction does not alter the ketoacid balance. This offered a means of selectively monitoring the elimination reaction. The results of steady state rate measurements are compared in table 1 with the parameters determined for the forward and the reverse redox reaction. It can be seen that in terms of k_{cat}/K_m, elimination is not much slower than the reverse reaction. For fluoropyruvate, no F^- loss could be observed; if it occurred

Table 1 STEADY STATE PARAMETERS FOR THE REACTIONS OF HALOGENOLACTATES AND HALOGENOPYRUVATES WITH FLAVOCYTOCHROME b_2

	Forward reaction[a]		Reverse reactive[a]		Elimination[b]	
Redox couple	K_m	k_{cat}	K_m	k_{cat}	K_m	k_{cat}
	mM	s^{-1}	mM	s^{-1}	mM	s^{-1}
bromolactate/ bromopyruvate	0.8 ± 0.1	105 ± 3	6.7 ± 0.4	3.6 ± 0.2	12 ± 2	3 ± 1
chlorolactate/ chloropyruvate	1.2 ± 0.1	34 ± 1	6.8 ± 0.9	1.5 ± 0.4	12 ± 1	0.4 ± 0.06
fluorolactate/ fluoropyruvate	1.1 ± 0.1	9.5 ± 0.2	7.2 ± 1.1	0.4 ± 0.1	not detected	

a) Taken from (13); b) taken from (14). All values were determined at 30°C in 0.1M phosphate buffer pH 7.

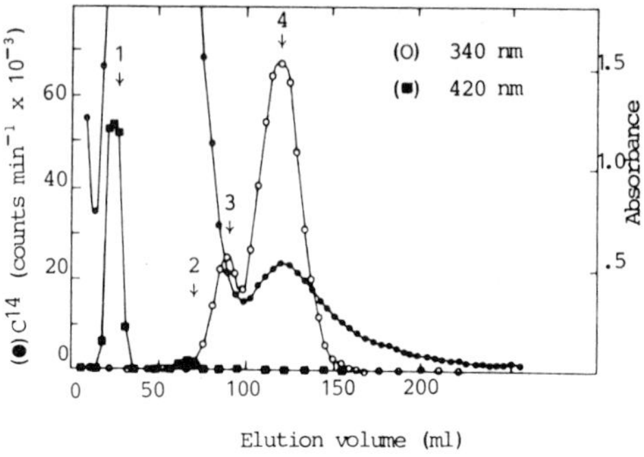

Figure 2. Reaction between oxydized flavocytochrome b_2 and {2-^{14}C}bromolactic acid: separation of reaction products.
At time zero, the reaction mixture contained 46.6 mM reduced glutathione, 25 mM L-{2-^{14}C}bromolactate (0.68 x 10^6 counts min^{-1} μmol^{-1}), 47.5 mM fluoropyruvate and 17 µM enzyme in a total volume of 0.88 ml standard buffer. After 45 min at 32°C under argon, more reagents were added : 25 µmols bromolactate, 41 µmols glutathione, 26 µmols fluoropyruvate and 17 nmols enzyme; the final volume was 1.56 ml. After an additional 65 min, 26.8 µmols carrier unlabelled pyruvate were added and the reaction was stopped by freezing. The separation was carried out as described in the legend of figure 1. Absorbance at 420 nm indicates the presence of hydroxyacid as assayed with flavocytochrome b_2, absorbance at 340 nm the presence of ketoacid as assayed with beef heart lactate dehydrogenase (13).
The arrows indicate the position expected for fluorolactic acid (1), bromolactic acid (2), fluoropyruvic acid (3), pyruvic acid (4).

at all, it must have been at least 150 times slower than dehydrochlorination, in view of our experimental detection limit.

If elimination was taking place from an intermediate on the redox pathway, then the reaction should also be observable starting from bromolactate and oxidized enzyme. The conditions of the experiment were devised so as to lead to the accumulation of a large number of turnovers (figure 2). Indeed, {2-^{14}C}pyruvate was again formed. The ratio of bromopyruvate to pyruvate formed was 488 compared to the ratio of 2 found in the reverse direction from the k_{cat}/K_m's ratio (14). These results are compatible with scheme A, where T is an

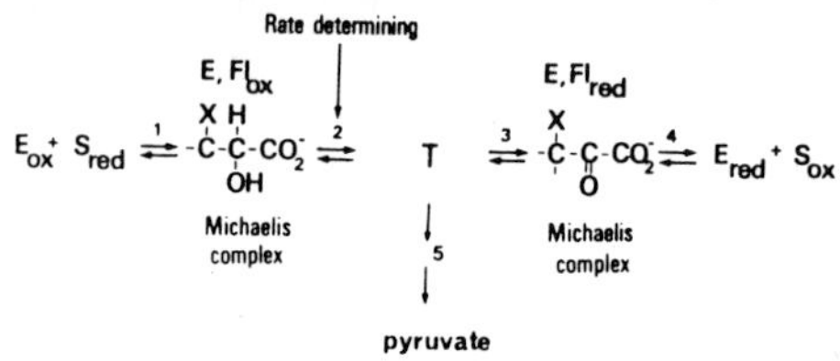

A

intermediate which has lost the α-hydrogen. According to the principle of microscopic reversibility, since elimination occurs both in the forward and the reverse reaction, it must be undergone by an intermediate on the redox pathway. Furthermore, since step 2 is rate determining in the forward reaction with a k_2^H/k_2^D = 7-8 (11), step 2 is rate determining in the reverse direction. This makes it most likely that it is T that undergoes elimination rather than either of the Michaelis complexes. This scheme also explains why flavocytochrome b_2 does not catalyze the non oxidative dehydrohalogenation in the absence of acceptor: after one turnover the halogenoketoacid formed in stoichiometric amount with the enzyme will dissociate and the reaction will stop. The 0.2% pyruvate formed will be undetectable.

Other lactate-oxidizing enzymes. Turning now our attention to lactate oxidase, I believe one can explain its catalysis of non oxidative dehydrohalogenation of chlorolactate in the absence of electron acceptor (6) by taking into account first the reversibility of the redox reaction, second the fact that halogenopyruvate dissociation from E_{red} may be very slow. Concerning the first point, Urban (unpublished experiments) has shown that under anaerobic conditions, reduced lactate oxidase can catalyze the transformation of halogenopyruvates (table 2). Ghisla has also observed the reverse reaction under rapid reaction conditions (S. Ghisla, personal communication). Concerning the second point, it has been determined that $k_4 = 0.04 s^{-1}$ when S_{ox} = pyruvate (12); a similar slow dissociation from E_{red} may be expected from halogenopyruvates. The qualitative explanation is then the following: because of slow S_{ox} dissociation, the $E, Fl_{red} \cdot S_{ox}$ complex can revert to T, leading to pyruvate formation. This suggestion is supported by the observation that during approach to the steady state under turnover conditions in the presence of chlorolactate, lactate

Table 2 REACTION OF REDUCED LACTATE OXIDASE WITH HALOGENOPYRUVATES

Substrate	Experimental conditions: Enzyme concentration	Substrate concentration	Duration	Products: Halogeno-lactate	Pyruvate
	µM	mM	min	%	%
Bromopyruvate	2.75	9.3	175	0	100
Chloropyruvate	7.60	53.0	90	9	91
Fluoropyruvate	3.30	44.0	120	100	0

Lactate oxidase came from Boehringer. Its concentration was taken to be that of the spectrophotometrically determined flavin concentration. Reactions were carried out in the presence of 90 to 100 mM L-lactate, under argon, at 30°C, in 0.1 M imidazole buffer pH 6.

oxidase appeared to be first reduced then partially reoxidized, and that hydroxylamine addition to the reaction mixture prevented this reoxidation (8). A similar qualitative explanation is probably also valid for D-amino acid oxidase, for which product dissociation from E_{red} is also rate limiting (7). Finally, for hydroxyacid oxidase, the steady state deuterium isotope effect of 8 (10) shows that step 2 is the rate limiting one, as observed for flavocytochrome b_2. The prediction is then that elimination, which was not observed when chlorolactate was oxidized by the enzyme in the absence or the presence of acceptors (10), should be observable starting from a halogenopyruvate and reduced enzyme. The prediction has now been borne out by experiment. I. Chirat has shown that reduced hydroxyacid oxidase can catalyze the disappearance of bromopyruvate under turnover conditions; when 21 μM enzyme was incubated with 30 mM L-lactate and 20 mM {2-^{14}C}bromopyruvate for 90 min at 30°C under argon in 0.1M phosphate buffer pH 7, the products formed were {2-^{14}C}bromolactate and {2-^{14}C}pyruvate in an approximate ratio of 1 to 10.

Carbanion versus other mechanisms. In scheme A, the intermediate T is missing the substrate α-hydrogen. It could be a carbanion, a radical, or a covalent intermediate of the type postulated to occur in normal flavodehydrogenase catalysis (1,2,7,15,16). One may expect these mechanisms to be differently affected by the nature of the halogen.

For a radical mechanism of the type proposed by certain authors (17,18) there exist no relevant well documented examples, a fact which may in itself indicate that it does not appear too probable.

A covalent intermediate of the type shown below

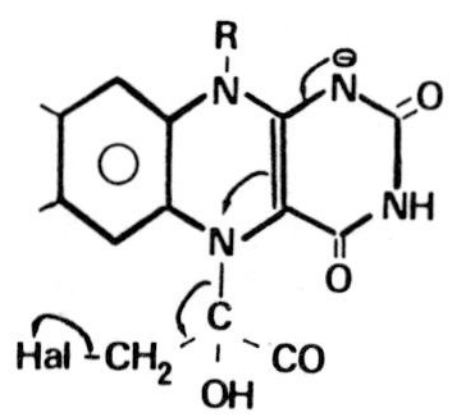

would be expected to expel a fluoride ion quite readily, since formation of the conjugated Fl_{ox} system could pay the energetic cost of breaking the C-F bond. Nevertheless, neither flavocytochrome b_2, nor lactate oxidase, nor D-amino acid oxidase (19) can catalyze fluoride elimination at an appreciable rate. Therefore T cannot be the covalent intermediate; this does not rule out the occurrence of this intermediate in the redox pathway (scheme A is a minimal scheme in terms of intermediates).

The only possibility left is that T be a carbanion. For an E1cB-type mechanism, with no carbon-halogen bond breaking in the transition state, one would expect loss of fluorine to take place even more easily than that of the other halogens. In scheme B, however, k_5 is probably of the same order of

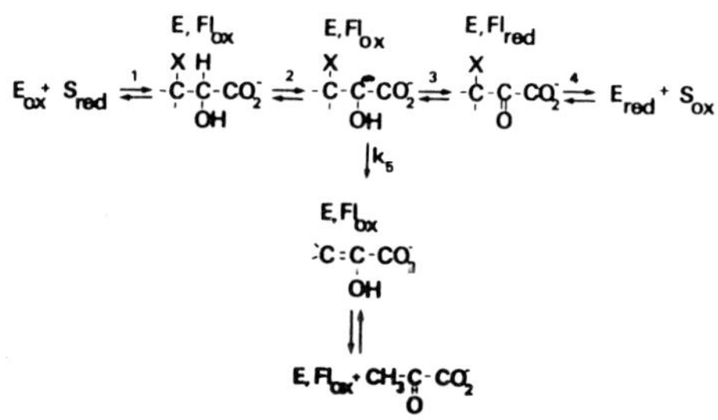

B

magnitude as k_{cat} for pyruvate formation and should therefore be the rate limiting step for pyruvate formation. In this case the relative elimination rates for the various halogens become understandable.

Intermolecular Hydrogen Transfer.

The question addressed was whether, under transhydrogenation conditions, the hydrogen picked up by the enzyme from Cα of one substrate could be transferred to Cα of the second substrate. An answer to this question may give clues as to the identity of the group(s) the hydrogen resides on and hence possibly on the reaction mechanism.

A review of the literature gives several examples of intramolecular transfers, in other words situations where the α-hydrogen was transferred to another carbon atom within the enzyme-intermediate complex. For some of these, tritium transfer was directly demonstrated. This is the case for the lactate-oxidase catalyzed elimination from chlorolactate (6), for the D-amino acid oxidase-catalyzed elimination from α-amino-β-chlorobutyrate, the stereospecificity of which was analysed (5,20) and for the glutaryl-CoA dehydrogenase reaction (21). In other cases, it could be concluded from kinetic studies that the protein-bound α-hydrogen was exchanged very slowly if at all during the life-time of the enzyme-intermediate complex, as for D-amino acid oxidase (22,23), for lactate oxidase with glycollate (16), for D-lactate dehydrogenase *from Megasphaera elsdenii* (23). It was also reported that the proton picked up by the lactate oxidase-oxalate complex exchanges only very slowly with the solvent (24). In all those cases, it is possible to invoke a physical shielding of the trapped proton, even though this may not be the right explanation.

Reported failures to observe intermolecular hydrogen transfer include the cases of mitochondrial NADH-cytochrome c reductase (25) D-amino acid oxidase (2) and glutamate synthase (26). Only two successful examples are to be found in the literature: those of the pyridine nucleotide transhydrogenase (27) and, most striking of all, of NADH-cytochrome b_5 reductase (28,29). Hydrogen transfer has also been observed between substrates and enzyme-bound 5-deazaflavins (30-35) but these examples are not relevant for this discussion.

Concerning our own results, two types of experiments will be described here which demonstrate the existence of flavocytochrome b_2- and lactate oxidase-catalyzed intermolecular hydrogen transfer.

Table 3 shows the result of a transhydrogenation reaction in which {2-^{3}H}lactate and bromopyruvate were incubated in the presence of flavocytochrome b_2. Clearly, a sizeable amount of transfer was observed. By enzymatic conversion of tritiated

Table 3 TRITIUM TRANSFER FROM LACTATE TO BROMOPYRUVATE

Radioactivity transferred to	Amount transferred	% Transfer	Specific radioactivity
	Counts min^{-1}		Counts min^{-1} μmol^{-1}
H_2O	2 700 x 10^3	94	116
Bromolactate	144 x 10^3	5	15 700
Pyruvate	28 x 10^3	1	2 840

Enzyme (10.2 μM) was incubated with 6.5 mM bromopyruvate and 25.7 mM L -{2-^{3}H}lactate (10^6 cpm/μmol) in 1 ml 0.1 M phosphate buffer. A total of 16.7 μmols bromopyruvate, 7 μmols lactate and 9 nmols enzyme were added at various time intervals. The reaction was stopped by 10 μl conc. HCl after 28 min. The final volume was 1.28 ml. Reaction products were separated by ion exchange chromatography as described in (13).

bromolactate to bromopyruvate, it was shown that the label was introduced stereospecifically. The transfer to pyruvate showed that the enzyme is capable of giving a hydrogen to the C3 position, as already observed with lactate oxidase (6) and D-amino acid oxidase (5). The lower specific radioactivity of pyruvate compared to that of bromolactate may mean that part of the pyruvate formed acquires its proton from the solvent after enol release from the protein. This could constitute another interesting piece of evidence in favour of elimination from a carbanion (scheme B). The lower specific radioactivity may however simply be due to irrelevant proton exchange at the level of pyruvate. Suitable controls experiments still have to be carried out. Similar observations were made with lactate oxidase.

The second set of results was obtained in transhydrogenation reactions between {α-^{3}H}lactate and fluoropyruvate. For both enzymes, the amount of tritium found in fluorolactate increased with fluoropyruvate concentration, as expected if exchange takes place essentially during the time the reduced enzyme is free from ligands. Extrapolation to infinite substrate concentration yielded a figure close to 1 (figures 3 and 4).

These experiments show that proton exchange between free reduced lactate oxidase or flavocytochrome b_2 and the solvent is slow, and furthermore that the α-proton never sees a heteroatom which would carry more than one hydrogen.

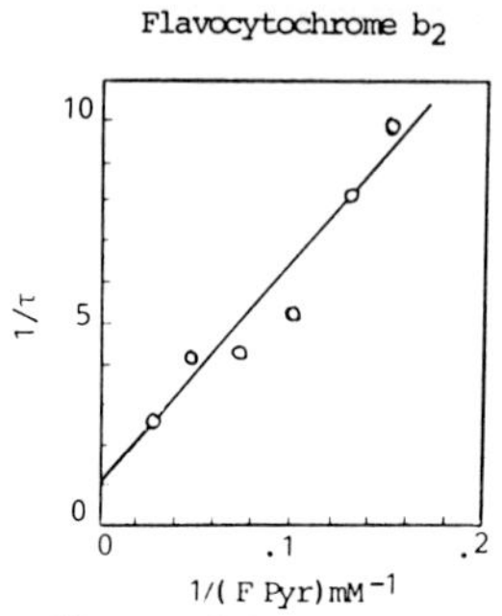

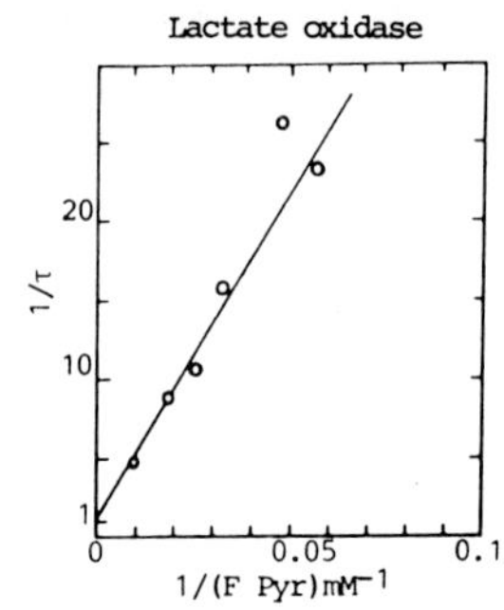

Figure 3 (left). Influence of fluoropyruvate concentration on flavocytochrome b_2-catalyzed intermolecular tritium transfer.
Each point corresponds to identical experimental conditions except for fluoropyruvate concentration: 28.6 mM L-{α-^{3}H}lactate (273×10^3 counts min^{-1} μmol^{-1}), 15 μM enzyme in 0.72 ml standard buffer; incubation for 20 min at 30°C under argon. The reaction was stopped by freezing. The reaction mixture was acidified just before being loaded onto a column of Dowex AG1 X10 (0.9 x 16 cm) eluted with a 400 ml gradient of 0 to 1.33 M.

Figure 4 (right). Influence of fluoropyruvate concentration on lactate oxidase-catalyzed intermolecular tritium transfer.
The reaction conditions are described in table 2. All experiments were carried out with 25 μM enzyme and 98 mM {α-^{3}H}lactate (98.2×10^3 counts min^{-1} μmol^{-1}).

It can be demonstrated that

$$1/\tau = \frac{v_e^*(k_r^* + k_{-1})}{k_r^* \, k_1} \cdot \frac{1}{S_{ox}} + 1$$

where

τ = amount of ^3H transfer to the acceptor divided by the total amount of ^3H labilized from lactate

k_1 and k_{-1} : on and off rate constants for S_{ox} binding

k_r^* : steady state rate constant for the reverse reaction of tritiated enzyme with the given ketoacid

v_{ex}^* : triton exchange rate of reduced enzyme under the given conditions.

For flavocytochrome b_2, we can make the reasonable assumptions that $k_1 \cong 10^5 M^{-1}s^{-1}$ and $k_{-1} \cong 10^3 s^{-1}$ (36,37) and $k_r^* = k_r/20$ because of a full isotope effect on carbanion protonation. From figure 3, we can then calculate, $v_{ex}^* \cong 0.1\ s^{-1}$ in 0.1 M phosphate buffer pH 7. This leads to $k_{ex} \cong 50\ M^{-1}s^{-1}$ for the exchangeable proton.

In buffered medium around pH 7, one would expect any freely accessible protein side chain to exchange at a rate of 10^7 to $10^8\ s^{-1}$ or faster (38). Hydrogen bonding will lower this rate by a factor of 10^3 to 10^4 in most cases (39). Therefore it appears likely that if the exchanged proton resides on the reduced enzyme active site base, it must enter into a very strong hydrogen bond.

Another way of looking at the results is to calculate the pK of the hydrogen donor from the equation

$$\log k_f - \log k_r = pK_A - pK_D$$

which applies to the equilibrium

$$DH + A^- \underset{k_r}{\overset{k_f}{\rightleftharpoons}} D^- + AH$$

where D and A stand for the donor and the acceptor respectively. If we take $k_r = 10^{10}\ M^{-1}s^{-1}$ (diffusion limited rate in the thermodynamically favoured direction) and $pK_A = 7.2$ (buffer pK_A), we can calculate $pK_D = 15$ to 16.

For lactate oxidase, the data at hand are less extensive and there are not yet sufficient kinetic data at hand to make a reliable estimate of the proton exchange rate. Highly preliminary experiments could indicate a pK_D at least as high as that calculated for flavocytochrome b_2.

These values appear too high for the pK of a protein side chain, even though environment can have an important influence on side chain pK's (it was shown for example by Ghisla and Massey that an active site group in lactate oxidase sees its pK change from 4.7 in the free enzyme to 9.8 in the oxalate-enzyme complex (24)). On the other hand, the N(5)-H pK in free Fl_{red} is thought to be higher than 20 (40).

The following suggestion can be made: the pK of 15-16 is that of Fl_{red} N(5)-H, which is perturbed by the protein environment. An attractive possibility would be that the α-proton be caught in a hydrogen bond linking the active site base and flavin N(5). This would obviate the problem of imagining how the α-proton can be transferred to the flavin. However, flavoproteins which, like cytochrome b_5 reductase, oxidize nicotinamide cofactors are supposed to be reduced by a hydride mechanism (41). In that case the substrate proton must reach the flavin directly, and a pK of 15 to 16 would be directly

attributable to an N(5)-H bond.
In conclusion, it appears that the phenomenon of intermolecular hydrogen transfer in flavoproteins deserves further study. Its dependence on hydrogen acceptor concentration can be used as an indicator of flavin N(5)-H pK. Data can be obtained concerning hydroxy acid oxidase. It would be interesting to see whether, under possibly more favourable conditions than those already used (5), D-amino acid oxidase still fails to catalyze intermolecular hydrogen transfer. It would be interesting also to obtain data about the N(5)-H pK of hydroxylases and oxygenases.
The results of such studies would help in deciding whether there is a correlation between the class a flavoprotein belongs to and the acidity of its flavin at N(5), an idea which is implicit in the hypothesis that hydrogen bonding at N(5) can be correlated with flavoprotein function (42). Whatever the case, the results of such studies will be useful, when combined with the results of physical methods and in particular of crystallographic determinations, in understanding the fine details of how protein structure influences reactivity.

References

1. Ghisla, S.: Flavins and Flavoproteins, Massey V. & Williams C.H. eds., Elsevier, Amsterdam 1982, pp. 133-142.

2. Walsh, C.T., Schonbrunn, A., Abeles, R.H.: J. Biol. Chem. 246, 6855-6866 (1971).

3. Voet, J.G., Porter, D.J.T., Bright, H.J.: Z. Naturforsch. 27b, 1054-1055 (1972).

4. Miyake, Y., Abe, T., Yamano, T.: J. Biochem. 73, 1-11 (1973).

5. Walsh, C.T., Krodel, E., Massey, V., Abeles, R.H.: J. Biol. Chem. 248, 1946-1955 (1973).

6. Walsh, C.T., Lockridge, O., Massey, V., Abeles, R.H.: J. Biol. Chem. 248, 7049-7054 (1973).

7. Bright, H.J., Porter, D.J.T.: The Enzymes, Boyer P.D. ed., Third edition, Academic Press, New York 1975, Vol. 12, pp. 421-505.

8. Massey, V., Ghisla, S., Ballou, D.P., Walsh, C.T., Cheung, Y.F., Abeles, R.H.: Flavins and Flavoproteins, Singer T.P. ed., Elsevier Scientific Publishing Co, Amsterdam 1976, pp. 199-212.

9. Lederer, F.: Eur. J. Biochem. 46, 393-399 (1974).

10. Cromartie, T.H., Walsh, C.T.: Biochemistry 14, 3482-3490 (1975).

11. Pompon, D., Iwatsubo, M., Lederer, F.: Eur. J. Biochem. 104, 479-488 (1980).

12. Lockridge, O., Massey, V., Sullivan, P.A.: J. Biol. Chem. 247, 8097-8106 (1972).

13. Urban, P., Alliel, P.M., Lederer, F.: Eur. J. Biochem. 134, 275-281 (1983).

14. Urban, P., Lederer, F.: Eur. J. Biochem., in press (1984).

15. Massey, V., Ghisla, S., Kieschke, K.: J. Biol. Chem. 255, 2796-2806 (1980).

16. Ghisla, S., Massey, V.: J. Biol. Chem. 255, 5688-5696 (1980).

17. Walling, C., Johnson, R.A.: J. Am. Chem. Soc. 97, 2405-2407 (1975).

18. Fendrich, G., Abeles, R.H.: Biochemistry 21, 6685-6695 (1982).

19. Dang, T.Y., Cheung, Y.F., Walsh, C.: Biochem. Biophys. Res. Comm. 72, 960-968 (1976).

20. Cheung, Y.F., Walsh, C.: Biochemistry 15, 2432-2441 (1976).

21. Gomes, B., Fendrich, G., Abeles, R.H.: Biochemistry 20, 1481-1490 (1981).

22. Fitzpatrick, P.F., Massey, V.: J. Biol. Chem. 257, 9958-9962 (1982).

23. Morpeth, F.F., Massey, V.: Biochemistry 21, 1307-1312 (1982).

24. Ghisla, S., Massey, V.: J. Biol. Chem. 252, 6729-6735 (1977).

25. Drysdale, G.R.: J. Biol. Chem. 234, 2399-2403 (1959).

26. Geary, L.E., Meister, A.: J. Biol. Chem. 252, 3501-3508 (1977).

27. Louie, D.D., Kaplan, N.O.: J. Biol. Chem. 245, 5691-5698 (1970).

28. Drysdale, G.R., Spiegel, M.J., Strittmatter, P.: J. Biol. Chem. 236, 2323-2328 (1961).

29. Drysdale, G.R.: Biochim. Biophys. Acta Library 8, 159-170 (1966).

30. Jorns, M., Hersh, L.: J. Am. Chem. Soc. 96, 4012-4014 (1974).

31. Fisher, J., Walsh, C.: J. Am. Chem. Soc. 96, 4345-4346 (1974).

32. Averill, B.A., Schonbrunn, A., Abeles, R.H.: J. Biol. Chem. 250, 1603-1605 (1975).

33. Hersh, L.B., Schuman Jorns, M.: J. Biol. Chem. 250, 8728-8734 (1975).

35. Pompon, D., Lederer, F.: Eur. J. Biochem. 96; 571-579 (1979).

36. Iwatsubo, M., Capeillère, C.: Biochim. Biophys. Acta, 146, 349-366 (1967)

37. Pompon, D., Iwatsubo, M., Lederer, F.: Eur. J. Biochem., 104, 479-488 (1980).

38. Fersht, A.: Enzyme Structure and Mechanism, Freeman & Co, Reading San Francisco 1977, pp. 127-128.

39. Eigen, M., Kruse, W.: Z. Naturforsch., 18b, 857-865 (1963).

40. Bruice, T.C. : Progress in Bioorganic chemistry, Kaiser E.T. & Kezdy F.J. eds., Wiley & Sons, New York 1976, pp. 1-87.

41. Powell, M.F., Bruice, T.C. : J. Am. Chem. Soc. 105, 1014-1021 (1983).

42. Massey, V., Hemmerich, P.: Biochemical Soc. Trans. 8, 246-257 (1980).

ON THE MECHANISM OF INACTIVATION OF FLAVOCYTOCHROME b_2 (BAKER's YEAST) BY ACETYLENIC SUBSTRATES.

Denis POMPON & Florence LEDERER, Centre de Génétique Moléculaire - C.N.R.S. 91190 GIF-SUR-YVETTE (France).

The inactivation of flavocytochrome b_2 by 2-hydroxy- and 2-keto-3-butynoate has been reinvestigated and the results were compared with published ones concerning lactate oxidase from Mycobacterium smegmatis.

No inactivation occurred upon incubation of the enzyme with saturating concentrations of hydroxybutynoate in the absence of electron acceptors. Introduction of either O_2 or ferricyanide caused substrate oxidation as well as a rapid loss of activity associated with a covalent flavin modification. The partition ratio between substrate oxidation and enzyme inactivation was 3 200. This ratio was independent of substrate and product concentration in the presence of excess ferricyanide.

Ketobutynoate had no effect on oxidized enzyme. However, under transhydrogenation conditions (Urban & Lederer (1983) Eur. J. Biochem. 134, 275) in the presence of excess lactate, ketobutynoate inactivated the enzyme while being reduced to hydroxybutynoate. The partition ratio between reduction and inactivation was 5.

Spectral and chromatographic comparisons showed that inactivation of the reduced enzyme by the ketoacid and of the oxidized enzyme by the hydroxyacid led to the same flavin modification. Two adducts were isolated. One of them (compound 1) was identical with the one exclusively obtained when lactate oxidase was treated with hydroxybutynoate (Schonbrunn et al. (1976) Biochemistry, 15, 1798) and processed under identical conditions. The structure of the major adduct (compound 2) has been determined. It arises, by a different decomposition mode while still enzyme-bound, from the same primary adduct to the flavin C(4a) position as in lactate oxidase. It is not clear whether the difference between the two enzymes is due to a different stereochemistry of the primary adduct or to differences in flavin-protein interactions in the N(10), C(10a), N(1) area.

These results strengthen the idea that inactivation is caused by reaction between Fl_{ox} and an allenic carbanion, the isomerisation product of a normal reaction intermediate, rather than by reaction between Fl_{red} and ketobutynoate. The widely different partition ratios for the forward and the reverse pathways can be easily explained if protonation of the carbanion is rate determining in the reverse direction, just as its formation is rate determining in the forward one. The same situation obtains for the reaction of flavocytochrome b_2 with other keto-and hydroxyacids (Urban & Lederer (1983) Eur. J. Biochem. 134, 275).

1

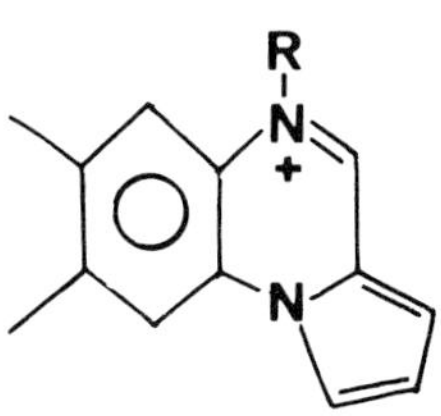

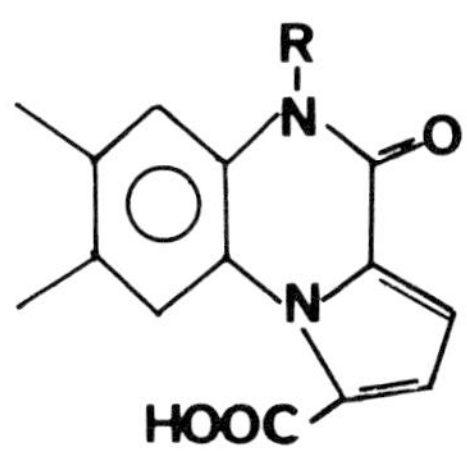

2

Flavins and Flavoproteins

INTERMOLECULAR HYDROGEN TRANSFER DURING TRANSHYDROGENATION CATALYZED BY FLAVOCYTOCHROME b_2 AND LACTATE OXIDASE.

Philippe URBAN & Florence LEDERER
Centre de Génétique Moléculaire, C.N.R.S., 91190 GIF-SUR-YVETTE (France)

We have shown before that flavocytochrome b_2 from baker's yeast can catalyze transhydrogenation reactions when incubated under anaerobic conditions in the presence of a hydroxyacid and a ketoacid. The same has now been shown for lactate oxidase from Mycobacterium smegmatis.

When these enzymes are incubated in the presence of excess L-lactate and a halogenoketoacid, the latter can not only be reduced but in some cases also undergo elimination. For example bromopyruvate is transformed into bromolactate and pyruvate by flavocytochrome b_2, and only into pyruvate by lactate oxidase. Fluoropyruvate is entirely converted to fluorolactate by both enzymes.

When the L-lactate used is tritiated in the α-position, a certain amount of label is found on the products halogenolactate and pyruvate, both with flavocytochrome b_2 and lactate oxidase. The rest of the label is found in solvent as tritiated water.

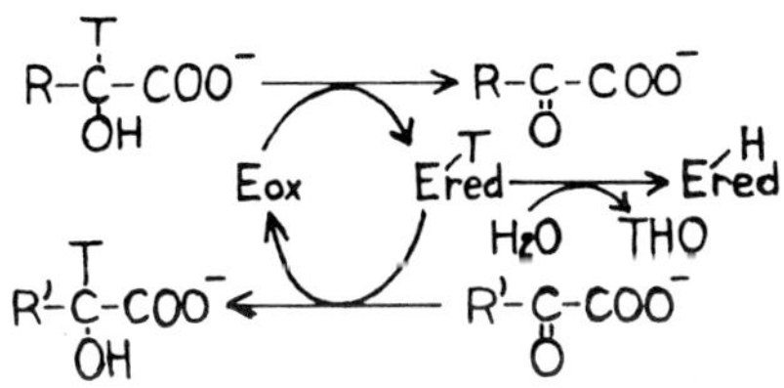

The transfer is stereospecific. The amount of transfer does not depend on the identity of the tritiated hydroxyacid. It depends on the initial concentration of acceptor ketoacid in a saturable fashion. Extrapolation to infinite acceptor concentration yields a value of transfer close to 100%. This shows that the α-proton never sees anything but a monoprotonic heteroatom. During flavocytochrome b_2 catalyzed transhydrogenation in $\{^3H\}$-H_2O, some label is also found incorporated in the products. No data have yet been obtained in this respect with lactate oxidase.

These results show that the free reduced enzyme exchanges the α-proton only slowly with the medium. The estimated value of the exchange rate constant yields a pK of 15 to 16 for the exchanging group in flavocytochrome b_2 and a possibly higher pK for lactate oxidase. As it seems unlikely that a protein side chain could have such a high pK, it appears probable that the α-proton has been transferred to the flavin N(5) position.

Flavins and Flavoproteins

PRODUCT BINDING AS MODULATOR OF FLAVIN REDOX PARAMETERS :
A MECHANISM OF ACTIVITY CONTROL IN DEHYDROGENASE-e^-TRANSFERASE ?

Françoise Labeyrie, Jean-Marc Janot and Mariella Tegoni,
Centre de Génétique Moléculaire du CNRS, 91190 Gif-sur-Yvette (France).

Introduction

Yeast L-lactate cytochrome c oxidoreductase or flavocytochrome b_2 is a complex two-domain enzyme that pertains, in the classification of flavoproteins defined by Hemmerich and Massey (1), to the class 4 (de)hydrogenase-e^-transferase. Indeed, the two electron-equivalents donated by lactate (yielding pyruvate), firstly accepted by flavin, are delivered one by one to the cytochrome moiety (preferential acceptor) with obligate formation of the semiquinone form, and hence to external acceptors (cf review (2)).

Equilibria concerning flavin-heme electron exchanges in flavocytochrome b_2 (from baker's yeast and Hansenula anomala) have already been thoroughly studied under conditions where no specific ligand other than the titrating reactant, ie L-lactate, was added to the system. The results (Blandin-Capeillère et al (3)) are presented in Table 1.

New redox determinations have been more recently carried out with Hansenula anomala flavocytochrome b_2 in the presence of pyruvate since this reaction product was suspected to modify flavin-heme redox equilibria (4). Correlations were established between the intensity of EPR semiquinone signal observed at the titration temperature and the level of heme reduction deduced from the absorbance level at 556 nm. It was thus shown that pyruvate considerably increases the proportion of semiquinone which reaches 90±10% during titrations (5,6): midpoint potential of the F_{ox}/F_{sq} couple was shown to be markedly altered (by +60 mV) and the flavin dismutation constant increased by a factor x 20 , as shown in table 1.

The present potentiometric/spectrophotometric study confirms the drastic modifications of flavin redox behavior in the presence of pyruvate.

Flavins and Flavoproteins

Results

The reductive anaerobic potentiometric titration of flavin and cytochrome moieties of flavocytochrome b_2 presents the profile shown in fig 1 when pyruvate is present at 10 mM. Midpoint potentials are respectively found equal to -25 mV for the cytochrome, +55mV for F_{ox}/F_{sq}, and -137mV for F_{sq}/F_{hq} . These direct potentiometric values confirm the order of magnitude of earlier estimates (cf table 1). As shown in fig 2, flavin spectra, do not vary in the whole interval 0 to -80 mV corresponding to the level of heme reduction between 35 and 80 % and are typical of red semiquinones as already underlined (5,6). ε values for bound flavin under all redox states are established here with uncertainty of ± 1 $mM^{-1}cm^{-1}$) as previously discussed (6). Best estimates are $\varepsilon_{490-570}$ = 4.6 $mM^{-1}cm^{-1}$ for the semiquinone and ε_{460} = 10 $mM^{-1}cm^{-1}$ for the oxidized form, both in the presence of pyruvate.

Table 1 : Midpoint potentials (mV) of heme and flavin of H-flavocytochrome b_2 at pH 7.00

[Pyruvate] ↓	t	$H_{ox/red}$	$F_{ox/sq}$	$F_{sq/hq}$	% F_{sq}	K_{dism} *1	Methods *2	Ref
Without	10°C	-25	-18	-42	40		P,S,E	*3
	18°C	(-25)	-20	-76	55	9	S,E	(5,6)
With 1 mM	"	(-25)	+14	-112	90	170	"	"
10 mM	"	(-25)	+35	-95	90	160	"	"
10 mM	"	-25	+55	-137	94	1000	P,E	this study

*1 : $K_{dism} = (F_{sq})^2/(F_{ox}) \times (F_{hq})$

*2 : Symbols : P for potentiometry, E for EPR estimate of F_{sq}, S for spectral study of F/H equilibria ; values for F are calculated from value for H in brackets.

*3 : Capeillère-Blandin & Bray : revised value (article in preparation).

Conclusions

Data in table 1 demonstrate that binding of the reaction product to the semiquinone form of such a dehydrogenase-e^-transferase, probably directly on the flavin (7), provokes a marked thermodynamic stabilization of the semiquinone, as shown by the K_{dism} value increasing by a factor of 20-100.

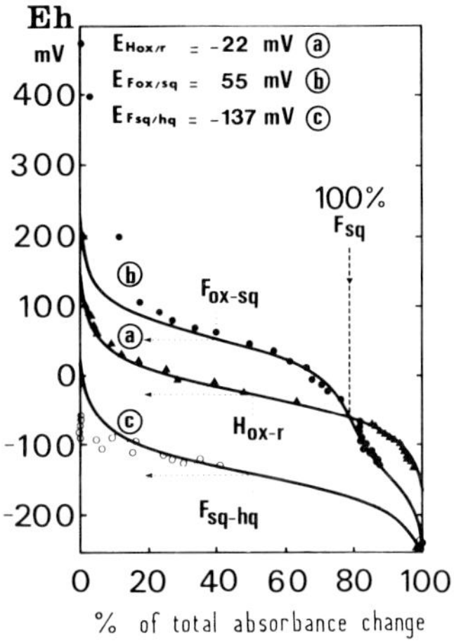

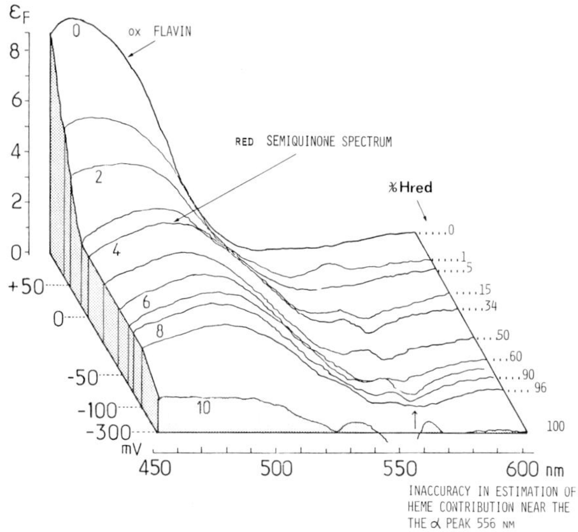

Fig 1: Spectrophotometric/potentiometric redox titration of flavin and cytochrome groups of flavocytochrome b_2 in the presence of 10 mM pyruvate (18°C).

Anaerobic redox titrations of flavocytochrome b_2 (35 μM in 100 mM phosphate buffer containing 10 mM pyruvate, final pH 7.00) are carried out, according to classical procedure, with the following mediators : Phenazine methosulfate: 3μM, Methylene blue: 1.4μM, Resorufine: 2μM, Naphtoquinone: 3μM. After an initial addition of ferricyanide (10 μM final), titration is carried out by addition of successive aliquots of L-lactate, each yielding a concentration increment of 2.5 μM, up to a total of 45 μM. Titration is completed by addition of aliquots of thiosulfite. Potentials and absorbance measurements at 6 wavelengths (λ-programm)(630,575,556,540,490,460nm) are performed after equilibration (1 to 4min). At critical phases during the titration, the whole spectrum is recorded and memorized. Flavin spectral contribution is calculated by substraction of heme contribution at each reduction level. Abscissa are for the different curves: a) % of absorbance variations at 556 nm, ie heme reduction level. b) % of flavin absorbance contribution at 460 nm ; 80% of the total variation corresponds to the F_{ox}/F_{sq} titration. c) % of flavin absorbance contribution at 490nm, near the isosbestic point of F_{ox}/F_{sq}. The curve shows therefore the titration F_{sq}/F_{hq} completed with thiosulfite. Solid lines are simulations based on Nernst formulation and values in table 1.

Fig 2: Evolution of flavin spectra during flavocytochrome b_2 anaerobic titration in the presence of 10 mM pyruvate.

Flavin spectra are computed from memorized partially reduced flavocytochrome b_2 spectra as described in (6). Baselines are taken as zero at 540 nm. Methylene blue contribution is visible at high E_h values between 550 and 600 nm. Intermediate spectra (35-90 % H_{red}) correspond to a red F_{sq}.

Since results at 1 and 10 mM pyruvate are not significantly different, the affinity of pyruvate for the flavosemiquinone enzyme must be higher than for the oxidized form (K_d = 5 mM)(unpublished) and similar to that of the substrate for the oxidized enzyme (K_m = 0.9 mM) .

The non-competitive inhibition exerted by pyruvate could be explained by the accumulation of an abortive flavosemiquinone-enzyme-pyruvate complex ; pyruvate, in this hypothesis, binds to an intermediate form of the enzyme, not to a site different from the substrate one.

This kind of thermodynamic control of the intermediate radical form and more generally of redox parameters governing the relative amounts of the three flavin species, in the presence of products at low and physiological concentrations, could be much more general and play an essential role in the retrocontrol of enzyme activity by reaction products.

Indeed, such an alteration of flavin redox parameters has been described by Gutman et al (8) for succinodehydrogenase in the presence of oxalacetate (not the direct product but a further keto-product in the citric cycle) and by Iyanagi et al (9) for NADH-cytochrome b_5 reductase in the presence of excess NAD^+.

References

1.Hemmerich, P & Massey, V. : Oxidases and related redox systems (King, ed) Pergamon Press, (1982) pp. 379-405.

2.Labeyrie, F. : Flavins and Flavoproteins (Massey, V. & Williams, C.H., eds) Elsevier/North Holland, Amsterdam, (1982) pp. 823-832.

3.Capeillère-Blandin, C., Barber, M.J. & Bray, R.C. ibidem, pp 838-843.

4.Tegoni, M., Silvestrini, M.C., Labeyrie, F. & Brunori, M. : Eur. J. Biochem. 140, 39-45 (1984).

5.Tegoni, M., Janot, J.M., Silvestrini, M.C., Brunori, M. & Labeyrie, F. : Biochem. Biophys. Res. Comm. 118, 753-759 (1984).

6.Tegoni, M., Janot, J.M. & Labeyrie, F. : Biochimie 66, 127-134 (1984).

7.Ghisla, S., Massey, V. & Choong, Y.S. : J. Biol. Chem. 254, 10662-10669 (1979).

8.Gutman, M., Bonomi, F.,Pagani, S., Cerletti, P. & Kroneck, P. : Biochim. Biophys. Acta 591, 400-408 (1980).

9.Iyanagi, T., Watanabe, S. & Anan, K.F. : Biochemistry 23, 1418-1425 (1984).

Tjump INVESTIGATION OF INTRAMOLECULAR ELECTRON EXCHANGES IN *Hansenula anomala* YEAST FLAVOCYTOCHROME b_2, L-LACTATE CYTOCHROME c OXIDOREDUCTASE.

Mariella Tegoni and Françoise Labeyrie,
Centre de Génétique Moléculaire du CNRS, 91190 Gif-sur-Yvette (France) and

Maria Chiara Silvestrini and Maurizio Brunori,
Istituto di Chimica, Facoltà di Medicina e Chirurgia, 00185 Roma (Italy).

Introduction

The study of intramolecular electron transfers taking place between the two functional domains of *Hansenula anomala* flavocytochrome b_2 is a necessary complement in understanding the complex reaction of this enzyme (review 1) and might help for a more general comprehension of the dehydrogenase electron-transfer flavoproteins mechanism.

Results

The redox potentials of the two prosthetic groups , flavin and heme, are not far from each other and, in partially reduced enzyme under anaerobic conditions, equilibrium is reached within a few secondc (2). After rapid (nsec) perturbation of the equilibrium (Tjump) we have followed relaxation kinetics corresponding to heme reduction. Flavin hydroquinone and flavin semiquinone are possible electron donors to oxidized heme. The reactions are:

$$F_{hq} + H_{ox} \rightleftarrows F_{sq} + H_{red} \qquad \text{reaction 1}$$

$$F_{sq} + H_{ox} \rightleftarrows F_{ox} + H_{red} \qquad \text{reaction 2}$$

Tjump experiments ($\Delta t = 5°C$, final temperature = 16°C) have been carried out at different enzyme concentration and different heme reduction level, at pH 7.0 and 6.0, in the presence or in the absence of pyruvate, product of lactate oxidation. All relaxations observed were independent of the enzyme concentration, as expected for intramolecular processes, and independent of observation wavelength. The identification of the couple donor-acceptor for each process has been done on the basis of: i) comparison between experimental amplitude of the relaxation at different wavelength and expected absorbance variation in the case of reaction 1 and 2 (fig.1) ;

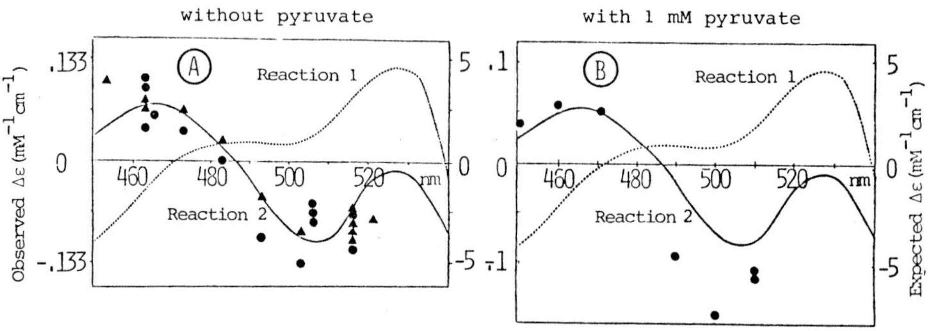

Fig.1 Comparison between Δ Absorbance expected for reaction 1 and 2 and experimental difference spectrum in the absence and in the presence of pyruvate. Experimental relaxation amplitudes (Abs.t=∞ - Abs.t=0) are taken as a function of λ at a constant % of heme reduction. Solid and dotted lines are the expected Δ A for reaction 1 and 2 (see text), calculated on the basis of known ε for each chromophore, as detailed in (3). Buffer solution was 100 mM phosphate, pH 7.0. A) Relaxation amplitudes at 45± 5% (●) and 65 ±10% (▲) heme reduction. B) Relaxation amplitudes at 25± 5% (●) heme reduction, in the presence of 1 mM pyruvate.

16°C	τ_1^{-1} (s^{-1})	τ_2^{-1} (s^{-1})	k_+ (s^{-1})	k_- (s^{-1})	REACTING SPECIES §
pH 6.0	~ 2 500		1 500	1 000	$F_{HQ} + H_{OX}$
		390	130	260	$F_{SQ} + H_{OX}$
pH 7.0		280 (σ = 70)	175	175	$F_{SQ} + H_{OX}$
pH 7.0 + 1 mM PYRUVATE		635 ± 85	140	500	$F_{SQ} + H_{OX}$

§ EPR AND DIFFERENTIAL SPECTROSCOPY RESULTS

Table 1: Reciprocal relaxation times (τ^{-1}, τ^{-2}) for the fast and slow phase are given for each experimental condition ; the corresponding values of k_+ and k_- are calculated on the basis of K_1 and K_2 given in fig. 2.

ii) comparison of the experimental amplitude at different level of heme reduction with the concentration of the possible donors during a titration (fig.2). Differential spectroscopy and EPR results have been used for both the comparison (Labeyrie et al., this book).

At pH 6.0 we have observed the more rapid process with $k_+ \sim 1500\ s^{-1}$ and $k_- \sim 1000\ s^{-1}$ at 16°C (Table 1). This exchange, visible only above 50% of heme reduction, has been assigned to the exchange between F_{hq} and H_{ox} (fig. 2A).

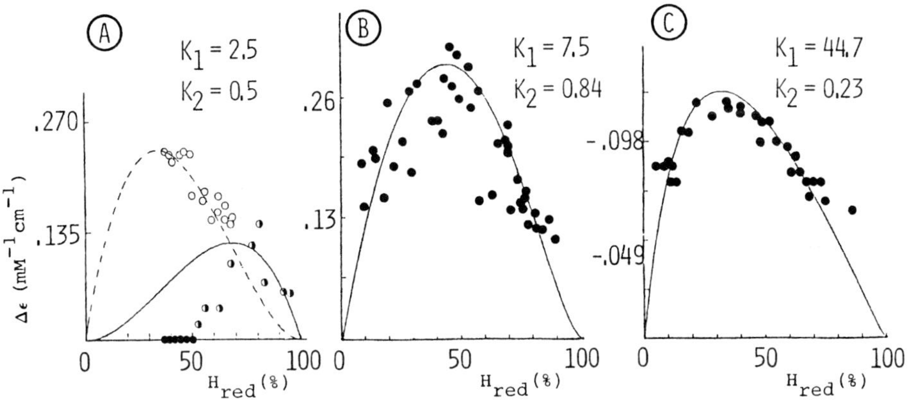

Fig.2 Dependence of the relaxation amplitude on the level of heme reduction. The amplitudes of relaxations ($\Delta\varepsilon\ mM^{-1}cm^{-1}$) are plotted as a function of estimated percentage of reduced heme (±10%). The solid lines represent the probability of finding in a same active center the couple donor-acceptor that is the product of the concentration of the reactants, calculated on the basis of K_1 and K_2, equilibrium constants for reaction 1 and 2, obtained by EPR measurements ; the values adopted for calculations are written on the corresonding part of the figure. A) (●) represent the amplitude of the fast phase $\tau^{-1} = 2500\ s^{-1}$, (O) is the total amplitude. The fast phase is visible above 50% heme reduction and is the only resolvable process above 75% heme reduction (◐). Solid line corresponds to $(F_{hq})\times(H_{ox})$ term ; dotted line is the product $(F_{sq})\times(H_{ox})$. Conditions are: 100 mM phosphate, pH 6.0; λ_{obs} = 557 nm. B) Solid line represents calculation of the term $(F_{sq})\times(H_{ox})$. Experimental points correspond to experiences in 100 mM phosphate, pH 7.0 ; λ_{obs} = 557 nm. C) Solid line is as in part B ; conditions are : 100 mM phosphate, pH 7.0, 1 mM pyruvate ; λ_{obs} = 510 nm.

A fast process has sometimes been noticed also at pH 7.0 in highly concentrated samples, but the amplitude was too small to allow a characterization. The rate constants found at pH 6.0 are compatible with estimates from stopped-flow studies at pH 7.0 (k_+ 600 s^{-1} reevaluated at 16°C (2)).

At pH 7.0 a slower reaction has been observed and assigned, on the basis of its amplitude at different level of heme reduction (fig. 2 B,C) and of its kinetic difference spectrum (fig. 1 A,B), to the exchange between F_{sq} and H_{ox}. The rate constants calculated for this process largely depend on the presence of pyruvate, the reaction product ; in particular k_- is 3 times greater than k_+ in the presence of 1 mM pyruvate (Table 1).This result suggest a possible mechanism of activity regulation of pyruvate by direct binding to the semiquinone form of the flavin. Similar kind of regulation has already been described in literature (4,5) and is discussed elsewhere in this book (Labeyrie et al.).

Conclusion

Previous stopped-flow results allowed the unambiguous evaluation of the rate constant in the electron exchange from F_{hq} to H_{ox} within active sites since this step was obligatory and partially rate limiting in enzyme reduction by lactate (2). This process at 16°C was too fast and of too small amplitude (ΔH) to be followed in the present Tjump study, except at pH 6.0. Contrastingly, the exchange between F_{sq} and H_{ox} has been well characterized : rate constants are much smaller (one order of magnitude) for this process and k_+ value is practically unaffected by the presence of pyruvate. Since we can assume that the relative position of the two prosthetic groups is independent of redox state of the flavin, the redox potential of the couple F_{hq}/F_{sq} in comparison with that of F_{sq}/F_{ox} couple can account for the greater reactivity of F_{hq}, as electron donor to oxidized heme .

References

1. Labeyrie, F. : Flavins and Flavoproteins (Massey, V. & Williams, C.H., eds.) pp 823-832, Elsevier/North Holland, Amsterdam,1982.
2. Capeillère-Blandin, C.,Barber, M.J. & Bray, R.C.: ibidem pp 838-843.
3. Tegoni, M., Silvestrini, M.C., Labeyrie, F. & Brunori, M.: Eur.J.Biochem. 140, 39-45 (1984).
4. Gutman, M.& Sinlan, N.: Flavins and Flavoproteins (Singer, T.P. ed.) pp. 537-545, Elsevier, Amsterdam, 1976.
5. Iyanagi, T.: Biochemistry, 16, 2725-2730, 1977.

A PULSE RADIOLYSIS STUDY OF FLAVOCYTOCHROME b_2 : DIFFERENCES IN REACTIVITY WITH CARBOXYLATE RADICALS BETWEEN FLAVIN AND HEME b_2.

Chantal Capeillère-Blandin
Centre de Génétique Moléculaire du CNRS, F-91190 Gif-sur-Yvette (France)

Christiane Ferradini
Laboratoire de Chimie Physique, 45 rue des Sts Pères, F-75270 Paris Cédex 06

Introduction

Reduction processes of flavocytochrome b_2 by a 2-electron donor, L-lactate, have already been studied at equilibrium and in the course of rapid kinetic studies (1,2). The electron transfer sequence starts by the 2-electron reduction of flavin by bound lactate within the flavodehydrogenase moiety followed by a reversible intramolecular 1-electron transfer between flavin hydroquinone and heme b_2, the latter belonging to the cytochrome b_2 domain with the concomitant formation of flavin semiquinone. Further reversible intramolecular electron transfer between semiquinone and cytochrome b_2 has been shown in T-jump studies on partially reduced enzyme (3). To bring additional information to the understanding of electron transfer processes inside such a multi-centered redox proteins we studied transient states of Hansenula anomala flavocytochrome b_2 obtained by reduction with a 1-electron donor of low potential, at a 1000 fold higher time resolution than with conventional rapid mixing techniques, using the radical $CO_2^{\cdot-}$ generated by radiolysis.

Results

1) Transient absorbance difference spectra. We carried out a detailed spectral investigation (280-600 nm) of the reduction products of oxidized

Flavins and Flavoproteins

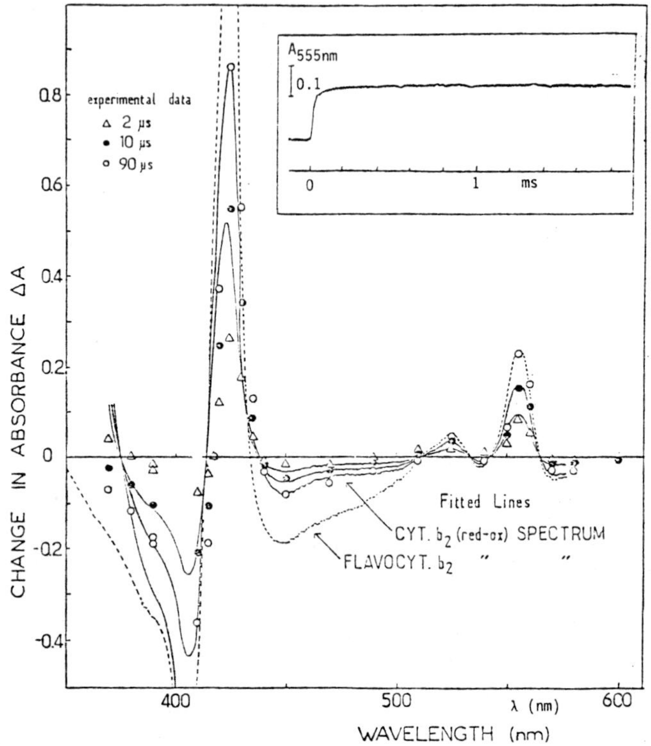

Fig.1. Total absorbance changes measured 2, 10 and 90 μs after the pulse in the course of the reaction between flavocytochrome b_2 and $CO_2^{\bar{\cdot}}$ in 2 mM phosphate buffer pH 7.0 and 0.16 M formate solution saturated with N_2O at 20°C. Solid lines passing through the points (600-400 nm) represent the (red-ox) spectra of cytochrome b_2 obtained by normalisation in the α band for the different reaction times. The dotted line corresponds to the (red-ox) spectrum of flavocytochrome b_2 normalized at 90 μs. Insert : absorbance recording of the heme b_2 reduction followed up to 2 ms at 555nm.

flavocytochrome b_2 (20 μM) by $CO_2^{\bar{\cdot}}$ produced at a (450∓50)μM concentration in less than 1 μsec (4). The transient absorbance difference spectra (Fig. 1) above 400 nm agreed quantitatively with the characteristic spectral features of heme b_2 as detected in the difference spectra of cytochrome b_2 core, isolated from flavocytochrome b_2. Below 400 nm the additional absorbance of the $CO_2^{\bar{\cdot}}$ prevents the fit. Over the whole spectrum no further changes were observed up to 2 ms : in particular at 0.3 ms when $CO_2^{\bar{\cdot}}$ has been consumed,heme b_2 was 47 % reduced and remained apparently at this level (Fig.1 insert). Nevertheless the formation of 20 % semiquinone could

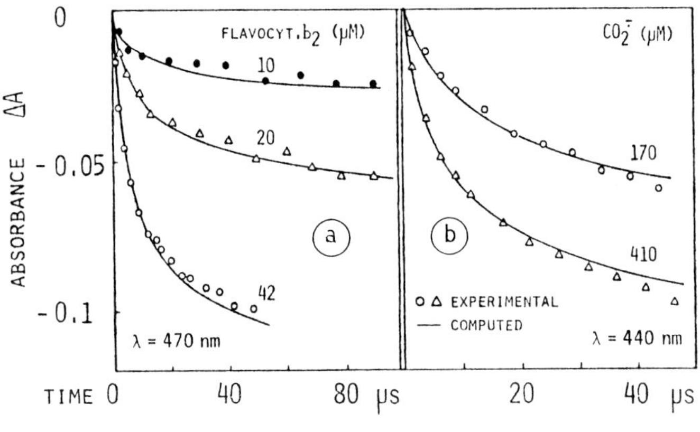

Fig. 2 : Dependence of the reaction time-course on the initial concentrations of oxidized flavocytochrome b_2 (a) and $CO_2^{\cdot-}$ (b). The solid lines have been computed using scheme I for the following best-fit parameters : $2k_1 = 1.1 \times 10^9\ M^{-1}s^{-1}$ and from top to bottom in (a) $k_2 = 2.2 \times 10^8\ M^{-1}s^{-1}$ with $\Delta\varepsilon = -5$, -5.5 and $-5.3\ mM^{-1}cm^{-1}$; in (b) $k_2 = 2.2 \times 10^8\ M^{-1}\ s^{-1}$ with $\Delta\varepsilon = -4.2\ mM^{-1}cm^{-1}$ and $k_2 = 2.4 \times 10^8 M^{-1}s^{-1}$ with $\Delta\varepsilon = -3.6\ mM^{-1}cm^{-1}$.

escape absorbance detection. Therefore it appears that the bound heme b_2 is preferentially reduced by $CO_2^{\cdot-}$.

2) Kinetic analysis of the reaction. The proportion of reduced flavocytochrome b_2 determined a constant time after the pulse ($t = 50\ \mu s$) were found, to be independent of the observation wavelength (fig. 1) but to vary with initial concentrations of $CO_2^{\cdot-}$ (170-500 μM) and of flavocytochrome b_2 (10-46 μM) (Fig. 2). These reaction time-courses were fitted using scheme I i.e taking into account two simultaneous second-order reactions: the self-recombination of $CO_2^{\cdot-}$ ($2k_1$) and the reduction of heme b_2 bound to flavocytochrome b_2 by carboxylate radical (k_2).

The fitting of the computed curves to the experimental data was obtained using absorbance difference coefficients within the experimental uncertainty limits for cytochrome b_2 : $\Delta\varepsilon_{440} = -3.2(\mp 0.7)\ M^{-1}cm^{-1}$; $\Delta\varepsilon_{470} = -4.6(\pm 0.5)\ M^{-1}cm^{-1}$ and reduction second-order rate constant $k_2 = (2.1 \pm 0.2) \times 10^8\ M^{-1}s^{-1}$. The close fit between experimental points and computed curves (Fig.2) strengthens the validity of such a kinetic model involving two second-order processes.

Discussion

In free enzyme molecules the protein-bound heme b_2 is directly reduced by $CO_2^{\bar{\cdot}}$ while the protein-bound flavin does not seem to be significantly reduced. These differences in reactivity with $CO_2^{\bar{\cdot}}$ between flavin and heme b_2 might indicate surface structural differences affecting the relative exposure of each prosthetic group to the solvent. Whereas the heme b_2 crevice would be accessible the flavin isoalloxazine ring would be buried, prevented from solvent contact by a compact folding of the polypeptide chain.
The mechanism of reduction of flavocytochrome b_2 by $CO_2^{\bar{\cdot}}$ which proceeds first by the heme b_2 attack, is different from that proposed for the catalytic electron transfer reaction where the reduction of flavin by bound lactate is the first step followed by an electron distribution among both prosthetic groups (1,2). The absence of detectable electron transfer between reduced heme b_2 and flavin, here observed on a short time scale $\leqslant$ 10 ms, points out that under our experimental conditions such a transfer is slowed down and no longer equal to 200 s^{-1} (3). The modification of the reactivity of flavin might be considered as an evidence of the existence of a possible mobility of the two structural domains of the molecule : in the free enzyme molecules random molecular motions could prevent the two electroactive groups from reaching the right position and orientation allowing electron tunneling. On the contrary when the substrate (or product) is bound the two domains may be subjected to constraints which would reduce the amplitude of such motions and thus might facilitate the electron transfer. However the use of high ionic strength conditions can be taken into consideration.

References

1. Capeillère-Blandin, C., Barber, M.J., Bray, R.C. : Flavins and Flavoproteins (Massey, V. & Williams, C.H., eds) Elsevier N.Y. (1982) 838-843.
2. Labeyrie, F. : ibidem, 823-832.
3. Capeillère-Blandin, C., Pucheault, J., Ferradini, C. : Biochim. Biophys. Acta, 786, 67-78 (1984).
4. Tegoni, M., Silvestrini, M.C., Labeyrie, F. Brunoni, M. : Eur. J. Biochem. 140, 39-45, (1984).

THE DISTRIBUTION OF REDUCING EQUIVALENTS AMONG SOME SPECIES IN SUCCINATE DEHYDROGENASE UPON REDUCTION OF SUCCINATE.

Silvia Pagani and Franco Bonomi

Department of General Biochemistry, University of Milan
I-20133 Milano

Introduction

The present communication reports the results of equilibrium studies on the redox state of covalently-bound flavin and of the Fe-S center S_1 in succinate dehydrogenase (SDH, E.C.1.3.99.1.), when low concentrations of substrate are interacting with the ligand-free, oxidized protein. The effect of the pH on the equilibria involving these species was followed in an attempt to get a deeper insight on the mechanism of the enzyme catalytic activity and of internal electron transfers.

Experimental

SDH was prepared in the presence of succinate and precipitated with ammonium sulphate at the gel eluate purification step (1). The succinate was removed by gel filtration on Sephadex G 25 column equilibrated with 50 mM Hepes-Tris pH 7.41 and the enzyme stored as 1.0 ml aliquots in liquid nitrogen. As judged by EPR and spectrophotometric measurements (2,3) the enzyme preparation is still 15% reduced. The enzyme activity was tested at 15°C by using PMS and DCIP as external acceptors: the catalytic parameters were affected by the ionic strenght of the Hepes-Tris buffer used, likely as a consequence of competitive inhibition by the Hepes-anion as described

Flavins and Flavoproteins

in (3). Constant ionic strenght buffers (0.025 M Hepes-) were therefore used. Fig. 1A shows the results of the kinetic experiments, fig. 1B and C

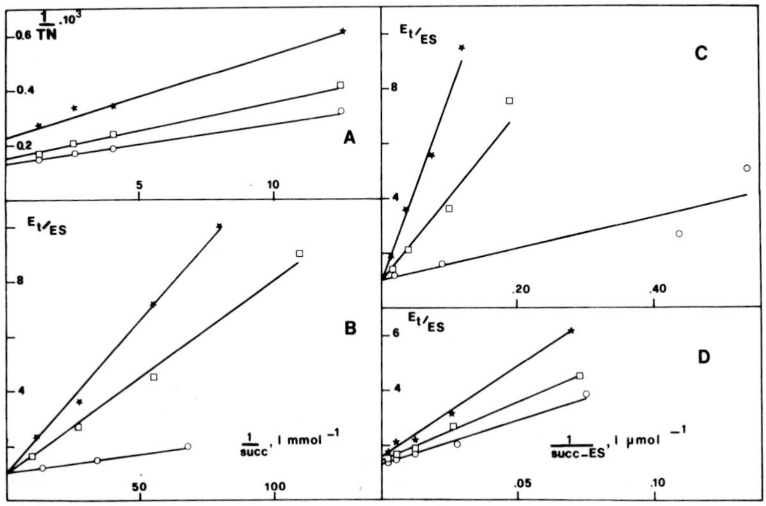

Fig. 1 - Double reciprocal plots of -A: catalytic activity of SDH with PMS and DCIP, -B: semiquinone accumulation, -C: S_1 reduction, -D: flavin reduction. ES/E_t is the molar fraction of the enzyme showing the investigated property. (✶) pH 7.00; (□) pH 7.50; (○) pH 8.00.

show the effect of succinate on the flavin semiquinone accumulation and on the reduction of S_1, as detected by the size of their EPR signals at liquid N_2 (4). Fig. 1D relates the decrease of absorbance in the 450-470 nm region (i.e. the disappearance of oxidized flavin) to increasing succinate concentration. Succinate concentration was kept as low as possible in order to exclude the involvement of the regulatory site (4). The titration of the visible absorbance of SDH with succinate is not monophasic. Tight isobestic points and a good linear correlation between $\frac{1}{red}$ vs $\frac{1}{S}$ were observed at all the pH values tested in the 0.0-0.2 mM succinate range. From the results summarized in fig. 1 two main parameters can be determined -i the value of the measured parameters at saturating succinate (P_{max}) and -ii the value of

K_d^{succ} (i.e. the apparent dissociation constant of the enzyme-succinate complex featuring the investigated properties). Table 1 shows the pH dependence of these values and reveals similarities between, respectively, flavin reduction and enzyme activity on one hand, and reduction of S_1 and

TABLE 1 - K_d (μM) for:

	activity	flavin reduction	flavin semiquinone accumulation	S_1 reduction
7.00	76	58	115	67
7.41			55	31
7.50	55	34		
7.85		21		
8.00	42		10	6

semiquinone accumulation, on the other hand. The acid-base properties of either succinate or fumarate are unlike to be involved in the decrease of K_d^{succ} at the pH values tested (5). Results can be explained taking into account dissociation of some acid group of SDH, the covalently-bound flavin being the most suitable one, as already suggested on a completely different approach (6). Accumulation of the flavin semiquinone appears to arise from the reaction $(E_{ox} + succ) \longleftrightarrow (E?sq) + H^+$ which is consistent with the stabilization of the anionic form of the flavin semiquinone in SDH (6,7) The scheme outlined for the pH dependence of the semiquinone formation also applies for the reduction of S_1. Since no evidence exists that the pH dependence of S_1 reduction is an intrinsic feature of the cluster (8,9), it must be attributed to some other reducing compounds, whose deprotonated forms are reducing S_1. Experimental observation are combined in the following scheme:

$$2e^- + E\text{-}F + 2H^+ \rightleftharpoons EFH_2 \rightleftharpoons E\text{-}FH^- + H^+$$

$$EF + EFH^- \rightleftharpoons 2EF^{\cdot -} + H^+ \text{(intermolecular equilibration)}$$

$$EFH^- + S_1^{ox} \rightleftharpoons S_1^{red} + EF^{\cdot -} + H^+ \text{(intramolecular equilibration)}$$

Acknowledgement

Work supported in part by grants from Italian Ministero della Pubblica Istruzione (MPI), Rome, project: "Struttura e funzione delle proteine: metodi di indagine".

References

1. Cerletti, P., Zanetti, G., Testolin, G., Rossi, C., Rossi, F., Osenga, G.: in Flavins and Flavoproteins (Kamin,H., ed.) University Park Press, Baltimore, pp.629-647 (1971).

2. Bonomi, F., Pagani, S., Cerletti, P.: in Flavins and Flavoproteins (Yagi,K. and Yamano, T., eds.) University Park Press, Baltimore, pp. 227-235 (1979).

3. Bonomi, F., Pagani, S., Cerletti, P.: Eur. J. Biochem. 119, 307-310 (1981).

4. Bonomi, F., Pagani, S., Cerletti, P., Giori, C.: Eur. J. Biochem. 134, 439-445 (1983).

5. Bonomi, F., Pagani, S.: this meeting.

6. Ohnishi, T., King, T.E., Salerno, J.C., Blum, H., Bowjer, J.R., Maida, T.: J. Biol. Chem. 256, 5577-5582 (1981).

7. Ackrell, B.A.C., McIntire, W., Edmonson, D.E., Kearney, E.B.: in Flavins and Flavoproteins (Massey, V. and Williams, C.Jr., eds.) Elsevier/Noth Holland, New York, pp. 488-491 (1982).

8. Ohnishi, T., Salerno, J.C., Winter, D.B., Lim, J., Yu, C.A., Yu, L., King, T.E.: Biol. Chem. 251, 2094-2104 (1976).

9. Beinert, H., Ackrell, B.A.C., Kearney, E.B., Singer, T.P.: Eur. J. Biochem. 54, 185-194 (1975

IONIC SPECIES OF THE FLAVIN AND THE CATALYTIC CYCLE OF SUCCINATE DEHYDROGENASE

Franco Bonomi and Silvia Pagani

Dept. of General Biochemistry, University of Milan, Via Celoria 2, I-20133 Milano, Italy.

Introduction

The redox properties of the electron transferring groups in succinate dehydrogenase (SDH; E.C.1.3.99.1) are modified when the protein interacts with a variety of positive and negative effectors (1). The values of E_m for the mono- and bielectronic transitions of the flavin in the ligand-free enzyme were also found to be pH-dependent (2) as a consequence of the acid-base properties of the reduced flavin and of the flavin semiquinone. The present study reports an investigation of the acid-base properties of the flavin in the complex between SDH and succinate or fumarate.

Experimental

Fig. 1 shows the effects of pH on the visible spectrum of SDH, at the gel-eluate stage of purification (3) in the presence of succinate only or of succinate and fumarate in a 1 to 10 ratio, a condition in which the enzyme is more than 90% oxidized at pH 7.5 (4). In both cases, a decrease in pH results in some oxidation of the flavin. This can be due to a decreased affinity of the enzyme for succinate, an increased affinity for fumarate or to modifications ensuing from the acid-base properties of either the flavin or its environment. The affinities of SDH for the two dicarboxylic acids were measured from activity experiments, which gave the

Flavins and Flavoproteins

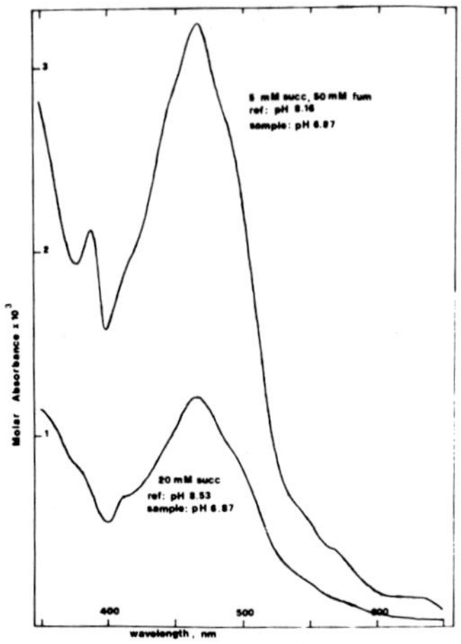

Fig.1 Difference spectra of 0.028 mM SDH in 180 mM Tris/Hepes in the presence of the given concentrations of dicarboxylic acids. The pH was adjusted to the given values with 1N acetic acid or NaOH.

results shown in Fig. 2 A and from which the following overall activity scheme can be drawn:

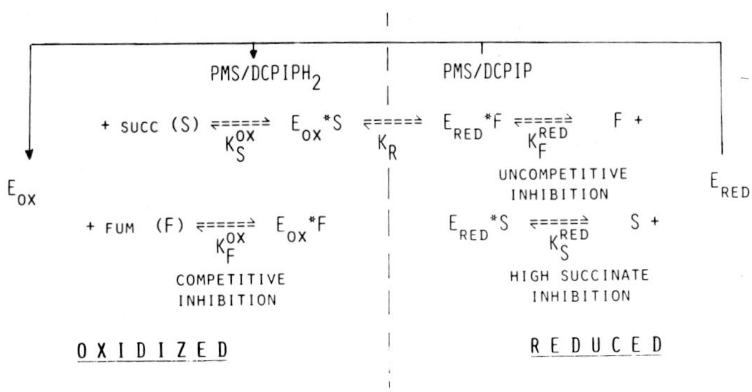

Spectrophotometric titrations of the visible absorbance spectrum of SDH were performed using the succinate/fumarate couple as the redox titratant in the pH range from 6.5 to 8.5 . Results were corrected for the affinity of SDH towards the acids and allowed to calculate $K_R = 4.02$ and $pK_a^R = 7.90$, as shown in Fig. 2 B.The exchange of reducing equivalents, which formally occurs between E-F and the reduced species ($E\text{-}FH_2$ + $E\text{-}FH^-$), produces the semiquinone, which could account for as much as 40% of the total flavin and is present in SDH only as the "red" $E\text{-}F^{\cdot -}$ anion (1,5) or its neutral

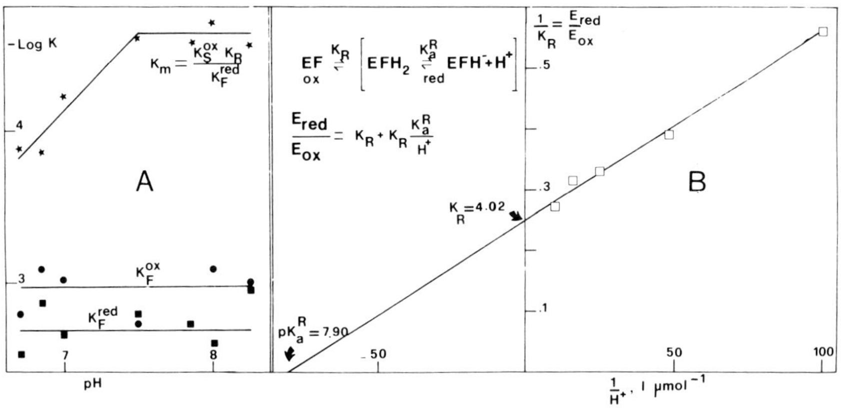

Fig.2 pH-dependence of some of the equilibria involved in SDH activity. A - Activity experiments were performed at 15°C with 2mM PMS and 0.08 mM DCPIP as acceptors. B - E^{red}/E^{ox} was determined by measuring at each pH value the fumarate/succinate ratio at which the flavin absorbance is half of the maximum. These raw values were then corrected for the relative affinities of the two dicarboxylic acids.

form (1,6). Succinate/fumarate redox titrations of the EPR-detectable semiquinone were performed at different pHs, as well as acid-base titrations of SDH in the presence of different fixed concentrations of the dicarboxylic acids. Results of these latter experiments are presented in Fig. 3. Semiquinone formation was observed at pH values where the ionized species $E\text{-}F^{\cdot -}$ is not expected to exist, and the semiquinone does not longer accumulate above pH 7.0 in the succinate-reduced enzyme or above pH 7.5 for the fumarate-oxidized protein. These results can be explained assuming: a- protonation of $E\text{-}F^{\cdot -}$ at low pH to give the "neutral red" semiquinone and b-deprotonation of some aminoacidic residues at high pH, where the lost positive charge on the protein results in the lack of stabilization of the $E\text{-}FH^-$ and $E\text{-}F^{\cdot -}$ anions. The calculated pKa for these groups are 6.55 and 9.20 respectively for the fumarate oxidized enzyme.

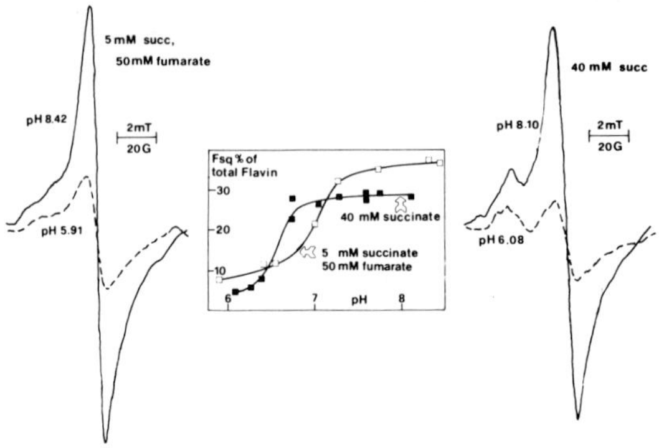

Fig.3 pH-dependence of the EPR signal of the flavin semiquinone in SDH. Liquid N_2 spectra of 0.033 mM SDH in the presence of the given concentrations of dicarboxylic acids and at the given pH values are reported. In the inset the titration curves for the semiquinone accumulation are given.

Acknowledgements. Work supported in part by grants from the MPI, Rome, Italy on the project "Struttura e funzione delle proteine: metodi di indagine".

References.

1) Bonomi,F., Pagani,S., Cerletti,P. and Giori,C. Eur.J.Biochem. 134, 439-445 (1983)

2) Ohnishi,T., King,T.E., Salerno,J.C., Blum,H., Bowjer,J.R., Maida,T. J.Biol.Chem. 256, 5577-5582 (1981)

3) Cerletti,P., Zanetti,G., Testolin,G., Rossi,C., Rossi,F., Osenga,G., in "Flavins and Flavoproteins"(Kamin,H. ed.) University Park Press, Baltimore, pp. 629-647 (1971)

4) Gutman,M., Bonomi,F., Pagani,S., Cerletti,P., and Kroneck,P.H. Biochim.Biophys.Acta 591, 400-408 (1980)

5) Ackrell,B.A.C., McIntire,W., Edmonson,D.E., Kearney,E.B. in "Flavins and Flavoproteins" (Massey,V. and Williams,C.H.,Jr eds.) Elsevier/NH, New York pp. 488-491 (1982)

6) Massey,V. and Hemmerich,P. Biochem.Soc.Trans. 8, 246-257 (1980)

ESR SPECTROSCOPIC STUDIES OF SUCCINATE DEHYDROGENASE AND FUMARATE REDUCTASE FROM ESCHERICHIA COLI

Richard Cammack, Daulat S. Patil

Department of Plant Sciences, King's College,
68, Half Moon Lane, London SE24 9JF, U.K.

Caro Condon, Peter Owen

Department of Microbiology, Moyne Institute,
Trinity College, Dublin 2, Ireland

Stewart T. Cole

Groupe de Génie Génétique, Institut Pasteur,
75724 Paris CEDEX 15, France

and Joel H. Weiner

Department of Biochemistry, University of Alberta,
Edmonton, Alberta, Canada T6G 2H7

Introduction

Succinate dehydrogenase is an enzyme that is produced by *Escherichia coli* during aerobic growth. Fumarate reductase is used as a terminal electron acceptor during anaerobic growth. Although they are distinct enzymes, succinate dehydrogenase and fumarate reductase have similar molecular properties. Both consist of two hydrophilic subunits, of M_r approx. 70 and 30 kD, and two smaller hydrophobic subunits of M_r approx. 15 kD. Both contain FAD bound covalently to the 70 kD subunit, and iron-sulphur clusters. In this paper we compare the types of iron-sulphur clusters in the two enzymes, and their redox potentials.

Materials and Methods

Succinate dehydrogenase was extracted from *E. coli* membranes with Lubrol PX, and precipitated with a specific antiserum (1). The preparation contained four subunits and included a cytochrome b. Fumarate reductase was purified from *E. coli* as described by Dickie and Weiner (2). The enzyme used for ESR spectroscopy was isolated as the two-subunit form by dissociation of the two small subunits (frdC and frdD) on a phenyl Sepharose column. ESR spectroscopy and redox titrations were as previously described (3).

Results and Discussion

The ESR spectroscopic properties of the intact membranes of succinate-grown cells were identical to those of the immunoprecipitated succinate dehydrogenase preparation, showing that succinate dehydrogenase is the predominant iron-sulphur protein. The membrane preparations were subsequently used for measurements of the redox potentials of succinate dehydrogenase. The ESR spectra of fumarate reductase were similar to those previously reported for anaerobically-grown *E. coli* cells (4), and in purified fumarate reductase from *Vibrio succinogenes* (5).

The ESR spectra of *E. coli* succinate dehydrogenase and fumarate reductase showed common features. In the oxidized state they consisted of a signal at g = 2.02, analogous to that of Centre 3 of mitochondrial succinate dehydrogenase, which is probably a $[3Fe\text{-}xS]^{3+}$ cluster (6); and a nearly-axial signal in the reduced state at g = 2.03, 1.94, analogous to that of Centre 1 which is a $[2Fe\text{-}2S]^{1+}$ cluster. Signals from the flavin semiquinone radical were relatively weak.

In both enzymes the g = 1.93 signal underwent a considerable increase in electron spin-lattice relaxation rate at low potentials, causing an apparent increase in signal amplitude when measured under saturating conditions. This change corresponds to the reduction of "Redox Centre 2". However the nature of this redox centre is controversial (7). If it was measured at non-saturating microwave power, the intensity of the g = 1.93 signal did not appear to increase significantly during the reduction of Centre 2. Therefore, as noted by Albracht (8), Centre 2 does not appear to be a normal [2Fe-2S] cluster.

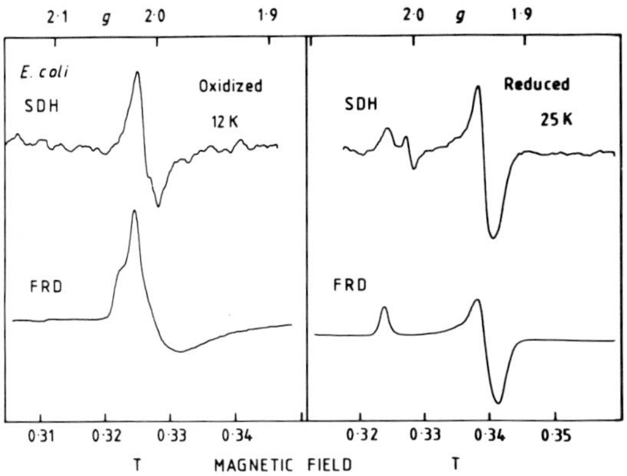

Fig. 1 ESR spectra of E. coli succinate dehydrogenase and fumarate reductase, oxidized (as prepared) and reduced with 2mM dithionite.

The midpoint potentials (E_m, pH 7.0) of the iron-sulphur proteins were measured by redox titrations in the presence of mediator dyes, monitored by ESR spectroscopy (3). The values obtained are summarized in Table 1, and compared with those of succinate dehydrogenase from beef heart and Micrococcus luteus. For succinate dehydrogenase, identical results were obtained from whole membranes and Lubrol extracts of membranes.

In fumarate reductase, the g = 1.93 signal showed an increase in electron spin relaxation rate at redox potentials between -20 and -70 mV, as well as below -330 mV (Fig. 2). This behaviour is similar to that observed in the succinate dehydrogenase of M. luteus (3), and might be attributed to a spin-spin interaction between Centres 1 and 3.

Table 1. Midpoint potentials of redox centres in succinate dehydrogenase (SDH) and fumarate reductase (FRD). Values are ±20mV

Enzyme	Centre 1	"Centre 2"	Centre 3
E. coli SDH	10	-175	65
E. coli FRD	-20	-330	-70
Beef heart SDH (9)	-5	-400	60
M. luteus SDH (3)	70	-295	10

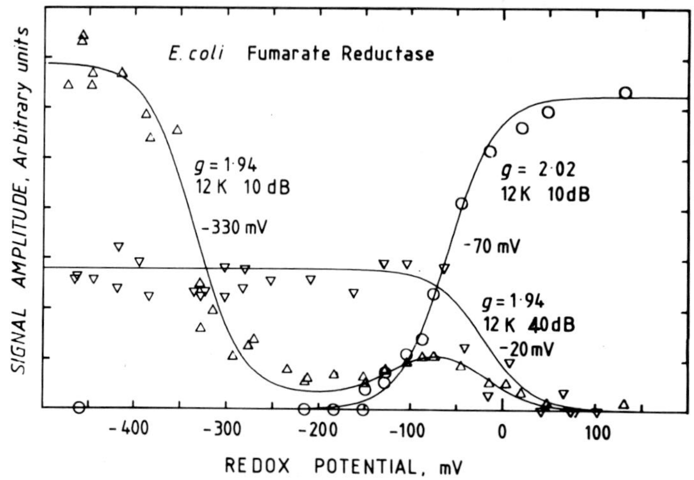

Fig. 2. Redox titration of E. coli fumarate reductase

The lower E_m of Centre 3 in fumarate reductase, compared with succinate dehydrogenase, may be related to the difference in function between the two enzymes, and possibly the use of menaquinone, which has a lower potential than ubiquinone.

Acknowledgements: This work was supported by a grant from the Science and Engineering Research Council.

References

1. Condon, C., Owen P.: FEMS Microbiol. Lett. 15, 109-113 (1982).
2. Dickie, P., Weiner, J.H.: Can. J. Biochem. 57, 813-821 (1979).
3. Crowe, B.A., Owen, P., Patil, D.S., Cammack, R.: Eur. J. Biochem. 137, 191-196 (1983).
4. Ingledew, W.J.: J. Gen. Microbiol. 129, 1651-1659 (1983).
5. Albracht, S.P.J., Unden, G., Kröger, A.: Biochim. Biophys. Acta 661, 295-302 (1981).
6. Ackrell, B.A.C., Kearney, E.B., Mims, W.B., Peisach, J., Beinert, H.: J. Biol. Chem. 259, 4015-4018 (1984).
7. Beinert, H., Albracht, S.P.J.: Biochim. Biophys. Acta 683, 245-277 (1982).
8. Albracht, S.P.J.: Biochim. Biophys. Acta 612, 11-28 (1980).
9. Ohnishi, T.: in Membrane Proteins in Energy Transduction, (Capaldi, R., ed.) pp. 1-87, Marcel Dekker, New York 1979.

FUMARATE REDUCTASE FROM *ESCHERICHIA COLI* REQUIRES THE *frd*C and *frd*D GENE PRODUCTS FOR QUINONE REDUCTASE ACTIVITY

Gary Cecchini, Brian A. C. Ackrell, and Edna B. Kearney

Department of Biochemistry and Biophysics, University of California, San Francisco, Calif. 94143 and Molecular Biology Division, VAMC, San Francisco, Calif. 94121

Robert P. Gunsalus

Department of Microbiology and Molecular Biology Institute, University of California, Los Angeles, Calif. 90024

Introduction

Fumarate reductase of *E. coli* is a complex flavoprotein bound to the inner surface of the cytoplasmic membrane, where it functions during anaerobic growth to reoxidize reduced menaquinone and reduce fumarate, the terminal electron acceptor. The enzyme will also oxidize succinate if appropriate electron acceptors are present. The *frd* operon encodes four polypeptides: two dissimilar subunits (A and B) of the enzyme and two small peptides, C and D of 15 kDa and 13 kDa, respectively, which are required for attachment of the enzyme to the membrane (1). The data reported here demonstrate a role for the C and D polypeptides in quinone reduction also, similar to that of peptides C_{II-3} and C_{II-4} in succinate-ubiquinone (Q) reductase of beef heart mitochondria (2).

Methods

The following plasmids were constructed. A 4.9-kb *Hind*III fragment from *E. coli* 1100 was ligated at the *Hind*III site of pBR322 to give the 9.2-kb plasmid pGC1002. This plasmid carries the genes coding for all four subunits of the fumarate reductase (FRD) complex. Plasmid pFRD23 contains the *Sal*I fragment of the *Hind*III insert ligated at the *Sal*I site of pACYC184;

Flavins and Flavoproteins

it carries the genes coding for the A and B subunits of the enzyme and also for a small part of the C peptide. Plasmid pFRD32 is derived by a *Sst*II reduction of pGC1002 to give a 7.8-kb plasmid coding only for peptides C and D. The succinate dehydrogenase mutant used in this study (TG414) was derived from *E. coli* 1100.

The FRD complex was extracted from pGC1002-amplified membranes with octyl-β-D glucopyranoside and purified by sucrose gradient centrifugation. Resolution of the complex was achieved by the method used with beef heart succinate-Q reductase, first extracting the enzyme with sodium perchlorate (3) and then treating the residue with Triton X-100 to solubilize together peptides C and D (2). Succinate-phenazine methosulfate (PMS) and succinate-DPB (2,3-dimethoxy-5-methyl-6-pentyl-1,4-benzoquinone) reductase activities were assayed as previously described (2).

Results and Discussion

A succinate dehydrogenase mutant (TG414) was transformed with plasmids carrying the genes for the complete FRD complex (subunits ABCD) or only for the enzyme (A and B) or peptides C and D, and grown aerobically on glycerol-fumarate medium (4) to minimize background due to chromosomally-encoded FRD complex. Cells were harvested ($OD_{600} \sim 1.0$), broken in the French press, and the membrane and cytoplasmic fractions prepared for assay.

Table I

Distribution of PMS and DPB Reductase Activities in Transformed Cells

	Membranes Reductase Activities*		Cytoplasm Reductase Activities*	
	PMS	DPB	PMS	DPB
TG414	8.1	5.3	0	0
TG414 + pGC1002	24.7	13.5	6.5	4.5
TG414 + pFRD23	2.9	1.5	42.6	0
TG414 + pFRD32	7.6	5.1	0.5	0

*Total units (μmol succinate oxidized/min at 38°). Values are normalized to 50 mg of cell protein before fractionation.

Transformation with pGC1002, which contains the entire *frd* operon, resulted in a 3-fold increase in the PMS and DPB reductase activities of the membrane fraction, as compared with the basal activities of the chromosomally-encoded complex in untransformed cells (Table I). Plasmid pFRD32, coding for peptides C and D alone, had no significant effect on the reductase activities. In cells transformed with the plasmid carrying genes for enzyme alone (pFRD23) and therefore unamplified in anchor peptides, the activity of the membrane fraction decreased markedly with both acceptors, with little change in the ratio (0.6) of the DPB and PMS reductase activities. Apparently the control level of expression of chromosomal FRD was suppressed in this experiment, possibly because of interference in membrane-assembly of the complex by the product of the partial C gene in this plasmid. The table shows that much enzyme was present in the cytoplasm of these cells, as measured by PMS reduction, but that it was devoid of DPB reductase activity. These data suggest that peptides C and D of the FRD complex serve not only to anchor the enzyme to the membrane (1) but that they also play a role in quinone reduction by the enzyme, like peptides C_{II-3} and C_{II-4} in succinate-Q reductase from beef heart (2). This was further investigated by recombination of resolved fractions enriched in either A and B or C and D. Table II shows the results of these experiments. It can be seen that the purified complex has a ratio of DPB:PMS reductase activity of about 0.6. This ratio was characteristic of several different preparations and is representative of the intact complex. When the complex is resolved by perchlorate the AB-enriched fraction loses DPB reductase activity. The residual DPB reductase activity in the AB- and CD-enriched fractions is

Table II

Succinate-PMS and-DPB Reductase Activities in the Reconstituted Complex

	PMS	DPB	DPB:PMS
FRD Complex (ABCD)	13.3	8.5	0.64
AB-enriched fraction	12.8	1.9	0.15
CD-enriched fraction	3.2	1.8	0.56
AB + CD fractions recombined*	11.6	6.9	0.60

The FRD complex was purified from membranes of *E. coli* 1100 transformed with pGC1002.

*Values given for the recombined fractions are those obtained after subtracting the activity of the AB and CD fractions measured separately.

due to some unresolved complex contaminating the fractions, as seen after SDS-PAGE (not shown). Recombination of the two fractions clearly restored the DPB reductase activity and the ratio of DPB:PMS reductase activities to ∿ 0.6. Thus functional reconstitution of the FRD complex was achieved on the recombination, and it may be concluded that peptides C and D function in full analogy with C_{II-3} and C_{II-4} of succinate-Q reductase in beef heart mitochondria. Moreover, just as FRD can function in reverse as succinate dehydrogenase, the 4-peptide complex could probably be either a reduced menaquinone-fumarate reductase or a succinate-Q reductase. It is noteworthy that in addition to having Fe/S redox centers and the same covalently bound flavin, the FRD from *E. coli* and the succinate dehydrogenase from beef heart cross-react immunologically: antibody to FRD peptide A reacts with the succinate dehydrogenase 70-kDa subunit and antibody to the succinate dehydrogenase 70-kDa and 30-kDa peptides reacted with the corresponding two subunits of FRD.

This investigation was supported by the Veterans Administration, by NIH Grant HL-16251 and NSF Grant PCM 81-10585, and by NSF Grant PCM 84-02974 (RPG).

References

1. Lemire, B.D., Robinson, J.J., Weiner, J.H.: J. Bacteriol. 152, 1126-1131 (1982).
2. Ackrell, B.A.C., Ball, M.B., Kearney, E.B.: J. Biol. Chem. 255, 2761-2769 (1980).
3. Baginsky, M.L., Hatefi, Y.: J. Biol. Chem. 244, 5313-5319 (1969).
4. Spencer, M.E., Guest, J.R.: Bacteriol. 114, 563-570 (1973).

GLUTAMATE SYNTHASE FROM *AZOSPIRILLUM BRASILENSE*

Sabina Ratti, Bruno Curti, Giuliana Zanetti

Dipartimento di Fisiologia e Biochimica Generali, Università di Milano, Milano, Italy

Introduction

Since the work of Tempest et al (1), glutamate synthase has been indicated as an enzyme partecipating in a new alternative pathway of ammonia assimilation. The enzyme (E.C.1.4.1.13) catalyzes the reductive transfer of the amino group from glutamine to α-oxoglutarate with the formation of two moles of glutamic acid, according to the following scheme:

$$\text{glutamine} + \alpha\text{-oxoglutarate} \;\xrightleftharpoons[\text{NAD(P)H} + \text{H}^{+} \;\rightarrow\; \text{NAD(P)}^{+}]{}\; 2\ \text{glutamate}$$

In prokariotes and in the yeast cells the enzyme, purified to a different degree of homogeneity, has been identified as an iron-sulfur flavoprotein which utilizes NAD(P)H as electron donor. It seemed to us interesting to study glutamate synthase also in a free-living diazotroph, the *Azospirillum brasilense*, a microorganism commonly associated with the roots of grasses. In diazotrophs, the glutamate synthase pathway is the preferred route for nitrogen fixation. In *Azospirillum* the level of glutamate synthase seems correlated to the level of

Flavins and Flavoproteins

nitrogenase expression (2), indicating a role for the glutamate synthase in the regulation of the nitrogen fixation cycle. Thus, we attempted the isolation of glutamate synthase from *Azospirillum brasilense Sp6*; some physico-chemical properties of the enzyme, purified almost to homogeneity, are reported and briefly discussed.

Results and discussion

In order to obtain cells with maximum specific activity of glutamate synthase, *Azospirillum brasilense Sp 6* was grown on a minimal medium (2) under limiting concentrations of ammonia. The purification procedure consisted of heat treatment, ammonium sulphate fractionations, ion exchange chromatography, gel filtration and affinity chromatography. The enzyme thus obtained was about 95% pure with a purification factor of 300-fold and a yield of 25%. The purified enzyme has a specific activity of 17 μmol NADPH oxidized.min^{-1}.mg^{-1}. If rapidly frozen in liquid nitrogen and stored at - 80°C, the enzyme is stable for several weeks. The molecular weight of the enzyme, as indicated by calibration of the Ultrogel AcA 22 column used at the fifth step of the purification procedure, was 740,000 + 10%. SDS gel electrophoresis of the enzyme gave two bands in equimolar amounts with a MW of 135,000 and 50,000, respectively. The glutamate synthase of *Azospirillum brasilense Sp6* thus seems an oligomer of four protomers, each being a dimer of non identical subunits.

HPLC analysis revealed the presence of one mole each of FAD and FMN for protomer; the iron and sulfur analyses gave about

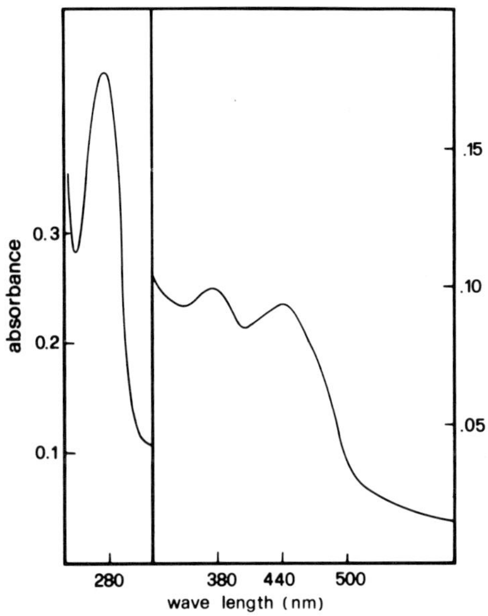

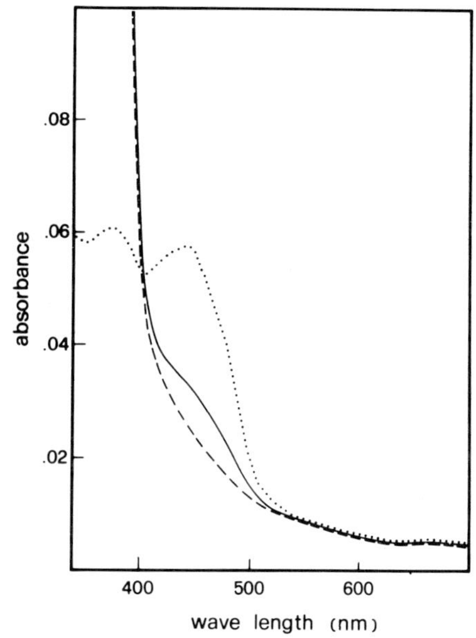

Fig. 1. - Absorption spectrum of glutamate synthase. Conditions: 0.388 mg.ml^{-1} enzyme in 25 mM HEPES, pH 7.5, containing 1 mM α-oxoglutarate, 1 mM EDTA, 5 mM β-mercapto ethanol.

Fig. 2. - Dithionite reduction of glutamate synthase. Conditions: (·····), oxidized enzyme, after the anaerobiosis was established; (——), immediately after Na_2SO_4 addition; (----), 30 min later.

8 atoms of Fe and 8 atoms of acid-labile S for protomer. The spectrum of the purified enzyme is reported in Fig. 1; the best preparations had a 278/440 absorbance ratio of 4.8. The enzyme is highly specific for NADPH as electron donor, as well as for glutamine, ammonia being less than 1% active as amino group donor. The pH optimum, with the standard substrates, is around 7.7. Catalytic activity was competiteve-

TABLE 1

KINETIC PARAMETERS OF GLUTAMATE SYNTHASE FROM *AZOSPIRILLUM BRASILENSE Sp6*

Substrate	Activity ($U.mg^{-1}$)	Substrate	K_M (μM)	K_I (mM)
NADPH + L-Glutamine	16.9	NADPH	6.04	
NADH + L-Glutamine	-	L-Glutamine	450.7	
NADPH + NH_4Cl	0.14	α-oxoglutarate	28.8	
		L-Methionine (a)		1.05

(a) Under standard substrates conditions.

ly inhibited by L-methionine with L-glutamine as substrate. A summary of the enzyme kinetic parameters is reported in Table 1: the data are very similar to values reported for glutamate synthase from other sources.

Earlier experiments of Miller et al (3) on anaerobic reduction of the *E. Coli* enzyme by NADPH, showed only partial reduction of the flavin spectrum, which was interpreted as the FAD being the species catalytically active. Using a NADPH regenerating system, we observed with our enzyme a complete reduction of the flavin spectrum (Fig. 3) similar to that observed with dithionite (Fig. 2) or by light irradiation

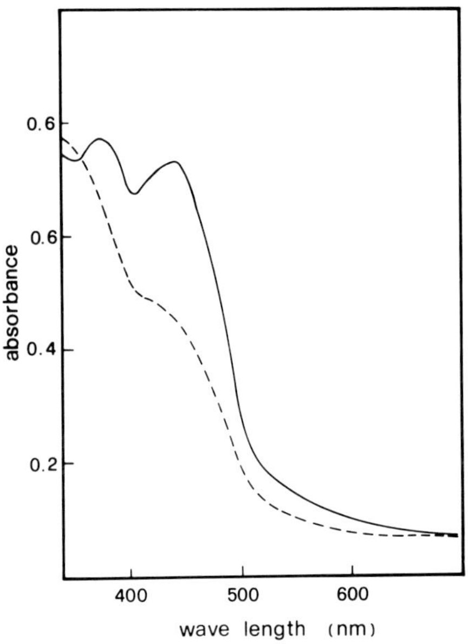

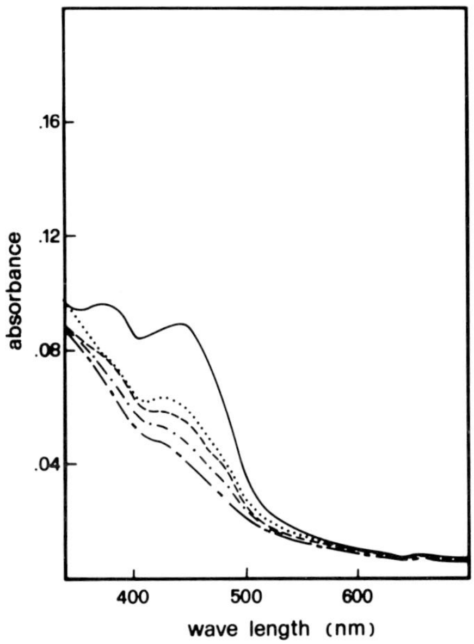

Fig. 3. - Reduction of glutamate synthase by a NADPH regenerating system. Conditions: 4.5 μM NADPH, 2 mM glucose--6-phosphate and 10 μg glucose-6-phosphate dehydrogenase. (——), oxidized enzyme; (- - -), after NADPH addition.

Fig. 4. - Photochemical reduction of glutamate synthase. Conditions: 8 mM EDTA, 0.5 μM 5-deazaflavin. (——), oxidized enzyme; (- -), 60 sec irradiation; (- · - · -), 90 sec; (- ·· -), 210 sec; (········), after adding 50 μl 0.1 mM $NADP^+$.

in the presence of EDTA + 5-deazaflavin (fig. 4). Thus NADPH is catalytically competent in reducing both the flavins. In the case of the NADPH reduction, full oxidized spectra were obtained only by addition of substrates after air admission to the spectrophotometric cell. In all cases, stable long wavelength intermediates were not observed along the reduction or the reoxidation path. Stopped flow and EPR experiments are required in order to ascertain the presence of semiquinone forms of the enzyme during catalysis.

References

1. Meer, J.L., Tempest, D.W., Brown, C.M.: J. Gen. Microbiol. 64, 187-194 (1970)

2. Barberio, C., Bazzicalupo, M., Federici, G., Gallori, E., Polsinelli, M.,: FEMS Microbiol. Letters 20, 361-364 (1983)

3. Miller, R.E., Stadtman, E.R.,: J. Biol. Chem. 247, 7407-7419 (1972)

OXIDATION-REDUCTION PROPERTIES OF GLYCOLATE OXIDASE

Charles Pace and Marian Stankovich

Department of Chemistry, University of Minnesota
Minneapolis, Minnesota 55455

Glycolate oxidase (GAO) catalyzes oxidation of α-hydroxy acids by O_2 to produce α-keto acids and H_2O_2. Previous studies of the isolated enzyme from pig liver have explored its physicochemical and steady-state kinetic properties (1-3). Particular attention was paid to the spectral and kinetic effects of mono- and dianions, e.g. Cl^-, oxalate. These anions are inhibitors except for P_i and arsenate which are activators (3).

The redox potential of GAO has not been reported. It has been suggested that the midpoint potential for native GAO should be more positive than -0.170 V, which is the midpoint potential of GAO in which FMN is replaced by 5-deazaFMN (4). Since the redox potential for the glycolate/glyoxalate couple is -0.086 V at pH 7 (5) and that for free FMN is -0.205 V (6), it seemed of interest to determine the oxidation-reduction properties of GAO and the effects of various anions upon these properties.

The spectroelectrochemical procedures have been described (7,8). Electrochemical reduction of GAO with methyl viologen (MV) as mediator of electron transfer produces the red radical form (9) of GAO (Figure 1). When absorbance data from experiments such as that shown in Figure 1 are plotted at selected wavelengths vs the amount of charge added to the system (inset to Figure 1), the intersections of the linear portions of these plots represent absorbance due to 100% radical formation. There is little difference in spectral characteristics of the radical formed in the presence of P_i or Cl^-. The maximum value obtained for the extinction coefficient at 375 nm is 18,000 M^{-1}-cm^{-1}.

The variation of the reduction potential of GAO as a function of pH is shown in Figure 2. The midpoint potential for the conversion of GAO radical (GAO_1) to fully reduced GAO (GAO_2) is more negative in 0.1 M P_i at

Flavins and Flavoproteins

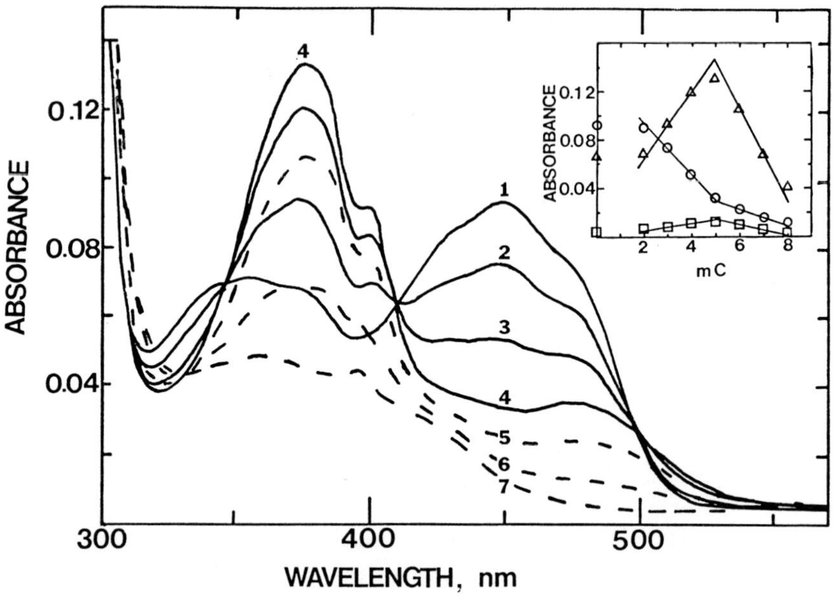

Figure 1. Absorption spectra from reductive coulometric titration of GAO (8.4 μM) at 10°C in 0.26 M triethanolamine-Cl at pH 9.2. Solid lines show radical increasing from the start of the titration (spectrum 1) to a maximum of 80% of the total FMN at 1.1 equivalents added per mole FMN (spectrum 4). Spectrum 2 corresponds to addition of a total of 3 mC. Dashed lines show the decrease in radical concentration which occurred between 1.1 and 2.1 equivalents added per mole FMN.
Inset. Absorbance vs charge at 375 nm (▲), 450 nm (●) and 520 nm (■).

pH 7 or pH 8 than in 0.1 M Cl^-, but the change from P_i to Cl^- makes no significant difference in the reduction potentials for the conversion of fully oxidized GAO (GAO_0) to GAO_1. From four experiments in 0.1 M P_i at pH 7.1, the reduction potential for conversion of GAO_0 to GAO_1 is -0.093 ± 0.013 V and that for conversion of GAO_1 to GAO_2 is -0.040 ± 0.009 V. From one experiment in 0.1 M imidazole-Cl, the values are -0.091 V and + 0.020 V, respectively.

The redox potential for conversion of GAO_0 to GAO_1 is more negative than that for conversion of GAO_1 to GAO_2 under all conditions tested. Radical stability increases with increasing pH (Figure 2), as indicated by the difference in potentials for the two one-electron steps (5). However, less than 30% radical is stable, even at pH 9. Higher percentage radical

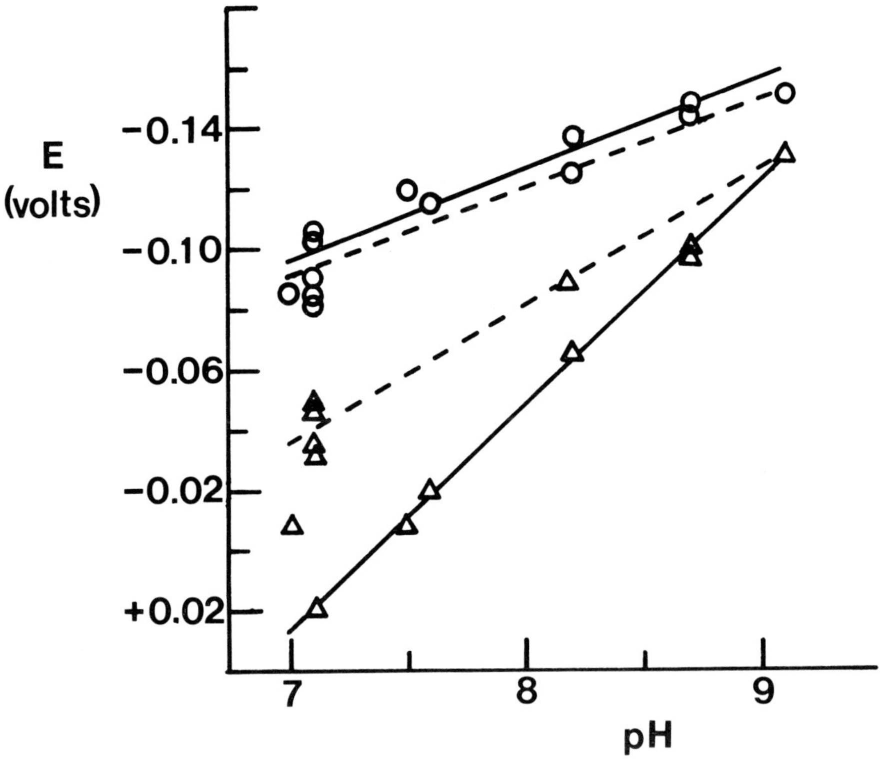

Figure 2. Reduction potentials for conversion of fully oxidized GAO to GAO radical (O) or GAO radical to fully reduced GAO (Δ) vs pH at 10°C. Lines are from least squares analysis of data from experiments in 0.1 M P_i (continuous lines) or 0.1 M Cl^- (dashed lines). The slopes of these lines in V/pH unit are 0.029 in P_i and 0.031 in Cl^- for the conversion of fully oxidized GAO to GAO radical. For the conversion of GAO radical to fully reduced GAO, they are 0.046 V/pH unit in P_i and 0.074 V/pH unit in Cl^-.

during coulometric titration is a kinetic phenomenon.

Both electron transfers in reduction of GAO are pH dependent. In 0.1 M P_i, the slope of the least squares plot indicates a stoichiometry for conversion of GAO_0 to GAO_1 of 0.52 proton per electron. In 0.1 M Cl^-, it is 0.54 proton per electron. For conversion of GAO_1 to GAO_2, the apparent stoichiometry is 0.82 proton per electron in 0.1 M P_i and 1.34 proton per electron in 0.1 M Cl^-. Transfer of one proton per electron during conversion of radical to fully reduced enzyme has been observed in this laboratory for L-amino acid oxidase (10) and D-amino acid oxidase (11). However the conversion of fully oxidized enzyme to radical was independent of pH for both amino acid oxidases. Variation of redox potential with pH is expected for interconversion of GAO_0 and GAO_1, especially at the alkaline end of the pH range examined here. The pK for dissociation of the 3-imino group of FMN is shifted from 10.35 for free FMN (6) to 8.0 for GAO (2). However, upward curvature of the pH dependence for interconversion of GAO_0 and GAO_1 is not apparent from the data in Figure 2.

Conclusions:

1. Upon binding to GAO, the redox potentials for FMN are shifted near that of the glycolate/glyoxalate couple and transfer of electrons from glycolate to FMN becomes thermodynamically feasible.
2. From pH 7 to pH 9, the reduction potential for conversion of GAO_0 to GAO_1 is more negative than that for conversion of GAO_1 to GAO_2 and the thermodynamic stabilization of radical is low.
3. Reduction of GAO in 0.1 M P_i exhibits no thermodynamic advantage over reduction of GAO in 0.1 M Cl^-.
4. Both electron transfers to GAO are pH dependent.

References:

1. Dickinson, F.M., Massey, V.: Biochem. J. 89, 53P (1963)
2. Schumann, M., Massey, V.: Biochim. Biophys. Acta 227, 500-520 (1971)
3. Schumann, M., Massey, V.: Biochim. Biophys. Acta 227, 521-537 (1971)
4. Jorns, M.S., Hersh, L.B.: J. Biol. Chem. 251, 4872-4881 (1976)
5. Clark, W. Mansfield: Oxidation-Reduction Potentials of Organic Compounds, Williams and Wilkins, New York 1960
6. Draper, R.D., Ingraham, L.L.: Arch. Biochem. Biophys. 125, 802-808 (1968)
7. Stankovich, M.T.: Anal. Biochem. New York 109, 295-308 (1980)
8. Stankovich, M, Fox, B.: Biochemistry 22, 4466-4472 (1983)
9. Massey, V., Palmer, G.: Biochemistry 5, 3181-3189 (1966)
10. Stankovich, M., Fox B.: Biochim. Biophys. Acta 786, 49-56 (1984)
11. Van den Berghe-Snorek, S., Stankovich, M.T.: submitted

STUDIES WITH THE FLAVIN DEPENDENT ALCOHOL OXIDASE FROM YEAST: PROPERTIES AND CATALYTIC MECHANISM

Johanna Geissler, Peter M.H. Kroneck and Sandro Ghisla
Fakultät für Biologie, Universität Konstanz, D-7750 Konstanz

Introduction

Alcohol oxidase (E.C. 1.1.3.13) from Candida boidinii (octamer, M_r subunit 74000, 1 FAD/subunit) catalyzes the oxidation of primary alcohols: $RCH_2OH + O_2 \rightarrow RCHO + H_2O_2$. The native enzyme stabilizes a FAD radical, $Fl^{\cdot -}$, even in the presence of oxygen and substrate (1,2). The amount of $Fl^{\cdot -}$ (approx. 30% of FAD of our enzyme) is increased 2.5 fold by hydroxylamine with concomitant quantitative inactivation of the enzyme (1). $Fl^{\cdot -}$ is considered to be catalytically inactive. Otherwise little is known on the catalytic mechanism: Cyclopropanol (2) and cyclopropanone (3,4) irreversibly inactivate the enzyme probably via radical pathways. In contrast to several other flavin oxidases (5) α-alkynoic substrates are not suicide inhibitors of alcohol oxidase: Upon turnover of propargyl alcohol inactivation is observed but is claimed to be due to reaction of the accumulating propynal with an active side residue (6).

Results and Discussion

The catalytic activity of alcohol oxidase is increased 1.7 fold in the presence of the redox dye radical $ABTS^{\cdot +}$ (fig. 1) which is produced via the peroxidase (POD) catalyzed oxidation of 2,2'-azino-bis-(3-ethylbenzthiazoline-sulfonic acid-6) ($ABTS_{red}$) by H_2O_2 (7). $ABTS^{\cdot +}$ might react with $Fl^{\cdot -}$ to give $ABTS_{red}$ and Fl_{ox}. The electron spectrum of the activated enzyme also indicates the loss of $Fl^{\cdot -}$ upon $ABTS^{\cdot +}$-treatment (fig. 1). This sup-

Flavins and Flavoproteins

ports the view of $Fl^{\cdot -}$ being catalytically inactive in alcohol oxidase. According to steady-state kinetic experiments (data not shown) the catalytic mechanism includes a ternary complex of alcohol oxidase with alcohol and oxygen. The K_M-values are 1.4 mM (8mM) for methanol ($[^2H_3]$-methanol) and 0.6 mM for oxygen. Steady-state and transient kinetics are accounted for by an overall kinetic scheme (scheme 1) as proposed for lactate oxidase (8). It is assumed that $k_{-2} << k_2$. As K_D (i.e. k_{-1}/k_1) is not very different from K_M(ROH) $k_2 << k_{-1}$ is implied (eq. III). The values for k_5 are calculated from eq. I. The observed rate constant k_{ox} for $E\text{-}Fl_{red} + O_2 \rightarrow E\text{-}Fl_{ox} + H_2O_2$ probably corresponds to k_6: Enzyme monitored turnover indicates a rate of approx. $7 \times 10^6\ M^{-1}min^{-1}$ for k_4 (9), which governs the reoxidation rate in catalysis (ternary complex mechanism).

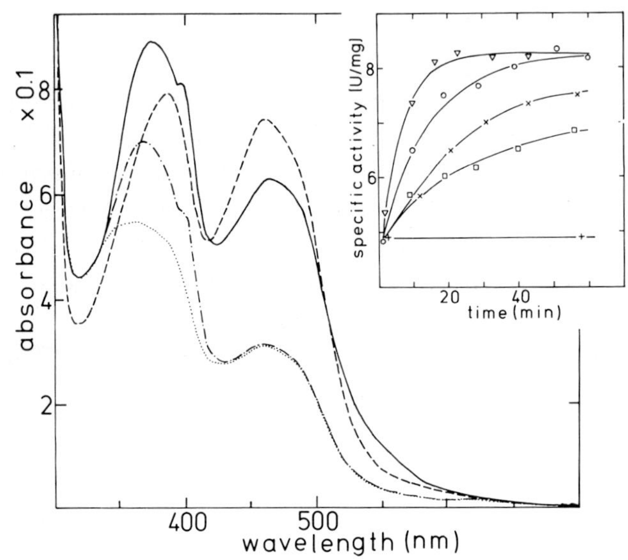

Fig. 1: Electron spectra of native and activated alcohol oxidase enzyme, 90µM, incubated for 19 h, 0°C, with $ABTS^{\cdot +}$, 2mM. After incubation the dye was removed by Sephadex G-25 gel filtration, 0.1M phosphate, pH 7.5. Activity (1 U = 1 µmole O_2/min) was assayed with an oxygen electrode, 0.1M methanol, air-saturated buffer, 25°C ($O_2 \sim 0.25$mM).
—— native enzyme, 90µM (4.9 U/mg); -·- native enzyme reduced with 0.1M methanol; --- enzyme after incubation and gel filtration (6.8 U/mg); ··· activated enzyme with 0.1M methanol.
Insert: Activation of alcohol oxidase by $ABTS^{\cdot +}$: 5µM enzyme, 1mM $ABTS_{red}$, 25µg POD in 1 ml; prior to the addition of enzyme $ABTS^{\cdot +}$ was produced with H_2O_2 (1 H_2O_2 yields 2 $ABTS^{\cdot +}$). (□) 1.1µM H_2O_2; (x) 2.1µM ; (o) 5.3µM; (▽) 21µM; (+) no H_2O_2 added.

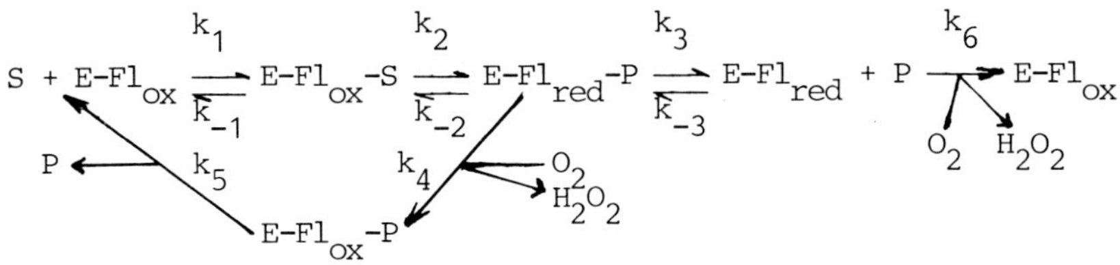

$$v_{Max} = \frac{k_2 \times k_5}{k_2 + k_5} \quad \text{(eq. I)}$$

$$K_M(O_2) = \frac{k_5(k_2 + k_{-2})}{k_4(k_2 + k_5)} \quad \text{(eq. II)}$$

$$K_M(ROH) = \frac{k_5(k_{-1} + k_2)}{k_1(k_2 + k_5)} \quad \text{(eq. III)}$$

k_{-1}/k_1 = 2 (8) mM

v_{max} = 4670 (3100) min^{-1}

k_5 = 12400 (11100) min^{-1}

k_2 = 7500 (4300) min^{-1}

k_{ox} = 3.6×10^6 $M^{-1}min^{-1}$

k_4 $\sim 7 \times 10^6$ $M^{-1}min^{-1}$

Scheme 1: Kinetic model of the turnover of methanol and [$^2H_3$]-methanol by alcohol oxidase (values in parentheses refer to deuterated methanol)

The enzyme is irreversibly inactivated by H_2O_2 in a process exhibiting saturation kinetics with K_I = 1.6 mM and k_{inact} = 0.55 min^{-1} (9). During methanol turnover the rate of inactivation increases as H_2O_2 accumulates. Methanol delays the inactivation as shown by decreasing rates of inactivation with increasing concentrations of methanol (fig. 2A). Even when H_2O_2 is eliminated by catalase prolonged alcohol turnover slowly inactivates the enzyme in a process showing (pseudo)first order saturation kinetics with all alcohols used (fig. 2B). With methanol k_{inact} = 0.005 min^{-1}, "K_I" = 0.6 mM, which is identical to K_{app} of methanol turnover under the same conditions. The characteristic ratio k_{cat}/k_{inact} varies for different alcohols but is always greater than 10^5:

$\frac{k_{cat}}{k_{inact}} \times 10^5$	3.00	2.80	2.66	2.18	2.11	1.70	1.57	1.43
R- (ROH)	CH_3-	C_2H_5-	C_3H_7-	C_4H_9-	CH_2=△CH_2-	$C_6H_5CH_2$-	$(CH_3)_2CH$-	HC≡CCH_2-

In the case of other flavin oxidases substrates with the methylenecyclopropyl or the acetylene function often act as efficient suicide substrates with k_{cat}/k_{inact} < 100 (5). With propargyl alcohol an inactivation mechanism via free propynal (6) is excluded by the (pseudo)first order kinetics of inactivation when catalase is present.

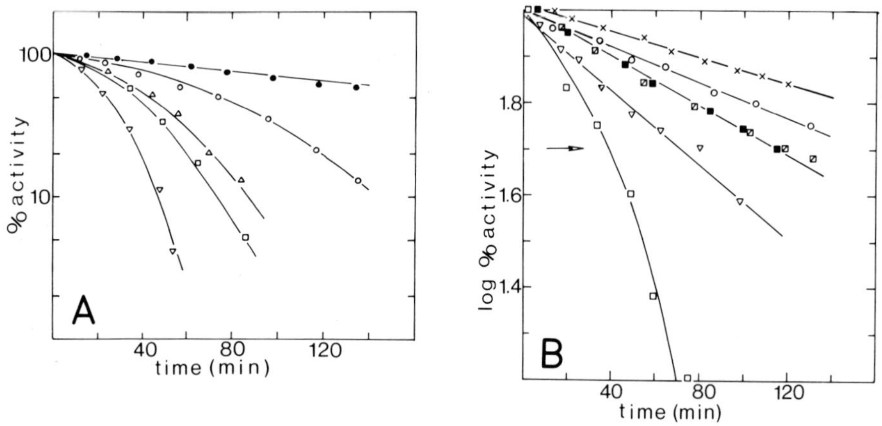

Fig. 2A: Inactivation by turnover of methanol
33nM enzyme and (●) 50mM methanol, 1.1 U/ml catalase; (o) 50mM methanol ; (▲) 10mM methanol; (■) 5mM methanol; (▽) 1mM methanol.
Fig. 2B: Inactivation by turnover of various alcohols with catalase present
17nM enzyme, 1.1 U/ml catalase, except (x) with 67nM enzyme; substrates are 0.1M (x) isopropanol, (o) butanol, (▨) methylenecyclopropyl alcohol, (■) methylenecyclopropyl alcohol, 1mM semicarbazide, (□) methylenecyclopropyl alcohol without catalase, (▽) propargyl alcohol.

References

1. Geissler, J. and Hemmerich, P.: FEBS Letters 126, 152-156 (1981).
2. Mincev, T., Tayrien, G., Mildvan, A.S. and Abeles, R.T.: Proc.Natl. Acad.Sci. USA 77, 7099-7101 (1980).
3. Cromartie, T.H.: Biochem.Biophys.Res.Commun. 105, 785-790 (1982).
4. Cromartie, T.H.: Flavins and Flavoproteins (Massey, V., Williams, C.H., eds.) Elsevier North Holland Inc., pp. 192-195, 1982.
5. Ghisla, S., Wenz, A. and Thorpe, C.: Enzyme Inhibitors (Brodbeck, U., eds.) Verlag Chemie GmbH, Weinheim, pp. 43-60, 1980.
6. Nichols, C.S. and Cromartie, T.: Biochem.Biophys.Res.Commun. 97, 216 - 221 (1980).
7. Werner, W., Riechmann, P., Rey, H., Wielinger, H., Kahle, K., Weiß, L., Klarwein, M. and Wieland, O.: Quad.Sclavo.Diagn. 7, 231-242 (1971).
8. Lockridge,O., Massey, V. and Sullivan, P.A.: J.Biol.Chem. 247, 8097 - 8106 (1972)
9. Geissler, J.: Ph.D. Thesis, Universität Konstanz, 1984.

PULSE RADIOLYSIS STUDIES ON THE FORMATION AND DECAY OF THE FLAVIN 4a-HYDROPEROXIDE SPECIES OF GLUCOSE OXIDASE

Robert F. Anderson

Cancer Research Campaign Gray Laboratory,
Mount Vernon Hospital, Northwood, Middlesex HA6 2RN, England.

Vincent Massey and Lawrence M. Schopfer

Department of Biological Chemistry, University of Michigan,
Ann Arbor, Michigan 48109, USA.

Introduction

It is known from the reactivity of substrates, substrate analogues and sulfite, that oxidases have the flavin N(5)-C(4a)-C(4) region exposed to solvent (1). It has thus been reasoned that a flavin hydroperoxido (HFOOH) is able to be formed during the oxidation of reduced flavin (FH_2) (2). Studies with chemically stablized N(5)-substituted C(4a)-flavin hydroperoxides have shown that such a species is stable only in aprotic solvents, breaking down rapidly in aqueous solution to yield H_2O_2 and oxidized flavin (F)(3). The spectral characteristics and rate constant for the decay of HFOOH formed by free flavins in aqueous solution have recently been determined using pulse radiolysis (4). We now report the use of the same technique to observe for the first time the formation and decay of the flavin 4a-hydroperoxide species of glucose oxidase (GOX-HFOOH).

Results and Discussion

1) Formation of glucose oxidase flavosemiquinone. GOX-FH$^{\bullet}$ species were produced by pulse radiolysis (2.5 Gy in 0.2 μs) of deaerated solution

Flavins and Flavoproteins

(pH 6.0) of GOX-F (35 μmol dm^{-3}) by i) reaction with e^-_{aq} ($k = 2.5 \times 10^{10}$ dm^3 mol^{-1} s^{-1}) in solution containing 2-methylpropan-2-ol (0.1 mol dm^{-3}) or ii) electron transfer from $CO_2^{\bullet -}$ species ($k = 2.5 \times 10^8$ dm^3 mol^{-1} s^{-1}) in solution containing formate ions (0.1 mol dm^{-3}) (5,6) figure 1. (Absorptions are presented as the product of the radiation chemical yield, G(mol J^{-1}) and extinction coefficient, ε(m^2 mol^{-1}) relative to the aerated KSCN dosimeter where $G((SCN)_2^-) = 0.29$ μmol J^{-1} and ε(472 nm = 758 m^2 mol^{-1} (7)). Using the literature ε(560 nm) = 414 m^2 mol^{-1} for GOX-FH$^\bullet$ (8) it is calculated that while electron transfer from $CO_2^{\bullet -}$ results in nearly quantitative formation of GOX-FH$^\bullet$ (<u>ca</u>. 0.26 μmol J^{-1}) a much lower yield (<u>ca</u>. 0.06 μmol J^{-1}) arises from e^-_{aq} scavenging. Also electron transfer from $CO_2^{\bullet -}$ results in the expected decrease in absorbance near 400 nm while e^-_{aq} scavenging does not. This is understood by the e^-_{aq} being scavenged by certain aminoacids which have known absorption bands <500 nm upon one-electron reduction (9).

2) <u>Formation of GOX-HFOOH</u>. When GOX-F is irradiated (60 Gy in 2 μs) in aerated solution competition occurs between GOX-F and O_2 for e^-_{aq} scavenging and electron transfer from $CO_2^{\bullet -}$. By <u>ca</u>. 10 μs after the pulse the maximum yield of GOX-FH$^\bullet$ produced can be calculated from the absorbance band >500 nm. The [GOX-FH$^\bullet$] is $\ll [O_2^{\bullet -}]$ and a fast first order decay of GOX-FH$^\bullet$ absorption is observed at all wavelengths. Both the blue GOX-FH$^\bullet$ (pH 6.0) and red GOX-F$^{\bullet -}$ (pH 8.7) react fast with $O_2^{\bullet -}$ (k = ca. 6×10^8 dm^3 mol^{-1} s^{-1}) to yield a similar intermediate figures 2 and 3. Correcting the spectra for loss of GOX-F absorption reveals the intermediate to have λ_{max} <u>ca</u>.380 nm with ε 1000 m^2 mol^{-1} (red GOX-F$^{\bullet -}$ was corrected on the basis of ε(525 nm) = 250 $m^2 mol^{-1}$ (8)). This is interpreted as $O_2^{\bullet -}$ reacting with GOX-FH$^\bullet$ in a similar way as with the flavosemiquinones of free flavins (4,10).

$$\text{GOX-FH}^\bullet + O_2^{\bullet -} + H^+ \longrightarrow \text{GOX-HFOOH} \qquad (1)$$

The absorption spectrum is similar to the flavin 4a-hydroperoxide of free flavin (4), bacterial lucerifase (11) and phenolic hydroxylases e.g. (12).

3) <u>Decay of GOX-HFOOH</u>. To study the decay of GOX-HFOOH at pH 6.0 it was found necessary to add a small amount of superoxide dismutase (SOD) (0.2 mg ml^{-1}) which did not interfere with the fast formation of GOX-HFOOH

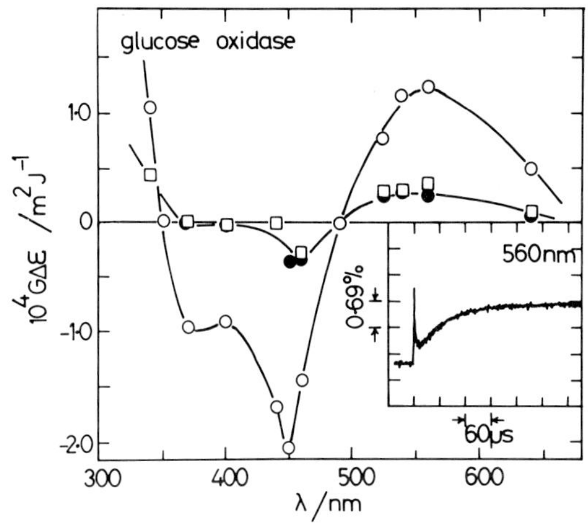

Figure 1. Absorption changes following pulse radiolysis (2.5 Gy in 0.2 µs) of deaerated solutions (pH 6.0) of glucose oxidase (35 µmol dm^{-3}) containing i) sodium formate (0.1 mol dm^{-3}). □, 6 µs. ○, 250 µs. (Insert: oscilloscope trace of absorbance vs. time) and ii) 2-methyl-propan-2-ol (0.1 mol dm^{-3}). ●, 4 µs.

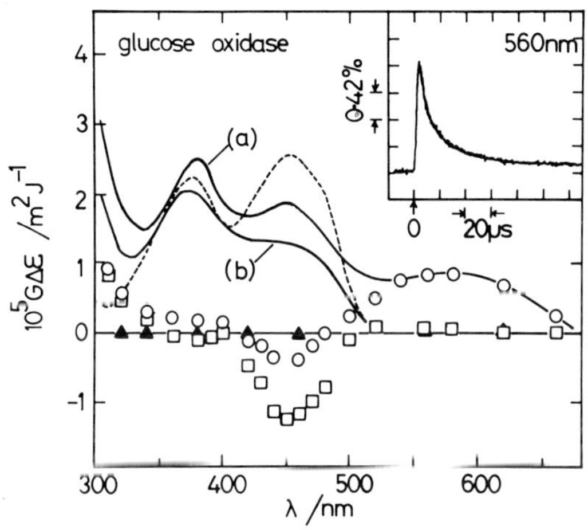

Figure 2. Absorption changes following pulse radiolysis (60 Gy in 2 µs) of aerated solutions (pH 6.0) of glucose oxidase (35 µmol dm^{-3}) containing sodium formate (0.1 mol dm^{-3}). ○, 6 µs; corrected spectrum (a), □, 120 µs; corrected spectrum (b). ▲, 8 ms. (Insert: oscilloscope trace of absorbance vs. time).

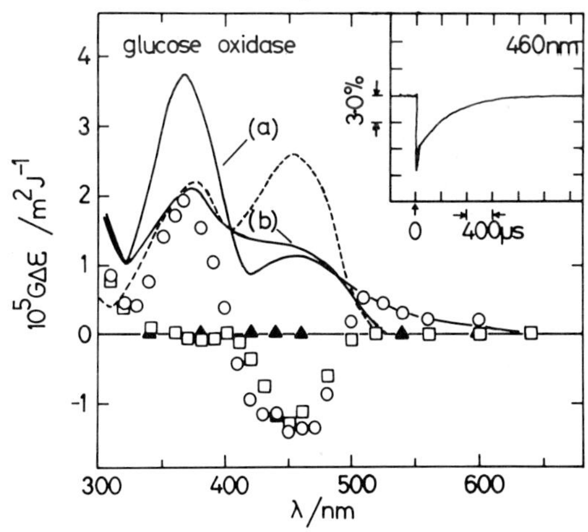

Figure 3. Absorption changes following pulse radiolysis (60 Gy in 2 µs) of aerated solutions (pH 8.7) of glucose oxidase (35 µmol dm^{-3}) containing sodium formate (0.1 mol dm^{-3}). ○, 4 µs, corrected spectrum (a). □, 120 µs (corrected for fast subsequent decay); corrected spectrum (b). ▲, 2 ms. (Insert: oscilloscope trace of absorbance vs. time).

but removed excess $[O_2^{\bullet -}]$ on a time scale shorter than the decay of GOX-HFOOH. This was because $O_2^{\bullet -}$ underwent a slow electron transfer reaction with GOX-F. The intermediate decayed to GOX-F with exponential kinetics independent of initial $[O_2^{\bullet -}]$. At pH 8.7 (where SOD addition was unnecessary) the intermediate decayed quicker.

$$\text{GOX-HFOOH} \longrightarrow \text{GOX-F} + H_2O_2 \quad (2)$$

$$\text{GOX-HFOOH} + OH^- \longrightarrow \text{GOX-F} + HO_2^-(H_2O_2) \quad (3)$$

The rate constants $k(2) = 3.4 \pm 0.2 \times 10^2\ s^{-1}$ and $k(3)$ = ca. $2.5 \times 10^8\ dm^3\ mol^{-1}\ s^{-1}$ are similar to those found for free flavin (4). It is clear from these rate constants why GOX-HFOOH was not observed by the relatively slow stopped-flow measurements (8) and its formation during the reoxygenation of fully reduced enzyme is most likely.

References

1. Massey, V., Hemmerich, P.: In Flavins and Flavoproteins, Massey, V., Williams, C.H. (eds.) pp 83-96, Elsevier North Holland, Amsterdam 1982.
2. Massey, V., Ghisla, S.: In Biological Oxidations, 34. Colloquium-Mosbach pp 114-139, Springer Verlag, Berlin Heidelberg 1983.
3. Kemal, C., Bruice, T.C.: Proc. Natl. Acad. Sci. USA 73, 995-999 (1976).
4. Anderson, R.F.: In Flavins and Flavoproteins, Massey, V., Williams, C.H. (eds.) pp 278-283, Elsevier North Holland, Amsterdam 1982.
5. Anderson, R.F.: Ber. Bunsenges. physik. Chem. 80, 969-972 (1976).
6. Land, E.J., Swallow, A.J.: Biochemistry 8, 2117-2125 (1969).
7. Schuler, R.H., Patterson,L.K., Janata, E.: J. Phys. Chem. 84, 2088-2089 (1980).
8. Stankovich, M.T., Schopfer, L.M., Massey, V.: J. Biol. Chem. 253, 4971-4979 (1978).
9. Bisby, R.H., Cundall, R.B., Redpath, J.L., Adams, G.E.: J. Chem. Soc. Faraday Trans. I 72, 51-63 (1976).
10. Anderson, R.F.: In Oxygen and Oxy-Radicals in Chemistry and Biology, Rodgers, M.A.J., Powers, E.L. (eds.) pp 597-600, Academic Press, New York 1981.
11. Hastings, J.W., Balny, C., Le Peach, C., Douzou, P.: Proc. Natl. Acad. Sci. USA 70, 3468-3472 (1973).
12. Detmer, K., Massey, V., Ballou, D.P., Neujahr, H.Y.: In Flavins and Flavoproteins, Massey, V., Williams, C.H. (eds.) pp 334-338, Elsevier North Holland, Amsterdam 1982.

E. COLI PYRUVATE OXIDASE: A HYSTERETIC ENZYME

Michael W. Mather, Robert B. Gennis
Department of Chemistry, University of Illinois
505 South Mathews Avenue, Urbana, Illinois 61801

Pyruvate oxidase is a flavoprotein dehydrogenase isolated from *E. coli* which catalyzes the oxidative decarboxylation of pyruvate to acetate plus CO_2 (1,2). Each subunit of the isolated enzyme contains a molecule of non-covalently bound FAD, and Mg^{2+}-thiamin pyrophosphate is a required co-factor for catalytic activity (2). Recent studies have shown that the natural electron acceptor for the flavoprotein is probably ubiquinone-8 in the *E. coli* cytoplasmic membrane (3). *In vitro* kinetics studies, however, have primiarly used a water-soluble electron acceptor such as ferricyanide. The enzyme is isolated as a water-soluble species, but is considered to be a peripheral membrane protein.

In the absence of pyruvate, the enzyme appears to exist in a conformation with little membrane affinity. Upon addition of pyruvate, a conformational change occurs which results in a dramatic alteration in the solution properties of the protein, and the enzyme binds to phospholipids or other amphiphiles with substantially greater affinity (4). Once bound to lipids, the V_{max} of the enzyme measured *in vitro* using water-soluble electron acceptors is increased (5). *In vivo*, the enzyme must be bound to the membrane in order to reduce the long chain quinone substrate (3). Controlled proteolysis can also be used to activate the enzyme *in vitro* apparently by cleavage at a single site at or near the lipid-binding domain (6,7). Hence, the enzyme can be locked into an "activated" form by either "lipid activation" or "protease activation". In this paper, recent kinetics studies on this enzyme are summarized. The motivation of the work was to determine the kinetic mecha-

Flavins and Flavoproteins

nism and, in particular, the step(s) primarily accelerated upon lipid or protease activation. The strategy was to carry out single turnover experiments to observe the rates of reduction and re-oxidation of the bound flavin by, respectively, pyruvate and ferricyanide, and to compare the activated and unactivated forms of the enzyme.

The traditional kinetic scheme for pyruvate oxidase incorporates the Breslow mechanism expected for thiamin pyrophosphate-dependent enzymes, in a ping pong scheme (see 8). In the absence of any activation, the turnover number is about 7 sec^{-1}, while the activated form of the enzyme turns over at about 200-350 sec^{-1} depending on the specific conditions (2,5). Earlier work (M. O'Leary and R. B. Gennis, unpublished) clearly showed that the actual carbon-carbon bond breaking step is not rate determining either for the unactivated or activated forms of the enzyme.

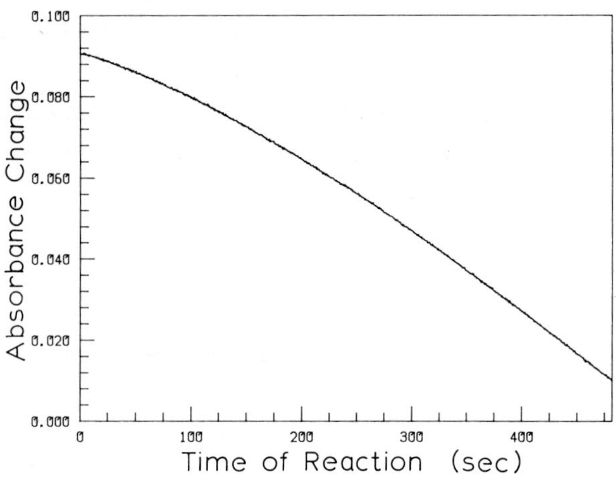

Figure 1: Progress curve for ferricyanide reduction by pyruvate oxidase at 25°C. Conditions are 6.25 mM pyruvate, 4 mM sodium ferricyanide, 0.15 mM thiamin pyrophosphate, 10 mM $MgCl_2$ 1% glucose, 0.1 M sodium phosphate at pH 6.0. Sufficient sodium sulfate was added to bring the ionic strength of 0.5. Protein concentration was 20 µg/ml.

Initial velocity studies of the ferricyanide reductase activity of the enzyme in the absence of an activator indicated that the enzyme is hysteric. The progress curves done both with manual mixing and by use of a stop flow apparatus showed reproducible evidence of an increase in enzyme specific activity during the course of the reaction. This was most evident at low pyruvate concentration (Figure 1).

The rates of flavin reduction (by pyruvate) and oxidation (by ferricyanide) were measured in stop flow experiments in the absence of an activitor. Ordinarily, for a non-hysteretic enzyme, the results of these single turnover experiments should be compatible with data obtained in "steady state" initial velocity experiments. Previous initial velocity studies done in the absence of activator were largely performed at high pyruvate concentrations and the non-Michaelis-Menton behavior was not readily apparent (2,9). Using values of K_m and k_{cat} from a previous study (9), clear gross disparities were noticed. For example the rate of reoxidation of the flavoprotein by 2 mM ferricyanide is predicted to be 7 sec^{-1} from previous initial velocity data (9), but is measured to be 0.012 sec^{-1}. Hence, in the single turnover experiments the enzyme is reoxidized almost 600 times slower than expected. A comparison of the rate of flavin reduction rates at 3.1 mM pyruvate yields values of 0.25 sec^{-1} (predicted) and 0.05 sec^{-1} (measured). When similar experiments are performed with protease activated enzyme there is no accelerating phase in the progress curve and the predicted and measured rates of flavin reduction are in good agreement.

One kinetic scheme which can account for the hysteretic behavior of the unactivated enzyme is that the enzyme exists in equilibrium between two states, with low and high turnover. Each state is presumed to operate _via_ a ping pong mechanism. The KINSIM kinetic programs (10) were used to simulate the kinetic behavior of the simplified two state ping pong system, and features of the unactivated pyruvate oxidase kinetics could be qualitatively duplicated. In single turnover experiments, the enzyme is largely in the unactivated state. However, in initial velocity measurements, as the system turns over, the amount of enzyme in the activated state progressively increases, resulting in the observed acceleration.

It is clear that lipid binding pulls the system into an activated state, as does protease activation. The stop flow results indicate that the rates of both flavin reduction and flavin reoxidation are accelerated upon activation. This is consistent with the general observation that the flavin active site "opens up" and appears to be more accessible when the enzyme is in the activated state (7). Further structural studies will be required to obtain detailed information about the exact nature of these alterations in the active site of this enzyme.

References

1. O'Brien, T. A., Schrock, H. L., Russell, P., Blake, R., II, Gennis, R. B.: Biochim. Biophys. Acta 452, 13-29 (1976).
2. Williams, F. R., Hager, L. P.: Arch. Bioch. Biophys. 116, 168-176 (1966).
3. Koland, J. G., Miller, M. J., Gennis, R. B.: Biochemistry 23, 445-453 (1984).
4. Schrock, H. L., Gennis, R. B.: Biochim. Biophys. Acta. 614, 215-220 (1980).
5. Blake, R., II, Hager, L. P., Gennis, R. B.: J. Biol. Chem. 253, 1963-1971 (1978).
6. Russell, P., Schrock, H. L., Gennis, R. B.: J. Biol. Chem. 252, 7883-7887 (1977).
7. Recny, M. A., Hager, L. P.: J. Biol. Chem. 258, 5189-5195 (1983).
8. Mather, M., Schopfer, L., Massey, V., Gennis, R. B.: Flavins and Flavo-proteins, V. Massey and C. H. Williams (ed.) Elsevier, New York, 869-872 (1982).
9. Blake, R., II: Ph.D. Thesis, University of Illinois, Urbana, IL (1977).
10. Barshop, B. A., Wren, R. F., Frieden, C.: Anal. Biochem. 130, 134-145 (1983).

MECHANISM AND FUNCTIONALITY OF FMN IN LIVER PYRIDOXINE (PYRIDOXAMINE) 5'-PHOSPHATE OXIDASE

Delores M. Bowers-Komro and Donald B. McCormick
Dept. of Biochemistry, Emory University, Atlanta, GA 30322

Background

The essential conversion of pyridoxine 5'-phosphate (PNP) and pyridoxamine 5'-phosphate (PMP) to pyridoxal 5'-phosphate (PLP) is accomplished in aerobic and facultative organisms by a cytosolic FMN-dependent oxidase (E.C.1.4.3.5) (1). Characterization of oxidase from rabbit liver (2) had revealed that the 54,000 MW enzyme is composed of two N-terminal-blocked, 27,000 MW chains (2,3) which associate ionically and hydrophobically to generate a catalytic cleft wherein the coenzyme is oriented with dimethylbenzenoid edge toward solvent (4), and substrate can bear a sizeable secondary amine function (2,5,6) instead of natural hydroxymethyl or aminomethyl group at the 4-position of the 5'-phosphopyridoxyl moiety. FMN is bound at the side chain by cationic interaction with the dianionic 5'-phosphate and by stereospecific hydrogen bonding to secondary hydroxyls (2,4,7); ring association occurs with a Try residue (8). Substrate and product are bound through their dianionic 5'-phosphate to a guanidinium function of an Arg; a second Arg is implicated in activity (9). Catalysis involves the base function of a His which serves to abstract a proton from the 4'-group of substrate (10), and a cationic group centered near N^1-O^2 of FMN (4), thereby enhancing its oxidative propensity by stabilization of the dihydroflavin anion generated. The oxidase proceeds through a binary (ping-pong) mechanism with PNP from which hydrogen abstraction is rate limiting, whereas a ternary complex (sequential) mechanism is followed with PMP from which hydrogen is abstracted more facilely but release of product PLP from the imine-enzyme complex is slower than O_2 reacting with reduced flavoprotein (11).

Flavins and Flavoproteins

Purpose

Two prime features needed elucidation: 1. The nature of the cationic residue which interacts at the N^1-O^2 region of the flavocoenzyme and is reflected by spectra of such analogs as the bound tautomeric 8-hydroxy-FMN (4). An Arg seems plausible, as modification of two are related to loss of activity yet only one serves as the counter-ion to the 5'-phosphopyridoxyl moiety of substrate or product (9). 2. The specific ionic mechanism involved in the heterolytic cleavage of a sigma bond from the 4'-hydrogen of the methylene group of substrate oxidized. A radical mechanism seems less likely as no semiquinone is spectrally observed during anaerobic reduction of the oxidase (2-4) which reflects a two-electron process. Either a base-catalyzed hydride ion mechanism, whereby a His is abstracting H^+ from the NH_3^+ or OH functions on the 4'-carbon of PMP or PNP, respectively, and flavocoenzyme is abstracting H^- from the 4'- CH_2 portion, or a carbanion mechanism, whereby His is abstracting H^+ from the 4'-CH_2 portion and the resultant carbanion attacks the flavin, probably at N^5 to yield an intermediate adduct, could be supposed.

Experimental

The probability that an Arg functions near the N^1-O^2 region of flavin was investigated by observing spectra of the oxidase with FMN or 8-hydroxy-FMN to which sulfite was added before or after modification of enzyme with cyclohexanone. Results demonstrate that the bound coenzyme generates the spectrum expected for the N^5-sulfite adduct only in unmodified enzyme. Exchange of tritium label from [4'-^{3}H-methylene]substrates to water with oxidase under anaerobic conditions would expectedly be greater with a carbanion mechanism in which $^3H^+$ is abstracted by a His rather than a hydride ion mechanism in which $^3H^-$ is directly transferred to form reduced flavocoenzyme, especially if the cofactor is the more stable 5-deaza analog.

Results shown in the table demonstrate that the ^3H_2O obtained by lyophilization of anaerobic, substrate-reduced oxidase is sufficient to allow a carbanion mechanism.

Exchange of ^3H from [4'-3H]substrates with oxidase to water

Coenzyme	O_2	mmol ^3H (H_2O)/mol oxidase[1]	
		PNP	PMP
FMN	+	118 (177)	216 (520)
	-	3 (24)	24 (22)
5-Deaza-FMN	+	212 (332)	70
	-	20 (35)	12

[1]Corrections made for exchange to buffer before (after) acid.

Phenol extraction of the flavocoenzyme from the [^3H]substrate-reduced 5-deaza-FMN-oxidase, followed by ether removal of phenol and HPLC of the flavin, revealed a ^3H-compound other than coenzyme and its dihydro form.

Probable Mechanism

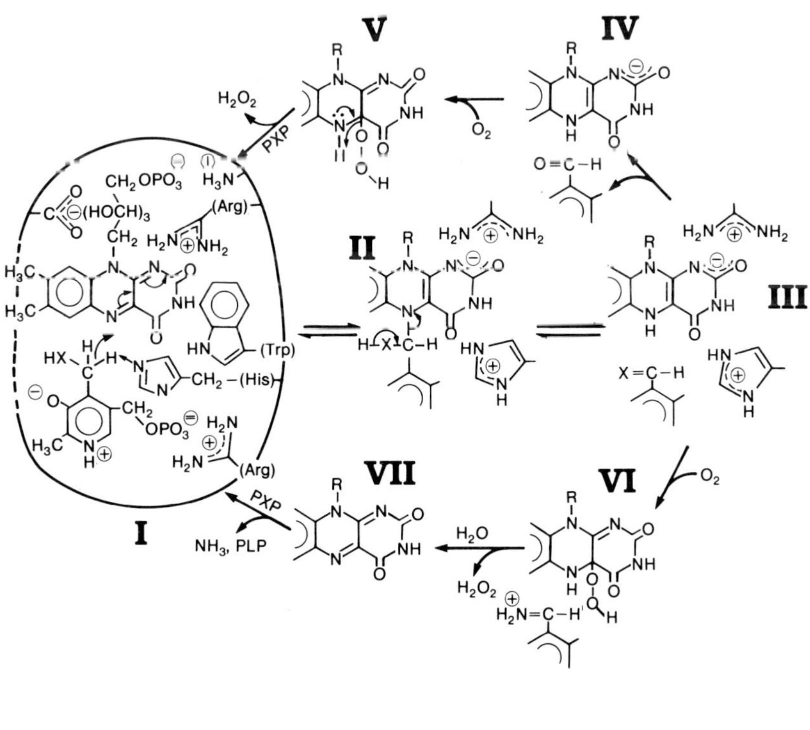

References

1. McCormick, D.B., Merrill, Jr., A.H.: Vitamin B_6 Metabolism and Role in Growth, G.P. Tryfiates (ed.), Food and Nutrition Press, Westpoint CT, 1980.

2. Kazarinoff, M.N., McCormick, D.B.: J. Biol. Chem. 250, 3436-3442 (1975).

3. McCormick, D.B., Kazarinoff, M.N., Tsuge, H.: Flavins and Flavoproteins, T.P. Singer (ed.), Elsevier, Amsterdam, 1976.

4. Merrill, Jr., A.H., Kasai, S., Matsui, K., Tsuge, H., McCormick, D.B.: Biochemistry 18, 3635-3641 (1979).

5. Kazarinoff, M.N., McCormick, D.B.: Biochem. Biophys. Res. Commun. 52, 440-446 (1973).

6. DePecol, M.E., McCormick, D.B.: Anal. Biochem. 101, 435 441 (1980).

7. Kazarinoff, M.N., McCormick, D.B.: Biochim. Biophys. Acta 359, 282-287 (1974).

8. McCormick, D.B.: Photochem. Photobiol. 26, 169-182 (1977).

9. Choi, J.D., McCormick, D.B.: Biochemistry 20, 5722-5728 (1981).

10. Horiike, K., Tsuge, H., McCormick, D.B.: J. Biol. Chem. 254, 6638-6643 (1979).

11. Choi, J.D., Bowers-Komro, D.M., Davis, M.D., Edmondson, D.E., McCormick, D.B.: J. Biol. Chem. 258, 840-845 (1983).

ENZYMATIC PROPERTIES OF YEAST PYRIDOXAMINE-P OXIDASE

Haruhito Tsuge, Toshitaka Okada, Izumi Nakane, Shinji Uchida, Reiko Sugiyama and Kazuji Ohashi

Department of Agricultural Chemistry, Gifu University, 1-1 Yanagido, Gifu 501-11, Japan.

Introduction

Pyridoxaminephosphate oxidase (EC 1.4.3.5) has been purified to an apparent homogeniety from baker's yeast by the authors (1-9). The enzyme is an FMN-dependent oxidase, whose molecular weight is about 55,000 dalton with two subunits and one FMN, and oxidizes the primary alcohol (pyridoxine 5'-P: PNP), the primary amine (pyridoxamine 5'-P: PMP) and the secondary amines (N-phosphopyridoxyl-amino acids: N-PP-A.A.). The purified enzyme altered its catalytic activity depending on the conditions used (1,3,5), for instance, aliphatic primary amines including Tris and spermine, enhanced the PMP oxidase over 4-fold, but not the PNP oxidase (5).

Thus, some disproportionate data concerning the catalytic reaction for some substrates are presented herewith.

Results and Discussion

Pyridoxal 5'-P (PLP), which was produced by this enzyme reaction, inhibited only the PMP oxidase competitively (K_i = 6 μM). However, PLP did not inhibit the PNP oxidase.

As shown in Fig. 1, Cd^{++} ion inhibited the PNP oxidase noncompetitively, regardless of the buffer species used. The PMP oxidase was also inhibited by H_2O_2, one of the products, while the PNP oxidase was not inhibited significantly. In general, the product being released finally from the enzyme-product complex should bring about the product inhibition from

Flavins and Flavoproteins

enzyme kinetic points of view. However, the PMP oxidase was inhibited by PLP and by H_2O_2. These facts are in conflict with the theoretical consideration.

Several pyridoxyl derivatives were synthesized from PLP and amino acids (10), and these synthetic N-PP-derivatives were used as substrate. Table I summarizes the Km and V_{max} values obtained throughout the experiments. The most efficient substrate was PNP in terms of V_{max}/Km in both Tris-HCl and Hepes-KOH buffer at pH 8.0. The enzyme changed its substrate specificity for PMP drastically when the buffer species were altered from Hepes-KOH (non-amine buffer) to Tris-HCl (aliphatic primary amine buffer)(5), and the value of PMP approached a comparative one to PNP's in Tris-HCl buffer at pH 8.0.

Furthermore, the inhibition experiments revealed that these substrates were not oxidized by the enzyme at one active site, because in the higher PMP concentration, the substrate analogue competed with PMP at the same active site. However, in the higher PNP concentration, these compounds seemed to work as mixed type inhibitor. In the lower concentration of PMP as well as PNP, the extent of inhibition (%) abruptly decreased. Thus, the biphasic double reciprocal plots were obtained as shown in Fig. 2.

Table I Substrate Specificity*

	in Tris-HCl buffer			in Hepes-KOH buffer		
	Km	V_{max}	V_{max}/Km	Km	V_{max}	V_{max}/Km
	(μM)	(%)		(μM)	(%)	
PNP	5.33	100	18.8	6.0	100	16.6
PMP	22.5	326	14.5	21.3	40.7	1.9
N-PP-L-Ala	20.0	100.7	5.0	14.1	21.6	1.5
N-PP-Gly	41.3	123	3.0	21.8	81.6	3.8
N-PP-GABA	109	150	1.4	198	217	1.1
N-PP-L-Ser	55.5	63.2	1.1	41.2	49.5	1.2
N-PP-EACA	98.2	41.4	0.4	22.3	7.8	0.35

* Specific activity of the enzyme preparation used was about 6 μmol of PLP formed/h/mg protein for PNP (285 μM) as substrate at pH 8. GABA, γ-aminobutyrate; EACA, ε-aminocaproate.

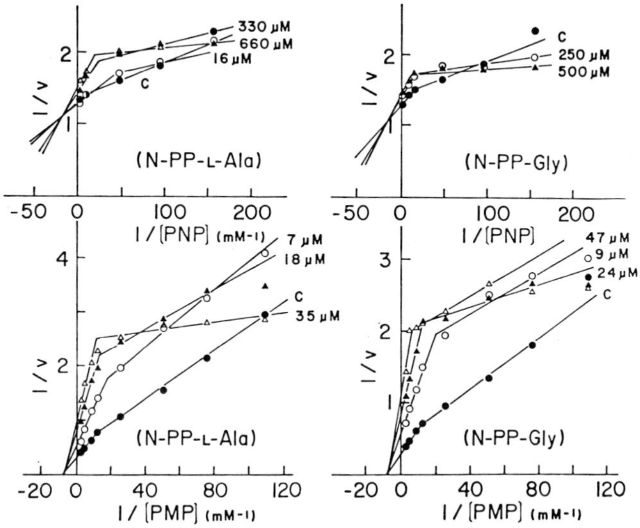

Fig. 1. Inhibitory effect of N-PP-analogues on pyridoxamine-P oxidase. Concentration of inhibitor was shown. C: control (= 0 M).

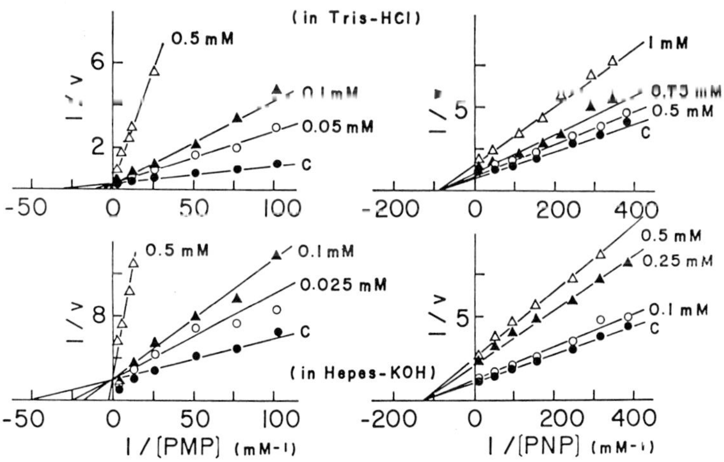

Fig. 2. Inhibition of pyridoxamine-P oxidase by Cd^{++}. Concentration of inhibitor was shown. C: control (= 0 M).

Based on the results shown above, yeast pyridoxamine-P oxidase was postulated to switch the substrate from PNP to PMP and _vice versa_ depending on the circumstances, and behaved as if the assembly of multi-enzymes with similar characteristics. However, more than two enzyme species could not be detected in the present preparation. A slow-transition from one conformation to the other might be suggested (4), but it was confirmed that HYSTERETIC change (11) by polyamine did not occur within the assay periods. Accordingly, it is herewith proposed that the yeast pyridoxamine-P oxidase has two interconvertible conformations (i.e., PMP- and PNP-type) or two different active sites. The above observation is not consistent with the description by Choi _et al_. (12) on the rabbit liver enzyme.

References

1. Tsuge, H., Hirose, A., Ohashi, K.: Res. Bull. Fac. Agric. Gifu Univ. (41) 71-77 (1977) (in Japanese).

2. Tsuge, H., Ozeki, K., Sen-maru, K., Ohashi, K.: Agric. Biol. Chem. _43_, 1801-1807 (1979).

3. Tsuge, H., Ozeki, K., Ohashi, K.: Agric. Biol. Chem. _44_, 2329-2335 (1980).

4. Tsuge, H., Itoh, K., Ozeki, K., Ohashi, K.: Agric. Biol. Chem. _46_, 1075-1077 (1982).

5. Tsuge, H., Itoh, K., Akatsuka, F., Okada, T., Ohashi, K.: Biochem. Int. _6_, 743-749 (1983).

6. Tsuge, H., Sen-maru, K., Ohashi, K.: FEBS Lett. _93_, 331-334 (1978).

7. Tsuge, H., Sakuma, J., Ozeki, K., Ohashi, K.: Vitamines, _55_, 437-443 (1981) (in Japanese).

8. Tsuge, H.: Vitamines, _57_, 67-79 (1983) (in Japanese).

9. Tsuge, H., Okada, T.: Biotech. Bioeng. _26_, 412-418 (1984).

10. Tschopp, J., Kirschner, K.: Biochemistry, _19_, 4514-4521 (1980).

11. Neet, K.E., Ainslie, G.R.: Methods in Enzymol. _64_, 192-226 (1980).

12. Choi, J-D., Bowers-Komro, D.M., Davis, M.D., Edmondson, D.E., McCormick, D.B.: J. Biol. Chem. _258_, 840-845 (1983).

A PHOTOLABELING REAGENT OF BRAIN PYRIDOXINE-5-P OXIDASE

J. E. Churchich
Department of Biochemistry

University of Tennessee Knoxville, TN USA

Introduction

Arylazide derivatives have been used as photolabeling reagents of the catalytic sites of several enzymes (1). Upon illumination of arylazide derivatives, a singlet nitrene characterized by a lifetime of about 100 ns is generated. The singlet nitrene reacts with amino acid residures of proteins to form covalent linkages. In our laboratory, we have synthesized an arylazide derivative, i.e. N-4-amino-2-nitrophenyl-pyridoxyl-5-P, which recognizes the binding sites of pyridoxal-5-P dependent enzymes, and it is an efficient photolabeling reagent of brain pyridoxine-5-P oxidase.

A characteristic feature of the arylazide derivative of pyridoxamine-5-P, i.e., its ability to recognize the substrate binding site of the oxidase, can be exploited to ascertain the nature of the amino acid residues located within the catalytic site domain.

Experimental Procedures

<u>Synthesis of N-4-azido-2-nitrophenyl-pyridoxyl-5-P (NANP)</u>

4-fluoro-3-nitrophenyl-azide (1 m mole) in 7 ml of ethanol was added to a solution of pyridoxamine-5-P (0.6 m mole) in

Flavins and Flavoproteins

5 ml of water containing sodium carbonate (10 m mole). Ethanol (14 ml) and another portion of water (6 ml) were added subsequently to enhance the homogeneity of the reaction mixture. The reaction was allowed to proceed at 50°C for 10 hours with an attached cooling condenser. At the end of the reaction period, the resulting dark red mixture was concentrated under vacuum to about 10 ml and then diluted with 10 ml of water. Two extractions with 50 ml of ether removed all of the excess starting azide.

The aqueous solution was concentrated to a volume of 5 ml, acidified with acetic cid to pH 3.2, and applied to a cation exchange resin (Bio-rad A G-50W-X_4) on a column of 24 x 2cm. The column was washed with NH_4-acetate (10 mM, pH 4.2) elute pyridoxamine-5-P. The red compound (N--4-azido-2-nitrophenyl-pyridoxyl-5-P) is eluted with NH_4-acetate (100 mM, pH 5.4). All the operations, synthesis and ion exchange chromatography, are performed in the dark to prevent photo decomposition. The arylazido derivative of pyridoxamine-5-P showed one spot when chromatographed on cellulose thin layer developed with the solvent system (water, acetone, ter-amylalcholol, acetic acid) (20/35/40/5). The red spot turns blue upon reaction with dichloroquinone chlorimide.

The absorption spectrum of N-4-azido-2-nitrophenyl-pyridoxyl-5-P shows two intense absorption bands centered at 340 and 475 mm. The infrared spectrum shows a band at 2130 cm-^{1}, characteristic of N_3 stretching vibrations. Upon irradiation with light of 360 nm, the absorption band at 475 nm is progressively decreased.

Preparation of brain pyridoxine-5-P oxidase

Brain pyridoxine-5-P oxidase was purified 2,250 fold from pig brain tissues (2). The rate of formation of pyridoxal-5-P was measured by following the increase in absorbance at 388 nm in a double beam amino spectrophotometer. A molecular weight of 60,000 for pyridoxine-5-P oxidase with a specific activity of 90 units/mg of enzyme at pH 8.4 was used in the calculation of molar enzyme concentrations.

Results and discussion

In the dark, the reaction of the oxidase with N-4-amino-2-nitrophenyl-pyridoxyl-5-P leads to the formation of pyridoxal-5-P. The rate of formation of PLP was measured by monitoring the increase in absorbance at 388 nm as a function of time. A $K_{cat} = 0.065\ S^{-1}$ and a $K_M = 5\mu M$ was determined for the arylazide derivative of PMP at pH 8.4. These results are interpreted to mean that the affinity label reagent undergoes a splitting reaction by which pyridoxal-5-P is generated in the presence of the oxidase.

Upon irradiation of mixture containing oxidase ($7\mu M$) and NANP ($60\mu M$) at pH 7, enzyme is irreversibly inactivated as directed by the results included in Fig. 1. As may be seen from the results included in Fig. 1, the rate of enzyme inactivation is paralleled by a change in the absorption band of the photolabeling reagent centered at 475 nm. This decrease in absorbance is due to decomposition of the arylazide derivative to generate singlet nitrene. Singlet nitrene reacts with amino acid residues of the oxidase to cause irreversible loss of catalytic activity. After dialysis, the inactive enzyme displays a strong absorption band over the range 360-540 nm due to incorporation of nitrophenyl residues (Fig. 2). Illumination of oxidase, in the absence of photolabeling reagent, has no effect on either the catalytic activity or the intensity of the absorption band due to FMN (Fig. 2).

The modified oxidase does not show any fluorescence due to

bound FMN (excitation 360, 450 nm) or to bound NANP (excitation 330 nm). However, reduction of the modified oxidase with sodium dithionite brings about an increase in the fluorescence emitted by excitation at 330 nm (Fig. 3).

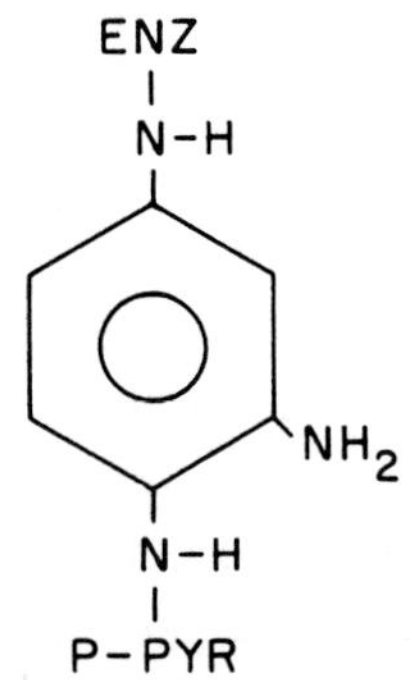

The finding that the emission spectrum of adduct I resembles that of P-pyridoxyl gives strong support to the contention that the P-pyridoxyl moiety of the photolabeling reagent has recognized the substrate binding site of the oxidase, whereas the nitrophenyl moiety has reacted with an amino acid residue located near the catalytic site. When tritium labeled N-4-azido-2-nitrophenyl-pyridoxyl-5-P was used in the photolabeling experiments, it was found that 1 mol of P-pyridoxyl is incorporated per mol of enzyme. Work is in progress to elucidate the amino acid sequence of the modified peptide.

References

1. Jeng, S.J. adn Guillroy, R.J. (1975). J. of Supramolecular Structure, 3, 448-468.
2. Churchich, J.E. (1984). Eur. J. Biochem. 138, 327-332.

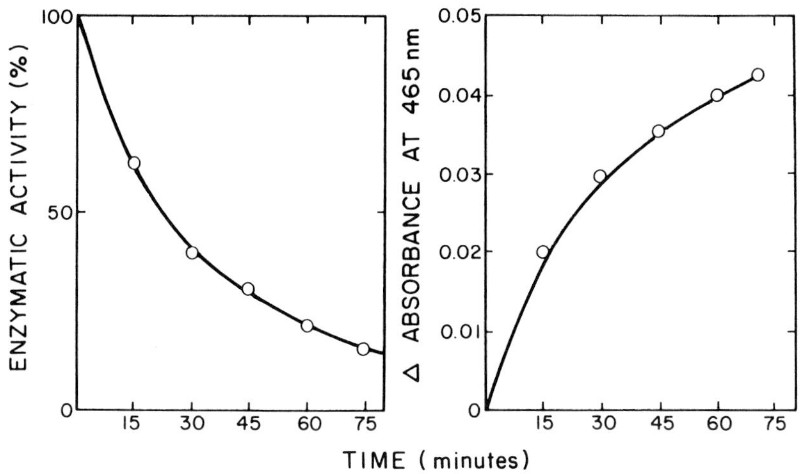

Figure 1. Left. Change in the catalytic activity of pyridoxine-5-P oxidase (7μ M) incubated with NANP (60 μM) and irradiated with light of 360 nm. Aliquots withdrawn from the incubation mixture at the indicated times were assayed for enzymatic activity using pyridoxine-5-P as a substate at pH 8.

Right. Decrease in absorbance at 465 nm of the Reaction mixture containing oxidase (7 μ M) and NANP (60 μ M) irradiated with light of 360nm.

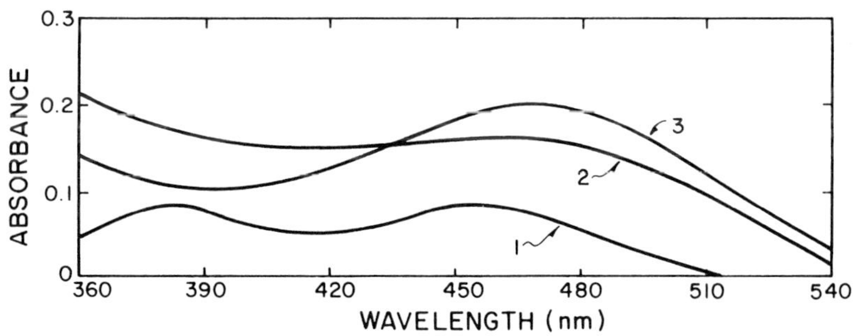

Figure 2. Absorption Spectra of Pyridoxine-5-P oxidase (7 μ M) in the absence (1) and presence (3) of NANP (60 μ M). Absorption spectrum of the sample containing oxidase (7μ M) + NANP (60 μM) (2) after irradiation (75 minutes), followed by Dialysis against o.1M phosphate buffer (pH 7).

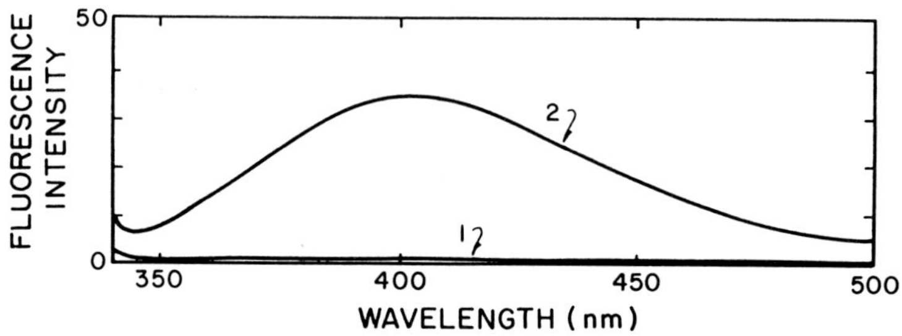

Figure 3. Emission spectrum of modified oxidase reacted with NANP (1). The same sample was reacted with sodium dithionite (1mM) and its emission spectrum recorded after excitation at 330 nm (2).

NON-STEREOSPECIFIC REDUCTION OF MONOAMINE OXIDASE FROM BOVINE LIVER BY ANALOGS OF AMPHETAMINE

Walter Weyler and James I. Salach

Molecular Biology Division, Veterans Administration Medical Center, San Francisco, California 94121; and the Department of Biochemistry and Biophysics, University of California San Francisco, California 94143

R.T. Coutts[1], and G.B. Baker[2]

[1])Faculty of Pharmacy and Pharmaceutical Sciences and [2])Department of Psychiatry, University of Alberta, Edmonton, Alberta, Canada T6G 2N8.

Introduction

The physiological action of amphetamine is complicated and not fully understood. Interactions with the re-uptake system of biogenic amines and inhibition of monoamine oxidase (1) are only two of its known actions. Recently Benderly et al., (2) reported that the optical isomers of the amphetamine analog 2-amino-3-fluoro-1-phenylpropane cause significant but small changes with regard to potentiation of release and inhibition of uptake of dopamine or 5-hydroxytryptamine in rat striatum when compared to the parent amphetamines. This result was surprising since there is a large shift in pK on substitution of fluorine for hydrogen and we, therefore, became interested in investigating these fluorinated analogs with respect to their interaction with monoamine oxidase. In this study we made the unexpected observation that both enantiomers of 2-amino-3-fluoro-1-phenylpropane are substrates for monoamine oxidase from bovine liver, a finding contrary to the notion that MAO is a stereospecific enzyme and that amphetamines are not substrates for this enzyme.

Methods

Monoamine oxidase B from bovine liver was prepared as previously reported (3,4). R- and S-2-amino-3-fluoro-1-phenylpropane were synthesized

Flavins and Flavoproteins

from R- and S-phenylalanine, respectively, as reported (5). Turnover of fluoroamphetamines was shown as follows. MAO B (16.8 nmol) was incubated with 20 nmol of the appropriate fluoroamphetamine in 0.1 mL, 10 mM NaPi buffer, pH 7.2, overnight at 23°C. Protein was precipitated, the supernatant was concentrated and analyzed by silica HPTLC (50% butanol, 25% acetic acid, and 25% water). Remaining amines were visualized with o-phthalaldehyde spray. The limit of detection was < 0.5 nmol.

Results and Discussion

Inhibition of MAO B by R- and S-2-amino-3-fluoro-1-phenylpropane was competitive with K_i^{app} values similar to those for amphetamine (Table I). Surprisingly, both enantiomers were slowly consumed when incubated aerobically with the enzyme, as demonstrated by HPTLC of the reaction mixtures. Further evidence that both enantiomers were substrates was obtained by single turnover experiments under anaerobic conditions monitoring flavin reduction. The semi-log plot of the rate of reduction was linear for at least five half-lives (Fig. 1). The minimum mechanism is shown in equation 1. With this mechanism, equation 2, and the assumption that $K_i = K_s$, we calculated k_2 from the relationship shown in equation 3. The true K_i values were determined by the use of the Henderson-Hasselbalch equation with the assumption that free amines are the substrate for MAO (6). The large decrease in binding energy on substitution of fluorine for hydrogen appears related to the change in pK. Since free amines are substrates, and alcohol

TABLE I. Kinetic parameters for the interaction of fluoroamphetamines with bovine monoamine oxidase.

Compound[a]	K_i^{app}(mM)[b]	k_{obs}(min^{-1})	k_2(min^{-1})	pK[c]	K_i[d]	ΔG kcal
R-FA	0.41	0.012	0.015	7.95	0.06 mM	-5.8
S-FA	1.46	0.018	0.021	7.95	0.22 mM	-5.1
S-A	0.42			9.58	1.8 μM	-8.0
R-A	0.74			9.58	3.1 μM	-7.6

[a]FA, fluoroamphetamine; A, amphetamine. [b]Determined in 50 mM NaPi buffer, pH 7.2, at 30°C, with benzylamine the substrate. [c]From reference 5. [d]Determined using the modified Henderson-Hasselbalch equation:
$K_i = (K_i^{app})/(1 + \text{antilog}(pK - pH))$

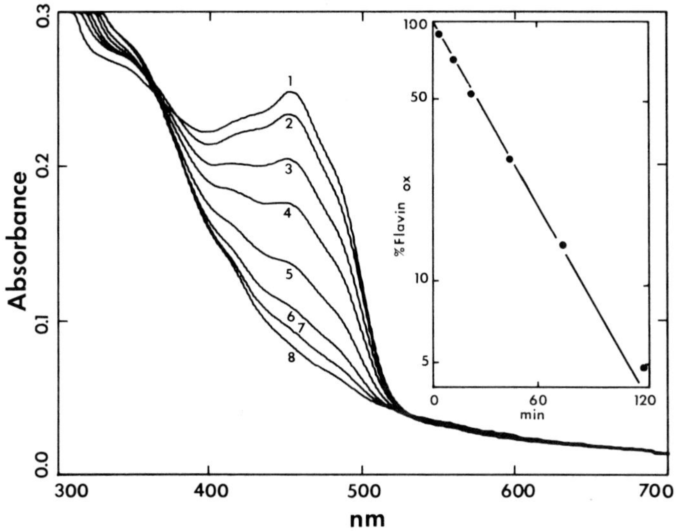

Figure 1. **Anaerobic reduction of MAO B from bovine liver mitochondria with fluoroamphetamine.** Reductions were carried out in 1 mL of a solution containing 50 mM NaPi, pH 7.2, 0.2% Triton X-100, 60 U of catalase, 2 U of glucose oxidase, 10 mM glucose, 14 µM MAO B and 0.25 mM R-2-amino-3-fluoro-1-phenylpropane or 1.82 mM S-2-amino-3-fluoro-1-phenylpropane. Solutions including all components except glucose and fluoroamphetamine were flushed with oxygen-free argon, glucose was added and cuvettes were allowed to incubate at room temperature for 20 min. The initial spectrum was taken, substrate was added, and spectral changes were followed as a function of time. The data shown are those for the S-isomer. 1, t=0 min; 2, t=2.6 min; 3, t=11 min; 4, t=22 min; 5, t=43 min; 6, t=73 min; 7, t=118 min; 8, endpoint, obtained by reduction with excess benzylamine. Inset, semi-log plot of absorbance changes at 450 nm versus time.

analogs of amines are potent competitive inhibitors of MAO (6), it is unlikely that ion-pairing interactions are involved in substrate binding. The decrease in K_i values observed for fluoroamphetamines compared to the parent compounds may be due to changes in hydrophobic interactions or hydrogen bonding as a result of decreased electron density on nitrogen.

$$E_{ox} + S \overset{K_s}{\rightleftharpoons} E_{ox}S \xrightarrow{k_2} E_{red} + P \qquad 1$$

$$E_{obs} = E_{ox} + E_{ox}S \qquad 2; \qquad k_2 = \frac{k_{obs} [K_i + (S)]}{(S)} \qquad 3$$

In stereochemical studies with partially purified MAO from rat liver, tyramine (7,8) and 1-aminoheptane (8) were oxidized with stereospecific removal of the pro-R-hydrogen and bovine MAO B oxidized d- but not l-amphetamine (unpublished observation, J.S. and W.W.). Therefore, we find it intriguing that 2-amino-3-fluoro-1-phenylpropane behaves in a non-stereospecific manner. Both the R- and S-enantiomers reduce the flavin chromophore of bovine liver MAO B with virtually identical rate constants. Thus, once the $E_{ox}S$ complex is formed reaction is equally probable with either enantiomer under anaerobic conditions.

Non-stereospecific behavior and similar rates of oxidation suggest there may be two productive binding modes as previously proposed for plasma amine oxidase with dopamine as substrate (10). It is well known that substrate specificity of MAO is wide and that inhibitors of quite unrelated structure are effective with sometimes very low K_i values. In light of these results a systematic investigation of the stereospecificity of MAO is needed to clarify this enzyme's role in normal metabolism (11).

References

1. Williams, C.H.: J. Pharm. Pharmacol. **34**, 386-387 (1982).
2. Benderly, A., Coutts, R.T., Mak A.L.C., and Baker, G.B.: Experientia **37**, 294-295 (1981).
3. Salach, J.I.: Arch. Biochem Biophys. **192**, 128 (1979).
4. Weyler, W. and Salach, J.I.: Arch. Biochem. Biophys. **212**, 147 (1981).
5. Coutts, R.T., Benderly, A., and Mak, A.L.C.: J. Fluorine Chem. **16**, 277-283 (1980)
6. McEwen, C.M., Saski, G., and Lenz, W.R.: J. Biol. Chem., **243**, 5217 (1968).
7. Belleau, B., Fang, M., Burba, J., and Moran, J.: J. Am. Chem. Soc. **82**, 5752-5754 (1960).
8. Battersby, A.R., Buckley, D.G., Stauton, J., and Williams, P.J. J. Chem. Soc., Perkin I **1979**, 2550-2558 .
9. Battersby, A.R., Chrystal, E.J.T., and Stauton, J.: J. Chem. Soc., Perkin I **1980**, 31-42 .
10. Summers, M.C., Markovic, F., and Klinman, J.P.: Biochemistry, **20**, 1969-1979 (1979).

11. We gratefully aknowledge support for this work by NIH grant # NS 14222-06

THE KINETICS OF NADPH DEPENDENT REDUCTION OF FAD AND CYTOCHROME b IN A SOLUBILISED PREPARATION OF THE SUPEROXIDE GENERATING OXIDASE OF NEUTROPHILS

Andrew Cross, John Parkinson, Owen Jones
Department of Biochemistry, University of Bristol, Bristol BS8 1TD. UK

Introduction

Polymorphonuclear leucocytes (neutrophils) form the major host defence mechanism against microbes. As they phagocytose bacteria there is a burst of oxygen uptake which is accompanied by the release of reduced oxygen radicals $O_2^{\overline{\cdot}}$ (superoxide) and its reaction products H_2O_2 and $OH^{\cdot}$. The electron donor of this oxidase is NADPH (1). Neutrophil plasma membranes contain a b-type cytochrome of very low redox potential (-245mV) (2) sufficiently low to reduce O_2 to $O_2^{\overline{\cdot}}$ ($E_{m\ 7}$ -160mV). The cytochrome is reduced when anaerobic suspensions of cells are stimulated and is rapidly oxidized when partly reduced cytochrome is mixed with air ($t\frac{1}{2}$ = 4.7ms). The cytochrome is absent from most patients with the X-linked form of chronic granulomatous disease who lack the ability to form $O_2^{\overline{\cdot}}$. In another group of patients the cytochrome is present but does not become reduced on stimulation and the ratio of FAD to cytochrome in the membranes is altered (3,4). The evidence suggests that the oxidase involves at least two proteins, cytochrome b_{-245} and a flavoprotein which are organised in an electron transfer chain :

$$\text{NADPH} \xrightarrow{e^-} \text{FAD protein} \xrightarrow{e^-} \text{cytochrome } \underline{b}_{-245} \xrightarrow{e^-} \left\{ \begin{array}{l} O_2^{\overline{\cdot}} \\ O_2 \end{array} \right.$$

Scheme 1

Flavins and Flavoproteins

Methods

Neutrophils were prepared from 4L of pig blood (5), suspended in Krebs-Ringer pH 7.4 and stimulated with phorbol for 5 min at 37°C. The cells were washed in saline and disrupted by sonication in 8.6% sucrose containing protease inhibitors. After centrifugation for 30 min at 100,000 x g the pellet was extracted by sonication with 50ml of buffer containing 0.25% Lubrol PX, 0.25% w/v sodium deoxycholate, 20mM glycine pH 8.0, 25% v/v glycerol, 1mM $MgSO_4$, 0.5mM $CaCl_2$, 1mM sodium azide and protease inhibitors. The suspension was centrifuged at 100, 000 x g for 1hr. to give a clear supernatant which contained 40% of the total $O_2^{\bar{\cdot}}$ generating activity of the postnuclear supernatant.

Results

The properties of the soluble oxidase are summarised below :

NADPH dependent $O_2^{\bar{\cdot}}$ generation	365 nmol min^{-1} mg^{-1} protein
Cytochrome <u>b</u>	331 pmol mg^{-1} protein
Flavin content (FAD)	421 pmol mg^{-1} protein

The rates of $O_2^{\bar{\cdot}}$ generation were higher than those of whole cells (130 nmol min^{-1} mg^{-1} protein). In extracts from unstimulated cells the rate of $O_2^{\bar{\cdot}}$ generation in the extract was only 2.0nmol min^{-1} mg^{-1} protein.
The absorption spectrum of the active oxidase is shown in Fig. 1. The absorption band of FAD at 450nm can be seen in the oxidised oxidase as well as the α,β and γ bands of the cytochrome in the dithionite reduced spectrum. The kinetics of reduction of the flavin and cytochrome by NADPH were studied under anaerobic conditions. NADPH was introduced from the side arm of a Thunberg cuvette to the oxidase preparation and the absorption spectrum recorded as rapidly as possible (Figs. 2 and 3. These Figures show that a substantial proportion of the cytochrome (24%) and of flavin (34%) is reduced as soon as measurements can be made (60 sec). This initial fast burst of reduction was followed by a slow phase which eventually resulted in almost complete reduction of flavin and about 60% reduction of the cytochrome. The reduction of these components was also

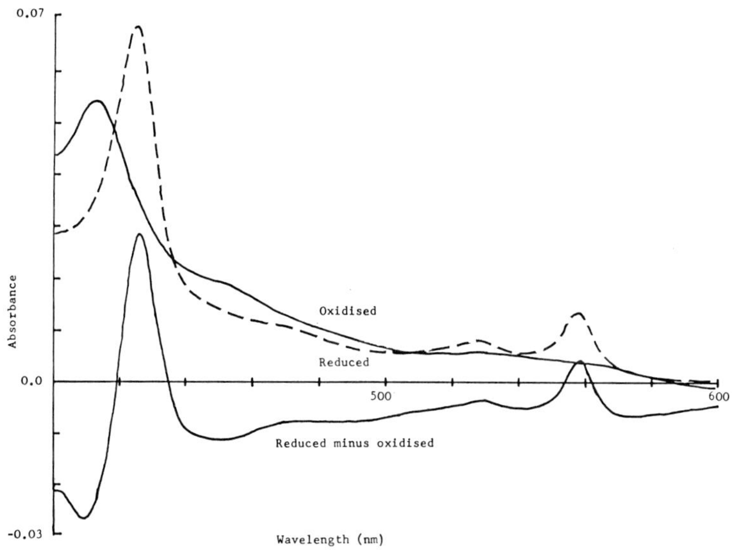

Fig. 1. Spectroscopic properties of the solubilised superoxide generating neutrophil oxidase.

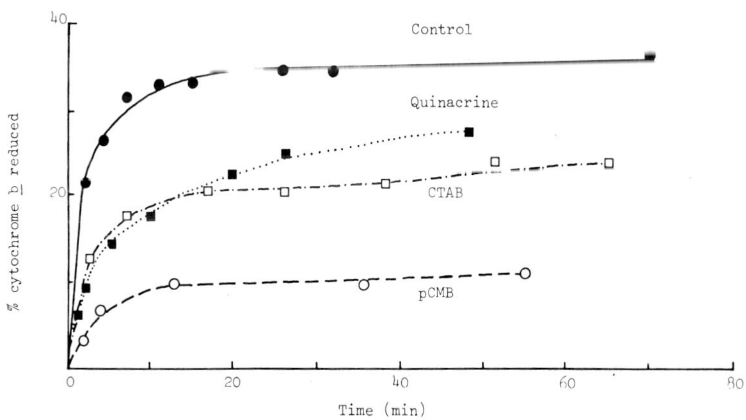

Fig. 2. Kinetics of the reduction of cytochrome b by NADPH

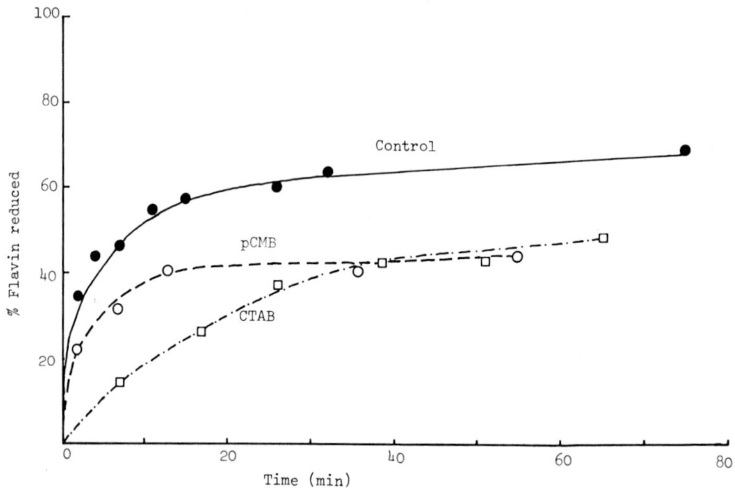

Fig. 3. Kinetics of the reduction of flavin by NADPH.

examined in the presence of the inhibitors of $O_2^{\cdot-}$ generation, p-chloromercuribenzoate (pCMB), quinacrine and cetyltrimethyl ammonium bromide (CTAB). pCMB is known to react with the cytochrome since it alters the redox potential (6). Quinacrine is an inhibitor of flavin-linked reactions, CTAB is a non-ionic detergent. Both CTAB and quinacrine strongly inhibited the rate of reduction of cytochrome b (Fig. 2). Flavin reduction could not be measured in the presence of quinacrine but was inhibited by CTAB. pCMB inhibited the extent and rate of cytochrome b reduction whilst having less effect on flavin reduction (Fig. 3). The kinetics of flavin and cytochrome b in the extract from unstimulated cells (Fig. 4) show decreased rates of reduction although eventually the extent of reduction is greater. The addition of a mobile electron carrier such as methyl viologen resulted in a rapid, almost complete reduction of both flavin and cytochrome.

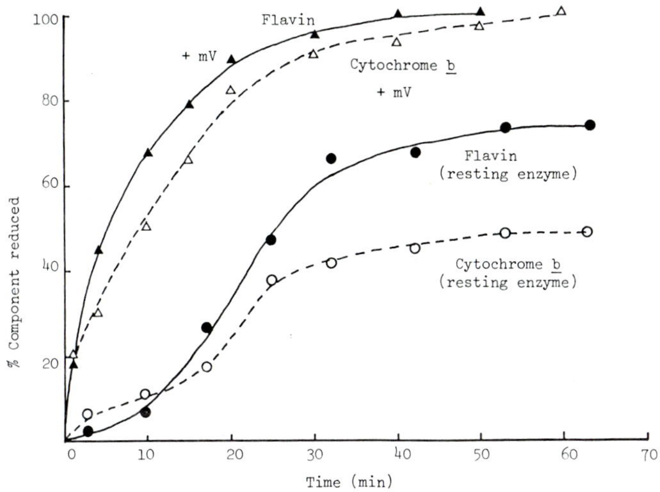

Fig. 4. Kinetics of the reduction of cytochrome b and flavin in the presence of methyl viologen and in enzyme prepared from resting cells.

Discussion

These results support the arrangement of electron carriers shown in Scheme 1. Both flavin and cytochrome b are present in the oxidase preparation and placing of FAD before cytochrome b is supported by the experiments which show that flavin becomes reduced before cytochrome b when the cytochrome is inhibited by pCMB. Not all the flavin or cytochrome was rapidly reduced even under strict anaerobiosis. It is possible that the slowly reacting components are not correctly associated, and the process of activation may be to alter the topological arrangement of the components to permit electron transfer between them but that in the preparation not all the components are activated. The stimulatory effect of methyl viologen on both flavin and cytochrome b reduction indicates that interaction between components is limited in a manner which can be overcome by a mobile mediator. An alternative explanation is that a reduced intermediate form may accumulate under anaerobic conditions which inhibits electron flow.

References

1. Rossi, F., Zatti, M. : Br. J. Exptl. Pathol. 45, 548-559 (1964)
2. Cross, A.R., Jones, O.T.G., Harper, A.M., Segal, A.W. : Biochem. J. 194, 599-606 (1980)
3. Cross, A.R., Jones, O.T.G., Garcia, R., Segal, A.W. : Biochem. J. 208, 759-763 (1982)
4. Gabig, T.G. : J. Biol. Chem. 258, 6352-6356 (1983)
5. Cross, A.R., Higson, F.K., Jones, O.T.G., Harper, A.M., Segal, A.W.: Biochem. J. 204, 479-485 (1982)
6. Bellavite, P., Cross, A.R., Serra, M.C., Davioli, A., Jones, O.T.G., Rossi, F. : Biochim. Biophys. Acta 746, 40-47 (1983)

FLAVOPROTEIN MONOOXYGENASES

David P. Ballou

Department of Biological Chemistry, University of Michigan, Ann Arbor Michigan, USA, 48109

Introduction

The flavoprotein monooxygenases comprise a diverse group of enzymes catalysing the following general reaction:

$$\text{Substrate} + \text{NAD(P)H} + H^{\oplus} + O_2 \xrightarrow[\text{Monooxygenase}]{\text{Flavoprotein}} \begin{matrix}\text{Oxygenated}\\ \text{Substrate}\end{matrix} + \text{NAD(P)}^{\oplus} + H_2O$$

One atom of molecular oxygen is incorporated into the substrate while the other is reduced to water. It is now 22 years since the first of these flavoprotein monooxygenases was reported by Katagiri et al.(1). Since then a great deal has been learned about the chemistries necessary for a) the efficient and controlled reduction of the flavin moiety by pyridine nucleotides, b) the formation of activated oxygen without the release of free radicals, and c) the specific oxygenation of a wide variety of compounds. Many of the essential features of these enzymes including biological relevance (2), enzymology (3, 6-8, 10,11), and model reactions (4,5,9) have been described. This paper will concentrate on some of the developments occurring since the Seventh Flavin Symposium in 1981. The catalytic cycle for all monooxygenases is described by scheme I. Reduction of the flavin (Step A) involves formation of a complex

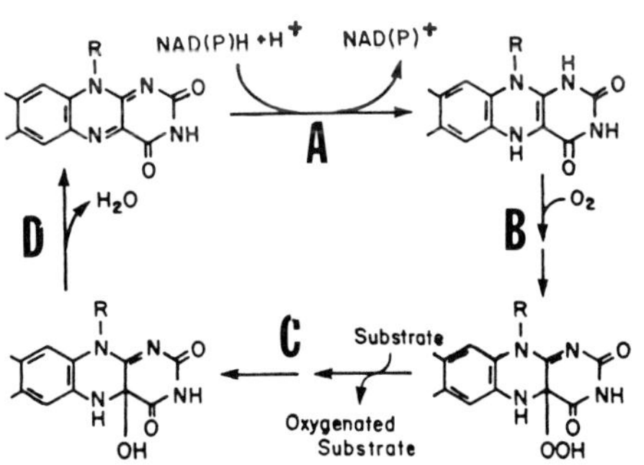

SCHEME I

Flavins and Flavoproteins

between NAD(P)H and the enzyme, frequently giving rise to charge transfer spectra. Upon hydride transfer (probably) to the N(5) of the flavin (12, 30) a second charge transfer spectrum can often be observed which decays if (and when) the NAD(P) is released (3 and references therein). Similar processes are observed for many NAD(P)H dehydrogenases.

The reaction of reduced flavins with oxygen (Part B) has been studied with many monooxygenases (13-20) and with model compounds (4, and references therein). The monooxygenases gave the first clear indication of oxygenated flavin species. Transient species were observed with absorbance maxima around 380 mm. (16,20,31). Subsequently, comparisons with spectra of model compounds (22, 23) and ^{13}C NMR studies (24) have confirmed that these species are C(4a)-hydroperoxides (Fl(4a)OOH) of the flavin (Species III in Scheme I).

Since the reaction of reduced flavin (Flred) with triplet O_2 to form the Fl(4a)OOH directly is spin forbidden, it is believed that electron transfer from Flred to O_2 first produces a flavin radical-superoxide pair with parallel spins. The flavin radical which is formed by this spin-allowed process is stabilized by electron delocalization on the flavin ring, thus enhancing its formation. After spin inversion, this radical pair rapidly collapses to form the Fl(4a)OOH. Model studies (25, 26) strongly support this hypothesis. The monooxygenases effect an extraordinary enhancement of this process. While free reduced flavins react with oxygen at a rate $\leq 50\ M^{-1}s^{-1}$ (33), frequently forming free radicals, the reduced oxygenases form Fl(4a)OOH stoichiometrically at rates typically between 10^4-$10^5\ M^{-1}s^{-1}$. The rate with cyclohexanone monooxygenases is $\geq 3X10^7\ M^{-1}s^{-1}$ (34).

Less is known of the mechanism of product hydroxylation (Step C), although this process has been subject to extensive research. This paper will address some of these features. Step D, the dehydration of the pseudobase (Fl(4a)OH), is common to all of the monooxygenases that have been studied in detail.

A useful classification for these monooxygenases is by the type of substrate oxygenated. Some enzymes catalyze the hydroxylation of activated aromatic compounds, a rather difficult reaction; the others catalyse reactions typical of classical organic peroxides. The latter group can be further subdivided into those enzymes acting upon nucleophilic substrates and those acting upon electrophilic substrates.

<u>Classical Peroxide Oxygenation of Nucleophilic Substrates</u> The prototype enzyme of this class is the mammalian FAD-containing monooxygenase (MFMO) originally isolated from hog liver microsomes by Ziegler and colleaques (see ref. 27). This enzyme has broad substrate specificity for hydrophobic nucleophiles containing nitrogen, sulfur, and selenium, as well as for iodide and thiocyanate anions. MFMO is a major protein component in hog and human liver (27); it probably acts in tandem with various cytochrome P-450's in the process of drug and xenobiotic detoxification. It may also have other functions (27) such as participation in the formation of protein disulfides. Recently, a form of MFMO induced during pregnancy and with slightly different properties has been isolated from rabbit lung (40).

In the presence of NADP, the reduced enzyme reacts quantitatively with oxygen to form a remarkably stable Fl(4a)OOH (27,32). The hydroperoxide intermediate is stabilized on the enzyme by NADP until a suitable substrate displaces the distal oxygen, thus resulting in the formation of the pseudo-base (Species IV of Scheme I-also see Fig. 1 of ref. 35). Conversion of the pseudobase to oxidized flavin is base catalyzed and is rate limiting in catalysis; it is largely unaffected by product or substrate. Thus the turnover rate with a wide variety of substrates is nearly constant (3). The oxygenation is well modeled by reactions with 4a-hydroperoxy-N(5)-ethyl-3-methyl-lumiflavin, first studied by Bruice (28). This model compound oxygenates essentially all the nucleophilic substrates of MFMO. converting the hydroperoxide to the pseudo-base, Fl(4a)OH (see fig. 3 of ref. 35).

Bruice has recently shown that the mechanism for sulfur-, nitrogen-, and iodide oxygenation reactions by peroxides is best described by a simple nucleophilic displacement of the β-peroxy oxygen (29). For a variety of peroxides including peracids, Fl(4a)OOH, and alkyl hydroperoxides, oxidation of $N\leftarrow$, $S\langle$, and I^- have similar sensitivities to the hydroperoxide structure, implying that there is a single reactivity series for oxygen donation from peroxides. The reactivity of this series of peroxides is directly related to the acidity of the leaving group (R-OH), as expected for a nucleophilic displacement. Thus the existence of special transition states for particular peracids and hydroperoxides constitutes mere perturbations on the general oxygen donation potential.

This work shows that N(5)-alkyl Fl(4a)OOH is only about 10^3-fold less reactive than m-chloro-perbenzoic acid, but 10^3-10^6-fold more reactive than alkyl hydroperoxides in these oxygenations. The Fl(4a)OH is reasonably acidic (pK 9) (29) as predicted by this reactivity. The reactivity of the Fl(4a)OOH towards nucleophiles is greater than originally predicted and its oxygenating potential may be better approximated by peracids than by hydroperoxides. MFMO behaves similarly towards these nucleophiles, but with enhanced rates (35). The enhancement may be due to a propinquity effect on the ES complex and/or a polarization of the hydroperoxide by interactions at the N(1) and O(2) positions of the flavin, as occurs with PHBH (36).

Classical Peroxide Oxygenation of Electrophilic Substrates The second class of monooxygenases catalyzes oxygenation of electrophilic carbonyl functions. Ketones are converted to esters (38) or to lactones (37), while aldehydes yield acids (8). Cyclohexanone monooxygenase (34 and refs. therein) was first isolated by Trudgill and colleagues (37) and has been the most thoroughly studied of the ketone monooxygenases. In this meeting, Itagaki and Katagiri will present studies on another ketone monooxygenase, while Trudgill will present work on one of the lactonizing enzymes of the camphor pathway.

Here, I will focus on the reactions of cyclohexanone monooxygenase. Upon mixing reduced enzyme with oxygen we observe very rapid formation of an intermediate ($k > 3 \times 10^7\ M^{-1}s^{-1}$); the intermediate has been tentatively identified as Fl(4a)OOH on the basis of its absorbance spectrum (Fig. 1 and ref. 34). When substrates are present, the Fl(4a)OOH decays to a second species which exhibits additional absorbance at about 530 nm. This absorbance property has not been observed with other monooxygenases. The rapid kinetics for this enzyme are somewhat complicated and suggest that slow hysteretic processes occur. Thus we may find that during the first turnover a portion of the enzyme is in a resting state. In subsequent turnovers when a different conformation predominates, the dynamics appear to be somewhat different. Comparison of the steady state kinetics (which show a burst) to single turnover experiments support this conclusion (44).

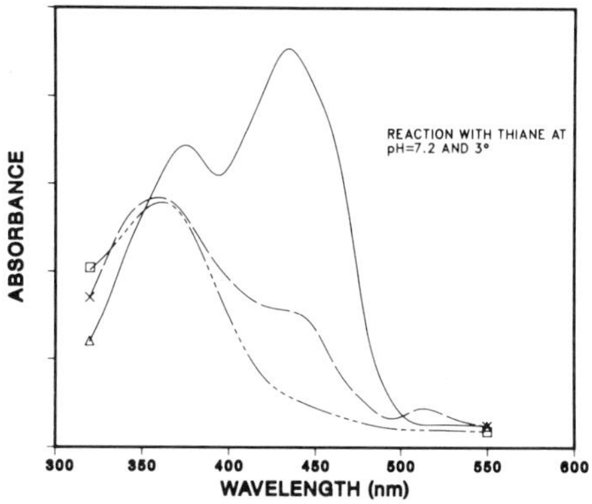

Figure 1. Reaction of cyclohexanone monooxygenase with oxygen and thiane in the stopped flow apparatus. Reduced enzyme (0.03 mM) was mixed with an equal volume of 1.25 mM oxygen. After mixing, buffer was KPi (21 mM), pH 7.2, 3 C, containing 10 mM mercaptoethanol, 0.015 mM NADP, and 0.12 mM thiane. (——) is final spectrum (Flox), (—·—·) is dead time spectrum (Fl(4a)OOH), and (- - -) is spectrum at 50 ms.

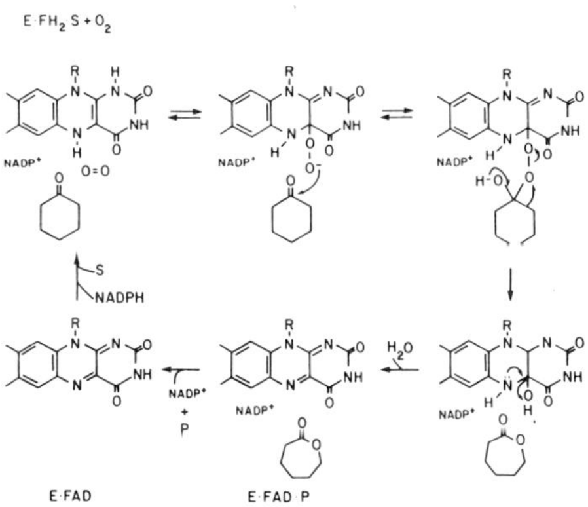

SCHEME II

Schwab (41) has elegantly shown that the enzyme-catalyzed reaction exhibits many characteristics of a Baeyer-Villiger reaction, including retention of configuration at the site of oxygen insertion, appropriate migrating tendencies of the alkyl substituents of the ketone, and retention of the substrate oxygen in the ketone functionality. Scheme II shows a postulated mechanism for cyclohexanone monooxygenase.

Branchaud and Walsh (42) have investigated a wide range of substrates for cyclohexanone monooxygenase. In general, it is found that the enzyme behaves like a peroxide-containing reagent. It readily oxygenates a variety of cyclic ketones, aldehydes and, interestingly, the highly electrophilic boronic acids. Branchaud and Walsh have also expanded earlier studies (34) on the reactivities with nucleophilic substrates. They find that sulfides, sulfoxides, selenides, triethyl phosphite and iodide are good substrates. It is clear that cyclohexanone monooxygenase has a broad specificity, hardly distinguishing between electrophilic and nucleophilic substrates. Boronic acids are also good substrates for MFMO (35), reinforcing the idea that individual oxygenases can deliver both nucleophilic and electrophilic oxygen atom equivalents typical of peracid (and hydroperoxide) chemistry.

The kinetic mechanisms of these "classical peroxide" oxygenases are quite different from the aromatic hydroxylases. In the latter systems oxidase activity (H_2O_2 production) is prevented since reduction of the enzyme is extremely slow in the absence of substrate (vide infra). With MFMO and cyclohexanone monooxygenase, reduction by NADPH in the absence of substrate readily occurs, but the subsequent reaction with oxygen produces a relatively stable Fl(4a)OOH which is discharged rapidly only after substrate binds. NADP, the specific product of the reduction reaction, remains tightly bound to reduced forms of these enzymes and stabilizes the Fl(4a)OOH, thus decreasing oxidase activity. Since these enzymes have such tolerant substrate requirements, strict substrate control of the reduction step as is found with the aromatic hydroxylases, is not practical, although substrate does control turnover.

The stability of the hydroperoxides of MFMO and cyclohexanone monooxygenases may allow these enzymes to carry out both electrophilic and nucleophilic oxygenations. Compared to the model flavin systems, these enzymes have stereoselectivities (43) and higher reaction rates, suggesting that propinquity effects are very important. It is also possible that an enzyme base which aids in making the Fl(4a)OOH more electrophilic could have a similar role in enhancing the electrophilicity of various substrates as the situation demanded. The Fl(4a)OOH could then be considered an opportunistic intermediate for these enzymes. It is notable that for each of these enzymes, electrophilic and nucleophilic substrates lead to similar spectral intermediates and rates (44). The Fl(4a)OOH forms rapidly, oxygen transfer occurs, and the rate limiting step is the breakdown of the resultant pseudobase. In the case of electrophilic substrates, the proposed Baeyer-Villiger rearrangement of the mixed peroxide must occur rapidly since no clear direct evidence for its existence can be found.

Bacterial Luciferase The production of light by bacterial luciferase has intrigued scientists from many different disciplines for years. Ziegler and Baldwin have reviewed in detail many of the interesting facets of luciferases occurring before 1981 (8). Recent progress has been substantial and strengthens the proposed reaction mechanism in Scheme III (from 10).

The luciferase-bound hydroperoxide (II) is well established (20,24) and like that of MFMO, is remarkably stable. Given the similarity of the overall reaction to that of cyclohexanone monooxygenase, the Baeyer-Villiger rearrangement shown is reasonable. Recently a Fl(4a)OH (V and VI) species has been demonstrated (45); simultaneous kinetic measurements of bioluminescence and absorbance showed that emission occurred prior to appearance of oxidized FMN (45). A species could be spectrally resolved from the kinetic studies and also isolated by chromatography (See Kurfürst et al. this volume) having a spectrum almost certainly ascribable to FL(4a)OH (see ref. 35, Figs. 1 and 3 - Note the slight blue shift relative to Fl(4a)OOH.) Moreover, it has fluorescence properties mimicking the bioluminescence (λ emission = 490 nm), making it the prime candidate for the ground state (VI) of the emitter, V. This species loses water to form stable FMN (VII). This is is consistent with studies where the bioluminescence and fluorescence properties of Fl(4a)OOH derived from different flavins correlated well (46).

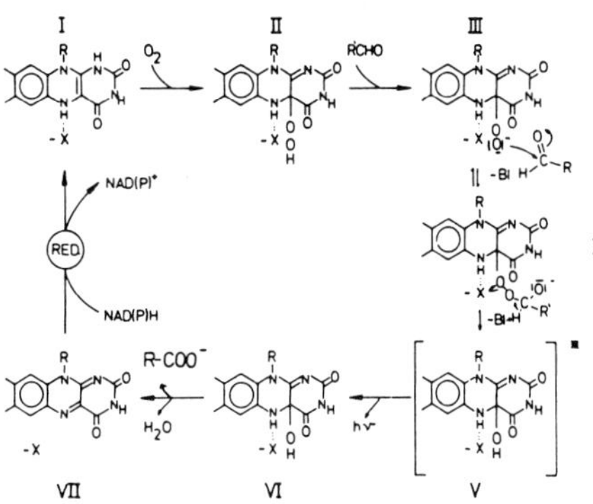

SCHEME III

Earlier work had shown the accumulation of a blue intermediate on time scales consistent with luminescence (47). This has since been demonstrated to be due to neutral semiquinone (48) which presumably forms via comproportionation of FMN and $FMNH_2$. It was also shown (49) that superoxide anion could recombine with FMNH• to form Fl(4a)OOH which could then decay, emitting light. This may account for the long lived light emission observed in earlier studies on luciferase.

While the mechanism of Scheme III would appear to be substantially correct, several questions remain. As normally studied, reactions are monitored in a single turnover mode; i.e. $FMNH_2$ is added to luciferase to initiate the reaction. It has been shown that $FMNH_2$ binds much more tightly to luciferase than does FMN (8). There is also evidence that the Kd for FMN to V. Harvei luciferase, just after the decay of Fl(4a)OOH, is approximately 10-fold lower than that for exogenously bound FMN (50). If this apparent hysteresis is real, then kinetic parameters may be different during steady state turnover than during a single turnover.

As shown in Scheme III, the reductase (RED) may actually be bound to the luciferase during steady state turnover. This needs to be investigated. The question of energy coupling to light emission is still not understood. Approximately 60 kcal/mole are required to generate the light. Whether the breakdown of the putative peroxy hemiacetal to the Fl(4a)OH is sufficient to release this energy is open for further experimentation.

Kemal and Bruice (22) have employed N(5)-ethyl Fl(4a)OOH to oxygenate aldehydes and produce chemiluminescence. Although many features of this reaction are similar to the luciferase reaction, several points of departure need be mentioned. The emission from the model system is maximal at ca. 550 nm vs. 490 nm for luciferase. The N(5)-ethyl-FL(4a)OH is nonfluorescent in solution, although the postulated FL(4a)OH on luciferase is fluorescent (47). N(5)-ethyl-Fl(4a)OOH on proteins (23) and in frozen glasses (22) is fluorescent. Thus, although the current model system offers important insight into the problem, it does not completely mimic the luciferase reaction.

The luciferase mechanism is typical of a peracid reaction. For example, McCapra and Hart (51) have reported the oxidation of long chain dialkyl sulfides by Fl(4a)OOH of luciferase. Thus at least three enzymes (see previous sections) are known in which Fl(4a)OOH can be either nucleophilic or electrophilic. This strongly suggests that these enzymes have many common active site characteristics.

Aromatic Hydroxylases The aromatic flavoprotein hydroxylases participate in the degradation of the aromatic breakdown products of lignin (one of the most abundant compounds in nature (2)) and of many man made pollutants. Most aerobic soil bacteria contain flavoprotein hydroxylases which can be genetically encoded by either chromosomal or plasmid DNA. Many of these flavoprotein hydroxylases have been studied in detail and several reports in this volume attest to the active interest in these enzymes. Information about para-hydroxybenzoate hydroxylase (PHBH) is by far the most complete of any of these enzymes. Extensive kinetic studies (3 and refs therein and this volume) and an X-ray structure (36) have made PHBH the paradigm for the aromatic hydroxylases.

All substrates for these enzymes are aromatic compounds with one or more activating substituents. Hydroxylation occurs in most cases either ortho or para to the substituents. Reduction of the flavin by NAD(P)H is tightly controlled by the presence of either the aromatic substrate or a closely similar effector molecule. Thus, reduction rates are 10^3-10^5-fold greater when the substrate is bound than when absent (3,6). Although reduction by NAD(P)H is greatly enhanced by the presence of the effector, the binding constant for NAD(P)H may be only slightly affected (52). Therefore, the presence of the effector must induce substantial conformational changes allowing rapid approach and an optimal orientation for NAD(P)H to reduce the flavin. The observation of charge-transfer intermediates (52-54) supports this idea. However, hydroxylases have different preferred methods for accomplishing this controlled reduction. In some cases the binding of substrate and NAD(P)H is random order (eg. 54) while in others, the substrate is bound first (eg 53). The ionization state of the bound substrate differs, depending upon the enzyme and the substrate. Thus a variety of protein-substrate interactions exist; hence simple generalizations may not be easily found.

The PHBH X-ray structure (36) is of the oxidized enzyme in complex with substrate. Correlation of the structure with that of glutathione reductase (56) has revealed many similarities for these two enzymes which interact with NADPH and also utilize the 4a-position of the FAD for covalent interactions. Notably, α-helices are present which point to the N-1,O-2α region (55) which could polarize the isoalloxazine ring. Attraction of negative charge towards these groups would enhance hydride transfer to N(5), as well as make the Fl(4a)OOH a better electrophile.

The reaction of oxygen with reduced forms of the aromatic hydroxylases has been of great interest for the past 12 years. Several papers in this volume describe the latest work on the important features of this set of reactions. The mechanism of Scheme I of reference 56, originally proposed by Entsch et al. (14), has been useful for discussion of PHBH. It explains most of the pertinent facts about PHBH. The Fl(4a)OOH (I in Scheme) and Fl(4a)OH (III in Scheme) are reasonably well established. However, the structure of intermediate II (pictured as a ring-opened species), has caused the most debate and is still "open" to question. The nature of this (and other species) has recently been investigated using both PHBH (56) and models of the ring-opened structure (57). The pH dependence of the spectra of II and of its formation and decay all show a pK of 7.8. Base catalysis is involved in formation of II and acid catalysis is involved in conversion of II to Fl(4a)OH (56). Similar results were obtained for PHBH modified both with 8-sulfonyl-FAD (56) and with 6-OH-FAD (58). Importantly, intermediate II has also been seen with a second aromatic hydroxylase, phenol hydroxylase from Trichosporon cutaneum (K. Detmer and V. Massey, personal communication). The spectra are somewhat different than those for PHBH, but have many similar features including the high extinction.

Bruice and colleagues (57) have synthesized an open ring lumiflavin model and have shown that its spectral properties do not fit those of II. Due to the lability of the model in water, it was not possible to convert it to the FAD level for binding to apo-PHBH. Such a test would be desirable to determine the effects the protein would have on its spectrum. The model is undoubtedly nonplanar. This may be reflected in the fact that it was also impossible to bind the model to the egg white riboflavin binding protein (L. Schopfer, A. Wessiak, V. Massey, and T. Bruice - personal communication), a protein which requires flat molecules for binding.

It is possible that the active site of PHBH presents a heterogeneous milieu not easily mimicked in solution. It might be this feature in conjunction with constraints on the mobility of the two rings, which results in the spectrum of II. Alternatively, this milieu might stabilize a flavin structure which is not ring-opened, but which has the intermediate II spectrum. This environment might even give rise to more unusual chemistry than we normally find in solution (59). Model flavin hydroperoxides in solution have never been found competent to effect aromatic hydroxylation, suggesting that with the aromatic hydroxylases specific interactions of flavin and protein (including dynamic processes) must play an extremely important role in catalysis.

References

1. Katagiri, M. Yamamoto, S. and Hayaishi, O.: J. Biol. Chem. 237, PC2413-PC2414 (1962).

2. Dagley, S.: in V. Massey, C.H. Williams (eds), Flavins and Flavoproteins, Elsevier, New York 311-317 (1982).

3. Ballou, D.P.: in V. Massey, C.H. Williams (eds), Flavins and Flavoproteins, Elsevier, New York 301-310 (1982).

4. Bruice, T.C.: in V. Massey, C.H. Williams (eds), Flavins and Flavoproteins, Elsevier, New York 265-277 (1982).

5. Mager, H.I.X.: in V. Massey, C.H. Williams (eds), Flavins and Flavoproteins, Elsevier, New York 204-293 (1982).

6. Massey, V. and Hemmerich, P.: The Enzymes, 3rd Ed. 12, 191-252 (1975).

7. Hastings, J.W. and Nealson, K.H.: Ann. Rev. Microbiol. 31, 549-595.

8. Ziegler, M.M. and Baldwin, T.O.: Current Top. Bioenergetics 12, 65-113 (1981).

9. Bruice, T.C.: Adv. Chem. Ser. 191, 88-118 (1980).

10. Massey, V. and Ghisla. S.: in Biological Oxidations (34. Colloquium, Mosbach) Springer-Verlag, Berlin, 114-139 (1983).

11. Walsh, C.: Acc. Chem. Res. 13, 148-155.

12. Walsh, C.: Enzymatic Reaction Mechanisms, Freeman, San Francisco, Chap. 11 and 12 (1979).

13. Detmer, K., Massey, V., Ballou, D.P., and Neujahr, H.: in V. Massey, C.H. Williams (eds), Flavins and Flavoproteins, Elsevier, New York 334-338 (1982).

14. Entsch, B., Ballou, D.P., and Massey, V.: J. Biol. Chem. 251, 2550-2563 (1976).

15. Presswood, R.P. and Kamin, H.: in T.P. Singer (ed), Flavins and Flavoproteins, Elsevier, NY, 145-154 (1976).

16. Strickland, S. and Massey, V.: J. Biol. Chem. 248, 2953-2962 (1973).

17. Entsch, B. Ballou, D.P., Husain, M., and Massey, V.: J. Biol. Chem. 251, 7367-7379 (1976).

18. Poulsen, L.L. and Ziegler, D.M.: J. Biol. Chem. 254, 6449-6455 (1979).

19. Ryerson, C.C., Ballou, D.P., and Walsh, C.: Biochemistry 21, 2644-2655 (1982).

20. Hastings, J.W., Balny, C., LePeuch, C., and Douzou, P.: Proc. Natl. Acad. Sci.(USA) 70, 3468-3472 (1973).

21. Bruice, T.C.: in V. Massey, C.H. Williams (eds), Flavins and Flavoproteins, Elsevier, New York 265-277 (1982).

22. Kemal, C. and Bruice, T.C.: Proc. Natl. Acad. Sci.(USA) 74, 405-409 (1977).

23. Ghisla, S., Entsch. B., Massey. V., and Husain, M.: Eur. J. Biochem. 76, 139-148 (1977).

24. Ghisla, S., Hastings, J.W., Favaudon, V., and Lhoste, J.M.: Proc. Natl. Acad. Sci.(USA) 75, 5860-5863 (1978).

25. Nanni, E.J., Sawyer, D.T., Ball, S.S., and Bruice, T.C.: J. Amer. Chem. Soc. 103, 2797-2802 (1981).

26. Anderson, R.F.: in V. Massey, C.H. Williams (eds), Flavins and Flavoproteins, Elsevier, New York 278-283 (1982).

27. Ziegler, D.M.: in Enzymatic Basis of Detoxification, Vol. 1, W.B. Jakoby (ed.), Academic Press, NY, 201-227 (1980).

28. Ball, S.S. and Bruice, T.C.: J. Amer. Chem. Soc. 102, 6498-6503 (1980).

29. Bruice, T.C., Noar, J.B., Ball, S.S., and Venkataram, U.V.: J. Amer. Chem. Soc. 105, 2452-2463 (1983).

30. Brustlein, M. and Bruice, TC.: J. Amer. Chem. Soc. 94. 6548-6549 (1972).

31. Spector, T. and Massey, V.: J. Biol. Chem. 247, 5632-5636 (1972).

32. Beaty, N. and Ballou, D.P.: J. Biol. Chem.: 256, 4619-4625 (1981).

33. Kemal, C., Chan, T.W., and Bruice, T.C.: J. Amer. Chem. Soc. 99, 7272-7286 (1977).

34. Ryerson, C.C., Ballou, D.P., and Walsh, C.: Biochemistry 21, 2644-2655 (1982).

35. Jones, K. and Ballou, D.P.: This volume.

36. Wierenga, R.K., Kalk, K.H., van der Laan, J.M., Drenth,J., Hofsteenge, J., Weijer, W.J., Jekel. P.A., Beintema, J.J., Muller, F., and van Berkel, W.J.: in V. Massey, C.H. Williams (eds), Flavins and Flavoproteins, Elsevier, New York 11-18 (1982).

37. Donoghue, N.A., Norris, D.B., and Trudgill, P.W.: Eur. J. Biochem. 63, 175-192 (1976).

38. Britton, L.N. and Markovetz, A.J.: J. Biol. Chem. 252, 8561-8566 (1977).

39. Ougham, H.J., Taylor, D.G., and Trudgill, P.W.: J. Bact. 153, 140-152 (1983).

40. Masters, B.S.S., Hale, S.E., Bedore, J.E., and Williams, D.E.: Fed. Proc. 43, 2062, #3749 (1984).

41. Schwab, J.M., Li, W., and Thomas, L.P.: J. Amer. Chem. Soc. 105, 4800-4808 (1983).

42. Branchaud, B.P. and Walsh, C.T.: Submitted (1984).

43. Light, D.P., Waxman, D.J., and Walsh, C.: Biochemistry 21, 2490-2498 (1982).

44. Jones, K., Ballou, D.P., Branchaud, B., and Walsh, C.T.: Unpublished (1984).

45. Kurfürst, M., Ghisla, S., and Hastings, J.W.: Proc. Natl. Acad. Sci.(USA) 81, 2990-2994 (1984).

46. Kurfürst, M., Ghisla, S., and Hastings, J.W.: in V. Massey, C.H. Williams (eds), Flavins and Flavoproteins, Elsevier, New York 353-358 (1982).

47. Presswood, R. and Hastings, J.W.: Photochem. Photobiol. 30, 93-99 (1979).

48. Kurfürst, M., Ghisla, S. Presswood, R., and Hastings, J.W.: Eur. J. Biochem. 123, 355-361 (1982).

49. Kurfürst, M., Ghisla, S., and Hastings, J.W.: Biochemistry 22, 1521-1525 (1983).

50. Becvar, J.E., Tu, S.C., and Hastings, J.W.: Biochemistry 17, 1807-1812 (1978).

51. McCapra, F. and Hart, R.: J. Chem. Soc., Chem. Commun. 273-274 (1976).

52. Shoun, H., Higashi, W., Beppu, T., Nakamura, S., Hiromi, K., and Arima, K.: J. Biol. Chem. 254, 10944-10951 (1979).

53. Schopfer, L.M. and Massey, V.: J. Biol. Chem. 254, 10634-10643 (1979).

54. Husain, M. and Massey, V.: J. Biol. Chem. 254, 6154-6166 (1979).

55. Wierenga, R., Drenth, J., and Schulz, G.E.: J. Mol. Biol. 167, 725-739 (1983).

56. Schopfer, L.M., Wessiak, A., and Massey, V.: This volume.

57. Wessiak, A., Noar, J.B., and Bruice, T.C.: Proc. Natl. Acad. Sci.(USA) 81, 332-336 (1984).

58. Entsch, B. and Massey, V.: This volume.

59. Visser, C.M.: Eur. J. Biochem. 135, 543-548 (1983).

THE NATURE OF THE 4a-HYDROPEROXYFLAVIN IN THE MAMMALIAN FLAVIN CONTAINING MONOOXYGENASE

Kenneth Jones and David P. Ballou

Department of Biological Chemistry, University of Michigan
Ann Arbor, MI 48109 USA

INTRODUCTION

The mammalian flavin containing monooxygenase (MFMO) is an excellent model enzyme for studying the reactions of the 4a-hydroperoxyflavin (4a-FlOOH). MFMO has a wide substrate specificity that includes many hydrophobic nitrogen and sulfur containing compounds (1) as well as the anions, iodide and thiocyanate. In the presence of $NADP^+$, but in the absence of substrate, MFMO stabilizes the 4a-FlOOH sufficiently to permit single turnover stopped-flow experiments (2). Addition of substrate to the 4a-FlOOH of MFMO causes a small, rapid blue shift in the spectrum which then decays more slowly to the oxidized flavin spectrum. Figure 1 shows the spectra of MFMO in the fully oxidized form as well as the 4a-FlOOH and the initial transient intermediate. The kinetics of these spectral changes yield information about the chemistry of the 4a-FlOOH.

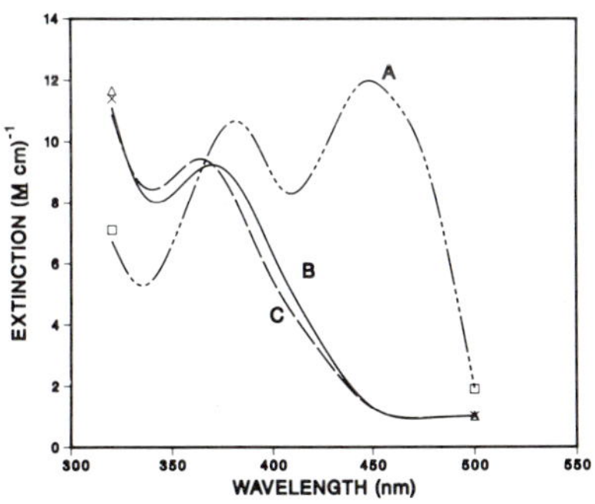

Figure 1 (A) Spectrum of MFMO fully oxidized. (B) Spectrum of MFMO immediately after reduction with stoichiometric NADPH in the presence of oxygen. (C) Spectrum of the first transient intermediate in the reaction of 4a-FlOOH with excess dimethylaniline. This spectrum was generated by computer analysis of kinetic traces at intervals of 10 nm (3).

Flavins and Flavoproteins

OXYGEN TRANSFER MONITORED WITH THIOBENZAMIDE

Thiobenzamide is oxygenated by MFMO to generate thiobenzamide S-oxide (4). Formation of the S-oxide gives a large increase in absorbance at 350 nm, where the enzyme shows very little spectral change. At 420 nm, the enzyme undergoes a small decrease in absorbance when substrate is mixed with 4a-FlOOH (figure 1). Thiobenzamide shows little absorbance change at this wavelength upon oxygenation. The two stopped-flow traces in Figure 2 show that oxygenation of thiobenzamide occurs concomitantly with the small blue shift of the enzyme that is seen with all substrates. This suggests that the spectral change is due to the oxygen transfer from the 4a-FlOOH to substrate resulting in the formation of oxygenated product and the 4a-flavin hydroxide (4a-FlOH).

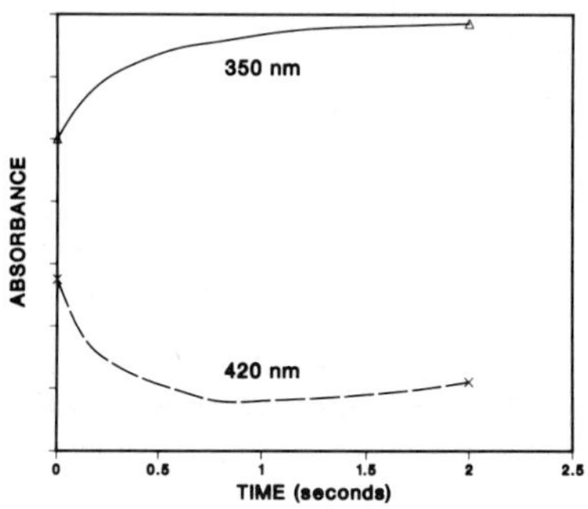

Figure 2 Traces from two stopped-flow experiments, both with 29 μM MFMO (of which approx. 80% is 4a-FlOOH and 20% fully oxidized) and 400 μM $NADP^+$ reacted with 130 μM thiobenzamide in 50 mM potassium phosphate, pH 7.2 at 3°. The absorbance scale for the 350 nm trace is eight times that of the 420 nm trace.

REACTION OF NUCLEOPHILES WITH 4a-HYDROPEROXY-5-ETHYL-3-METHYL LUMIFLAVIN

A model compound for the 4a-FlOOH of MFMO is the 4a-hydroperoxide of 5-ethyl-3-methyl lumiflavin. This was first synthesized in the laboratory of Dr. T. Bruice (5). It oxygenates a wide variety of nucleophiles including all of those tested that are substrates for MFMO. The oxygenation reaction converts the model hydroperoxide to the 4a-hydroxy flavin in a manner quite similar to that of the 4a-FlOOH of MFMO. Figure 3 shows the spectral change that occurs upon oxygenation of a nucleophile (in this case, dimethylbenzylamine) by the model peroxide.

That the hydroperoxide of lumiflavin free in solution performs the same chemistry as the 4a-FlOOH of MFMO, implies that the enzyme is not altering the chemistry of the flavin hydroperoxide in a dramatic way. However, the rate of oxygenation by the enzyme is enhanced by a large factor over the rate using the lumiflavin hydroperoxide.

Two explanations may be used to describe the enhancement of rate for the enzyme bound flavin hydroperoxide oxygen transfer. The protein may serve to activate the hydroperoxide functional group by placing strain on the flavin to result in an electron withdrawing effect. This would be necessary in making the nucleophilic substrates more susceptible to attack. Second, the enzyme has a hydrophobic binding pocket which serves to orient the nucleophilic moiety of the substrate into the proper configuration.

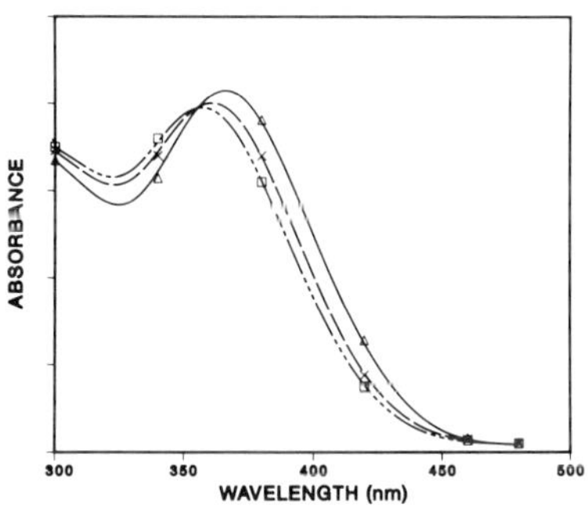

Figure 3 (Δ——Δ) Spectrum of 4a-hydroperoxy 5-ethyl-3-methyl lumiflavin in dimethylformamide at 25°. (x——x) Spectrum 67 minutes after addition of 1.5 mM dimethylbenzylamine. (□——□) Spectrum 138 minutes after addition of dimethylbenzylamine.

OXYGENATION OF ELECTROPHILIC COMPOUNDS BY MFMO

Luciferase and cyclohexanone monooxygenase are flavin monooxygenases that utilize a 4a-FlOOH intermediate. The method of oxygenation that is believed to occur is an attack of a nucleophilic hydroperoxide on an electrophilic center of the substrate (6).

Cyclohexanone monooxygenase has also been shown to be an effective oxygenating agent for sulfur containing nucleophiles (7). The ability of MFMO to react with both electrophiles and nucleophiles has been investigated. Butyl boronic acid is an electrophile that reacts with the 4a-FlOOH of MFMO and gives spectral changes that are similar for the reaction of nucleophiles with the 4a-FlOOH.

Peroxide anions are known to oxidize boronic acids as in scheme 1.

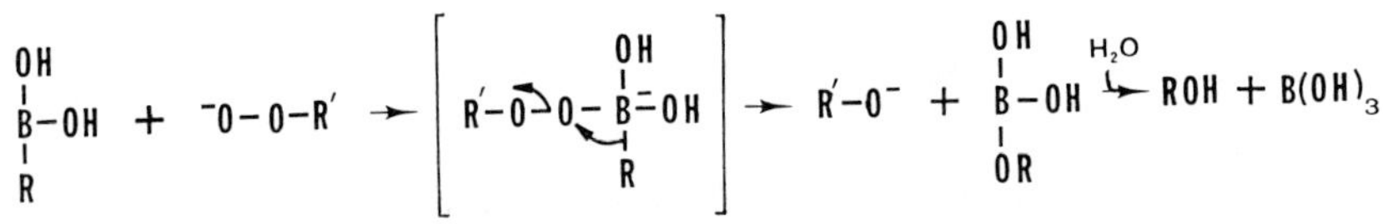

SCHEME 1

For MFMO, the alcohol product of the peroxide (depicted as $R\text{-}O^-$) is the 4a-FlOH just as it is for the oxygenation of nucleophiles. The rate of decay of the 4a-FlOH generated in oxygenation of the electrophilic boronic acids is the same as the decay of the 4a-FlOH resulting from the oxygenation of nucleophilic substrates.

REFERENCES

1. Ziegler, D.M.: Enzymatic Basis of Detoxication, Vol. 1, W.B. Jakoby (ed.) Academic Press, New York 201-227 (1980).

2. Beaty, N.B., Ballou, D.P.: J. Biol. Chem. 256 4619-4625 (1981).

3. Entsch, B., Ballou, D.P., Massey, V.: J. Biol. Chem. 251 2550-2563 (1976).

4. Cashman, John R., Hanzlik, Robert P.: Biochem. Biophys. Res. Commun. 98 147-153 (1981).

5. Kemal, C., Bruice, T.C.: Proc. Nat. Acad. Sci. USA 73 995-999 (1976)

6. Walsh, C.: Flavins and Flavoproteins, V. Massey and C.H. Williams Jr. (eds.) Elsevier, New York 121-132 (1982).

7. Ryerson, C.C., Ballou, D.P., Walsh, C.: Biochemistry 21 2644-2655 (1982).

THE EFFECTS OF PH AND IONIC STRENGTH ON THE BINDING OF NADPH AND NADPH ANALOGUES TO p-HYDROXYBENZOATE HYDROXYLASE FROM *PSEUDOMONAS FLUORESCENS*: THE IMPORTANCE OF MONOPOLE-MONOPOLE AND MONOPOLE-DIPOLE INTERACTIONS.

Robert A. Wijnands, Johan W. van Leeuwen, Willem J.H. van Berkel and Franz Müller

Department of Biochemistry, Agricultural University, De Dreijen 11, 6703 BC, Wageningen, The Netherlands

Introduction

p-Hydroxybenzoate hydroxylase from *P.fluorescens* catalyzes the conversion of p-hydroxybenzoate into 3,4-dihydroxybenzoate using NADPH as an electron donor. The substrate and several analogues were found to increase the reduction of protein-bound FAD by NADPH (1,2). It has been suggested for the enzyme from *P.desmolytica* that the substrate also shifts the pH optimum of NADPH binding to more alkaline pH values (3). This suggestion seems plausible as optimal NADPH binding to the free enzyme was found at pH 6, while optimal activity occurs at pH 8. In addition it was suggested that the pK_a value of a histidine residue is shifted to a higher pH value upon substrate binding. It was proposed that this pK_a shift is responsible for the shift of optimal NADPH binding to higher pH values (4). In later studies, however, it was shown that no histidine residues are involved in substrate binding (5,6). In order to clarify this apparent discrepancy we decided to investigate the interaction between the enzyme and NADPH in dependence of the pH and the ionic strength.

Theory

When two particles interact the interaction forces involved are either electrostatic or non-electrostatic. Only the former ones are mainly affected by

Flavins and Flavoproteins

the ionic strength of the solution. At zero ionic strength they are maximal and at infinite ionic strength they are virtually absent. The observed (overall) dissociation constant of the complex can be expressed as

$$K_d = K_{d(\infty)} \cdot K_{d(I)}$$

where $K_{d(\infty)}$ is the ionic strength independent term or the dissociation constant in the absence of electrostatic interactions and $K_{d(I)}$ is the ionic strength dependent term. $K_{d(I)}$ is given by the electrostatic energy at the encounter radius, normalized to kT, as has been described elsewhere (7). For the sake of simplicity it is sufficient to note that the electrostatic energy depends on the term Z, related to the charges of both particles, the term P, related to the dipole moment of the particles, and the term f(κ), a function of the ionic strength.

Results and Discussion

The pH and ionic strength dependence of the dissociation constant of p-hydroxybenzoate hydroxylase and NADPH is shown in figure 1. Optimal NADPH binding occurs at pH 6.4 and the dissociation constant increases with increasing

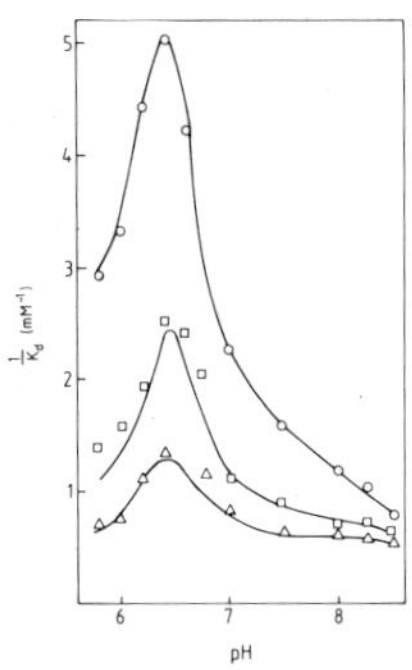

Fig. 1. The dissociation constant of the p-hydroxybenzoate hydroxylase-NADPH complex as a function of pH and ionic strength. The experimentally determined values of $1/K_d$ are plotted against pH at three different ionic strengths: I=45mM (o), 56 mM (□) and 115 mM (Δ). The solid curves represent calculated data. The enzyme concentration was 5 μM.

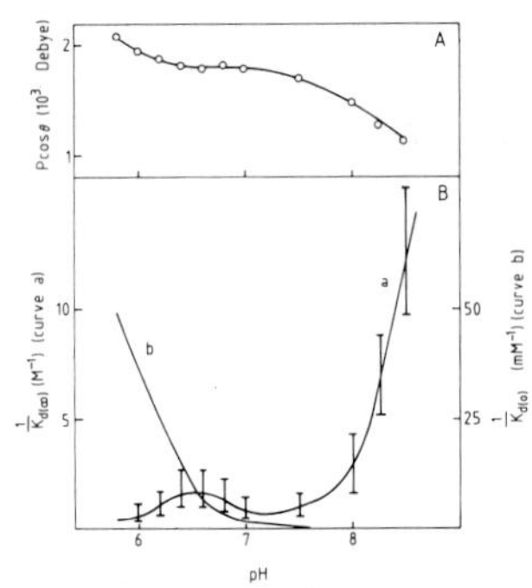

Fig. 2. Component of the dipole moment of p-hydroxybenzoate hydroxylase in the direction of the NADPH binding site (A) and dissociation constants at infinite and zero ionic strength (B) as a function of the pH (for further details see ref. (8).

ionic strength. As both the enzyme and NADPH are negatively charged at pH > 5.8 (8,9) the decrease in affinity can only be explained when asymmetry in the charge distribution over the enzyme, i.e. the dipole moment, is taken into consideration. Using suitable equations (7) the dipole moment of the enzyme in the direction of the NADPH binding site can be calculated from the data in figure 1 and the net charge of the enzyme (figure 2A) (7) Once the dipole moment is known, the dissociation constant of the p-hydroxybenzoate hydroxylase-NADPH complex can be calculated at any ionic strength (figure 2B). The theoretical results clearly show that the enzyme possesses a higher affinity for NADPH under alkaline conditions when electrostatic interactions are minimal, and under acidic conditions when electrostatic interactions are maximal. Under experimental conditions an optimum is found in between (figure 1).

The dipole moment of the enzyme also allows us to calculate the dissociation constants of p-hydroxybenzoate hydroxylase complexes with NADPH analogues, provided the non-electrostatic interactions are the same. A comparison of calculated and experimental values showed that the difference in the affinity of the enzyme for NADPH, 2',5'-ADP and NADH can be fully accounted for by the difference in charge between the three nucleotides.

Binding of 2',5'-ADP to the enzyme-substrate complex did not reveal the proposed shift of the optimal binding to higher pH values (3). A similar result was found using 1,4,5,6-tetrahydronicotinamide adenine dinucleotide 2'-phosphate, instead of 2',5'-ADP. Therefore it must be concluded that the proposed shift of the optimal binding of NADPH in the presence of substrate does not occur and that the corresponding results (3) must be ascribed entirely to the pH dependence of the ionic strength of the buffer used, as proven previously by our results (figure 3C, ref. 7).

Conclusions

The apparent discrepancy of the observations that optimal binding of NADPH occurs at more acidic and optimal activity is observed at more alkaline pH values can now be rationalized by the theory presented in this paper. Theoretically the favoured electrostatic interactions under acidic conditions also hold for the binding of $NADP^+$. It can therefore be expected that optimal activity is observed at pH values where $NADP^+$ binding is less favourable. This is indeed the case for the enzyme from P.fluorescens showing an optimal pH at about 8. The same holds for the enzyme from P.desmolytica as shown by the data presented above. This statement is supported by the facts that the physical, chemical and enzymatic properties of the two enzymes are very similar, if not identical. Previously we have mentioned (8) that attempts to purify the enzyme by a 2',5'-ADP affinity column were unsuccessful. These studies were carried out at pH values 7.5-8.5. It is now obvious that such an affinity column could still be useful if applied at pH 6.5.

Acknowledgements

We thank Mr. M.M. Bouwmans for preparing the figures, Lyda Verstege for typing the manuscript and Dr. C.A.H. Rasmussen for carefully reading the manuscript. This study was carried out under the auspices of the Dutch Foundation for Chemical Research (SON) with financial aid from the Dutch Organization for the Advancement of Pure Research (ZWO).

References

1. Hosokawa, K. & Stanier, R.Y.: J.Biol.Chem. 241, 2453-2460 (1966).
2. Spector, T. & Massey, V.: J.Biol.Chem. 247, 4679-4687 (1972).
3. Shoun, H., Higashi, N., Beppu, T., Nakamura, S., Hiromi, K. & Arima, K.: J.Biol.Chem. 254, 10944-10951 (1979).
4. Shoun, H., Beppu, T. & Arima, K.: J.Biol.Chem. 254, 899-904 (1979).
5. Shoun, H. & Beppu, T.: J.Biol.Chem. 257, 3422-3428 (1982).
6. Wijnands, R.A. & Müller, F.: Biochemistry 21, 6639-6646 (1982).
7. Wijnands, R.A., Van der Zee, J., Van Leeuwen, J.W., Van Berkel, W.J.H. & Müller, F.: Eur.J.Biochem. 139, 637-644 (1984).
8. Müller, F., Voordouw, G., Van Berkel, W.J.H., Steennis, P.J., Visser, S. & Van Rooijen, P.J.: Eur.J.Biochem. 101, 235-244 (1979).
9. Dawson, R.M.C., Elliot, D.C., Elliot, W.H. & Jones, K.M.: Data for Biochemical Research, p. 204, Oxford University Press, Oxford 1969.

KINETIC MECHANISM OF THE REDUCTIVE HALF OF REACTION CATALYZED BY SALICYLATE HYDROXYLASE

Lee-Ho Wang, Shiao-Chun Tu

Department of Biochemical and Biophysical Sciences, University of Houston-University Park, Houston, Texas 77004, U.S.A.

Introduction

Salicylate hydroxylases have been isolated from *Pseudomonas putida* (1), *Pseudomonas* sp. 29351 (2), *P.* sp. 29352 (3), and *Pseudomonas cepacia* (4,5). All four species of this hydroxylase exhibit some differences in structural and/or kinetic properties. The kinetic mechanisms for the first three species have been partially resolved (2,6,7). In the present study, we have examined the kinetic mechanism of the *P. cepacia* enzyme, with a particular emphasis on the reductive half of reaction.

Results and Discussion

Initial velocities of salicylate hydroxylase were examined at various levels of the substrates salicylate, NADH, and O_2. Double reciprocal plots of initial rates versus salicylate concentrations at a constant level of oxygen and several fixed concentrations of NADH yielded a set of linear lines that converged to a common point. Using the same data, double reciprocal plots of initial rates versus varying NADH concentrations at a constant level of O_2 and several constant concentrations of salicylate again yielded a family of converging lines. These results clearly indicate that the hydroxylase is capable of forming a ternary complex containing salicylate and NADH. Double reciprocal plots of initial rates versus NADH concentrations at a fixed concentration of salicylate and several constant levels of oxygen, however, produced a set of parallel lines. Similarly, parallel lines were also obtained from double

Flavins and Flavoproteins

reciprocal plots of initial rates versus varying salicylate concentrations at a constant level of NADH and several fixed concentrations of oxygen. These results indicate that, subsequent to the binding of salicylate and NADH, a product is released and then oxygen binds to an enzyme form which is different from the original oxidized enzyme. Our results are consistent with a reaction mechanism involving, sequentially, the binding of salicylate and NADH to form a ternary complex, the reduction of the enzyme-bound FAD by NADH, the release of NAD^+ as a product, the binding of O_2 to the reduced enzyme:salicylate complex, the formation and release of products, and the regeneration of the oxidized holoenzyme.

Results described thus far, however, do not distinguish a random from a fixed-order binding of salicylate and NADH. This was resolved by comparisons of dissociation constants for salicylate (K_{Sal}) and NADH (K_{NADH}) determined by equilibrium measurements with those deduced from the steady-state kinetic results (8). At 23°C and pH 7.6, K_{Sal} and K_{NADH} were calculated to be 12.5 µM and 0.4 mM, respectively, based on the above described steady-state measurements. Using equilibrium techniques, K_{Sal} and K_{NADH} were found to be 12 µM and 0.45 mM, respectively, corresponding well with those deduced from initial rate determinations. Following the arguments detailed previously (8,9), these observations indicate a random binding of salicylate and NADH to the enzyme.

Such a conclusion is further supported by the examination of tritium isotope effect. Salicylate hydroxylase catalyzes a slow oxidation of NADH to form H_2O_2 in the absence of any benzenoid substrate. Benzoate greatly stimulates the rate of NADH oxidation and leads to stoichiometric formation of H_2O_2 without being hydroxylated itself. Using (4R)-[4-^{3}H] NADH and benzoate as cosubstrates, a tritium isotope effect on V_m/K_m ($^TV/K$) was observed. Such an isotope effect was dependent upon the level of benzoate used (Fig. 1). Based on the analysis described previously (10), $^TV/K$ should be independent of benzoate concentration if benzoate binds to enzyme prior to NADH. On the other hand, $^TV/K$ should be reduced to unity as the benzoate concentration approaches infinity if NADH binds to the enzyme prior to benzoate. Only when benzoate and NADH bind to the hydroxylase in a random order, $^TV/K$ will decrease at higher con-

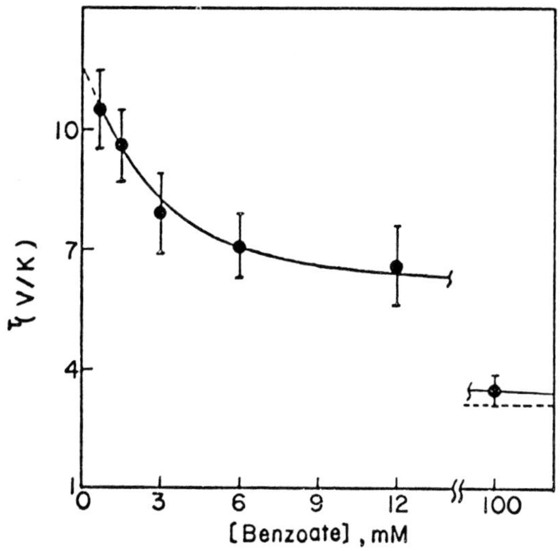

Fig. 1. Dependence of the $^{T}V/K$ effect of (4R)-[4-^{3}H]NADH on benzoate concentration. Isotope effects were determined at 23°C in 0.02 M KPi, pH 7.6, containing 18 nM enzyme, 0.1 mM tritiated NADH, and various amounts of benzoate as indicated.

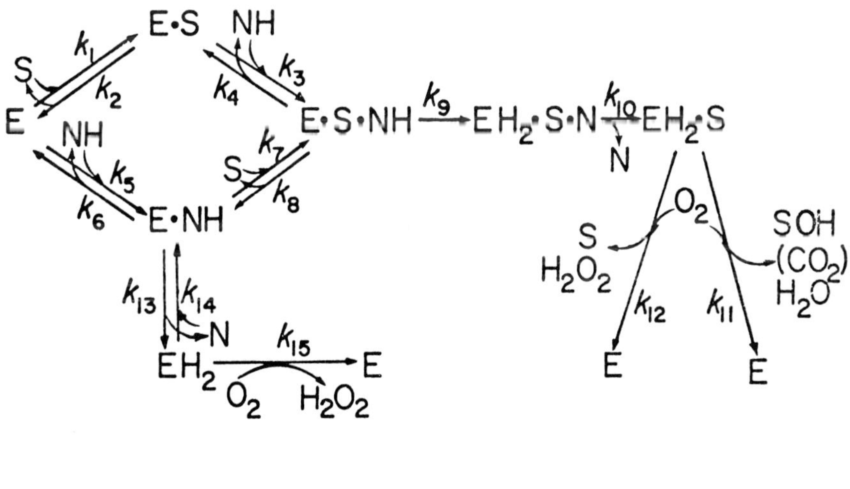

Fig. 2. Kinetic mechanism of salicylate hydroxylase. E and EH_2 are oxidized and reduced enzyme, respectively; N and NH are NAD^+ and NADH, respectively; S and SOH are substrate (or effector) and hydroxylated product, respectively. The isotope-sensitive step is indicated by k_9.

centrations of benzoate but reaches a final level significantly larger than 1 at infinite concentration of benzoate. Our results thus clearly indicate a random binding for NADH and benzoate. At 0.14 mM salicylate, (4R)-[4-^{3}H]NADH also exhibited a $^TV/K$ effect of 3.36 ± 0.37.

A mechanism is thus proposed (Fig. 2) to depict both the substrate hydroxylation and H_2O_2 formation activities of salicylate hydroxylase.

Acknowledgment

This work was supported by grants GM25953 and K04 ES00088 from National institutes of Health and by a Robert A. Welch Foundatin grant E-738.

References

1. Katagiri, M., Yamamoto, S., Hayaishi, O.:J. Biol. Chem. 237, 2413-2414 (1962).

2. Presswood, R. P., Kamin, H.:in Flavins and Flavoproteins (Singer, T. P., ed) pp. 145-154, Elsevier, Amsterdam (1976).

3. White-Stevens, R. H., Kamin, H.:J. Biol. Chem. 247, 2358-2370 (1972).

4. Tu, S.-C., Romero, F. A., Wang, L.-H.:Arch. Biochem. Biophys. 209, 423-432 (1981).

5. Wang, L.-H., Tu, S.-C.:J. Biol. Chem. in press (1984).

6. Takemori, S., Nakamura, M., Suzuki, K., Katagiri, M., Nakamura, T.:Biochim. Biophys. Acta 284, 382-393 (1972).

7. White-Stevens, R. H., Kamin, H., Gibosn, Q. H.:J. Biol. Chem. 247, 2371-2381 (1972).

8. Cleland, W. W.:in The Enzymes (Boyer, P. D., ed) 3rd Ed, Vol. 2, pp. 1-65, Academic Press, New York (1970).

9. Husain, M., Massey, V.:J. Biol. Chem. 254, 6657-6666 (1979).

10. Klinman, J. P., Humphries, H., Voet, J. G.:J. Biol. Chem. 255, 11648-11651 (1980).

THE pH-DEPENDENCE OF ENZYME-PHENOL COMPLEXES OF PHENOL HYDROXYLASE

Halina Y Neujahr
Dept of Biochemistry, The Royal Institute of Technology, S-100 44 Stockholm, Sweden

The flavoprotein phenol hydroxylase (EC 1.14.13.7; phenol, NADPH: oxygen oxidoreductase, 2-hydroxylating) contains 2 FAD and 2 subunits per 148 000 dalton. FAD is non-covalently attached. Its attachment varies (reversibly) depending on proton equilibria and temperature, as indicated by the significant changes in the absorption spectra with change in pH and temperature (cf Fig 1). FAD is released from the enzyme by chaotropes (1) and by SH-blocking reagents (2).

Phenol acts as both, effector and substrate, bringing about a conformation with greatly increased affinity towards the co-substrate NADPH (3). The conformation change caused by phenol is reflected in decreased fluorescence of the enzyme-bound FAD and in a characteristic perturbation of the absorption spectrum in its 450-480 nm region, indicating a change in the environment of FAD. From such perturbations, the dissociation constants of the enzyme-phenol complexes at different pH were determined. The data were also analyzed for cooperativity of the phenol binding sites by Scatchard and Hill plots, nomenclature as in ref 4.

Fig 2 illustrates the perturbation of the absorption spectrum by phenolat pH 7.6. There is a sharp increase of the perturbation with sub-saturating concentrations of phenol, followed by a comparatively narrow interval of saturation and a rather flat decrease of the perturbation with excess phenol. The concentration of the enzyme-phenol complex is calculated from the extent of the perturbation, $\Delta A_{478-458}$. "Enzyme" denotes here

Flavins and Flavoproteins

enzyme-bound FAD. The concentrations of free enzyme and free phenol are calculated.

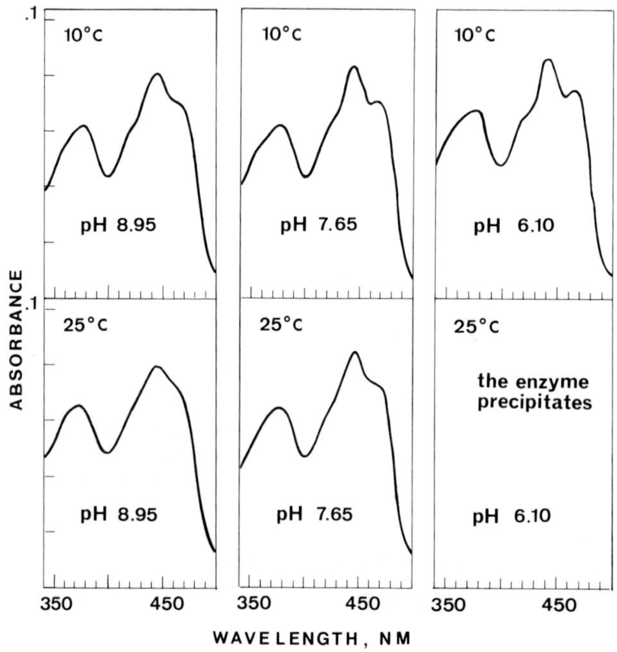

Fig 1 (left) Absorption spectra of phenol hydroxylase.

Fig 2 (right) Perturbation of the absorption spectrum of phenol hydroxylase by phenol at pH 7.6 (10°C). Difference spectra are in the inset. Enzyme-FAD was initially 12.34 µM.

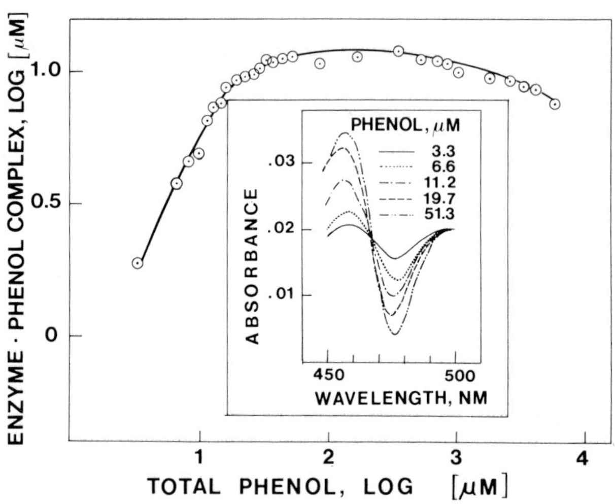

Dissociation constants of enzyme-phenol complexes are defined as

$$[\text{enzyme-phenol}] \underset{}{\overset{K_d(E)}{\rightleftharpoons}} [\text{enzyme}] + [\text{phenol}] \quad \text{and}$$

$$[\text{enzyme-phenol}] - [\text{phenol}]_x \overset{K_d(I)}{\rightleftharpoons} [\text{enzyme-phenol}] + x[\text{phenol}]$$

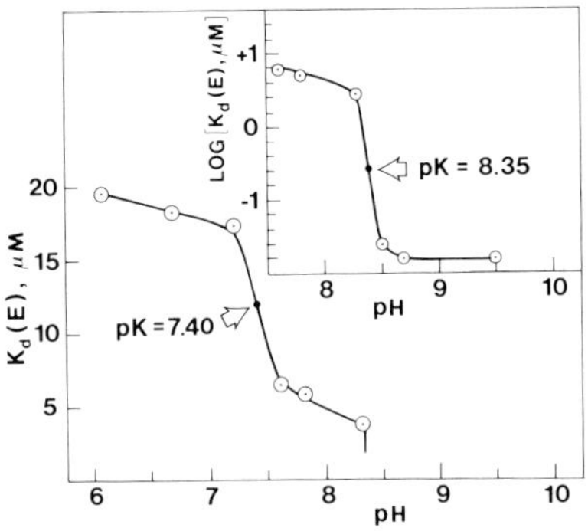

Fig 3 The pH-dependence of $K_d(E)$. E:effector

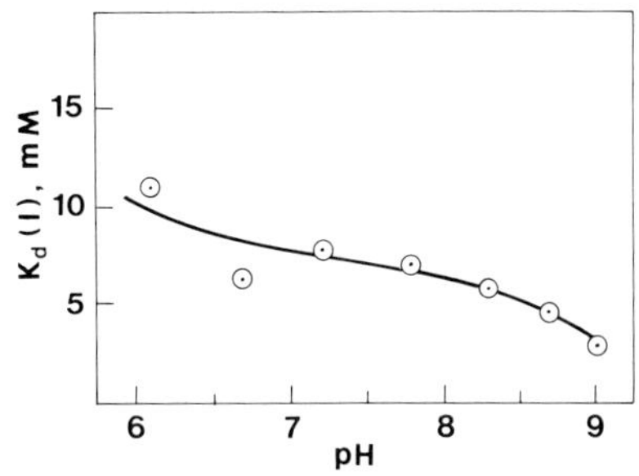

Fig 4 The pH-dependence of $K_d(I)$. I:inhibitor

The pH-dependence of $K_d(E)$ (Fig 3) shows inflexion points at pH 7.40 and pH 8.35. The pK 8.35 is easily correlated to the pK_a of cystein (8.33) and its possible involvement in the attachmont of FAD, observed earlier (2). The present results add the information that this involvement may be through proton abstraction steps, necessary for proper attachment of FAD in the conformation(s) brought about by the binding of phenol. The pK 7.4 could be interpreted by the involvement of histidyl residues in an interaction between phenol and imidazole, leading to the formation of phenolate and protonated imidazole. Allthough pK 7.4 does not seem very close to the pK_a of imidazole (6.0), the occurrence in phenol hydroxylase of essential histidyl residues with unusually high pK_a (close to 7.0) has been observed in chemical modification studies (Sejlitz & Neujahr, in preparation).

Also $K_d(I)$ decreases with increasing pH, but much less so (Fig 4). The values of $K_d(I)$ are at least three orders higher than those of $K_d(E)$. Gel size-exclusion chromatography suggests the occurrence of oligomers of the enzyme at high concentrations of phenol (4-5 mM), a tetramer and a dodecamer. Thus, the lower perturbation of the absorption spectra at high concentrations of phenol could be accounted for by partial aggregation of the enzyme.

Analysis of the binding data by Scatchard plots indicates interacting phenol sites, with positive cooperativity below pH 8.3 (Fig 5). Above pH 8.3, the pattern shows negative cooperativity. This correlates with the sharp decrease of the $K_d(E)$. A similar pattern of cooperativity is indicated by the values of the Hill coefficient: h=1.2-1.6 at pH 6.1-8.3, but h≤1.0 above pH 8.3.

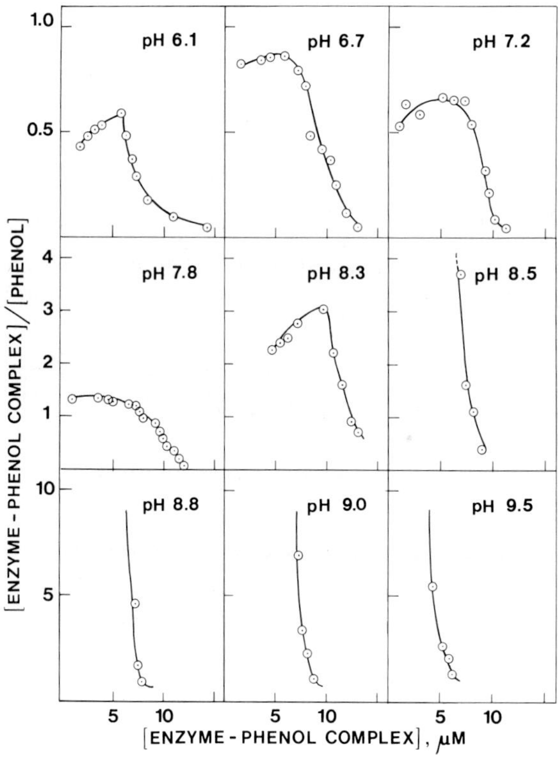

These indications of positive cooperativity suggest the occurrence of two phenol binding sites. One of them may bind phenol as effector, the other one - as substrate.

Fig 5 Scatchard plots of enzyme-phenol (effector) complexes at different pH (10°C). Note the difference of scale on y-axis.

References

1. Neujahr, H.Y.: Biochemistry, 22, 580 (1983).
2. Neujahr, H.Y., Gaal, A.: Eur.J.Biochem., 58, 351 (1975).
3. Neujahr, H.Y., Kjellén, K.G.: J.Biol.Chem., 253, 8835 (1978).
4. Hammes, G.G., Enzyme Catalysis and Regulation, pp 163-166, AP (1982).

DIFFERENTIAL INDUCTION AND CHARACTERIZATION OF FLAVOPROTEIN HYDROXYLASES IN *PSEUDOMONAS CEPACIA*

Riyad Y. Hamzah, Shiao-Chun Tu

Department of Biochemical and Biophysical Sciences, University of Houston-University Park, Houston, Texas 77004, U.S.A.

Introduction

Previously, orcinol and resorcinol hydroxylases have been shown to be differentially induced in *Pseudomonas putida* ORC cells (1). In recent years, *Pseudomonas cepacia* cells were used by us for the induction and isolation of salicylate hydroxylase (Sal HOase) (2,3,4). In this report, we demonstrate that, in addition to Sal HOase, *P. cepacia* can also be differentially induced to synthesize *m*-hydroxybenzoate-6-hydroxylase (*m*-HB HOase) and *p*-hydroxybenzoate-3-hydroxylase (p-HB HOase).

Results and Discussion

When (0.2%) 3-methylsalicylate, 4-aminosalicylate, 2,3-dihydroxybenzoate, 2,4-dihydroxybenzoate, 2,5-dihydroxybenzoate, and 2,6-dihydroxybenzoate were each used in a minimal medium (5) as the sole carbon source for *P. cepacia*, no significant cell growth was detected. However, *P. cepacia* can utilize salicylate, *m*-hydroxybenzoate, or *p*-hydroxybenzoate for growth, all with similar growth rates. The crude lysates prepared from these cells were all active in stimulating the rates of NAD(P)H oxidation in the presence of their respective inducers as cosubstrates. Sal HOase from *P. cepacia* has already been previously purified and characterized. The *m*-hydroxybenzoate- and *p*-hydroxybenzoate-induced NAD(P)H oxidation activities have now been purified 23- and 19-fold, respectively, using similar procedures involving sonication extraction, protamine sulfate treatment, ammonium sulfate fractionation, DEAE-Sephadex column chroma-

Flavins and Flavoproteins

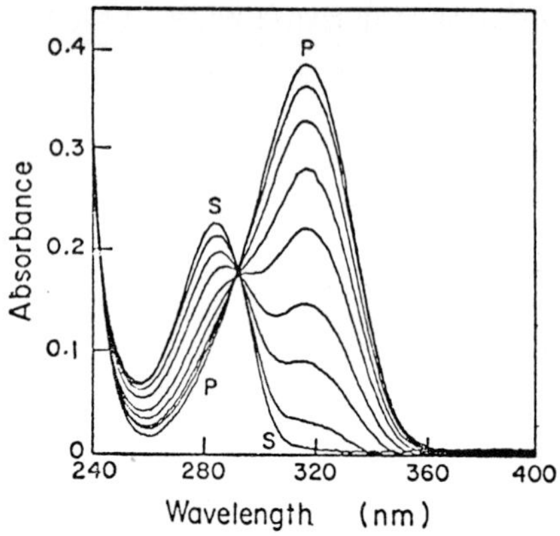

Fig. 1. Conversion of m-hydroxybenzoate (S) to gentisate (P) by the m-HB HOase. The spectrum of the final product is identical to that of a gentisate (2,5-dihydroxybenzote) standard.

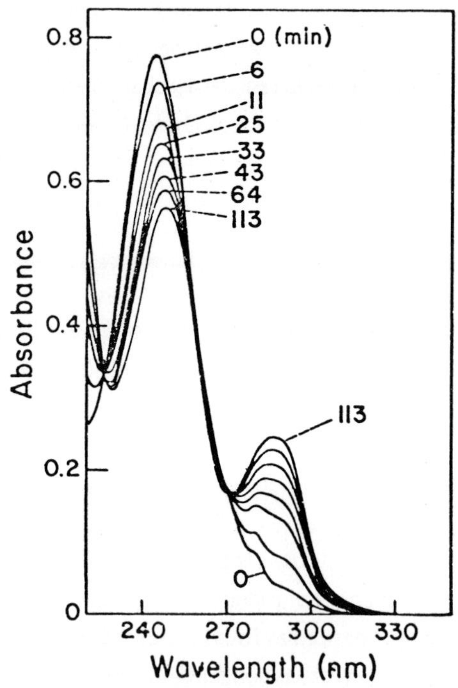

Fig. 2. Conversion of p-hydroxybenzoate (at time zero) to protocatechuate (at 113 min) by p-HB HOase. The spectrum of the final product is identical to that of a protocatechuate (3,4-dihydroxybenzoate) standard.

tography, and Sephadex G-100 column chromatography. Based on patterns of SDS polyacrylamide gel electrophoresis, the m-hydroxybenzoate-induced enzyme so obtained was apparently homogeneous whereas the p-hydroxybenzoate-induced enzyme showed a few minor bands in trace quantities.

The identities of the m-hydroxybenzoate- and the p-hydroxybenzoate-induced enzymes are determined to be m-HB HOase and p-HB HOase, respectively, by analyses of the reaction products (Figs. 1,2). This is further supported by the observations that neither the m-hydroxybenzoate-stimulated NAD(P)H oxidation by m-HB HOase nor the p-hydroxybenzoate-stimulaed NADPH oxidation by p-HB HOase form any H_2O_2.

Some general properties of the newly identified m-HB HOase and p-HB HOase are summarized in Table 1 along with that of Sal HOase for comparison.

Table 1. Comparison of properties of flavo-hydroxylases differentially induced in *Pseudomonas cepacia*

Enzyme	Cofactor	NAD(P)H requirement	Subunit structure	M.W. per monomer
Salicylate HOase	FAD	NADH > NADPH	Dimer	45,000
m-Hydroxybenzoate HOase	FAD	NADH > NADPH	Dimer	47,000
p-Hydroxybenzoate HOase	FAD	NADPH	N.D.	49,000

The m-HB HOase was also found to exhibit a tritium isotope effect on V_m/K_m ($^TV/K$) using (4R)-[4-^{3}H]NADH as an external reductant (Fig. 3). The $^TV/K$ effect was determined to be 5.0 ± 0.5. Moreover, it was independent of variations of m-hydroxybenzoate concentration (Fig. 3).

Following the kinetic isotope models described previously (6), the observed independence of $^TV/K$ on m-hydroxybenzoate concentration suggests that m-HB HOase assumes an ordered sequential binding of m-hydroxybenzoate and NADH.

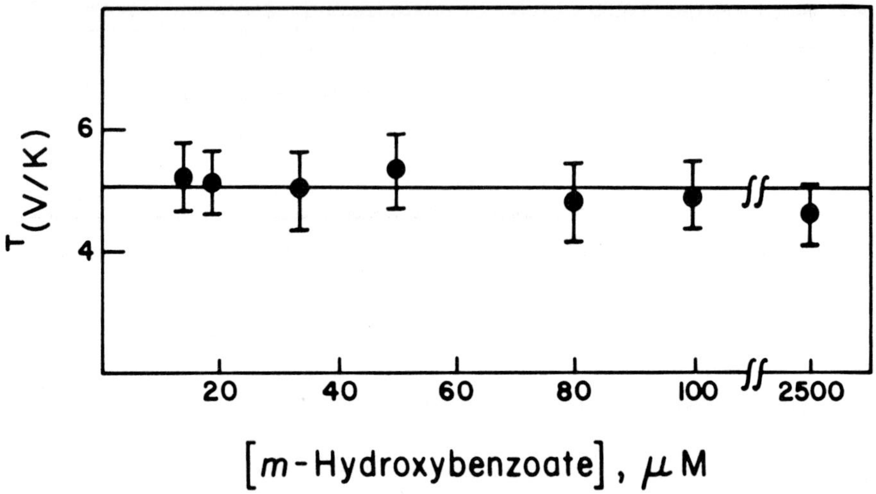

Fig. 3. Effect of m-hydroxybenzoate concentration on $^{T}V/K$ of (4R)-[4-^{3}H]-NADH. Isotope effects were determined at 23°C in 20 mM KPi, pH 7.6, containing 30 nM m-HB HOase, 1 μM FAD, 0.22 mM tritiated NADH, and m-hydroxybenzoate as indicated.

Acknowledgment: This work was supported by grants GM25953 and K04 ES 00088 from National Institutes of Health.

References

1. Ohta, Y., Ribbons, D. W.:Eru. J. Biochem. 61, 259-269 (1976).

2. Tu, S.-C., Romero, F. A., Wang, L.-H.:Arch. Biochem. Biophys. 209, 423-432 (1981).

3. Wang, L.-H., Tu, S.-C., Lusk, R. C.:J. Biol. Chem. 259, 1136-1142 (1984).

4. Wang, L.-H., Tu, S.-C.:J. Biol. Chem. in press (1984).

5. White-Stevens, R. H., Kamin, H.:J. Biol. Chem. 247, 2358-2370 (1972).

6. Klinman, J. P., Humphries, H., Voet, J. G.:J. Biol. Chem. 255, 11648-11651 (1980).

STEROID MONOOXYGENASE FROM *Cylindrocarpon radicicola*

An FAD-containing Baeyer-Villiger Type oxygenase.

Eiji Itagaki and Masayuki Katagiri

Department of Chemistry, Faculty of Science,
Kanazawa University, Kanazawa 920, Japan

Introduction

A few enzymes have been known to catalyze the insertion of an oxygen atom into a bond between carbon atoms of the substrate. Steroid monooxygenase, [ketosteroid,NADPH:oxygen oxidoreductase (oxidative esterifying, lactonizing)] from *Cylindrocarpon radicicola* ATCC 11011, reported by Rahim and Sih (1), is one of such enzymes performimg oxidation of progesterone to form testosterone acetate. Androstenedione has also been reported to be oxidized to testololactone by micro-organisms as described by Prairie and Talalay (2), but the enzyme catalyzing the latter reaction has not been studied to date (Refer Scheme I). In this study, we purified the steroid monooxygenase of *C. radicicola* to homogeneity by affinity chromatography on a pregnenolone - ligated gel, elucidating the enzyme as an FAD-monooxygenase. We also disclosed that the same enzyme catalyzed the conversion of androstenedione to testololactone.

Results

Isolation of steroid monooxygenase. The organism was grown under the conditions as described (1). Crude extract from the frozen cells was applied to DEAE-cellulose. The enzyme was eluted with 0.3M NaCl-0.03M Tris-HCl buffer, pH7.4, and then

Flavins and Flavoproteins

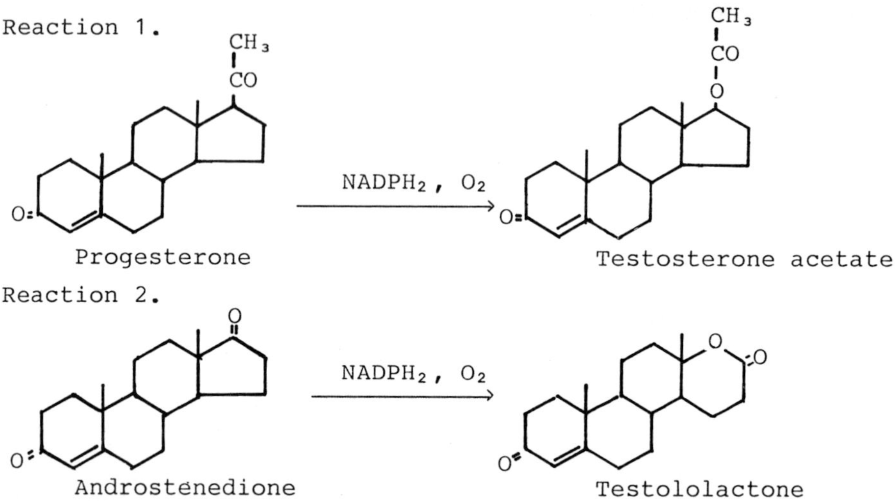

Scheme I. Enzymic reactions catalyzed by steroid monooxygenase.

applied to the affinity gel column, which had been prepared by linking the 3α-hydroxyl group of pregnenolone to Sepharose gel through N-carboxymethyl ethylenediamine bridge. After washing the gel, the enzyme was eluted with 0.03M Tris-HCl buffer, pH 7.4, containing NaCl, progesterone, NADPH, and dithiothreitol at 25°C, and was collected at 4°C. The pooled, active fractions were dialyzed against 0.03M Tris-HCl buffer, pH 7.4, and chromatographed on a DEAE - Sephacel column. This isolation procedure was shown to be very efficient in terms of both the yield and specific activity of the active enzyme giving a 1,900 fold purification from the crude extract. Assay for the enzyme activities. The enzyme is able to catalyze the two types of monooxygenase reactions shown in Scheme I. Reaction 1, the oxidative esterifying reaction of C_{21}-20-keto steroids, was termed progesterone monooxygenase reaction (PGO), and reaction 2, the oxidative lactonization of 17-ketosteroids, the androstenedione monooxygenase reaction (ADO). PGO activity was estimated by determining $[^{3}H]$ testosterone produced from $[^{3}H]$progesterone with the aid of testosterone acetate esterase in 0.03M Tris-HCl buffer, pH7.4, and ADO activity assay, by measuring the rate of

$[^3H]$testololactone production from $[^3H]$androstenedione in 0.1M potassium phosphate buffer, pH 6.5, both using NADPH as the electron donor.

Molecular properties of steroid monooxygenase. The purified enzyme preparation was in homogeneity with the criterion of gel electrophoretic analysis and exhibited an absorption spectrum typical of flavoproteins but the maximum in visible region shifted hypsochromically to 438nm. The molecular size of the enzyme was estimated to be 115,000 by sedimentation equilibrium analysis, gel filtration on Sephadex G-200 column, and gel electrophoresis. The size of subunit, 56,000, was obtained by SDS-gel electrophoresis with the reduced, denatured protein, indicating that the oxygenase is composed of two subunits of an identical size. The prosthetic group was identified to be FAD by TLC-analysis and by reconstruction experiments of the active enzyme with the apo-enzyme and flavin.

Catalytic properties. The reactions depicted in Scheme I are the two representative reactions catalyzed by the enzyme. The product of PGO was confirmed to be testosterone acetate by TLC and GLC after hydrolysing with purified testosterone acetate esterase of the same fungus. The product of ADO was identified as D -homo - 17α-oxa - androst-4-ene - 3,17-dione, (testololactone), by mass-spectrometry and GLC by comparison with the authentic compound. The enzyme also catalyzed conversion of 17α- hydroxy-20-ketosteroids to the 17-keto-androgens by cleaving the pregnane side chain without the aid of esterase. Stoichiometries of these reactions were determined to be 1:1:1 for the produced steroid : oxidized NADPH : consumed oxygen, respectively. NADPH was not oxidized by the enzyme in the absence of steroid. NADH was inactive as the electron donor for the reactions. These lines of evidence indicate clearly that the enzyme catalyzing these reactions belongs to the typical monooxygenase of EC14 category.

Reaction 1 in Scheme I is the oxidative estrification of an aliphatic ketone and reaction 2 is the oxidative lactonization of an alicyclic ketone. Both of these reactions are so-called biological Baeyer-Villiger type oxidations of ketones. The specific activity of oxygenation for progesterone and the molecular size gave the molecular activity of about 82 mol per min per mol of enzyme under the standard assay conditions. The pH-optima for the activity of PGO were pH 7.8 and of ADO, pH 6.5: PGO was more active than ADO. Substrate affinities of the enzyme were extremely strong toward progesterone and weak toward androstenedione. For oxygen and NADPH, Km values were similar to those reported for many other oxygenases. The enzyme exhibited a wide range of specificity for the substrate steroids. Results examined with several steroids having 20-keto, 20-hydroxyl, 17-keto or 17-hydroxyl group indicate that the ketosteroids, but not steroid alcohols, are the substrate for the steroid monooxygenase.

Conclusion

Steroid monooxygenase from Cylindrocarpon radicicola is an FAD-containing oxygenase of the external electron donor type, which catalyzes the oxidative esterification and lactonization of 20- and 17-ketosteroids, respectively, with the stoichiometric consumptions of NADPH and molecular oxygen.

Acknowledgments: We are pleased to acknowledge Dr. D. N. Kirk, Steroid Reference Collection, Chemistry Department, Westfield College, University of London, for the donation of steroids.

References

1. Rahim,M.A., Sih, C.J.: J.Biol.Chem. 241, 3615-3623 (1966).
2. Prairie, R.L., Talalay, P.: Biochemistry 2, 203-208 (1963).

UNUSUAL PROPERTIES OF THE FLAVOCYTOCHROME *P*-CRESOL METHYLHYDROXYLASE

William S. McIntire, Steven C. Koerber, Craig W. Bohmont, and Thomas P. Singer

Department of Biochemistry and Biophysics, University of California, San Francisco, CA 94143 and Molecular Biology Division, Veterans Administration Medical Center, San Francisco, CA 94121.

Introduction

This paper presents three aspects of the interesting and unusual flavocytochrome *p*-cresol methylhydroxylase (PCMH). These are 1) the reversible resolution into subunits and the factors affecting the association-dissociation reaction, 2) some unexpected findings on its electron-transfer mechanism which came to light during spectrophotometric titration with the substrate, and 3) the intermediates detected during rapid reaction studies with *p*-cresol and its trideuteromethyl analog.

Results and Discussion

PCMH is comprised of a single flavoprotein subunit and a single *c*-type cytochrome subunit. Table I lists the six variants of this enzyme isolated to date, two of which are produced by *P. putida*, strain 9869. The table also presents the molecular weights of the flavocytochromes and of their subunits as well as the potentials of their heme components.

Flavins and Flavoproteins

Table I: MOLECULAR WEIGHTS AND E'_o VALUES OF

p-CRESOL METHYLHYDROXYLASES[a]

source	molecular weights: flavo-cytochrome	molecular weights: subunits	E'_o (mV): flavo-cytochrome	E'_o (mV): cytochrome subunit
P. putida				
9866[b]	115,000	56,000	+ 250	+ 180
9869A[b]	114,000 (108,000)	58,000	+ 248	+ 187
9869B	99,000 (100,000)	56,000	+ 230	--
P. testosteroni				
8955	123,000	--	+ 226	--
P. alcaligenes[b]				
9867	129,000 (115,000)	--	+ 250	--
Alcaligenes sp.				
	129,000 (120,000)	--	--	--

[a] Data from references 1-3. Molecular weights were determined by ultracentrifugation for the flavocytochromes and by SDS gel electrophoresis for the subunits. Values in parentheses were determined by gel filtration.

[b] The flavocytochromes from these species dissociate on isoelectric focusing.

Table II lists some of the unusual properties of PCMH and also describes the catalytic reaction involved in the oxidation of *p*-cresol by PCMH. The enzyme also oxidizes other 4-alkylphenols besides *p*-cresol, some of which are substituted at the 2 or 3 position of the benzene ring (8).

Table II: UNUSUAL PROPERTIES OF *P*-CRESOL METHYLHYDROXYLASE[a]

(1) The enzyme oxidizes a methyl group without molecular O_2 or other activating agents. The incorporated oxygen derives from H_2O (4).

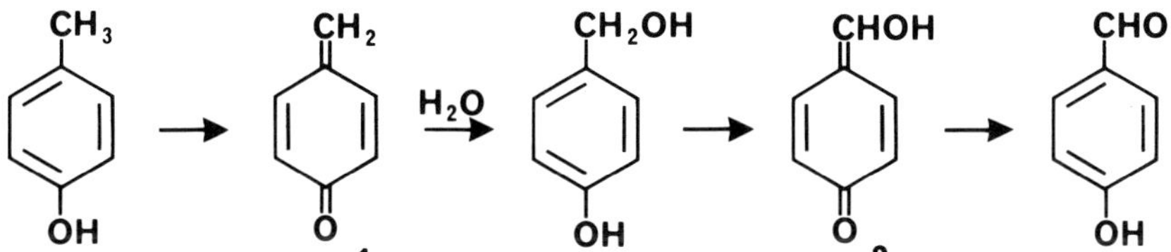

1 2

(2) The product *p*-hydroxybenzyl alcohol is also a good substrate (5).

(3) FAD is attached to the protein via an 8α-O-tyrosyl bond. This is the only enzyme known to contain this type of covalent flavin linkage (6).

(4) The flavoprotein and cytochrome subunits are separated easily by isoelectric focusing (7). On reconstitution, all properties of the native flavocytochrome are recovered (i.e., steady-state kinetic parameters, E'_o of the cytochrome component, MW, etc.).

(5) The free flavoprotein dimerizes, has only 2% of the maximal activity of the flavocytochrome, and appears to follow a different steady-state mechanism (7,8).

(6) On titration with *p*-cresol an unusual event, namely inter-enzymic electron transfer, occurs (8).

(7) The deazaflavin radical generated via flash photolysis reduces only the heme in the flavocytochrome and reduces the flavin in the flavoprotein with the formation of the *neutral* flavin radical. The deazaflavin radical generated via continuous irradiation reduces both the heme and the flavin in the flavocytochrome and reduces the flavin in the flavoprotein with the production of the *anionic* flavin radical (9).

(8) Stopped-flow studies (5) involving the reduction of the

Table II: (continued)

enzyme by *p*-cresol indicate the formation of an unknown "*flavin radical species*" which converts to the anionic flavin radical.

(9) There are 3 potentially stereospecific steps in the oxidation of 4-ethylphenol; 1) the pro-*R* or pro-*S* proton can be abstracted; 2) water can add to either the *Re* or *Si* side of the quinone methide; or 3) there can be preferential oxidation of either the *R*- or *S*-1-(4'-hydroxyphenyl)ethanol.

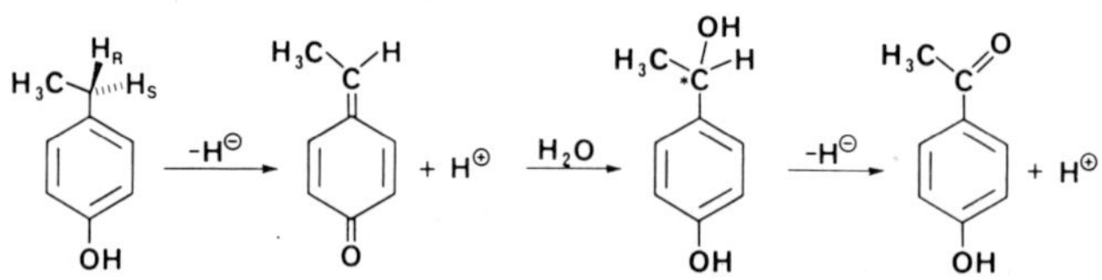

The first experiment in the analysis indicates 70% *S* and 30% *R* alcohol are formed (10).

[a] Unless otherwise stated, the properties to be discussed in this paper are for the A enzyme isolated from *P. putida* strain 9869.

We have selected from this array of interesting properties three aspects for presentation here: the process of reversible subunit dissociation, events occuring during titration of the flavocytochrome with substrate, and evidence for the occurence of two different forms of flavin radical formed sequentially after rapid mixing.

Isoelectric focusing experiments. The flavoprotein and cytochrome subunits of the three types of PCMH indicated in Table I are easily separated by isoelectric focusing (IEF) using an agarose gel support (9). During IEF separation three bands were visualized, two of which represented the subunits with the third tentatively assigned to the undissociated flavocytochrome. As the experiment progressed, this last band decreased in intensity with a concomitant increase of the other two bands. When *p*-cresol was included in the agarose matrix these enzymes no longer dissociated. The reason for this will be discussed later. When substrate was added to the agarose gel following IEF of the subunits, further focusing resulted in a small but measurable decrease in the pI value of the flavoprotein subunit (Table III). Because of technical problems the pI values of the reduced cytochromes were measured by chromatofocusing in the presense of ascorbate. The pI values are presented in Table III for the various oxidized and reduced holoenzymes and the constituent subunits studied.

Table III: pI VALUES[a]

source	subunits				flavoproteins	
	Cyt_{ox}	Cyt_{red}	Fl_{ox}	Fl_{red}	ox.	red.
P. alcaligenes						
9867	4.67 5.06[b]	-- 4.98[b]	5.47	5.40	5.06 (?)	4.97
P. putida						
9866	4.77 5.14[b]	-- 4.91[b]	5.22	5.15	5.00 (?)	4.80
9869A	4.55 4.70[b]	-- 4.46[b]	5.25	5.16	4.99 (?)	4.73 4.55[b]
9869B	--	--	--	--	5.73[c] 5.83[c]	5.58
P. testosteroni						
8955	--	--	--	--	5.0 - 5.6	5.0 - 5.3

[a] Determined by isoelectric focusing on an agarose support except where otherwise noted. [b] Determined by chromatofocusing. [c] Two major bands were observed.

In many of the experiments large quantities of the pure subunits were needed. The use of Sephadex G-200 Superfine as the IEF support medium (10) on a 7x12x0.3 cm plate permitted the processing of 3 to 50 mg of pure protein in a single run. The subunits were freed of ampholytes prior to use.

The IEF technique was further extended to allow the determination of the relative net charge on the protein from pH 3 to 9. First, a pH gradient was developed on an agarose thin-layer plate, then the protein was applied on a thin strip of filter paper transversing the gradient. Finally, an electric field was applied in a perpendicular direction for a period of time short enough to be compatible with the stability of the gradient. As a result the protein migrated a distance in the second direction proportionally to its charge, in effect producing a charge *vs.* pH titration curve (11). From pH 6 to 9 there is a small constant (but not necessarily equivalent) negative charge on each of the variant flavocytochromes, flavoproteins subunits, and cytochrome subunits regardless of the oxidation state. This is an indication that histidine residues do not contribute significantly to the net charge of these proteins (12). As the pH was lowered from a value of 6, the net charge rapidly changed from negative to positive.

<u>Dynamics of the dissociation-association process.</u> The association-dissociation of PCMH at pH 7.6 and 30 ^{o}C was studied by two techniques. One involved removal of aliquots from a dilute solution of the enzyme at known time intervals and measurement of the remaining activity. Since the free flavoprotein subunit is only 2% as active as the flavocytochrome (7), and the dissociation is slow compared with the time needed for sampling and assay, the activities accurately reflect the amount of undissociated enzyme. The rate constant for dissociation, k_{off}, was determined by a computer fit of the data to a single exponential curve. The second technique

required incubation of several solutions, each containing the same concentration of the flavoprotein subunit but different concentrations of the cytochrome subunit. Incubation was continued until a constant activity was reached. The relationship between the activity and the cytochrome concentration is described by a quadratic binding equation, and computer analysis yielded the dissociation constant (K_D). The rate constant for association, k_{on}, was calculated from the ratio k_{off}/K_D. Table IV summarizes the constants obtained.

It is important to note that the K_D of the dithionite-reduced enzyme is much lower than that of the oxidized enzyme, and that when the enzyme is reduced by excess substrate the K_D becomes even smaller, i.e., so low that it cannot be measured (Table IV). This implies that in addition to the oxidation state, the occupancy of the active site also modulates the affinity of the subunits for each other. This last point was

Table IV: OXIDATION STATE DEPENDENCE FOR THE SUBUNIT ASSOCIATION-DISSOCIATION REACTION

enzyme form	k_{off} ($\times 10^{4}$ sec^{-1})	k_{on} ($\times 10^{-3}$ M^{-1} sec^{-1})	K_d (nM)
oxidized	9.02	8.09	112
+ dithionite	2.16	8.47	25.5
+ *p*-cresol	----	----	< 5
+ *o*-cresol	----	----	< 5
+ *p*-bromophenol	----	----	< 5

10 mM phosphate, pH 7.60, I=0.03, T= 25 °C.

confirmed by the use of the competative inhibitors *o*-cresol and *p*-bromophenol which also decrease the K_D below a measurable level (Table IV). The tight binding of the reduced subunits was also observed in IEF experiments in which *p*-cresol was present (Table III), since the enzyme did not dissociate at a pH equal to its pI value. At this pH the reduced subunits were also virtually uncharged, suggesting that local charges and/or hydrophobic interactions are important for binding.

Both the k_{on} and k_{off} values increase as a function of ionic strength (Fig. 1) indicating the subunits have like charges. The IEF experiments indicate that these charges are negative. In the dithionite-reduced enzyme virutally no ionic strength dependence of the k_{on} and k_{off} values was noted. This is puzzling since there is a net negative charge on the free reduced subunits at pH 7.6.

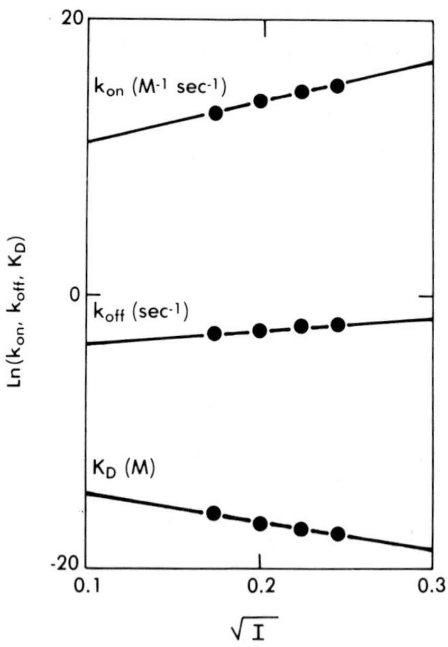

Figure 1. Plots of the natural logarithm of K_D, k_{on} (association rate constant), and k_{off} (dissociation rate constant) *vs.* the square root of the ionic strength. Conditions: 10 mM potassium phosphate; ionic strengths = 0.03, 0.04, 0.05, and 0.06 (adjusted with KCl); 25 °C.

It order to show that the results thus obtained are valid, it was required to demonstrate that the subunits generated by IEF are unaltered and that native enzyme is fully reconstituted from these subunits. To this end the steady-state kinetic properties of the reassociated enzyme were measured. The V_{MAX} and K_M values for both *p*-cresol and phenazine methosulfate (the reoxidizing substrate) were identical with those of the untreated flavocytochrome. Further, using the method of Job (13) a 1:1 stoichiometry for the subunits was observed, which is identical to that in the untreated flavocytochrome (1,2,8). Finally, Hopper (3) has found the same potentials for the heme in the unresolved and reconstituted PCMH's from *P. putida* strain 9866 and 9869 (the A enzyme).

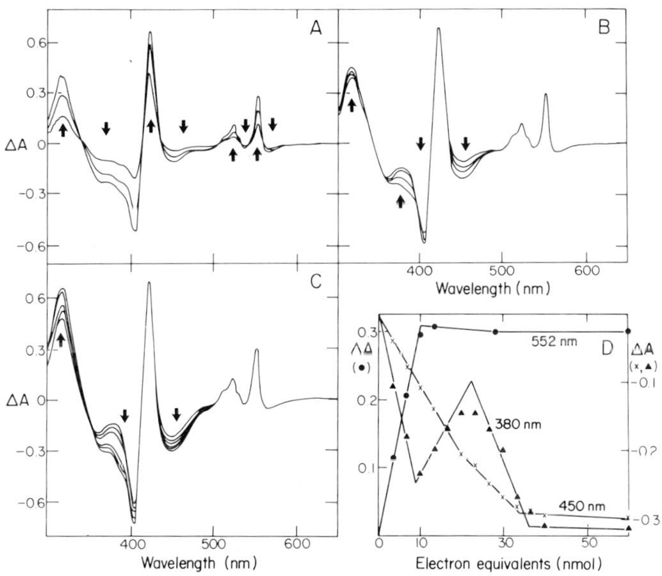

Figure 2. Reductive titration of the enzyme with *p*-cresol. The difference spectrum resulted from the subtraction of the spectrum of the oxidized enzyme from the spectrum of each reduced enzyme species. (A) heme reduction spectra; (B) formation of the anionic flavin radical; (C) formation of the fully reduced flavin; (D) absorbance change *vs*. electron equivalents derived from *p*-cresol. For (A)-(C) the arrows indicate the direction of absorbance change during reduction. Conditions: [PCMH]= 12 μM; 50 mM sodium pyrophosphate, pH 8.0; 20 °C.

<u>Reaction of the enzyme with the substrate.</u> The usual approach to investigating the redox chemistry of flavoenzymes involves spectrophotometric monitoring of absorbance changes during reductive titrations. When the flavocytochrome was titrated with dithionite (6) or *p*-cresol (Fig. 2) three distinct $1\ e^-$ phases were recorded. The first phase represents the reduction of the heme (Fig. 2A), the second formation of the anionic (red) flavin radical (Fig. 2B) which was confirmed by electron spin resonance spectroscopy (14), and the final phase full reduction of the flavin (Fig. 2C). The titration with *p*-cresol is surprising, since *p*-cresol is an obligatory $2\ e^-$ donor. The explaination of these observations must include inter-enzymic electron transfer (steps A, B, and C in the following scheme):

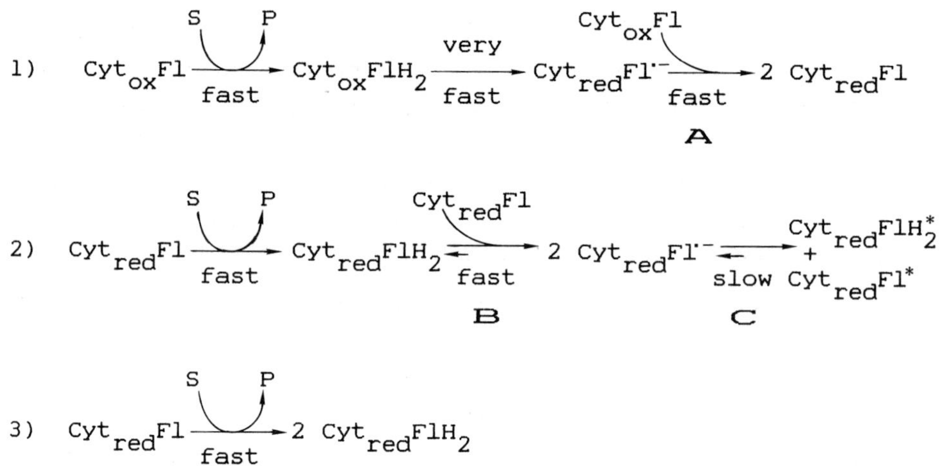

A necessary step in this scheme is the disproportionation of the flavin radical (step C). Without this step the flavoquinone form of the enzyme, the only form that can be reduced by *p*-cresol, would not be produced and thus step C would not occur. Also, when the titration was halted at any stage during the second phase, the radical disappeared via disproportionation. This seems to create a thermodynamic dilemma in that the radical is rapidly formed by a reaction between $Cyt_{red}FlH_2$ and $Cyt_{red}Fl$ (step B) and the radical then apparently converts back to these species by a different route (step C). If nothing else, this seemingly violates the principle of microscopic reversibility. A possible explaination which circumvents this dilemma would require that one or both of the species finally formed, i.e., $Cyt_{red}FlH^*_2$ and $Cyt_{red}Fl^*$, be different from those formed initially, possibly by conversion to a new conformational state. Another possibility would involve tight binding of the product so that disproportionation would occur only after its release. The underlying assumption in this discussion is that a substrate-radical mechanism is not operating. If this were not assumed and a radical mechanism were functioning, inter-enzymic electron transfer would still occur, but the flavin radical would not necessarily form by comproportionation, and the ensuing disproportionation would not create a dilemma.

These studies were extended to monitoring the reduction of the flavocytochrome by excess *p*-cresol or its trideuteromethyl analog with stopped-flow spectroscopy (5). The experiments were conducted at 6 ^{o}C because the reaction was too fast to measure at 25 ^{o}C. The fastest reaction directly observed on mixing the substrates and the enzyme was the formation of the reduced heme as monitored at 552 nm (Fig. 3). This reaction is concentration-independent and sensitive to the isotopic substitution. These facts indicate that the fast 2 e^- reduction of the flavin is followed by a much faster 1 e^- transfer to the heme. The isotope effects for this reaction are 7.05 +/- 0.22 at pH 7.6 and 5.44 +/- 0.10 at pH 6.5.

When the reaction with *p*-cresol was monitored at 385 nm, at pH 6.5, the rate constant for the apparently monophasic formation of the red flavin radical was lower than the rate constant for

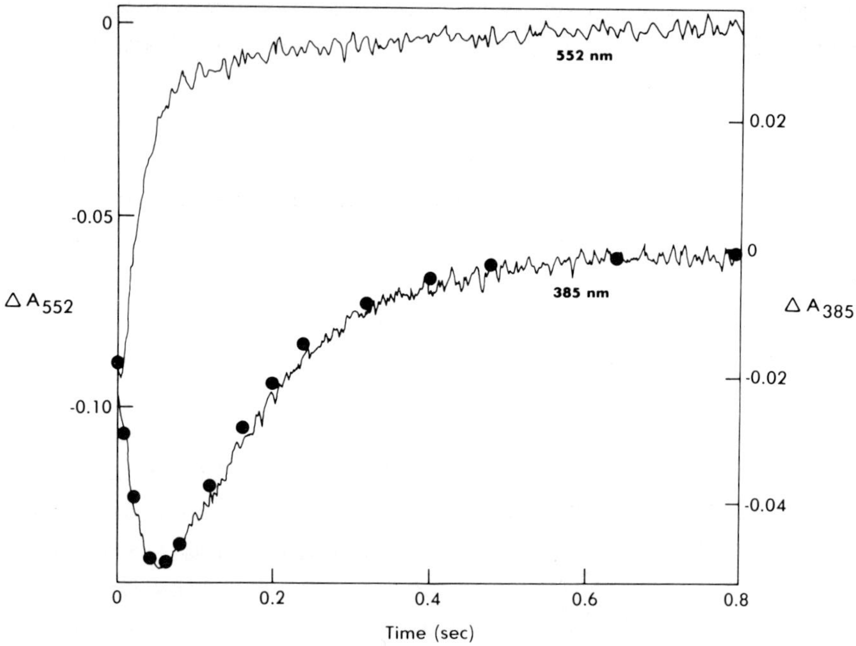

Figure 3. Stopped-flow time courses at 385 nm and 552 nm of the reduction of PCMH by *p*-cresol-α,α,α-d_3. The change at 552 nm is due predominantly to the reduction of the heme. The filled circles represent values obtained from the computer analysis described in the text. Conditions: [PCMH]= 19 μM; 25 mM sodium phosphate, pH 6.5; I=0.075; 6 ^{o}C.

the reduction of heme by a factor of 25. *p*-Cresol-α,α,α-d_3, on the other hand, produced a biphasic change at 385 nm (Fig. 4). The first phase, represented by a *decrease* in absorbance, had the same rate constant as that found for the heme reduction monitored at 552 nm (Fig. 3). In the second phase the absorbance *increased*. The rate constant for this increase was equal to that produced by *p*-cresol at 385 nm and was only a factor of 4.5 lower than the rate constant for heme reduction. The slower phases for both substrates represent the formation of the red radical. Since the formation of this radical occured more slowly than the 1 e^- reduction of the heme, and since the substrates are 2 e^- donors, it is concluded that an "unidentified flavin radical" must be formed during the fast phase and converted to the red radical in the slower phase.

The points on the curve in Fig. 3 represent the values derived from computer fitting to the data (5). The computer analysis also gave the extinction coefficients for the unknown radical and the flavin red radical at 385 nm; 26.9 and 30.6 mM^{-1} cm^{-1} respectively. These values are numerically similar to that estimated for the extinction coefficient of the red radical formed during dithionite titration of the pure flavoprotein subunit (~ 26 mM^{-1} cm^{-1}; ref. 5).

It would seem that the transient species is not the neutral flavin (blue) radical because of the large extinction at 385 nm. Also, the formation of a blue radical was not observed when the reaction was monitored at 600 nm. One may also envisage the formation of a flavin radical-product complex (15) or covalent intermediate as the transient species. The spectrum of the neutral flavin red radical is similar to that of the anionic radical (16) and could preceed the formation of the latter. This would be in accord with the large absorbance associated with the intermediate. Freeze-quench EPR studies are planned to shed light on the identity of this unknown intermediate.

References

1. Keat, M.H., and Hopper, D.J.: Biochem. J. 175, 649-658 (1978).
2. Hopper, D.J., and Taylor, D.G.: Biochem. J. 167, 155-162 (1977).
3. Hopper, D.J.: FEBS Lett. 161, 100-102 (1983).
4. Hopper, D.J.: Biochem J. 175, 345-347 (1978).

5. McIntire, W.S.: Ph.D. dissertation, University of California, Berkeley (1983).

6. McIntire, W.S., Edmondson, D.E., Hopper, D.J., and Singer, T.P.: Biochemistry 20, 3068-3075 (1981).

7. McIntire, W.S., and Singer, T.P.: FEBS Lett. 143, 316-318 (1982).

8. McIntire, W.S., Hopper, D.J., and Singer, T.P.: J. Biol. Chem., (submitted) (1984).

9. Tollin, G., McIntire, W.S., Bhattacharrya, A., and Singer, T.P.: Biochem. J., (submitted) (1984).

10. Radola, B.J.: Biochem. Biophys. Acta 386, 181-195 (1975).

11. Troungos, C., Khrishnamoorthy, R., Elion, J., and Lubie, D.: J. Chrom. 250, 73-79 (1982).

12. Righetti, P.G., Krishnamoorthy, R., Gianazza, E., and Lubie, D.: J. Chrom. 166, 455-460 (1978).

13. Job, P.: Ann. Chim. 9, 113 (1928).

14. Ackrell, B.A.C., McIntire, W.S., Edmondson, D.E., and Kearney, E.B.: Flavins and Flavoproteins (Massey, V. and Williams, C.H., eds.) pp. 488-491, Elsevier/North Holland, New York (1982).

15. Yagi, K., Sugiura, N., Okamura, K., and Kotaki, A.: Biochem. Biophys. Acta 151, 343-352 (1968).

16. Muller, F., Hemmerich, P., and Ehrenberg, A.: Flavins and Flavoproteins (Kamin, H., ed.) pp. 107-122, University Park Press, Baltimore and Butterworth, London (1971).

Acknowlegements

These studies were supported by Program Project HL 16251 from the National Institutes of Health, Grant PCM 81-19609 from the National Science Foundation, and the Veterans Administration.

IDENTIFICATION OF THE LUCIFERASE-BOUND FLAVIN-4A-HYDROXIDE AS THE PRIMARY EMITTER IN THE BACTERIAL BIOLUMINESCENCE REACTION

Manfred Kurfürst, J. Woodland Hastings

Department of Cell & Developmental Biology, Harvard University
Cambridge, Massachusetts, USA

Sandro Ghisla and Peter Macheroux

Fakultät Biologie der Universität Konstanz
D-7750, FRG

Summary

The luciferase light-emitting reaction was carried out at 1°C by mixing purified luciferase-FMN-4a-hydroperoxide with long chain aldehyde (decanal). Simultaneous kinetic measurements of bioluminescence and absorbance showed that light emission decayed more rapidly than oxidized FMN appeared, indicative of a transient intermediate species subsequent to light emission. The same species was found in reaction mixtures examined immediately after light emission was completed. Both its absorption spectrum (λmax, 360 nm) and its fluorescence emission (λmax, 490 nm) are consistent with the hypothesis that the chromophore is the luciferase-bound flavin-4a-hydroxide, the ground state of the primary emitter in the reaction. It has a relatively short lifetime (7 min at 9°C) and decays to the stable product, FMN, by losing water. The activation energy for this step was determined to be 83 kJ mol^{-1}.

Introduction

In the bacterial bioluminescent reaction the electronically excited light emitting species has not been previously identified (1,2,3). FMN, a product and a highly fluorescent molecule, could be the emitter in the

Flavins and Flavoproteins

reaction, but its fluorescence emission is centered at 530 nm while the luciferase-mediated bioluminescence peaks around 490 nm (4). Moreover, luciferase-bound FMN is non-fluorescent (5,6). Nevertheless, from studies with active flavin analogs having different fluorescence properties, it was concluded that the emitter is a flavin species (3,4,7).

The luciferase-bound flavin 4a-hydroperoxide (8,9,10) has a half lifetime of about 1 hour at 2°C in the absence of aldehyde (11, 12) and, under appropriate conditions, exhibits a high quantum yield fluorescence with an emission closely matching that of the bioluminescence (13). However, this peroxyflavin cannot be the emitter either, since it has not yet reacted with aldehyde in the catalytic pathway. The luciferase flavin-4a-hydroxide should have a similar fluorescence and, as suggested earlier (1,14), could be formed in the singlet excited state. However, this species might be elusive, since it would be expected to lose water and form FMN. The experiments described below provide evidence for the occurrence and emitter role of the hydroxyflavin chromophore (15).

Materials and Methods

Luciferase was extracted and purified from Vibrio harveyi, M-17 (16). A commercial spectrophotometer (Kontron, Uvikon, model 820) was modified so as to measure absorbance and bioluminescence simultaneously, with automatic corrections (50 times/sec) for any changes in real or apparent (e.g., bioluminescence) dark current. Fluorescence measurements were made with a Perkin-Elmer MFP-44 fluorescence spectrofluorometer. The luciferase flavin hydroperoxide intermediate was prepared and purified by molecular sieve column chromatography at 1°C (11). Spent reaction mixtures were prepared by running the luminescent reaction on an aldehyde (0.01% v/v sonicated decanal)-equilibrated Sephadex G-25 column (1 x 15 cm) in 0.35 M phosphate buffer at 1°C, pH 7.0 (15). Light emission had disappeared prior to the elution of the protein fraction (0.4 ml), which was collected (elution time, 10 min) after the void volume (4.7 ml) and transferred to a cuvette for measurement of either fluorescence or absorbance spectra.

Results and Discussion

In the absence of added aldehyde the decay of the flavin hydroperoxide is exponential and the spectral absorbance changes occur isosbestically with the appearance of oxidized FMN, much of it still enzyme bound (11). This decay (as monitored by absorbance increase at 440 nm) is paralleled by the decay of a very weak "endogenous" light emission from the preparation as well as by the loss of its capability to emit bioluminescence upon the addition of long chain aldehyde.

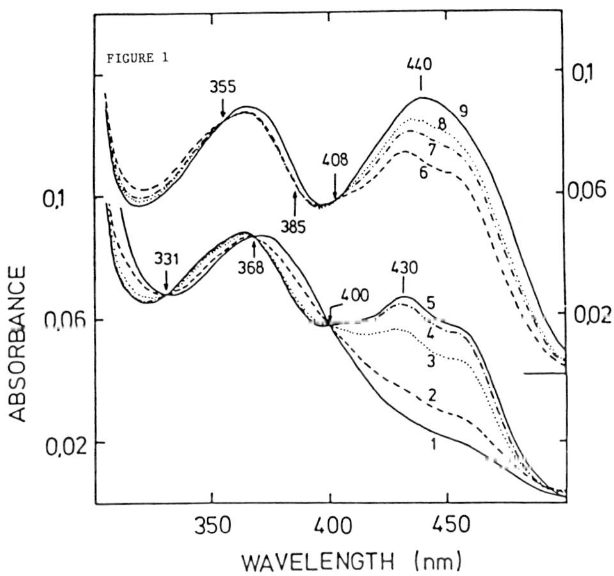

Figure 1
ABSORBANCE CHANGES WITH TIME OCCURRING UPON REACTION OF THE PURIFIED LUCIFERASE FLAVIN HYDROPEROXIDE WITH DECANAL
The hydroperoxide (8×10^{-6} M) in 0.35 M phosphate buffer pH 7.0 (trace 1, already corrected for dilution) was mixed (mixing time, <10 s) with decanal (1.6×10^{-4} M) at 1°C. Recording of trace 2 was started immediately after mixing; the scan (500 nm/min) required 55 sec for one cycle. The subsequent spectra of (traces 3-8) were started at 2, 4, 5, 10, 20 and 30 min respectively. The final spectrum (trace 9) was recorded at 1°C after warming the sample to 25°C and recooling. Arrows are used to indicate the isosbestic points.

In the present study, we measured simultaneously the changes in absorbance and in bioluminescence intensity following the addition of decanal to the isolated luciferase-flavin hydroperoxide at 1°C (Figs. 1 & 2). The absorbance changes (at 440 nm) are indicative of the appearance of oxidized FMN, but they may be divided into an early phase (traces 1-5; 0-300 sec), and a late phase (traces 6-9), characterized also by different isosbestic points.

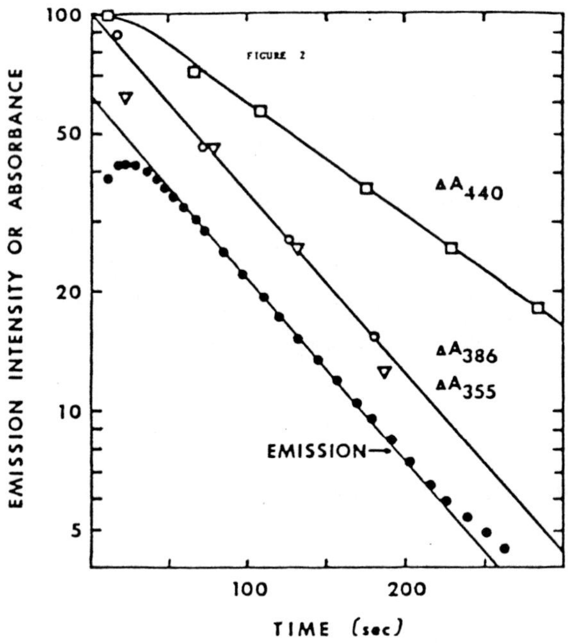

Figure 2
COMPARISON OF THE KINETICS OF SPECTRAL CHANGES AND BIOLUMINESCENCE EMISSION FOLLOWING REACTION OF LUCIFERASE HYDROPEROXIDE WITH DECANAL
The data from the experiment of Fig. 1 plotted on a semilogarithmic scale. Open symbols, absorbance changes at 440 nm (squares), 386 nm (circles), and 355 nm (triangles). The decay of bioluminescence emission (solid circles; $k = 0.62\ min^{-1}$) parallels the spectral conversions at 355 nm and 386 nm whereas the changes in absorbance at 440 nm (reflecting formation of oxidized flavin) proceed at a slower rate ($k = 0.39\ min^{-1}$). Absorbance changes were normalized so that they all extrapolated to about the same value at zero time. Emission intensity in arbitrary units.

In the earlier phase the isosbestic points (331, 368 and 400 nm) are similar to those previously reported (329, 368 and 401 nm) for the dark decay of the V. harveyi flavin hydroperoxide (10). Also, the increase in absorbance is more rapid at 355 nm ($t_{1/2}$ = 67 s; k = 0.62 min^{-1}) than at 440 nm ($t_{1/2}$ = 106 s; k = 0.39 min^{-1}), with the decay in bioluminescence paralleling the former (Fig. 2). At the same time, only about 55% of the final absorbance at 440 nm (thus oxidized FMN) has developed during this first phase even though the major fraction (70%) of the light has been emitted.

The later phase, with isosbestic points at 355 and 408 nm, involves only small absorbance changes in the 300-400 nm range, but continued changes at 440 nm. This phase may therefore involve an intermediate flavin species and its conversion to oxidized FMN. A spectrum for this species (postulated to be the luciferase-FMN-4a-hydroxide; see below) can be obtained (Fig. 3) by subtracting from an intermediate spectrum (trace 5) the appropriate amounts of both oxidized FMN (the endpoint spectrum) and the amount of still unreacted flavin-4a-hydroperoxide (trace 1), as deduced from the amount of light still to be emitted.

A preparation of the hydroxy FMN species without the peroxy species was obtained by running the reaction at 1°C on a Sephadex G-25 column preequilibrated and eluted with decanal-containing buffer. Two forms of flavin were bound to eluted luciferase, as deduced by absorbance and fluorescence spectra. A major part was in the oxidized form, with absorbance maxima in the 370 nm and 440 nm regions and fluorescence emission peaking at 530 nm. A second form was evident from its fluorescence emission in the 490 nm range (Fig. 4). Its absorbance (Fig. 3, trace C), determined by correcting the absorbance of the freshly eluted reaction mixture for the amount of luciferase-bound oxidized FMN calculated to be present, is similar to that obtained by the previous method. Its fluorescence emission spectrum (λmax, 495 nm; Fig. 4, trace F) is similar to the bioluminescence emission and to that of the fluorescence of the luciferase flavin hydroperoxide (13, 17).

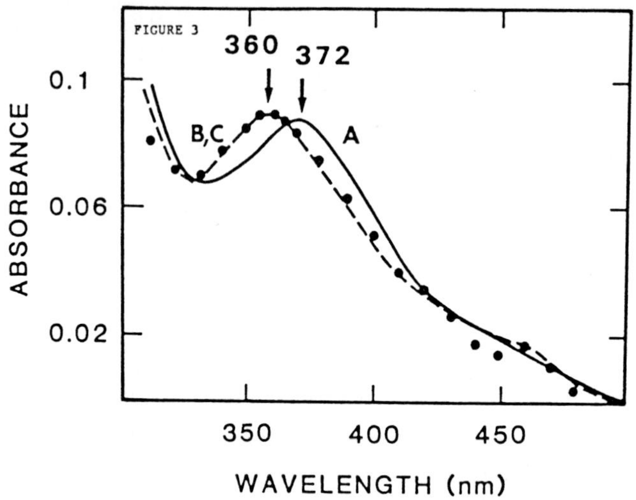

Figure 3
ABSORBANCE SPECTRA FOR THE LUCIFERASE-PEROXY AND -HYDROXY FLAVINS
Curve A (solid line) shows the spectrum of the luciferase flavin-4a-hydroperoxide (same as trace 1, Fig. 1). Curve B (dashed line) is the spectrum of the postulated hydroxy flavin, obtained by subtracting from trace 5 (Fig. 1) 55% of trace 9 (the amount of oxidized FMN already formed, calculated from the amount of peroxy and hydroxy flavin still present), 30% of trace 1 (the amount of flavin peroxy remaining in the mixture, calculated from the amount of light still to be emitted), and normalizing the obtained spectrum by a factor of 5.8 in order to compare with the flavinhydroperoxide. Curve C (solid circles) is also a spectrum calculated for the luciferase-hydroxyflavin, measured by the absorbance of a spent reaction mixture taken immediately after its elution from a column, subtracting the absorbance of the equilibration buffer (phosphate-aldehyde) and the luciferase-bound FMN present. The spectrum was normalized at 360 nm by multiplying by 2.35.

A minimal mechanism which, in our opinion, best fits the present data is shown in Scheme 1. On the assumption that k_2 and k_4 are the limiting steps, and about equal in value, the kinetic behavior of ΔA_{355} will be first order but that for ΔA_{440} will exhibit a lag. The data (Fig. 2) are consistent with this.

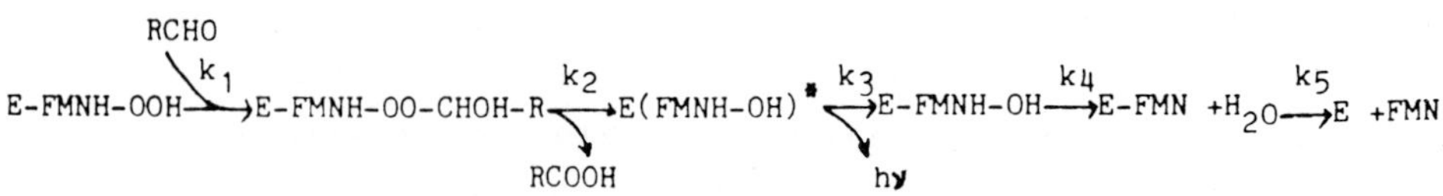

Scheme 1

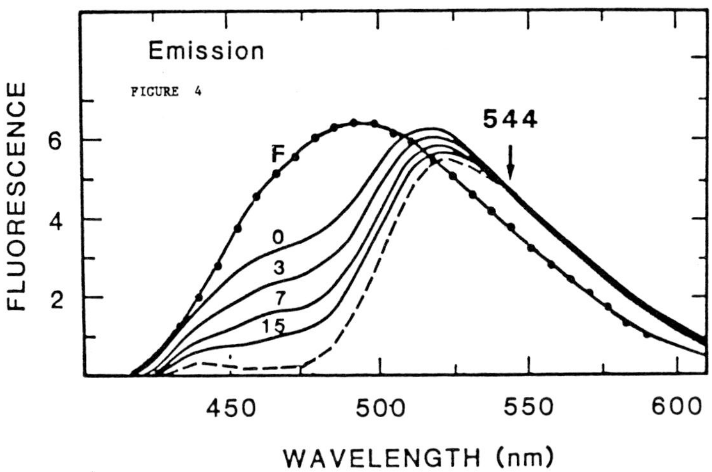

Figure 4
TIME DEPENDENT CHANGES OF THE FLUORESCENCE EMISSION SPECTRA OF THE FLAVIN INTERMEDIATE AFTER THE COMPLETED LIGHT REACTION
A solution of 50 µl luciferase (3.1×10^{-4} M), 25 µl FMN (4.5×10^{-3} M) was reduced by sodium dithionite and applied to a G-25 column, equilibrated with 0.01% v/v decanal in 0.35 M phosphate buffer, pH 7.0. The protein fraction (0.4 ml) was eluted and fluorescence spectra were recorded at 9°C. Curve 0 shows the first emission spectrum (excitation at 380 nm) immediately after the elution. The other curves, as marked, were recorded after 3, 7 and 15 min. A final spectrum (dashed line) was recorded also at 9°C after warming the sample to 25°C. The corrected fluorescence emission spectrum for the intermediate (solid points) was calculated by subtracting the fluorescence attributable to oxidized FMN and then applying the corrections for the fluorimeter. Excitation, 380 nm; sensitivity, 1; slit, 6 nm; scan speed, 120 nm/min.

The proposed catalytic cycle for the bacterial luciferase reaction is shown in Fig. 5. The conversion of the peroxyhemiacetal to the flavin-4a-hydroxide and RCOOH (Step k_2) may be written as a hydride shift (Baeyer-Villiger; 18) or equally well as a proton loss (15). Although the proposed reaction scheme does not in itself demand the formation of the hydroxide in an excited state, it accounts for the energetic requirements of light emission: the labile O-O bond disappears in the reaction step in which the postulated luminophore is created. Moreover this mechanism is in agreement with the studies of Bruice and coworkers (19) with flavin-4a-hydroperoxide model systems, and the spectra are consistent with previous work (20). The identity of the new intermediate as the hydroxide also follows from comparisons with studies of p-hydroxybenzoate hydroxylase (21), phenol hydroxylase (22), and N, S-monooxygenase (23). The spectrum of the postulated luciferase-4a-hydroxide is also consistent with the absorbance of authentic FAD-4a-hydroxide bound to p-hydroxybenzoate hydroxylase (24).

The flavin-4a-hydroxide intermediate is a clear candidate as the primary excited state and emitter in the bacterial luciferase reaction (14, 15). Such a species would also argue against proposed flavin structural rearrangements in the reaction (25, 26, 27). While the exact steps responsible for populating the excited state have not been established, both the proposed Baeyer-Villiger mechanism (18) and a postulated free-radical mechanism (28) are compatible with the formation of the flavin-4a-hydroxide.

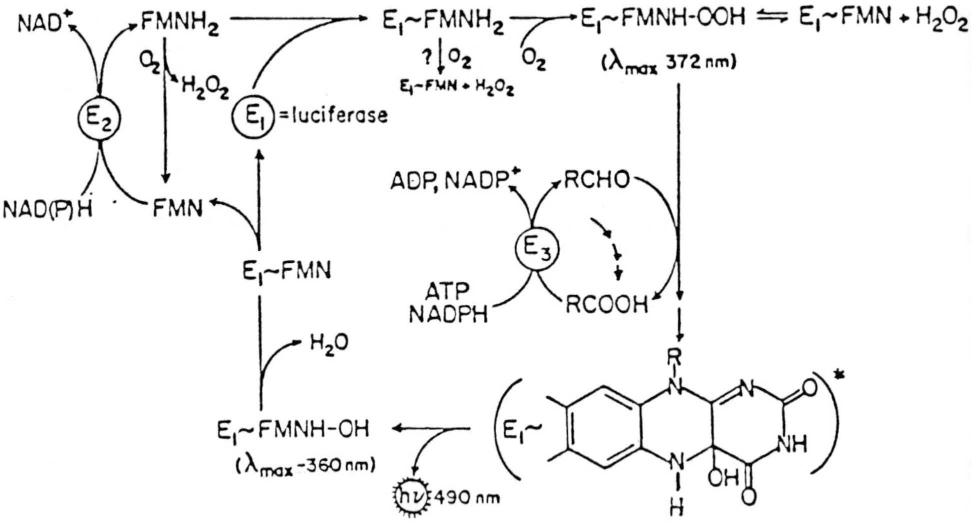

Figure 5
Scheme representing the catalytic pathway of bacterial luciferase. The reaction of the flavin peroxy with aldehyde results in the formation of the peroxyhemiacetal (not shown), followed by its conversion to the excited state hydroxide. E_1 represents luciferase; E_2, FMN reductase and E_3, fatty acid reductase.

The intermediate reported here may be the same as a transient fluorescent species reported by Matheson and Lee (29). Although they did not identify it as a flavin, they noted its occurrence during the luciferase reaction initiated by mixing $FMNH_2$ with aldehyde and oxygen, and proposed a model in which the fluorescent species is formed in the reaction, but not initially as an excited state. In our model, by contrast, the transient which exhibits fluorescence in the 490 nm region is the reaction product, namely the luciferase-bound FMN-4a-hydroxide, formed in the reaction initially in the excited state, and therefore the emitting species in the bioluminescent reaction. Emission from secondary chromophores (ones not involved in the chemical reaction *per se*) has been shown to occur in several bioluminescent systems, including certain coelenterates (30) and bacteria (31). In the bacterial system chromophores such as lumazine (32) and oxidized flavin (33), bound to distinct separate proteins, are postulated to be secondary emitters, populated by energy transfer from a primary excited state. The identity of the primary excited state as the FMN-4a-hydroxide in our scheme should be contrasted to the model of Matheson and Lee (29, 34), which assumes that the primary excited species is still unidentified and, in particular, that the flavin does not play any role whatsoever in this respect.

Supported in part by grants from the Deutsche Forschungsgemeinschaft (Project Gh 24/4) to S.G. and of the "Wissenschaftsausschuss der NATO", distributed by DAAD to M.K. and the U.S. National Science Foundation (PCM 80-19504 and 83-09414) to J.W.H.

References

1. Hastings, J.W., Wilson, T.: Photochem. Photobiol. 23, 461-473 (1976).
2. Shimomura, O.: In Chemical and Biological Generation of Excited States, pp. 249-276, eds. Adam, W. & Cilento, G., Academic Press, New York (1982).
3. Ziegler, M.M., Baldwin, T.O.: In Current Topics in Bioenergetics, ed. Sanadi, D.R., Vol. 12, pp. 65-113, Academic Press, New York (1981).
4. Mitchell, G., Hastings, J.W.: J. Biol. Chem. 244, 2572-2576 (1969).
5. Baldwin, T.O.: Biochem. Biophys. Res. Comm. 57, 1000-1005 (1974).

6. Baldwin, T.O., Nicoli, M.Z., Becvar, J.W., Hastings, J.W.: J. Biol. Chem. 250, 2763-2768 (1975).

7. Hastings, J.W., Ghisla, S., Kurfürst, M., Hemmerich, P.: In Proceedings of the Second International Congress of Chemiluminescence and Bioluminescence, pp. 97-201, eds. DeLuca, M. & McElroy, W.D., Academic Press, New York (1981).

8. Hastings, J.W., Balny, C., Le Peuch, C., Douzou, P.: Proc. Natl. Acad. Sci. USA 70, 3468-3472 (1973).

9. Ghisla, S., Hastings, J.W., Favaudon, V., Lhoste, J.M.: Proc. Natl. Acad. Sci. USA 75, 5860-5863 (1978).

10. Hastings, J.W., Balny, C.: J. Biol. Chem. 250, 7288-7292 (1975).

11. Becvar, J.E., Tu, S.-C., Hastings, J.W.: Biochemistry 17, 1807-1812 (1978).

12. Tu, S.-C.: J. Biol. Chem. 257, 3719-3725 (1982).

13. Balny, C., Hastings, J.W.: Biochemistry 14, 4719-4723 (1975).

14. Hastings, J.W., Nealson, K.H.: Ann. Rev. Microbiol. 31, 549-595 (1977).

15. Kurfürst, M., Ghisla, S., Hastings, J.W.: Proc. Natl. Acad. Sci. USA 81, 2990-2994 (1984).

16. Hastings, J.W., Baldwin, T.O., Nicoli, M.Z.: Methods in Enzymology 57, 135-152 (1978).

17. Tu, S.-C.: Biochemistry 26, 5940-5945 (1979).

18. Eberhard, A., Hastings, J.W.: Biochem. Biophys. Res. Commun. 47, 348-353 (1972).

19. Muto, S., Bruice, T.C.: J. Am. Chem. Soc. 104, 2284-2290 (1982).

20. Ghisla, S.: Methods in Enzymology 66, 360-373 (1980).

21. Entsch, B., Ballou, D.P., Massey, V.: J.Biol. Chem. 251, 2550-2563 (1976).

22. Massey, V., Ghisla, S.: In Coll. der Ges. fuer biol. Chem. Mosbach, eds. Sund, H. & Ullrich, V., Springer Verlag, Berlin, (in press, 1984).

23. Beaty, N.B, Ballou, D.P.: J. Biol. Chem. 256, 4619-4625 (1981).

24. Ghisla, S., Entsch, V., Husein, M.: Eur. J. Biochem. 76, 139-148 (1977).

25. Mager, H.J., Addink, R.: Tetrahedron Lett. 37, 3545-3548 (1979).

26. McCapra, F., Hysert, D.W.: Biochem. Biophys. Res. Commun. 52, 298-304 (1973).

27. Wessiak, A., Trout, G.E., Hemmerich, P.: Tetrahedron Lett. 21, 739-742 (1980).

28. Kosower, E.M.: Biochem. Biophys. Res. Commun. 92, 356-364 (1980).

29. Matheson, I.B.C., Lee, J.: Photochem. Photobiol. 38, 231-240 (1983).

30. Morin, J.G., Hastings, J.W.: J. Cell. Physiol. 77, 305-312 (1971).

31. Gast, R., Lee, J.: Proc. Natl. Acad. Sci. USA 75, 833-837 (1978).

32. Koka, P., Lee, J.: Proc. Natl. Acad. Sci. USA 76, 3068-3072 (1979).

33. Leisman, G., Nealson, K.H.: In Flavins and Flavoproteins, pp. 383-386, eds. Massey, V. & Williams, C., Elsevier, Amsterdam (1982).

34. Matheson, I.B.C., Lee, J.: Biochem. Biophys. Res. Commun. 100, 532-536 (1981).

STUDIES ON THE BACTERIAL LUCIFERASE REACTION: ISOTOPE EFFECTS ON THE LIGHT EMISSION. IS A "CIEEL" MECHANISM INVOLVED?

Peter Macheroux, Sandro Ghisla
Fakultät für Biologie der Universität Konstanz
D-7750 Konstanz, FRG

Manfred Kurfürst and J. Woodland Hastings
The Biological Laboratories, Harvard University
Cambridge, MA 02138, USA

Introduction

Bacterial luciferase, an FMN dependent enzyme catalyses the conversion of a long chain aldehyde to the corresponding fatty acid, with the concomitant emission of light. In the last years progress was achieved in the elucidation of the mode of activation of O_2, which involves formation of a flavin C(4a)-hydroperoxide (1,2). In a parallel presentation (3) it is shown, that emission of light goes along with formation of a flavin C(4a)-hydroxide, the latter is proposed to be the primary emitting chromophore. Several important questions still await elucidation: What is the mechanism of oxidation of the aldehyde? What is the mechanism of population of the excited emitter chromophore, i.e. how is the energy resulting from oxidation of the aldehyde converted into excitation?

Results and Discussion

Studies with deuterated aldehydes

We have reinvestigated the effects of 1-deuterated aldehydes on the rate of the luciferase reaction. Contrary to previous results (4), we have found nearly identical values for the isotope effects with octanal, decanal and dodecanal (Table 1). Very littel isotope effect (∼5 %) was found on the quantum yield using either 1-protio or 1-deutero aldehydes. Similar to the

Flavins and Flavoproteins

earlier report, the onset of light in the bioluminescence reaction is distinctly biphasic, with a rapid rise at the onset followed by a slower increase (Fig. 1); the rapid onset is strongly dependent on the concentration of aldehyde.

Table 1: Rates and isotope effects in the bioluminescence reaction

Aldehyde	k_{decay} (s^{-1})	k_{rise} (s^{-1})	$\frac{k_{decay}(H)}{k_{decay}(^2H)}$	$\frac{k_{rise}(H)}{k_{rise}(^2H)}$
Octanal	0.00146	n.d.	1.45	n.d.
(1-^{2}H)-Octanal	0.0010	n.d.		
Decanal	0.010	0.143	1.5	1.5
(1-^{2}H)-Decanal	0.0067	0.095		
Dodecanal	0.0015	0.037	1.55	1.4
(1-^{2}H)-Dodecanal	0.00097	0.026		

The rates were obtained using purified luciferase flavin-hydroperoxide (1). The values for k_{rise} and k_{decay} were obtained at 2° and at -4° respectively under the conditions described in the Legend of Fig. 1.

Schuster (5) has proposed a CIEEL (Chemically Induced Electron Exchange Luminescence) mechanism for the chemical generation of excited states. Such a mechanism can be formulated (Scheme 1, see also Mager and Addink (6)) for the breakdown of the proposed flavin peroxyhemiacetal intermediate to yield an electronically excited product upon electron transfer back to the flavin. According to Schuster one can expect a dependence of the rate of light production and decay on the redox potential of an activator, which in this case would be the reduced 4a,5-dihydroflavin hydroperoxide. To test this we have investigated the rates of decay of the bioluminescence as related to the redox potential of different FMN analogues used in the reaction. Fig. 2 shows that a correlation does exist and that, as required by the CIEEL theory, with low potential flavins a higher rate of the process is observed. In the case of bacterial luciferase, mechanisms can be formulated in which the 4a,5-dihydro-flavin nucleus acts as an activator thus initiating the reaction. If, alternatively, the bioluminescence reaction is governed primarily by a Bayer-Villiger type mechanism (7),

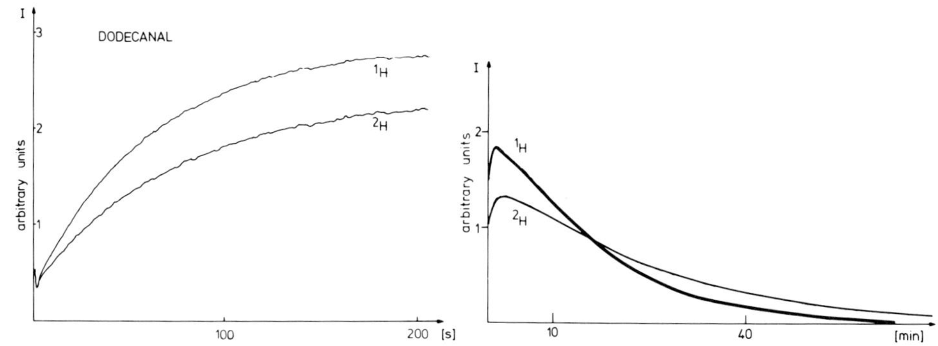

Fig. 1) Course of luminescence rise (left hand side), and of luminescence decay (right hand side) observed upon reaction of luciferase flavin-C(4a)-hydroperoxide with 1(^{1}H)- or with 1(^{2}H)-dodecanal

Sephadex purified luciferase hydroperoxide (1) in 0.01 M phosphate buffer, pH 7.0, and 0.35 M NaCl was reacted at 2° in the stopped-flow spectrophotometer with dodecanal (left). The fluorescence decay (right) was measured in a conventional cuvet at -4° under the same conditions.

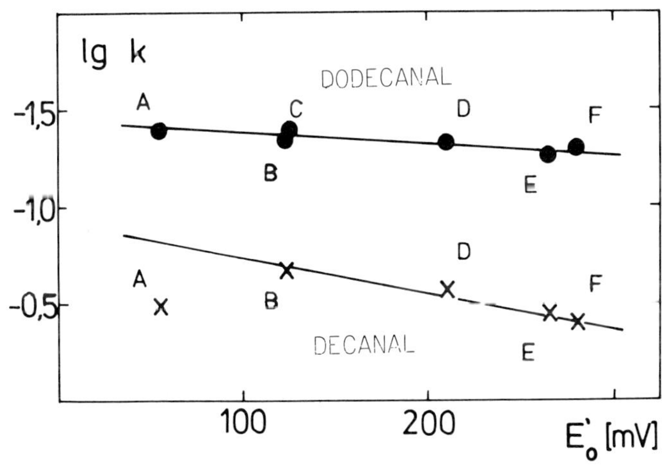

Fig. 2) Dependence of Luminescence decay rate from the FMN redox potential

The luminescence emission was measured in a ditionite assay (8) at aldehyde final concentrations of 10^{-5} M, and with 6.7×10^{-7} M luciferase. The data shown are the average of at least 5 measurements. The FMN derivatives used were (redox potential in V): A) 4-Thio-FMN (-0.055); B) 2-Thio-FMN (-0.122); C) 7,8-Dichloro-FMN (-0.125); D) FMN (-0.210); E) 6-OH-FMN (-0.265); and F) 1-Deaza-FMN (-0.280)

Scheme 1) CIEEL Mechanism for the bacterial luciferase reaction

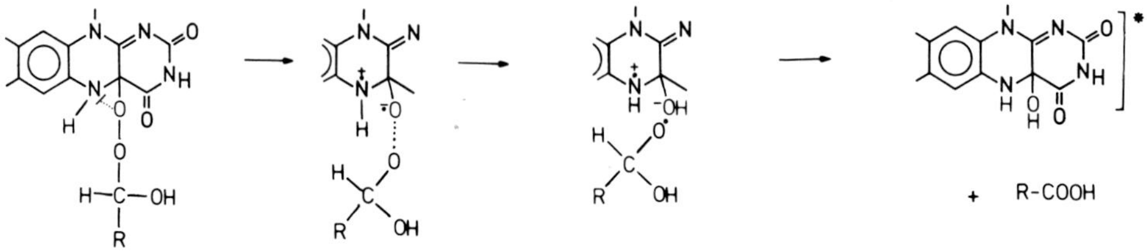

then a reverse dependence of the reaction rate from the flavin redox potential might be expected (flavins with high redox potential should yield "stronger" peroxides and with them aldehyde oxidation might be expected to be faster).

References

1. Hastings, J.W., Balny, C., Le Peuch, C. and Douzou, P.: Proc. Natl. Acad. Sci. USA 70, 3768-3472.

2. Ghisla, S., Hastings, J.W., Favaudon, V., and Lhoste, J.M.: Proc. Natl. Acad. Sci. USA 75, 5860-5863 (1978).

3. Kurfuerst, M., Ghisla, S., Macheroux, P., Hastings, J.W.: this volume.

4. Presswood, R., Hastings, J.W., Photochem. Photobiol. 30, 93-99 (1977).

5. Koo, J.Y., Schuster, G.B.: J. Amer. Chem. Soc. 99, 6107 (1977).
Schuster, G.B.: Acc. Chem. Res. 12, 366-373 (1979).

6. Mager, H.X. Addink, R.: this volume.

7. Hastings, J.W., Nealson, K.H.: Ann. Rev. Microbiol. 31, 549-595 (1977).

8. Hastings, J.W., Baldwin, T.O., Nicoli, M.Z.: Methods in Enzymology 57, 135-152 (1978).

9. Wilson, Th.: "Singlet Oxygen" (Frimer, A., ed.) CRC Press, Inc. (1984) in press.

PROBES FOR THE ACTIVE SITE OF BACTERIAL LUCIFERASE

Frank McCapra, Christopher S. J. Walpole and Ian P. Street
School of Chemistry and Molecular Sciences, University of Sussex, Brighton BN1 9QJ, U.K.

Introduction

In spite of many years of intensive effort, the mechanism of the light emitting step which occurs during the oxidation of long chain aldehyde by bacterial luciferase has not been defined. However the main reactants and products are known largely as the result of the work by Hastings and his collaborators[1]. The accepted scheme is shown.

$$FMNH_2 + E \quad \underset{I}{FMNH_2 \cdot E} \xrightarrow{O_2} \underset{II}{FMNH(OOH) \cdot E}$$

$$FMNH(OOH) \cdot E + RCHO \longrightarrow FMNH(OH) \cdot E + RCO_2H + \text{light.}$$

It has recently been confirmed (this volume) that flavin 4a-hydroxide is the immediate product and the likeliest emitter in the unperturbed system.

However no mechanistic explanation for the generation of the excited state is available and model compounds, so successful in the elucidation of the mechanism of light emission in the firefly and coelenterates[2], do not appear to be relevant. Other observations make identification of the primary excited state problematical. Lee[3] has shown that a blue fluorescent protein with ribityl lumazine as its chromophore accepts energy from bacterial luciferase with a shift in the emission maximum from 494 to 476 nm. The overlap between the emission spectrum of flavin 4a hydroxide (λmax 490nm) and the absorption spectrum of the lumazine chromophore (λmax 414 nm) is negligible.

Flavins and Flavoproteins

Transfer of energy from the lower flavin excited state to the more energetic lumazine is therefore improbable. An excited state sufficiently energetic to populate both chromophores may exist, and we describe preliminary experiments designed to examine this point.

We also report an interesting observation concerning a co-operative effect between the substrate-related sulphydryl inhibitor 2-bromodecanal and $FMNH_2$.

Materials and Methods

Luciferase was extracted from freshly grown cells of Photobacterium phosphoreum by the method of Hastings et al.[4]. For the inhibition studies luciferase (3.97 mg ml^{-1}) was incubated with 2-bromodecanal alone or in the presence of $FMNH_2$ obtained by the reduction of FMN by $Na_2S_2O_4$ or by EDTA and light in oxygen free solution. Reactions were carried out in 0.2 M phosphate buffer at pH7 and 25 ± 0.2^{o}. Assays of enzyme activity, measured as light intensity were made by injecting $FMNH_2$ into an aerated solution of ensyme in the presence of decanal. Energy transfer experiments were carried out using the reductase associated with the luciferase preparation and NADH as reductant (turn-over assay). 2-Bromodecanal was synthesised by a modification of the method used in the synthesis of 2-bromopentanal[5]. The synthesis of the fluorescent aldehydes will be described elsewhere. Analytical and spectral data were in agreement with the structures shown.

Fluorescent Aldehydes as Substrates

The specificity of the enzyme for its aldehyde substrates is high, but there is evidence that modification of the chain sufficiently remote from the aldehydic group preserves

Reaction of fluorescent aldehydes with luciferase

Aldehyde	I_o(rel)	R[a]	R[b]	R[c]
Decanal	1.0	6.5	6.5	6.5
(1)	0.7	5.0	3.6	3.5
(2)	0.1	5.1	4.1	6.5
(3)	0.02	-	-	-

(a) (1) at 2×10^{-5}M, (2) at 1.5×10^{-5}M

(b) (1) at 4×10^{-5}M, (2) at 3×10^{-5}M

(c) Total solution (b) diluted 10 fold

$$R = \frac{I_o\ 490\text{nm}}{I_o\ 535\text{nm}}$$

activity[6]. Three aldehydes have been synthesised, and in spite of the large end groups, shown to be substrates with varying degrees of activity in a turn-over assay. The chain length may not yet be optimal but since for light emission the aldehyde must be at the active site, the fluorescent acceptor is ideally placed for energy transfer by the Förster mechanism. The results in the table were obtained using a specially constructed spectrometer which delivers a ratio of light emission at two wavelengths (490 and 535 nm in this experiment). Its construction will be described elsewhere. The flavin-substituted aldehyde is an effective substrate, giving about 70% of the initial intensity obtained using decanal. The lumazine derivative(3) on the other hand is a poor substrate. Alteration of the structure and chain length may improve the light yield. The nitrobenzoxadiazole derivative(2) appears also to bind non-specifically and to dissociate more readily than the flavin derivative as shown by the change in ratio on dilution. Other aldehydes covering the range 350-530 nm are being synthesised and the results will be reported shortly.

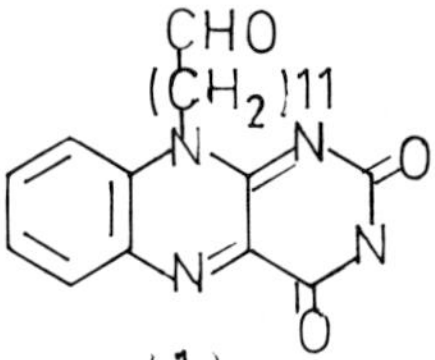

(1)

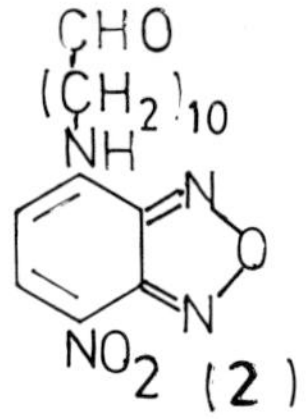

(2)

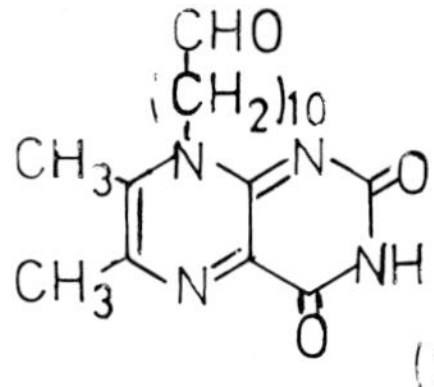

(3)

2-Bromodecanal as an Inhibitor

Various maleimides are inhibitors of the luciferase, by reaction with an essential sulphydryl group[7]. A closer substrate analogue, 2-bromodecanal, has been synthesised and found to be a more effective inhibitor. Although FMN protects against inhibition[7], we find that $FMNH_2$ causes a remarkable <u>increase</u> in the rate of inactivation (figure). Such a syncatalytic effect on the binding of a second substrate has been discussed and ascribed to changes in enzyme conformation[8]. Perhaps the binding of $FMNH_2$ exposes the thiol to the reagent. Titration showed that the group inhibited had a pKa of between 9.0 and 9.5.

Inactivation of Luciferase

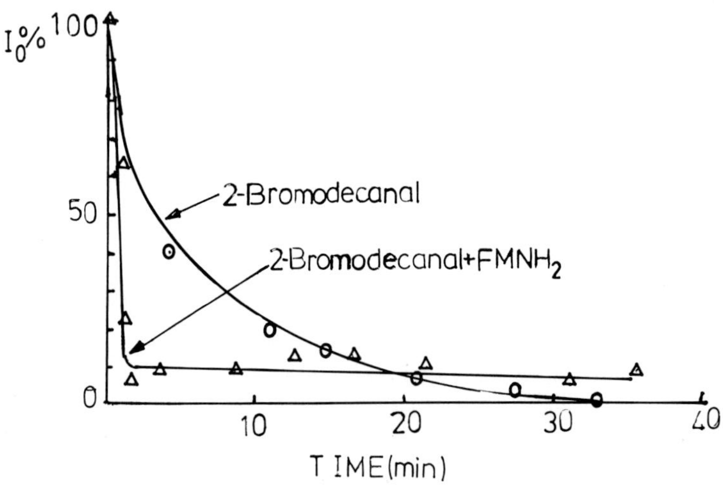

References

1. Nealson, K.H. and Hastings, J.W.: Microbiol.Revs. 43, 496 (1979).

2. McCapra, F.: Proc.R.Soc.Lond. B215, 247 (1982).

3. Lee, J., Carreira, L.A., Gast, R., Irwin, R.M., Koka, P., Small, E.D. and Visser, A.J.W.G.: Bioluminescence and Chemiluminescence, M. DeLuca and W.D. McElroy (eds.), Academic Press, 1981, p.103.

4. Hastings, J.W., Baldwin, T.O. and Nicoli, M.Z.: Methods in Enzymology 57, 135 (1978).

5. Erlenmeyer, H. and Jung, J.P.: Helv.Chim.Acta 32, 35 (1949).

6. Hastings, J.W., Gibson, Q.H., Friedland, J. and Spudich,J.: Bioluminescence in Progress, Johnson, F.H. and Haneda, Y. (eds.), Princeton University Press 1966.

7. Nicoli, N.Z., Meighen, E.A. and Hastings, J.W.: J.Biol. Chem. 249, 2385 (1974).

8. Christen, P. and Gehring, H.: Methods Biochem.Anal. 28, 151 (1982).

THE CELLOBIOSE OXIDOREDUCTASES OF SPOROTRICHUM PULVERULENTUM

Fraser F Morpeth

National College of Food Technology, University of Reading, Whiteknights, Reading, Berks, RG6 2AP, U.K.

Introduction

Enzymes capable of oxidising cellobiose to the corresponding lactone have been found in the extracellular medium of a number of fungi. The "white rot" fungi *Sporotrichum pulverulentum* produces three isoenzymes (1). Two of these enzymes have been purified and partially characterised, one was found to be a flavoprotein (2) and one a flavocytochrome (3). The flavoprotein isoenzyme has been named cellobiose dehydrogenase and the flavocytochrom cellobiose oxidase. This was on the basis of cellobiose dehydrogenase having the ability to reduce quinones and cellobiose oxidase, oxygen. Eriksson has claimed on this basis that the dehydrogenase is involved in lignin biodegredation and the oxidase in the oxidative stimulation of cellulose degredation. The cellobiose oxidising isoenzymes of *S. pulverulentum* and the ability of several fungi to produce these activities have been examined to test these hypothesis.

Materials and Methods

S. pulverulentum was grown in a minimal salt medium (4) with absorbant cotton wool as the sole carbon source. Less enzyme activity was released by the fungi when other forms of cellulose with lower degrees of crystallinity were substituted. Other fungi were grown on the same medium, plus biotin if required, except *Aspergillus niger* where soluble carboxymethyl cellulose was the carbon source. Cellobiose oxidoreductase activity was monitored at 25°C and 600 nm in 2.5 ml of a solution containing 0.125 mmol potassium phosphate buffer, pH 6, 1 μ mol of cellobiose and 60 nmol of

Flavins and Flavoproteins

DCPIP. The three cellobiose oxidoreductases were purified and separated as follows. Extracellular (cell free) growth media (32L) was concentrated to 2L. The pH of the media was adjusted to 6 and it was clarified by centrifugation. Ammonium sulphate was added to a final concentration equal to 60% saturation and the precipitate was desalted by passage through a column of Bio-Gel P2 equilibrated in 5 mM potassium phosphate buffer pH 7.6. The elutant was concentrated to 50 ml by ultrafiltration and loaded onto a DEAE-Sephadex column (100 cm x 2.5 cm) equilibrated in the above buffer. The column was washed with 5 mM potassium phosphate buffer 7.6 and then with a linear gradient from 5 mM to 80 mM phosphate buffer pH 7.6.

Results and Discussion

Table 1 shows the maximal level of cellobiose oxidoreductase activity produced by a number of fungi.

TABLE 1

Organism	Activity	moles of cellobiose oxidised min^{-1} ml^{-1} of culture medium
Sporotrichum pulverulentum	4×10^{-9}	
Trichoderma reesi	0	
Trichoderma koningii	0	
Coriolus versicolor	3×10^{-9}	
Penicillium funiculosam	1.1×10^{-9}	
Fusarium solani	1.4×10^{-9}	
Aspergillus niger	0	

All those fungi which had cellobiose oxidoreductase activity showed multiple bands on activity staining of native gels. *T. reesi* and *T. koningii* are both very efficient at cellulose degredation yet have only a very limited ability to degrade lignin. This suggests that cellobiose oxidase does not play an essential role in crystalline cellulose degredation as has been suggested. All the fungi in Table 1 which have cellobiose oxidoreductase activity are also capable of biodegrading lignin. Lignin and cellulose are always found together in nature and liginolysis cannot proceed in the absence of an energy source which is usually cellulose (5). Thus it is not surprising that en enzyme using a cellulose breakdown product is involved in lignin biodegredation.

FIGURE 1 CHROMATOGRAPHY ON DEAE SEPHADEX

THE AMMONIUM SULPHATE PRECIPITATE OF THE EXTRACELLULAR MEDIUM WAS DESALTED AND LOADED ONTO A DEAE SEPHADEX COLUMN EQUILIBRATED WITH 5 MM POTASSIUM PHOSPHATE PH 7.6. ISOENZYME I WAS ELUTED BY WASHING WITH 5 MM BUFFER AND THE OTHER ISOENZYMES BY A 2L LINEAR GRADIENT FROM 5 TO 80 MM POTASSIUM PHOSPHATE BUFFER PH 7.6. THE ARROW INDICATES WHERE THE GRADIENT COMMENCED. EACH FRACTION CONTAINED APPROXIMATELY 12 ML.

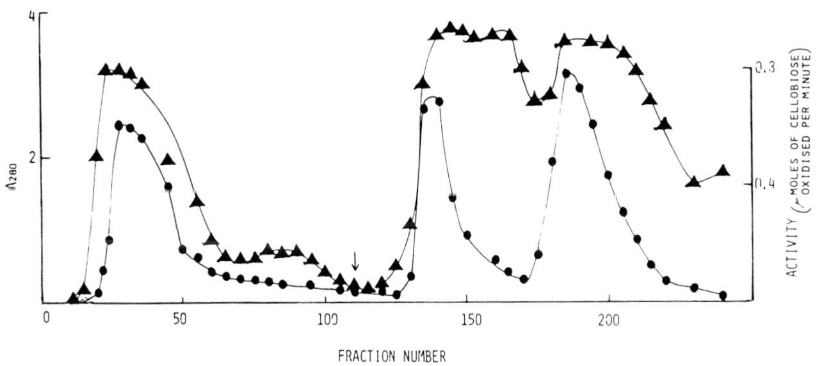

In Figure 1 the elution profile of the DEAE-Sephadex column is shown. The three isoenzymes can thus be readily separated. Peak 1 seems to correspond to the cellobiose dehydrogenase purified previously by Eriksson (2). It contains in addition to FAD the 6-OH and 8-OH derivatives. This enzyme elutes from a calibrated Sephadex column with a molecular weight corresponding to 97,000 daltons. On reaction with oxygen it produces hydrogen peroxide and when DCPIP is the electron acceptor it has a Km for cellobiose of 35 μM. The enzyme is strongly substrate inhibited by high concentrations of cellobiose.

The enzyme which elutes in peak II seems to be very similar to that in peak 1 in all its properties having the same apparent molecular weight and a Km for cellobiose of 33 µM. The spectral properties of this enzyme also suggest it contains hydroxylated flavin derivatives.

Cellobiose oxidase elutes in peak III. This enzyme may be purified to homogenity by a further ammonium sulphate precipitation and chromatography on Sephacryl S-200 and reactive red. The enzyme elutes from a calibrated Sephacryl column with a MW corresponding to 84,000 daltons. It is a glycoprotein containing 10% carbohydrate. The amino acid analysis of the protein shows it has only a low level of basic amino acids and DTNB titrations suggest that there are no thiols. The enzyme has a Km for cellobiose of 50 µM at pH 6 and as with the other isoenzymes, is substrate inhibited. With cytochrome c as an electron acceptor the reaction is strongly inhibited by superoxide dismutase. This along with the observation that cellobiose oxidoreductase produces 10 to 20% hydrogen peroxide suggests that the reduced oxygen product of the enzyme is the superoxide anion. Apart from cellobiose this enzyme oxidises a large range of polysaccharides (but not glucose) providing they have a B 1-4 glycosidic bond next to the sugar residue to be oxidised. The precise role of these enzymes in cellulose lignin biodegredation is not clear. However, activated forms of oxygen are known to be involved and important in the disruption of the lignin polymer. Thus production of hydrogen peroxide and the superoxide anion could be one of the functions of these cellobiose oxidoreductases. The fact that all three isoenzymes are strongly substrate inhibited supports this, since lignin degredation only commences when carbohydrate is diminished (6).

References

1. Westermark, U. and Eriksson, K-E.: Acta Chemica Scand. B28, 209-214 (1974).
2. Westermark, U. and Eriksson, K-E.: Acta Chemica Scand. B29, 419-424 (1975).
3. Ayers, A.R., Ayers, S.B. and Eriksson, K-E.: Eur. J. Biochem. 90, 171-181 (1978).
4. Eriksson, K-E. and Pettersson, B.: Eur. J. Biochem. 51, 193-206 (1975).
5. Kirk, T.K., Connors, W.J. and Zeikus, J.G.: Appl. Environ. Microbiol. 32, 192-194 (1976).
6. Jeffries, T.W., Choi, S. and Kirk, T.K.: Appl. Environ. Microbiol. 42, 290-296 (1981).

THE EQUILIBRATION OF REDUCING EQUIVALENTS WITHIN MILK XANTHINE OXIDASE

Russ Hille
Department of Biological Chemistry, University of Michigan, Ann Arbor, MI

Introduction

Milk xanthine oxidase is a complex molybdoflavoprotein containing a molybdenum center, FAD, and a pair of 2Fe/2S centers in each of its two catalytically independent subunits. The enzyme is reduced by xanthine at the molybdenum center, and oxidized by molecular oxygen at the FAD. Thus the intramolecular transfer of reducing equivalents from molybdenum to FAD is an integral part of the catalytic cycle. It has been shown (1, 2) that the thermodynamic properties of xanthine oxidase are well-described by a model in which it is assumed that the equilibration of reducing equivalents among the several redox-active sites of the enzyme is rapid and that the distribution at equilibrium is determined solely by the relative oxidation-reduction potentials of the sites (1). Successful kinetic simulations of both the reductive (1) and oxidative (3, 4) half-reactions of xanthine oxidase using parameters based on the rapid equilibrium model also lend support to this concept, and suggest a lower limit for the rate of internal equilibration of approximately 100 s^{-1} (3). It has been recently concluded from flash photolysis experiments, however, that at least at low pH and low levels of enzyme reduction the transfer of reducing equivalents from molybdenum to FAD can be no faster than some 11 s^{-1} (5). This rate is barely sufficient to support the established enzyme turnover under the conditions of the flash photolysis experiment, and cannot account for the rapid oxidation of the iron-sulfur and molybdenum centers in the course of the oxidative half-reaction. We have therefore undertaken an examination of the rate of internal electron equilibration using the pH jump technique, taking advantage of the known pH-dependence of the distribution of reducing equivalents within partially reduced enzyme. Our results indicate that the rate of reequilibration is invariably rapid relative to enzyme turnover under the same conditions. At pH 8.5 in 0.05 M pyrophosphate (after mixing) the observed rate is approximately 330 s^{-1} (V_{max} being 17 s^{-1}).

Flavins and Flavoproteins

Results and Discussion

It has been previously shown that the spectral properties of partially reduced xanthine oxidase exhibit a pronounced pH dependence (2). From the room-temperature oxidation-reduction potentials of the several sites in the enzyme at low and high pH (6), the distributions of reducing equivalents within the partially reduced enzyme may be quantitated (Table 1). It may be

Table 1. The oxidation-reduction potentials of xanthine oxidase at low and high pH (ref. 4; fraction reduced calculated at 50% enzyme reduction).

	pH 6.1		pH 9.2	
	E_o' (mV vs NHE)	Fraction Reduced	E_o' (mV vs NHE)	Fraction Reduced
Fe/S II	-150	0.90	-255	0.96
Fe/S I	-261	0.17	-345	0.43
FAD/FADH·	-203	0.12	-365	0.02
FADH·/$FADH_2$	-168	0.82	-252	0.83
Mo(VI)/Mo(V)	-239	0.28	-430	0.00
Mo(V)/Mo(IV)	-301	0.00	-350	0.00

seen that the net effect of rapidly lowering the pH is the formation of FADH· and Mo(V) at the expense of the two iron-sulfur centers. Shown in Figure 1 are the kinetic time courses observed on jumping the pH from 9 to 6, and the reverse, following the absorbance change at 600 nm due primarily to the formation or loss, respectively, of FADH·. The reactions are monophasic and quite rapid, exhibiting rate constants of 155 s^{-1} and 330 s^{-1}, respectively. When followed at 450 nm, the observed rates were somewhat more rapid (but never more than some 30%); the absorbance changes at 450 nm were in the same direction as those seen at 600 nm, as expected given the partial reduction of Mo(V) at low pH. In the presence of alloxanthine, which binds essentially irreversibly to Mo(IV) removing the molybdenum center from the redox equilibrium, the spectral changes observed at 450 and 600 nm are in opposite directions, but the observed rates are essentially unchanged. The observation of such rapid rates lends strong support to the rapid equilibrium model, as originally proposed (1). The observed in the above experiments were found to be independent of the level of enzyme reduction between 20% and 80% reduced, the buffer, and the ionic

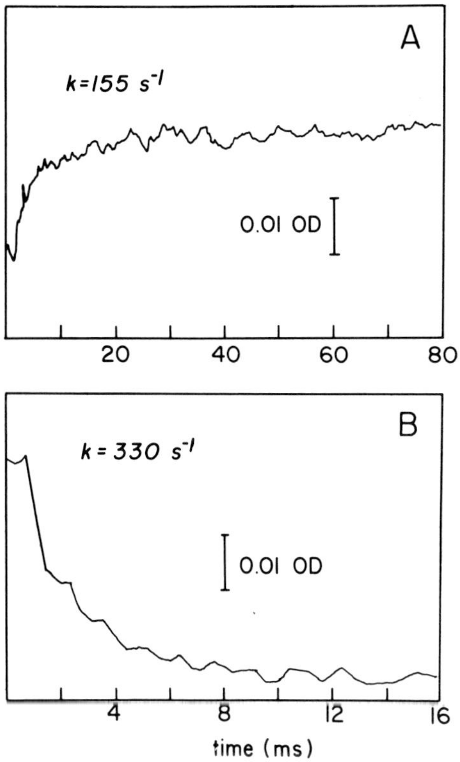

Figure 1. Time courses observed at 600 nm on jumping the pH of partially reduced xanthine oxidase. Panel A, the time course observed on mixing 56 uM enzyme (approximately 70% reduced) in 10 mM glycine, 0.1 N NaCl, pH 9.0 with 0.1 M MES, 0.1 N NaCl, pH 6.0 (final pH=6.1) at 25°C. Panel B, the time course observed on mixing 41 uM enzyme (approximately 65% reduced) in 10 mM MES, 0.1 N NaCl, pH 6.0 with 0.1 M pyrophosphate, pH 8.5 (final pH=8.5) at 25°C.

strength (0.1 N NaCl was added to stabilize the enzyme at low pH and low buffer concentration). The rate was dependent on the final pH, however, doubling over the pH range 6 to 9. While this dependence is significant, it should be noted that rates faster than 100 s^{-1} are always observed.

The temperature-dependence of the rates observed following upward and downward jumps are shown in Figure 2. It may be seen that the activation energy thus determined at a given pH is independent of the buffer used, but is approximately twice as great at pH 6.1 as at pH 8.5. While this correlates well with the observed ratio of rates observed at the two pH values, the significance of this observation is not clear. While the pH-jump reaction may be thought of as the conversion of enzyme from one state to another, it must be kept in mind that these states are not discrete chemical species but aggregations of microstates consisting of each of the possible electron distributions at a given level of enzyme reduction. These

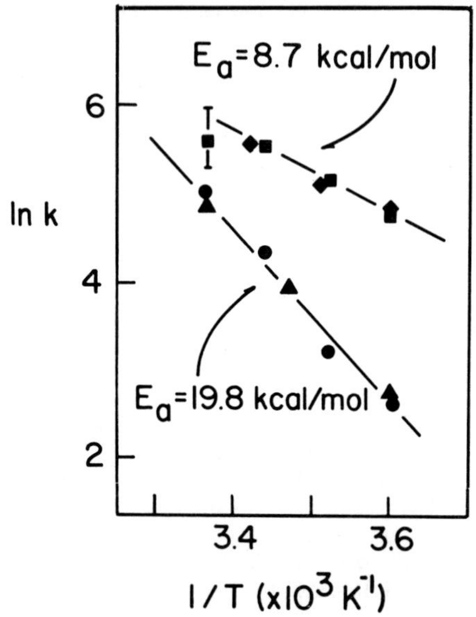

Figure 2. The temperature-dependence of the observed rates subsequent to a pH-jump. Circles, glycine pH 9.0 to MES pH 6.1; diamonds, glycine pH 9 to phosphate pH 6.1; squares, MES pH 6.0 to pyrophosphate pH 8.5; diamonds, MES pH 6.0 to glycine pH 8.8. Enzyme for these experiments was approximately 40 uM and 60% reduced.

microstates would be expected to interconvert at microscopic rates which would in all likelihood depend on the two redox-active sites involved in the electron-transfer reaction in question. It would not be intuitively expected that the conversion of one entire set of microstates to another, by means of a finite but large number of microscopic steps, would be characterized by a single, macroscopic activation energy barrier. This matter is currently being examined by computer simulation.

References

1. Olson, J.S., Ballou, D.P., Palmer, G., Massey, V.: J. Biol. Chem. 249, 4363-4382 (1974).

2. Hille, R., Fee, J.A., Massey, V.: J. Biol. Chem. 256, 8933-8940 (1981).

3. Hille, R., Massey, V.: J. Biol. Chem. 256, 9090-9095 (1981).

4. Porras, A.G., Olson, J.S., Palmer, G.: J. Biol. Chem. 256, 9096-9103 (1981).

5. Battacharrya, A., Tollin, G., Davis, M., Edmondson, D.E.: Biochem. 22, 5270-5279 (1983).

6. Porras, A.G., Palmer, G.: J. Biol. Chem. 257, 11617-11626 (1982).

REACTIONS BETWEEN XANTHINE OXIDASE AND 4-HYDROXY-7-AZAPTERIDINE

Richard C. Stewart and Vincent Massey
Department of Biological Chemistry, University of Michigan, Ann Arbor, MI

Introduction

Milk xanthine oxidase (XO) catalyzes oxidation of numerous classes of heterocyclic compounds including pteridines (1). Steady state turnover of most pteridines by XO is relatively slow (1); however, turnover of 4-hydroxy-7-azapteridine by XO has been reported to be significantly more rapid than xanthine turnover (2). We have investigated the interaction of this substrate with XO by characterizing the steady state kinetics, the reductive half-reaction, and the reaction between reduced enzyme and substrate. The optical spectral change associated with oxidation of 4-hydroxy-7-azapteridine to 2,4-dihydroxy-7-azapteridine and the chemical structures of these compounds can be seen in Fig. 1. In addition to the expected oxidation reaction, the azapteridine can also be reduced to the corresponding dihydro-azapteridine by sodium dithionite or by reduced XO. A midpoint potential of -180mV (vs. SHE) at pH 8.3 has been determined for this couple. Reduction of the azapteridine results in bleaching of the visible absorbance spectrum of this compound (Fig. 1). The reduced azapteridine is rapidly reoxidized by oxygen. All experiments reported here were performed in 0.1M Na Bicine buffer at pH 8.3 and 25°C.

Results

Results of steady state experiments monitoring formation of 2,4-dihydroxy-7-azapteridine indicate $V_{max}=55s^{-1}$ compared to a value of $18s^{-1}$ for xanthine (3). While the observed rate of turnover exhibits a hyperbolic dependence on the azapteridine concentration (observed $K_m=40\mu M$), the rate is essentially independent of the oxygen concentration. In fact,

Flavins and Flavoproteins

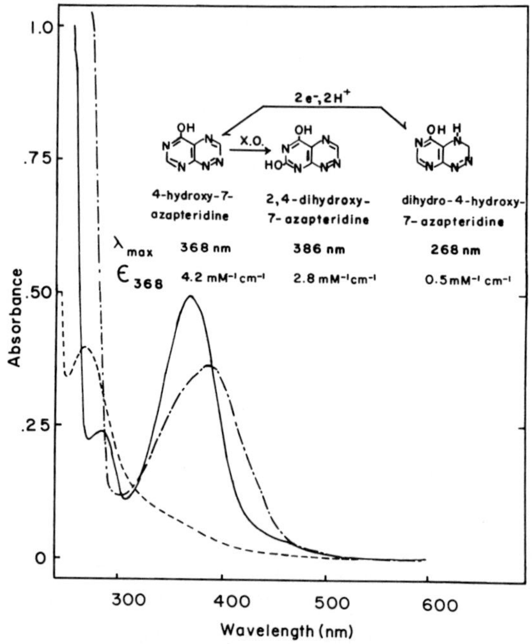

Figure 1. Optical absorbance spectra of 120µM 4-hydroxy-7-azapteridine (———), 2,4-dihydroxy-7-azapteridine (—·—), and dihydro-4-hydroxy-7-azapteridine (-----) in 0.1M Na Bicine, at pH 8.3 and 25°C.

anaerobic assays at several concentrations of azapteridine indicate that turnover is not diminished in the complete absence of oxygen. Thus the azapteridine appears to serve as both a reducing and oxidizing substrate for XO. This finding is corroborated by the ability of iodoacetamide-alkylated XO (which has a non-functional C(4a)-alkylated FAD) to turn the azapteridine over just as rapidly as does the native enzyme (data not shown). Neither oxygen nor an oxygen-reactive center on the enzyme are required for turnover of this substrate.

The anaerobic reaction between active XO (AFR 195) and 4-hydroxy-7-azapteridine is complex. At relatively low concentrations of substrate, absorbance changes such as those shown in Fig. 2 are observed. Because the azapteridine serves as both reducing substrate and oxidizing substrate, the anaerobic reaction shown in Fig. 2 is actually a steady state situation. The three phases of the reaction observed at 500nm (rapid absorbance

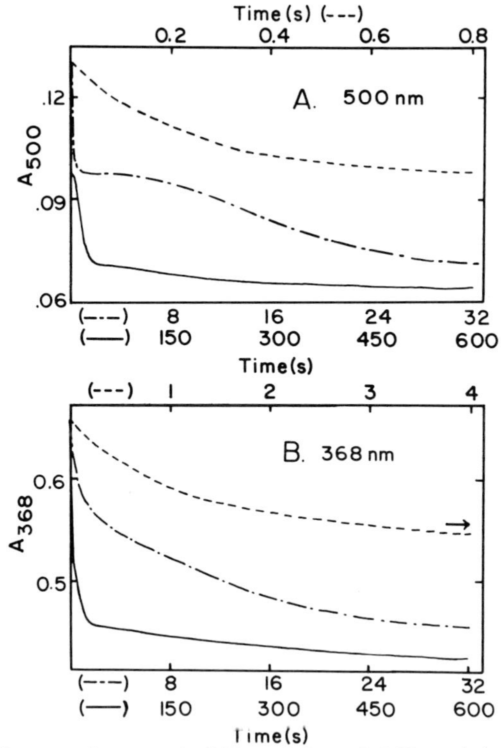

Figure 2. Absorbance change at 500nm (A) and 368nm (B) observed during anaerobic reaction of 3.3μM XO (AFR 195) with 100μM 4-hydroxy-7-azapteridine. The arrow in (B) indicates the final A_{368} expected for complete reduction of the enzyme by three equivalents of azapteridine.

decrease, level-off, and slow absorbance decrease) represent the approach to steady state, steady state, and decay from steady state, respectively. At 368nm (B) both the enzyme and the azapteridine contribute to the multiphasic absorbance decrease. The observed ΔA_{368} is more than twice the change expected for complete reduction of the enzyme by three equivalents of azapteridine. This can be accounted for by the spectral change arising from reduction of the substrate to the corresponding dihydro-4-hydroxy-7-azapteridine by the reduced XO.

Under anaerobic conditions, 4-hydroxy-7-azapteridine quantitatively reoxidizes the reduced forms of desulfo XO (Fig. 3), XO-arsenite, alloxanthine-complexed XO, and iodoacetamide-alkylated desulfo XO. The kinetics of the reoxidation reaction are altered by each of these

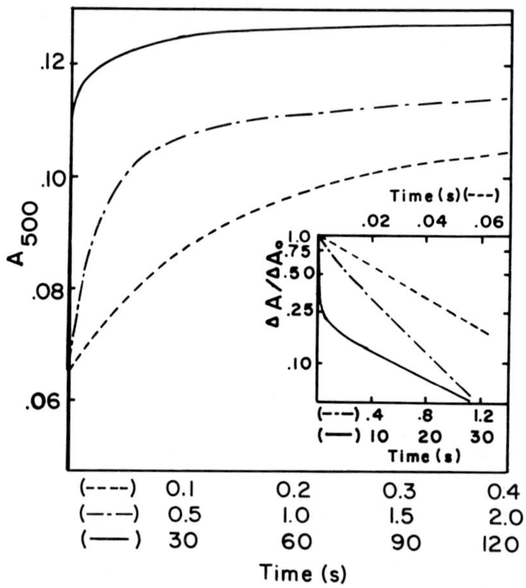

Figure 3. Absorbance change observed at 500nm during the reaction of 3.3 µM desulfo XO with 250µM azapterdine. The reaction is triphasic. Shown in the inset are semilog plots of the original data (——), the data after subtracting out the contribution of the slowest phase (—·—), and the data after subtracting out the contributions of the two slower phases (----).

modifications, suggesting that more than one of the XO redox centers may participate in this reaction.

Acknowledgements

This research was supported by NSF Grant PCM-8208240 to VM and a predoctoral NSF fellowship to RCS.

References

1. Bergmann, F., Levene, L., Tamir, I., Rahat, M.: Biochim. Biophys. Acta 480, 21-38 (1977).

2. Bergmann, F., Levene, L., Nieman, Z., Brown, D.J.: Biochim. Biophys. Acta 222, 191-196 (1970).

3. Olson, J.S., Ballou, D.P., Palmer, G., Massey, V.: J. Biol. Chem. 249, 4350-4362 (1974).

"TYPE 1" AND "TYPE 2" RAPID AND SLOW E.P.R. SIGNALS FROM THE MOLYBDENUM CENTRES OF MOLYBDENUM-CONTAINING HYDROXYLASES AND THEIR SIGNIFICANCE

R.C. Bray, G.N. George, S. Gutteridge, F.F. Morpeth, N. Turner
School of Chemistry and Molecular Sciences, University of Sussex, Brighton, BN1 9QJ, U.K.

The molybdenum atoms of the molybdo-flavoproteins xanthine oxidase and aldehyde oxidase exist in active and desulpho forms, the former containing in the oxidized state a sulphido ligand of Mo, which in the latter is replaced by an oxo group (1,2). Reduced forms of the active and desulpho enzymes give rise, respectively, to Mo(V) e.p.r. signals known as Rapid and Slow. Bray and Vänngård (3) showed that when xanthine oxidase is reduced by xanthine, two Rapid signals, designated type 1 and type 2, are produced. The former has two non-equivalent, and the latter two equivalent (or nearly equivalent) protons coupled to Mo. These signals have different $\underline{\underline{g}}$-tensors but almost identical values of g_{av}. Later work (4) has shown that whether xanthine oxidase, on reduction, gives rise to a Rapid type 1 or a type 2 signal, depends on the pH and on the presence of additives which form complexes with the Mo centre. However, the nature and significance of the type 1/type 2 transformation has remained obscure.

The features of the xanthine oxidase molecule causing it to give two spectrally distinct reduced forms are not dependent on the presence of the sulphido group. This is shown by our new finding that the Slow signal, hitherto observed only as a type 1 form (Fig. 1g), can exist also in a type 2 form. Borate, which causes the Rapid signal to assume the type 2 form (5), has a comparable effect on the Slow signal (Fig. 1i). Type 1 and type 2 Slow signals have almost identical values

Flavins and Flavoproteins

of g_{av}, in contrast to the 6-fold increase in coupling of the second proton in the type 2 species (Table 1).

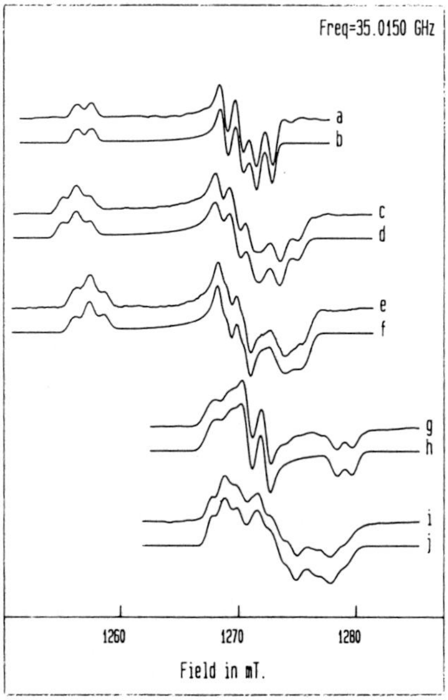

Fig. 1. E.p.r. spectra at 35 GHz of some type 1 and type 2 Rapid and Slow signals

All spectra were recorded in Bicine buffer pH 8.2; where no substrate was present the enzyme was reduced with dithionite. (a) and (c) are signals from active xanthine oxidase, (e) is from active aldehyde oxidase and (g) and (i) are for desulpho xanthine oxidase. (a) is the formamide Rapid type 1 signal, (c) the borate Rapid type 2 and (e) is also Rapid. (g) is the Slow type 1 signal and (i) the borate Slow type 2. Simulations of (a), (c), (e), (g) and (i) are shown respectively in (b), (d), (f), (h) and (j). Parameters were (b) as ref.(6), (d) as ref.(5), (f) $g_{1,2,3}$ 1.9895, 1.9700, 1.9622; $A(^1H)_{1,2,3}$ (1st) 1.38, 1.50, 1.50; (2nd) 1.13, 0.65, 0.71, (h) as ref. (1), (j) $g_{1,2,3}$ 1.9713, 1.9643, 1.9570; $A(^1H)_{1,2,3}$ (1st) 1.45, 1.45, 1.50; (2nd) 0.85, 0.85, 1.10. (Note that the combinations of hyperfine splittings here and in Table 1 have been chosen arbitrarily, so as to maximize the ratio of A_{av} values for the two protons).

Table 1. Summary of parameters of some type 1 and type 2 Rapid and Slow signals

XO_a, XO_{ds} and AO are active and desulpho xanthine oxidase and aldehyde oxidase, respectively.

Enzyme	Additive	Signal	g_{av}	$A(^1H)_{av}$ 1st	2nd	Ratio	Ref.
XO_{ds}	-	Slow type 1[a]	1.9647	1.63	0.16	10.2	(1)
XO_{ds}	Borate	Slow type 2[a]	1.9642	1.47	0.93	1.58	Present[b]
XO_a	Formamide	Rapid type 1	1.9760	1.27	0.26	4.9	(6)
AO	-	Rapid[a]	1.9739	1.46	0.83	1.76	Present[b]
XO_a	Pyridine-3-aldehyde	Rapid type 2	1.9726	1.46	1.03	1.42	Present[c]
XO_a	Borate	Rapid type 2[a]	1.9745	1.52	1.16	1.31	(5)
XO_a	Xanthine	Rapid type 2	1.9760	1.01	1.01	1.00	(7)

[a] The enzyme was reduced with dithionite.
[b] For other parameters, see Fig. 1.
[c] Parameters were $g_{1,2,3}$ 1.9861, 1.9695, 1.9623; $A(^1H)_{1,2,3}$ (1st) 1.35, 1.57, 1.45; (2nd) 1.35, 0.74, 0.99.

As shown in Fig. 1 and Table 1, clear distinction between Rapid type 1 and 2 signals breaks down when new data on pyridine-3-aldehyde, and more particularly on active aldehyde oxidase, are considered. Thus, there is a progression of the ratio of A_{av} values for the two protons in Rapid signals, from 4.9:1 (xanthine oxidase, type 1, in the presence of formamide ref. 6) though 1.76:1 (aldehyde oxidase) to 1.00:1 (xanthine oxidase, type 2, in the presence of xanthine, ref. 7). Note particularly that for aldehyde oxidase the protons are almost equally coupled in the A_1 direction (Fig. 1e).

Since g_{av} values remain constant for all Rapid signals (Table 1) we think it unlikely (despite information from ^{17}O splittings, ref. 8) that type 1 and 2 species differ in the nature and number of ligands to Mo. More likely, there is simply a change of ligand geometry, probably with the additives not coordinated to Mo, but binding in the active site in specific ways that cause the coordination geometry of the metal to vary. The significance of such structural changes in relation to catalysis remains uncertain.

Acknowledgements

The work has been supported by the MRC, the SERC and the Wellcome Trust. N.A.T. is in receipt of a studentship from the MRC.

References

1. Gutteridge, S., Tanner, S.J., Bray, R.C.: Biochem.J. 175, 997-897 (1978).
2. Cramer, S.P., Wahl, R., Rajagopalan, K.V.: J.Am.Chem.Soc. 103, 7721-7727 (1981).
3. Bray, R.C., Vänngård, T.: Biochem.J. 114, 725-734 (1969).
4. Bray, R.C.: Advances in Enzymology and Related areas of Molecular Biology, 51, 107-165 (1980).
5. Malthouse, J.P.G., Gutteridge, S., Bray, R.C.: Biochem.J. 185, 767-770 (1980).
6. Morpeth, F.F., George, G.N., Bray, R.C.: Biochem.J. 220, 235-242 (1984).
7. Bray, R.C., Barber, M.J., Lowe, D.J.: Biochem.J. 171, 653-658 (1978).
8. Bray, R.C., Gutteridge, S.: Biochemistry 21, 5992-5999 (1982).

Present addresses:

S. Gutteridge: Rothamsted Experimental Station, Harpenden, AL5 2JQ, U.K.

F.F. Morpeth: Department of Food Technology, University of Reading, Reading, RG1 5AQ, U.K.

RE-SULPHURATION OF XANTHINE OXIDASE

A. Ventom and R.C. Bray
School of Chemistry and Molecular Sciences, University of Sussex, Brighton, BN1 9QJ, U.K.

Active xanthine oxidase (containing a -Mo=S group) is converted to inactive desulpho xanthine oxidase (containing -Mo=O) by treatment with cyanide or, more slowly, on storage. Partial reversal of this reaction is obtained on treatment of the desulpho enzyme with sulphide (1,2,3). We report a systematic study of conditions affecting the reconstitution reaction and related experiments.

We found optimum re-sulphuration at pH 8.5 (Fig. 1), with an optimum temperature of about 45°C. Increasing Na_2S concentrations led to increasing re-activation, up to an optimal range of 7-12 mM, with a slight fall off at higher sulphide concentrations. The extent of reactivation showed a clearly defined dependence on enzyme concentration (Fig. 2). The limiting extent of re-activation, at the highest enzyme concentration (about 1 mM) corresponded to about 50%. Re-activation fell off as the enzyme concentration was decreased, and at the lowest levels (about 2 nM) amounted to 10 to 15%. No doubt incorporation of sulphide competes in some way with inactivation processes dependent on the concentration of the enzyme.

Contrary to claims in the literature (4) we found that the extent of re-activation of desulpho oxidase under optimum conditions was not increased by reduction of the enzyme with dithionite. In fact we found, under anaerobic conditions, that sulphide in concentrations stoichiometric with the

Flavins and Flavoproteins

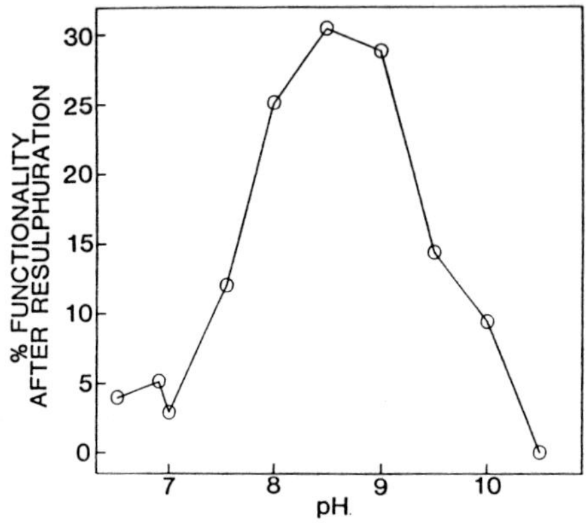

Fig. 1. Effect of pH on re-sulphuration

Samples of desulpho xanthine oxidase (27 μM) were incubated at various pH values with 7 mM Na_2S for 160 min at 45°C, then xanthine oxidase activity was measured spectrophotometrically. Buffers (250 mM) were as follows: PIPES (to pH 6.9), MOPS (to pH 7.55), BICINE (to pH 8.4), CHES (to pH 9.5) and CAPS.

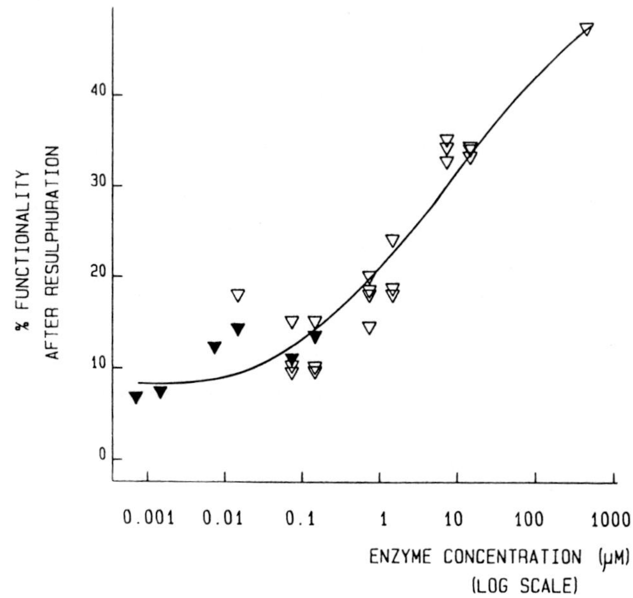

Fig. 2. Effect of enzyme concentration on re-sulphuration

Samples of desulpho xanthine oxidase, at different concentrations, were incubated with 7 mM Na_2S for 150 min at 45°C at pH 8.45 (200 mM Bicine with 1 mM EDTA). Activity was then measured spectrophotometrically with xanthine (open symbols) or fluorimetrically with 2-amino-4-hydroxypteridine (filled symbols).

enzyme caused partial reduction of the enzyme.

Reversible interconversion of active and desulpho xanthine oxidase by a system consisting of rhodanese, thiosulphate and dithiothreitol has been reported (5). In agreement with these workers, our investigations provided no evidence for direct interaction between xanthine oxidase and rhodanese. On the contrary, sulphide liberated (6) by the system was estimated (7) and found to be sufficient to account for the observed re-sulphuration. In confirmation, substantial re-sulphuration was observed in the presence of thiosulphate and dithiothreitol, even when desulpho xanthine oxidase and rhodanese were separated by a dialysis membrane.

An indication of as yet unexplained complexities in the de-sulphuration/re-sulphuration processes of the enzyme was provided by experiments utilizing enzyme enriched (8) in the CN^--labile site with ^{33}S. Using this enzyme, we were able to use e.p.r. specifically to follow exchange, in the presence of unlabelled sulphide, of sulphur out of the active site of the enzyme. A fast phase, in which about half of the ^{33}S exchanged ($t_{1/2}$ approx. 10 min, with 10 mM Na_2S at pH 8.45 and 45°C), was followed by much slower exchange.

Acknowledgements

A.V. in supported by a studentship from the S.E.R.C. Facilities for e.p.r. work have been provided by the M.R.C.

References

1. Massey, V., Edmondson, D.: J.Biol.Chem. 245, 6596-6598 (1970).

2. Gutteridge, S., Tanner, S.J., Bray, R.C.: Biochem.J. 175, 887-897 (1978).

3. Cramer, S.P., Wahl, R., Rajagopalan, K.V.: J.Am.Chem.Soc. 103, 7721-7727 (1981).

4. Wahl, R.C., Rajagopalan, K.V.: J.Biol.Chem. 257, 1354-1359 (1982).

5. Nishino, T., Usami, C., Tsushima, K.: Proc.Nat.Acad.Sci. U.S.A. 80, 1826-1829 (1983).

6. Pecci, L., Pensa, B., Costa, M., Cignini, P.L., Cannella, C.: Biochim.Biophys.Acta 245, 104-111 (1976).

7. Bray, R.C.: Analyst 83, 379-381 (1961).

8. Malthouse, J.P., George, G.N., Lowe, D.J., Bray, R.C.: Biochem.J. 199, 626-637 (1981).

REACTIVATION OF DESULFO XANTHINE OXIDASE BY AN ENZYMATIC SYSTEM

Takeshi Nishino, Chikako Usami and Keizo Tsushima
Department of Biochemistry, Yokohama City University School of Medicine, Yokohama 232, Japan

The inactive enzyme of desulfo form is known to be present in the enzyme preparation of the molybdenum-containing enzymes xanthine oxidase(1), xanthine dehydrogenase(2,3) and aldehyde oxidase(4). The desulfo form has been considered to be merely preparation or storage artifacts(5). However almost half of enzyme molecule in supernatant solution of rat liver homogenate was found to be inactive form(6). Furthermore the molecular activity of chicken liver xanthine dehydrogenase changes with protein content in diet(7). These suggest that a mechanism exists in vivo whereby interconversion between sulfo and desulfo forms may be effected. The successful separation of sulfo and desulfo forms by affinity chromatography(8) prompted us to investigate the interconversion between sulfo and desulfo forms by an enzymatic system. We found that desulfo enzyme could be reactivated by incubation with rhodanese, thiosulfate and dithiothreitol(9). Evidence was given for reversible interconversion between sulfo and desulfo forms by this system. However, studies of intracellular distribution of rhodanese in rat liver have revealed that the enzyme is confined to the mitochondria(10). We could not detect rhodanese activity in cow's milk. These suggest that rhodanese may not have physiological role in trans sulfur reaction in xanthine oxidase.

We have found that the alternate enzyme system of mercaptopyruvate sulfur transferase (3-mercaptopyruvate: cyanide sulfurtransferase, EC 2.8.1.2) can also reactivate

Flavins and Flavoproteins

desulfo enzyme. This enzyme activity could be detected in cow's cream. This enzyme having fairly high activity in 100,000 x g supernatant solution of rat liver homogenate could be purified to homogeneity by the procedure including ammonium sulfate fractionation, DEAE-cellulose column chromatography, Bio-gel HTP column chromatography, ACA 44 gel filtration and HPLC using TOYO TSK-gel column. The system of the purified mercaptopyruvate sulfurtransferase reactivates desulfo enzyme. Sulfhydryl reagent activated this reactivation. The enzyme activity reactivated by this system was about 30% of full activity in the adopted experimental condition and the same system inactivated the full active enzyme down to the same level, suggesting that this reaction is also reversible.

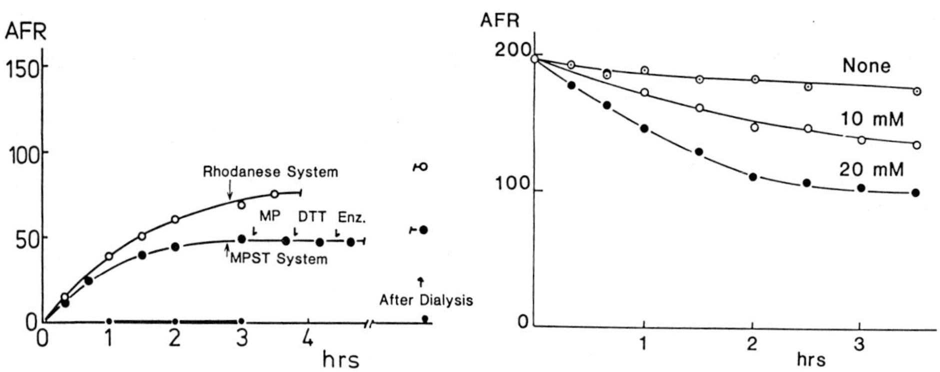

Fig. 1. (left) Reactivation of desulfo xanthine oxidase by the system of mercaptopyruvate sulfurtransferase.
Desulfo xanthine oxidase (final A_{450}=0.201) was incubated with the system containing 1.5 mM mercaptopyruvate, 5mM DTT and 14.3 U of mercaptopyruvate sulfurtransferase.

Fig. 2. (right) Inactivation of sulfo xanthine oxidase by Na_2S. Sulfo enzyme (AFR 196, A_{450}=0.0803) was incubated at 30°C in 0.1 M pyrophosphate (pH 8.5) with Na_2S of the indicated concentrations.

The reversible reaction between sulfo and desulfo forms catalyzed by these systems can be explained by the involvement of a common intermediate. As we mentioned previously(9,11) possible reactive sulfur species such as S^{--} or $R\text{-}S\text{-}S^{-}$ may be involved in the reaction.

Rhodanese system:

$$Na_2S_2O_3 + RSH \longrightarrow Na_2SO_3 + RSSH$$

$$RSSH + RSH \longrightarrow RSSR + H_2S$$

Mercaptopyruvate sulfurtransferase system:

$$HSCH_2COCOOH + RSH \longrightarrow CH_3COCOOH + RSSH$$

$$RSSH + RSH \longrightarrow RSSR + H_2S$$

Kinetic analysis of reactivation by the rhodanese system revealed that $Na_2S_2O_3$ and dithiothreitol were real substrate of rhodanese, but xanthine oxidase was not. Reactivation velocity was simply proportional to the concentration of xanthine oxidase. Furthermore Zn^{++}ion, that should trap S^{--} as a ZnS precipitate, inhibited these reactivation, while this ion did not inhibit mercaptopyruvate sulfurtransferase. These suggest that sulfide ion formed in these systems is incorporated into desulfo xanthine oxidase. On the contrary, in the desulfuration reaction of sulfo enzyme by the sulfurtransferase systems, sulfide at the molybdenum of xanthine oxidase is assumed to be released by the attack of the sulfide anion. In fact sodium sulfide was found to inactivate sulfo enzyme. Coughlan et al(12,13) also reported the sulfide inactivation of turkey liver xanthine dehydrogenase. A mechanism of desulfuration of sulfo enzyme by HS^- is considered to be similar to that of desulfuration by CN^-, because both HS^- and CN^- are soft bases as classified by Pearson(14) and calculated E_{HOMO} value of HS^- is very close to that of CN^-(15). In physiological conditions, xanthine oxidase is assumed to be equilibrated with sulfide anion formed by an enzymatic system.

And this may explain why desulfo enzyme exists in vivo.

Acknowledgement

This work was supported by a Grant-in-Aid for Scientific Research from Japanese Ministry of Education, Science and Culture. (No. 58480142).

References

1. Massey, V. and Edmondson, D.: J. Biol. Chem. 245, 6595-6598 (1970).
2. Cleere, W. F. and Coughlan, M.P.: Biochem.J. 143, 331-340 (1974).
3. Nishino, T., Itoh, R., and Tshushima, K.: Biochim. Biophys. Acta 403, 17-22 (1975).
4. Branzoli, V. and Massey, V.: J. Biol. Chem. 249, 4346-4349 (1974).
5. Bray, R. C.: The Enzymes, 3rd Ed. 12, 299-419 (1975).
6. Ikegami, T., Nishino, T. and Tsushima, K.: (This volume)
7. Itoh, R. Nishino, T., Usami, C. and Tsushima, K.: J. Biochem. (Tokyo) 84, 21-26 (1978).
8. Nishino, T., Nishino, T., and Tsushima, K.: FEBS Lett. 131, 369-372 (1981).
9. Nishino, T., Usami, C, and Tsushima, K.: Proc. Natl. Acad. Sci., USA, 80, 826-1829 (1983).
10. De Duve, C., Pressman, B. C., Gianetto, R., Watteaux, R and Appelmans, F. Biochem. J. 60, 604-617 (1955).
11. Nishino, T., Nishino, T. and Tsushima, K.: Flavins and Flavoproteins, (Massey, V. and Williams, C. H. eds) pp792-795 (1982).
12. Cleere, W. F. and Coughlan, M. P.: Biochem. J. 143, 331-340 (1974).
13. Coughlan, M. P. and Cleere, W. F.: Flavins and Flavoproteins, (Singer, T. P. eds), pp576-581 (1976).
14. Klopman, G. : J. Amer. Chem. Soc. 90, 223 (1968).
15. Pearson, R. G.: J. Amer. Chem. Soc. 85, 3533-3539 (1963).

THE EXISTENCE OF DESULFO XANTHINE DEHYDROGENASE IN RAT LIVER

Tadashi Ikegami, Takeshi Nishino and Keizo Tsushima
Department of Biochemistry, Yokohama City University School of Medicine, Yokohama 232, Japan

Introduction

It is known that avian xanthine dehydrogenase (1) or milk xanthine oxidase (2) contains an inactive desulfo form. This inactive form has been considered to be merely a preparation or storage artifact (3). However, some evidence suggest the existence of an inactive enzyme in vivo (4). We examined the existence of an inactive enzyme in both crude and purified enzyme from rat liver.

Materials and Methods

Rat liver xanthine dehydrogenase was purified by the method including 0-40% acetone fractionation, 40-60% $(NH_4)_2SO_4$ fractionation, hydroxyl apatite column chromatography, Ultrogel column chromatography and first Sepharose 4B/folate affinity chromatography (5). ^{14}C-allopurinol was synthesized by the method of Elion et al. (6). Anti rat liver xanthine dehydrogenase antibody was purified from the serum obtained from rabbits immunized with purified rat liver enzyme. All reagents used were of analytical grade.
Determination of the amount of inactive enzyme in crude extract is as follows. 2ml of 100,000xg supernatant from rat liver was incubated at 25°C with 6.5 mM ^{14}C-allopurinol till complete inactivation. Then the solution was passed through Sephadex

Flavins and Flavoproteins

G-25 to remove excess ^{14}C-allopurinol. The enzyme fractions were collected and the immunoprecipitate was obtained by excess specific antibody. The immunoprecipitate was washed and dissolved in 2ml of 5M guanidine-HCl. 1ml each of this solution was subjected to the determination of radioactivity and FAD content (7). For determination of non-specific binding, the remaining portion of the collected enzyme fraction was reactivated by incubation with 50μM ferricyanide and 1mM salicylate for 10 hours. Then the same procedure as above was applied and the radioactivity and FAD content in the immunoprecipitate were determined.

Results and Discussions

Rat liver xanthine dehydrogenase was purified in good yield (40-50%). The AFR value of the prepared enzyme is usually about 120 and the ratio of absorbance at 280nm to that at 450nm is usually between 5.15-5.50.

With purified enzyme preparation, (A) reduction of the enzyme by xanthine under anaerobic condition proceeded in two phases. This biphasicity is considered to be due to the presence of inactive desulfo enzyme (1,2). (B) The enzyme was inactivated by cyanide. The release of sulfur atom resulted in this inactivation and was protected by pre-treatment of the enzyme with allopurinol. These results show that a labile sulfur atom exists close to molybdenum. (C) The active and inactive enzymes were able to be resolved with affinity chromatography on Sephrose 4B/folate gel as milk xanthine oxidase (4). These results provide evidence that purified enzyme preparation contains inactive desulfo form.

Inactive form in the enzyme preparation has been considered to be desulfo form caused by elimination of sulfur atom from the enzyme during preparation or storage (3). And even when the enzyme is purified in good yield (40-50%), 40-50% of the inactive enzyme was found in the purified enzyme preparation. Furthermore, the molecular activity of the enzyme purified from livers of chickens fed a high protein diet has been reported to be high and changes with the protein content of the diet (4). These suggest that the inactive enzyme originally exists in vivo. And so we examined the existence of inactive form in crude extract from rat liver homogenate. The amount of ^{14}C-allopurinol per FAD in the immunoprecipitates of crude and purified enzymes were shown in Table 1.

Table 1. Allopurinol bound to purified and crude enzyme

	Purified enzyme	Crude enzyme
AFR	118	—
^{14}C-allopurinol bound (mol/mol of FAD)	0.54	0.48 ± 0.04

The value in crude enzyme is the average of triplicate experimental values with different rats.

The corrected value by subtracting non-specific binding were the same level of 0.48-0.54 in both samples. This indicates that about 50% of the enzyme exist as an inactive form in purified and even in crude enzyme preparations suggesting that the inactive form is not merely an artifact and significant amount of this form exists in vivo.

Possible regulatory system of sulfuration and desulfuration has been suggested to exist in several enzyme systems (8). Recently reversible interconversion between sulfo and desulfo forms of xanthine oxidase was known to occur by the systems of rhodanese (9) and mercaptopyruvate sulfurtransferase (10). This reversible reaction was mediated by sulfide anion formed by the enzymatic systems. In physiological condition, xanthine dehydrogenase is assumed to be equilibrated with sulfide anion formed by the enzymatic system. This may explain the existence of desulfo enzyme.

In chicken liver xanthine dehydrogenase a system may be considered to exist to stabilize sulfur atom at molybdenum so that the content of desulfo form might be effected.

Acknowledgement

This work was supported by a Grant-in-Aid for Scientific Research from the Japanese Ministry of Education, Science and Culture (No. 58480142).

References

1. Cleere, W. F. and Coughlan, M. P.: Biochem. J. 143, 331-340 (1974).
2. Massey, V. and Edomondson, D.: J. Biol. Chem. 245, 6595-6598 (1970).
3. Bray, R. C.: The Enzymes, 3rd edition, Academic Press, New York, Vol. 12, 299-419 (1975).
4. Itoh, R., Nishino, T., Usami, C. and Tsushima, K.: J. Biochem. (Tokyo) 84, 21-26 (1978).
5. Nishino, T., Nishino, T. and Tsushima, K.: FEBS Lett. 131, 369-372 (1981).
6. Elion, G. B., Kovensky, A. and Hitchings, G. H.: Biochem. Pharmacol. 15, 863-880 (1966).
7. Yagi, K.: Methods in Enzymology, Vol. 8 part B, 290-296 (1971).
8. Coughlan, M. P.: Biochem. Soc. Trans. 9, 307-308 (1981).
9. Nishino, T. Usami, C. and Tsushima, K.: Proc. Natl. Acad. Sci. USA 80, 1826-1829 (1983).
10. Nishino, T. Usami, C. and Tsushima, K.: Flavins and Flavoproteins, this volume.

THE STRUCTURES AND CATALYTIC MECHANISMS OF MOLYBDENUM CENTRES IN ENZYMES STUDIED BY e.p.r. AND X-RAY SPECTROSCOPY

R.C. Bray

School of Chemistry and Molecular Sciences, University of Sussex, Brighton, BN1 9QJ, U.K.

Summary

The usefulness and limitations of e.p.r. spectroscopy in studies of the structures of molybdenum centres in enzymes is discussed. For the ten ligands of the metal so far identified evidence, regarded as definitive in determining the structure, has been provided by EXAFS in three cases and by e.p.r. in seven. In large measure the two methods are complementary. As a result of such studies, in conjunction with kinetic e.p.r. work, the mechanism of interaction of reducing substrates with xanthine oxidase now seems relatively well understood. Future applications of the methods in this field are briefly considered.

Introduction

Molybdoenzymes have the special properties of involving in their reactions coupled transfer of electrons and protons, and of having the ability to transform 2-electron transfer sequences into single-electron transfer steps. This latter ability is a property that molybdoenzymes share only with flavoproteins and quinones. Since, furthermore, molybdoenzymes frequently contain flavin centres, a full-length paper (1) devoted to molybdenum centres may not be entirely out of place in this volume.

Flavins and Flavoproteins

Molybdoenzymes (see Table 1) perform a number of vital roles in biological redox processes. These enzymes include nitrate reductase and nitrogenase, from the biological nitrogen cycle, as well as sulphite oxidase and xanthine oxidase which are involved, respectively, in sulphur and purine metabolism, in mammals and elsewhere. The biological versatility of molybdenum is shown by the wide span of the redox potentials of the reactions catalysed by the enzymes, which range (cf. 2) from about +590 to -580 mV.

Though many molybdoenzymes have been well characterised by standard biochemical procedures, nevertheless none has so far proved amenable to structure determination by X-ray crystallography. To obtain structural information, we are therefore limited to spectroscopic methods. Of these, ^{95}Mo n.m.r. has so far not had adequate sensitivity, and contributions from the metal to electronic absorption spectra are masked by the larger contributions from other constituents of the enzymes (flavin, haem and iron-sulphur centres). Thus, only e.p.r. and X-ray spectroscopy have proved useful in studies of the structures of molybdenum centres.

Comparison of the e.p.r. and X-ray Spectroscopic Methods

E.p.r. of Mo(V) in molybdoenzymes, and EXAFS and X-ray edge spectra, have generally provided complementary information (6, 38) on the structures of molybdenum centres in enzymes. Thus, we know more about the enzymes from a combination of studies by the two methods, than we would have gained from studies by either method on its own. Both methods give information about ligands of molybdenum. With appropriate use of substitutions with stable isotopes, e.p.r. can give unambiguous information about the nature and number of these ligands. It also gives some information about ligand geometry (particularly if

Table 1. Reactions catalysed by molybdoenzymes

Reactions are listed in approximate order of decreasing redox potentials at pH 7.

reaction			enzyme	reference
chlorate + $2e^- + 2H^+$	$\rightleftharpoons$	chlorite + H_2O	chlorate reductase[b]	(35)
nitrate + $2e^- + 2H^+$	$\rightleftharpoons$	nitrite + H_2O	nitrate reductase	(6)
tetrathionate + $2e^-$	$\rightleftharpoons$	2 thiosulphate	tetrathionate reductase[b]	(35)
thiosulphate + $2e^-$	$\rightleftharpoons$	sulphite + sulphide		
biotin + H_2O	$\rightleftharpoons$	biotin sulphoxide + $2e^- + 2H^+$	biotin sulphoxide reductase[b]	(3)
nitrogen + $6e^- + 6H^+$	$\rightleftharpoons$	2 ammonia	nitrogenase	(6)
formate + H_2O	$\rightleftharpoons$	bicarbonate + $2e^- + 2H^+$	formate dehydrogenase[b]	(6)
xanthine + H_2O	$\rightleftharpoons$	uric acid + $2e^-$ + $2H^+$	xanthine oxidase[a]	(6)
sulphite + H_2O	$\rightleftharpoons$	sulphate + $2e^-$ + $2H^+$	sulphite oxidase	(6)
carbon monoxide + H_2O	$\rightleftharpoons$	carbon dioxide + $2e^- + 2H^+$	carbon monoxide oxidase[a][b]	(36)
acetaldehyde + H_2O	$\rightleftharpoons$	acetate + $2e^-$ + $2H^+$	aldehyde oxidase[a]	(6)

(a) Molybdoflavoprotein

(b) Enzyme not yet fully characterized.

relations between the g- and A-axes can be established (7-10), but no information about distances of ligand atoms from the metal. Conversely, EXAFS is only moderately good at identifying ligands, rather poor at determining ligand numbers and normally provides no information about geometry. However, once the ligands are known, EXAFS can provide very precise information on bond lengths.

The first e.p.r. studies on a molybdoenzyme (4) preceded those by EXAFS (5) by almost twenty years. Successful EXAFS studies of these enzymes have so far proved possible only by using the synchrotron at Stanford CA, USA and, to a lesser extent, that in Hamburg, Germany. At the time of writing (see ref (6) for summary of earlier work) up to perhaps two hundred papers on molybdenum in enzymes, based on e.p.r. information, have appeared in the literature, compared with less than ten papers based on EXAFS information. In addition to advantages of availability of experimental facilities, e.p.r. has considerable advantages over EXAFS in terms of sensitivity, since excellent e.p.r. spectra can usually be obtained with around 0.02 μmole of a molybdoenzyme, whereas EXAFS requires around 1 μmol. However, a basic disadvantage of the e.p.r. studies is that they give information only about the metal present in the Mo(V) oxidation state. In enzymes, this occurs almost always accompanied by, and often in equilibrium with, the diamagnetic Mo(IV) and Mo(VI) oxidation states. EXAFS on the other hand gives information about the metal in any oxidation state.

The most serious disadvantage of EXAFS is apparent when molybdenum in a sample does not occur in only one type of chemical environment, but is present instead, as a number of distinct chemical species. The resultant EXAFS spectrum is then the average of the contributions of each species. That this is a very serious limitation is shown by e.p.r. results. Here, individual contributions to an e.p.r. spectrum from several different Mo(V) species in a single sample, can frequently be deconvoluted by the use of appropriate techniques (difference spectra, computer simulations and the use of two microwave frequencies). E.p.r. studies (6) of this type, indicate that it is quite usual to have two, three or even more molybdenum(V) species coexisting in a single sample. Furthermore, frequently only perhaps 30% of the molybdenum is present in the Mo(V) state. Thus, presumably it is reasonable to assume that multiple molybdenum(IV) and molybdenum(VI) species are also present in some samples. Therefore, if meaningful results are to be obtained, EXAFS should be applied to molybdo-enzymes, only under conditions where everything possible has been done to minimize the number of species in the sample. In practice, this has unfortunately not always been done.

Interpretation of e.p.r. involves extraction of parameters (g-values, A-values [for ^{95}Mo and for ligand nuclei] and angles of non-coincidence). These parameters are obtained from the spectra, normally via simulations. The weakness is that structural information has to be deduced from comparison with theoretical predictions, or from data on model compounds, or both. The parameters are very sensitive to ligand geometry and bonding, with effects of these two factors closely inter-related. Thus, full interpretation in terms of structures is rarely possible. Nevertheless, if a magnetic nucleus shows hyperfine coupling to molybdenum, we can say unambiguously that this atom must be a ligand of the metal, or nearby.

Extraction of information from EXAFS spectra generally depends (cf. 38) on assuming the nature of the ligands, then from this simulating the spectrum. Ligand distances and to some extent their nature, are varied until the simulation gives an optimum fit to the experimental spectrum. The weakness would seem to be that quite different initial assumptions about the nature of the ligands might conceivably give comparable simulations. As indicated above, the strength of the EXAFS method is however, that very precise ligand distance information can, in favourable circumstances, be forthcoming.

Identification of Ligands

One major goal, both of e.p.r. and of EXAFS, is the identifica -tion of ligands of molybdenum in enzymes. Table 2 summarizes information on the twelve or thirteen different ligands of molybdenum which have so far been identified in particular forms, or states of different molybdoenzymes. The Table lists references from EXAFS (X), e.p.r. (E) or from chemical models or chemical analogies (C), in relation to each ligand. The columns correspond to a somewhat subjective categorization of the references as presenting, for the ligand in question, "first indications", "reasonable evidence", or "definitive evidence".

Clearly according to Table 2, at the present time e.p.r. has yielded considerably more definitive information on ligands of molybdenum in molybdoenzymes than has EXAFS. (This is particularly so if it is borne in mind that some of the work on model compounds e.g. ref (15) has depended on e.p.r.). It should be noted, though, that definitive information on ligands has in the main emerged from e.p.r. work relatively recently (Table 2), as improved experimental and computing techniques have made it possible to derive maximum information from isotopic substitution experiments.

Probably the greatest (and the first) triumph of EXAFS of molybdoenzymes was the identification (5), with the help of model compounds (24), of -S-Fe-ligands in nitrogenase, providing evidence that molybdenum in this enzyme is present in the form of some type of cluster containing several iron and sulphur atoms per molybdenum atom. (In this respect, nitrogenase differs from all other molybdoenzymes). Other definitive information from EXAFS concerns the presence, in

Table 2. Evidence for various ligands of Mo in enzymes

References (preceded by the year of publication) to work depending on chemical models or analogy are indicated by a letter C, those from e.p.r. by E and those for X-ray spectroscopy by X. Assignment to the different columns is of course somewhat arbitrary.

Ligand	Enzymes[a]	First indications	Reasonable evidence	Definitive evidence
=O	XO, SO, NR	C 1974(11)	E 1984(12)	X 1981(13), 1984(39)
=S	XO	C 1974(11)	X 1980(14), E 1981(7)	X 1981(13)
-OH	XO, SO, NR	E 1966(16)	E 1978(17), X 1981(13)	C 1981(15)
			X 1984(39)	
-SH	XO	E 1966(16)	X 1981(13)	E 1983(9)
-SR	XO, SO, NR	E 1966(18)	X 1981(13), 1984(39)	
-Cl, -F	SO, NR		E 1983(19)	E 1983(20)
-O-C≡	XO		E 1978(21)	E 1980(22)
$-O-PO_3H_2$	SO	E 1974(23)		E 1983(19)
-S-Fe-	N_2		X 1978(5)	C 1978(24)
-S-As=	XO		X 1983(25)	E 1983(8)
-S-Hg-	XO	X 1984(26)		E 1983(9)
-N=	XO		E 1984(27)	
Unknown	NR	X 1984(39)		

[a] Enzymes: XO, xanthine oxidase or dehydrogenase; SO, sulphite oxidase; NR, nitrate reductase; N_2, nitrogenase

oxidized forms of certain enzymes, of =O ligands (in xanthine oxidase, sulphite oxidase and nitrate reductase) and of =S ligands (in xanthine oxidase). There had been many indications for such ligands previously (11,17) (particularly in the case of =O) but the definitive information came from EXAFS (13,14, 39). For the =S ligand this was provided by the dramatic improvement of fit of the simulated to the experimental EXAFS spectrum (13) when a sulphur was included, at the optimum distance of 2.15 Å. This ligand is present only in the active form of xanthine oxidase and related enzymes.

Of the ligands identified by e.p.r., the strongly coupled protons on -OH and -SH ligands were detected at a very early stage from their strong hyperfine coupling (16). However, the nature of the intervening oxygen or sulphur atoms was not properly clarified until substitutions with ^{17}O ref (28) and with ^{33}S ref (7) had been performed and the couplings from these atoms evaluated. Important in this work was the first, and relatively recent, appearance (15) of a model compound having an -OH ligand with the proton coupled strongly to molybdenum, as in the enzyme species.

Detection of halide ligands of molybdenum by e.p.r. is a relatively new development (19,20). Since EXAFS work (13,29) on sulphite oxidase failed to take into account the possible ligation of molybdenum to either chloride or phosphate (19) ions from buffer constituents, some reinterpretation of the published work (13) may be required. (Chloride would presumably be confused with sulphur and the 'unknown' ligand in nitrate reductase (39) might conceivably be a phosphate group).

The -As= and -Hg-ligands of molybdenum, in complexes of inhibitors with xanthine oxidase, are remarkable in that their e.p.r. spectra (8,9) show not only strong hyperfine, but also strong quadrupole coupling of ^{75}As or ^{201}Hg to molybdenum. Estimation of the quadrupole coupling parameters for the e.p.r.

spectra has provided additional information bearing on the structures of these complexes.

The -O-C≡ ligand of molybdenum is particularly important in relation to the mechanism of action of xanthine oxidase (as is discussed below). This ligand was identified from ^{17}O and ^{13}C substitution studies. In the appropriate enzyme species, the carbon atom is rather weakly coupled (21) and the oxygen atom strongly and isotropically coupled (22).

Origins of Multiple Enzyme Species

So far different ligands of molybdenum in enzymes have been discussed in general terms, with little reference to the enzymes themselves. The most general conclusion from extensive e.p.r. work (6) over many years has been that the molybdenum centre of each enzyme in the Mo(V) oxidation state can exist in a variety of different chemical forms. Much work by e.p.r. has gone, first into distinguishing these, then, on the one hand into obtaining structural information on them by identifying ligands as discussed above, and on the other hand in elucidating relationships among the species by investigating their appearance, disappearance and interconversions.

In general, multiple species of the enzymes in the Mo(V) oxidation state (and presumably in the Mo(IV) and Mo(VI) states also) arise from causes such as equilibria with protons and complex formation with anions or substrate or product molecules (40). Additional species represent relatively stable complexes with inhibitors (41), and one [the species from xanthine oxidase known as Very Rapid (37)] represents a transient intermediate in substrate turnover. Finally, normal preparations of xanthine oxidase contain not only the active enzyme but also a catalytically inactive degradation product, the desulpho form of the enzyme. Only the active form of the enzyme contains the =S ligand of molybdenum (Table 2), originally supposed (30) from biochemical data to be a persulphide group [see also (42)].

Mechanism of Action of Xanthine Oxidase

Xanthine oxidase is the most studied and best understood of the molybdoenzymes. Understanding its mechanism of action (6,31, 37) has come about over a long period of time, as a result of some of the e.p.r. studies summarized above, in conjunction with kinetic studies. The generally accepted catalytic mechanism, and essential evidence (17,21,22,32) for it, are summarized in Scheme 1.

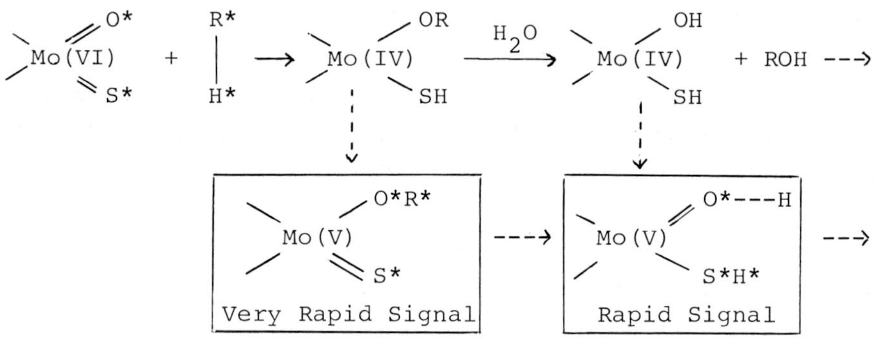

Scheme 1: Mechanism of reaction of xanthine oxidase with a reducing substrate, RH (refs.9,12,17,21,22,32,37). Atoms marked in the oxidized enzyme or the substrate with asterisks, have been substituted with stable isotopes (^{17}O, ^{33}S, ^{13}C, or ^{2}H) and can then, under appropriate conditions be detected from their hyperfine coupling to molybdenum, in the e.p.r. signal-giving species in the boxes.

The mechanism involves reaction of =O and =S ligands of Mo(VI) in the enzyme with the substrate molecule, RH, to give the Mo(IV) state, with these ligands transformed to -OR and -SH respectively. In further stages of the catalytic cycle this species breaks down, with fission of the Mo-O-bond, to liberate the product anion, RO^-, while Mo(IV) is re-oxidized by intramolecular electron transfer to the flavin and iron-sulphur centres of the enzyme.

As indicated above, finally establishing presence of =O and =S ligands in oxidized xanthine oxidase depended on EXAFS. However, the mechanistic conclusions of Scheme 1 otherwise depend entirely on observations on two Mo(V) e.p.r. signals, indicated in the boxes. These signals are known as Very Rapid and Rapid (type 1), respectively (6). The definitive observations were made in fast kinetic studies employing the rapid freezing method (33). The Very Rapid signal is, with the substrate xanthine, a transient species detectable only by rapid freezing. Its structure has been established (37) by identification, by the use of isotopic substitution, of three atoms, marked in Scheme 1 by asterisks. ^{17}O from water can be incorporated into the oxidized enzyme and then appears in the Very Rapid signal (22). ^{13}C in the 8-position of xanthine also shows (21) hyperfine coupling in this signal, as does ^{33}S, initially incorporated into =S of the oxidized enzyme, by reactivation of the desulpho form with $^{33}S^{2-}$ (ref. 7). Coupling of this sulphur is particularly strong and anisotropic, indicating retention of the =S ligand in the Very Rapid species, while coupling of O is isotropic, in keeping with the Mo-O-C≡ structure.

Information from the Rapid signal on the mechanism depends on the observation of direct transfer of hydrogen from the substrate RH to xanthine oxidase, first detected (34) in 1968, but not interpreted, quantitatively, until ten years later (32). If the deuterated substrate is used, then under the appropriate conditions, the coupled hydrogen of the Rapid signal appears as deuterium. This atom then exchanges relatively rapidly into the solvent, so that rapid freezing is required for observation of the direct hydrogen transfer phenomenon. The oxygen and sulphur ligands are also detected in isotopic substitution experiments, as for the Very Rapid signal. However, now it is coupling of oxygen which is strong and anisotropic, indicating (12,28) the structure of the Rapid species to be as shown in Scheme 1.

It should be noted that the signal-giving species in Scheme 1 are depicted, not as on the 'primary' catalytic pathway, which is assumed to involve only the Mo(VI) and Mo(IV) oxidation states, but instead as being on a parallel path, resulting from alternative reaction sequences, involving Mo(V). The evidence seems (37) to be that the parallel path is kinetically closely related to the main pathway, so that the validity of conclusions derived from e.p.r. is in no way weakened.

Confirmation of the mechanism of Scheme 1 has come from several studies with inhibitors (31). Particularly significant, is reaction (9) of xanthine oxidase with p-mercuribenzoate to give the Mo-S-Hg-species. This reaction is fully in keeping with the presence of a Mo-SH group in reduced enzyme species such as Rapid. The fact that the mercury in the inhibitory complex and the proton in the Rapid species are both strongly coupled to molybdenum, provides evidence that the proton is a sulphur. This, in turn, confirms that the =S, and not the =O, ligand in the oxidized enzyme, is the proton acceptor (32) in catalytic turnover. In additional work, the mechanism of reaction with the potent inhibitor, alloxanthine, has recently been elucidated (27) by e.p.r. This compound is not oxidized by the enzyme, but nevertheless binds strongly to a reduced form of it, to form a species with a structure analogous to that of the Very Rapid species, but having, in place of the Mo-O-C≡, a Mo-N= bond, believed to be formed from the N-2 atom of alloxanthine by elimination of the original oxygen ligand.

Prospects

Though much has been achieved, clearly further work is required, both by e.p.r. and by EXAFS. Further biochemical work is required on several of the enzymes listed in Table 1, to be followed by physical studies. E.p.r. is clearly capable of yielding more information, e.g. on ligands of molybdenum in nitrate reductase. It might be speculated, since proton transfer in xanthine oxidase catalysis may be observed directly, that a similar phenomenon might apply, and perhaps be observable by kinetic e.p.r. work, e.g. in sulphite oxidase. Transfer of oxygen, from molybdenum of the enzymes to product molecules, though widely assumed to take place in catalysis, has so far not actually been demonstrated, and could, and should, be confirmed. More work is needed on the relevance to catalysis of the anion complexes of Mo(V) in sulphite oxidase and nitrate reductase. For all the enzymes, full interpretation (cf. ref. 10) of ^{95}Mo and where possible of ^{97}Mo (cf. 37) couplings, in terms of angles of non-coincidence of the q- and A-axes and of quadrupole coupling constants, should provide substantial new information.

For EXAFS, priority should be given to studies of states of the enzymes anticipated from e.p.r. work to correspond to single chemical species. Both the As and Hg complexes of xanthine oxidase seem good candidates for such studies, since, in contrast to the vast majority of the enzyme species, approaching 100% conversion of these complexes to the Mo(V) state can be achieved. Thus, direct comparison of EXAFS and e.p.r. data on single species should, for the first time, be possible. The comparative ease with which =O and =S ligands may be detected, makes it desirable that these should be looked for in the Very Rapid (37) and Inhibited (12) species. Additionally, comparison of the structures of corresponding enzyme species in the Mo(V) and Mo(VI) oxidation states should be attempted. In general, much slow and painstaking EXAFS work is needed in relation to the nature and numbers of ligands of molybdenum not detected by e.p.r.

Completion of the above e.p.r. and EXAFS work will confirm the molybdoenzymes as the best understood group of metalloenzymes, for which structural information is not available from X-ray crystallography.

Acknowledgements

The author thanks his many past collaborators and particularly, his present collaborator, Dr. G.N. George, for their parts in work from this laboratory, as listed in the references. E.p.r. work has been supported by grants from the MRC and work by EXAFS, by the SERC.

References

1. This paper is based in part on a lecture by the author presented in March 1984, which is to be published in Biological Systems: Structure and Analysis: Proceedings of the Daresbury Study Weekend, Ed. Garner, C.D., SERC, Daresbury, U.K.
2. Claassen, V.P.: Doctoral Thesis, University of Amsterdam (1983).
3. Del Campillo-Campbell, A., Campbell, A.: J.Bacteriology, 149, 469-478 (1982).
4. Bray, R.C., Malmström, B.G., Vänngård, T.: Biochem.J., 73, 193-197 (1959).
5. Cramer, S.P., Hodgson, K.O., Gillum, W.O., Mortenson, L.E.: J.Am.Chem.Soc., 100, 3398-3407 (1978).
6. Bray, R.C.: Advances in Enzymology and Related Areas of Molecular Biology, 51, 107-165 (1980).
7. Malthouse, J.P.G., George, G.N., Lowe, D.J., Bray, R.C.: Biochem.J., 199, 629-637 (1981).
8. George, G.N., Bray, R.C.: Biochemistry, 22, 1013-1021 (1983).

9. George, G.N., Bray, R.C.: Biochemistry 22, 5443-5452 (1983).

10. George, G.N.: D.Phil. Thesis, University of Sussex (1983).

11. Williams, R.J.P., Wentworth, R.A.D.: in Chemistry and Uses of Molybdenum, ed Mitchell, P.C.H. (Climax Molybdenum Company, London, 1974) pp.212-215.

12. Morpeth, F.F., George, G.N., Bray, R.C.: Biochem.J. 220, 235-242 (1984).

13. Cramer, S.P., Wahl, R., Rajagopalan, K.V.: J.Am.Chem.Soc., 103, 7721-7727 (1981).

14. Bordas, J., Bray, R.C., Garner, C.D., Gutteridge, S., Hasnain, S.S.: Biochem.J., 191, 499-508 (1980).

15. Yamase, T., Sasaki, R., Ikawa, I.: J.Chem.Soc. Dalton, 628-634 (1981).

16. Bray, R.C., Meriwether, L.S.: Nature (London) 212, 467-469, (1966).

17. Gutteridge, S., Tanner, S.J., Bray, R.C.: Biochem.J., 175, 887-897.

18. Meriwether, L.S., Marzluff, W.F., Hodgson, W.G.: Nature (London) 212, 465-467 (1966).

19. Bray, R.C., Gutteridge, S., Lamy, M.T., Wilkinson, T.: Biochem.J., 211, 227-236 (1983).

20. Morpeth, F.F., Boxer, D.H., Bray, R.C., George, G.N.: in Recent developments in the chemistry of chromium, molybdenum and tungsten, ed. Dilworth, J.R., Lappert, M.F. (Royal Soc. Chemistry, Dalton, London, 1983) p.P10; also unpublished work.

21. Tanner, S.J., Bray, R.C., Bergmann, F.: Biochem.Soc.Trans., 6, 1328-1330 (1978).

22. Gutteridge, S., Bray, R.C., Biochem.J., 189, 615-623 (1980).

23. Kessler, D.L., Rajagopalan, K.V.: Biochim.Biophys.Acta, 370, 389-398 (1974).

24. Wolff, T.E., Berg, J.M., Warrick, C., Hodgson, K.O, Holm, R.H.: J.Am.Chem.Soc., 100, 4630-4632 (1978).

25. Cramer, S.P.: in Recent developments in the chemistry of chromium, molybdenum and tungsten, ed. Dilworth, J.R., Lappert, M.F. (Royal Soc. Chemistry, Dalton, London, 1983) p.L27.

26. Cramer, S.P., George, G.N., Bray, R.C.: unpublished work.

27. Hawkes, T.R., George, G.N., Bray, R.C.: Biochem.J., 218, 961-968 (1984).

28. Bray, R.C., Gutteridge, S.: Biochemistry, 21, 5992-5999 (1982).

29. Cramer, S.P., Gray, H.B., Rajagopalan, K.V.: J.Am.Chem. Soc., 101, 2772-2774 (1979).

30. Massey, V., Edmondson, D.: J.Biol.Chem., 245, 6595-6598 (1970).

31. Bray, R.C., George, G.N., Hawkes, T.R., Morpeth, F.F.: in Recent developments in the chemistry of chromium, molybdenum and tungsten, ed, Dilworth, J.R., Lappert, M.F. (Royal Soc. Chemistry, Dalton, London, 1983) p.L7.

32. Gutteridge, S., Tanner, S.J., Bray, R.C.: Biochem.J., 175, 869-878 (1978).

33. Bray, R.C.: Biochem.J., 81, 189-195 (1961).

34. Bray, R.C., Knowles, P.F.: Proc.Roy.Soc.(London) A302, 351-353 (1968).

35. Oltmann, L.F., Claassen, V.P., Kastelein, P., Reijnders, W.N.M., Stouthamer, A.H.: FEBS Letters, 106, 43-46 (1979).

36. Meyer, O.: J.Biol.Chem., 257, 1333-1341 (1982).

37. Bray, R.C., George, G.N.: Biochem.Soc. Transactions (in press).

38. Cramer, S.P.: in Inorganic and Bioinorganic Mechanisms Vol.2 (1983) ed Sykes, A.G. (Academic Press, New York) pp.259-316.

39. Cramer, S.P., Solomnson, L.P., Adams, M.W.W., Mortenson, L.E.: J.Am.Chem.Soc., 106, 1467-1471 (1984).

40. Bray, R.C., George, G.N., Gutteridge, S., Morpeth, F.F., Turner, N.: (this volume).

41. Morpeth, F.F., Bray, R.C.: Biochemistry, 23, 1332-1338 (1984).

42. Ventom, A., Bray, R.C.: (this volume).

BIOCHEMISTRY OF NATURALLY OCCURRING DEAZAFLAVIN COENZYMES

Christopher Walsh

Departments of Chemistry and Biology,
Massachusetts Institute of Technology
Cambridge, Massachusetts 02139

INTRODUCTION

Flavin coenzymes with their tricyclic isoalloxazine redox moiety are perhaps the most versatile of the redox cofactors in biochemistry. A critical feature is the ability of flavins to undergo either $2e^{\ominus}$ or $1e^{\ominus}$ redox steps, allowing flavoenzymes thereby to sit at the crossroads of $2e^{\ominus}$ and $1e^{\ominus}$ electron transfer processes in biology (1). The kinetic and thermodynamic accessibility of the flavin semiquinone radical also appears to permit flavoenzymes to function in reductive metabolism of O_2 either to superoxide and/or hydrogen peroxide along with monooxygen transfer capacity. The nitrogen atoms at positions one and five (N^1,N^5) of the isoalloxazine have been analyzed for their contributions to flavin redox vesatility by synthetic substitution with carbon to yield 1-carba (2,3)(1-deaza) and 5-carba (5-deaza) flavin analogs (4-6). The 1-deazaflavin, at the FMN

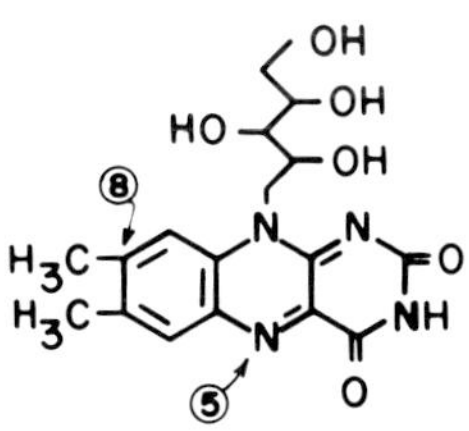

Riboflavin

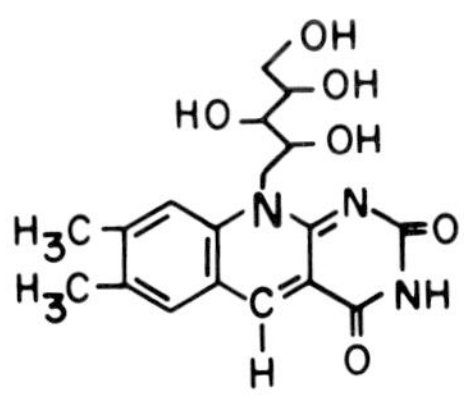

5-Deazariboflavin (5-carba)

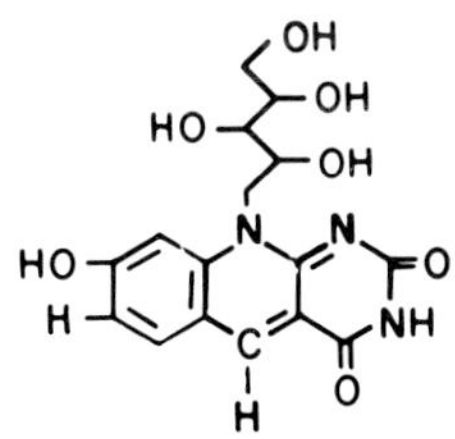

8-Hydroxy-7-demethyl-5-deazariboflavin (Fo)

Flavins and Flavoproteins

and FAD analog, has a more negative redox potential (-280 mV) than FMN or FAD (-210 mV) but still retains one electron redox capacity and at least some O_2 reactivity. The 5-deazaflavin has a central pyridine ring rather than the pyrazine ring in the parent flavin tricycle; in the dihydro forms this dihydropyridine for dihydropyrazine replacement is consequential. The 5-deazaflavin system reacts only by $2e^{\ominus}$ steps in ground state chemistry as the pyridine/dihydropyridine system dominates the chemical reactivity and the analogy to NAD/NADH cofactors is very close indeed. We (3,7) and Hemmerich and Massey (8) have elsewhere noted that 5-deazaflavins are, in reactivity terms, more like nicotinamides in flavin's clothing.

Factor 420: 8-Hydroxy-7-demethyl-5-deazaflavins

The 5-deazaflavin system first synthesized by man in 1970 (4) turns out to have been produced in nature throughout the past two billion years as the key redox locus in the yellow, highly blue-green fluorescent Factor 420 of methanogenic bacteria, a family of the Archaebacterial kingdom. Wolfe and his colleagues detected, isolated and identified F_{420} as the first example of an 8-hydroxy-7-demethyl-5-deazariboflavin (9,10). The side chain is a ribityl-5-phosphate in phosphodiester to a lactyl diglutamate, a bit reminiscent of a folate side chain. The structure was confirmed by synthesis of the 8-OH-7-demethyl-5-deazariboflavin, Fo, by Ashton and Brown in 1979 (11,12). Under some growth conditions, Fo, is

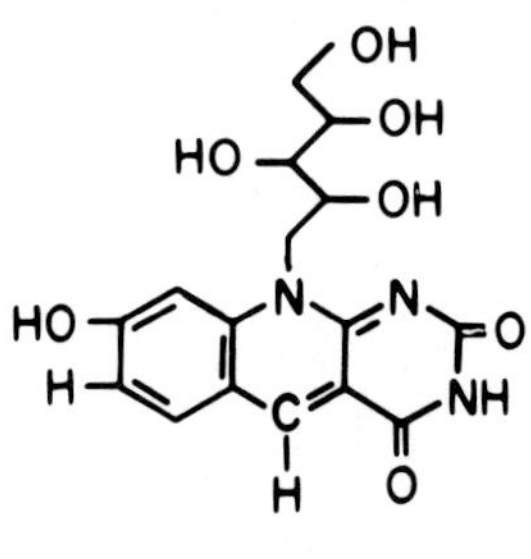

Fo

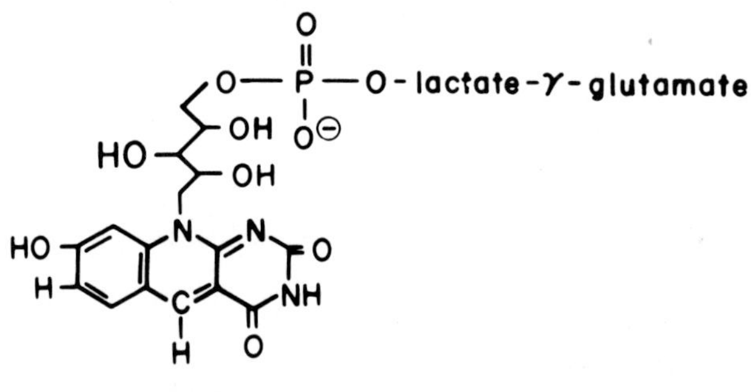

F_{420}

(up to 100mg/kg of methanogenic bacteria)

excreted into the media by the methanogenic bacteria. The rest of this chapter deals with biochemical roles for F_{420} elucidated over the past five years.

Chemical Features of F_{420} relevant to its Biochemical Roles

The effect of the 8-hydroxy group on the 5-carbaflavin system was presaged by Ghisla and Mayhew in 1976 (13) in their study of 8-hydroxy FAD released from D-lactate dehydrogenase from Megasphera elsderii. The phenolic pKa is 4.8 in that system and analogously we (14) and Eirich et al (10) found a value of 5.8 for Fo. The absorbance at 420 nm is due to

8-Hydroxyriboflavin

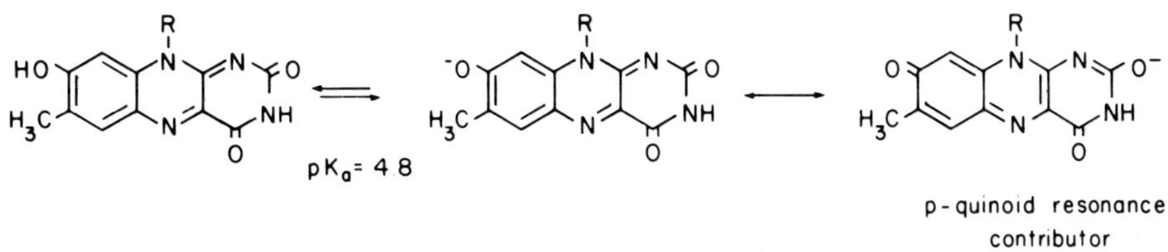

8-Hydroxy-5-deazariboflavin

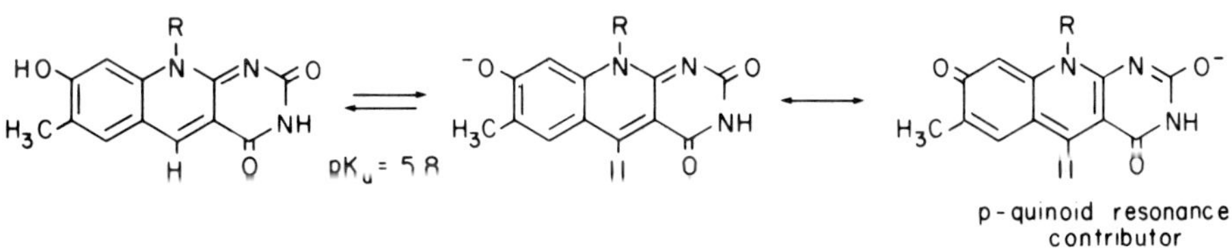

the 8-phenoxy anion which has quinonoidal character. In fact above this pKa, at physiological pHs, the electron rich Fo or F_{420} anion is very difficult to reduce and specific proteins may alter this pKa on binding (e.g. flavodoxin (14)).

The dihydro Fo or F_{420} is the 1,5-dihydro isomer and this, like NADH, is kinetically stable to autoxidation and useful as a low potential general redox currency in methanogenic gells. The $E^{0\prime}$ for the Fo/FoH_2 couple is -350 to -370 mV (14) so the equilibrium for electron transfer to NADP, catalyzed by a specific methanogen oxidoreductase is 10-100/1 in favor of NADPH generation. Also, like the C_4 locus of NADPH, the C_5 locus

$$FoH_2 + NADP \rightleftharpoons NADPH + Fo$$

of FoH_2 and $F_{420}H_2$ is prochiral and not readily exchangeable with solvent. This allows inquiry into the origin of the transferring C_5 hydrogen and the stereochemical course of transfer, inquiries characteristic for NADH and NADPH but not possible for $FADH_2$ with its nitrogen atom at position 5 from which hydrogen equilibrates rapidly with solvent. We note below experiments on these issues with methanogen enzymes.

Roles for F_{420} in Methanogen Metabolism

Scheme one proposes several enzymic roles for F_{420} in methanobacterial redox metabolism. F_{420} is reduced to $F_{420}H_2$ by the two enzymes proceeding down the left of the scheme, hydrogenase and formate dehydrogenase, while $F_{420}H_2$ may be reoxidized to drive the equilibria proceeding up the right side of the scheme, including postulated roles in three crucial sequences, CH_4 generation (15), NADPH production (16), and reductive acetyl CoA carboxylation to pyruvate (17) in cell carbon fixation.

BIOLOGICAL FUNCTIONS OF FACTOR 420

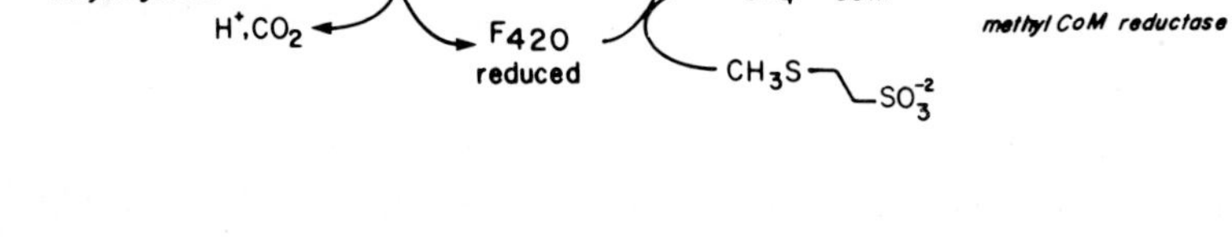

Studies in crude extracts (9) suggested H might reduce F_{420} but only when we suceeded in purifying a F_{420}-reducing hydrogenase from Methanobacterium thermoautotrophicium ΔH was this conclusively demonstrated (18).

The $F_{420}H_2$ase is an intriquing enzyme. We could purify it to ~90% homogeneity in an inactive aerobic scheme then reductively reactivate it. In the course of purification we separated out and purified a distinct hydrogenase which could not reduce deazaflavins (19). The physiological roles of both hydrogenases are under investigation. The F_{420} hydrogenase is an $\alpha_2\beta_2\gamma$ multi subunit enzyme which contains not only the anticipated Fe/S clusters but also stoichiometric FAD and inorganic nickel ion (18,19). We anticipated a $2e^{\ominus}.1e^{\ominus}$ redox switch would be required to interface between the one e fe/S clusters and the obligate two $e^{\ominus}$ acceptor F_{420} and therefore looked for and detected enzyme-bound FAD (18). This proposed role seems validated by recent work from Ferry's laboratory where FAD dissociates from the hydrogenase purified anaerobically from M. formicicum. Only when FAD is added back can that purified hydrogenase reduce F_{420}. Here then is a case of a flavoprotein catalyzing net hydride transfer to the natural deazaflavin.

We have analyzed for substrate deuterium isotope effects on Vmax and on Vmax/Km with D_2 and the pure $F_{420}H_2$ase from M. thermo ΔH and do not see values distinct from 1.0 (21). Analysis of deuterium content at C_5 of FoH_2 reveals no deuterium. The transferring deuterium atom washes out completely into solvent H_2O before transfer to the exchange stable C_5 locus of FoH_2. The complementary study with H_2 in D_2O validated this premise since 5-D_1-dihydro Fo is detected. Again V_H/V_D values are essentially 1.0 suggesting hydrogen transfer from H_2 to enzyme out to solvent and from solvent back into enzyme and on to Fo do not occur in kinetically significant transition states (21). The V/K value for H_2 is in fact $5 \times 10^7\ M^{-1}sec^{-1}$ with a Km of 2μM,, below the likely ambient concentration in the environment of these bacteria. By this criterion the reductively reactivated F_{420} H_2ase is close to a perfect catalyst (22). Finally, we note studies by Daniels, Spencer et al (23,24) that in vivo D_2 gives only CH_4, consistent with this H ase being a site for the quantitative exchange with solvent hydrogen.

The most striking and surprising feature about the purified F_{420} hydrogenase was the tightly bound nickel (18,19). Subsequent to detection of nickel in this and a M. thermoautotophicum Marburg strain hydrogenase (25), nickel has been detected as a stoichiometric component in uptake hydrogenases in sulfate reducers (26), *Chromatium* (27), *Alcaligines* (28), Vibrio (29), and Rhizobium (30) species, suggesting nickel is a general constituent of hydrogenases. The role of nickel is under intense scrutiny and may require ligand identification as structural prelude to functional analysis. We have recently reported initial nickel EXAFS (31) and nickel electron spin echo studies (32) suggesting sulfur ligands and a more distant nitrogen ligand to nickel in the F_{420} hydrogenase.

The hydrogen electrode potential, $H_2/2H^+ + 2e^{\ominus}$, and the formate/CO_2 potential are both at -420mV and have the thermodynamic driving force to generate $F_{420}H_2$ from F_{420}. Ferry and his colleagues have purified a three subunit F_{420}-reducing formate dehydrogenase to homogeneity (33) and find a molybdopterin cofactor, fe/S clusters, and bound FAD. Again they have demonstrated that FAD is the redox outlet to the cosubstrate F_{420}, delineating again the redox versatility of FAD to sit in a niche between $1e^{\ominus}$ centers (fe/s) and $2e^{\ominus}$ centers (in this case F_{420}).

On the ascending arm of Scheme one only the NADP oxidoreductase has yet been purified to homogeneity and shown to use F_{420} as direct cosubstrate (16). Of special interest will be whether $F_{420}H_2$ is a *direct* reductant in the conversion of methyl-S-CoM to methane, the last reductive step in methanogenesis.

The stereochemical analysis of F_{420} (Fo) redox enzymes in methanogens is just beginning. In the NADH, NADPH systems, stereochemical studies on *A* side and *B* side specificity, later shown to correlate with *re* and *si* absolute chirality, were both classical and very important questions in enzyme stereochemistry. We have shown the F_{420} hydrogenase (14) and the NADP oxidoreductase (34) recognize the same prochiral H (or chiral ^3H in enzymically generated 5-^3H FoH_2 samples) which we have defined as *A* side (14). The opposite *B* side recognition is demonstrated by an NADH -dependent oxido reductase from the marine bacterium *Benekea harveyii* (14).

Non-Methanogenic Roles for F_{420}

Subsequent to the detection of F_{420} by Wolfe and colleagues as a ubiquitous marker of methanogenic bacteria, the 8-hydroxy-7-demethyl-5-deazaflavin molecules have been detected in other organisms including several strains of streptomycetes, Anacystis nidulans (35) and most recently Mycobacterium avium. Thus F_{420} and congener seem much more widely distributed in nature than just in methanogens. There are currently two distinct types of metabolic roles in the non-methanogenic organisms detected for F_{420} in Streptomycetes as will be noted below.

Terminal Stages of Tetracyline Biosynthesis

The biosynthesis of the tetracycline family of antibiotics, e.g. chlortetracycline and oxytetracycline were carried out almost thiry years ago by the use of blocked mutants of streptomycetes, allowing the accumulation and detection of intermediates. Studies by McCormick and colleagues at Lederlie Laboratories resulted in a publication in 1960 (37) noting that a mutant blocked in the last step, conversion of 5, 11a, 7-dehydrochlortetracycline, inactive as an antibiotic, to the active chlortetracycline required a small molecule, cosynthase I, from an S.

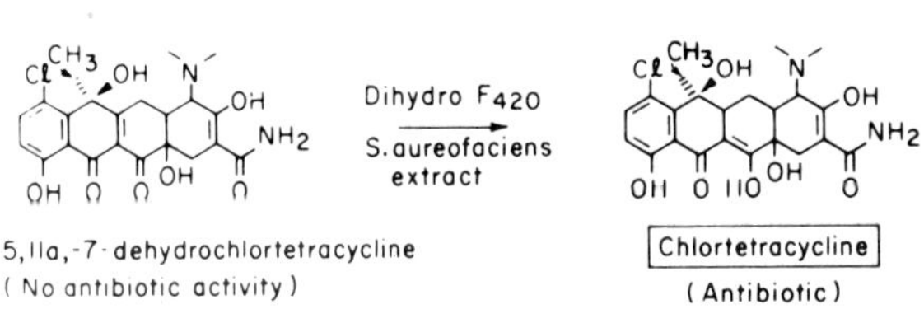

aureofaciens extract. That small molecule, they noted 22 years later in 1982, (38) is F_{420}. The conversion in question, shown below, is a formal reduction of an α,β-enone and, although experiments have not been

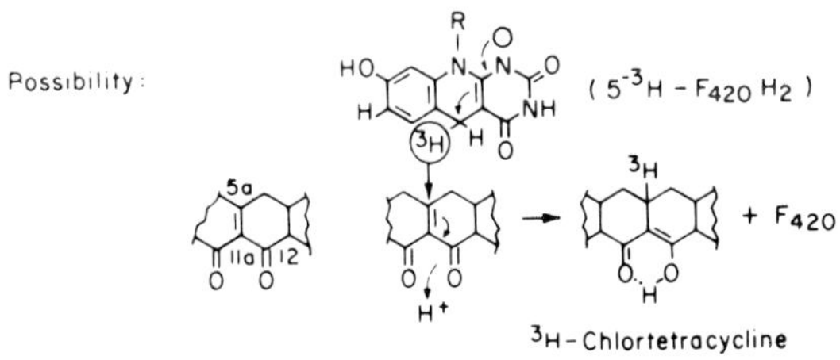

reported, it is likely that $F_{420}H_2$ is serving as a hydride donor to C5a in a conjugate addition sequence. $F_{420}H_2$ appears to fulfill a similar crucial role in oxytetracycline biosynthesis (39). How general a metabolic reducing currency $F_{420}H_2$ is in the antibiotic-producing streptomycetes will be of interest.

Photoreactivation of UV-damaged DNA

DNA damaged by UV-light contains as predominant lesion the cis-syn-cyclobutane dimer of two adjacent intrachain thymine residues. These thymine dimers can be repaired enzymically either by dark processes, e.g. excision repair, or by photoreactivation with visible light (350-500 nm) (35). The photoreactivation spectrum differs in various organisms and

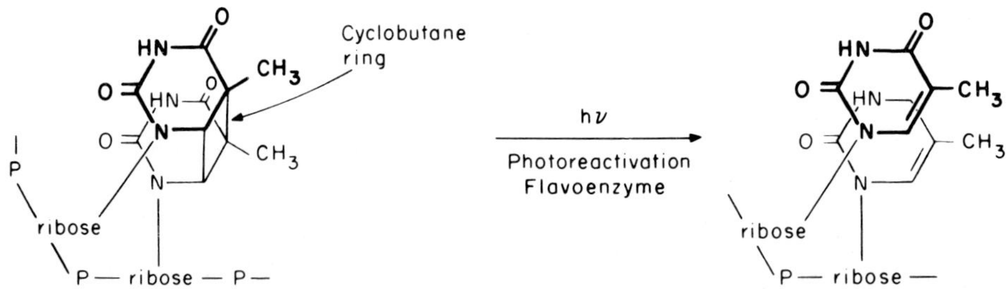

cis-syn-intrachain
thymine dimer

reflects the cofactor requirements of specific photorepair enzymes or DNA photolyases, as they are also known. Two types of purified DNA photolyases are now known. The yeast enzyme and the E. coli enzyme contain, as isolated, a reduced FAD species, possibly the 4a,5-isomer of the flavin (40,41). Sancar and colleagues (42) have cloned the Eco phr gene and amplified it 15,000-fold such that photolyase goes from 10-20 copies per cell to 15% of the cell protein and can be purified in quantity for study.

The other type of photorepair enzyme has a distinct action spectrum at 420-430 nm (35) and stems from F_{420} as the light-absorbing cofactor of the DNA photolyase. Eker (43,44) has purified the photolyase to homogeneity from Streptomyces griseus and the alga Anacystis nidulans (35) and has proven stoichiometric content of F_{420}. Work can now begin on the mechanism for DNA cyclobutane reversion by the photosensitizing flavin and 5-deazaflavin coenzymes.

There is evidence that the photoreversion can occur in model studies with thymine dimer. Thus, Eker, et al (44) could detect 7% monomerization by N^{10}-ethyl-7-demethyl-8-hydroxy-5-deazaflavin but Lamola (45) reported that at pH 7 lumiflavin is inactive as a sensitizer for thymine dimer cleavage. We (46) have recently conducted model studies using HPLC separation of [^{14}C]-thymine dimer and product monomer and detect catalytic behavior at high pH (>10) for flavins such as lumiflavin and for 5-deazariboflavin and 8-methoxy-5-deazariboflavin; action spectra are

Action Spectrum of Thymine Dimer Cleavage vs. Absorbance Spectrum of 8-Methoxy-5-deazaflavin

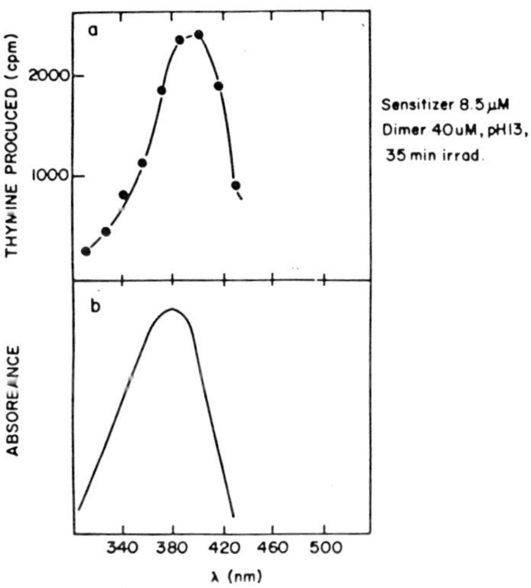

Action Spectrum of Thymine Dimer Cleavage vs. Absorbance Spectrum of Lumiflavin

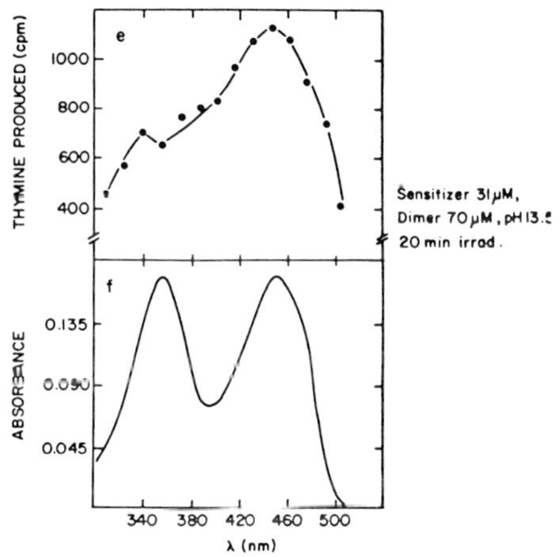

shown below. In this model system the 8-hydroxy-5-deazaflavin is inactive, indicating that the ionization at C_8, blocked in the methoxy, is deleterious. The S. griseus enzyme may alter this pKa in bound F_{420}.

Mechanistic studies on the model system (46) show saturation kinetics indicative of complexation prior to energy transfer. It is likely that the flavin or 5-deazaflavin absorbs light, and goes to the excited state triplet in the (deaza)flavin-thymine dimer complex. Then transfer of an electron or H· may occur, for example from dimer to flavin to yield the dimer radical cation and (deaza)flavin semiquinone. The dimer radical cation could undergo facile cleavage of the cyclobutane to yield thymine and a thymine radical which would regain the electron from the waiting

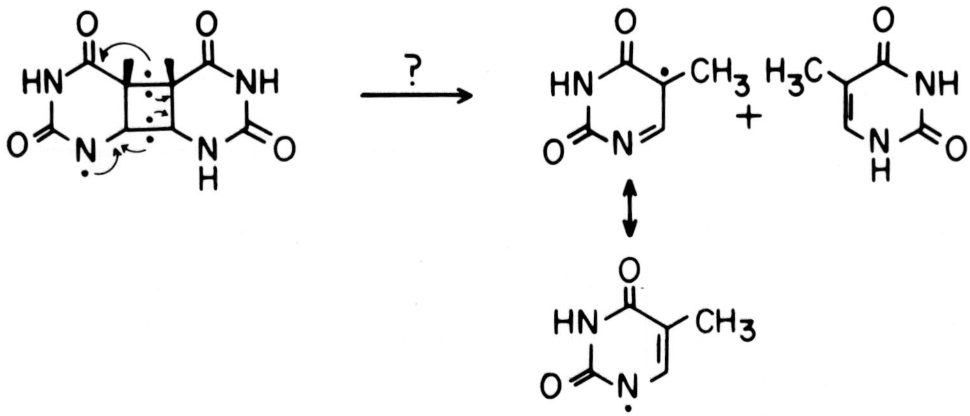

flavin semiquinone. The availability of pure photolyases and a good model for thymine dimer photorepair should facilitate progress in the near future.

Conclusion

At this juncture, a half dozen years after the initial characterization of F_{420} as a methanogen natural product 8-hydroxy-5-deazaflavin, the chemistry open to this redox cofactor has begun to be delineated and a wide range of biological niches both in methanogens and in other organisms has been detected. The remarkable roles of F_{420} in tetracycline biosynthesis and in DNA photorepair suggest additional novel biochemical functions for this deazaflavin coenzyme may yet be unraveled.

References

1. Walsh, C.: Accts. Chem. Res. 13, 148 (1980).

2. Walsh, C., Fisher, J., Graham, D., Spencer, R., Ashton, W., Brown, J., Brown, R., Rodgers, E.: Biochemistry 17, 1942 (1978).

3. Spencer, R., Fisher, J., Walsh, C.: Biochemistry 16, 3586-3594 (1977).

4. O'Brien, P., Weinstock, L., Cheng, C.: J. Heterocyclic Chem. 7, 99 (1970).

5. Edmondson, D., Barman, B., Tollin, G.: Biochemistry 11, 1133 (1972).

6. Fisher, J., Spencer, R., Walsh, C.: Biochemistry 15, 1043-1054 (1976).

7. Walsh, C., Jacobson, F., Ryerson, C.: Biomimetic Chemistry, ACS Symposium Series vol. 34, p. 119 (1980).

8. Massey, V., Hemmerich, P.: Biochemistry 17, 1 (1978).

9. Tzeng, S., Wolfe, R., Bryant, M.: J. Bacteriol. 121, 184 (1975).

10. Eirich, D., Vogels, G., Wolfe, R.: Biochemistry 17, 4583 (1978).

11. Ashton, W., Brown, R.: J. Heterocyclic Chem. 17, 1709 (1980).

12. Ashton, W., Brown, R., Jacobson, F., Walsh, C.: J. Am. Chem. Soc. 101, 4419 (1979).

13. Ghisla, S., Mayhew, S.: Eur. J. Biochem. 63, 373 (1976).

14. Jacobson, F., Walsh, C.: Biochemistry 23, 979 (1984).

15. Ellefson, W., Wolfe, R.: J. Biol. Chem. 255, 8388 (1980).

16. Yamazaki, S., Tsai, L.: J. Biol. Chem. 255, 6462 (1980).

17. Zeikus, J., Fuchs, G., Kenealy, W., Thauer, R.: J. Bacteriol. 132, 604 (1977).

18. Jacobson, F., Daniels, L., Fox, J., Walsh, C., Orme-Johnson, W.: J. Biol. Chem. 257, 3385 (1982).

19. Kojima, N., Daniels, L., Fox, J., Hausinger, R., Orme-Johnson, W., Walsh, C.: Proc. Nat. Acad. Sci. 80, 378 (1983).

20.

21. Fox, J.: Ph.D. dissertation, MIT, Chemistry Department (1984).

22. Fersht, A.: Enzyme Mechanisms, W. H. Freeman, San Francisco (1978).

23. Daniels, L. Fulton, G., Spencer, R., Orme-Johnson, W.: J. Bacterol. 141, 694 (1980).

24. Spencer, R., Daniels, L. Fulton, G., Orme-Johnson, W.: Biochemistry 19, 3678.

25. Graf, E., Thauer, R.: FEBS Letts 136, 165 (1981).

26. Commack, R., Patil, D., Aguirre, R., Hatchikian, E.: FEBS Lett 142, 289 (1982).

27. Albracht, S., Grafe, Thauer, R. (1982) FEBS Lett 140, 311 (1982).

28. Friedrich, B., Cornelius (1982).

29. Under, G., Bacher, R., Knecht, J., Kroger, A.: FEBS Lett 145, 230 (1982).

30.

31. Lindahl, P., Kohima, N., Hausinger, R., Fox, J., Teo, B., Walsh, C, Orme-Johnson, W.: J. Am. Chem. Soc. 106, (1984).

32. Tan, S., Fox, J., Kojima, N., Walsh, C., Orme-Johnson, W.: J. Am. Chem. Soc. 106, (1984).

33. Schauer, N., Ferry, J.: J. Bacteriol. 150, 1, (1982).

34. Yamazaki, S., Tsai, L. Stadtman, T., Jacobson, F., Walsh, C.: J. Biol. Chem. 255, 9025 (1980).

35. see Eker, A. M.: in Molecular Models of Photoresponsiveness, Montagndi, G., and Erlanger, B., editors, Plenum Ptress, p. 109, for summary.

36. Naraoka, T., Momoi, K., Fukasawa, K., Goto, M.: Biochim. Biophys. Acta 797, 379 (1984).

37. Miller, P., Sjolander, N., Nalesnyk, A., Arnold, N., Johnson, S., Doerschuk, A., McCormick, J.: J. Am. Chem. Soc. 82, 5002 (1960)/

38. McCormick, J. and Morten, G.: J. Am. Chem. Soc. 104, 4014 (1982).

39. Rhodes, P., Winskill, N., Friend, E., Warren, M.: J. Gen. Microbiol. 124, 329 (1981).

40. Iwatsuki, N., Joe, C., Werbin, H.: Biochemistry 18, 1172 (1980).

41. Sancar, A., Smith, F., Sancar, B.: J. Biol. Chem. 259, 6028 (1984).

42. Sancar, G., Smith, F., Lorence, M., Rupert, C., Sancar, A.: J. Biol. Chem. 259, 6033 (1984).

43. Eker, A.: Photochem. Photobiol. 32, 593 (1980).

44. Eker, A., Dekker, R., Berends, W.: Photochem. Photobiol. 33, 65 (1981).

45. Lamola, A.: Mol Photochem. 4, 107 (1972).

46. Rokita, S., Walsh, C.: J. Am. Chem Soc., in press.

INTERACTION OF THE HERBICIDE SULFOMETURON METHYL WITH ACETOLACTATE SYNTHASE: A SLOW-BINDING INHIBITOR

John V. Schloss

Central Research & Development Department
E. I. du Pont de Nemours & Company
Wilmington, Delaware 19898 USA

Sulfometuron methyl (SM) (Fig. 1), the active component of the commercial herbicide Oust®, is a potent and selective inhibitor of acetolactate synthase (ALS) (E.C. 4.1.3.18) in various bacteria (1,2), yeast (3), and plants (4,5). ALS is the first common enzyme in the biosynthesis of the branched-chain amino acids. This enzyme has an absolute requirement for FAD (1,6,7) although an ALS involved in the biosynthesis of acetoin from *Aerobacter aerogenes* does not require FAD (8). Since the reaction catalyzed by ALS involves no net redox, the requirement for FAD is unusual. Spectral changes in ALS-bound FAD upon addition of SM, or upon initiating the enzymic reaction, together with the kinetic details of ALS's inhibition by SM are reported here.

Results and Discussion

Inhibition of *Salmonella typhimurium* acetolactate synthase isozyme II (ALSII) by SM was biphasic. Figure 2 illustrates assay progress curves (obtained by continuously monitoring pyruvate consumption at 333 nm, $\varepsilon = 17.5\ cm^{-1}\ M^{-1}$) in the presence of various concentrations of SM. From the data in Fig. 2, an initial K_i of 1.7 µM, a final steady-state K_i of 82 nM, and a maximal rate constant for transition between initial and final inhibition of 0.15 min^{-1} were obtained, values that are in reasonable agreement with those previously

Flavins and Flavoproteins

determined by use of a fixed-time assay (1). Inhibition of ALSII by SM was competitive with pyruvate (Fig. 3). Initial and final inhibition constants of 0.35 μM and 39 nM were obtained, respectively. Since the apparent inhibition constants determined by varying SM (Fig. 2) were obtained in the presence of 50 mM pyruvate (K_m = 6 mM; pyruvate concentration = 8 x K_m), these values are somewhat smaller than expected (only 5 and 2-fold greater than the pyruvate-independent values, respectively). This apparent discrepancy may be due to the previously observed (1) facilitation of tight complex formation by pyruvate. Competing effects of the first and second pyruvate on SM inhibition would explain similar assay progress curves obtained at various pyruvate concentrations(1).

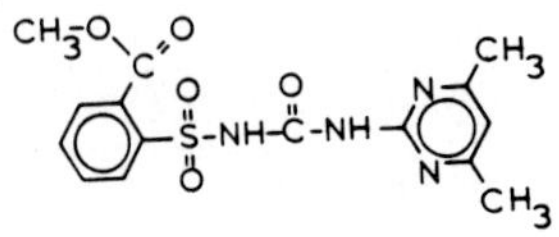

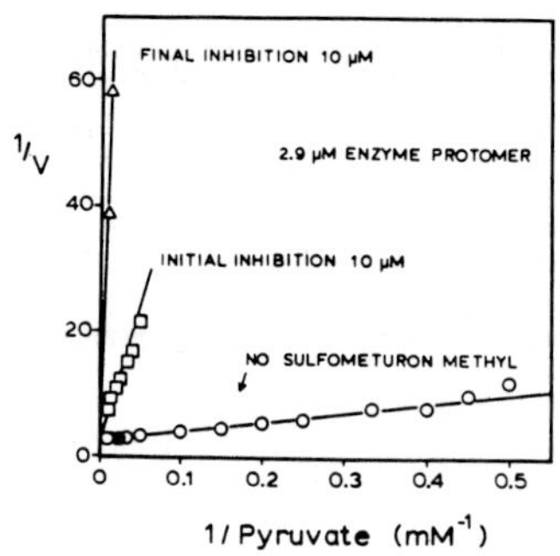

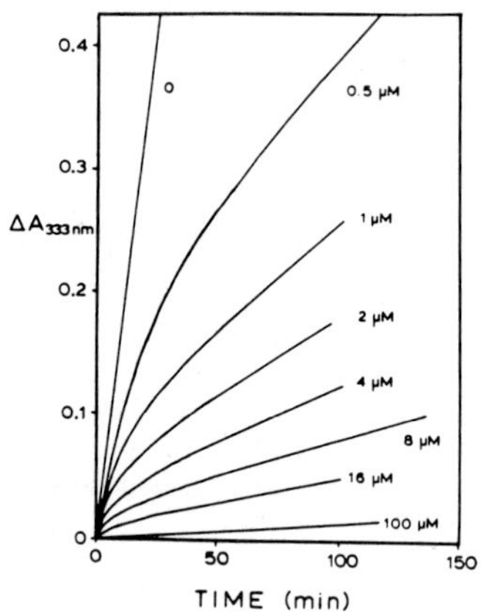

Figure 1. Sulfometuron methyl.
Figure 2. Assay progress curves (pH 7.8, 37°) in the absence and presence of various concentrations of SM.
Figure 3. Inhibition of 2.9 μM (protomer) ALSII by 10 μM SM. ALSII was activated prior to assay.

Addition of thiamine pyrophosphate, pyruvate, or $MgCl_2$ to ALSII-FAD, or any combination of two of these ligands, did not result in perceptible changes in the absorbance spectrum of enzyme-bound FAD. However, upon addition of all three and initiation of the enzymic reaction, there was an immediate loss of absorbance by enzyme-bound FAD between 370 and 530 nm

(Fig. 4). Addition of SM to ALSII-FAD resulted in the spectral perturbation illustrated in Figure 5. Initiation of the enzymic reaction in the presence of SM (Fig. 5) resulted in a similar loss of FAD absorbance, however, of about half of the magnitude of that observed in the absence of SM. Consumption of pyruvate under the conditions of Figure 5, proceeded at 6 % of the uninhibited rate.

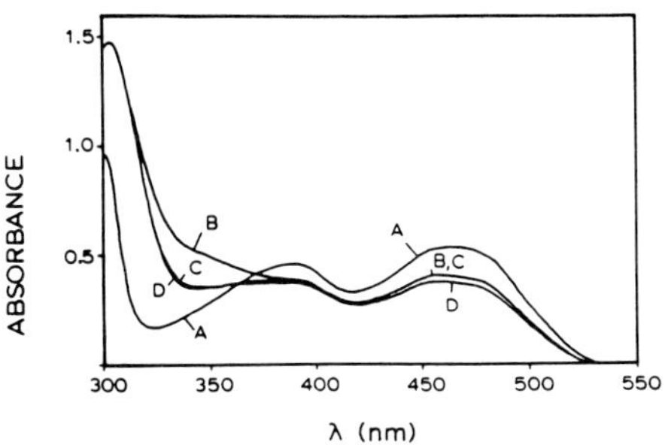

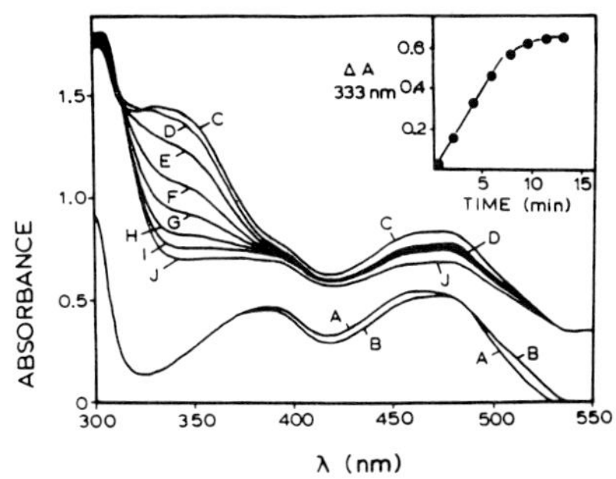

Figure 4. Absorbance spectra of 58 μM ALSII-FAD (4 mg/mL). A. In the presence of 0.1 M Tricine-NaOH, pH 7.8, 0.1 mM thiamine pyrophosphate, and 2 mM $MgCl_2$; B. 30 sec; C. 2 min; D. 10 min after addition of 50 mM pyruvate.
Figure 5. Absorbance spectra of 58 μM ALSII-FAD (4 mg/mL). A. In the presence of 0.1 M Tricine-NaOH, pH 7.8; B. 0.25-10 min after the addition of 0.1 mM SM; C. 0.25-5 min after the addition of 0.1 mM thiamine pyrophosphate and 50 mM sodium pyruvate, the spectra were offset for clarity; D. 0.5 min; E. 2 min; F. 4 min; G. 6 min; H. 8 min; I. 10 min; J. 30 min after addition of 2 mM $MgCl_2$. The change in absorbance at 333 nm (consumption of pyruvate) after initiating the enzymic reaction is shown as an insert.

Since FAD may have no direct role in catalysis, a possible structural or allosteric role for FAD with this enzyme has been suggested (9). However, the loss of absorbance by FAD upon initiation of the ALS reaction suggests it participates in the reaction. Further, SM, which is competitive with pyruvate, also perturbs the absorbance spectrum of enzyme-bound FAD. A potential role for FAD in the ALS reaction is illustrated on the next page. The isoalloxazine ring of FAD could serve as an electrophile to provide additional stability to the carbanionic intermediate formed by decarboxylation of the first pyruvate. Attack on the isoalloxazine ring by the carbanion would result in the observed loss of absorbance by FAD. SM appears to bind most tightly to the ALS-FAD-TPP-Mg^{2+}-(decarboxylated)pyruvate complex, at the site of accommodation for the second pyruvate. However, the reason for its rather high affinity for this site ($1.5x10^5$-fold more tightly than pyruvate) is still obscure.

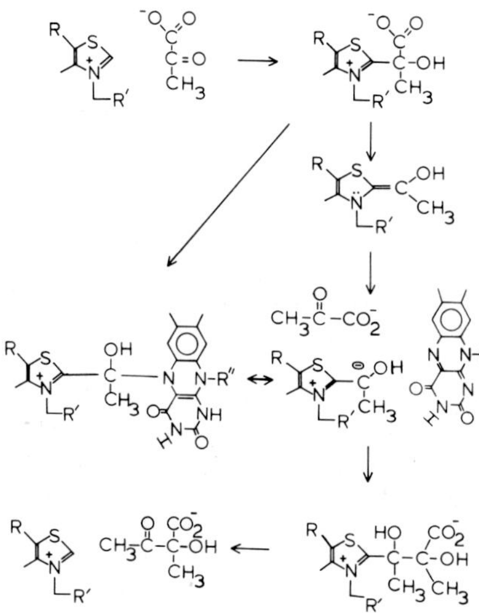

References

1. LaRossa, R. A., Schloss, J. V.: *J. Biol. Chem., in press.*
2. LaRossa, R. A., Smulski, D. R.: *J. Bacteriol., in press.*
3. Falco, S. C., Dumas, K. S.: *Genetics, in press.*
4. Ray, T. B.: *Plant Physiol., in press.*
5. Chaleff, R. S., Mauvais, C. J.: *Science, in press.*
6. Stormer, F. C., Umbarger, H. E.: *Biochem. Biophys. Res. Commun. 17,* 587-592 (1964).
7. Grimminger, H., Umbarger, H. E.: *J. Bacteriol. 137,* 846-853 (1979).
8. Stormer, F. C.: *J. Biol. Chem. 242,* 1756-1759 (1967).
9. Walsh, C.: *Enzymatic Reaction Mechanisms, p. 689,* W. H. Freeman and Company, San Francisco 1979.

THE RATE DETERMINING STEP IN PTERIDINE DEPENDENT MONOOXYGENASES IS NOT RING CLEAVAGE

David Pike, Mary Hora, Steven Bailey, and June Ayling
University of South Alabama, Mobile, Alabama 36688, USA

The tetrahydrobiopterin and dihydroflavin dependent monooxygenases have in common the formation of an adduct between cofactor and molecular oxygen. In both cases all but three analogous structures have been eliminated as the reagent which finally transfers oxygen to substrate: C4a hydroperoxide (a), C4a carbonyl oxide (b) (or its cyclic trioxide tautomer), and C4a-C8a (10a in flavin) perepoxide (c) (1,2,3,4,5).

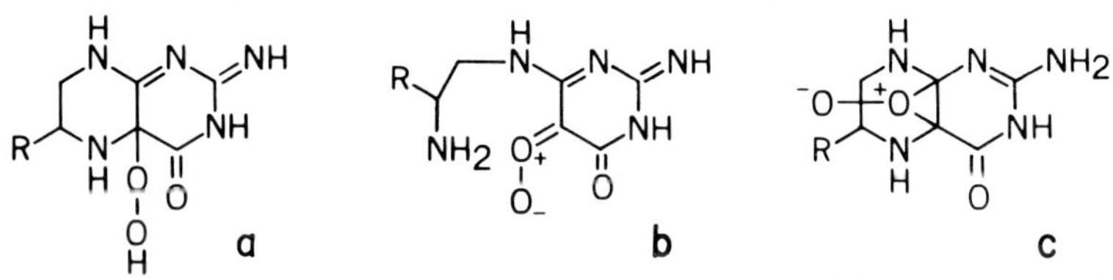

Hamilton has proposed that the carbonyl oxide results from an initial hydroperoxide by cleavage of the C4a-N5 bond (2). The following describes cofactor analogs designed to test whether this cleavage might be rate limiting.

It has been suggested that entropy of activation does not play a major role, in comparison to strain, in determining the relative rates of ring opening in a series of varying ring size. For example, the rate of cleavage of cyclic dimethylammonium ions by methoxide follows the pattern of ring strain in the corresponding cycloalkanes (6). Although counter examples have been observed, the above generalization is probably applicable to the opening of a cofactor hydroperoxide, since rearrangement in the transition state is minimal. A pyrazine ring expanded cofactor homolog,

Flavins and Flavoproteins

6-methyl-tetrahydropyrimidodiazepine (6-Me-PDH_4), was synthesized in order to observe the effect of increased strain on the rate of phenylalanine hydroxylase.

Three lines of evidence indicate that the additional methylene group increases strain. (1) The ring opened products of both 6-methyl-tetrahydropterin (6-Me-PH_4) and 6-Me-PDH_4 were synthesized (6-(2-aminopropyl)-quinoid divicine and 6-(3-aminobutyl)-quinoid divicine, respectively). The yield and rate of cyclization of the former to quinoid 6-Me-PH_2 was pH dependent consistent with that expected of Schiff's base condensation. An equilibrium is established with a ΔG of -4 Kcal M^{-1} in favor of quinoid pterin. In contrast, cyclization of the higher homolog is not observed. This is probably due either to (a) a reversal of the equilibrium by the additional strain of a seven-membered ring (7) and/or (b) transition state strain and entropy decreasing the rate of closure to the point where decomposition of starting material is dominant.

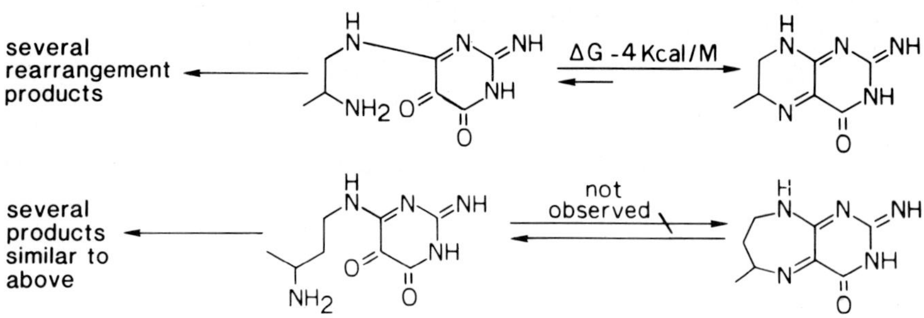

(2) In contrast to the stability of 6-Me-7,8-PH_2, 6-Me-7,8-PDH_2 is in equilibrium with a ring opened form due to hydrolysis of the N5-C6 bond. In air, the latter further decomposes by autooxidation.

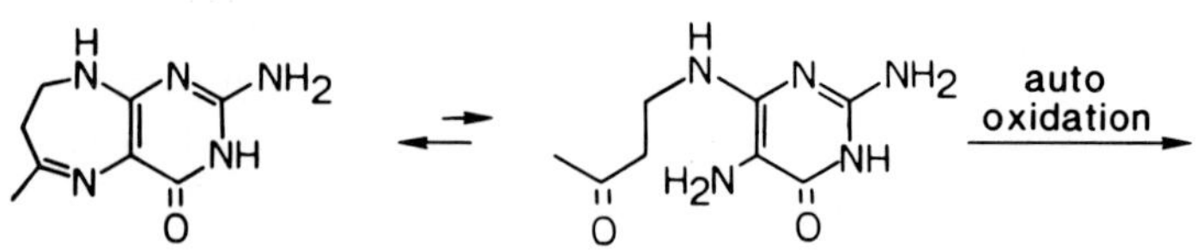

(3) Quinoid 6-Me-PH_2 tautomerizes quantitatively to 6-methyl-7,8-dihydropterin. On the other hand, quinoid 6-Me-PDH_2 (generated by oxidation of 6-Me-PDH_4) is only half converted to 6-methyl-7,8-dihydropyrimidodiazepine; the other half appears as the same decomposition products that result from the decay of the above ring-opened compound.

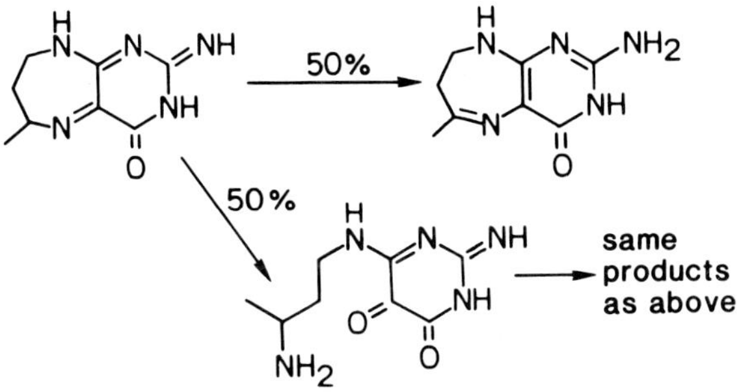

Considering the above indications of extra strain in a pyrimidodiazepine over that in the corresponding pteridine, acceleration of carbonyl oxide formation from the hydroperoxide of the former would be expected. However, although 6-Me-PDH_4 is a cofactor of phenylalanine hydroxylase with a Km (0.15 mM) that is only twice that of 6-Me-PH_4, V_{max} is decreased by a factor of 20. Thus the cleavage of the C4a-N5 bond, if it occurs at all, is not rate limiting (8).

Another step which is unique to the mechanism proposed by Hamilton is the transfer of oxygen from the carbonyl oxide to substrate. It has been shown that 2,5,6 triamino-4-pyrimidinone (TP) is the minimal structure embodying cofactor activity for phenylalanine hydroxylase (9). The loss of the 5 amino group in this compound would leave the carbonyl oxide of quinoid divicine regardless of any substitution of that amine.

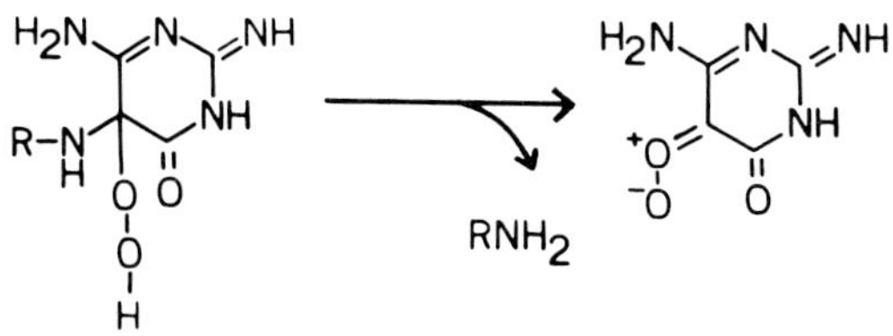

If utilization of this oxene reagent is rate limiting, the maximum rate of hydroxylation should not be dependent on R. With cofactor, phenylalanine, and oxygen extrapolated to saturating concentrations, V_{max} with R=benzyl is 2.6 fold faster and with R=methyl 5.6 fold slower than that of the parent TP (R=H). This fourteen fold range of rates demonstrates that the hypothesized transfer of oxygen from a carbonyl oxide to phenylalanine is also not rate limiting. Further, the critical step cannot occur later in Hamilton's sequence. Thus, the continued viability of the carbonyl oxide as an intermediate in pteridine dependent monooxygenases requires that the rate limiting step occur earlier than cleavage of the C4a-N5 bond (8).

Supported by NIH grants CA-31852 and GM-30368.

1. Bailey, S.W., Weintraub, S.T, Hamilton, S.M., and Ayling, J.E.: J. Biol. Chem. 257, 8253-8260 (1982).
2. Hamilton, G.A.: In Molecular Mechanisms of Oxygen Activation, Hayaishi, O. (ed.) Academic Press, New York (1974) pp. 405-451.
3. Bruice, T.C.: In Flavins and Flavoproteins, Massey V., and Williams, C.H. (eds.) Elsevier/North-Holland, New York. Amsterdam (1982) pp. 265-277.
4. Dmitrienko, G.I., Snieckus, V., and Viswanatha, T.: Biorganic Chemistry 6, 421-429 (1977).
5. Entsch, B., Ballou, D.P., and Massey, V.: J. Biol. Chem. 251, 2550-2563 (1976).
6. Cerichelli, G., Illuminati, G., and Lillocci, C.: J. Org. Chem. 45, 3952-3957 (1980).
7. Capon, B. and McManus, S.P.: Neighboring Group Participation, Vol. 1, Plenum Press, New York. London (1976).
8. Pike, D.C., Hora, M.T., Bailey, S.W., and Ayling, J.E.: in press.
9. Bailey, S.W. and Ayling, J.E.: Biochem. Biophys. Res. Comm. 85, 1614-1621 (1978).

THE ONE ELECTRON REDUCTION OF RIBOFLAVIN BINDING PROTEIN

M. Faraggi and Michael H. Klapper

Departments of Chemistry, Nuclear Research Center - Negev, Beersheva, Israel and The Ohio State University, Columbus, Ohio 43210

Two questions concerning the one electron reduction of flavoproteins interest us: how and from where proton and electron reach the flavin group when the neutral semiquinone is the stable product. Pulse radiolysis permits the observation of radical transients and of radical initiated protein reactions. Here we present data on the hydrated electron (e^-_{aq}) and the CO_2^- reduction of chicken egg white riboflavin binding protein (RBP). On the basis of these and previous results (1,2), we feel that pulse radiolysis is useful for dissecting complex flavoprotein reductions. Because flavin reduction in RBP may involve intramolecular electron transfer, this system could also provide convenient entry into the generally important subject of electron migration within proteins.

Results And Discussion

1. **e^-_{aq}** Our previous studies on the reaction of e^-_{aq} with RBP (1) serve as background for the CO_2^- results to follow. The techniques of e^-_{aq} and CO_2^- production and reaction detection have been described elsewhere (1-3). We observed the one electron reduction of RBP by e^-_{aq} over the time and spectral ranges of 1 μsec to 1 sec, and 300 to 600 nm, respectively. The e^-_{aq} to protein concentration ratio was kept low to minimize double electron reduction of the RBP molecule (3). The final product of the overall reaction is RBP-FlH, the protonated flavin semiquinone form of RBP. The minimal scheme we

Flavins and Flavoproteins

proposed to explain the reaction kinetics is:

$$e^-_{aq} + RBP \begin{array}{l} \nearrow RBP\text{-}XSSX^- \longrightarrow \text{products} \\ \searrow RBP\text{-}Fl^- \xrightarrow[H^+]{} RBP\text{-}FlH \end{array} \qquad (1)$$

The reaction as monitored at 350 and 540 nm is shown in fig. 1. Formation of RBP-Fl$^-$, the anionic semiquinone, is concurrent with the disappearance of e^-_{aq}, seen as the positive absorbance at 540 nm immediately after the fast e^-_{aq} decay. The protonation of RBP-Fl$^-$ to form RBP-FlH is then observed as an absorbance increase ($k_{app} = 4.8 \times 10^3\ s^{-1}$ at pH 9.0). However, a second transient, the one electron reduced disulfide (XSSX$^-$), is also formed during the e^-_{aq} reaction ($k_{app} = 1.4 \times 10^{10}\ M^{-1}\ s^{-1}$). Two decays are seen at 350 nm. The faster is the RBP-Fl$^-$ protonation also observed at 540 nm; the slower ($k_{app} = 9.7\ s^{-1}$) is the decay of XSSX$^-$ to products that do not absorb at this wavelength. The yields of XSSX$^-$ and RBP-FlH are each 20-25% of the reacted e^-_{aq}. We have presented (1) the spectral analysis that supports the scheme of eq. 1.

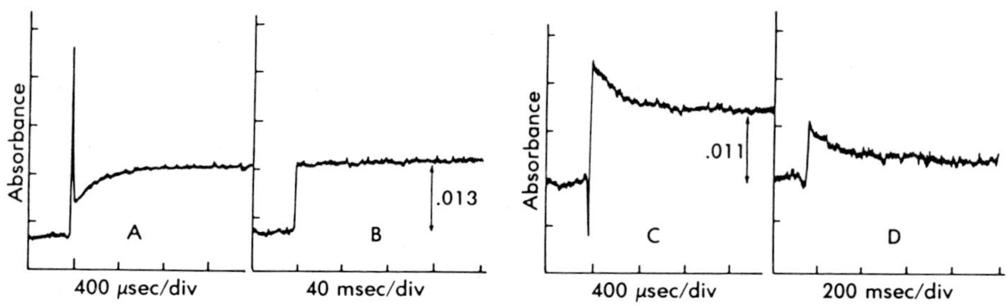

Fig 1: Absorbance changes during the reaction of e^-_{aq} with RBP at 540 nm (A,B) and 350 nm (C,D) at pH 9.0: RBP, 5×10^{-5}M; e^-_{aq}, 8×10^{-6}M.

2. **CO_2^-** XSSX$^-$ and RBP-FlH are also products of the reaction between CO_2^- and RBP. The spectrum of the neutral semiquinone is seen after the decay of XSSX$^-$ (fig 2). However, there is no evidence for RBP-Fl$^-$ as an intermediate, or for the protonation reaction seen in the e^-_{aq} reduction. (fig 3; 352 nm is

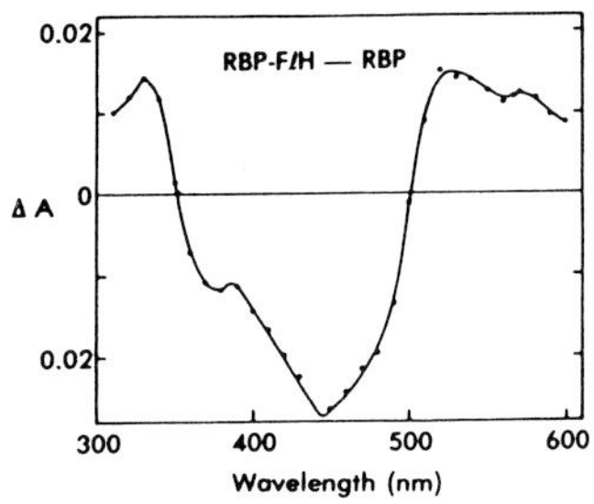

Fig 2: Difference spectrum .5 sec after the reaction of CO_2^- with RBP. Experimental conditions are identical with those of fig 2.

an isosbestic point between RBP and RBP-FlH) The rate of $XSSX^-$ formation from CO_2^- ($k_{app} = 1.7 \times 10^4\ s^{-1}$) is almost an order of magnitude faster than the rate of RBP-FlH formation measured at 540 nm ($k_{app} = 2.3 \times 10^3\ s^{-1}$). The yields of $XSSX^-$ and RBP-FlH were both between 35 and 40%. The rate constant for the slow decay of the disulfide radical is 9.7 sec^{-1}, the value observed after e_{aq}^- reduction.

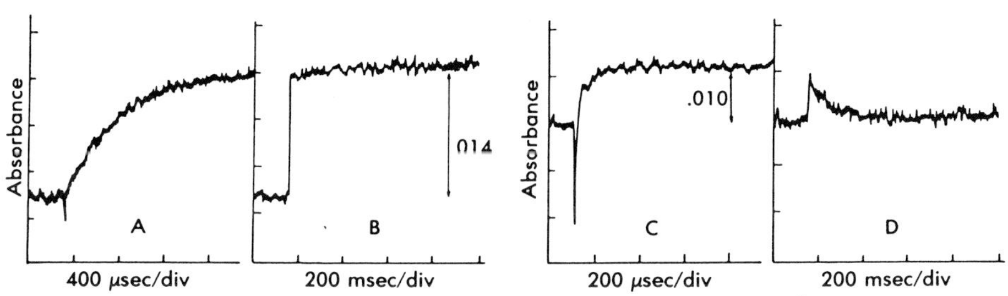

Fig 3: Absorbance changes during the reaction of CO_2^- with RBP at 540 nm (A,B) and 352 nm (C,D) at pH 9.0: RBP, 5.5×10^{-5}M; CO_2^- 3.5×10^{-6}M.

We propose at least two primary reductive processes.

$$CO_2^- + RBP \begin{cases} \nearrow \ RBP\text{-}XSSX^- \longrightarrow \text{products} \\ \searrow \ Z \longrightarrow RBP\text{-}FlH \end{cases} \qquad (2)$$

In contrast with the e_{aq}^- reaction, direct reduction of the flavin moiety cannot be competitive with $XSSX^-$ formation. Different reductive reactions of CO_2^- with protein must be linked. Thus, formation of the first protein radical products must have the same apparent 1st order rate constant (protein in excess over radical), with the ratio of product yields equal to the ratio of primary rate constants. Since the apparent rate constant for RBP-FlH formation is smaller than that for $XSSX^-$, RBP-FlH cannot be a primary reaction product, but must arise from a precursor that we designate as Z in eq 2. Since the yields of $XSSX^-$ and RBP-FlH are comparable, the rates of $XSSX^-$ and Z formation are probably similar. Although we do not know the nature of Z, this intermediate is not $RBP\text{-}Fl^-$, since the unprotonated species is not seen. Nor is $XSSX^-$ a precursor of reduced flavin, since the formation of RBP-FlH is faster than the decay of $XSSX^-$. The unknown intermediate may be either a CO_2^--protein complex, or an unrecognized radical formed at still another site on the protein.

The scheme of Equation 2 requires transfer of an electron from a radical Z to flavin. As seen in figure 4, the apparent 1st order RBP-FlH formation rate constant increases with both decreasing pH and increasing protein concentration. In the latter dependence, linear extrapolation of k_{app} to zero concentration results in a non-zero intercept. Electron transfer from Z to flavin thus has both a first and second order component, so that reduction of flavin may occur, in part, by intramolecular electron migration.

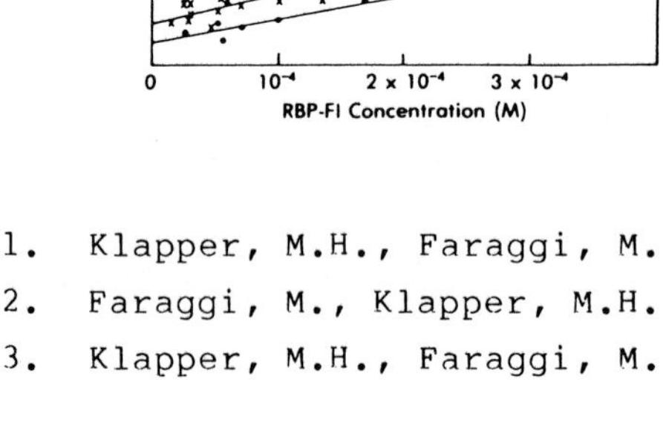

Fig 4: The RBP concentration dependence of the k_{app} for RBP-FlH formation in the CO_2^- reaction at pH 9.0 (A - intercept $2.4 \times 10^3\ s^{-1}$, slope $2.1 \times 10^7\ M^{-1}s^{-1}$), pH 7.0 (B), and pH 5.2 (C)

1. Klapper, M.H., Faraggi, M.: Biochemistry 22, 4067 (1983)
2. Faraggi, M., Klapper, M.H.: J. Biol. Chem. 254, 8139 (1979)
3. Klapper, M.H., Faraggi, M.: Q. Rev. Biophys. 12, 465 (1979)

Modified Flavins in Flavoproteins

8-AZIDOFLAVINS: PHOTOAFFINITY LABELS FOR FLAVOPROTEINS

Sandro Ghisla
Faculty of Biology, University of Konstanz, D-7750 Konstanz, FRG

Paul F. Fitzpatrick, and Vincent Massey
Dep. of Biological Chemistry, Univ. of Michigan, Ann Arbor, Mi. 48109, USA

Introduction

The photoreactivity of flavins has, surprisingly, never been used as a tool for the specific labelling of active centers. In contrast, an FAD analog modified in the adenine moiety, 8-azido-adenosine-FAD has been used for labelling studies with D-amino acid oxidase and glucose oxidase (1). In the past decade arylazides have been studied intensively as photoactivity labels (2), they react covalently with protein residues upon photochemical generation of highly reactive nitrene intermediates. We have synthesized flavin coenzymes carrying the azido function at position 8 of the flavin since they were expected to be photolabelling agents. Here we describe some of their properties, and the labelling of three typical flavoproteins.

Results and Discussion

General Properties of 8-Azido-flavins

8-Azidoflavins can be synthesized conveniently by two routes: Diazotization of 8-aminoflavins and reaction of the (crystalline) diazonium perchlorate with N_3^- lead to formation of $(8)N_3$-flavins (Scheme 1). This method is convenient for the synthesis of $(8)N_3$-flavin models and of larger amounts of derivatives. For the preparation of $(8)N_3$-flavin coenzymes the method of choice is the direct conversion of 8-F-FMN or -FAD using N_3^- as a nucleophile (Scheme 1). $(8)N_3$-flavins are extremely photolabile; in the presence of various solvent molecules (8)N-R substituted flavins are formed probably

Flavins and Flavoproteins

by nitrene insertion. These products have been identified by comparison with derivatives obtained by direct substitution of the corresponding 8-F-flavin (Scheme 1). Thus 8(CH_3-O-NH-)riboflavin, 8(HO-NH-)riboflavin, and 8(CH_3-NH-NH-)riboflavin were obtained by reaction with (O)methylhydroxylamine, hydroxylamine, and monomethyl hydrazine respectively. These flavins are characterized by a strong absorption band in the 480-500 nm region ($\varepsilon \sim 20000$ to $35000\ M^{-1}cm^{-1}$). 8(R-O-NH-)Riboflavins have an unusually low pK of ~ 5.9, which does not depend on the nature of the substituent R (R = -H or CH_3-), indicating that this pK reflects deprotonation at N(1)-H of a p-quinoid flavin, as is observed in the case of the 8-OH-flavins. All of these 8(R-N-)flavins are easily and essentially quantitatively converted to 8-NH_2-flavins upon reduction with dithionite and subsequent reoxidation with oxygen. The reaction of (8)N_3-flavins with sulfite leads to formation of 8-aminoflavins via a spectrally distinct intermediate. The latter probably results from addition of sulfite to the azido group. (8)N_3-flavins also show a pronounced solvatochromy, suggesting their use as active site probes for lypophilic/lypophobic environments.

Scheme 1:

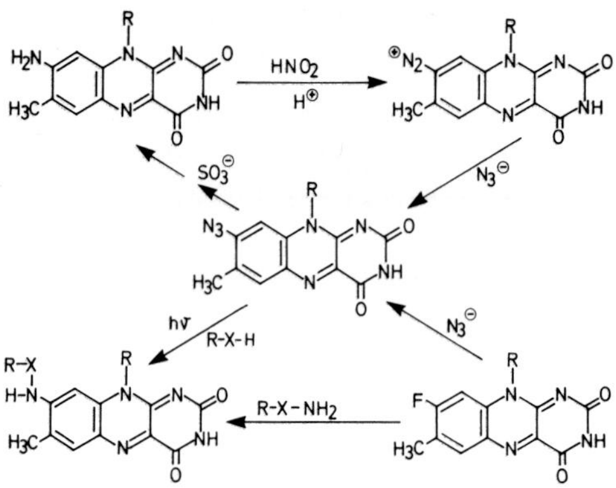

Interaction with Flavodoxins

(8)N_3-FMN is bound tightly by M. elsdenii apoflavodoxin (3), with spectral shifts similar to those observed when other flavins are bound to the apoprotein, i.e. with a decrease in extinction at the wavelength maximum in the visible (from ~ 22000 to ~17900 $M^{-1}cm^{-1}$) and the development of a pronounced shoulder at 470 nm. The complex is very poorly photoreactive being orders of magnitude less sensitive than free 8-N_3-flavin, but with similar spectral changes. Precipitation of the irradiated protein with 5% trichloroacetic acid and dissolving in 8 M guanidine.HCl revealed approximately 5 % covalently labelled protein, consistent with the X-ray crystallographic studies, which show that the 8-CH_3 moiety of the flavin is exposed to solvent (4). It should be noted that photoreaction of the (8)N_3-FAD-glutathione reductase complex, in which the flavin 8-CH_3 position also is in contact with solvent, does not lead to covalent modification of the protein (5).

Interaction with Riboflavin Binding Protein

Riboflavin binding protein (6) forms a stoichiometric complex with (8)N_3-riboflavin. The complex is much less photosensitive than free (8)N_3-flavin, and irradiation must be carried out at 0-4° in order to minimize dissociation, and subsequent photoreaction of free (8)N_3-flavin. Approx. 30% of the flavin present after illumination is covalently bound to the protein.

Interaction with D-Amino Acid Oxidase

(8)N_3-FAD is bound tightly by the apoenzyme (7) of this flavoprotein. In contrast to riboflavin binding protein, it is almost as light sensitive as free (8)N_3-flavin (Fig. 1). The extent of covalent binding of the chromophore to the protein upon illumination was $\sim$17%.

Conclusions

These studies demonstrate that the photoreactivity of (8)N_3-flavocoenzymes turns the latter into potent affinity labels. In the three cases examined up to 50% labelling was obtained per cycle of irradiation. The present studies cannot tell which amino acid residue might have reacted, however a correlation with the spectral properties of appropriately substituted 8-amino-flavins might provide such information. Mapping studies with enzymes of known primary sequence should also provide direct information on which protein region is located near the flavin benzene moiety.

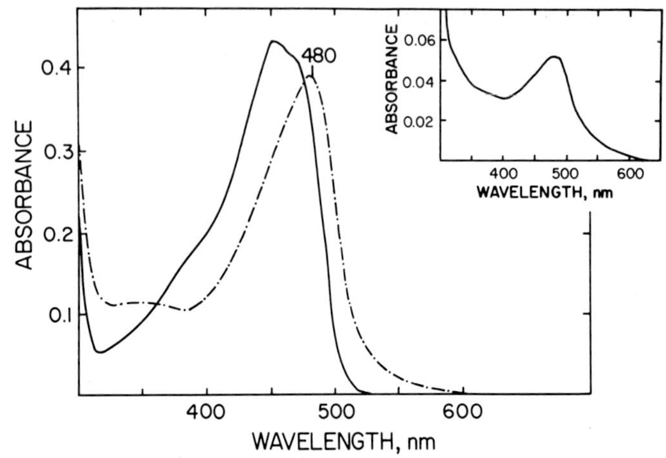

Fig. 1) Spectral changes occurring during the irradiation of $(8)N_3$-FAD-D-amino acid oxidase

Spectrum (——) is that of apoenzyme (50 μM) and $(8)N_3$-FAD (30 μM) in 20 mM pyrophosphate buffer, pH 8.5 at 4°. Curve (-·-·) was obtained after 8 s irradiation with white light (~ $8x10^6$ erg $cm^{-2}s^{-1}$). The inset shows the spectrum of the chromophore covalently bound to the protein, obtained after exhaustive dialysis against 2 M KBr, 20 mM pyrophosphate pH 8.5 (which removes non-covalently bound flavin), and then against pyrophosphate minus KBr.

References

1. Koberstein, R.: Eur. J. Biochem. 67, 223-229 (1976).

2. Bayley, H., and Knowles, J.R.: Methods in Enzymology 46, 69-114 (1977).

3. Mayhew, S.G.: Biochim. Biophys. Acta 235, 289-302 (1971).

4. Mayhew, S.G., and Ludwig, M.L.: The Enzymes, Vol. 12b, 57-109 (1975).

5. Krauth-Siegel, L., Ghisla, S., and Schirmer, H.: to be published.

6. Blankenhorn, G., Osuga, D.T., Lee, H.S., and Feeney, R.E.: Biochim. Biophys. Acta 386, 470-478 (1975).

7. Massey, V., and Curti, B.: J. Biol. Chem. 241, 3417-3423 (1966).

GLUTATHIONE REDUCTASE SPECIES CONTAINING FAD ANALOGUES

R.L. Krauth-Siegel, R.H. Schirmer and S. Ghisla*
Institut für Biochemie II, Universität Heidelberg
*Fakultät für Biologie, Universität Konstanz

INTRODUCTION

Glutathione reductase (GR) (1,2,3) is a suitable enzyme for correlating spectroscopic properties and chemical reactivities of protein-bound flavin analogues with structural data. As reported here FAD, the prosthetic group of the enzyme, was replaced by analogues which carried modifications at the positions 8,6,4,2 and 1 of the isoalloxazine ring.

RESULTS AND DISCUSSION

Binding of 8-mercapto-FAD to apoglutathione reductase causes a shift of λ_{max} from 530 to 560 nm. Within 12h this intermediate changes to a final stable spectrum with absorption maxima at 575 and 455 nm (Fig.) The spectral changes probably reflect a very slow protein rearrangement subsequent to a primary binding step. 8-Mercapto-FAD is bound to apo·GR predominantly in the blue p-quinoid form which carries a negative charge in the pyrimidine subnucleus (4). This charge is probably neutralized by the positive pole of helix 338-354 (5,6). 8-Mercapto-FAD-GR possesses 40% of the enzyme activity of native FAD·GR. It reacts readily with thiol reagents such as methylmethane thiosulfonate (MMTS) (k_2= 1.4 x 10^5 $M^{-1}cm^{-1}$) to give a spectrum characteristic of an 8-S-"alkylated" flavin (7). The conversion is complete with a stoichiometric amount of MMTS so that side reactions with cystein residues of the enzyme in the oxidized form (E(1,8)) can be excluded.

Flavins and Flavoproteins

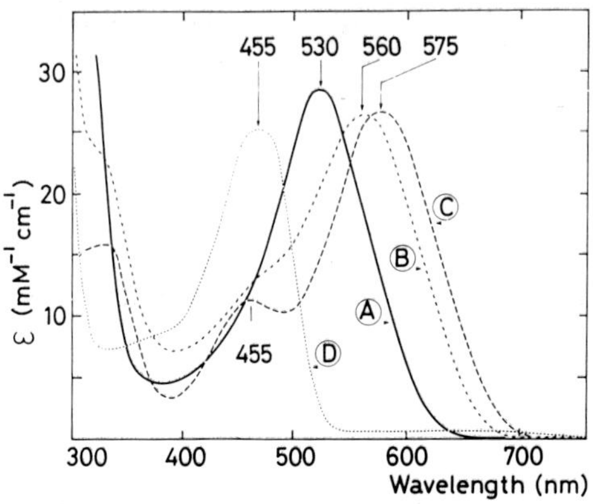

Fig.: 8-Mercapto-FAD in phosphate buffer, pH 6.9 (curve A) was titrated with apoglutathione reductase at 25°C. The spectrum was recorded immediately after titration (curve B) and 12h later (curve C). Curve D resulted from the reaction of 8-mercapto-FAD·GR (curve C) with one equivalent of methylmethane thiosulfonate.

The reaction is completely reversible by using 10 mM 2-mercaptoethanol. Reduction of 8-mercapto-FAD·GR with one equivalent NADPH leads to a slight hypsochromic shift of the long wavelength maximum to 568 nm and a decrease of the intensity of the 568 nm and 444 nm bands by 10%. No absorption increase at 340 nm is observed, so that this species is likely to represent 8-mercapto-FAD·EH_2. The spectrum does not show the characteristic "charge transfer" absorption of native EH_2 (1,8). Reduction of the flavin is achieved by a large excess of NADPH: the resulting species is 8-mercapto-$FADH_2$·EH_2 (=EH_4). Enzyme bound <u>8-Cl-FAD</u> or <u>8-F-FAD</u> reacted only poorly with sulfide to give 8-mercapto-FAD·GR. At most, 50% conversion was obtained by incubating 8-F-FAD·GR with 20 mM HS^- at pH 8.8 for 20h at 25°C.

The chemical reactivities at position 8 can probably be explained in structural terms. In native GR the xylene ring of FAD around the 7α-methyl group sticks in a very hydrophobic pocket. C(8) is in van-der-Waals contact with the oxygen atom of the peptide bond of Val61. The 8α-methyl group is at a distance of 4 Å to N_η of Arg291. These steric constraints may explain why, in the cases of 8-Cl-FAD·GR and 8-F-FAD·GR, a nucleophilic substitution by HS^- is hindered whereas the reaction of the 8-mercapto-FAD enzyme with MMTS is fast.
Addition of an equimolar amount of apo·GR to 6-OH-FAD does not alter the spectrum of the chromophore appreciably. A pH titration of the complex leads to spectral changes which however, cannot be described by a single ionisation process. The estimated pK for the changes at 330 and 680 nm is 7.2 ± 0.3. In contrast to the other GR species tested, 6-OH-FAD·GR has no detectable catalytic activity. This is not surprising in light of the three-dimensional structure of native FAD·GR. In native GR three amino acid residues, namely Gly62, Cys63 and Lys66, are in van-der-Waals contact to position C(6)(G.E.Schulz, personal communication). Introduction of an OH group at this position must disturb the protein structure, and/or the modified flavin is likely to be bound weakly. Poor binding is consistent with the observation that ultrafiltration of 6-OH-FAD·GR leads to the loss of the chromophore.
Reconstitution of apo·GR with 4-thio-FAD leads to a shift of the absorption maximum from 486 to 504 nm. This enzyme species shows 24% catalytic activity when compared with native FAD·GR. Reduction of the flavin in 4-thio-FAD·GR was achieved with a 1.9 fold molar excess of NADPH under anaerobic conditions. In this respect the 4-thio-FAD enzyme differs from all other GR species tested. The native enzyme, 8-Cl-FAD·GR and 1-deaza-FAD·GR for instance are reduced by NADPH to EH_2 species, which are characterized by (re)oxidized flavin and a dithiol group at the catalytic site (1,3). 4-Thio-FAD·GR does not react with methylmethane thiosulfonate (MMTS) and it reacts very

slowly with H_2O_2. Excess H_2O_2 leads to isosbestic formation of a new enzyme species with maxima at 452 and 364 nm which shows no catalytic activity. This species possibly contains FAD covalently bound to a side chain of the protein. Lys66 with its ε-NH_2 group at a distance of 3 Å to position O(4α) in the native enzyme, would be a good candidate; the long side chain of this residue lines the flavin around C(6)/N(5)/O(4α) (6).

When 2-thio-FAD binds to apoglutathione reductase, the absorption maxima shift from 486 to 504 nm and from 316 to 324 nm, respectively. 2-Thio-FAD·GR possesses 17% of the activity of native FAD·GR. The 2-thio-FAD enzyme was reacted with MMTS in order to test the accessibility of the region N(1)-C(2)-N(3) in the bound flavin. Even drastic conditions (10mM reagent, 24h incubation time, pH 9 at 25°C) did not lead to the expected spectral changes, which indicates that the pyrimidine subnucleus is unaccessible to solvent borne reagents. This is consistent with structural data. The peptide NH group of Thr339 forms an H-bond with N(1)/O(2α). In addition some other residues of the α-helix 338-354 are in close contact to N(1)/O(2α)/N(3) of the flavin. The strong hydrogen bond between N(3) and the carbonyl group of His467 is also important (6).

Upon incubation of 1-deaza-FAD with apo·GR, the long wavelength band of the chromophore is shifted from 535 to 560 nm and the intensity of the 365 nm band increases by 70%. The complex is stable at 2°C but tends to precipitate at room temperature, and also upon aerobic addition of NADPH at 2°C. Anaerobic addition of NADPH leads to a species resembling EH_2 of the native enzyme. Evidence for this is the disappearance of the absorption around 340 nm in the absence of flavin reduction. λ_{max} is shifted from 560 to 540 nm but the spectrum lacks the long wavelength absorption band of native EH_2. 1-Deaza-FAD·GR was found to have about 22% activity when compared with native FAD·GR. This means that an H-bond between protein and atom-1 of flavin is not essential for enzyme activity.

In conclusion, FAD-analogues are promising tools for mechanistic and drug designing studies on glutathione reductase. As to the latter point, inhibitors of GR are used clinically in the chemotherapy of malignancies and experimentally in malaria research. The structurally unstable 1-deaza-FAD·GR, for instance, is of interest because it resembles the GR species found in hereditary glutathione reductase deficiency of human erythrocytes (9). This condition, like favism, is believed to provide some protection against Plasmodium falciparum malaria.

Acknowledgments

We thank Georg E. Schulz, Freiburg, and Emil F. Pai, Heidelberg, for communicating unpublished data and for stimulating discussions. Our work is supported by the Deutsche Forschungsgemeinschaft and by the Fond der Chemischen Industrie.

References

1. Williams,C.H.Jr. (1976) in The Enzymes (Boyer, P.D., ed.) 3rd Ed, Volume 13, 89-173, Academic Press, New York
2. Pai,E.F., Horn,E. and Schulz,G.E. This volume
3. Schulz,G.E. This volume
4. Massey,V., Ghisla,S. & Moore,E.G. (1979) J.Biol.Chem. 254, 9640-9650
5. Hol,W.G.J.& Wierenga,R.K. (1984) In: X-ray crystallography and drug action (Horn,A.S. & De Ranter,C.J. eds) pp 151-168,Clarendon Press, Oxford
6. Schulz,G.E., Schirmer,R.H. & Pai,E.F. (1982) J.Mol.Biol. 160, 287-308
7. Moore,E.G., Ghisla,S. & Massey,V. (1979) J.Biol.Chem. 254, 8173-8178
8. Pai,E.F. & Schulz,G.E.(1983) J.Biol.Chem. 258, 1752-1757
9. Loos,H., Roos,D., Weening,R. & Houwerzijl,J. (1976) Blood 48, 53-62

FLAVIN ANALOG STUDIES OF PIG KIDNEY ELECTRON TRANSFERRING FLAVOPROTEIN

Robert J. Gorelick and Colin Thorpe

Department of Chemistry, University of Delaware,
Newark, Delaware 19716

Introduction

Mammalian electron transferring flavoprotein (ETF) mediates the flow of reducing equivalents not only from the acyl-CoA dehydrogenases of fatty acid oxidation, but also from several other mitochondrial dehydrogenases involved in amino acid and 1-carbon metabolism. ETF is a dimer of dissimilar subunits containing only 1 molecule of FAD (1-3). This communication describes the use of flavin analogues to examine the flavin binding site and the mode of interflavin electron transfer between pig kidney general acyl-CoA dehydrogenase and ETF.

Results and Discussion

ETF-apoprotein was prepared essentially as described previously (4) and was immediately reconstituted with one equivalent of 8-Cl-FAD yielding a highly resolved spectrum closely similar to that of native ETF (1). The accessibility of the 8-locus of the bound flavin can then be probed on the addition of appropriate sulphur nucleophiles (5,6). Thiophenol (1 mM at pH 8.5) reacts with bound flavin yielding 8-(arylmercapto)-bound-FAD ($t_{\frac{1}{2}}$ = 6 min). Sodium sulphide under similar conditions also effects the displacement of chloride giving a spectrum typical of the paraquinoid form of 8-mercapto-FAD (5). This is in accord with the stabilization of the red semiquinone on reduction

Flavins and Flavoproteins

of native ETF (1). These data suggest that the 8-position of the bound flavin is accessible to solvent. This was further tested by preforming 8-(alkylmercapto)-FAD-derivatives containing substituents of various sizes, and examining their interaction with ETF-apoprotein. Analogues were generated by incubating 8-Cl-FAD with methanethiol (6), 2-mercaptoethanol, thiophenol, or CoA-SH, and purified by gel filtration. All of these flavin derivatives are bound effectively by ETF-apoprotein yielding very similar spectra (see e.g. Fig. 1A, curve 1).

The derivatives prepared above were reduced anaerobically by catalytic levels of substrate reduced general acyl-CoA dehydrogenase. In all cases, data similar to those shown for 8-((methylmercapto)-ETF (Fig. 1A) were obtained. Reduction is biphasic with accumulation of the red semiquinone (Fig. 1A,

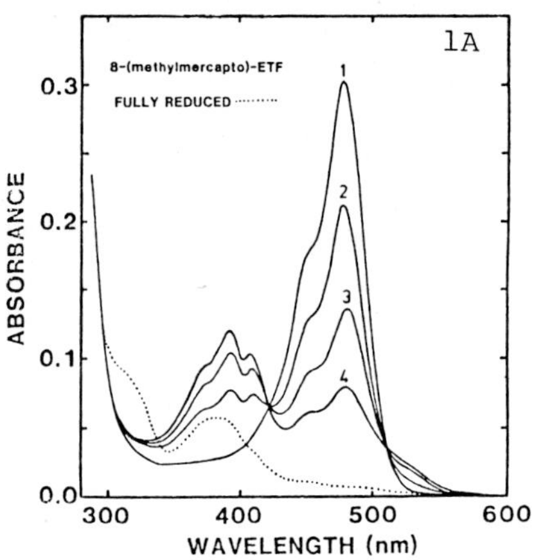

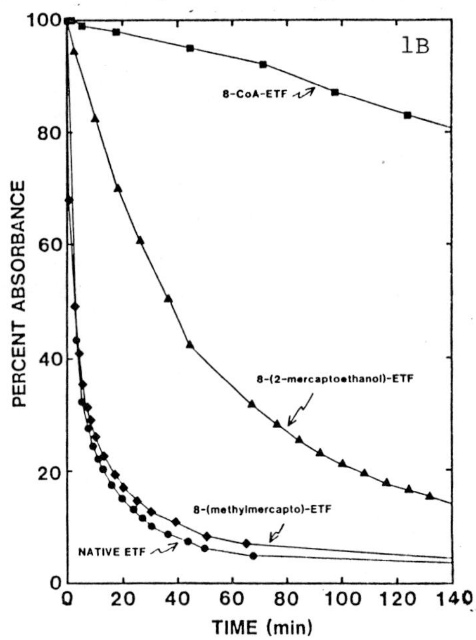

Figure 1A. Anaerobic reduction of 8.5 uM 8-(methylmercapto)-ETF (curve 1) by 6.4 nM general acyl-CoA dehydrogenase, 60 uM octanoyl-CoA in 15 mM Tris/HCl buffer, pH 8.5 at 4°C. Curves 2-4; 0.83, 4.23, and 15.0 min after adding octanoyl-CoA respectively.

Figure 1B. The ETF's indicated were reduced as in Fig. 1 and followed at their respective absorbance maximum.

curve 4), as is observed with native ETF (1), followed by slower reduction to the dihydroflavin species. The rate of reduction of ETF derivatives is shown in figure 1B. It should be noted that the redox potentials of the 8-(alkylmercapto)-FAD's are very similar to those of FAD(6).

Two electron reduced general acyl-CoA dehydrogenase is rapidly and extensively oxidized by one equivalent of native ETF (7,8). However, when the dehydrogenase flavin is replaced by the more strongly reducing 1,5-dihydro-5-deaza-FAD, no reduction of ETF occurs over 5 h. The reduced 5-deaza-dehydrogenase is nevertheless competent in rapidly reducing crotonyl-CoA with transfer of a hydride from the 5-position to the C-3 of the bound thioester (9). Thus reduction of ETF by the dehydrogenase is unlikely to involve direct transfer of a hydride equivalent between the N-5 positions of the reactants. Rather, in agreement with rapid reaction studies (7,8), interflavin electron transfer probably occurs via discrete 1-electron steps thermodynamically inaccessible to 1,5-dihydro-5-deaza-FAD (10). In view of the inhibition of electron transfer by bulky 8-substituents (Figures 1A & 1B), a favourable geometry for this reaction may require approach of the dehydrogenase flavin to the exposed dimethylbenzene edge of ETF. It is interesting to note that this region of the dehydrogenase flavin appears solvent inaccessible by the criteria mentioned in this paper. However, electron transfer e.g. via outer sphere or tunneling processes does not necessarily require molecular contact of donor and acceptor. Thus it is possible that electron transfer occurs between flavin rings aligned such that their dimethylbenzene rings approach most closely.

Another interesting aspect is the spatial relationship of the bound FAD to the two dissimilar subunits of ETF (1). 8-Cl-FAD under certain conditions reacts covalently with apoprotein yielding an 8-SR-flavin spectrum. Our preliminary data suggest that this flavin is bound in a non-native state to the smaller subunit. However, the very low recovery of labeled subunit on chromatofocusing makes this conclusion a very tentative one.

Acknowledgement

This work was supported in part by NIH grant GM 26643.

References

1. Gorelick, R.J., Mizzer, J.P., Thorpe, C.: Biochemistry 21, 6936-6942 (1982).

2. Husain, M., Steenkamp, D.J.: Biochem. J. 209, 541-545 (1983).

3. McKean, M.C., Beckmann, J.D., Frerman, F.E.: J. Biol. Chem. 258, 1866-1870 (1983).

4. Mayer, E.J., Thorpe, C.: Anal. Biochem. 116, 227-229 (1981).

5. Massey, V., Hemmerich, P.: Biochem. Soc. Trans. 8, 246-257 (1980).

6. Moore, E.G., Ghisla, S., Massey, V.: J. Biol. Chem. 254, 8173-8178 (1979).

7. Hall, C.L., Lambeth, J.D.: J. Biol. Chem. 255, 3591-3595 (1980).

8. Gorelick, R.J., Schopfer, L.M., Ballou, D.P., Massey, V., Thorpe, C.: Unpublished.

9. Ghisla, S., Thorpe, C., Massey, V.: Biochemistry, in press.

10. Blankenhorn, G.: Eur. J. Biochem. 67, 67-80 (1976).

CHEMICALLY MODIFIED FLAVINS AS PROBES OF PHENOL HYDROXYLASE STRUCTURE AND FUNCTION

Kristina Detmer and Vincent Massey

Department of Biological Chemistry, The University of Michigan, Ann Arbor, Michigan, 48109 USA

The flavin prosthetic group of phenol hydroxylase can be replaced by chemically modified FAD derivatives. Reactions of these modified enzymes provide information about phenol hydroxylase structure and function.

Accessibility of the 8-position to solvent. The results of the reaction of 8-mercapto-FAD phenol hydroxylase with excess iodoacetamide are shown in figure 1. Conversion to 8-SCH_2CONH_2-FAD proceeds by a monophasic reaction exhibiting pseudo first order kinetics. A plot of the observed rate constant as a function of iodoacetamide concentration gives a straight line with slope equal to the second order rate constant of 97 $\underline{M}^{-1}$ min^{-1}. When the enzyme is saturated with phenol, the rate constant for the iodoacetamide reaction is 160 $\underline{M}^{-1}$ min^{-1}. The rate constant for the reaction of free 8-mercapto-FAD is 47 $\underline{M}^{-1}$ min^{-1} (1). These data

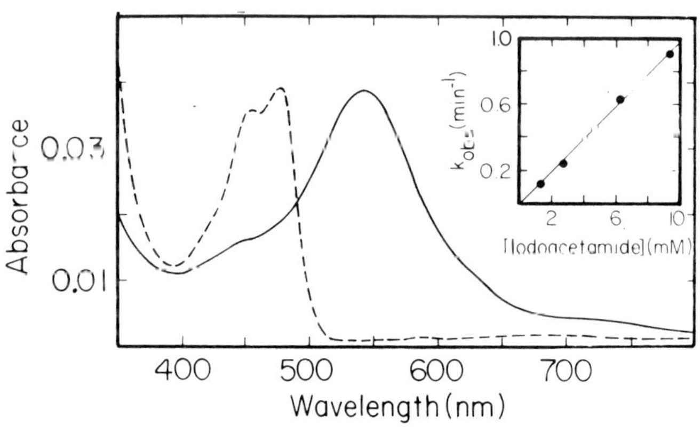

reaction of 8-mercapto-FAD phenol hydroxylase with excess iodoacetamide are shown in figure 1. Conversion to 8-SCH_2CONH_2-FAD proceeds by a monophasic reaction exhibiting pseudo first order kinetics. A plot of the observed rate constant as a function of iodoacetamide concentration gives a straight line with slope equal to the second order rate constant of 97 $\underline{M}^{-1}$ min^{-1}. When the enzyme is saturated with phenol, the rate constant for the iodoacetamide reaction is 160 $\underline{M}^{-1}$ min^{-1}. The rate constant for the reaction of free 8-mercapto-FAD is 47 $\underline{M}^{-1}$ min^{-1} (1). These data

Flavins and Flavoproteins

indicate the 8-position of phenol hydroxylase is exposed to solvent.

The spectral properties of flavoenzymes with 8-mercapto-flavins have been correlated with the protein structure around the flavin (2). The absorbance maximum of 8-mercapto-FAD phenol hydroxylase, 530 nm, is that of a thiolate species as opposed to a thioquinoid species. The thiolate appearance suggests that there is no positive charge in the N(1)C(2)O region of the flavin, at least in the oxidized form of the enzyme.

Reactivity of the pyrimidine subnucleus. Flavins containing modifications in the pyrimidine subnucleus are unreactive when bound to phenol hydroxylase. The reaction of 2-thio-FAD phenol hydroxylase with methyl methane thiolsulfonate (MMTS) at 4° and pH 7.0 is very slow (k_{obs}=0.004 min^{-1}, independent of MMTS concentration) and does not vary with the presence of phenol. The corresponding free flavin rate is 0.11 min^{-1} (3). After the reaction is complete, the 2-thio-FAD spectrum can be regenerated by the addition of dithiothreitol. However, the flavin is no longer bound to the enzyme. The observed rate for the reaction with MMTS may represent the off rate for the flavin; comparable rates for the reaction of 2-thio-enzymes in which $k_{obs}= k_{off}$ have been reported (3).

4-Thio-flavins react with stoichiometric quantities of H_2O_2 to produce 4-S-oxides; further reaction with H_2O_2 results in the formation of 4-oxo-flavins (4). The reaction of 4-thio-FAD phenol hydroxylase is shown in figure 2. No 4-S-oxide is detectable during the course of the reaction.

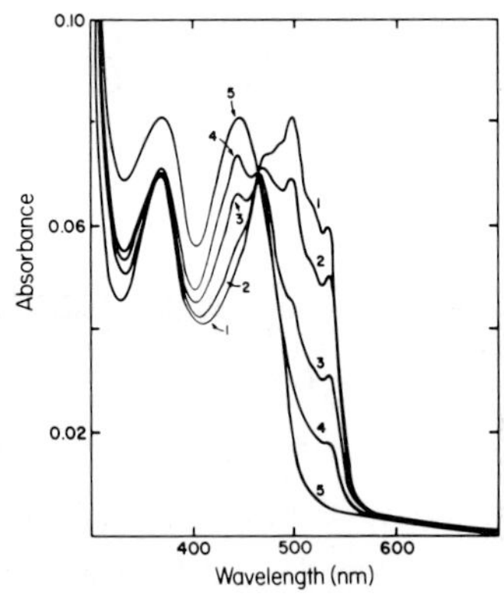

Fig. 2. Reactions of 4-thio-FAD phenol hydroxylase with H_2O_2. Curve 1, 5 uM 4-thio-FAD phenol hydroxylase in 50 mM KPi, pH 7.0; curve 2, 50 min after addition of 8 mM H_2O_2 at 4°; curve 3, 416 min; curve 4, 850 min; curve 5, after 2 hr at 25°.

4-Oxo-flavin is formed at a rate of 1.5×10^{-3} min^{-1} at 4° compared to an expected rate of 1.4×10^{-2} min^{-1} for the free FAD reaction (4). Similarily, the rates of reaction of the 4-thio-enzyme with sulfite or hydroxylamine are slower than the corresponding free flavin rates.

In single turnover experiments, neither 2-thio- nor 4-thio- phenol hydroxylase formed catechol from phenol.

The unreactivity of the 2-thio- and 4-thio- enzymes indicates the pyrimidine subnucleus is inaccessible to solvent. The highly resolved spectrum of the 4-thio-enzyme is consistent with a hydrophobic binding pocket.

Reactions of phenol hydroxylase containing FAD substituted in the benzene ring. Native phenol hydroxylase forms 3 transient intermediates in the oxidative half reaction, a flavin C(4a)-hydroperoxide, a second intermediate of unknown structure, and a flavin C(4a)-hydroxide (5). Monovalent anions retard the conversion of the flavin C(4a)-hydroxide to oxidized enzyme sufficiently that in the presence of an NADPH generating system and azide the flavin C(4a)-hydroxide can be stabilized for several hours. This property is retained by phenol hydroxylase containing certain flavins modified in the benzene ring. Figure 3 shows the absorbance and emission spectra of the flavin C(4a)-hydroxide obtained with 8-F-FAD phenol hydroxylase. Table I summarizes the absorbance and emission maxima for a

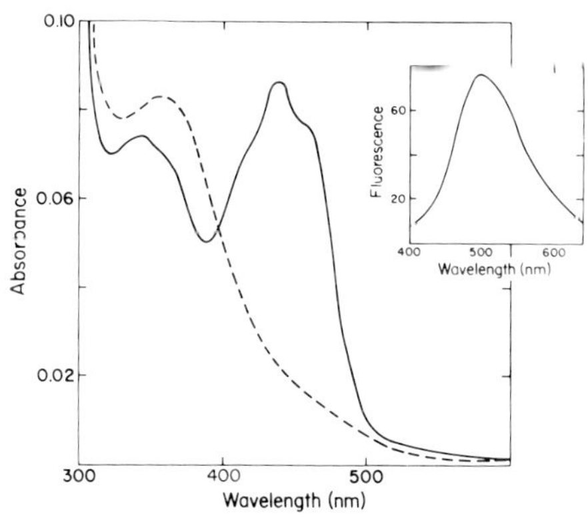

Fig. 3. Spectra of 8-F-FAD phenol hydroxylase, oxidized and C(4a)-hydroxide. Solid line, 7 uM 8-F-FAD phenol hydroxylase in 1 ml 50 mM KPi, pH 7.6, containing 2 mM phenol, 0.2 M KN_3, 40 mM glucose-6-phosphate, and 12 ug glucose-6-phosphate dehydrogenase at 4°. Dashed line, 15 min after addition of 8 uM NADPH, corrected for NADPH absorbance. Inset, fluorescence emission of the C(4a)-hydroxide. Excitation was at 350 nm.

TABLE I

Absorbance and emission maxima of the phenol hydroxylase flavin C(4a)-hydroxide for native and modified flavins.

flavin	absorbance maximum (nm)	emission maximum (nm)
native	375	505
8-F	355	495
$8\text{-}SO_3$	350	485
8-Cl	360	not detected
7-Br	360	not detected

number of flavin C(4a)-hydroxides. All of the flavins have electron withdrawing substituents on the benzene ring, and the absorbance and emission maxima are blue shifted relative to native flavin.

It is perhaps significant that the catalytically competent, modified phenol hydroxylases contain modifications in the benzene ring of the flavin. Since 2-thio-, 4-thio-, and 1-deaza- (6) phenol hydroxylase do not form product, the electronic configuration of the pyrimidine ring appears crucial for oxygen transfer. Experiments are planned to examine the effect of substitution in the benzene ring on the rate of oxygen transfer.

References

1. Schopfer, L. M., Massey V., and Claiborne, A.: J. Biol. Chem. 256, 7329-7337 (1981).
2. Massey, V.: Flavoproteins: Structure and Function Relationships in Biochemical Systems, Bossa, F., Chiancone, E., Finazzi-Agro, A., and Strom, R., eds., Plenum Press, NY, pp. 295-308 (1982).
3. Claiborne, A., Massey., V., Fitzpatrick, P. F., and Schopfer, L. M.: J. Biol. Chem. 257, 174-182 (1981).
4. Massey, V., Claiborne, A., Biemann, M., and Ghisla, S.: J. Biol. Chem., submitted.
5. Detmer, K. and Massey, V.: J. Biol. Chem., in press.
6. Detmer, K., Schopfer, L. M., and Massey, V.: J. Biol. Chem. 259. 1532-1538 (1984).

Supported by grants GM-11106 and 5-T32-GM07767 from the USPHS.

4-THIOFLAVINS AS ACTIVE SITE PROBES OF FLAVOPROTEINS: REACTION WITH SULFITE AND FORMATION OF 4-HYDROXY-4-SULFONYLFLAVINS

Al Claiborne, Vincent Massey
Department of Biological Chemistry, The University of Michigan, Ann Arbor, MI 48109, USA

Monika Biemann, and Sandro Ghisla
Fakultät für Biologie der Universität Konstanz, D 7750 Konstanz

Introduction

4-Thioflavincoenzymes are useful as active site probes for a variety of flavoproteins (1). This results from their high reactivity at C(4)=S with either nucleophiles or electrophiles, which has allowed investigation of the interaction of the flavin N(3)-C(4)=O regions with the apoprotein and its accessibility from the solvent. Significantly different reactions with sulfite have been observed with different flavoenzyme classes reconstituted with 4-Thio-FAD or -FMN. 4-Thio-FAD-D-amino acid oxidase and 4-thio-FMN-lactate oxidase react rapidly to yield "classical" N(5)sulfite adducts (Fig. 1, Curve 4) as do the native enzymes (2). 4-Thio-FAD-p-hydroxybenzoate hydroxylase on the other hand, reacts slowly and biphasically to yield a product (Fig. 1, Curve 3) the spectral properties of which are typical of 3,4-dihydroflavins (5). In view of possible relevance for classifying different types of flavoenzymes, we studied the reaction of free 4-thioflavins with sulfite and elucidated the structure of the products as 4-hydroxy-4-sulfonyl flavins.

Flavins and Flavoproteins

Results and discussion

4-Thio-FAD-D-amino acid oxidase reacts with sulfite rapidly and reversibly to yield a fluorescent species with an absorption maximum at 360 nm (Fig. 1, Curve 4). This species can reasonably be ascribed to a flavin-N(5)-sulfite adduct, which in the case of native enzyme has been shown to exist in a reversible equilibrium (3). The reaction of 4-thio-FAD-p-hydroxybenzoate hydroxylase with sulfite on the other hand is slow and biphasic. A first intermediate is formed in a sulfite concentration dependent reaction (Fig. 1, Curve 2), which is converted to a final blue fluorescent species with an absorption maximum at 415 nm (Fig. 1, Curve 3).

Figure 1: Effect of sulfite on 4-thio-FAD reconsituted enzymes

Curve 1; 4-thio-FAD-p-hydroxybenzoate hydroxylase in 0.05 M phosphate, pH 7, 25°C. Curve 2; 2.5 min after addition of 10 mM sodium sulfite. Curve 3; after incubation for 20 hr at 25°C. Curve 4; N(5)-sulfite adduct of 4-thio-FAD-D-amino acid oxidase in 0.1 M phosphate, pH 7.0, 4°C.

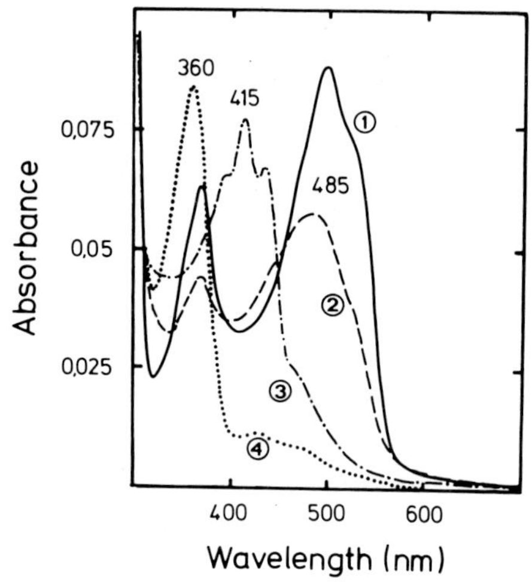

When 4-thioriboflavin is mixed aerobically with sodium sulfite, a blue fluorescent species with absorption maximum at 410 nm (Fig. 2. Curve 3) is formed. The reaction is biphasic and sulfite concentration dependent, and proceeds via formation of a transient species with absorption maximum at 465 nm (Fig. 2, Curve 2). Under anaerobic conditions the same intermediate is formed, but the final product is a typical N(5)-sulfite flavinadduct (Fig. 2, Curve 3). Addition of riboflavin binding protein to the first intermediate restores 84 % of the starting material, demonstrating that the

C(4) sulfur is not lost at this stage of reaction. In the final product, however no sulfur at the S^{-2} oxidation level is present at C(4). N(3) blocked and unblocked 4-thioflavins react similarly. This excludes a flavin-4-sulfonyl structure for the sulfite adducts. Their chromatographic properties indicate the presence of a negative charge on the flavin. The flavin sulfite adducts are extremely photolabile, they yield normal oxidised flavin and sulfite. They are stable if kept in the dark. Anaerobic or aerobic photolysis leads quantitatively to normal oxidized flavins and not to the reduced one as it should be expected from a 3,4-dihydroflavin.

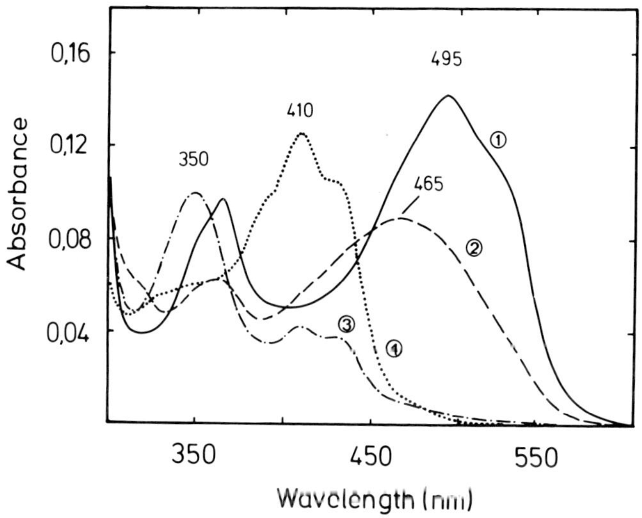

Figure 2: Effect of sulfite on 4-thioriboflavin under aerobic and anaerobic conditions.
Curve 1; 4-thioriboflavin in oxygen free 0.1 M phosphate, pH 7.0, 0 - 1°C in an anaerobic cuvette at the same concentrations as the remaining curves. Curve 2; 80 - 160 sec after addition of 0.18 M sodium sulfite. Curve 3; 12 hr later. Curve 4; 4-thioriboflavin in aerobic phosphate, pH 7.0, 12 hr after addition of 0.18 M sodium sulfite.

In the presence of 5 % H_2O_2, in 1 N acetic acid, or with meta-chlorperbenzoate or methylmethanthiolsulfonate at pH 7, the sulfite adducts are stable. When irradiated in the presence of amines they undergoe nucleophilic substitutions to yield the corresponding 4-amino- or 4-imino-flavins similar to those observed with 4-thioflavins (1,4). While the sulfite adducts have similar UV-VIS and fluorescence properties as 3,4-dihydroflavins (5), unlike them they show no ^{1}H-NMR resonance which could be attributed to a C(4)-H function. These properties indicate that the flavin sulfiteadducts have an sp^3 center at the C(4) position, and deduce that they have a 4-hydroxy-4-sulfonyl structure. 4-Thio-FAD reconsituted p-hydroxybenzoate hydroxylase forms 4-hydroxy-4-sulfonylflavins with sulfite like the free coenzyme under aerobic conditions (Scheme below, "hydroxylase" reaction), while 4-thio-FAD / -FMN reconstituted oxidases or 4-thioflavins under anaerobic conditions form N(5)-sulfite adducts (Scheme below, "oxidase" reaction).

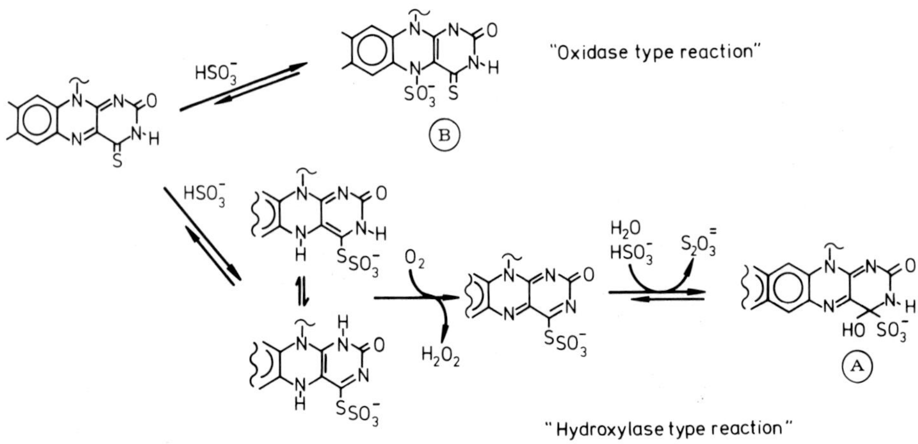

References

1. Massey, V., Claiborne, A., Biemann, M., and Ghisla, S.: J. Biol. Chem. in press.
2. Massey, V., Müller, F., Feldberg, R., Schuman, M., Sullivan, P. A., Howell, L. G., Mayhew, S. G., Matthews, R. G., and Foust, G. P.: J. Biol. Chem. 244, 3999 - 4006 (1969).
3. Müller, F., and Massey, V.: J. Biol. Chem. 244, 4007 - 4016 (1969)
4. Müller, F., and Hemmerich, P.: Helv. Chim. Acta 49, 2352 - 2364
5. Müller, F., Massey, V., Heizmann, C., Hemmerich, P., Lhoste, J-M., and Gould, D. C.: Eur. J. Biochem. 9, 392 - 401 (1969)
6. Hemmerich, P., and Erlenmeyer, H.: Helv. Chim. Acta 40, 181 - 186 (1957)

OXYGEN REACTIVITY OF 4-THIO-FAD *p*-HYDROXYBENZOATE HYDROXYLASE

Al Claiborne

Department of Biochemistry, Wake Forest University Med. Ctr.
Winston-Salem, North Carolina 27103

Lawrence M. Schopfer and Vincent Massey

Department of Biological Chemistry, University of Michigan
Ann Arbor, Michigan 48109

Introduction

Previous studies have shown (1) that *p*-hydroxybenzoate hydroxylase (PHBH) from *Pseudomonas fluorescens* catalyzes the S-oxygenation of the substrate analog *p*-mercaptobenzoate. Though never actually demonstrated, this reaction presumably proceeds via direct oxygen transfer from the flavin-C(4a)-hydroperoxide (PHBH intermediate I; Ref. 2) to the sulfur atom of the substrate analog (1). More recent studies with free 2- and 4-thioflavins (sulfur replacing oxygen substituents at positions 2 and 4, respectively) have shown (3, 4) that these modified flavins undergo facile oxidative conversions to the corresponding flavin-S-oxides. *p*-Hydroxybenzoate hydroxylase has thus been reconstituted with 4-thio-FAD, in order to probe the reactivity of the 4-thioflavin-C(4a)-hydroperoxide toward the sulfur at the adjacent 4-position.

Results and Discussion

Figure 1 shows the spectral changes accompanying the loss of sulfur from enzyme-bound 4-thio-FAD during a single turnover of the enzyme in the presence of the effector 6-hydroxy-

Flavins and Flavoproteins

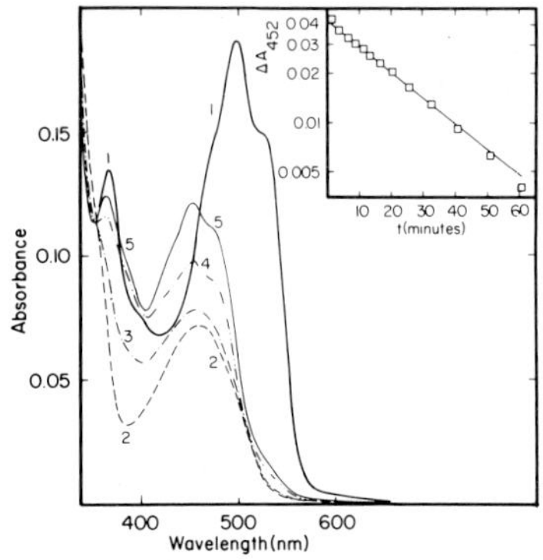

Figure 1. Spectral changes on photoreduction and air oxidation of 4-thio-FAD PHBH in the presence of 6-hydroxynicotinate (6HNA) and azide. 4-Thio-FAD enzyme plus 6HNA (1); reduced enzyme-6HNA plus azide (2); and 40 seconds (3), 16 minutes (4), and 2 hours (5) after air oxidation. INSET: Semilog plot of absorbance increase at 452 nm; $t_{1/2}$ = 19.4 minutes.

nicotinate (6HNA) and 0.1 M azide. Spectrum 5 contains less than 10% of the original 4-thio-FAD; the elimination of sulfur during the oxidation of reduced 4-thio-FAD PHBH leads to the appearance of native oxidized enzyme.

During the reoxidation of the 4-thio-FAD enzyme-6HNA complex, a fluorescent species appears (Figure 2). The excitation

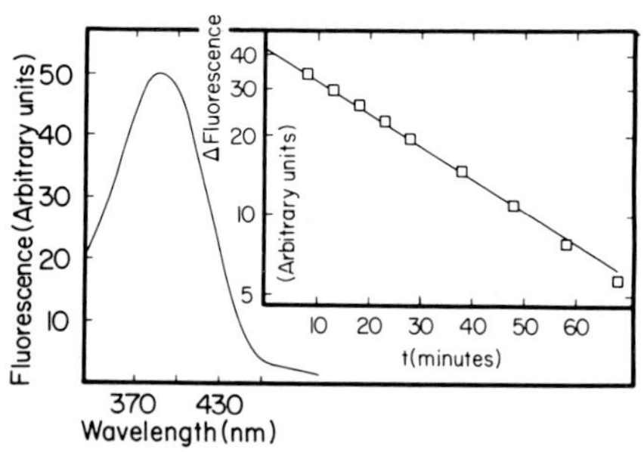

Figure 2. Fluorescent species observed during oxidation of reduced 4-thio-FAD enzyme-6HNA complex in the presence of azide. Fluorescence excitation spectrum (λ_{max}^{EX} = 385 nm, λ_{max}^{EM} = 505 nm) measured on opening reduced enzyme-6HNA to air. INSET: Semilog plot of fluorescence decay at 385 nm; $t_{1/2}$ = 24.7 minutes. Conditions as in Figure 1.

wavelength maximum is 385 nm, when monitored at the emission maximum of 505 nm. This species slowly decays (k_{obs} = 0.028 min^{-1} at 4°C) to native oxidized enzyme; the pseudo-first order rate compares favorably with that measured for the slow absorbance changes shown in Figure 1 (k_{obs} = 0.036 min^{-1} at 4°C). These data indicate that this fluorescent species is an intermediate in the conversion of 4-thio-FAD enzyme to native enzyme.

In contrast to the nearly quantitative desulfurization observed for 4-thio-FAD PHBH in the presence of 6HNA, virtually no sulfur elimination occurs during turnover in the presence of 2,4-dihydroxybenzoate (24DHB). Less than 10% conversion to native FAD enzyme occurs during single turnover of the 4-thio-FAD enzyme-24DHB complex.

Scheme 1 summarizes these results in terms of a single primary 4-thio-FAD PHBH oxygen adduct, the 4-thioflavin-C(4a)-hydroperoxide. Similar peroxyflavin intermediates have been observed in studies of PHBH reconstituted with other modified flavins as well, such as 2-thio-FAD (5) and 1-deaza-FAD (6).

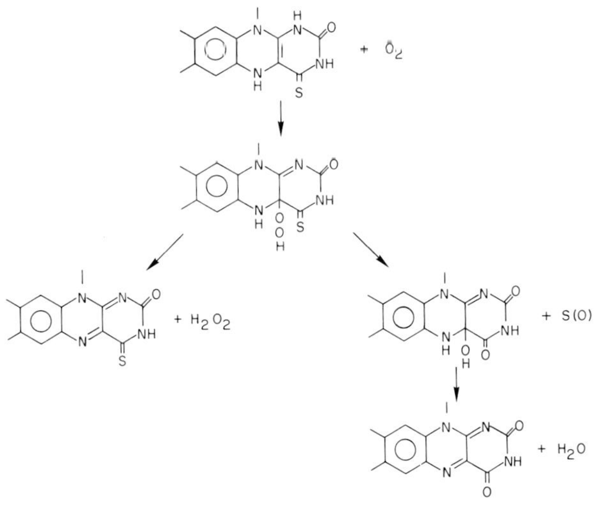

Scheme 1

In both these cases the flavin hydroperoxide breaks down to yield free hydrogen peroxide and the oxidized flavin. Such a process would account for the reappearance of oxidized 4-thio-FAD PHBH after single turnover in the presence of 24DHB, and is also consistent with the absence of 24DHB hydroxylation observed with the modified 4-thio-FAD enzyme.

The identification of the fluorescent intermediate in the conversion of 4-thio-FAD enzyme as the native C(4a)-hydroxyflavin is primarily supported by the comparison of its fluorescence properties with those of the enzyme-bound N(5)-ethyl-C(4a)-hydroxy-FAD (λ_{max}^{EX} = 372 nm, λ_{max}^{EM} = 505 nm; Ref. 7). This conclusion also implies that one oxygen atom is transferred from the 4-thioflavin-C(4a)-hydroperoxide during the desulfurization process. Both the fate of this oxygen atom and the form in which the 4-thioflavin sulfur is eliminated are currently under investigation.

Acknowledgment. This work was supported by NIH Grant No. GM 11106 from the U.S. Public Health Service to V.M.

References

1. Entsch, B., Ballou, D.P., Husain, M., Massey, V.: J. Biol. Chem. 251, 7367-7379 (1976).
2. Entsch, B., Ballou, D.P., Massey, V.: J. Biol. Chem. 251, 2550-2563 (1976).
3. Biemann, M., Claiborne, A., Ghisla, S., Massey, V., Hemmerich, P.: J. Biol. Chem. 258, 5440-5448 (1983).
4. Massey, V., Claiborne, A., Biemann, M., Ghisla, S.: J. Biol. Chem. 259, in press (1984).
5. Claiborne, A., Massey, V.: J. Biol. Chem. 258, 4919-4925 (1983).
6. Entsch, B., Husain, M., Ballou, D.P., Massey, V., Walsh, C.: J. Biol. Chem. 255, 1420-1429 (1980).
7. Ghisla, S., Entsch, B., Massey, V., Husain, M.: Eur. J. Biochem. 76, 139-148 (1977).

OXYGEN REACTIONS OF PARA-HYDROXYBENZOATE HYDROXYLASE CONTAINING 6-HYDROXY-FAD

Barrie Entsch

Department of Biochemistry and Nutrition, The University of New England, Armidale, 2351, Australia

Vincent Massey

Department of Biological Chemistry, The University of Michigan, Ann Arbor, 48109, USA

Introduction

It is now recognized that the reaction mechanisms of flavoprotein oxygenases are known in some detail, principally because of the research with p-hydroxybenzoate hydroxylase (EC 1.14.13.2) from Pseudomonas fluorescens (1). However, major gaps remain in our knowledge. For example, during catalysis by aromatic hydroxylases, the transfer of oxygen to the substrate involves an unidentified transient state of flavin. Modified flavins provide one experimental tool to probe flavoprotein mechanisms, by perturbing the active site. With the naturally occurring flavin, 6-hydroxy-FAD (2), apo-para-hydroxybenzoate hydroxylase forms an active enzyme which hydroxylates p-hydroxybenzoate (A. Claibourne, unpublished result), in contrast to some other modified flavins (3). We chose to examine the oxygen reactions of the 6-OH-FAD enzyme in some detail, as the 6-hydroxy group changes the chemistry of the isoalloxazine ring.

Results and Discussion

Dr L.M. Schopfer kindly prepared 6-OH-FAD using enzymes from

Flavins and Flavoproteins

Brevibacterium ammoniagenes. Four compounds which are hydroxylated efficiently by native enzyme were tested with 6-OH-FAD enzyme. The same products were formed as with native enzyme. Enzyme containing ≤ 0.2% FAD was used to quantify product formation. Rates of NADPH oxidation in turnover were approximately an order of magnitude lower than with native enzyme at pH 7, 25°. The ratio of moles product formed to NADPH oxidised was calculated - 0.69 for p-hydroxybenzoate, 0.25 for p-aminobenzoate, 0.29 for 2,4-dihydroxybenzoate, and 0.52 for p-mercaptobenzoate. Thus, 6-OH-FAD enzyme was an effective catalyst, but not as efficient as native enzyme.

The oxygen half-reaction of enzyme was studied as described in the Table and (4). Complex reaction traces were interpreted in terms of transient chemical species of flavin. The most

Figure. Spectra of flavin species in the reaction E_R-p-aminobenzoate + O_2 (see Table) - reduced enzyme (—); intermediate I (□—□); intermediate II (O-O); intermediate III (●-●); oxidised enzyme (— — —).

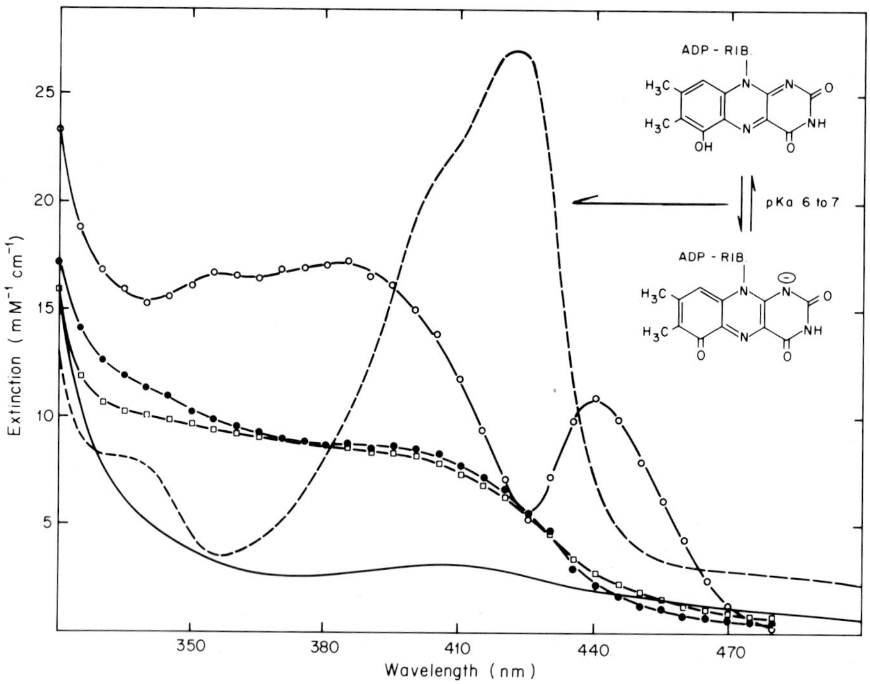

Table. RATE CONSTANTS AND INTERMEDIATES IN REACTIONS OF REDUCED ENZYME WITH OXYGEN

The 6-OH-FAD enzyme was mixed with substrate to saturate the reduced enzyme, based upon known dissociation constants for native enzyme. This assumption was justified for p-hydroxybenzoate and p-aminobenzoate, but probably not for 2,4-dihydroxybenzoate. After the solution was made anaerobic, the enzyme was reduced with buffered sodium dithionite. The reduced enzyme complex was reacted with oxygenated buffer in a stopped-flow spectrophotometer with computerized data aquisition. The reactions were followed at many wavelengths and different oxygen concentrations. Reaction solutions finally contained 16μM enzyme, substrate, and oxygen in 50mM phosphate (K salt) and 10mM EDTA (Na salt), pH6.55 at 3 to 4°, or 20mM phosphate (K salt with Tris) and 4mM EDTA, pH8.5 at 3 to 4°.

Substrate	Minimum Scheme for each Reaction from Observed Kinetics
p-hydroxybenzoate (pH6.55)	$E_RS + O_2 \xrightarrow[\mathit{2.6 \times 10^{5b}}]{3.5 \times 10^{5a}} I \xrightarrow[\mathit{48}]{14} (II)^c \xrightarrow[\mathit{Fast}]{Fast} III \xrightarrow[\mathit{14.5}]{2.0} E_{OX}S + P + H_2O$ $I \xrightarrow[\mathit{<1}]{7} E_{OX}S + H_2O_2$
p-aminobenzoate (pH6.55)	$E_RS + O_2 \xrightarrow[\mathit{3.3 \times 10^{5}}]{2.9 \times 10^{5}} I \xrightarrow[\mathit{4.9}]{1.4} II \xrightarrow[\mathit{.031}]{.63} III \xrightarrow[\mathit{Fast}]{.020} E_{OX}S + P + H_2O$ $I \xrightarrow[\mathit{<.2}]{5} E_{OX}S + H_2O_2$
2,4-dihydroxybenzoate (pH8.5)	$E_RS + O_2 \xrightarrow[\mathit{5.0 \times 10^{5}}]{4.5 \times 10^{4}} (I) \xrightarrow[\mathit{123}]{} II \xrightarrow[\mathit{.145}]{} (III) \xrightarrow[\mathit{Fast}]{} E_{OX}S + P + H_2O$ $(I) \xrightarrow[\mathit{<5}]{Fast} E_{OX}S + H_2O_2$

[a] Numbers in this column are second order rate constants ($M^{-1}s^{-1}$), and in other columns are first order constants (s^{-1}).

[b] Numbers in italics are constants previously obtained for native (FAD) enzyme.

[c] Parentheses indicate that an intermediate is kinetically invisible.

important spectral data is illustrated in the Figure, and appropriate rate constants in the reactions are compared to native enzyme under the same conditions in the Table.

By analogy with native enzyme, intermediate I (Figure and Table) is C(4a)-peroxyflavin and intermediate III is C(4a)-hydroxyflavin. The 6-hydroxy group has changed these spectra relative to native enzyme (4). The C(4a)-peroxyflavin decays partly to H_2O_2 (Table). A third transient species (II) was clearly observed between intermediates I and III with p-aminobenzoate as substrate. This species has not been detected yet in the hydroxylation of other substrates. On the basis of known spectral data, there does not appear to be any known flavin structure responsible for intermediate II. An inspiration is needed to establish the chemical change in C(4a)-peroxyflavin which results in the observed spectrum.

The interaction between 2,4-dihydroxybenzoate and 6-OH-FAD enzyme shows little similarity to the native enzyme. No transient intermediates were detected upon reaction with oxygen (Table). Our tentative explanation is that the Kd for reduced enzyme with substrate is very large ($\geq$ 0.1M), compared to 1.8×10^{-4}M for native enzyme (4). Results suggest that the 6-hydroxy group on the flavin interferes with the binding of this substrate, which places the substrate below and appressed to the flavin ring in the reduced conformation.

References

1. Malmström, B.G.: Ann. Rev. Biochem. 51, 21-59 (1982).
2. Mayhew, S.G., Whitfield, C.D., Ghisla, S., Schuman-Jörns, M.: Eur. J. Biochem. 44, 579-591 (1974).
3. Entsch, B., Husain, M., Ballou, D.P., Massey, V., Walsh, C.: J. Biol. Chem. 255, 1420-1429 (1980).
4. Entsch, B., Ballou, D.P., Massey, V.: J. Biol. Chem. 251, 2550-2563 (1976).

THE EFFECT OF PH AND MODIFICATIONS IN POSITION 8 ON THE OXIDATION OF REDUCED P-HYDROXYBENZOATE HYDROXYLASE

L.M. Schopfer, Albert Wessiak and Vincent Massey
Department of Biological Chemistry, University of Michigan, Ann Arbor, Michigan, USA, 48109

Introduction

Though the mechanism of flavin dependent aromatic hydroxylation is well understood, the structure of the second intermediate in the oxidative half reaction is still unknown. Scheme I illustrates the mechanism, emphasizing the structures of the intermediates. The ring open structure for II was originally suggested by Entsch et.al. (1) and is the structure with which our data are most consistent.

Scheme I

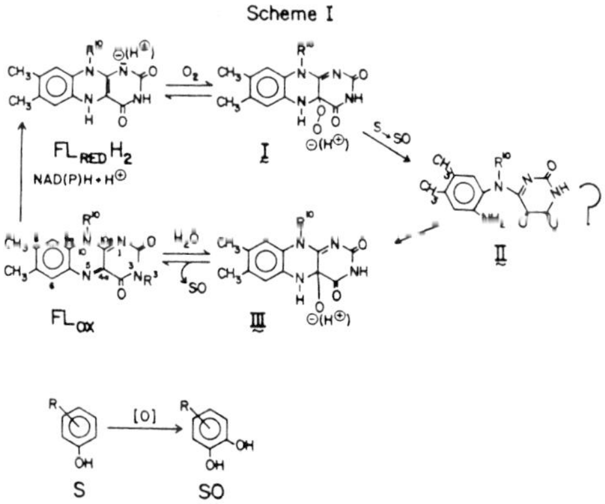

Using p-OH-benzoate hydroxylase (PHBH), plus 2,4-diOH-benzoate (2,4-D) and NaN_3, we could resolve the rates for the conversion of I to II, II to III and III to Flox. Furthermore, we could directly measure the spectrum of II using a fast scan, stopped-flow spectrophotometer. In order to obtain more information on the chemistry of II, we have examined the oxidative half reaction for this system: A) as a function of pH, and B) using enzyme reconstituted with modified flavins.

Flavins and Flavoproteins

Results and Discussion

The oxidation of native, reduced PHBH·2,4-D complex was examined at pH values ranging from 6.6 to 9.2, both in the presence (Figure 1) and in the absence of NaN_3.

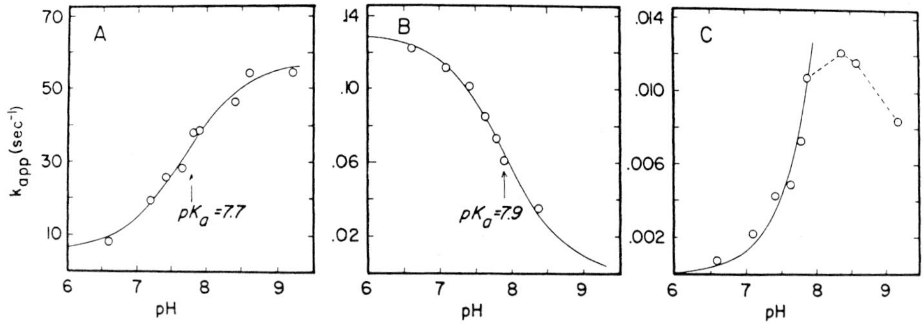

Figure 1. pH dependence of the oxidation of reduced PHBH·2,4-D complex by O_2, in the presence of NaN_3. Conditions: 12.3 micromolar PHBH in 10 mM KPi buffer, pH 6.6 containing 10 mM 2,4-D, 100 mM NaN_3 and 0.3 mM EDTA was reduced with sodium dithionite, and then mixed at 4° in a stopped-flow spectrophotometer with an equal volume of a 200 mM buffer pH (variable) containing 10 mM 2,4-D, 100 mM NaN_3, 0.3 mM EDTA and 1.95 mM O_2. The final pH was measured at the end of the reaction. Panel A, the apparent rate (kapp) for formation of II. Panel B, kapp for conversion of II to III. Panel C, kapp for conversion of III to Flox. The points are data and the lines are calculated.

Three intermediates could be distinguished. The conversion of $FlredH_2$ to I was complete in the dead-time of the stopped-flow apparatus. The conversion of I to II was base catalyzed (pKa=7.7). Conversion of II to III was acid catalyzed (pKa=7.9). Conversion of III to Flox was directly dependent on $[OH^-]$ from pH 6.6 to 7.9 ($k=1.4\times10^4\ M^{-1}sec^{-1}$), at higher pH values, the preceeding step became limiting. Azide reduced kapp for each step in the oxidative half reaction, consistent with earlier reports (1), but did not significantly alter the nature of the pH catalysis or the observed pKa values. This is at variance with a proposal by Shoun et.al. that monovalent anions exert their influence on the oxidative half reaction by perturbing the pKa of an essential group in the active site (2). Our data are more consistent with the proposal (3) that monovalent anions effect a conformational change in the enzyme active site. We suggest further that this change adversely affects the orientation of the catalytically essential groups but not the fundamental chemistry of catalysis. The spectrum of II was also pH dependent, as previously described (1). The peak shifted from 388 nm (extinction 15.3 $mM^{-1}cm^{-1}$) at pH 6.6 to 423 nm (extinction 16.9 $mM^{-1}cm^{-1}$) at pH 9.2. The pKa was 7.8.

An analogous set of experiments was performed with 8-sulfonyl-FAD reconstituted PHBH. Essentially identical results were obtained. Conversion of I to II was base cataylzed, though only about 2-fold. Conversion of II to III was acid catalyzed (pKa=7.9) and conversion of III to Flox was directly dependent on hydroxide ion concentration from pH 6.7 to 7.9 ($k=3x10^4\ M^{-1}sec^{-1}$). The spectrum of II (Figure 2) shifted from a single peak at 378 nm (extinction 14.5 $mM^{-1}cm^{-1}$) at pH 6.7 to a double banded species with peaks at 395 nm (extinction 12.9 $mM^{-1}cm^{-1}$) and 430 nm (extinction 11.3 $mM^{-1}cm^{-1}$) at pH 9.2. The pKa was 7.8.

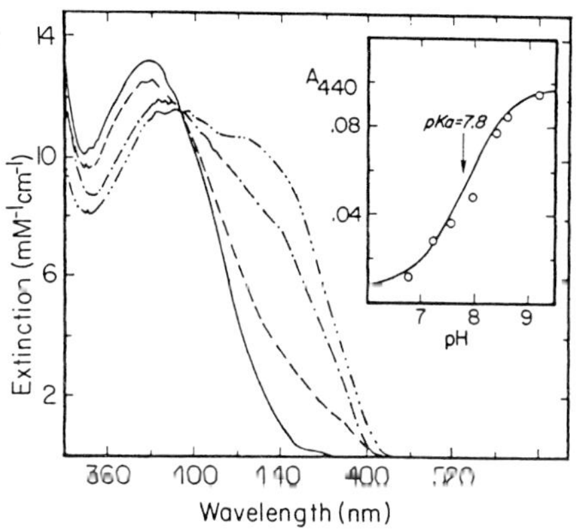

Figure 2. The pH dependence of II spectra for 8-SO_3-FAD PHBH. Conditions: same as Figure 1 except that 10.9 micromolar 8-SO_3-FAD PHBH was used in place of native enzyme. Spectra: pH 6.75 (solid line), pH 7.55 (dashed line), pH 8.40 (dashed-dot line) and pH 9.17 (dashed-dot-dot line). Inset is a plot of pH versus absorbance at 440 nm. The points are data and the line is calculated for a pKa of 7.8.

The reactions involved in the formation and decay of II are base and acid catalyzed, respectively. Changing the redox potential of the flavin by 90 mV has no significant effect on these rates. It appears that the chemistry of II is dominated by acid/base catalysis. Based on the fact that the same pKa is measured for each reaction, we suggest that the same group may be responsible for catalysis of both steps. This group would most likely be a phenol moiety deriving either from the substrate/product, or one of the three tyrosines found in the active site (4).

Chemistry of this type is consistent with the reversible reactions of carbonyls and amines to form carbinolamines as formulated by Sayer et.al. (5). Such a mechanism for the formation and decay of II would require that II be the 4a,5-ring open form shown in Scheme I. Of the structures proposed for II, the 4a,5-ring open form can be most readily incorporated into a mechanism which is consistent with the pH dependences observed in this work.

If the structure of II is ring opened as shown in Scheme I, the visible absorbance cannot be due simply to a sum of the absorbances of the two halves. We suggest that it is the result of a charge-transfer complex, wherein the benzene half is the donor and the uracil half is the acceptor. This is consistent with the effect on the spectrum of substituting an electron withdrawing sulfonyl group for the methyl at position 8. The reduced electron donating capacity of the benzene is reflected by a 10 nm blue shift in the spectrum (at pH 6.6) of II. The perturbation of the spectrum of II at high pH might then result from a shift in the relative orientation of the components of the charge transfer complex, in response to the appearance of a new anionic species in the active site. Alternately, it might reflect the appearance of a new charge transfer donor in the form of a phenolate, for the uracil acceptor.

Supported by GM-11106 of the U.S. Public Health Service.

References

1. Entsch, B., Ballou, D.P. and Massey, V.: J. Biol. Chem. 251, 2550-2563 (1979).

2. Shoun, H., Arima, K. and Beppu, T.: J. Biochem. (Tokyo) 93, 169-176 (1983).

3. Steenis, P.J., Cordes, M.M., Hilkens, J.G.H. and Müller, F.: FEBS Lett. 36, 177-180 (1973).

4. Weijer, W.J., Hofsteenge, J., Beintema, J.J., Wierenga, R.K. and Drenth, J.: Eur. J. Biochem. 133, 109-118 (1983).

5. Sayer, J.M., Pinsky, B., Schonbrunn, A. and Wastitien W.: J. Am. Chem. Soc. 96, 7998-8009 (1974).

REACTION OF BACTERIAL LUCIFERASE FROM VIBRIO HARVEYI WITH 8 - SUBSTITUTED FLAVINS

Lorenzo Chen and Thomas O. Baldwin

Department of Biochemistry and Biophysics and the Texas Agricultural Experiment Station, Texas A&M University, College Station, TX 77843

Introduction

There are two observations leading to the conclusion that the emitter species in the bioluminescence reaction of bacterial luciferase consists of an enzyme-flavin complex. Altering the flavin structure results in a change in color of bioluminescence (1) and changes in the enzyme through mutations also result in changes in bioluminescence emission (2). In addition, the presence of the 5' phosphate on the ribityl moiety of $FMNH_2$ is crucial for good binding as well as for high quantum yield (3). Anion binding also causes structural changes in the enzyme (4). Studies with *Photobacterium phosphoreum* using 8 - substituted flavins have suggested that the imino group at N(5) is involved in the binding of $FMNH_2$ to luciferase (5). Using 8-mercaptoflavins as probes of the active centers of flavoproteins, Schopfer *et al.* have suggested that upon binding to luciferase, the 8 position of FMN may be oriented away from the protein and possibly exposed to the solvent (6). Our studies with *Vibrio harveyi* luciferase demonstrate this more conclusively, by showing that the enzyme binds 8 - substituted flavins despite variations in substituent charge and size.

Procedures and Results

A number of 8 - substituted flavins were synthesized from 8-Cl FMN. This analogue was converted to the 8-mercapto- and 8-thiopropyl Sepharose derivatives. The 8-mercapto-FMN was then treated with various alkylating

Flavins and Flavoproteins

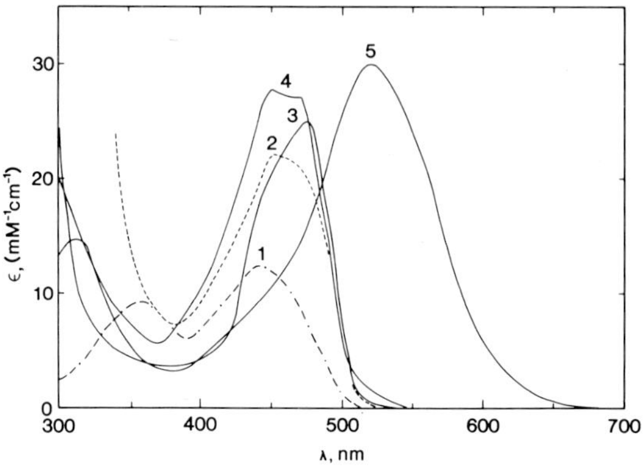

Figure 1. Spectra of 8 - substituted FMN analogues. 1) 8-chloro-FMN; 2) 8-S-(N-ethylsuccinimido)-FMN; 3) 8-S-carboxymethyl-FMN; 4) 8-S-carboxamidomethyl-FMN; 5) 8-S-FMN. Extinction values were normalized to 8-chloro-FMN (12,400 M^{-1} cm^{-1} at 444 nm).

reagents. The progress of the reactions was monitored spectrophotometrically. Separation of the derivatized flavins from other reagents was achieved by chromatography over a BioGel P2 column equilibrated in water. Figure 1 shows the spectra of the final products.

The binding affinity of these analogues to luciferase was determined through competition experiments with $FMNH_2$ as substrate. All the analogues were found to bind tightly to luciferase despite large differences in charge and size of the 8 substituent. The premise that the 8 position is oriented away from the protein was most dramatically illustrated by the ability of luciferase to bind the thiopropyl Sepharose derivative. As seen from the representative set of data for 8-S-carboxamidomethyl-$FMNH_2$ (Figure 2), the binding of the analogues was competitive with $FMNH_2$, indicating that changes in size and charge of the 8 substituent do not alter the ability of luciferase to recognize and bind the analogues. The tighter binding observed with 8-S-(N-ethylsuccinimido)-$FMNH_2$ could be correlated with the position of the active center in or near a hydrophobic pocket (7).

SPECTRAL PROPERTIES OF $FMNH_2$ ANALOGUES

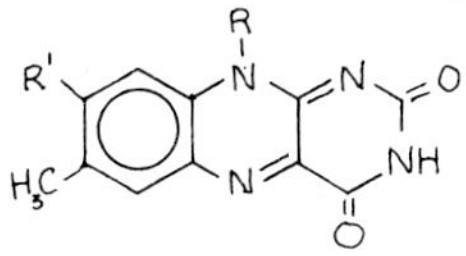

R'	K(μM)	Absorbance λ_{max}(nm)	ε mM^{-1} cm^{-1}	Fluorescence ex λ_{max}(nm)	Fluorescence em λ_{max}(nm)
$-CH_3$*	0.3	450	12	452	519
-Cl*	0.1	444	12	450	516
-S	3.0	520	30	-	-
-S-CM*	1.0	475	25	469	520
-S-CAM	1.8	450	27	N.D.	N.D.
-S-NES	0.8	450	22	468	517

* These analogues showed bioluminescence activity
Abbreviations: -S-CM, S-carboxymethyl-; -S-CAM, S-carboxamidomethyl-; ε NES, S-(N-ethylsuccinimido)-.

While substitutions at the 8 position did not significantly alter the ability of the enzyme to bind the analogues, there was a dramatic effect on the bioluminescence activity with these substitutions. As seen in Table 1, only 8-chloro- and 8-S-carboxymethyl-$FMNH_2$ were capable of inducing light emission upon binding to luciferase. Although light emission by these analogues was only 2-3% of the maximal intensity seen with $FMNH_2$, the 8-S-carboxymethyl derivative showed a peculiar time course for bioluminescence decay. The peak intensity (I_o) with this derivative occurred several seconds after aldehyde injection in the dithionite assay, compared with less than one second with $FMNH_2$, indicating that some step subsequent to formation of the 4a-peroxydihydroflavin intermediate is altered, perhaps resulting in a change in the rate-limiting step. Emission of light requires a stable 4a-peroxydihydroflavin intermediate, and the inability of most of these analogues to induce light emission with luciferase could be due to alterations in the formation and/or stability of this intermediate. Thus, substitutions at the 8 position may affect oxygen binding to form the functional intermediate II, or once formed, this complex may be unstable and/or decay by some dark pathway.

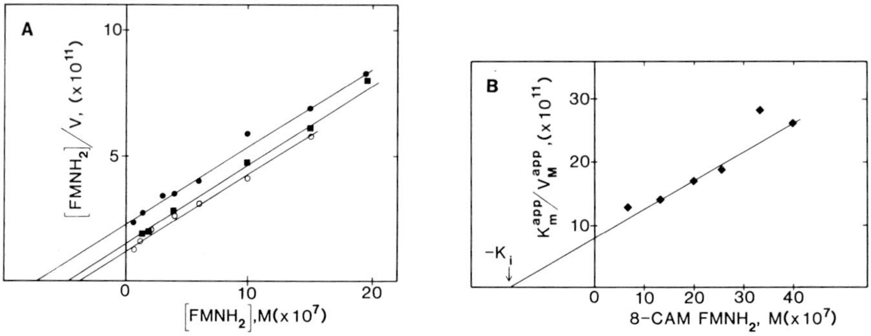

Figure 2. Representative data for K_d determination of 8-S-carboxamido-methyl-FMN with V. harveyi luciferase. (A) Final luciferase concentration: 60 nM; final 8-S-CAM FMN concentrations: (●), 3.25 mM; (■), 2.5 mM; (○), 1.3 mM. (B) Secondary plot of data determined from (A).

In summary, the binding of 8-substituted flavin analogues to luciferase was not significantly perturbed by charge and size variations in the 8 substituent, but these changes did affect bioluminescence, apparently by altering the formation and/or stability of the intermediate in the rate determining step.

(This work was supported by grants from NSF, NIH and the Welch Foundation.)

References

1. Mitchell, G., Hastings, J.W.: J. Biol. Chem. 244, 2572-2576 (1969).
2. Cline, T.W., Hastings, J.W.: J. Biol. Chem. 249, 4668-4669 (1974).
3. Meighen, E.A., Mackenzie, R.E.: Biochemistry 12, 1482-1489 (1973).
4. Baldwin, T.O., Riley, P.L.: Flavins and Flavoproteins (Yagi, K., Yamano, T., eds.), 139-147 (1980).
5. Watanabe, T., Matsui, K., Kasai, S., Nakamura, T.: J. Biochem. 84, 1441-1446 (1978).
6. Schopfer, L.M., Massey, V., Claiborne, A.: J. Biol. Chem. 256, 7329-7337 (1981).
7. Nicoli, M.Z., Hastings, J.W.: J. Biol. Chem. 249, 2393-2396 (1974).
8. Ziegler, M.M., Baldwin, T.O.: Curr. Top. Bioenerg. 12, 65-113 (1981).

Biomedical Aspects

INTESTINAL ABSORPTION OF RIBOFLAVIN

Sabu Kasai, Hideko Nakano, Hirokazu Matsukawa,
Yoshihiro Miyake*, and Kunio Matsui

Research Institute for Atomic Energy, Osaka City University, Sumiyoshi-ku, Osaka 558, Japan and *National Cardiovascular Center, Research Institute, Suita, Osaka 565, Japan

Introduction

Many reports about intestinal absorption of riboflavin have been already published. But the precise absorption mechanism is not clear yet. After a passive diffusion mechanism was suggested in the studies using everted sacs (1,2), a saturable mechanism was added to that, using improved methods with everted intestine (3,4). Experiment with perfused jejunal segment suggests that the absorption is a "biological process" and not passive diffusion (5). For the saturable mechanism, a phosphorylation-dephosphorylation process was suggested (6).

Materials and Methods

The analogues used were 5'-deoxyriboflavin (5'-dRF), [2-^{14}C] 5'-dRF (^{14}C-5'-dRF), 2'-deoxyriboflavin (2'-dRF), [2-^{14}C]2'-dRF (^{14}C-2'-dRF), diacetone riboflavin (DARF), lumiflavin (LF), 8-methylamino-8-demethylriboflavin (MAF), 8-dimethylamino-8-demethylriboflavin (roseoflavin,RoF), 8-diethyl-8-demethylriboflavin (DEF), 8-hydroxyl-8-demethylriboflavin (HF), 3-methylriboflavin (3-MRF), 3-isobutylriboflavin (3-IBRF), 3-carboxymethylriboflavin (3-CMRF) and lumichrome (LC) (7,8).
In vivo study: ^{14}C-RF, ^{14}C-2'-dRF or ^{14}C-5'-dRF was administered to male Wistar rat stomach and after 1.5 hr the gastrointestinal tract was washed with saline. Radioactivity of the

Flavins and Flavoproteins

washings was measured and the incorporated value was calculated by subtracting that remaining from that administered.
Closed in situ circulation system: Male Wistar rats were fed on RF deficient diet for 6 days and fasted for 24 hr before the experiment. Two jejunal segments ($\sim$20 cm) were used, one is for a test and the other is for a control. The segment was connected with a vial by two silicone rubber tubes and the loop was perfused with flavin soln., by the aid of peristaltic pump. Samples were taken from the vial every 10 min.

Results and Discussion

In vivo study: ^{14}C-RF, ^{14}C-5'-dRF or ^{14}C-2'-dRF was administered in the range of 0.05 to 0.2 µmole and the absorption rates were compared. The rate of absorption of ^{14}C-5'-dRF or ^{14}C-2'-dRF decreased to 69% or 88% of ^{14}C-RF's but t-test shows that there is practically no difference among those of the three flavins. As the phosphorylation occurs at the position 5' of RF, ^{14}C-5'-dRF was expected to remain in the tract, but these three flavins were absorbed almost equally. Then Phosphorylation-dephosphorylation mechanism was contradicted.
Site specificity of the absorption: Fig 1A shows that two jejunal segments absorb RF almost equally. The same absorption was observed using even ileal segment. These results don't coincide with the conclusion of early in vivo studies, that RF is absorbed from upper gastro-intestinal tract (9,10), but coincide with the result of Hegazy and Schwenk (11).
Saturation of the absorption: The absorption of ^{14}C-RF was observed in the presence of non-labeled RF. Fig 1B shows that the absorption can not be observed, when the concentration of non-labeled RF is above 100 µM. These results show that the absorption is fully saturable and don't coincide with the results of the studies using everted intestine (3,4), of which condition seems to be non-physiological.
Effect of flavin analogues on the absorption of labeled RF and

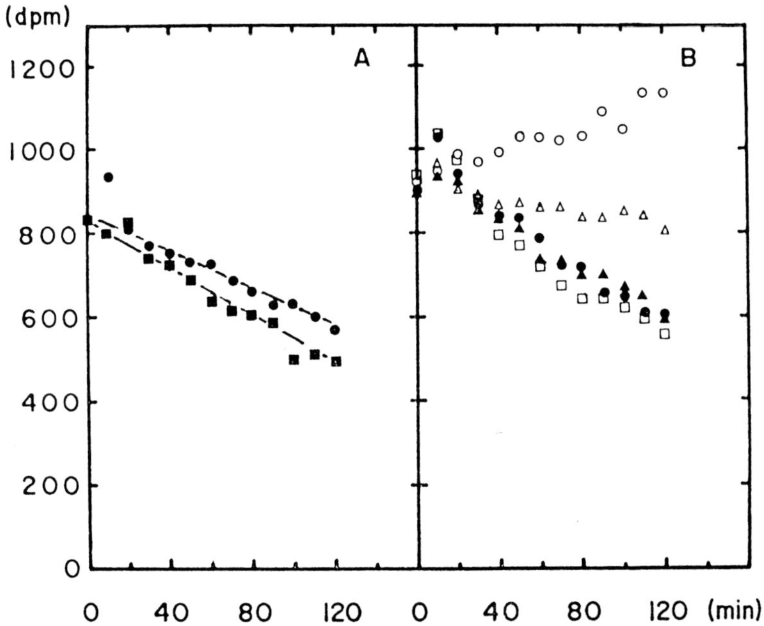

Figure 1. A. Comparison of the absorption by two jejunal segments. (■) Upper segment. (●) Lower segment.
B. Saturation of the absorption of RF. (□) ^{14}C-RF, 140nM + RF, (▲) 100 nM, (●) 1 μM, (Δ) 10μM or (O) 100μM.

Table 1. Inhibition of the absorption of ^{14}C-RF (140 nM) by flavin analogues.

Flavins	Substituents at Positions; 3	8	10	Concentration (μM) 1	10	100	1000
RF	H	CH_3	$CH_2(CHOH)_4H$	-	±	+	
5'-dRF	H	CH_3	$CH_2(CHOH)_3CH_3$	-	±	+	
2'-dRF	H	CH_3	$(CH_2)_2(CHOH)_3H$	-	±	+	
DARF	H	CH_3	$CH_2(CHO·CHO)_2H$	-	+		
LF	H	CH_3	CH_3	-	±	+	
MAF	H	$NHCH_3$	$CH_2(CHOH)_4H$	-	+		
RoF	H	$N(CH_3)_2$	$CH_2(CHOH)_4H$	-	+		
DEF	H	$N(C_2H_5)_2$	$CH_2(CHOH)_4H$	-	-	+	
HF	H	OH	$CH_2(CHOH)_4H$	-	-	±	+
3-MRF	CH_3	CH_3	$CH_2(CHOH)_4H$		-	-	+
3-IBRF	$CH_2CH(CH_3)_2$	CH_3	$CH_2(CHOH)_4H$			-	-
3-CMRF	CH_2COOH	CH_3	$CH_2(CHOH)_4H$		-	-	-
LC	H	CH_3	H		-	-	-

a profile of the carrier: Non-labeled RF was replaced with RF analogues and results were summarized in Table 1. In the table, the plus sign (+) means complete inhibition of the absorption of ^{14}C-RF. These results suggest that the transport of RF involves solely carrier mechanism and not simple diffusion. Comparing the extent of the inhibition by each analogue, we can image the flavin binding site of the carrier. The ribityl group has no meaning for the carrier, as all 10-N analogues, even LF, inhibited equally to non-labeled RF. This might be consistent with the result of Hegazy and Schwenk, no inhibition of RF uptake by ribose (11), and with the result of Daniel et al, reduction of the transport rate by LF (3). Isoalloxazine ring has great meaning for the carrier, as a slightly modified ring, alloxazine ring, didn't inhibit and HF, which has a charge at pH 7, inhibited weaker. Space surrounding 8-position of isoallxazine ring is not very wide, as DEF inhibited weaker. Space surrounding 3-position is very narrow, as 3 substituted analogues inhibited very weakly or didn't.

References

1. Spencer, R.P. & Zamcheck, N.: Gastroenterology 40, 794-797, (1961)
2. Turner, J.B. & Hughes, D.E.: Q.J.Exp.Physiol. 47, 107-123, (1962)
3. Akiyama, T., Selhub, J. & Rosenberg, I. H.: J. Nutr. 112, 263-268, (1982)
4. Daniel, H., Wille, U., & Rehner, G.: J. Nutr. 113, 636-643, (1983)
5. Rivier, D. A.: Experimentia 29, 1443-1446, (1973)
6. Jusko, W.J. & Levy, G.: in Riboflavin, Rivlin R. ed, pp. 99-152, Plenum Press, New York (1975)
7. Kasai, S., Kubo, Y., Yamanaka, S., Hirota, T., Sato, H., Tsuzukida, Y. & Matsui, K.: J. Nutr. Sci. Vitaminol. 339-350, (1978)
8. Kasai, S., Yamanaka, S., Wang, S-C. & Matsui, K.: J. Nutr. Sci. Vitaminol. 25, 289-298, (1979)
9. Campbell, J.A. & Morrison, A.B.: Amer. J. Clin. Nutr. 12, 162-169, (1963)
10. Levy, G. & Jusko, W.J: J. Pharm. Sci. 55, 285-289, (1966)
11. Hegazy, E. & Schwenk, M.: J. Nutr. 113, 1702-1707, (1983)

TRANSFER OF HYDROGEN ATOMS FROM PENTOSE PHOSPHATE TO THE XYLENE RING OF RIBOFLAVIN.

Keller, P.J., Floss, H.G.
Department of Chemistry, The Ohio State University
Columbus, Ohio, USA

Le Van, Q., Neuberger, G., Nielsen, P., Bacher, A.
Lehrstuhl für Organische Chemie und Biochemie,
Technische Universität München
D-8046 Garching, FRG

The xylene ring of riboflavin (III) is formed by a dismutation of two molecules of 6,7-dimethyl-8-ribityllumazine (II). Formally, this is a contribution of two biogenetically identical 4-carbon units consisting of carbon atoms 6α, 6, 7 and 7α of the lumazine II to riboflavin. This reaction is catalyzed by the enzyme riboflavin synthase which has been studied in considerable detail (1). The ultimate origin of the 4-carbon unit is less well understood. Lumazine II forms from a pyrimidine precursor (I) by condensation with a compound of unknown structure which contributes the nascent 4-carbon unit. Strong evidence has accumulated from *in vivo* studies with ^{13}C-labeled precursors (2) and from *in vitro* studies (3,4) that all four carbon atoms in question are derived from a pentose phosphate or a pentulose phosphate by elimination of C-4 of the sugar. However, neither the structures of intermediates nor the mechanism of the C-4 elimination is known, i.e., the transfer of carbon atoms is known only in a rather formal sense.

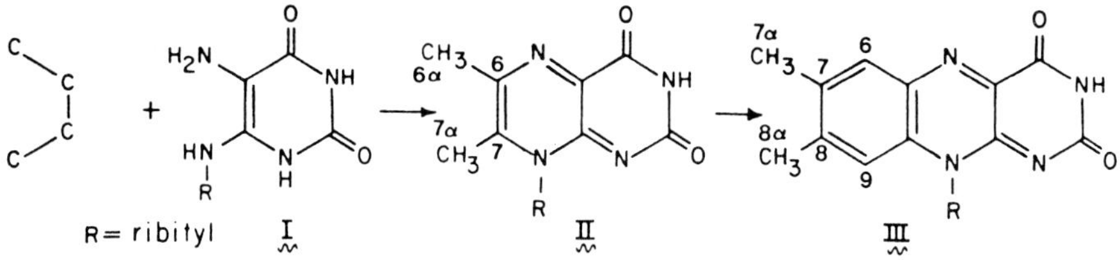

Flavins and Flavoproteins

Results

We have approached the question of the transfer of hydrogen atoms from the pentose/pentulose precursor into riboflavin by in vitro and in vivo studies. It has been shown that riboflavin can be formed from I or its 5'-phosphate plus ribose phosphate or ribulose phosphate in the presence of cell extract from *Candida guilliermondii* (3,4). It was also demonstrated by isotope labeling studies that C-4 of the pentose was not incorporated into the product riboflavin, in accord with our earlier in vivo results (4). In the present work the transfer of tritium from C-1, C-3, C-4 and C-5 of ribulose-5-phosphate to III was examined. Results are shown in the Table. Tritium transfer was rather efficient from C-1 and C-5 of the sugar phosphate, but little or no incorporation of label was detected for [3-^{3}H] or [4-^{3}H]ribulose phosphate. However, the absolute incorporation rates suggest that a substantial amount of the label was washed out in the transfer of tritium from C-1 of the carbohydrate. This was confirmed by double label experiments using mixtures of ribulose phosphate samples appropriately labeled with ^{3}H and ^{14}C, respectively (data not shown).

Table

Ribulose 5-phosphate	dpm/nmol	Incorporation (%)
[1-^{3}H]	7151	33
[3-^{3}H]	1567	3.1
[4-^{3}H]	3686	0.1
[5-^{3}H]	2520	27

Incorporation of tritium from ribulose-5-phosphate into the isoalloxazine moiety of riboflavin by *C. guilliermondii* cell extract. Assay mixtures contained ribulose-5-phosphate, pyrimidine I, Mg^{++}, and protein.

With the above results in hand, it was of interest to determine whether only one or both of the protons at C-1 and C-5 of a pentulose are retained in the methyl groups of riboflavin. The observed loss of tritium could be due to (a) stereospecific elimination of one of two hydrogen atoms bound to C-1 or C-5 of the precursor pentulose phosphate or, (b) a random washout mechanism. We sought to obtain more detailed information by supplying 2R-[1-$^2H_2$]glycerol to the flavinogenic ascomycete *Ashbya gossypii*. Primary metabolism through the glycolytic and pentose phosphate pathways could yield pentulose phosphates labeled with deuterium at position 5, and to a lesser extent at position 1 (for a discussion of the metabolic sequence, see Ref. 2).

Riboflavin was isolated from the culture fluid, and sites of labeling in the molecule were determined by several independent NMR techniques. Deuterium was shown to be present at C-7α and C-5' by ^{2}H-NMR, however possible C-8α labeling was masked by the solvent signal.

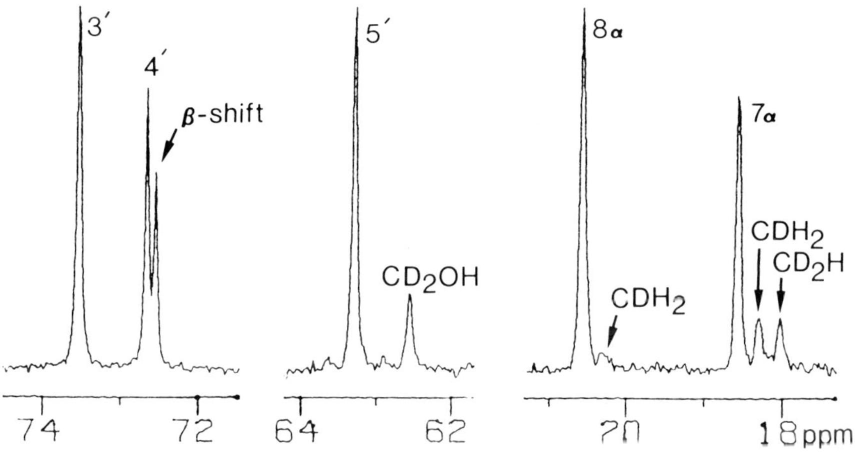

SCHEME. Portions of the ^{13}C-{^{1}H,^{2}H}-NMR spectrum of riboflavin from 2R-[1-$^2H_2$]glycerol

Further information was obtained by observation of the ^{13}C-NMR spectrum under the condition of simultaneous ^{1}H and ^{2}H broadband decoupling. Here the presence of deuterium is indicated by an isotope induced chemical shift change for a carbon bearing deuterium. It was found (Scheme) that the 7α methyl group gave three signals indicating the presence of CH_3, CDH_2 and CD_2H species. In contrast, only CH_3 and a minor amount of CDH_2 was detected for C-8α. The C-5' signal as anticipated displayed the presence of CH_2OH and CD_2OH but no CDHOH species. A quantitative measure of 5' labeling was obtained from the two bond isotope shift effect (also denoted as a secondary or β shift) observed at C-4'. Integration of the total C-4' signal indicated 40% CD_2OH and 60% CH_2OH at C-5' of the molecule. Quantitative information for the methyl groups was obtained through the observation of β-shifts for the methyl signals in the ^2H broadband decoupled ^1H-NMR spectrum. Integration yielded the following values: C-7α = CH_3 (64%), CDH_2 (15%), CD_2H (21%); C-8α = CH_3 (96%), CDH_2 (4%).

It follows that both protons bound to C-5 of the pentulose can be transferred to C-7α of riboflavin, although some washout yielding CDH_2 occurs in agreement with the tritium results. The transfer of only one hydrogen atom from C-1 of the pentulose phosphate was observed, however, the evidence for C-8α is as yet inconclusive due to the low incorporation rate at this position.

1. Plaut, G.W.E. and Beach, R.L. in Flavins and Flavoproteins, T.P. Singer, Ed., (1976) Elseveir Scientific, Amsterdam, pp. 737-746.
2. Bacher, A., Le Van, Q., Keller, P.J., Floss, H.G.: J. Biol. Chem. 258, 13431-13437 (1983).
3. Logvinenko, E.M., Shavlovskii, G.M., Zakal'skii, A.E., and Zakhodylo, I.V.: Biokhimiya 47, 931-936 (1982).
4. Nielsen, P., Neuberger, G., Floss, H.G., Bacher, A.: Biochem. Biophys. Res. Commun.: 118, 814-820 (1984).

BIOSYNTHESIS OF RIBOFLAVIN.

Enzymatic formation of 6,7-dimethyl-8-ribityllumazine

A. Bacher*, P. Nielsen*, G. Neuberger*, and H. G. Floss#

*Lehrstuhl für Organische Chemie und Biochemie, Technische Universität München, 8046 Garching, West Germany, and #Department of Chemistry, The Ohio State University, Columbus, Ohio 43210

Riboflavin (**3**) is ultimately formed by a dismutation of 6,7-dimethyl-8-ribityllumazine (**2**) catalyzed by riboflavin synthase. The formation of **2** from the pyrimidine precursor **1** requires the addition of 4 carbon atoms whose origin has remained elusive (for review see Ref. 1). We have shown that the lumazine atoms arbitrarily designated as 1^*-4^* are biosynthetically equivalent to carbon atoms 1, 2, 3, and 5 of pentoses, whereas C-4 of the pentose precursor is lost by an intramolecular skeletal rearrangement (1, 2). Logvinenko et al. have recently reported on the formation of **2** from an enzymatically generated pyrimidine precursor and ribose phosphate by cell extract from *Candida guilliermondii* (3). We have studied this reaction in closer detail.

Results

The pyrimidines **1** and **4** can both serve as precursors for the formation of riboflavin by cell extract of *C. guilliermondii*. Various pentose phosphates and ribulose phosphate can serve as cosubstrate. Hexose phosphates are poorly utilized, and unphosphorylated carbohydrates are no substrates (4).

Flavins and Flavoproteins

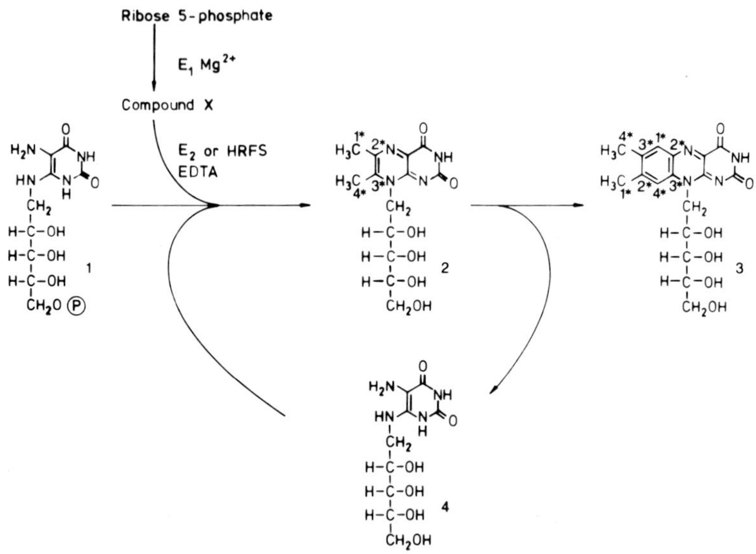

Fig. 1
Biosynthesis of riboflavin. E_1, cell extract from *C. guilliermondii* class 1 mutants; E_2, cell extract from *C. guilliermondii* class 2 mutants; HRFS, heavy riboflavin synthase from *B. subtilis*

The incorporation of various ^{14}C-ribulose phosphate samples into riboflavin in the cell-free system is shown in Table I. The incorporation is well in line with the in vivo experiments, i. e. carbon atoms 1, 3, and 5, but not 4 are transferred to riboflavin from the pentulose phosphate precursor.

Label from [1-^{13}C]ribose 5-phosphate was exclusively transferred to the positions marked as 1^* of the riboflavin molecule as shown by ^{13}C-NMR analysis of the product in agreement with earlier findings *in vivo* (1). Isotope from [1'-^{14}C]5-amino-6-ribitylamino-2,4(1H,3H)-pyrimidinedione (**4**) was incorporated into the ribityl side chain of **2** and **3**, but not into the heterocyclic moiety.

Table I
Incorporation of isotope from ribulose 5-phosphate into the isoalloxazine moiety of riboflavin.

Ribulose 5-phosphate	Relative incorporation (%)
$[1-^{14}C]$	125
$[3,5-^{14}C]$	109
$[4-^{14}C]$	3.1
$[5-^{14}C]$	113

Genetic data had suggested the involvement of at least 2 enzymes in the formation of **2** (4, 5). We have therefore isolated and studied mutants of *C. guilliermondii* which are unable to form **2** from **1**. Cell extracts of these mutants were unable to form **2** under standard conditions. However, enzyme activity could be restored by pairwise combination of certain mutant cell extracts. Thus the mutants could be divided into two groups which show *in vitro* complementation.

It was subsequently shown that mutants of class 1 produce an intermediate X from ribose phosphate by an Mg^{II} dependent reaction. X can be converted to riboflavin by cell extract from class 2 mutants in the presence of the pyrimidines **1** or **4**. This reaction is not inhibited by EDTA.

X can also be converted to riboflavin by pure heavy riboflavin synthase from *Bacillus subtilis* in the presence of the pyrimidines **1** or **4**. It follows that this enzyme is a bifunctional complex of riboflavin synthase and lumazine synthase as suggested earlier on the basis of genetic evidence (6).

We have shown that the intermediate X is destroyed by phosphatase. We conclude that compound X is a phoshoric acid ester. Work on the structure of this material is in progress.

We thank the Deutsche Forschungsgemeinschaft and the National Science Foundation for support.

References

1. Bacher, A., Le Van, Q., Keller, P. J., Floss, H. G.: J. Biol. Chem. 258 13431-13437 (1983)

2. Floss, H. G., Le Van, Q., Keller, P. J., Bacher, A: J. Amer. Chem. Soc. 105, 2493-2494 (1983)

3. Logvinenko, E. M., Shavlovskii, G. M., Zakal'skii, A. E., Zakhodylo, I. V.: Biokhimiya 47, 931-936 (1982)

4. Nielsen, P., Neuberger, G., Floss, H. G., Bacher, A.: Biochem. Biophys. Res. Comm. 118, 814-820 (1984)

5. Oltmanns, O., Bacher, A., Lingens, F., Zimmermann, F. K.: Molec. Gen. Genetics 105, 306-313 (1969)

6. Shavlovskii, G. M., Sibirnyi, A. A., Kshanovskaya, B. V., Koltun, L. V., Logvinenko, E. M.: Genetika 15, 1561-1568 (1979)

7. Bacher, A., Mailänder, B.: J. Bacteriol. 134, 476-482 (1978)

CELL-FREE BIOSYNTHESIS OF LIPOAMIDE DEHYDROGENASE

Kichiko Koike, Akihiko Tsuji, Yoshishige Urata, Matsuko Moriyasu, Masahiko Koike

Department of Pathological Biochemistry, Atomic Disease Institute, Nagasaki University School of Medicine, Nagasaki, Japan

Introduction

Lipoamide dehydrogenase (NADH-lipoamide oxidoreductase, EC 1. 6. 4. 3) is an essential component of the pyruvate dehydrogenase and 2-oxoglutarate dehydrogenase multienzyme complexes (1). Previously, we demonstrated that the two forms of the lipoamide dehydrogenase separated from the pig heart mitochondrial multienzyme complexes are functionally interchangeable, and identical in molecular weight, dimeric structure, amino acid composition and immunochemical properties (2). These results suggest that the mammalian lipoamide dehydrogenase is also the product of a single structural gene just like Escherichia coli.(3). We are interested in the regulation of biosynthesis of this common component enzyme of two species of the multienzyme complexes. In this report, we describe the purification of the lipoamide dehydrogenase mRNA by polysome immunoadsorption method, and the cell-free biosynthesis of the lipoamide dehydrogenase in the reticulocyte lysate system.

Experimental Procedures

Enrichment of Lipoamide Dehydrogenase mRNA---Total polysomes isolated from the pig liver by the Mg-precipitation method (4), were incubated with monospecific anti-lipoamide dehydrogenase IgG (2.5mg/1,000A_{260}) purified by antigen-Sepharose

Flavins and Flavoproteins

chromatography (5), at 0 °C for 3h. The antibodies bound to polysomes were adsorbed to the protein A-Sepharose column (6). After washing, enriched polysomal RNA was eluted, deproteinized by protease K treatment and poly(A) RNA was isolated by oligo(dT)-cellulose chromatography.

Cell-Free Translation and Immunoprecipitation---Staphylococcal nuclease-treated rabbit reticulocyte lysate was prepared as described by Pelham and Jackson (7), and in vitro translation was carried out by the method of Mori et al. (8). Incubation was carried out at 25 °C for 90 min. Immunoprecipitation of the translation product was performed by using anti-lipoamide dehydrogenase IgG and protein A-coated Staphylococcus aureus cells (9). SDS-polyacrylamide gel electrophoresis was carried out by the method of Laemmli (10) by using 10% gel and fluorographed by using Amplifier (Amersham).

Results and Discussion

Lipoamide dehydrogenase mRNA was found in the pig liver at a very low level (i.e. less than 0.01% of the total cellular mRNA as measured by an in vitro translation assay). Therefore, it was necessary to concentrate by several enrichment procedures to obtain sufficient amounts of mRNA for translation or cDNA cloning. The lipoamide dehydrogenase mRNA was purified by immunoadsorption of polysomes isolated from the one month old pig liver. Translation product with the purified mRNA was precipitated by antibody. Analysis of the immunoprecipitate by SDS-gel electrophoresis revealed a 53K band and an additional 55K band which was co-migrated with the mature lipoamide dehydrogenase as shown in Fig. 1, lanes 1 and 3. These bands were not observed in the translation products with the crude liver RNA, however, they were observed in the products with the purified mRNA (lane 2). Furthermore these bands were not detected in the precipitates by using the control IgG from non-immune rabbit.

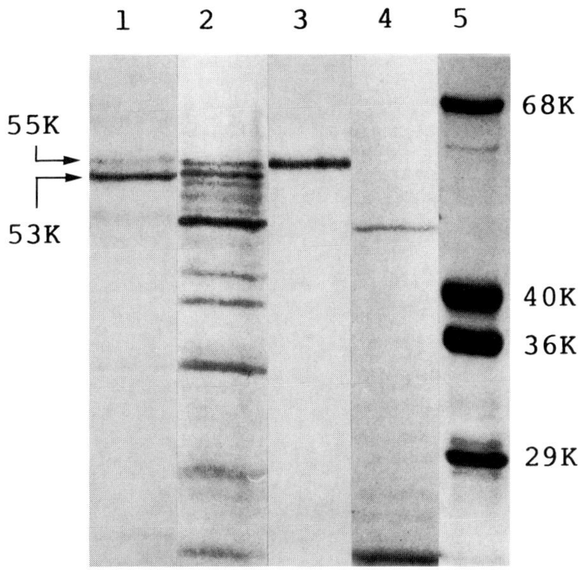

Fig. 1. Electropherograms of the lipoamide dehydrogenase subunits synthesized in vitro by the purified mRNA. The purified mRNA (0.5 µg) was translated at 25 °C for 90 min in a final volume of 10 µl of the reticulocyte lysate system containing 2 µCi of [^{35}S]methionine. Translation products were analyzed by the immunoprecipitation and SDS-gel electrophoresis. Lane 1 shows the immunoprecipitated products; lane 2, total products; lane 3, mature ^{14}C-methylated lipoamide dehydrogenase subunit; lane 4, products without exogenous mRNA; and lane 5, ^{14}C-methylated marker proteins (bovine serum albumin, aldolase, glyceraldehyde 3-phosphate dehydrogenase, and carbonic anhydrase). Arrows indicate the lipoamide dehydrogense subunits (55K & 53K) synthesized.

Recently, it is thought that most of the mitochondrial protein coded by nuclear gene, are synthesized as a larger precursor on cytoplasmic polysomes. However, the size of the lipoamide dehydrogenase synthesized in vitro was not larger than the mature form. This could be due to a number of factors such as premature termination of translation, cleavage of the mRNA either during isolation or translation by nuclease, partial degradation of the primary translation product and an abnormal behavior of the product on SDS-gel electrophoresis.

The comparison of radioactivities of the immunoprecipitated translation products by using both crude RNA and the purified mRNA indicated about 50 to 100-fold enrichment. We believe that this enriched lipoamide dehydrogenase mRNA would be sufficient to prepare corresponding cDNA for cloning or screening purposes.

Ackowledgement

This work was supported in parts by the grants from the Ministry of Education, Science and Culture of Japan and from Vitamin B. Research Committee of Japan.

Reference

1. Koike, M., Koike, K.: Advances in Biophysics, University of Tokyo Press, Vol.9, pp.187-227 (1976)
2. Sakurai, Y., Fukuyoshi, Y., Hamada, M., Hayakawa, T., Koike, M.: J. Biol. Chem. 245, 4453-4462 (1970)
3. Guest, J.R., Creaghan, I.T.: J. Gen. Microbiol. 81, 237-245 (1974)
4. MacGillivray, R.T.A., Chung, D.W., Davie, E.W.: Eur. J. Biochem. 98, 477-485 (1979)
5. Koike, M., Koike, K., Tanabe, S., Ichinose, H.: Flavins and Fravoproteins, Yagi, K. and Yamano, T. (eds.) Tokyo : Japan Scientific Societies Press, pp. 551-555 (1980)
6. Kraus, J.P., Rosenberg, L.E.:Proc. Natl. Acad. Sci. USA 79, 4015-4019 (1982)
7. Pelham, H.R.B., Jackson, R.J.: Eur. J. Biochem. 67, 247-257 (1976)
8. Mori, M., Miura, S., Tatibana, M., Cohen, P.P.: Proc. Natl. Acad. Sci. USA 76, 5071-5075 (1979)
9. Kessler, S.W.: J. Immunol. 117, 1482-1490 (1976)
10. Laemmli, U. K.:Nature (London) 227, 680-685 (1970)

COVALENTLY ^{14}C-RIBOFLAVIN-LABELED PROTEINS IN ARTHROBACTER OXIDANS AND THEIR POSSIBLE RELATIONSHIP TO 6-HYDROXY-D-NICOTINE OXIDASE

Karl Decker, Henry Reeves, Roderich Brandsch

Biochemisches Institut der Universität Freiburg
D-7800 Freiburg i.Br., F.R.G.

Introduction

6-Hydroxy-D-nicotine oxidase (6-HDNO) (EC 1.5.3.6) is an inducible flavoprotein of A. oxidans (1). It is involved in the degradation of D-nicotine and catalyzes the opening of the pyrrolidine ring in a stereospecific manner. The cofactor FAD is bound to the monomeric polypeptide chain of 6-HDNO (M_r = 47 500) by an 8α-N_τ (his) linkage (2). Antisera raised against highly purified 6-HDNO did not crossreact with 6-hydroxy-L-nicotine oxidase or with any other protein of A. oxidans. Using a riboflavin-requiring mutant (rf^-) of A. oxidans (3) the dependence of the synthesis of this enzyme on the intracellular concentration of freely dissociable FAD was shown (4). Earlier studies on the cell-free synthesis of 6-HDNO produced evidence for cotranslational addition of FAD (5).

Results

Growth of rf^- mutant cells of *A.oxidans* in the presence of ^{14}C-riboflavin (15 µCi/ml) but in the absence of the inducer D-nicotine revealed one labeled protein of M_r = 60000 in the soluble fraction (Fig. 1). Succinate dehydrogenase of Arthrobacter is a tightly membrane-associated protein that does not yield radioactivity to the supernatant of a sonicated cell homogenate. The 60 kD covalent flavoprotein was not observed in extracts of other *Arthrobacter* species. It could not be precipitated from the soluble extract by antisera raised against 6-HDNO and it did not reveal enzymatic activity with 6-hydroxy-D-nicotine. This riboflavin-containing polypeptide is considered not to belong to the set of proteins induced in *A. oxidans* by nicotine.

After growth in the presence of D-nicotine, two additional ^{14}C-riboflavin-labeled polypeptides were seen in SDS PAGE of cell-free extracts (Fig. 1). Both could be precipitated by the anti-6-HDNO antisera bound to protein A-Sepharose and both could be displaced from the immunocomplexes by authentic unlabeled 6-HDNO. One of them showed a M_r of about 48000 and appeared to be identical with 6-HDNO. The M_r of the other was about 72000. Both of them displayed enzymatic activity when PAGE was run under non-denaturing conditions and the gel incubated with 6-hydroxy-D-nicotine, peroxidase and dianisidine (Fig. 2). The 72 kD polypeptide was also observed in SDS PAGE of ^{35}S-methionine-labeled cells. The intensity of this band declined in pulse-chase experiments while that of the 48 kD band increased.

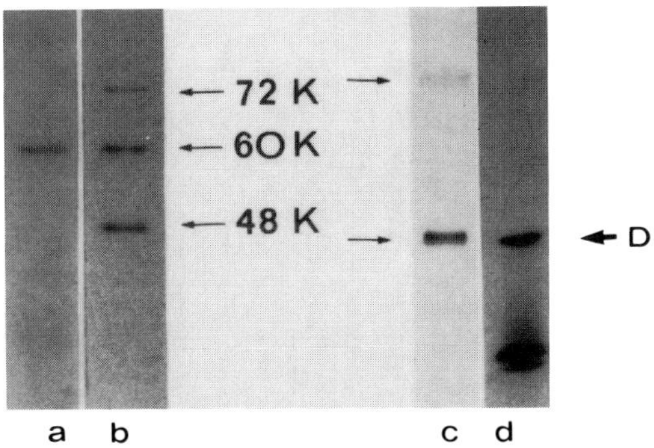

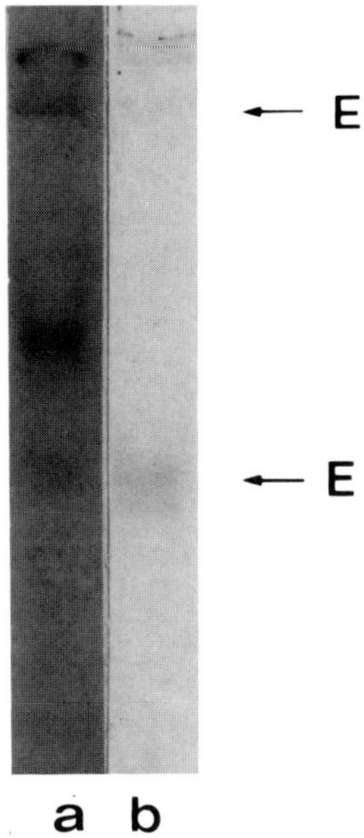

Fig. 1 Flavinylated polypeptides of A. oxidans

Rf^- cells were grown in the presence of ^{14}C-riboflavin. Soluble proteins were analyzed by SDS PAGE (7.5 %) and fluorography.
Lane a: acetone-precipitated protein from uninduced cells;
b: same from induced cells;
c: proteins precipitated by 6-HDNO-specific antiserum from 18 h-induced cells;
d: same as c, but 72 h-induced cells.
D marks position of 6-HDNO.

Fig. 2 Enzymatic activity of flavinylated polypeptides

The soluble proteins of ^{14}C-riboflavin-labeled, induced cells were analyzed by PAGE (7.5 %) under non-denaturing conditions.
Lane a: fluorogram of the dried gel;
b: enzymatic stain of the same gel as in a.
E marks position of 6-HDNO activity.

The extracts of cells kept in culture beyond the active growth phase of the organism revealed a number of ^{14}C-riboflavin- and ^{35}S-methionine-labeled polypeptides of decreasing molecular weight. The most prominent of these was a piece of about 18 kD.The patterns of labeled proteins observed after extended periods of incubation seem to indicate a progressive proteolytic degradation that is most conspicuous with 6-HDNO, but is also seen with the remaining 72 kD species.

The data presented here strongly suggest that the 72 kD polypeptide is a precursor of 6-HDNO: besides 6-HDNO itself, it is the only soluble protein that is induced by D-nicotine and contains covalently bound riboflavin. It crossreacts with specific anti-6-HDNO antisera and displays enzymatic 6-HDNO activity. Its decrease with time precedes that of 6-HDNO.

References

1. Decker, K.: Trends in Biochem. Sciences 1, 184-185 (1976).

2. Möhler, H., Brühmüller, M., Decker, K.: Europ. J. Biochem. 29, 152-155 (1972).

3. Hamm, H.-H., Decker, K.: Arch. Microbiol. 119, 65-70 (1978).

4. Hinkkanen, A., Decker, K.: In Flavins and Flavoproteins (Massey, V., Williams, H.C., eds.) pp. 478-482, Elsevier-North Holland Biomedical Press, New York. Amsterdam 1982.

5. Hamm, H.-H., Decker, K.: Europ. J. Biochem. 92, 449-454 (1978).

ELECTRON IMMUNOCHEMICAL LOCALIZATION OF 6-HYDROXY-NICOTINE OXIDASES IN ARTHROBACTER OXIDANS

James R. Swafford and Henry C. Reeves

Department of Botany and Microbiology
Arizona State University
Tempe, Arizona 85287, U.S.A.

Roderich Brandsch and Karl Decker

Biochemisches Institut der Albert-Ludwigs-Universität
7800 Freiburg, i Br, Fed. Rep. Germany

Introduction

During growth in the presence of D,L-nicotine, *Arthrobacter oxidans* synthesizes the flavin enantiozymes, 6-hydroxy-L-nicotine oxidase (6-HLNO) and 6-hydroxy-D-nicotine oxidase (6-HDNO) (1). The 6-HLNO consists of two identical subunits with two FAD molecules non-covalently linked (2), while the 6-HDNO is a single protein chain containing covalently linked FAD (3). The induction of 6-HLNO occurs during the logarithmic phase of the growth cycle, while the cells are rod-form, and continues into the stationary phase (2). In contrast, the 6-HDNO is induced in the stationary phase of growth when the cells are in a coccoid-form (1, 4).

Although these two enzymes have been assumed to be localized in the cytoplasm of *Arthrobacter oxidans*, this has never been demonstrated. In the experiments to be described, the intracellular localization of 6-HLNO and 6-HDNO was determined by electron immunochemical techniques with ultrathin sections of the bacterium. The thin sections, which were obtained by ultracryomicrotomy, were incubated first with antiserum specific for the enzyme and then with a protein-A gold complex

Flavins and Flavoproteins

as previously described (5). The sections were subsequently examined by transmission electron microscopy.

Methods

Arthrobacter oxidans was grown in 150 ml flasks containing 10 ml of citrate medium with 0.1% D,L-nicotine at 31°C with vigorous aeration. Aliquots of the culture were harvested during the stationary phase, washed with 100 mM sodium phosphate buffer at pH 7.5 and then fixed with 0.1% glutaraldehyde for 30 min at 4°C. After fixation, the cells were washed again in the same buffer and then enrobed in 10% gelatin. Small pieces of this cell mass were infused with 1.6 M sucrose in 100 mM sodium phosphate buffer, pH 7.5, for 1 hr at 4°C, then mounted on copper specimen pegs and rapidly frozen in Freon-22 slush. Ultrathin frozen sections were obtained at -90°C on glass knives (6) with an ultramicrotome (Porter-Blum MT-2 B) equipped with a cryokit (DuPont).

Carbon-stabilized, support-filmed grids with the sections adhering to them were treated with 2% gelatin prepared in 100 mM glycine phosphate buffer for 10 min to inhibit nonspecific antibody binding. The grids were floated on 10 µl of antibody, raised in rabbits against either the 6-HLNO or the 6-HDNO, for 10 min at 23°C. Control sections were incubated in an identical manner with purified IgG obtained from preimmune serum. After being thoroughly rinsed, each grid was floated for 10 min at 23°C on 10 µl of protein A-gold particles (10 mg/ml; diameter 10 nm) diluted 1:10 with 5% bovine serum albumin in 100 mM sodium phosphate buffer. The immunocomplexes were washed and then stabilized with 2% glutaraldehyde, and sections were stained and embedded as described (6).

Results and Conclusions

Samples were examined with a transmission electron microscope at an acceleration of 100 KeV. Fig. 1A and 1B are experimental and control sections, respectively, examined for 6-HLNO. Fig. 2A and 2B are experimental and control sections, respectively, examined for 6-HDNO. The pattern of gold labelling clearly demonstrates that both the 6-HLNO and the 6-HDNO are localized in the cytoplasm of Arthrobacter oxidans.

Acknowledgments

These studies were supported in part by a grant from the National Science Foundation (PCM 7921714) to HCR and by grants to KD from the Deutsche Forschungegemeinshaft, Bonn, through SFB 206, and by the Fonds der Chemischen Industries, Frankfurt.

References

1. Gloger, M., Decker, K.: Z. Naturforsch 24b, 1016-1025 (1969).
2. Decker, K., Dai, V.D.: Eur. J. Biochem. 3, 132-138 (1967).
3. Möhler, H., Brühmüller, M., Decker, K.: Eur. J. Biochem. 29, 152-155 (1972).
4. Brühmüller, M., Schimz, A., Messmer, L., Decker, K.: J. Biol. Chem. 250, 7747-7751 (1975).
5. Swafford, J.R., Malloy, P.J., Reeves, H.C.: Science 221, 295-296 (1983).
6. Tokuyasu, K.T., Singer, S.J.: J. Cell Biol. 71, 894-898 (1976).

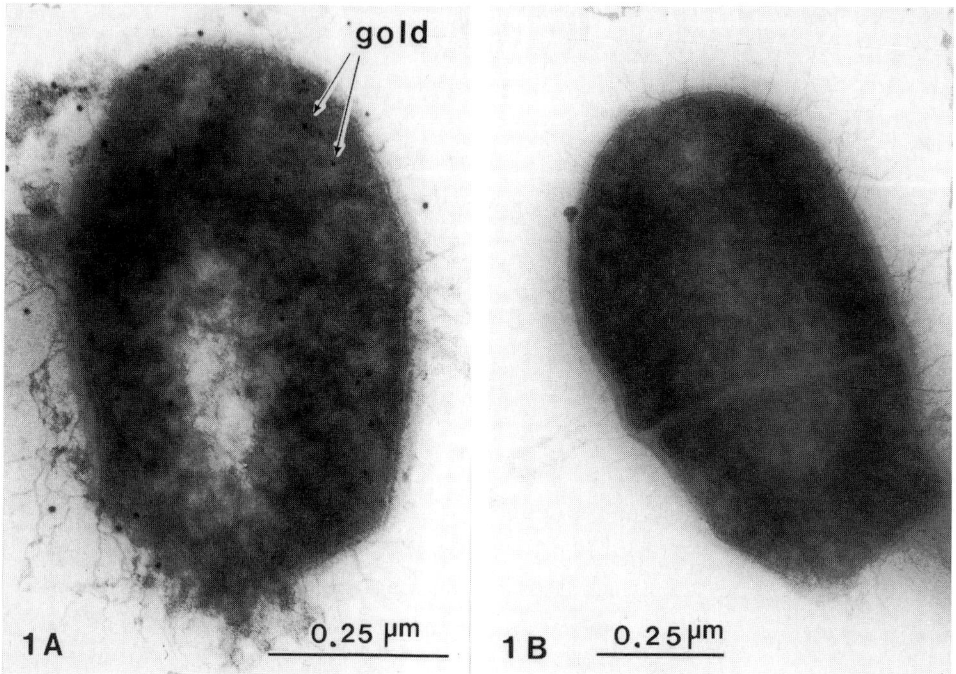
gold
1A
0.25 μm
1B
0.25 μm

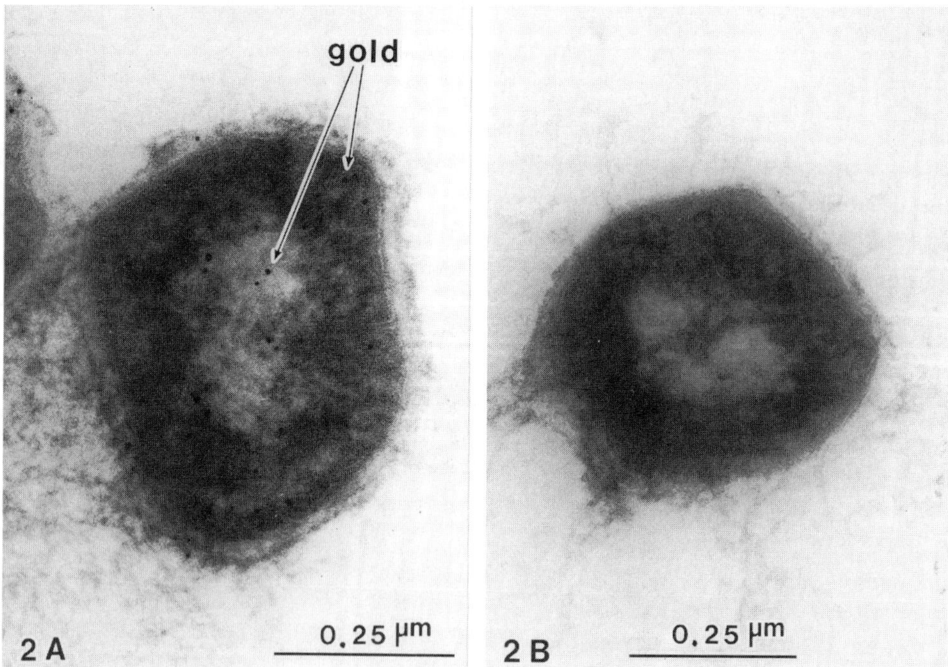
gold
2A
0.25 μm
2B
0.25 μm

CLONING IN *ESCHERICHIA COLI* OF THE 6-HYDROXY-D-NICOTINE OXIDASE GENE FROM *ARTHROBACTER OXIDANS*

Roderich Brandsch

Biochemisches Institut der Universität Freiburg
D-7800 Freiburg i.Br., F.R.G.

Introduction

The Gram-positive bacterium *Arthrobacter oxidans* has the ability to metabolize D,L-nicotine. One of the enzymes of this pathway is 6-hydroxy-D-nicotine oxidase (6-HDNO) which contains the cofactor FAD covalently linked to the apoprotein. We undertook to clone the gene for this enzyme on a multicopy plasmid in *E. coli* in order to make it available for structural analysis and "in vivo" and "in vitro" studies on the flavinylation process.

Methods

Isolation of pA01 from *A. oxidans* wild-type cells was done as described (1). Cloning in *E. coli* with the aid of multicopy plasmid vectors was done according to published methods (2). The spectrophotometric assay of 6-HDNO was performed according to Gloger and Decker (3) and the enzymatic stain according to Tsuge and Nakanishi (4).

Flavins and Flavoproteins

Results

The 6-HDNO structural gene was associated with the presence in A. oxidans of a catabolic plasmid, pA01, of approximately 160 kb (5). Subcloning of DNA fragments obtained by partial Sau3AI digestion of pA01 into the single BamHI site of pUK230 (6) resulted in two recombinant clones (pRB1) showing 6-HDNO activity. Both contained the same 12.5 kb DNA insert. Subcloning of the PstI generated fragments of this insert into the PstI site of pBR322 gave 6-HDNO-positive clones associated with the presence of a 5.7 kb insert (pRB11). The 6-HDNO activity conferred to E. coli cells by pRB1 and pRB11 as compared to that of A. oxidans is presented in Table 1.

Table 1. Specific activity of 6-HDNO in E. coli strain HB101 cells harbouring the vector plasmid pUK and pRB322 or the recombinant plasmids pRB1 and pRB11 as compared to A. oxidans cells.

Strains	Inducers (0.02 %)	Specific activity (mU/mg protein)
A. oxidans	none	1.5
	D,L-nicotine	85
E. coli K12 HB101/pUK	none	0
	D,L-nicotine	0
E. coli K12 HB101/pRB1	none	6 mU/mg
	D,L-nicotine	5 mU/mg
E. coli K12 HB101/pBR322	none	0
	D,L-nicotine	0
E. coli K12 HB101/pRB11	none	10 mU/mg
	D,L-nicotine	8 mU/mg

Expression of 6-HDNO in E coli was not inducible by nicotine and was not as copious as in A. oxidanns. The orientation of the 5.7 kb insert within the vector plasmid pBR322 had no effect on enzyme expression suggesting that the A. oxidans promotor is used for the transcription of the gene.

The cloned enzyme shows the same characteristics as 6-HDNO from A. oxidans: methylene blue as electron acceptor stimulates its activity; it is competitively inhibited by the stereoisomer 6-hydroxy-L-nicotine; it is not inactivated by dialysis against 4 M KBr as is the enantiozyme 6-hydroxy-L-nicotine oxidase with noncovalently attached FAD.

Immunologically it is indistinguishable from authentic 6-HDNO and shows the same specific color reaction in the enzyme assay (Fig. 1).

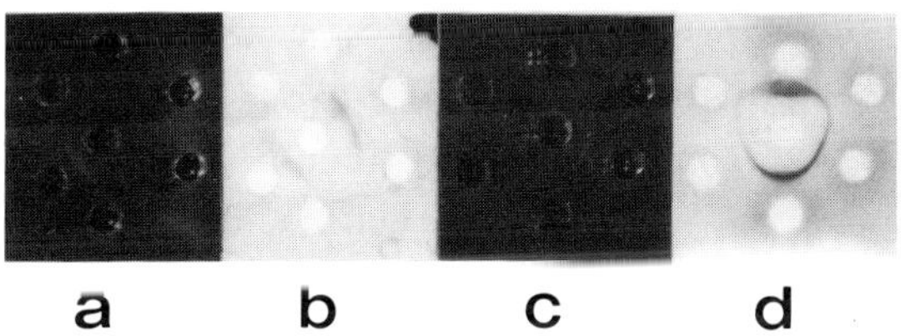

Fig. 1. Immunodiffusion assay with E.coli cell-free extracts and authentic 6-HDNO against 6-HDNO-specific antiserum.

a. Central well: 6-HDNO antiserum; outer wells, clockwise from top : E. coli HB101; E. coli HB101/pRB1; E. coli HB101/pUK230; E. coli HB101; E. coli HB101/pRB11; E. coli HB101/pRB322.
b. As a, stained specifically for 6-HDNO activity.
c. Central well: 6-HDNO antiserum; outer wells, clockwise from top: authentic HDNO; E. coli HB101/pRB1; E. coli HB101/pRB11; 1:10 dilution of HDNO; E. coli HB101/pRB1; E. coli HB101/pRB11.
d. As c, stained specifically for 6-HDNO activity.

Acknowledgement

I wish to thank Prof. Decker for his interest in this work. The technical assistance of Waltraud Faller is gratefully acknowledged.

References

1. Brandsch, R., Decker, K.: Arch. Microbiol. 138, 15-17 (1984).
2. Maniatis, T., Fritsch, E.F., Sambrook, J.: Molecular cloning. Cold Spring Harbor Laboratory (1982).
3. Gloger, M., Decker, K.: Z. Naturforsch. 24b, 1016-1025 (1969).
4. Tsuge, H., Nakanishi, Y.: in Methods in Enzymology, vol. 66, p. 344-350, Academic Press, New York, 1980.
5. Brandsch, R., Hinkkanen, A.E., Decker, K.: Arch. Microbiol. 132, 26-30 (1982).
6. Konen, M., Rüther, U., Müller-Hill, B.: Embo J. 1, 509-512 (1982).

HYDROXYLATION OF RIBOFLAVIN 7- AND 8-METHYL GROUPS IN MAMMALS

Kunio Yagi

Institute of Applied Biochemistry, Yagi Memorial Park, Mitake, Gifu 505-01, Japan

Nobuko Ohishi

Institute of Biochemistry, Faculty of Medicine, University of Nagoya, Nagoya 466, Japan

Hiroshi Ohkawa

Research Laboratories, Tokyo Tanabe Co., Ltd., Tokyo 115, Japan

By detailed survey of metabolites of flavins in the urine of rats and humans, new flavin derivatives were found. 7-Carboxylumichrome and 8-carboxylumichrome were found in rat urine; and 7α-hydroxyriboflavin and 8α-hydroxyriboflavin, in human urine. The latter two compounds were produced from riboflavin by the microsomal electron-transfer system of rat liver. These hydroxylation reactions are considered to be the first step in a newly discovered pathway of the metabolism of riboflavin in mammals.

Introduction

The general features of the absorption, metabolism, and excretion of riboflavin in mammals have been well elucidated. Flavins are absorbed as free riboflavin, which is converted to FMN and FAD as well as covalently bound flavins in cells. As for its excretion into the urine, it was believed that the major flavin excreted is free riboflavin, with minor ones being lumichrome, lumiflavin, and 10-hydroxyethylflavin (1,2). Recently,

Flavins and Flavoproteins

we have contributed further information on particular aspects of riboflavin metabolism in mammals by performing an extensive survey of metabolites in the urine of rats and humans and confirmed the occurrence of a new metabolic pathway initiated by hydroxylation of methyl groups of the isoalloxazine ring (3-6).

Metabolites Appearing in Rat Urine

Up to this time, many studies on the excretion of metabolites of riboflavin or its derivatives in rat urine have been made with radioactive riboflavin. We also examined extensively flavin derivatives in the urine of rats orally administered physiological amounts of [2-^{14}C]riboflavin (37.3 µg per rat of 140-160 g body weight) (4). After separation of chloroform-soluble compounds, flavins in the urine were extracted with phenol, transferred to the water layer, and subjected to paper chromatography (PPC) using n-butanol/glacial acetic acid/water (4:1:5, v/v, upper phase) as solvent. In a typical radiochromatogram two major radioactive peaks, I and II, were found; and these two spots exhibited greenish-yellow fluorescence under 365 nm-illumination. When the eluate of peak I was further subjected to PPC with another solvent system, benzene/n-butanol/methanol/water (1:2:1:1, v/v, upper phase), a spot of R_f 0.30, corresponding to riboflavin, and another unknown spot of R_f 0.12 were found. The latter spot was further separated into two radioactive spots by silica gel thin-layer chromatography (TLC) with a solvent system of n-butanol/glacial acetic acid/water (4:1:5, v/v, upper phase; R_f's of 0.49 and 0.46). After purification, each of these compounds, termed A and B, respectively, gave a single radioactive and fluorescent spot on TLC chromatograms with various solvent systems. From their chromatographic behavior, they did not correspond to any known flavin derivative.

To obtain large amounts of compounds A and B, the pooled urine of rats administered non-radioactive riboflavin was sub-

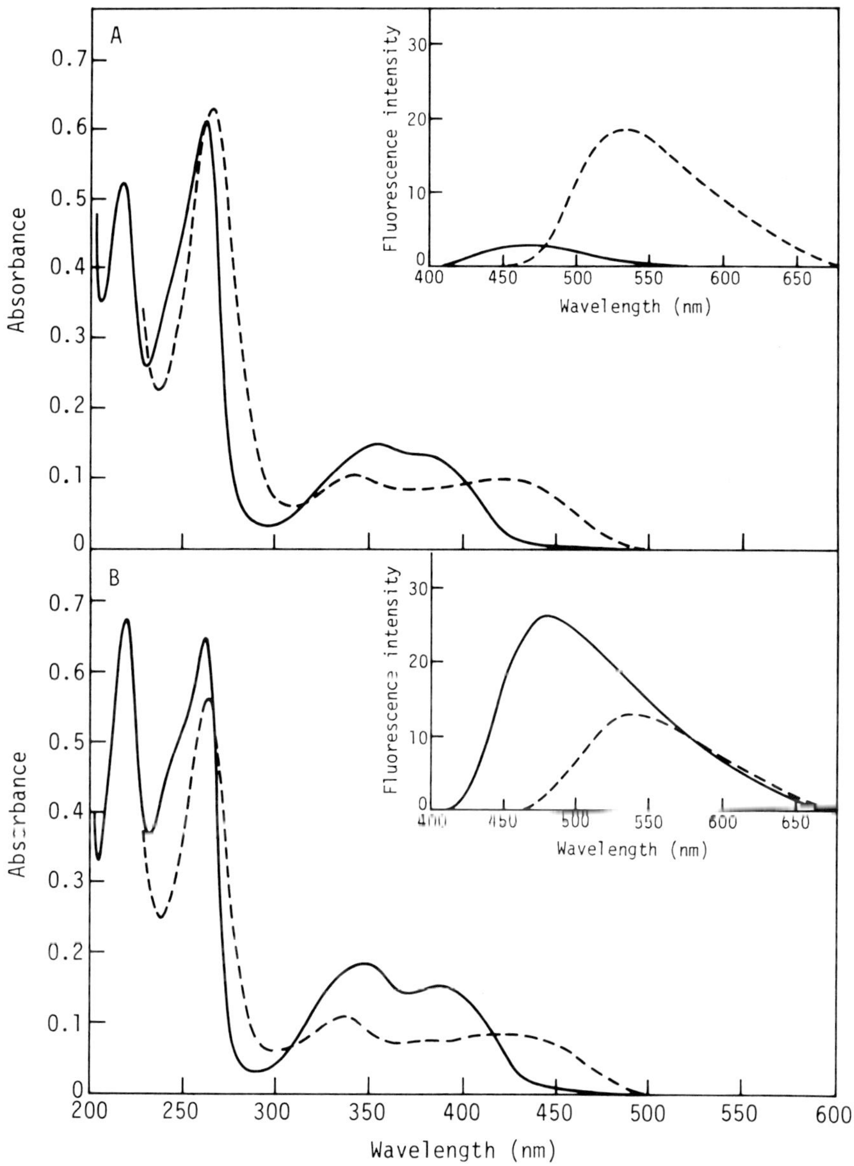

Figure 1. Absorption and fluorescence (inset) spectra of compounds A and B. Solid line, in 0.1 M sodium phosphate buffer (pH 7.0); dashed line, in 0.1 M NaOH. A, compound A; B, compound B.

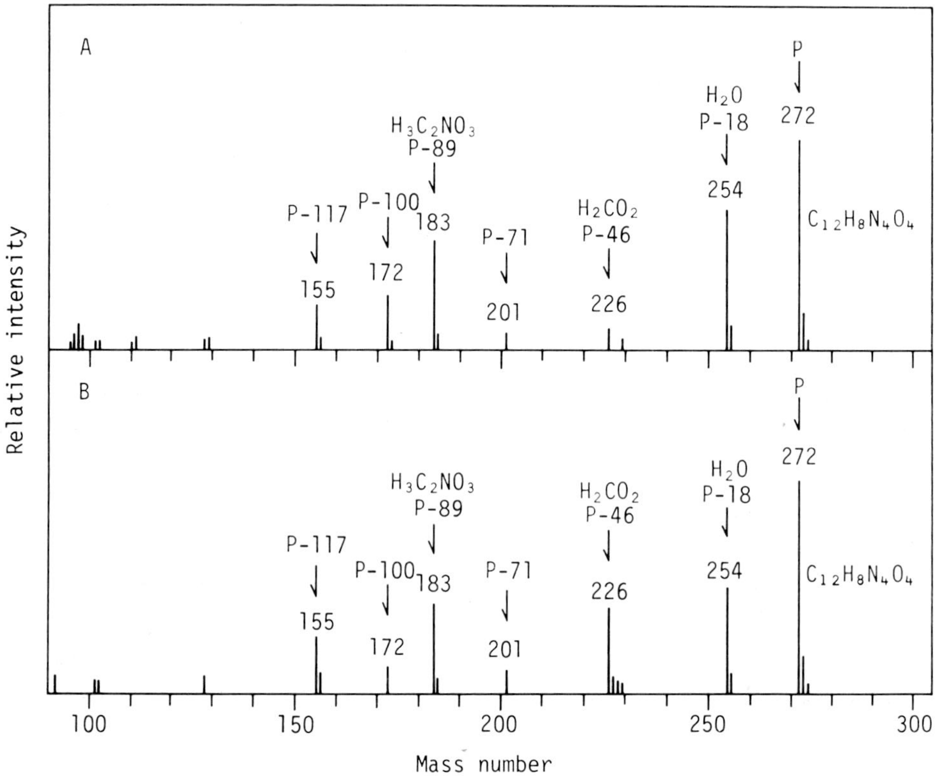

Figure 2. Mass spectra of compounds A and B. A, compound A; B, compound B.

jected to DE-32 column and paper chromatography after phenol extraction. The absorption and fluorescence spectra of the purified compounds A and B are shown in Figure 1A and B. The characteristic changes in these spectra provoked by changing the pH indicate the occurrence of a tautomeric structure as observed in lumichrome (7) and the possibility that these two compounds are derivatives of alloxazine.

In Figure 2, mass spectra of compounds A and B are shown, indicating that their molecular weight is m/e 272. By the analysis of fragment ions, it was confirmed that both compounds are alloxazine derivatives having a carboxyl group.

In ^{1}H-NMR spectra, compound A exhibited signals having chemical shift values of 7.76 and 7.94 ppm (aromatic protons) and 2.58 ppm (methyl protons); and compound B, those having 7.75 and 7.90 ppm (aromatic protons) and 2.52 ppm (methyl protons). The signal of methyl protons of compound B was in a slightly higher field than that of compound A. The occurrence of only one methyl group was confirmed in each compound, and proton resonance signals of the ribityl side chain observed in riboflavin and those of the methyl group attached to N(10) of lumiflavin were not detected.

From all these results as well as from those reported on alloxazine, we conclude that compounds A and B are 7-carboxylumichrome and 8-carboxylumichrome, respectively. Further, the chromatographic behavior and physico-chemical characteristics of compound B coincided completely with those of authentic 8-carboxylumichrome (4).

The substance in peak II was also purified to homogeneity and identified as riboflavinyl glucoside (8), suggesting the metabolic pathway of riboflavin to its glucoside in mammals. The enzyme which catalyzes the formation of this glucoside has already been found in rat liver (9) and pig liver (10).

Metabolites Appearing in Human Urine

Flavin derivatives in human urine were investigated by high-performance liquid chromatography (HPLC) using an ODS-type column (5), in which authentic samples including FAD, FMN, riboflavin, riboflavinyl glucoside, 10-hydroxyethylflavin, and lumiflavin could be separated completely. In a typical chromatographic pattern obtained from 15 μl of injected raw human urine, seven major peaks having retention times of 3.3, 4.2, 4.8, 5.9, 6.7, 9.5, and 10.4 min were detected. Among them, the peak at 9.5 min was identified as riboflavin; the peak at 10.4 min, as 10-hydroxyethylflavin; and the purified substance with a retention time of 3.3 min, as a derivative of 10-hydroxyethylflavin (5). Upon oral administration of 50 mg of riboflavin, the total intensity of fluorescence at 530 nm excited at 440 nm on HPLC in the urine increased, and this increase was mostly due to riboflavin and the substances with 4.2 and 4.8 min-retention times. The latter two substances were unknown in nature and detected in all of the raw urine tested.

Their structures were examined with samples purified from the

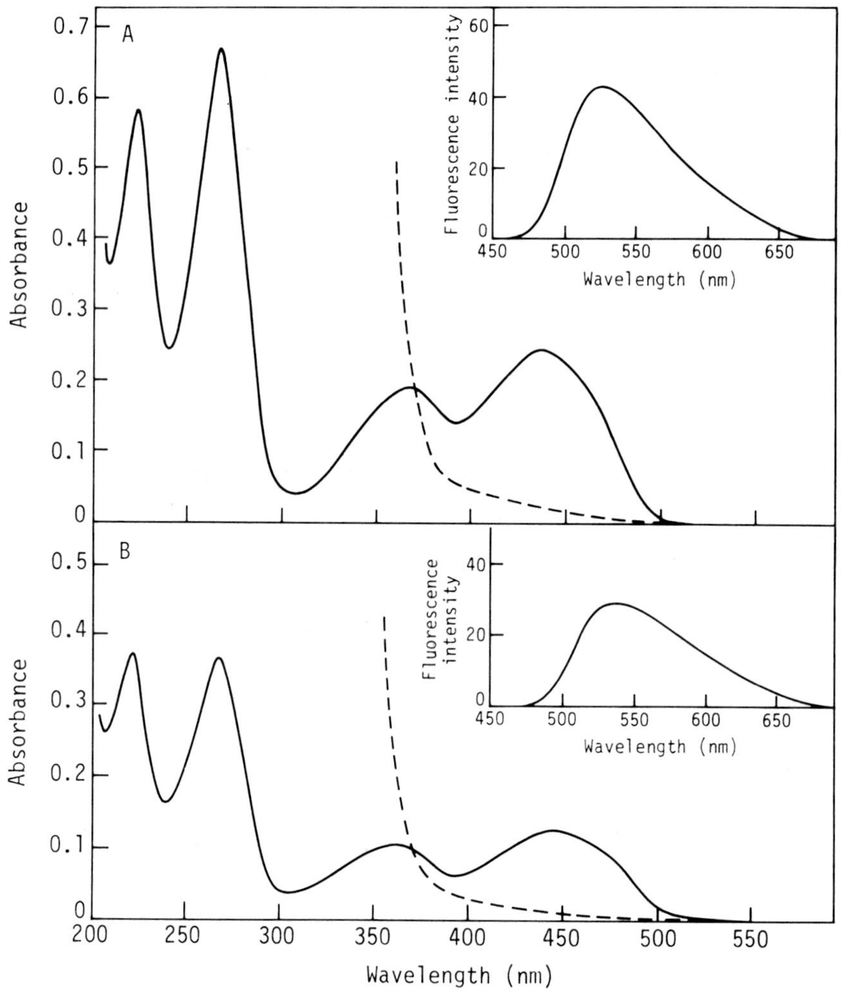

Figure 3. Absorption and fluorescence (inset) spectra of 4.2 min- and 4.8 min-compound. Solvent, 0.1 M sodium phosphate buffer (pH 7.0). Dashed line, after reduction with dithionite. A, 4.2 min-compound; B, 4.8 min-compound.

pooled urine obtained following the oral administration of 50 mg of riboflavin. A series of column chromatographic steps using DE32, cellulose, florisil, CM-Sephadex C-25, and Sephadex LH-20, together with PPC were employed for their purification (5).

Chromatographic behaviors of these two compounds were different from each other and from those of riboflavin, riboflavinyl glucoside, and lumiflavin as judged by silica gel TLC with various solvent systems. Their products of alkaline photolysis also differed from each other and were not identical with lumiflavin.

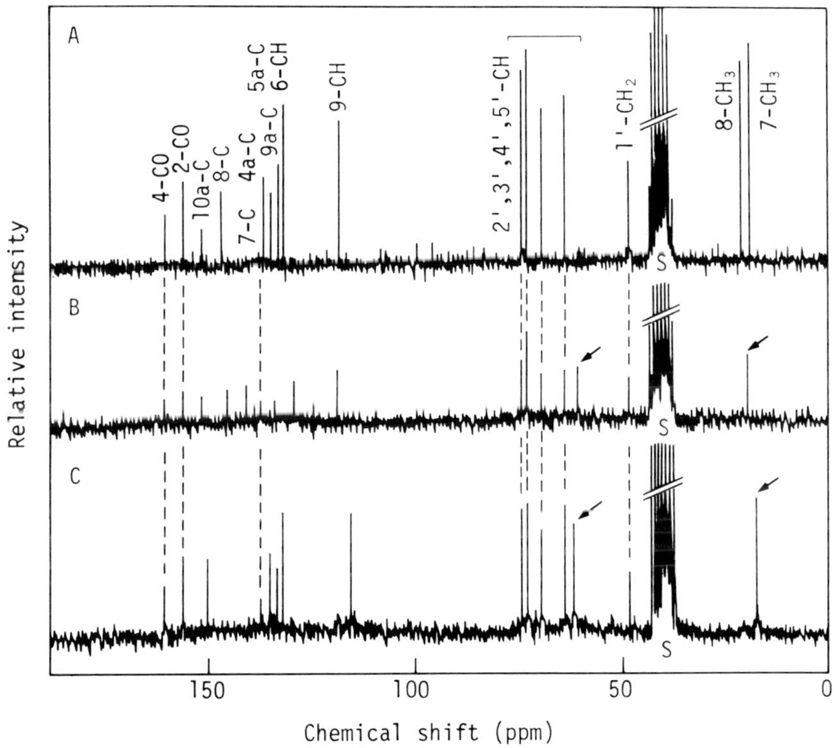

Figure 4. Proton-decoupled ^{13}C-NMR spectra of 4.2 min- and 4.8 min-compound. Arrows, in the higher and lower fields, the signals of the methyl and methine carbons, respectively. S, the signals due to dimethyl sulfoxide as solvent. A, riboflavin; B, 4.2 min-compound; C, 4.8 min-compound.

The absorption and fluorescence spectra of the compounds at 4.2 min and 4.8 min are shown in Figure 3A and B. One characteristic of the spectra of these compounds was the blue shift of the second absorption maximum compared with that of riboflavin. As for the 4.2 min-compound, the blue shifts of the first absorption band and fluorescence maximum were also observed.

Mass spectra revealed that the molecular weight of both compounds is m/e 392.

In ^{1}H-NMR spectra of these two compounds (5), new chemical shifts appeared at 4.81 ppm for the 4.2 min-compound and at 4.89 ppm for the 4.8 min-compound along with the signals of the ribityl side chain; and one of the two signals of the two methyl groups at positions 7 and 8, resonating between 2.3 and 2.6 ppm (11), disappeared. The new signals seem to be due to the proton of the methine group of the $-CH_2OH$ residue. The existence of a methine group and a ribityl side chain and the disappearance of one of the two methyl groups were also confirmed by proton-decoupled ^{13}C-NMR spectra (Figure 4).

From all of these results, we conclude that the compound appearing at 4.2 min is 7α-hydroxyriboflavin and that at 4.8 min, 8α-hydroxyriboflavin. The latter compound was definitely identified as authentic 8α-hydroxyriboflavin (5) from its R_f value by TLC, and from its absorption, fluorescence, mass, ^{1}H-NMR, and ^{13}C-NMR spectra.

Hydroxylation of 7- and 8-Methyl Groups of Riboflavin

The finding of 7- and 8-carboxylumichrome in rat urine and 7α- and 8α-hydroxyriboflavin in human urine led us to presume the occurrence of an oxidation reaction of riboflavin at its 7- and 8-methyl groups in its metabolism in higher animals, with the hydroxylation of the methyl groups being the initial event. We tested this presumption using rat liver, and clearly demonstrated (6) that 7- and 8-methyl groups were hydroxylated to form 7α- and 8α-hydroxyriboflavin by a liver microsomal elec-

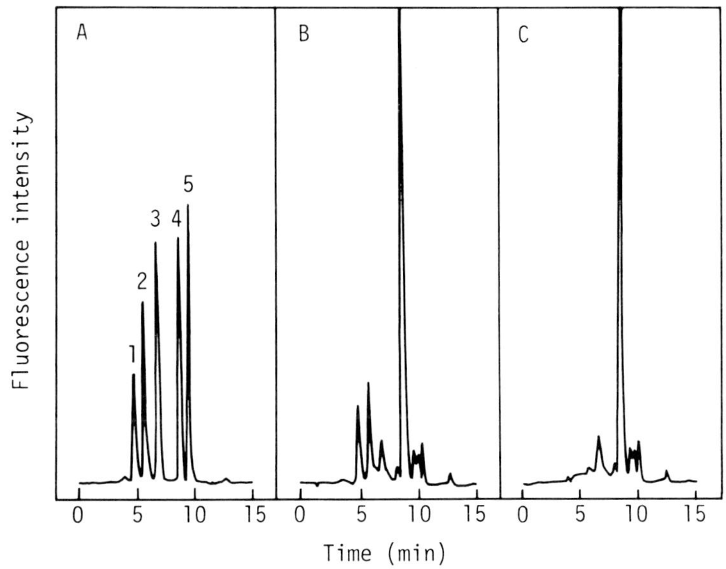

Figure 5. HPLC-Pattern of an alkaline hydrolysate of the hydroxylation products of riboflavin tetrabutyrate. A, authentic samples: 1, 7α-hydroxyriboflavin; 2, 8α-hydroxyriboflavin; 3, FMN; 4, riboflavin; 5, lumiflavin. B, alkaline hydrolysate of the reaction products of riboflavin tetrabutyrate after incubation with rat liver microsomes and NADPH with its generating system. C, the same as B except for the absence of NADPH and its generating system.

tron-transfer system.

Riboflavin or riboflavin tetrabutyrate was incubated with rat liver microsomes, NADPH, and an NADPH-generating system at pH 7.4 and 37°C for 60 min with continuous shaking. In the case of riboflavin tetrabutyrate, butyryl esters were hydrolyzed in 1.0 M NaOH before HPLC.

In the HPLC with an ODS-type column, the peaks corresponding to 7α-hydroxyriboflavin and 8α-hydroxyriboflavin were found (Figure 5), and further identification was made by HPLC and TLC after collection of each product.

A stoichiometric relation was observed between the sum of 7α- and 8α-hydroxyriboflavin and the amount of riboflavin or riboflavin tetrabutyrate which disappeared, indicating that other side reactions did not occur under the experimental conditions

Table 1. Effect of NADPH and Oxygen on Hydroxylation of the 7- and 8-Methyl Groups of Riboflavin Tetrabutyrate by Rat Liver Microsomes

	Relative formation	
	7α-Hydroxyriboflavin	8α-Hydroxyriboflavin
NADPH	100 %	100 %
NADH	60.1	17.3
$NADP^+$	ND	1.9
NAD^+	4.8	4.9
N_2/O_2 (90:10, v/v)	100 %	100 %
N_2/O_2 (100:0, v/v)	11.9	ND
CO/O_2 (90:10, v/v)	18.4	6.7

Concentrations: NADPH, $NADP^+$, and NAD^+, 2 mM; NADH, 4 mM; riboflavin tetrabutyrate, 40 nmol/ml; rat liver microsomes, 2 mg protein/ml. ND, not detected.

employed.

The formation of 7α-hydroxyriboflavin increased with increasing amounts of microsomes from 0.5 mg to 4 mg of protein per ml of reaction mixture, and that of 8α-hydroxyriboflavin leveled off at 2 mg of protein per ml. This hydroxylation required NADPH as electron donor, but NADH could replace it to some extent for the formation of 7α-hydroxyriboflavin as shown in Table 1.

Upon changing the gas phase of the reaction from a mixture of N_2 and O_2 (90:10, v/v) to 100% N_2 or to a mixture of CO and O_2 (90:10, v/v), the hydroxylation was markedly decreased, being almost negligible in the case of 8α-hydroxyriboflavin (Table

Table 2. Apparent K_m Values for Hydroxylation of the 7- and 8-Methyl Groups of Riboflavin and Riboflavin Tetrabutyrate

	7α-Hydroxyriboflavin	8α-Hydroxyriboflavin
Riboflavin	22.2 μM	28.6 μM
Riboflavin tetrabutyrate	14.3	33.3

Table 3. Effect of Pretreatment of Rat with Phenobarbital and 20-Methylcholanthrene on Hydroxylation

	7α-Hydroxy-riboflavin (nmol/h)	8α-Hydroxy-riboflavin (nmol/h)	Cytochrome P-450 (nmol/mg protein)
Phenobarbital			
Control			0.98
Riboflavin	0.11	0.12	
Riboflavin tetrabutyrate	2.96	2.36	
Treated			2.90
Riboflavin	0.18	0.11	
Riboflavin tetrabutyrate	4.46	1.80	
20-Methylcholanthrene			
Control			0.90
Riboflavin	0.12	0.16	
Riboflavin tetrabutyrate	2.18	1.80	
Treated			2.12
Riboflavin	0.25	0.11	
Riboflavin tetrabutyrate	3.44	1.17	

1).

The optimal pH value for formation of 7α-hydroxyriboflavin was found to be 7.4; and that for 8α-hydroxyriboflavin, 7.5.

Apparent K_m values are listed in Table 2. The rate of hydroxylation of riboflavin tetrabutyrate was about 15 times higher than that of riboflavin, probably due to its hydrophobic property. On the other hand, 7α- and 8α-hydroxy derivatives of FMN and FAD were not detected under the above conditions.

The formation of these two compounds was completely inhibited by 1.5 mM p-chloromercuribenzoate, an inhibitor of microsomal NADPH-cytochrome c reductase, and was also inhibited to ca. 50% by 1 mM NaCN or 10 mM ethylisocyanide, suggesting that the involvement of the cyanide-sensitive factor (12) is partial.

Pretreatment of rats with phenobarbital or 20-methylcholanthrene resulted in a marked increase in the formation of 7α-hydroxyriboflavin and increased the amount of microsomal cytochrome P-450, while hydroxylation of the 8-methyl group tended to be decreased compared with that of control rats (Table 3).

The differences in the results of pretreatment of rats with inducers and in those of NADH replacement suggest that the mechanism of 7-methyl hydroxylation is different from that of 8-methyl hydroxylation. This seems to be at least partly due to the multiplicity of cytochrome P-450. Considering that the amount of 7α-hydroxyriboflavin excreted was 31% of the total flavin in human urine, being larger than that of 8α-hydroxy-riboflavin (5%), as well as taking into account the above results, there is a possibility that the hydroxylation of 7-methyl group is caused by a general drug-metabolizing enzyme system, while that of 8-methyl group is related to a specific cytochrome P-450.

From the data described above, it is easily deduced that 7α- and 8α-hydroxyriboflavin found in human urine are mainly synthesized from riboflavin by microsomal enzymes in some organs and that these two compounds are further metabolized to 7- and 8-carboxylumichrome, which have been detected in the urine of

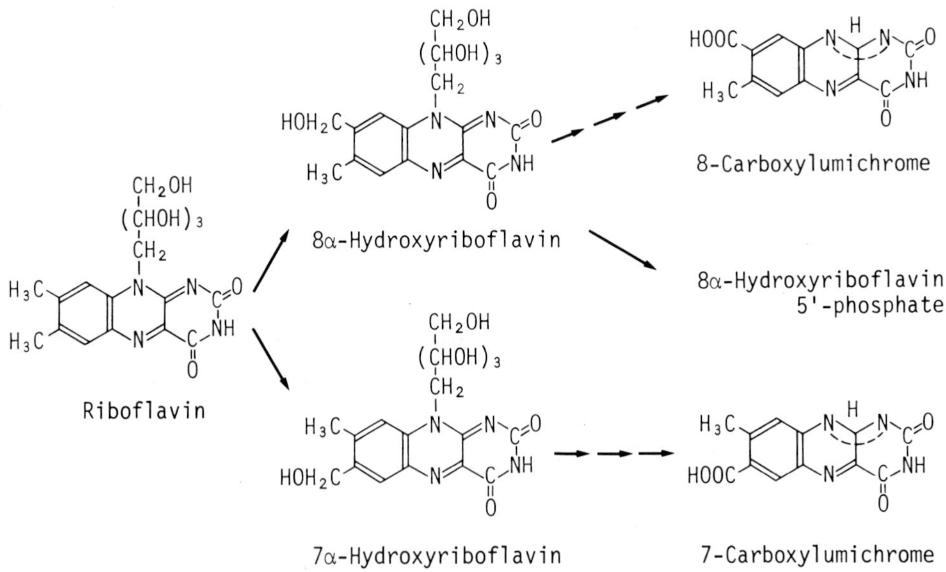

Figure 6. New metabolic pathway of riboflavin in mammals.

rats, mice, hamsters, guinea pigs, and rabbits (4). We are thus led to suspect that a definite metabolic pathway initiated by hydroxylation of the methyl groups of the isoalloxazine ring occurs in the animal body and that some part of the process is common in mammals. In relation to this, the sum of the radioactivities of 7- and 8-carboxylumichrome was found to be about 46% of the total radioactivity in rat urine, being comparable to that of percentages of 7α- and 8α-hydroxyriboflavin in human urine (ca. 36%). Furthermore, we detected 8α-hydroxyriboflavin and its 5'-phosphate in some rat tissues (13).

The mechanism for the conversion of hydroxy derivatives of riboflavin to carboxylumichromes is now under investigation.

In Figure 6, the possible new metabolic pathway of riboflavin in mammals is depicted on the basis of our results. This is the first case demonstrating the occurrence of 7-carboxylumichrome, 8-carboxylumichrome, 7α-hydroxyriboflavin, 8α-hydroxyriboflavin, and 8α-hydroxyriboflavin 5'-phosphate in nature, though the latter two compounds were synthesized chemically by Zhilina and Berezovskii (14,15).

Further we should point out the possibility that 8α-hydroxyriboflavin participates in the coupling of flavin with amino acid residues in covalently bound flavoproteins.

It is supposed that in vivo hydroxylation of the 7- and 8-methyl groups of flavins occurs at the riboflavin level rather than at the FMN or FAD level, because FMN or FAD could not be converted directly to their hydroxy derivatives in vitro (6). On the other hand, 7α- and 8α-hydroxyriboflavin were easily converted to their 5'-phosphate by partially purified rat liver flavokinase (16). Their coenzyme forms combined with the apoprotein of old yellow enzyme, and the enzymatically active holoenzyme exhibited characteristic absorption spectra and changes in spectra upon binding with phenol derivatives. In the case of 8α-hydroxyriboflavin 5'-phosphate-old yellow enzyme, a rather stable anionic radical was observed upon reduction (16). 8α-Hydroxyriboflavin 5'-phosphate could be further converted to its FAD level by the FAD synthetase complex partially purified from Brevibacterium ammoniagenes. This compound could then bind to the apoproteins of FAD enzymes (17). As active site probes, we feel that derivatives hydroxylated at the 8α-position are promising analogues to give further information along these lines.

References

1. Yang, C., McCormick, D. B.: J. Nutr. 93, 445-453 (1967).
2. West, D. W., Owen, E. C.: Br. J. Nutr. 23, 889-898 (1969).
3. Ohkawa, H., Ohishi, N., Yagi, K.: Biochem. Int. 2, 447-453 (1981).
4. Ohkawa, H., Ohishi, N., Yagi, K.: Biochem. Int. 6, 239-247 (1983).
5. Ohkawa, H., Ohishi, N., Yagi, K.: J. Biol. Chem. 258, 5623-5628 (1983).
6. Ohkawa, H., Ohishi, N., Yagi, K.: J. Biol. Chem. 258, 5629-5633 (1983).
7. Koziol, J., Koziolowa, A.: In Flavins and Flavoproteins, Physicochemical Properties and Function. Ostrowski, W. (ed.) Polish Scientific Publishers, Cracow, pp. 81-107 (1977).
8. Ohkawa, H., Ohishi, N., Yagi, K.: J. Nutr. Sci. Vitaminol. 29, 515-522 (1983).
9. Whitby, L. G.: Biochem. J. 50, 433-438 (1952).
10. Uchida, K., Suzuki, Y.: Agric. Biol. Chem. 38, 195-206 (1974).
11. Grande, H. J., van Schagen, C. G., Jarbandhan, T., Müller, F.: Helv. Chim. Acta 60, 348-366 (1977).
12. Oshino, N., Imai, Y., Sato, R.: Biochim. Biophys. Acta 128, 13-28 (1966).
13. Ohkawa, H., Ohishi, N., Yagi, K.: unpublished results.
14. Berezovskii, V. M.: In Flavins and Flavoproteins, Physicochemical Properties and Function. Ostrowski, W. (ed.) Polish Scientific Publishers, Cracow, pp. 63-73 (1977).
15. Zhilina, T. A., Berezovskii, V. M.: J. Nutr. Sci. Vitaminol. 23, 265-271 (1977).
16. Ohishi, N., Ohkawa, H., Yagi, K.: unpublished results.
17. Ohishi, N., Ohkawa, H., Yagi, K.: unpublished results.

MAMMALIAN METABOLISM OF FLAVINS

D.B. McCormick, W.S.A. Innis, A.H. Merrill, Jr., and S.-S. Lee
Dept. of Biochemistry, Emory University, Atlanta, GA 30322

General Introduction and Purpose

There has been a considerable increase in knowledge concerning the total metabolic fate of flavins in mammals. In particular, we have gained information on the diversity of natural flavins ingested, those circulatory carrier proteins which vector the vitamin to cells, the nature of cellular uptake and metabolic trapping as coenzyme, control of flavocoenzyme synthesis and turnover including effects of development, riboflavin status, and thyroid hormone regulation, additional aspects of catabolism, and more clearly, how flavin deprivation affects redox balance in the animal. Our input to current understanding, including most recent findings, will be outlined in the following in context with work reported by others.

Digestion, Absorption, and Overview of Metabolic Fate

The processes by which not only riboflavin, but also lesser amounts of natural derivatives are released by digestion of complexes with food proteins and then absorbed has received some attention experimentally (1) and noted in broader reviews (2,3). Salient features are that coenzyme forms of the vitamin (mainly FAD and less FMN) are released from noncovalent attachment to proteins as a consequence of gastric acidification. Nonspecific action of pyrophosphatase and phosphatase on the coenzyme forms occurs in the upper gut (4). Several percent of 8α-(amino acid)riboflavins originally in covalent attachment to certain enzymes, e.g. mitochondrial succinate dehydrogenase or monoamine oxidase, and traces of other ring

Flavins and Flavoproteins

and side-chain substituted flavins are also released by these actions following proteolysis. The vitamin is absorbed in the small intestine by a saturable transport system that is rapid and proportional to dose before leveling off at about 25 mg in the human (5). Bile salts appear to facilitate the uptake, and a modest amount of the vitamin circulates via the entero-hepatic system. Active transport at lower levels of intake appears to be Na^{+}-dependent as indicated from studies on up-take in rat intestine in vivo (6) and in vitro (7,8) and by isolated enterocytes from guinea pigs (9). Following release of the vitamin from the serosal side of the enterocyte, its subsequent fate can be schematized as in Fig. 1.

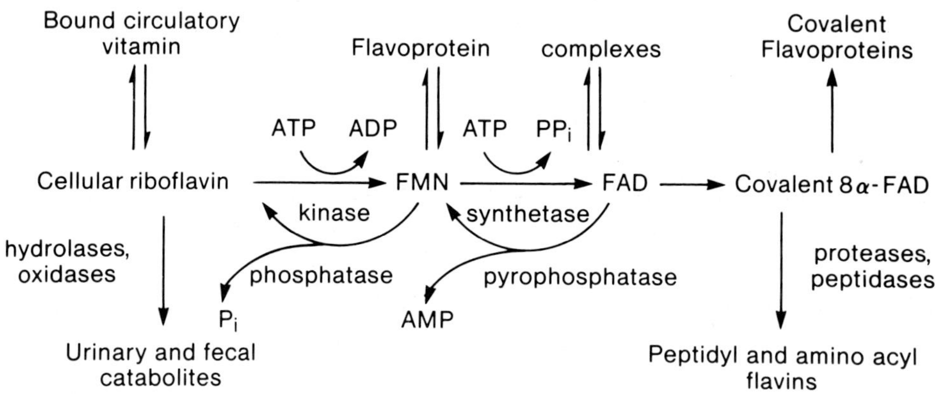

Fig. 1. Metabolic fate of flavins in mammals.

Transport in circulation

The vitamin is vectored in circulation not only as a weak complex with plasma albumin, presumed earlier to be the main carrier (10), but also in tighter associations. Our previous work described the identification and isolation of a subclass of human immunoglobulins which bound riboflavin moderately tightly (11). This has been expanded in more recent work

(12). A chromatographic profile of a human plasma sample which distinguishes the predominant binding of riboflavin by immunoglobulins as opposed to albumin is shown in Fig. 2.

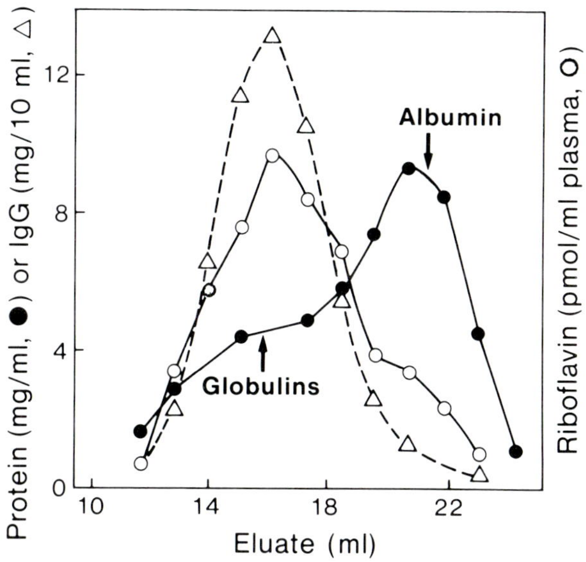

Fig. 2. Chromatography of human plasma binding 0.611 pmol/mg protein (39.1 pmol/ml plasma) on Sephacryl S-200 with binding determined by use of [2-^{14}C]riboflavin and IgG's measured nephelometrically.

A high-affinity riboflavin-binding IgG has been isolated from the serum of a patient with multiple myeloma (13); furthermore, a monoclonal IgA produced by a mouse plasmacytoma MOPC-315 binds riboflavin (14). Moreover, pregnancy-induced riboflavin-binding protein may be a feature of transport systems in the human (15), as had been observed in bovine (16) and rat (17) blood. These latter carrier proteins reflect the modest need for increased supply of vitamin to a fetus remaining in circulatory contact with the mammalian maternal host and are less dramatic in quantity compared to the reservoir proteins deposited by avian species within the egg (18). Correspondingly less information is available on the binding properties of the proteins in mammals than, for example, the chicken egg-white riboflavin-binding protein (19-21). However, a fuller appreciation of the importance of different riboflavin-binding proteins in mammalian circulation, significant in decreasing renal loss through rapid filtration of the free vitamin (22) and perhaps in the special case of fetal supply, has been gained. This information, including our recent investigation of the variation of binding among individuals (12), is summarized in Table I.

TABLE I. Riboflavin-binding proteins in circulation

Protein	K_D	Plasma level,%	Circulating riboflavin bound,%
Albumin	$\gtrless$4-10 mM	3.5 - 5.0	13-15
Immunoglobulins	$\geq$4 μM	1	1-68
Pregnancy-specific[1]	0.2 nM		

[1]Bovine and rat types have K_D's of $\leq$1 and 0.4 μM, respectively.

Cell Uptake and Metabolic Trapping

In addition to Na^+-dependent uptake by cells of the small intestine, the ability of other tissues to accumulate riboflavin has received some attention. The reports that aldosterone, a mineralocorticoid hormone, stimulates riboflavin incorporation into rat renal flavocoenzymes (23), and certain riboflavin analogs inhibit sodium reabsorption so influenced (24), suggest a Na^+ interrelationship in kidney. Riboflavin uptake has been measured in rabbit kidney slices where a relation to organic acid transport has been claimed (25). Active transport has also been reported for the central nervous system (26) where studies have been done on choroid plexus (27) and brain slices (28). It is clear, however, that uptake of the vitamin by parenchymal liver cells (29) is neither sensitive to Na^+ flux (oubain insensitive) nor exhibits characteristics expected for common organic acid anion carriers (bilirubin insensitive). With the isolated rat hepatocyte, entry of the free vitamin by facilitated diffusion is followed by rapid metabolic trapping which is most dependent upon flavokinase-catalyzed phosphorylation. This latter is sensitive to competitive flavin substrate inhibitors, reagents that decrease supply of cosubstrate ATP, and temperature. Findings are in Table II.

TABLE II. Effectors of ^{14}C-riboflavin uptake at pH 7.4 and 37°C by hepatocytes isolated from livers of normal rats

Treatment (with 3 μM riboflavin)		Initial uptake rate (as % of control at 100)[1]
Flavin analogs		
Lumiflavin, 23 μM		67
2'-Hydroxyethylflavin,	3 μM	64
	9 μM	48
7,8-Dichlororiboflavin,	3 μM	62
	6 μM	46
ATP depletors		
FCCP, 0.8 μM		62
1.0 μM		50
Ethionine, 5 mM		74
10 mM		52
Temperature		
20°		38
0-2°		2
Compounds found inactive:		
oubain, choline (for Na^+), bilirubin		100

[1]Initial rate of controls averaged 13.2 pmol/(10^6 cells·min).

Interconversions of Coenzymes

Conversion of riboflavin to coenzymes occurs within the cellular cytoplasm of most tissues, but particularly in the small intestine, liver, kidney, and heart (2,3). The obligatory first step is the ATP-dependent phosphorylation of the vitamin catalyzed by flavokinase. The FMN product can be complexed with specific apoenzymes to form several functional flavoproteins, but the larger quantity is further converted to FAD in a second ATP-dependent reaction catalyzed by FAD synthetase (pyrophosphorylase). The biosynthesis of flavocoenzymes, particularly at the flavokinase step, is tightly regulated. Several factors such as age, substrate supply, product utilization, and hormonal status contribute to alteration in levels of activities and often interrelate. One example of this is provided by our study of the developmental change in flavokinase in chickens (30). Specific activity in embryonic organs is lower in comparison with activity of this enzyme in mature organs of the adult. This is in agreement with earlier measurements of flavokinase activity in livers of pre- and postnatal rats (31). Moreover, the specific activity of the chick kinase remains relatively constant within a given tissue during the primary development phase, but the amount (total activity) increases with production of more tissue (protein) as probably reflects an adequate level of FMN formation that parallels growth. The increase in specific and total activities seen in the adult must indicate a greater need for flavocoenzyme supply as would be expected for this more metabolically active and more oxygen-dependent stage of life. Typical results are given in Table III.

TABLE III. Development of flavokinase activity in chickens

		Activity units	
Organ	Age	Specific, per mg[1]	Total, per organ
Liver	12-d embryo	0.52 ± 0.10	6.9
	18-d embryo	0.35 ± 0.05	17.6
	adult	1.68	4115
Heart	12-d embryo	0.50 ± 0.08	2.7
	18-d embryo	0.49 ± 0.15	4.3
	adult	2.00	777
Intestine	12-d embryo	0.83 ± 0.21	3.1
	18-d embryo	0.79 ± 0.26	14.8
	adult	1.59	2137

[1]Standard deviations from 6 to 21 embryos per group.

Riboflavin status also affects level of flavocoenzyme biosynthesis. As shown previously (32), the lower activities of flavokinase measured in small intestine and liver from riboflavin-deficient rats is increased within hours following a single per os dose of vitamin.

TABLE IV. Effect of riboflavin status on specific activities of flavocoenzyme biosynthesizing and degrading enzymes from livers of weanling rats maintained for 5 weeks on diet

Enzyme	Riboflavin, μg/15g diet	Activity, units/mg[1]
Flavokinase	0	3.1 ± 1.4
	30	8.0 ± 1.3
FAD synthetase	0	34.3 ± 5.6
	30	65.9 ± 5.9
FMN phosphatase	0	45.1 ± 10.7
	30	36.5 ± 8.9
FAD pyrophosphatase	0	141.0 ± 15.0
	30	98.0 ± 19.0

[1]Standard deviations from 3 animals per group.

Our more recent work has extended these findings in a comparison of changes in activities of hepatic enzymes directly involved in flavocoenzyme interconversions. Typical results are given in Table IV. There is greatest effect during deficiency on flavokinase, less on FAD synthetase, and only age-related changes seen with nonspecific FMN phosphatase and FAD pyrophosphatase (33).

Other investigators have noted the stimulation of flavocoenzyme biosynthesis elicited by thyroid hormone (34). The extent to which this is attributed to formation of more active enzyme catalyzing formation of FMN and/or FAD has now been secured. By examining the correspondence of activity with immunologically quantitated flavokinase in hypo-, eu-, and hyperthyroid rats, we have demonstrated that the thyroid response is caused by an increased amount of enzyme. Interestingly, another flavin-binding protein decreases upon thyroid hormone stimulation. This is illustrated in Fig. 3.

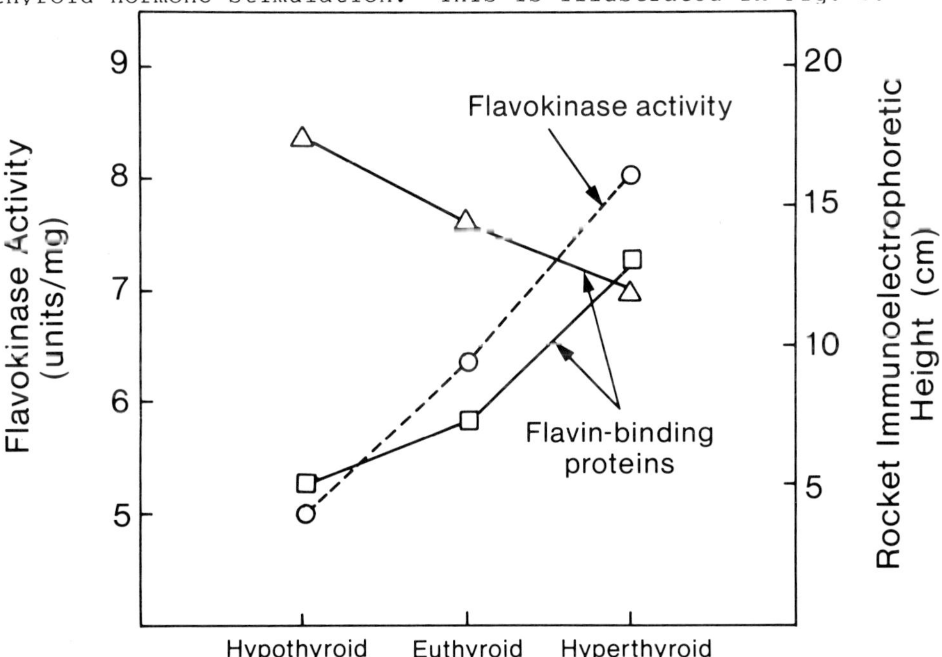

Fig. 3. Correspondence of immuno-reactive flavokinase with liver cytosolic activity from hypo (propylthiouracil), hyper (triiodothyronine) and cuthyroid rats.

Comparison of the changes in both biosynthetic and degradative activities in livers from such animals reveals greater effect of thyroid hormone on flavokinase than FAD synthetase. These results are given in Table V.

TABLE V: Effect of thyroid state on specific activities of flavocoenzyme biosynthesizing enzymes from liver of rats

	Activity, units/mg[1]		
Enzyme	Hyperthyroid	Euthyroid	Hypothyroid
Flavokinase	8.03 ± 0.85	6.31 ± 1.58	5.02 ± 0.66
FAD synthetase	68.4 ± 11.9	59.1 ± 15.5	51.4 ± 12.5

[1] Standard deviations from 4 to 6 animals per group.

Whether directly or indirectly, the biosynthesis of covalently bound FAD, demonstrated earlier in proteins from mitochondria of rats administered ^{14}C-riboflavin (35,30), is also regulated by thyroid hormones (37). Again the interplay of other factors should be noted. An age dependence of thyroxine stimulation of riboflavin incorporation into flavin coenzymes in liver of newborn or adult, but only in brain of newborn rats has been found (38). Also aldosterone and other corticosteroids affect flavocoenzyme biosynthesis in kidney and liver (23,39). One report has been made for induction of the FAD synthetase system in rat liver by phenobarbital (40).

Catabolism and Excretion

The findings that a portion of urinary flavin is represented by such flavin catabolites as have undergone oxidation in the original flavin benzenoid methyls, e.g. 7α- and 8α-hydroxyriboflavins (41), and even loss of the D-ribityl chain, e.g. 7- and 8-carboxylumichromes (42), amplifies our understanding of the catabolic fate of the vitamin in mammals. It has been shown that a cyt P_{450}-dependent microsomal oxidase system is responsible for oxidation of the methyls (41), just as generally known for demethylation reactions with steroids and similar aromatics. Though not all intermediates have been secured, it is probable that oxidation proceeds from methyl to hydroxymethyl to formyl to carboxyl and even conceivably with some decarboxylation to form demethyl compounds. We had reported earlier that catabolism of 8α -(amino acid)flavins, such as derived in turnover of covalently bound FAD, leads to a number of urinary products, some compatible with 8α-hydroxy and 8-formylriboflavins (43). A chromatographic profile of such urinary flavins is given in Fig. 4.

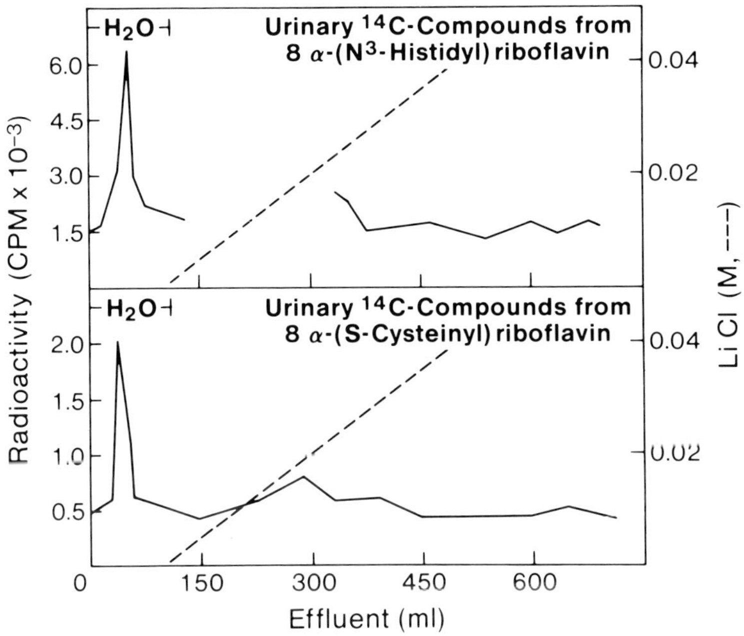

Fig. 4. Elution patterns from column chromatography on DEAE-cellulose (Cl^-) of ^{14}C-compounds concentrated by phenol extraction of pooled urine from rats which had received the stipulated compounds orally.

It was also reported by others that since putatively germ-free rats excrete carboxylumichromes, there are tissue enzymes responsible for complete removal of the flavin side chain (41). Though such activity may exist, there is no question that photochemical degradation contributes and is difficult to exclude from even germ-free animals and that intestinal microflora contribute significantly to ribityl side-chain cleavage in normal rats. This latter fact is easily emphasized by results from our ealier work that demonstrated increased excretion of chloroform-soluble catabolites from [2-^{14}C]riboflavin when extra vitamin is administered and a curtailment when antibiotics such as a sulfa drug is administered to decrease gut bacteria (44). Synoptic results are given in Table VI. Hence, further work should quantitate the extent to which tissue enzymes may contribute to side-chain cleavage.

TABLE VI. Effects of amount of riboflavin and intestinal microflora on rat excretion of side-chain-degraded flavin

Diet (for 2 weeks before 110 µg [2-^{14}C]riboflavin given i.p.)	$CHCl_3$-extracted % ^{14}C within 50h post ^{14}C-riboflavin injection[1]	
	Urine	Feces
Laboratory ration	0.25	3.10
Ration + riboflavin (15 µg/g diet)	0.32	9.91
Ration + riboflavin + sulfa (1%)	0.21	4.45

[1]Numbers are from 3 animals per group.

Consequences of Deficiency

An important consideration in the global economy of riboflavin is the extent the symptomatology of deficiency relates to stringencies of need at the biochemical level. There are both general and specific consequences of inadequate vitamin intake. With the fairly recent findings that both glutathione peroxidase (45) and catalase (46) are decreased in riboflavin deficiency, the decreased protection against toxic peroxides (47) including increase of lipid peroxides observed in ariboflavinosis is more comprehensible. The relatively greater sensitivity of mammalian catalase than other hemoproteins (46) to flavin status is probably the result of less substrate induction, since a bacterial catalase has been shown to be specifically induced by hydrogen peroxide (48). Hence, a major sequence of events now documented is that less vitamin → less coenzymes → less oxidases → less H_2O_2 → less catalase as well as glutathione reductase and peroxidase → less protection against peroxidative damage. The rate and extent of decrease and increase of peroxisomal enzymes during riboflavin deficiency and repletion, respectively, in rats is shown in Fig. 5 where change in catalase is compared to the insensitive Cu-dependent urate oxidase and the sensitive FMN-dependent glycolate oxidase and FAD-dependent D-amino acid oxidase (46).

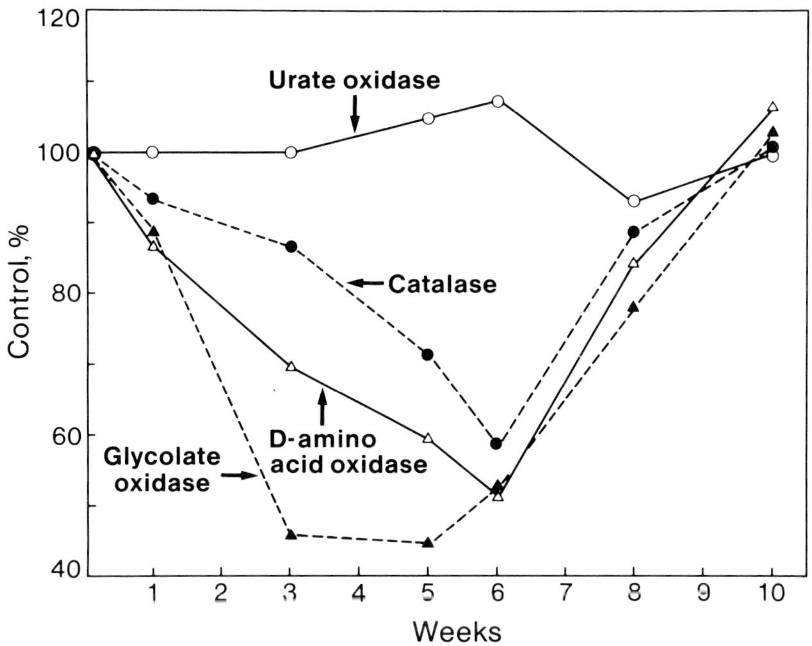

Fig. 5. Mean changes of liver peroxisomal enzymes in rats (3 to 6 each) fed a riboflavin-deficient diet which was supplemented after 5 weeks (2 μg/g diet).

Another relatively sensitive site afflicted by riboflavin deficiency is the FMN-dependent pyridoxine (pyridoxamine) 5'-phosphate oxidase necessary for utilization of vitamin B_6 (49,50). One of the complicating aspects of severe riboflavin deficiency is a secondary, tissue-level deficit of functional pyridoxal 5'-phosphate. This is but one of several instances where a deficit in a given vitamin leads to inadequate formation or utilization of another; hence, it is one contributing factor to commonly encountered multi-vitamin deficiencies. Decrease in level of liver pyridoxine (pyridoxamine)5'-phosphate oxidase with continuing riboflavin deficiency in the rat is compared in Fig. 6 with the reciprocal changes in the activity coefficient of erythrocyte glutathione reductase (49). The former FMN-dependent oxidase is at least as sensitive to flavin status in the mammal as is the latter FAD-dependent reductase which is routinely followed as a clinical chemical indicator for the human (51).

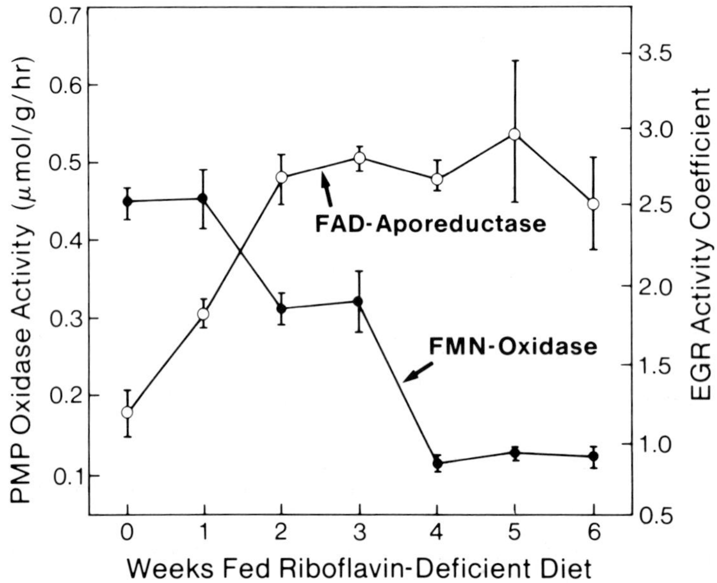

Fig. 6. Changes in oxidase and reductase during riboflavin deficiency.

References

1. McCormick, D.B., Addison, R.: Mammalian Metabolism of Flavins, Xth Int. Cong. Nutr., Kyoto, 1975.

2. Merrill, Jr., A.H., Lambeth, J.D., Edmondson, D.E., McCormick, D.B.: Ann. Rev. Nutr. 1, 281-317, 1981.

3. McCormick, D.B.: Modern Nutrition in Health and Disease, M.E. Shils and V.R. Young (ed.), 7th edition, Lea & Febiger, Philadelphia, 1984.

4. Daniel, H., Binninger, E., Rehner, G.: Int. J. Vit. Nutr. Res. 53, 109-114, 1983.

5. Jusko, W.J., Levy, G.: Riboflavin, R.S. Rivlin (ed.), Plenum Press, New York, 1975.

6. Rivier, D.A.: Experientia 29, 1443-1446, 1973.

7. Meinen, M., Aeppli, R., Rehner, G: Nutr. Metab. 21 (Suppl. 1), 264-266, 1977.

8. Daniel, H., Wille, U., Rehner, G.: J. Nutr. 113, 636-643, 1983.

9. Hegazy, E., Schwenk, M.: J. Nutr. 113, 1707, 1983.

10. Jusko, W.J., Levy, G.: J. Pharm. Sci. 58, 58-62, 1969.

11. Merrill, Jr., A.H., Froehlich, J.A., McCormick, D.B.: Biochem. Med. 25, 198-206, 1981.

12. Innis, W.S.A., McCormick, D.B., Merrill, Jr., A.H.: Biochem. Med., submitted, 1984.

13. Farhangi, M., Osserman, E.F.: N. Engl. J. Med. 294, 177-183, 1976.

14. Eisen, H.N., Michaelides, M.C., Underdown, B.J., Schulenberg, E.P., Simms, E.S.: Fed. Proc. 29, 78, 1970.

15. Murthy, C.V.R., Adiga, P.R.: Biochem. Int. 5, 289-296, 1982.

16. Merrill, Jr., A.H., Froelich, J.A., McCormick, D.B.: J. Biol. Chem. 254, 9362-9364, 1979.

17. Muniyappa, K., Adiga, P.R.: Biochem. J. 187, 537-540, 1980.

18. Froelich, J.A., Merrill, Jr., A.H., Clagett, C.O., McCormick, D.B.: Comp. Biochem. Physiol. 66B, 397-401, 1980.

19. Choi, J.-D., McCormick, D.B.: Arch. Biochem. Biophys. 204, 41-51, 1980.

20. Becvar, J., Palmer, G.: J. Biol. Chem. 257, 5607-5617, 1982.

21. Miura, R., Kasai, S., Horiike, K., Sugimoto, K., Matsui, K., Yamano, T., Miyake, Y.: Biochem. Biophys. Res. Commun. 110, 405-411, 1983.

22. Merrill, Jr., A.H., Shapira, G., McCormick, D.B.: Flavins and Flavoproteins, V. Massey and C.H. Williams, Jr. (ed.), Elsevier North Holland, New York, 1981.

23. Tan, E.L., Trachewsky, D.: J. Steroid Biochem. 6, 1471-1475, 1975.

24. Trachewsky, D.: J. Clin. Invest. 62, 1325-1333, 1978.

25. Spector, R.: J. Pharmacol. Exp. Ther. 221, 394-398, 1982.

26. Spector, R.: J. Clin. Invest. 66, 821-831, 1980.

27. Spector, R., Boose, B.: J. Biol. Chem. 254, 10286-10289, 1979.

28. Spector, R.: J. Neurochem. 34, 1768-1771, 1980.

29. Aw, T.Y., Jones, D.P., McCormick, D.B.: J. Nutr. 113, 1249-1254, 1983.

30. Zak, Z., McCormick, D.B.: Comp. Biochem. Physiol. 73B, 341-345, 1982.

31. Rivlin, R.S.: Amer. J. Physiol. 216, 979-982, 1969.

32. Merrill, Jr., A.H., Addison, R., McCormick, D.B.: Proc. Soc. Exp. Biol. Med. 158, 572-574, 1978.

33. Lee, S.-S., McCormick, D.B.: J. Nutr. 113, 2274-2279, 1983.

34. Rivlin, R.S.: Riboflavin, Plenum Press, New York, 1975.

35. Sato, M., Ohishi, N., Nishikimi, M., Yagi, K.: Biochem. Biophys. Res. Commun. 78, 868-873, 1977.

36. Addison, A., McCormick, D.B.: Biochem. Biophys. Res. Commun. 81, 133-138, 1978.

37. Pinto, J., Rivlin, R.S.: Arch. Biochem. Biophys. 194, 313-320, 1979.

38. Fazekas, A.G., Pinto, J., Huang, Y.P. Chandhuri, R., Rivlin, R.S.: Endocrinol. 102, 641-648, 1978.

39. Fazekas, A.G., Sandor, T.: J. Steroid Biochem. 7, 29-32, 1976.

40. Hamajima, S., Ono, S., Hirans, H., Obara, K.: Int. J. Vit. Nutr. Res. 49, 59-63, 1979.

41. Ohkawa, H., Ohishi, N., Yagi, K.: J. Biol. Chem. 258, 5623-5628 and 5629-5633, 1983.

42. Ohkawa, H., Ohishi, N., Yagi, K.: Biochem. Int. 6, 239-247, 1983.

43. Chia, C.P., Addison, R., McCormick, D.B.: J. Nutr. 108, 373-381, 1978

44. Yang, C., McCormick, D.B.: J. Nutr. 93, 445-453, 1967.

45. Yagi, K.: 12th Int. Cong. Biochem., Abst. PLE 003-1, Perth, Western Australia.

46. Lee, S.-S., Ye, J., Jones, D.P., McCormick, D.B.: Biochem. Biophys. Res. Commun. 117, 788-793, 1983.

47. Chance, B., Sies, H., Boveris, A.: Physiol. Rev. 59, 527-605, 1979.

48. Clayton, R.K.: J. Cell. Comp. Physiol. 55, 1-7 and 9-14, 1960.

49. Rasmussen, K.M., Barsa, P.A., McCormick, D.B.: Proc. Soc. Exp. Biol. Med. 161, 527-530, 1979.

50. Rasmussen, K.M., Barsa, P.A., McCormick, D.B., Roe, D.A.: J. Nutr. 110, 1940-1946, 1980.

51. McCormick, D.B.: Clinical Clinical Chemistry, N.W. Tietz (ed.), W.B. Saunders Co., Philadelphia, 1984.

FLAVOENZYMES AS DRUG TARGETS

R. Heiner Schirmer, Florian Lederbogen, R. Luise Krauth-Siegel, Gerhard Eisenbrand

Institut für Biochemie II der Universität und Deutsches Krebsforschungszentrum, Im Neuenheimer Feld 328/280, D-6900 Heidelberg, FRG

Georg Schulz

Institut für Organische Chemie und Biochemie der Universität, Albertstr. 21, D-7800 Freiburg, FRG

Albrecht Jung

Institut für Physiologische Chemie, Hoppe-Seyler-Str. 1, D-7400 Tübingen, FRG

Introduction

"For someone starting his career as a chemist, it was a distinct handicap in those years, viz. the 50' s, that B_2, in contrast to vitamins that were discovered later, was a 'breadless' vitamin, i.e. it had no lucrative pharmacological aspects to it. It was found ubiquitously, it was not expensive to produce, so that the producing company... could not realize much more out of it than chicken feed and an, in every respect, harmless food coloring... Pharmacologically active derivatives of B_2 were not known and have barely been developed until now, so that interest in B_2 was at a low point, particularly in medicine. This situation is only changing now... : we have an involvement of B_2 even in human medicine, namely in the oxygen metabolism of cells."

(From the memoirs of Peter Hemmerich dictated in August 1981)

Flavins and Flavoproteins

Flavins in practical and experimental medicine

The use of riboflavin and riboflavin analogues in clinical and experimental medicine - e.g. in hypertension research - have been reviewed recently (1,2). Another aspect, the structural similarity of flavin with drugs like tricyclic antidepressants has also attracted the attention of clinical pharmacologists (3).
Here we should like to comment on the enzymatic test for the detection of riboflavin (= B_2) deficiency (4,5). In our opinion, this test is compulsory for malnourished women prior to pregnancy. The B_2-status of a person is assessed by measuring the activity of glutathione reductase in hemolysate before and after addition of FAD; in this way the proportion of apoglutathione reductase lacking the flavin-containing cofactor is determined. The basic assumption underlying this test, namely the biological and physical stability of apoglutathione reductase of red blood cells (6,7), was recently confirmed by X-ray diffraction analysis. We found that the apoenzyme has a defined geometric structure; it even closely resembles the holoenzyme from which the structure of the prosthetic group has been subtracted (6,8). The occurrence of the stable apoenzyme *in vivo*, as well as the similarity of apo- and holoenzyme are points which have to be accounted for also in drug-designing studies (9).

Flavoproteins as targets of drugs and toxic compounds

As reviewed by Merrill et al. (1) inhibitors of flavoproteins play a prominent role in pharmacology. Points in case are allopurinol and other inhibitors of xanthine oxidase which are used in the treatment and prevention of gout, or inhibitors of monoamine oxidases which serve as antidepressant drugs.

A number of toxic compounds found in foodstuffs inhibit flavoenzyme activities. Examples are 3-nitropropionate, a suicide substrate of succinate dehydrogenase and hypoglycin A, the causing agent of Jamaican vomiting sickness; hypoglycin A is metabolized to methylene-cyclopropyl-acetyl-CoA, an inhibitor of acyl-CoA dehydrogenase. The interactions of dioxygen, the most abundant and possibly the most toxic foodstuff, with flavoenzymes is discussed below. The cited examples illustrate that flavin enzymologists have greatly contributed to the elucidation of the mode of drug action. Consequently one may hesitate to use the term 'flavin pharmacology' in a narrow sense as it is done in the next section.

Flavoenzymes in OSVAC processes, a challenge for drug research

An adult person consumes appr. 1000g O_2 per day. Some 3% to 5% of this oxygen, that is 30 to 50g O_2, is not directly reduced to water by cytochrome oxidase but metabolized to $^1O_2$, $O_2^{\bullet -}$, H_2O_2, $OH^{\bullet}$, OCl^- and other cytotoxic compounds. The sum of these reactive oxygen species represents the <u>oxidant stress</u> (10-16). Flavoenzymes play a predominant role in the production of oxidant stress (Table 1). Points in case are the $O_2^{\bullet -}$-producing NADPH-oxidase (16-19) of neutrophilic leucocytes and macrophages and the xanthine oxidase of endothelial and sinusoidal cells (20). These enzymes reduced O_2 directly; other flavoenzymes such as NADPH-cytochrome P-450 (c) oxidoreductase do so indirectly by reducing so-called redox-cyclers (=X in the Scheme), which in turn spontaneously react with O_2:

---

*OSVAC = Oxidant Stress versus Antioxidant Capacity

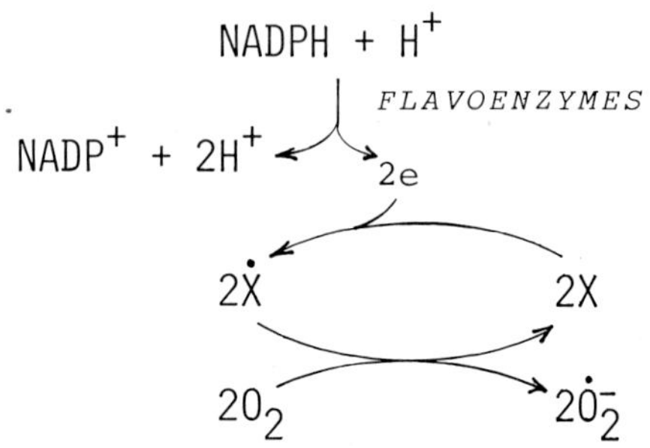

Scheme I

$\dot{O}_2^-$ is the educt for a number of other reactions (10-12,16) some of which are sketched in Fig.1

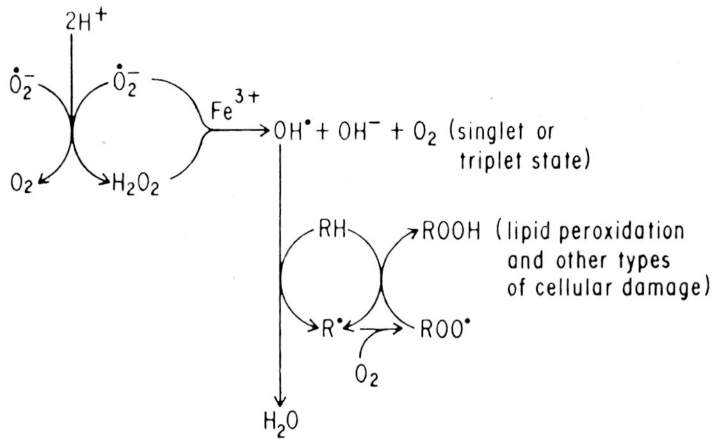

Fig. 1 Oxidant stress leading to damage of macromolecules(=R)
As radical-initiated processes are conservative and propagative they may lead to the production of secondary and tertiary free radicals derived from lipids, amino acids, glutathione or nucleic acids (after Docampo and Moreno (12)).

Examples for the increasing number of redox-cycling compounds are catechols and related substances, iron chelators, aromatic nitrocompounds such as nitrofurantoin or nifurtimox, adriamycin and related cytostatic agents, the herbicide paraquat, and many naturally occurring pesticides (10-14). It should be noted that the reactions discussed do not only cause oxidant stress; at the same time they consume precious cellular compounds such as NADPH and O_2. The hypoxic condition induced by the redox-cycling of adriamycin, for instance, is believed to contribute to the cardiotoxicity of this compound (10).

As shown in Table 1, reactive oxygen species are not only harmful for the human organism; indeed they are vital for the destruction of pathogenic microorganisms and tumour cells.

There exist powerful defense systems against oxidant stress the sum of which represents the *antioxidant capacity* of a cell or tissue. A cornerstone of the antioxidant capacity is the FAD-enzyme glutathione reductase (6-9,21). Thus there are flavoenzymes on both sides in OSVAC processes. However, in the presence of trinitrobenzene sulfonate *in vitro* (22) and probably in the presence of paraquat *in vivo* (23), glutathione reductase turns into an NADPH oxidase which means that the enzyme can change sides in OSVAC processes. This behaviour is believed to contribute to the damaging effects of herbicides on plant cells as well as on lung cells or on fibres of the eye lens (23,24).

In conclusion, if OSVAC-processes are really as important in medicine as it is currently believed (Table 1 and ref. 10-20), the flavoenzymes involved will play a central role in drug research. *As a rule, oxidative damage to cell constituents and tissues is preceded by a reduction step which is catalysed by a flavoenzyme.*

TABLE 1

ROLE OF FLAVOENZYMES IN OSVAC-PROCESSES (OSVAC = OXIDANT STRESS VERSUS ANTIOXIDANT CAPACITY)

SYSTEM	OXIDANT STRESS	REASON FOR IMPAIRED ANTIOXIDANT CAPACITY	CONSEQUENCES
Tumours treated with the cytostatic agent BCNU (25,28)	Respiratory burst of macrophages, catalyzed by the FAD-enzyme NADPH-oxidase ($NADPH+2O_2 \rightleftharpoons NADP^++H^++2\dot{O}_2^-$	Inhibition of the FAD-enzyme GSSG reductase by BCNU in the malignant cells	Death of the tumour cells
Heart muscle cells suffering from the side effects of cytostatic drugs(26,27)	$\dot{O}_2^-$-Production catalyzed by flavoenzymes and adriamycin(10)	A) Low activity of antioxidant enzyme systems in heart myocytes (27) B) Inhibition of GSSG-reductase by BCNU (28)	Cardiomyopathy
Bacterial infections (16-19)	Respiratory burst of neutrophilic leucocytes catalyzed by the FAD-enzyme NADPH-oxidase	Lack of antioxidant enzyme systems in the bacterial coats	Death of the microorganisms
Infections in patients having chronic granulomatous disease (CGD) (31)	No $\dot{O}_2^-$-production by leucocytes because of NADPH-oxidase deficiency		Frequent and serious bacterial infections
Chronic rheumatic diseases (10,11)	Respiratory burst of neutrophilic leucocytes in joint cavities	Lack of antioxidant enzyme systems in extracellular spaces	Chronic disease
Chagas disease (12)	$\dot{O}_2^-$-production catalyzed by flavoenzymes plus drugs	Low activity of antioxidant enzyme systems in <u>Trypanosoma cruzi</u>	Death of trypanosomes
Alveolar cells in the lung of patients treated with nitroaromatic drugs (10-12,29)	A) High O_2-tension in the lung B) $\dot{O}_2^-$-production, catalyzed by flavoproteins plus drugs	The FAD-enzyme glutathione reductase is inhibited or changes sides under the influence of the drug	Lung fibrosis
Malaria-infected erythrocytes in patients having favism * (30,32-35)	H_2O_2 production, catalyzed by isouramil and divicine, two active principles of fava-beans (Fig.2)	Inborn deficiency of glucose-6-phosphate dehydrogenase	Death of Plasmodium falciparum

* This system is listed here because in can be imitated by inhibiting the FAD-enzyme GSSG-reductase of parasitized erythrocytes (see text)

Inhibitors of glutathione reductase as potential antimalarial drugs

Recent biochemical studies (32) corroborate the hypothesis that favism confers some protection against malaria, and suggested to us that the biochemistry of favism might be mimicked by inhibiting the enzyme glutathione reductase in human erythrocytes (Fig.2) The threedimensional structure of this flavoenzyme which catalyzes the reaction NADPH + GSSG + $H^+ \rightleftharpoons NADP^+$ + 2GSH is known (8,9,21,36), so that the interactions of glutathione reductase and inhibitors can be studied in atomic detail and the design of new inhibitors with desirable properties is facilitated. -We have studied two compounds which inhibit glutathione reductase as antimalarials: the clinically used antineoplastic agent carmustine (BCNU) and its newly developed water-soluble and less toxic analogue, 1-(2-chloroethyl)-1-nitroso-3-(2-hydroxyethyl)urea = HECNU (37). Both drugs inhibited the growth of Plasmodium falciparum in culture. In addition, HECNU was shown to have a curative effect on ro-

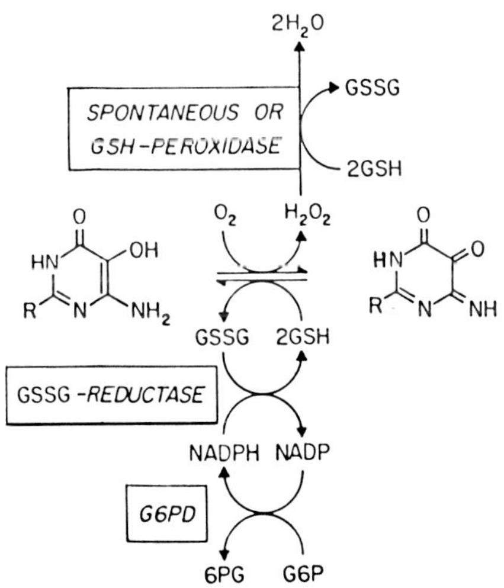

Fig. 2

Favism: The action of broad bean pyrimidines in the red blood cell. One of the 2e-redox-cycling pyrimidines is isouramil (R=OH). When isouramil reacts with O_2, H_2O_2 representing oxidant stress is formed. Both H_2O_2 and oxidized isouramil are reduced at the expense of (four) GSH molecules. The oxidized glutathione is reduced by the flavoenzyme glutathione reductase. G6PD, glucose-6-phosphate dehydrogenase; 6PG, 6-phosphogluconolactone; G6P, glucose 6-phosphate (after Chevion et al.(38)).

Table 2 Effect of BCNU and HECNU on rodent malaria

All animals were infected with $5 \cdot 10^8$ parasitized erythrocytes given intraperitoneally on day zero. From day 1 onwards the drugs were administered as two intraperitoneal injections (7.5 mg/kg body weight at 9 a.m. and 7.5 mg/kg at 1 p.m.).- Results analogous to those shown here were obtained with NMRI-mice after infection with *Plasmodium vinckei*, with the chloroquine-resistant strain K65 of *Plasmodium berghei* and with the chloroquine-sensitive strain of *Plasmodium berghei*, respectively.

Parasitemia on day 3 after infection with *Plasmodium vinckei*

Control rats (n = 5)	BCNU-treated rats (n = 5)	HECNU-treated rats (n = 5)
70%	8%	1%
60%	25%	1%
32%	20%	3%
60%	3%	6%
90%	8%	3%
all animals died before or on day 6	all animals were alive on day 6 but suffered from side effects of BCNU and the treatment was stopped. The animals died of malaria before day 11.	all animals remained apparently healthy; on day 8 their blood appeared to be free of parasites. On day 10-when the treatment was stopped-it was no longer infectious.

dent malaria being equally effective against chloroquine-resistant and chloroquine-sensitive strains of P.vinckei (Table 2). It remains to be studied to what extent the in vivo effects of HECNU are indeed caused by the inhibition of glutathione reductase, but our results are consistent with the notion that parasitized erythrocytes have a critical OSVAC balance. Increasing the oxidant stress (13) or decreasing the reducing capacity by inhibiting glutathione reductase (34) can lead to the selective destruction of parasitized cells. Being carcinogenic, BCNU and HECNU might never be considered for treating human malaria. Therefore a series of other compounds inhibiting glutathione reductase from human erythrocytes and from P. vinckei (39-41) were designed and synthesized; most of these inhibitors are carbamoylimidazolides and carbamoyloxosuccinimides. 1-Deaza-flavin is also of interest because 1-deaza-FAD glutathione reductase resembles the enzyme species found in congenital glutathione reductase deficiency of red blood cells (ref. 42 and Krauth-Siegel et al., this volume).

Some considerations on the development of flavoenzyme-directed drugs

a) Experimenta naturae including diseases (31,42-44) might be used as guidelines in drug research. For instance, the properties of the NADPH-oxidase in CGD (Table 1) or the properties of glutathione reductase in congenital glutathione reductase deficiency (42) should be studied and then be imitated by using inhibitors. Attempts in this direction have been mentioned above.
b) Specific inhibitors may be obtained by drug-tailoring in light of the three-dimensional structure of an enzyme (Fig. 3). Alternatively, one may try to link flavin with another ligand of a given flavoenzyme: If the binding sites do not overlap, the dissociation constant for the pluriligand inhibitor is expected to correspond the product of the K_i-

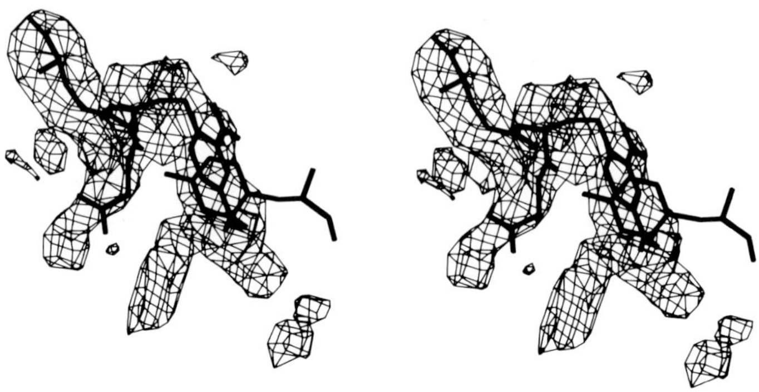

Fig. 3

Example of a drug derivative as bound to glutathione reductase at the GSSG-binding site (46). 1-Chloro-2,4-dinitrobenzene - which is used in studies on oxidant stress in liver (and in the treatment of warts) - is conjugated in vivo with GSH to give S-(2,4-dinitrophenyl)-glutathione (DNPG). As shown by X-ray crystallography, DNPG binds to glutathione reductase in an unexpected manner. This example illustrates that drugtailoring on the basis of a three-dimensional structure of a target enzyme must be accompanied by many structural analyses.
The figure shows a stereo view of DNPG as bound to glutathione reductase which is represented by Tyr-114 at the upper left hand side. The difference electron density is shown in bird cage representation. Details are given in ref. 46.

values of the individual ligands (45).

c) The effect of the drug may depend on the presence of the prosthetic group FAD or FMN. The cytostatic agent carmustine (BCNU), for instance, irreversibly modifies hologlutathione reductase in vivo but not the FAD-free apoenzyme which remains as a potential apoenzyme. This fact complicates clinical and experimental studies with inhibitors of glutathione reductase.

d) Inhibitors targeted to flavoenzymes of pathogenic bacteria or protozoa are of special interest. This is even more the case if the inhibitor can change the flavoenzyme into an oxidase (22-24); many pathogens possess poor antioxidant capacity (12,13).

e) In general - but particularly in the case of flavin analogues - the free compound, its precursors and catabolites may have pharmacologic effects (1).

Finally one should not forget the rule of the drug industry that rational approaches to drug design are doomed to fail.

Conclusions

OSVAC processes which reflect the versatility of oxygen metabolism contribute to many pathophysiological conditions. In most OSVAC mechanism flavoenzymes play dominant roles on both sides (9-20,33,34). Consequently flavoprotein-directed compounds may become an important field of drug research. To quote again from the memoirs of Peter Hemmerich: "The chemistry of B_2 will see considerable changes and will, in the long run, be anything but a 'breadless' undertaking."

Acknowledgments: We thank Dr. Renate Untucht-Grau, Universitätskinderklinik Heidelberg, for valuable discussions. Our work is supported by the Deutsche Forschungsgemeinschaft and the Fonds der Chemischen Industrie.

References

1. Merrill,A.H.Jr.,Lambeth,J.D.,Edmondson,D.E. & McCormick,D. B.:Annu.Rev.Nutr. 1, 281-317 (1981)

2. Borzenko,I.A.:Vrach Delo (10) 4-8 (1983)

3. Pinto,J.,Huang,Y.P.,Pellicione,N.,Rivlin,R.S.: Biochem. Pharmacol. 31(21),3495-3499 (1982)

4. Glatzle,D.,Weber,F. & Wiss,O.: Experientia 24, 1122 (1968)

5. Bates,C.J.,Prentice,A.M.,Paul,A.A.,Sutcliffe,B.A.,Whitehead, R.G.: Trans.R.Soc.Trop.Med.Hyg. 76(2), 253-258 (1982)

6. Fritsch,K.G.: Diplomarbeit, Freie Universität Berlin (1982)

7. Williams,C.H.Jr.: The Enzymes (Boyer,P.D.ed.), 3rd ed., vol.13,pp.89-172, Academic Press, New York (1976)

8. Schulz,G.E.,Schirmer,R.H. & Pai,E.F.: J.Mol.Biol. 160, 287-308 (1982)

9. Schirmer,R.H. & Schulz,G.E. in: Biological Oxidations, 34th Colloquium Mosbach (Sund,H. & Ullrich,V. eds) pp. 93-113, Springer Verlag,Berlin (1983)

10. Kappus,H. & Sies,H.: Experientia 37, 1233-1241 (1981)

11. Ames,B.N.:Science 221, 1256-1264 (1983)

12. Docampo,R. & Moreno,S.N.J.:Rev.Infect.Diseas. 6, 223-238 (1984)

13. Clark,I.A. & Hunt,N.H.: Infect.Immun. 39, 1-6 (1983)

14. Harisch,G.,Stegmann,H.,Scheffler,K.,Philippens,K.M.H., Schole,J.:Chronobiologia VII, 181-187 (1980)

15. Harlan,J.M.,Levine,J.D.,Callahan,K.S.,Schwartz,B.R. & Harker,L.A.: J.Clin.Invest. 73(3), 706-713 (1984)

16. Babior,B.M.: New Engl.J.Med. 298, 721-725 (1978)

17. Light,D.R.,Walsh,C.,O'Callaghan,A.M.,Goetzl,E.J. & Tauber, A.I.: Biochemistry 20(6), 1468-1476 (1981)

18. Schirmer,R.H.,Schulz,G.E. & Untucht-Grau,R.: FEBS-Lett.154, 1-4 (1983)

19. Borregaard,N. & Tauber,A.I.: J.Biol.Chem. 259(1), 47-52 (1984)

20. Jarasch,E.D.,Grund,C.,Bruder,G.,Heid,H.W.,Keenan,T.W. & Franke,W.W.: Cell 25, 67-82 (1981)

21. Pai,E.F. & Schulz,G.E.: J.Biol.Chem. 258(3),1752-1757(1983)

22. Carlberg,I. & Mannervik,B.: FEBS. Lett. 115, 265-268 (1980)

23. Richmond,R. & Halliwell,B.: J.Inorg.Biochem. 17, 95-107 (1982)

24. Pirie,A.,Rees,J.R. & Holmberg, N.J.: Exptl.Eye Res. 9, 204-

218 (1970)

25.Nathan.C.F.,Arrick,B.A., Murray,H.W.,Desantis,N.M. & Cohn, Z.A.: J.Exp.Med. 153(4), 767-782 (1981)

26.Babson,J.R.,Abell,N.S. & Reed,D.J.: Biochem.Pharmacol. 30 (16), 2299-2304 (1981)

27.Dershwith,M. & Novak,R.F.: Biochem.Pharm. 30, 676-678(1981)

28.Frischer,H. & Ahmad,T.: J.Lab.Clin.Med. 89, 1080-1091(1977)

29.Lederbogen,F.,Krauth-Siegel,R.L.,Schulz,G.E. & Schirmer,R. H.: Hoppe-Seyler's Z.Physiol.Chem. 363,966 (Abstr.)(1982)

30.Lederbogen,F.: MD-thesis Heidelberg (1983)

31.Gabig,T.G. & Lefker,B.A.: J.Clin.Invest. 73(3), 701-705, (1984)

32.Golenser,J.,Miller,J.,Spira,D.T.,Navok,T. & Chevion,M.: Blood 61, 507-510 (1983)

33.Kosower,N.S. & Kosower,E.M.: Lancet ii, 1343-1345 (1970)

34.Schirmer,R.H.,Lederbogen,F., Eisenbrand,G.,Königk,E. & Jung,A.: Parasitology (1984) in the press

35.Huheey,J.E. & Martin,D.L.: Experientia 31, 1145-1147 (1975)

36.Untucht-Grau,R.,Blatterspiel,R.,Krauth-Siegel,R.L.,Saleh,M., Schiltz,E.,Schirmer,R.H. & Wittmann-Liebold,B. in: Developments in Biochemistry vol.21 (Massey,V. & Williams,C.H.Jr., eds) pp 38-43, Elsevier North-Holland, Amsterdam (1982)

37.Eisenbrand,G.,Fiebig,H.H. & Zeller,W.J.: Z.Krebsforsch. 86, 279-286 (1976)

38. Chevion,M.,Navok,T.,Glaser,G. & Mager,J.: Eur.J.Biochem. 127, 405-409 (1982)

39.Picard-Maurcau,A.,Hempelmann,E.,Krämmer,G.,Jackisch,R. & Jung,A.: Tropenmed.Parasit. 26, 405-416 (1975)

40.Eckman,J.R. & Eaton,J.W.: Nature 278, 754-756 (1979)

41.Fritsch,KG.,Bayer,B.,Dicckmann,A.,von Cunow,D.,Schirmer,R. H. & Jung,A.: Hoppe-Seyler's Z.Physiol.Chem. (Abstr.)(1984) in the press

42.Loos,H.,Roos,D.,Weening,R. & Houwerzijl,J.: Blood 48(1), 53-62 (1976)

43.Varma,R.N.,Mankad,V.N.,Phelps,D.D.,Jenkins,L.D.,Suskind,R. M.: Am.J.Clin.Nutr. 38(6), 884-887 (1983)

44.Swarup-Mitra,S.,Bhattacharya,J. & Mukherji,A.: Indian J. Med.Res. 76, 668-673 (1982)

45.Lienhard,G.E. & Secemski,J.J.: J.Biol.Chem.248,1121 (1973)

46.Bilzer,M., Krauth-Siegel,R.L.,Schirmer,R.H.,Akerboom,T.P.M. Sies,H. & Schulz,G.E.: Eur.J.Biochem. 138, 373-378

STUDIES ON ESCHERICHIA COLI DNA PHOTOLYASE: ROLE OF FLAVIN IN DNA REPAIR

Marilyn Schuman Jorns

Hahnemann University School of Medicine
Philadelphia, Pennsylvania 19102, U.S.A.

Gwendolyn B. Sancar and Aziz Sancar

University of North Carolina School of Medicine
Chapel Hill, North Carolina 27514, U.S.A.

Introduction

Exposure of DNA to ultraviolet light results in the formation of dimers between adjacent pyrimidine bases in a photochemically reversible reaction. The dimers can cause mutation, cancer and death (1). They can be monomerized with shorter wavelenths of ultraviolet light or by DNA photolyase (E) in an enzymic reaction (equation 1) which requires visible light.

$$E + DNA\ (UV) \underset{k_2 \atop \text{dark}}{\overset{k_1}{\rightleftharpoons}} E \cdot DNA\ (UV) \xrightarrow[\text{light}]{k_3} E + DNA \qquad (1)$$

Since direct excitation of the dimers does not occur with visible light, it has been proposed that the photolyase reaction is a photosensitized process involving an enzyme-bound chromophore which absorbs in the visible region. This hypothesis has been difficult to evaluate since the enzyme is typically present at very low levels in most tissues. A major breakthrough in the study of DNA photolyase was provided by cloning techniques which have generated a strain of E. coli where photolyase

Flavins and Flavoproteins

constitutes 15% of the total protein in contrast to the parent strain which produces only 10-20 copies of the enzyme per cell (2). In this chapter we report studies (3) on the identification of the chromophores in the E. coli enzyme and compare the results with data obtained for photolyase from other sources. We also describe studies (4) on the substrate specificity of the E. coli enzyme which provide insight regarding the nature of the interaction of the enzyme with DNA.

Results

E. coli DNA photolyase is a blue protein exhibiting absorption maxima at 580, 475 and 384 nm and prominent shoulders at 625, 505 and 450 nm (Figure 1). The spectrum of the enzyme, particularly in the 500-700 nm range [a region overlooked in preliminary studies (5) when the spectrum of the enzyme was zeroed at 500 nm] is similar to that observed for blue neutral flavoprotein radicals (6). However, relative to absorbance at 580 nm, the absorbance of the photolyase at 384 nm is about twice the intensity expected for a neutral flavoprotein radical. Heat denaturation of the enzyme results in the disappearance of the long wavelength absorption band and the appearance of a typical oxidized flavin spectrum without any indication of a second chromophore in the 380 nm region (data not shown). TLC studies show that the heat extract contains a single fluorophore which comigrates with FAD. Evidence for a second chromophore was obtained by studying the denaturation of the enzyme with sodium dodecyl sulfate (SDS). Addition of SDS (0.77%) at 25°C results in the immediate disappearance of the absorption band around 580 nm and the appearance of oxidized flavin as judged by the increase in absorption at 450 nm (Figure 1). The initial spectrum after SDS addition also shows an intense band at 360 nm which decays in a relatively slow first order reaction ($k = 5.59 \times 10^{-2}\ min^{-1}$) to yield a typical flavin spectrum ($A_{450}/A_{375} = 1.17$ versus $A_{450}/A_{375} = 1.22$ with FAD), similar to that observed after heat denaturation. The results indicate that photolyase contains a second chromophore which decomposes upon denaturation at neutral pH to product(s) which do not absorb significantly

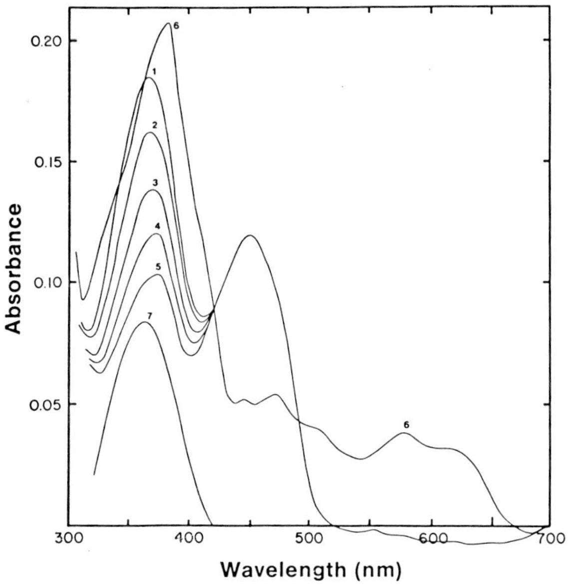

Figure 1 - Denaturation of photolyase with SDS. Curve 6 is the spectrum of native enzyme at pH 7.4. Curves 1-5 were recorded 50 seconds, 7, 16, 30 and 101 minutes, respectively, after addition of 0.77% SDS at 25°C. Curve 7 is the difference spectrum of curve 1 minus curve 5. Spectra are not corrected for a small negative deviation in the instrument baseline which is superimposible with curve 6 at λ > 515 nm.

in the visible. The spectrum of the protein-free second chromophore can be calculated by correcting the spectrum observed immediately after SDS addition for the contribution due to oxidized flavin. The calculated spectrum (Figure 1, curve 7) exhibits a peak at 360 nm and an intensity, relative to the initial native enzyme, which suggests that about 50% of the absorbance of the enzyme in this region is attributable to the second chromophore. The stability of the second chromophore is pH-dependent. Unlike the slow decomposition observed at pH 7.4, denaturation with SDS at pH 10.1 results in immediate decomposition whereas at pH 3.0 the chromophore is stable for at least 7 hours.

The amount of oxidized FAD released in denaturation experiments corresponds to 0.70 mole flavin/mole enzyme and can be used to calculate an extinction coefficient for the enzyme at 580 nm (ε = 3.6 x 10^{3} M^{-1} cm^{-1}) which falls with the range (3.4 - 5.4 x 10^{3} M^{-1} cm^{-1}) (6)

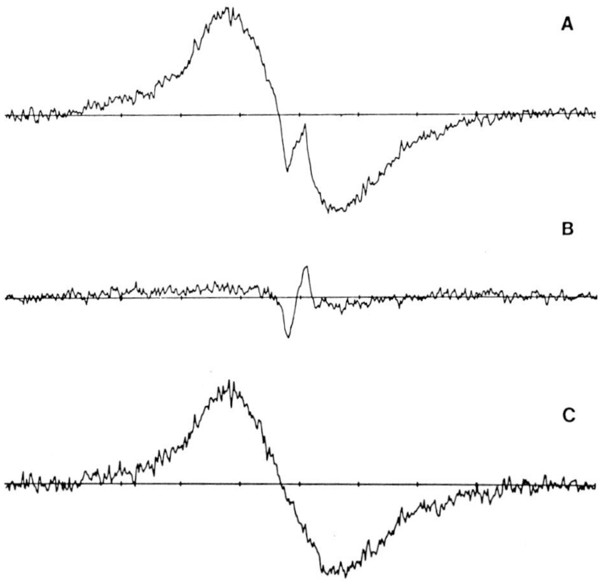

Figure 2 - The ESR spectrum of photolyase in Panel A was recorded in 50 mM Tris pH 7.4 containing 50 mM NaCl, 1.0 mM EDTA, 10 mM DTT and 50% glycerol. The signal observed for buffer alone is shown in Panel B. Panel C shows the difference spectrum of enzyme minus buffer. Each division on the X-axis represents 10 gauss.

observed with other blue flavoprotein radicals. Assuming a 1:1 stoichiometry of flavin and the second chromophore, the data can also be used to estimate an extinction coefficient for the protein-free second chromophore ($\varepsilon_{360} = 7.8 \times 10^3\ M^{-1}\ cm^{-1}$).

The ESR spectrum observed for DNA photolyase (Figure 2) exhibits a prominent signal with a line width of 19 gauss, similar to that observed for neutral flavoprotein radicals whereas red anionic flavin radicals show a band width of 14-15 gauss (7). The spectrum of the enzyme also contains an unexpected narrow signal superimposed on the broader flavin spectrum. This narrow signal is also observed with buffer alone (possibly due to sulfur radical from DTT) and is completely eliminated in the difference spectrum of enzyme minus buffer. The radical concentration in the sample used for ESR studies was estimated at 1.09×10^{-4} M using potassium nitrosodisulfonate as standard. This is in excellent agreement with that predicted from visible absorption studies (1.08×10^{-4} M).

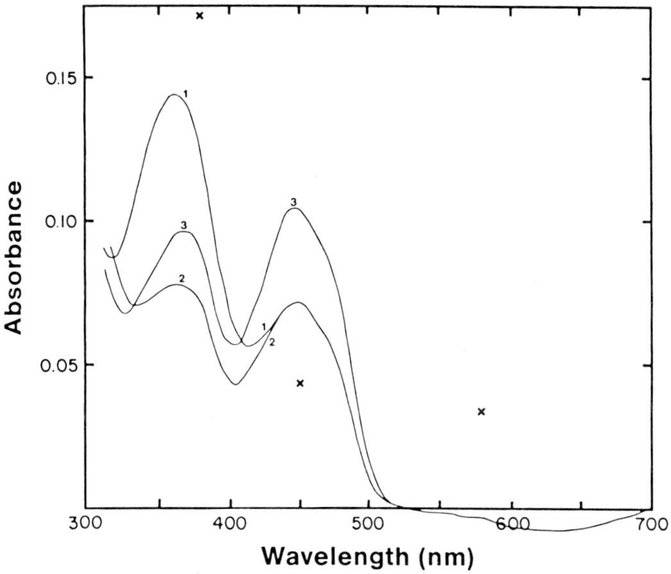

Figure 3 - Anaerobic denaturation of photolyase. After degassing at 0^oC, the absorbance of photolyase (pH 7.4) was quickly checked at several wavelengths (shown by X) to verify that no denaturation had occurred. The sample was warmed to 25^oC and SDS (0.77%) was added from a side arm. Curves 1 and 2 were recorded 2 and 60 minutes, respectively, after mixing. The sample was made aerobic after 64 minutes and curve 3 was recorded.

Unlike the air-stable radical observed with NADPH cytochrome P_{450} reductase (8), in tests with DNA photolyase no evidence for radical oxidation was obtained even with a 10-fold excess of $K_3Fe(CN)_6$. The air stability of photolyase was unaffected by addition of phenazine methosulfate as a potential mediator to oxygen. On the other hand, the flavin radical in DNA photolyase is readily reduced by dithionite. The fully reduced enzyme is not stable towards air oxidation which results in the return of the original radical spectrum. Radicals stable towards disproportionation can be formed with free flavin derivatives containing an alkyl substituent at position N(5). Since DNA photolyase contains a second chromophore along with flavin radical, it is conceivable that the second chromophore might be attached to the 5 position of the radical and that the protein-free radical might be stable in the absence of oxygen. To test this hypothesis, the enzyme was denatured in an anaerobic experiment under conditions where the radical is immediately oxidized when oxygen is present. As shown in Figure 3, anaerobic denaturation results in the immediate decay of the radical as judged by the disappearance of the long wavelength absorption band. An immediate increase in absorbance at 450 nm is also observed but the magnitude of this change is less than that observed in aerobic experiments. No further changes are observed at 450 nm upon incubation under anaerobic conditions. Slow decomposition of the second chromophore (monitored by decreases in absorption at 360 nm) does occur under anaerobic conditions at a rate ($k = 5.54 \times 10^{-2}$ min^{-1}) which is similar to that observed in aerobic experiments ($k = 5.59 \times 10^{-2}$ min^{-1}). After complete decomposition of the second chromophore, the sample was made aerobic which caused a 32% increase in absorbance at 450 nm. The results show that the protein-free radical is not stable under anaerobic conditions but rather disproportionates to a mixture of oxidized and reduced flavin as would be expected for an unsubstituted FAD radical. The percent of oxidized flavin recovered upon air oxidation is 18% less than expected based on the stoichiometry of the disproportionation reaction. This might reflect a limited oxidation of the reduced flavin by small amounts of disulfide or sulfur radical associated with the DTT (1 mM) component of the buffer. No oxygen ($< 10^{-6}$ M) was detected in control tests of the anaerobic system.

E. coli DNA photolyase is fluorescent, exhibiting an emission maximum at 470 nm (Figure 4) and an excitation maximum at 398 nm. That the latter does not precisely match the absorption maximum at 384 nm is not unexpected, since aside from the fact that fluorescence spectra are uncorrected, the absorption of the enzyme in this region reflects contributions from two chromophores. An unlikely fluorescent flavin radical can be excluded because the observed emission band occurs at higher energy than the lowest energy band in the radical absorption spectrum. The enzyme fluorescence is tentatively attributed to the second chromophore. Denaturation with heat or SDS results in the disappearance of the 470 nm emission band and the appearance of a band at 519 nm, as expected for free FAD. The heat extract also shows a weak band at 455 nm which is not seen with authentic FAD. The band at 455 nm slowly appears in samples denatured with SDS at pH 7.4 but is not detected at pH 3.0

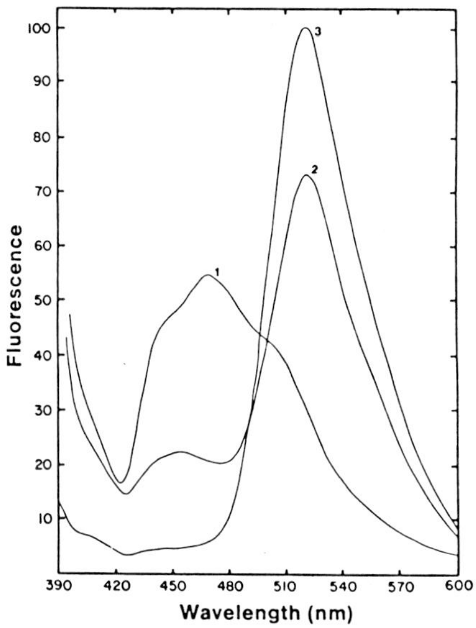

Figure 4 - Emission spectra of photolyase (excitation at 380 nm) were recorded at 5.5°C before (curve 1) and after (curve 2) heat denaturation at pH 7.4. Curve 3 is the emission spectrum recorded at 25°C after denaturation at pH 3.0 with 0.77% SDS.

(Figure 4) where the protein-free second chromophore is stable. This suggests that the 455 nm band arises from the decomposition of the second chromophore. (The decomposition product is not detectable in excitation spectra which are very similar to that observed with FAD.) The loss of the emission band at 470 nm when the sample is denaturated at pH 3.0 suggests that the fluorescence of the second chromophore is quenched in acidic solution.

In order to facilitate studies on the nature of the interaction of photolyase with DNA and on the role of the chromophores in catalysis, we initiated a search for simple DNA analogues which might function as substrates and also provide the basis for a new activity assay. Existing assays (2, 9) use DNA as substrate, are quite sensitive but are slow and/or require time-consuming preparation of radioactive substrates. Studies to determine whether the enzyme might repair dimers in oligodeoxythymidylates (oligo$(dt)_n$) were prompted by a report by Setlow and Bollum (10) who showed that UV-irradiated oligo$(dt)_n$ ($n \geq 9$) was bound to yeast DNA photolyase. We initiated our studies with oligo(dt)$_{18}$ since this was the shortest chain oligomer that exhibited binding properties with the yeast enzyme similar to that observed with polydeoxythymidylate (poly(dt)). In comparison with thymine, thymine dimers do not exhibit significant absorption at 260 nm. Irradiation of oligo(dt)$_{18}$ with light at 254 nm results in the rapid formation of dimers as evidenced by decreases in absorption at 260 nm. At a maximum, 37% conversion to dimers was observed which compares favorably with results obtained by Deering & Setlow (11) with poly(dt) (39% dimer) and oligo(dt)$_2$ (33% dimer). In contrast, a maximum of 7% dimer is formed with DNA under similar conditions (12). If the oligomer functioned as a substrate for photolyase, the high dimer content and the absence of interference due to purine absorption could be useful properties in developing a rapid spectrophotometric assay. We find that illumination of UV-irradiated poly(dt)$_{18}$ with black light in the presence of photolyase results in increases in absorption at 260 nm to a level similar to that observed before UV-irradiation. Studies with varying amounts of photolyase show that the rate of dimer monomerization is directly proportional to enzyme concentration. The reaction is catalytic as evidenced by the fact that the enzyme is able to repair dimers present in 400-fold excess. Absorbance changes at 260 nm in the range 0.01 - 0.10 are routinely measured in a 250 ul assay containing 2.36×10^{-6} M poly(dt)$_{18}$ (34-37% dimer). This corresponds to 180-1800 picomoles of dimer repaired. With 5 picomoles of enzyme the lower detection limit is reached in 5 minutes. While less sensitive than other photolyase assays (1 picomole detectable) the spectrophotometric assay requires only modest amounts of enzyme and offers the advantage of speed and convenience since the substrate is commercially available.

A turnover number for the enzyme with oligo(dt)$_{18}$ was calculated based on the rate observed with 2.38×10^{-6} M poly(dt)$_{18}$ (7.86×10^{-6} M dimers). That this substrate concentration is saturating is evidenced by the fact that the rate is linear for 90% of the reaction and is unaffected by doubling or tripling the initial substrate concentration. A value of 3.4 min^{-1} is obtained when enzyme concentration is based on its flavin content or 2.4 min^{-1} when calculated based on protein since the preparation contains 0.7 mole flavin/mole protein. The latter calculation is comparable to that used in previous studies where a turnover number of 2.4 min^{-1} was obtained (2) for the enzyme with plasmid DNA as substrate under otherwise similar reaction conditions. The results indicate that oligo$(dt)_{18}$ is remarkably comparable to natural DNA as a substrate for the enzyme. This means that turnover is not significantly affected by the fact that DNA is double-stranded. That the reaction rate observed with oligo$(dt)_{18}$ is constant until the dimer concentration falls to about 8×10^{-7} M suggests that the K_m for oligo$(dt)_{18}$ is very small. A K_m value for the E. coli enzyme with DNA as substrate has not been reported but a value around 10^{-8} to 10^{-9} M can be estimated for the dissociation constant of the enzyme-DNA complex (13).

With double-stranded DNA the introduction of a dimer results in denaturation of about 4 base pairs. The ability of photolyase to seek out and repair dimers in DNA might reflect the ability of the enzyme to recognize the particular deformation in DNA structure caused by the dimer. Alternatively, the critical factor may be recognition of the dimer itself. The latter hypothesis is supported by the fact that the enzyme exhibits similar activity with single-stranded oligo$(dt)_{18}$ and duplex DNA. Further evidence was sought by examining the effect of chain length on activity. If dimer recognition is a critical factor, the enzyme might also be expected to repair dimers in short chain oligodeoxythymidylates. The results of a survey show that oligomers containing 7 or more thymine residues appear to be as good as substrates as the higher homologues, including polydeoxythymidylate. Starting with oligo$(dt)_6$, a progressive decrease in activity is observed (activity with oligo$(dt)_n$, expressed as a percent of that observed with oligo$(dt)_{18}$ = 76%, 55%, 49% and 13% when n = 6, 5, 4 and 3, respectively) until, with oligo$(dt)_2$, no activity was detectable. This survey was conducted at a constant thymine residue concentration (4.21×10^{-5} M) and a fairly constant thymine dimer concentration except for a relatively modest decrease (at maximum, 38% lower with oligo$(dt)_2$) with oligomers containing 7 or less thymine residues. The dimer concentrations used in this survey are saturating in the case of oligomers containing 7 or more thymine residues but not for the shorter homologues where an increase in activity is observed when the dimer concentration is doubled or tripled. The data at higher substrate concentrations show that the decrease in activity observed with $oligo(dt)_n$ when n = 4-6 is due to a K_m effect and not to a decrease in V_{max}. No evidence for saturation was obtained with $oligo(dt)_3$ where tripling the substrate concentration results in an increase in activity from 13% to 43% of the rate observed with $oligo(dt)_{18}$. No activity was detectable with $oligo(dt)_2$ even at higher substrate concentrations.

Nearly quantitative repair (90-100%) is observed with $oligo(dt)_n$ when $n \geq 9$. A progressive decrease in the extent of repair is observed with the lower homologues. Table I compares the observed extent of repair with the expected distribution of dimers at internal positions and at the 3' and 5' ends for the case of short chain oligomers containing one dimer per molecule. The latter condition is reasonable, based on the number of residues per molecule and the percent of residues present as dimers. The observed extent of repair agrees reasonably well with that expected for repair of internal dimers plus selective repair of dimers involving a residue at one of the two terminal positions. With the oligomers used in these studies $(p(dt)_n)$, a thymine residue at the 5' end is attached to a

TABLE I

Extent of Dimer Repair with Short Chain Oligodeoxythymidylates

n	Expected Dimer Distribution (%): Internal (a)	5' end (b)	3' end (c)	a + b	Extent of Enzymic Repair (%)
7	67	17	17	83	78
6	60	20	20	80	81
5	50	25	25	75	71
4	33	33	33	67	54
3	0	50	50	50	36

ribosyl moiety containing phosphate groups as both the 3' end and 5' positions and, in this sense, more closely resembles an internal residue as compared with a thymine at the opposite end where the corresponding ribosyl moiety contains an unsubstituted 3'-hydroxyl group. That this difference might provide the basis for selective repair of dimers at the 5' end is supported by studies with dephospho-oligo$(dt)_4$ $((dt)_n$ where the observed extent of repair (35%) agrees fairly well with the expected content of internal dimers (33%), in contrast to the considerably higher repair observed with oligo$(dt)_4$ (54%). The results suggest that the enzyme can repair dimers at the 5' end only when it is phosphorylated and cannot repair dimers at an unphosphorylated 3' end. The latter can explain why oligo$(dt)_2$ is not a substrate since the only dimer possible in oligo$(dt)_2$ involves an unphosphorylated 3' end.

Discussion

E. coli DNA photolyase contains two chromophores, a neutral FAD radical and a second chromophore which is probably responsible for enzyme fluorescence. Assuming the enzyme repairs DNA via a photosensitized reaction, it is reasonable to suppose that the light required for catalysis is absorbed by one of two chromophores. Action spectra could be used to discriminate between the two chromophores but available data, obtained with crude enzyme or in in vivo studies (14) are not useful since they show a maximum in the region (360-380 nm) where both chromophores absorb and were not extended into the 500-700 nm region where only the flavin radical absorbs. Evidence in favor of the second chromophore is provided by the fact that no dimer repair is observed for "dark" controls when assay mixtures are prepared under yellow light which would be absorbed only by the flavin radical. While speculations regarding the mechanism of the E. coli enzyme are clearly premature, the presence of a flavin radical as a potential electron donor or acceptor is particularly tantalizing since it is known (15,16) from model reactions that 1-electron reduction or 1-electron oxidation results in monomerization of thymine dimers.

Flavin has been found in photolyase from other sources. The enzyme from Streptomyces griseus contains a 7,8-didemethyl-8-hydroxy-5-deazaflavin derivative (17). Photolyase I from yeast releases FAD upon denaturation. The yeast enzyme exhibits absorption (λ_{max} = 380 nm) and fluorescence (emission λ_{max} = 475-480 nm, excitation λ_{max} = 375-380 nm) properties similar to the E. coli enzyme (18,19). Based primarily on the absorption band at 380 nm it was suggested that the yeast enzyme might contain a 4a,5-reduced FAD derivative (18). A similar proposal was made for the E. coli enzyme in preliminary studies (5) when the absorbance of the enzyme in the 500-700 nm region was not known. The absorbance of the yeast enzyme in the long wavelength region is currently unknown. The photolyase from E. coli described in this chapter (M_r = 53994) has been referred to as photolyase F (2) in order to distinguish it from an entirely different RNA-containing protein (photolyase R, M_r = 36800) also isolated from E. coli (20,21). Photolyase F appears to be the major enzyme in E. coli since it is 1000-fold more active than photolyase R and cells mutated at the gene which codes for the apoenzyme of photolyase F have very little photoreactivating activity (22).

E. coli photolyase can efficiently repair dimers in compounds which correspond to short chain, single-stranded analogues for DNA. This provides strong evidence that recognition of the dimer itself is likely to be a critical factor in catalysis with native DNA. With UVRABC excision nuclease [a key enzyme in an alternate pathway for dimer repair in E. coli (23)] recognition of the deformation in the DNA structure caused by the dimer is important in catalysis. This is evidenced, in part, by the fact that the nuclease requires double-stranded DNA as substrate (A. Sancar, unpublished). In addition, the nuclease responds to a broad variety of DNA modifications that have no resemblance to pyrimidine dimers. That the photolyase and the nuclease have very different binding sites is supported by the fact that photolyase causes a 2-fold stimulation of nuclease activity (measured in the dark) when dimer-containing DNA is used as substrate but not with nuclease substrates containing other DNA modifications. We interpret this to mean that, by binding the pyrimidine dimer, photolyase increased the distortion caused by the dimer and thus made it a better substrate for the nuclease.

Flavin-containing photolyase enzymes may be grouped with several other flavoenzymes that also catalyze reactions that do not involve a net oxidation-reduction. In oxynitrilase, flavin functions as a structural component at the active site (24). Flavin is not bound at the active site in glyoxylate carboligase where it apparently functions to maintain the gross conformation of the enzyme (25). Although the oxidation-reduction properties of flavin are not directly involved in catalysis with either enzyme, these properties may be important in regulation since the enzymes are inhibited when the flavin is reduced. N-Methylglutamate synthetase may represent the closest model for the photolyase reaction in the broad sense that catalysis with the synthetase involves a cyclic oxidation-reduction process where flavin functions as a typical redox catalyst (26).

References

1. Rupert, C.S.: J. Gen. Physiol. 45, 703-724 (1962).

2. Sancar, A., Smith, F.W., Sancar, G.B.: J. Biol. Chem. 259, 6028-6032 (1984).

3. Jorns, M.S., Sancar, G.B., Sancar, A.: Biochemistry, in press (1984).

4. Jorns, M.S., Sancar, G.B., Sancar, A.: submitted.

5. Sancar, A., Sancar, G.B.: J. Molec. Biol. 172, 223-227 (1984).

6. Müller, F., Brüstlein, M., Hemmerich, P., Massey, V., Walker, W.H.: Eur. J. Biochem. 25, 573-580 (1972).

7. Palmer, G., Müller, F., Massey, V.: in Flavins and Flavoproteins (Kamin, H., Ed.), pp. 123-137, University Park Press, Baltimore (1971).

8. Vermilion, J.L., Coon, M.J.: J. Biol. Chem. 253, 8812-8819 (1978).

9. Sutherland, B.M.: in The Enzymes, Volume XIV, pp. 481-515, Academic Press, New York (1981).

10. Setlow, J.K., Bollum, F.J.: Biochim. Biophys. Acta 157, 233-237.

11. Deering, P.A., Setlow, R.B.: Biochim. Biophys. Acta 68, 526-534 (1963).

12. Patrick, M.H., Rahn, P.O.: in Photochemistry and Photobiology of Nucleic Acids, Volume 11 (Wang, S.Y., ed.), p. 57, Academic Press, New York (1976).

13. Harm, H., Rupert, C.S.: Mutation Res. 6, 355-370 (1968).

14. Setlow, J.K.: Curr. Top. Radiat. Res. 2, 195-248 (1966).

15. Helene, C., Charlier, M.: Photochem. Photobiol. 25, 429-434 (1977).

16. Roth, H.D., Lamola, A.A.: J. Am. Chem. Soc. 94, 1013-1014 (1972).

17. Eker, A.B.M., Dekker, R.H., Berends, W.: Photochem. Photobiol. 33, 65-72 (1981).

18. Iwatsuki, N., Joe, C.O., Werbin, H.: Biochemistry 19, 1172-1176 (1980).

19. Madden, J.J., Werbin, H.: Photochem. Photobiol., in press.

20. Snapka, R.M., Sutherland, B.M.: Biochemistry 19, 4201-4208 (1980).

21. Sutherland, B.M., Chamberlin, M.J., Sutherland, J.C.: J. Biol. Chem. 248, 4200-4205 (1973).

22. Sancar, A., Rupert, C.S.: Gene 4, 294-308 (1978).

23. Sancar, A., Rupp, W.D.: Cell 33, 249-260 (1983).

24. Jorns, M.S.: J. Biol. Chem. 254, 12145-12152 (1979).

25. Cromartie, T.H., Walsh, C.T.: J. Biol. Chem. 251, 329-333 (1976).

26. Jorns, M.S., Hersh, L.B.: J. Biol. Chem. 250, 3620-3628.

ELECTRON-TRANSFER TO NITROGENASE IN *K. PNEUMONIAE*. *nifF* GENE CLONED AND THE GENE PRODUCT, A FLAVODOXIN, PURIFIED AND PARTIALLY CHARACTERISED

J. Deistung, S. Hill, R.N.F. Thorneley
A.F.R.C. Unit of Nitrogen Fixation, University of Sussex, Brighton, BN1 9RQ, U.K.

F. Cannon and M. Cannon
Biotecnica International Inc., 85 Bolton Street, Cambridge, Massachusetts 02140, U.S.A.

The nitrogen fixation (*nif*) gene cluster of *K. pneumoniae* comprises 17 genes which are required for the *in vivo* synthesis and activity of nitrogenase. *K. pneumoniae* nitrogenase comprises the MoFe protein (Kp1, M_r = 220000) and the Fe protein (Kp2, M_r = 67000). The *nifF* gene product is a flavodoxin that is involved in the electron transfer sequence shown in Eqn. (1).

$$\text{Pyruvate} \rightarrow \underset{\substack{\text{oxido-reductase} \\ nifJ}}{\text{Pyruvate-Flavodoxin}} \rightarrow \underset{nifF}{\text{Flavodoxin}} \rightarrow \text{Kp2} \rightarrow \text{Kp1} \rightarrow \underset{(N_2,\ H^+)}{\text{substrate}} \qquad \text{Eqn. (1)}$$

An elevated synthesis of the *nifF* gene product *in vivo* was achieved by cloning the *nifF* gene on a multicopy plasmid. The transformant, UNF5112, a *K. pneumoniae* strain deleted for chromosomal *nif* genes, contained the constructed plasmid pBCC13, a derivative of pBR328 carrying the *nifF* gene with part of *nifL* and *nifM* on a *Bgl*II/*Xho*I fragment, and the *Eco*RI/*Hind*III fragment of *nifH* and *nifJ*. Because maximal transcription of *nifF* requires the *nifA* gene product, the transformant also contained a plasmid (pMC71a) carrying the nifA gene under the control of a constitutive promoter.

The nifF gene product was purified using a method based on that described by Shah et al. (1) with the assay procedure developed by Hill & Kavanagh (2). Cells were ruptured by French pressing and a crude extract supernatant obtained by centrifugation. Chromatography with DEAE cellulose, gel filtration with Sephadex G-50 and finally preparative gel electrophoresis yielded a yellow protein (M_r = 18000±500) with ε_{450} = 9.3 mM cm^{-1}, $\varepsilon_{450}:\varepsilon_{280}$ = 0.212, which was homogenous as judged by SDS-PAGE.

References

1. Shah, V.K., Stacey, G., Brill, W.J.: J. Biol. Chem. 258, 12064-12068 (1983).
2. Hill, S., Kavanagh, E.P.: J. Bact. 141, 470-475 (1980).

Practical Applications of Flavoprotein Studies

IN VITRO SYNTHESIS OF (BIO)CHEMICALS USING FLAVIN-CONTAINING ENZYMES

Colja Laane*, Riet Hilhorst, Ruud Spruijt, Koen Dekker and Cees Veeger

Department of Biochemistry, Agricultural University, De Dreijen 11,
6703 BC Wageningen, The Netherlands.
*To whom correspondence should be addressed.

Abstract

We will demonstrate two systems in which flavin-containing enzymes together with others are used for the synthesis of various (bio)chemicals of commercial interest.
The first system consists of the enzymes hydrogenase, the FAD-containing enzyme lipoamide dehydrogenase and 20β-hydroxysteroid dehydrogenase. Upon enclosure of this multi-enzyme system in a reversed micellar medium, hydrogenase and lipoamide dehydrogenase regenerate NADH which is consumed by 20β-hydroxysteroid dehydrogenase during the stereo- and sitespecific reduction of apolar ketosteroids to their corresponding 20β-hydroxyform.
We will also demonstrate that in this system NADH can be regenerated electrochemically with lipoamide dehydrogenase in the absence of H_2 and hydrogenase.
In the second system the FAD-containing enzymes glucose oxidase and xanthine oxidase are used in combination with other enzymes to produce various (bio)chemicals in a bioelectrochemical fuel cell according to the following equations:

$$\text{glucose} \xrightarrow{\text{glucose oxidase}} \text{gluconic acid} + H_2 + \text{electricity}$$

$$\text{cellulose} \xrightarrow[\text{glucose oxidase}]{\text{cellulases}} \text{gluconic acid} + H_2 + \text{electricity}$$

$$\text{xanthine} + \tfrac{1}{2}O_2 \xrightarrow{\text{xanthine oxidase}} \text{product(s)} + H_2O + \text{electricity}$$

$$\text{glucose} + O_2 + \text{barbituric acid} + HCl \xrightarrow[\text{chloroperoxidase}]{\text{glucose oxidase}} \text{gluconic acid} + \text{5-chlorobarbituric acid} + 2H_2O$$

Flavins and Flavoproteins

Introduction

Enzymes possess several properties that make them more skillful in synthesizing (bio)chemicals than ordinary chemical catalysts. For example, enzymes catalyze, under mild conditions, site- and/or stereo-specific reactions that are otherwise difficult or impossible to perform and furthermore produce pure compounds. Despite these obvious merits less than 1% of all known enzymes are used in industry. This has both scientific and practical causes. First, enzymes are relatively labile catalysts and are usually expensive to isolate. Second, enzymes preferentially function in aqueous environments, while in many important chemical reactions non-watersoluble compounds are involved. Third, certain classes of enzymatic reactions of general interest involve cofactors which are expensive and are consumed during catalysis. Hence, the cofactor has to be regenerated from inexpensive reagents to make an enzymic process economically feasible.
Recently we designed a system in which these problems were overcome [1,2]. It consists of a solution of tiny water droplets stabilized by surfactants in a bulk water-immiscible organic solvent. We were able to show that in such a reversed micellar solution poorly water-soluble steroids can be converted enzymatically using an in situ NADH-regenerating enzyme system and H_2 or electricity as ultimate reductants [1,2].
The first part of this paper summarizes our present state of knowledge on reversed micellar enzymology. In the second part we will demonstrate that the flavoenzymes glucose oxidase and xanthine oxidase have potential for the synthesis of various (bio)chemicals in a bioelectrochemical fuel cell.

Materials and Methods

Hydrogenase (EC 1.12.2.1) from Desulfovibrio vulgaris strain Hildenborough NCIB 8303 was purified as in [3]. Pigheart lipoamide dehydrogenase (EC.1.6.4.3), 20β-hydroxysteroid dehydrogenase (EC 1.1.1.53) from Streptomyces hydrogenans, Caldariomyces fumago chloroperoxidase (EC 1.11.1.10) were obtained from Sigma. Aspergillus niger glucose oxidase (EC 1.1.3.4), cow milk xanthine oxidase (EC 1.2.3.2) and all enzymes required for analytical purposes were purchased from Boehringer. The industrial enzyme preparations

Maxazym CL (mainly endoglucanase (EC 3.2.1.4) and exoglucanase (EC 3.2.1. 91) , deficient in β-glucosidase (EC 3.2.1.21)) and Rapidase C-80 (mainly β-glucosidase, deficient in endo- and exoglucanase) were obtained from Gist-Brocades, The Netherlands for the breakdown of phosphoric-acid-swollen Avicel cellulose. All other (bio)chemicals were of analytical grade and obtained from commercial sources.

Reversed micellar media were prepared and enzyme activities were measured exactly as described in [1,2]. A typical reversed micellar medium contained: 0.2 M CTAB in octane, 50 mM HEPES pH 7.6 and the desired amount of hexanol. At a w_0 of 10 this is 54 μl 50 mM HEPES, pH 7.6, 0.11 g CTAB, 1.5 ml octane and 0.16 ml hexanol (10% w/w).

The bioelectrochemical cell described in [4] was used. The experiment performed with the xanthine oxidase containing cell (Fig. 7) and the combined enzymatic fuel cell (Fig. 8) are described in [4].

The conditions for the H_2-producing fuel cells were as follows: The anaerobic anode compartment contained a solution (4 ml) of 0.1 M Tris/HCl, 1 mM N,N'-di(γ-aminopropyl)dipyridinium bromide hydrobromide (DAPV) and 10 mg glucose oxidase, final pH 8.2. DAPV was synthesized according to [5]. To this solution either glucose (10 μmoles), or 3.5 mg (9 μmole glucose units) phosphoric-acid-swollen Avicel cellulose together with 18 mg of Maxazym CL plus 0.5 mg of glucose-free Rapidase C-80 were added. The cathode compartment contained a solution (4 ml) of 0.5 M sulphuric acid, final pH 1.0. Both compartments were stirred rapidly with a magnetic stirrer and kept at 25°C. Reactions were started by adding the appropriate substrate to the anode compartment.

H_2 production was determined by gas chromatography (Pye Unicam GCD chromatograph equipped with a catharometer detector device) and O_2 concentrations were measured with a Clark O_2-electrode mounted in the cathode compartment. The concentrations of glucose, fructose, sucrose, and gluconic acid were determined as described in refs. 6, 7, 8 and 9, respectively. The current produced by the cell was measured as described in refs. 4 and 10.

Results and discussion

Cofactor regeneration and conversion of apolar steroids in reversed micellar media

Fig. 1 shows our NADH regenerating steroid converting reversed micellar model system. The medium consists of 4 components *e.g.* the cationic surfactant cetyltrimethylammonium bromide (CTAB), the cosurfactant hexanol, an organic solvent usually octane, and an aqueous solution of HEPES buffer.

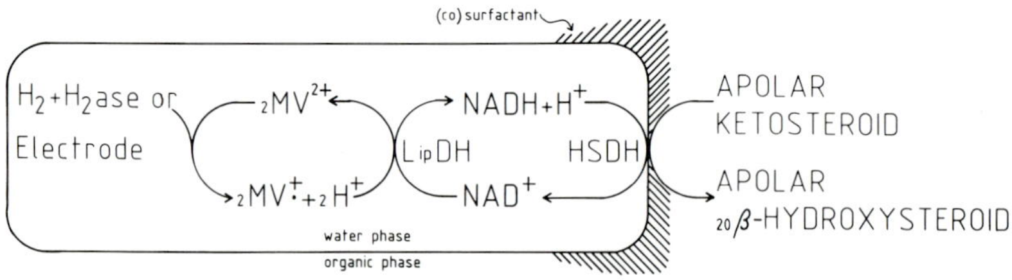

Fig. 1. Scheme for the H_2- or electrochemically driven regeneration of NADH and the subsequent reduction of an apolar steroid in a reversed micellar medium. Abbreviations: H_2-ase, hydrogenase; lipDH, lipoamide dehydrogenase; HSDH, hydroxysteroid dehydrogenase; MV, methylviologen.

In analogy to a system for aqueous media [11] hydrogenase (H_2-ase) and lipoamide dehydrogenase (lipDH) are used for the *in situ* regeneration of NADH which is consumed during the site- and stereospecific reduction of an apolar ketosteroid to its corresponding 20β-hydroxyform. Alternatively NADH can be regenerated electrochemically, in the absence of H_2 and hydrogenase.

Fig. 2 shows the time course of progesterone reduction using either H_2 (Fig. 2A) or electricity as ultimate reductant (Fig. 2B). In both cases the reaction rate declines when the conversion is almost complete. In the H_2-driven system not more than 95% conversion was achieved, whereas in the electrochemical system the conversion is >99.5%. This difference stems from the fact that the reducing power generated with the H_2-H_2ase couple declines in time due to inactivation of H_2ase, whereas such difficulties

do not occur when reducing power is imposed by the potentiostat.
Fig. 2 also shows that at higher concentrations of progesterone the reaction rate increases and that the reaction proceeds steadily for >10 hrs., indicating that the system is stable for a considerable length of time. Determination of turnover numbers (moles of product formed/mole of enzyme or cofactor) in the H_2-driven system after 28 h reaction time gave values of 50000, 13000 and 11000 for H_2ase, lipDH and HSDH respectively, whereas

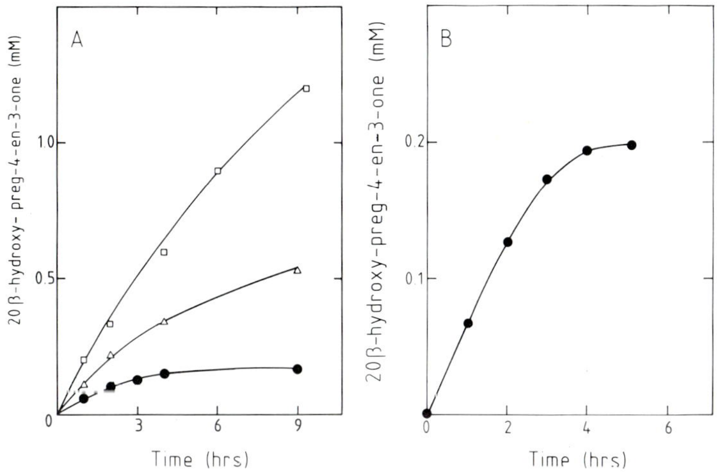

Fig. 2. Time course of 20β-hydroxy-pregn-4-en-3-one formation at different concentrations of progesterone. A. H_2-driven system; B. Electrochemically-driven system. In A, experiments were performed as in [1]. In B, experiments were performed at pH 6.0 in a 3 ml gas-tight bioelectrochemical cell in which the compartments were separated by a dialysis membrane. The Pt-cathode potential was set at -450 mV *vs.* the Ag/AgBr reference electrode with a Gerhard Bank potentiostat. In the anode compartment an Fe-counter electrode was used in combination with bathophenanthroline. In both systems CTAB-hexanol-octane reversed micelles were applied as described in [1], except that in B 0.1 M tetraheptylammonium bromide was added as a conductivity salt. (●-●) 0.2 mM; (Δ-Δ) 1.0 mM; (□-□) 5.0 mM.

for NADH a value of 100 was obtained. As a result the cofactor costs are reduced from ~$ 6000/mol for NADH to $ 60/mol using cheap starting materials. Although higher turnover number (>1000) are required to make such an enzymatic conversion economically feasible, our results are encouraging and future research will be focused on the stability of NADH regenerating systems in reversed micellar media. A system presently under investigation contains the FAD-enzyme H_2ase from <u>Alcaligenes eutrophus</u>. This enzyme reduces NAD^+ directly with H_2 [12] and as a result only one enzyme is required for NADH-regeneration instead of two.

In evaluating the pros and cons of NADH-regenerating systems in reversed micellar media several remarks must be made. A beneficial property of the electrochemical system is that the electrode potential can be adjusted to any desired value, whereas the potential of the H_2-H_2ase couple is predetermined by the conditions in the aqueous environment. Consequently, in an electrochemical system both oxidation and reduction reactions can be performed. Yet, the electrochemical system is much more laborious and seems therefore less suitable for practical purposes. An exception might be a system in which an oxidation reaction is coupled to a reduction reaction (Fig. 3).

The electrons generated at the oxidation site are pumped *via* an external circuit to the reduction site and the protons liberated during the oxida-

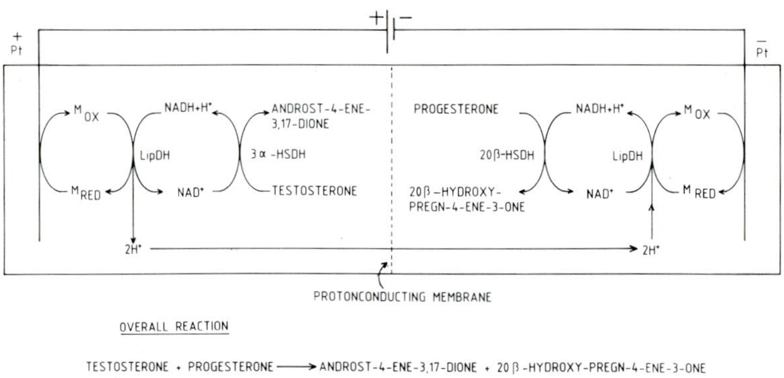

Fig. 3. Schematic representation of an electrochemical system for the oxidation and reduction of apolar steroids in a reversed micellar medium. See for abbreviations Fig. 1. M denotes mediator.

tion are consumed during the reduction. Hence, two products are synthesized simultaneously and in different compartments.

Synthesis of (bio)chemicals in a bioelectrochemical cell

Fig. 4 shows an enzyme-containing bioelectrochemical fuel cell. To date, the main objective of such devices is the generation of electricity [13], or the development of sensitive detectors for (bio)chemicals [10,14]. In addition to these objectives we have demonstrated recently that a bioelectrochemical cell can also be applied for the synthesis of various (bio)-chemicals [4].

For example, in case the cell contains glucose oxidase the reactions are as follows: in the anaerobic anode compartment glucose oxidase oxidizes glucose to glucono-1,5-lactone which is non-enzymatically hydrolyzed to gluconic acid, a valuable industrial chemical currently produced by fermentation [15]. In the absence of O_2 the reduced enzyme in turn donates its reducing equivalents to an electron mediator, which is subsequently oxidized at the anode.

The electrons thus generated flow through the external circuit to the cathode, where an acceptor is reduced provided the redox potential of the mediator in the anode compartment is lower than the potential of the acceptor in the cathode compartment. O_2 usually serves as an electron acceptor

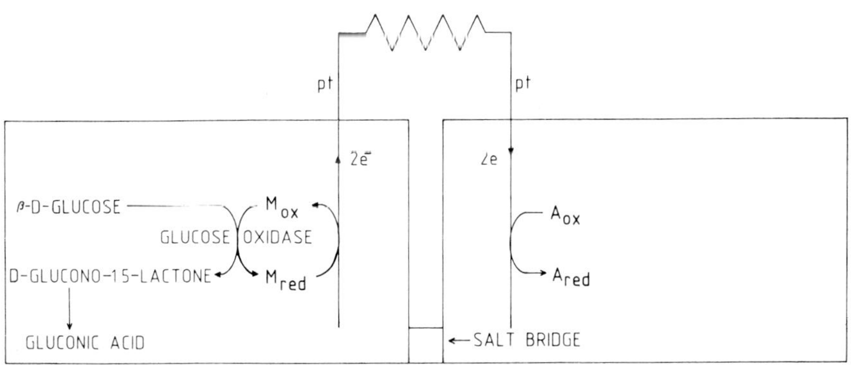

Fig. 4. Generalized scheme of reactions taking place in a glucose oxidase containing bioelectrochemical cell. For explanation see text. M denotes mediator and A acceptor.

In that case the cell can operate using mediators such as 2,6-dichlorophenolindophenol (E_0' = +0.217 V vs. NHE), N,N,N',N'-tetramethyl-4-phenylenediamine (E_0' = +0.26 V) or phenazine (m)ethosulfate (E_0' = +0.08 V) since the driving force is sufficient to reduce O_2 (E_0' = +0.82 V in case H_2O is produced or E_0' = +0.28 V in case H_2O_2 is produced). However, for the production of H_2 the reducing power of these mediators is insufficient. Redox mediators are required with an E_0' between 0 and the glucose-gluconic acid redox couple (-0.38 V). Of all the compounds tested only DAPV (E_0' = -0.355 V) and to a lesser extent benzyl viologen (E_0' = -0.359 V) were capable of mediating electrons between glucose oxidase and the platina electrode.

Fig. 5 shows the results obtained with our glucose-driven, H_2-producing bioelectrochemical cell. DAPV is used as a mediator. In the absence of O_2, H_2 is produced in the cell in almost stoichiometric amounts with gluconic acid (80-90%). Furthermore, the results indicate that glucose is converted completely and exclusively according to equation (1) and that the total amount of current passing through the circuit is directly proportional to the amount of glucose added.

$$\text{glucose} \xrightarrow{\text{glucose oxidase}} \text{gluconic acid} + H_2 + \text{electricity} \qquad (1)$$

The current output of the cell and thus the rate of gluconic acid and H_2 formation depends strongly on the pH differences between the two compartments. The pH of the anode compartment was fixed at pH 8.2 since at this pH the rate of DAPV reduction was highest. Routinely the pH of the cathode compartment was kept at pH ~1. Under these circumstances the current output is ~50 μA and the operational potential difference of the cell is 0.25 V. With O_2 as acceptor currents of 600 μA were easily obtained which is not surprisingly given the larger driving force [4].

As was shown by Van Bekkum and co-workers reaction (1) can also be performed chemically by a Pt/C or Rh/C catalyzed dehydrogenation of glucose [16]. However, in the course of time the rate of dehydrogenation decreases due to co-adsorption of the product on the catalyst surface. In our bioelectrochemical cell such phenomena do not occur and hence constant conversion rates are obtained.

Fig. 6 shows the results obtained with cellulose as fuel. Here, glucose is produced _in situ_ by the combined action of cellulases _i.e._ endoglucanase,

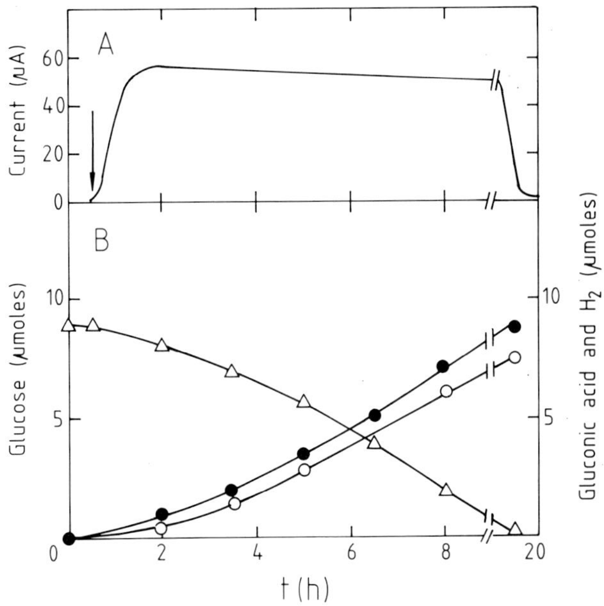

Figure 5. Performance of a bioelectrochemical cell using glucose as a fuel. Experiments were performed as described in Materials and Methods. At the time indicated by an arrow glucose was added to the anode compartment. (A) Time course of current output. (B) Time course of: O, H_2-production; ●, gluconic-acid production; Δ, glucose consumption.

exoglucanase and β-glucosidase. The enzyme activities were tuned in such a way with respect to each other that the cellobiose concentration is very low and the steady-state concentration of glucose in the system remains at ~3 mM. Under these circumstances cellulolytic enzymes are hardly hindered by product inhibition and as a result steady currents and continuous production rates are observed as long as glucose derived from cellulose is available. The fact that the current remains constant implies that the Pt-electrode is invariant with respect to product accumulation, to the presence of cellulose and to the industrial enzyme preparation.

As is shown in Fig. 6 cellulose is converted completely and exclusively into gluconic acid and almost completely into H_2 according to reaction (2):

$$\text{cellulose} \xrightarrow[\text{glucose oxidase}]{\text{cellulases}} \text{gluconic acid} + H_2 + \text{electricity} \qquad (2)$$

Other interesting fuels are sucrose and starch. In the presence of invertase sucrose is hydrolyzed to a mixture of glucose and fructose. In the cell glucose is subsequently converted into gluconic acid and H_2. Hence 4

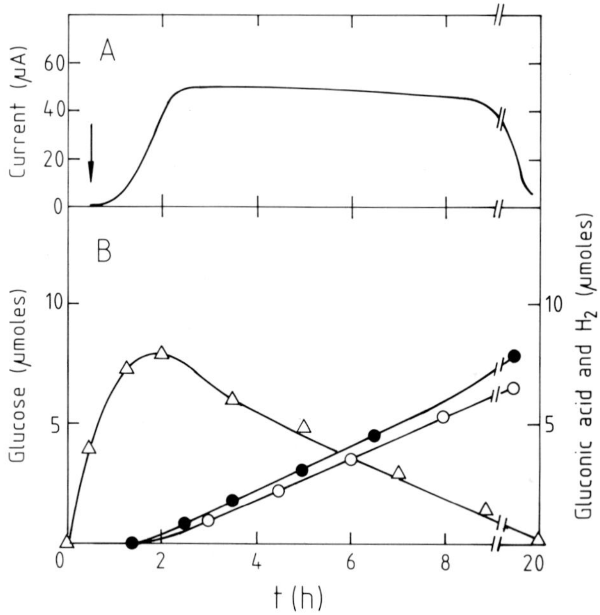

Figure 6. Performance of a cellulose containing bioelectrochemical cell. Experiments were performed as described in Materials and Methods. At the time indicated by an arrow cellulose was added to the anode compartment. Δph = 6. (A) Time course of current output. (B) Time course of: O, H_2-production; ●, gluconic-acid production; Δ, glucose concentrations.

products are produced. Likewise, starch can be degraded to glucose units by the combined action of α-amylase and amyloglucosidase. Work is in progress to investigate these possibilities.

Besides glucose oxidase other oxidases or related enzymes can be used as well. For example amino acid oxidase [17], laccase [18], hyaluronidase [19], hydrogenase [20], methanol dehydrogenase [10] and alcohol dehydrogenase. Recently, we investigated the possibilities of xanthine oxidase in our biofuel cell [4]. Xanthine oxidase was chosen since this enzyme shows a notably wide substrate-specificity for many azoheterocyclics like pyrimidines, purines pteridines and related compounds. Fig. 7 shows the performance of a xanthine oxidase containing fuel cell with O_2 as ultimate electron acceptor. In addition to the enzymatic conversion of xanthine into uric acid, it appeared that uric acid is further oxidized electrochemically. Initially uric acid is oxidized in an n=2 reaction to a diimine intermediate which

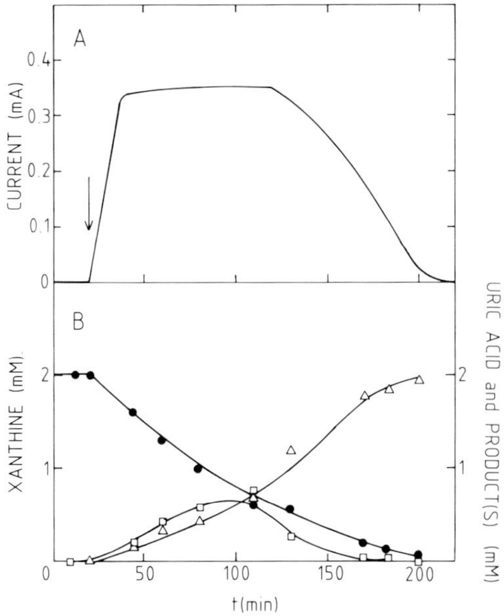

Figure 7. Performance of a xanthine oxidase containing bioelectrochemical cell. Experiments were performed as in [4]. At the time indicated by an arrow xanthine was added to the anode compartment. (a) Time course of current output. (b) Time course of: •, xanthine consumption; □, uric acid formation; Δ, product(s) formation.

reacts further to give products that are not clearly identified [22].
So far, the anode compartment either produces H_2O or H_2. However, by using a gold electrode instead of a platinum electrode H_2O_2 can be generated and be subsequently scavenged by chloroperoxidase. In the presence of Cl^- this enzyme halogenates a variety of compounds such as steroids [23], β-dicarbonyl systems [24], alkenes [25], alkynes [26], cyclopropanes [26] and many heterocyclics [27]. Figure 8 shows the reactions taking place in a glucose oxidase and chloroperoxidase containing fuel cell. Barbituric acid was taken as a model substrate [27].
With glucose as fuel 3 products are formed; gluconic acid, 5-chlorobarbituric acid and electricity. After 3 days our small scale model had produced ~10 mg gluconic acid and 8 mg of 5-chlorobarbituric acid. Determination of turnover numbers (moles of product formed per mole of enzyme) gave values of 1.8×10^4 and 10^7 for glucose oxidase and chloroperoxidase respectively.

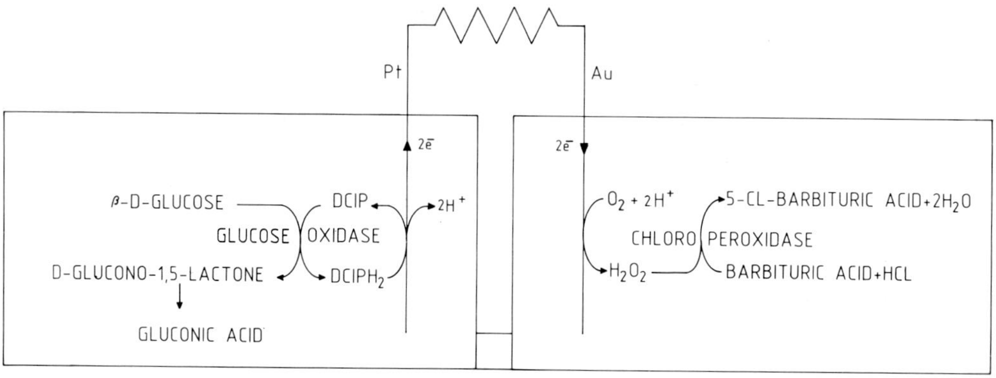

Figure 8. Reactions taking place in a bioelectrochemical cell containing glucose oxidase in the anode compartment and chloroperoxidase in the cathode compartment. For explanation see text and for details ref. 4.

The results clearly demonstrate that the enzymes in the redox cycle of Fig. 8 operate catalytically. These results, although preliminary, are encouraging and future developments such as enzyme stabilization and improvement of the electrode and cell design should facilitate much higher currents and higher production yields.

Normally the product of chloroperoxidase action is a mixture of mono- and dihalogenated compounds [25,26]. Surprisingly, in our cell only the monohalogenated compound could be detected. We found evidence that the enzyme produces the dihalogenated form of barbituric acid, but that the second chloride is removed by a reductive electrochemical reaction. Other (bio)-chemicals than those shown above can also be prepared simply by using different oxidases, dehydrogenases or peroxidases together with their respective substrates. The fact that the chemicals produced in different compartments facilitates their isolation. Hence, in a biofuel cell optimum use of a fuel can be made due to a combination of chemical, biochemical and electrochemical reactions.

An essential difference between the "classical" fuel cells and our cell is that the former converts chemical energy mainly into electrical energy, whereas our cell converts chemical energy into other forms of chemical energy and just a little bit of electrical energy. The reactions taking place in our cell are chosen in such a way that chemical energy released by the fuel is stored sequently as it travels through the system into chemicals of interest.

Acknowledgements

We thank Mr. M.M. Bouwmans for the preparation of the figures, Mrs. J.C. Toppenberg-Fang for typing the manuscript and the pharmaceutical industry Duphar BV (Weesp) for financing part of this investigation.

References

1. Hilhorst, R., Laane, C. and Veeger, C. (1983) FEBS Lett. 159, 225-228.
2. Hilhorst, R. (1984) Thesis, Agricultural University, Wageningen, The Netherlands.
3. Van der Westen, H.M., Mayhew, S.G. and Veeger, C. (1978) FEBS Lett. 86, 122-126.
4. Laane, C., Pronk, W., Franssen, M. and Veeger, C. (1984) Enzyme Microb. Technol. 6, 165-168.
5. Simon, M.S. and Moore, P.T. (1975) J.Polym.Sci., 13, 1-16.
6. Bergmeyer, H.U., Bernt, E., Schmidt, F. and Stork, H. (1970) in Methode der Enzymatischen Analyse (Bergmeyer, H.U. ed.) Verlag Chemie, Weinheim and Ac. Press, New York and London, Vol. II, pp. 1163-1172.
7. Bernt, E. and Bergmeyer, H.U. ibid, pp. 1266-1269.
8. Bergmeyer, H.U. and Bernt, E., ibid, pp. 1143-1146.
9. Möllering, H. and Bergmeyer, H.U. (1974) in: Methods in Enzymatic Analysis (Bergmeyer, H.U., ed.) Verlag Chemie, Weinheim and Ac. Press, New York and London, Vol. 3, pp. 1243-1292.
10. Plotkin, E.V., Higgins, I.J. and Hill, H.A.O. (1981) Biotechnol. Lett. 3, 187-192.
11. Wong, C-H., Daniels, L., Orme-Johnson, W.H. and Whitesides, G.M. (1981) J.Am.Chem.Soc. 103, 6227-6228.
12. Egerer, P., Simon, H., Tanaka, A. and Fukui, S. (1982) Biotechn. Lett. 4, 489-494.
13. Bockris, J.O'm., Conway, B.E., Yeager, E. and White, R.E. in: Comprehensive Treatise of Electrochemistry, Plenum Press, New York, 1981, vol. 3.

14. Davis, G., Hill, H.A.O., Aston, W.J., Higgins, I.J. and Turner, A.P.F. (1983) Enzyme Microb.Technol., 5, 383-388.
15. Kominek, J. (1983) in Biotechnology: A Comprehensive Treatise (Rehm, H.J. and Reed, G., eds.) Verlag Chemie, Weinheim, pp. 355-465.
16. De Wit, G., De Vlieger, J.J., Kock-Van Dalen, A.C., Kieboom, A.P.G. and Van Bekkum, H., Tetrahedron Lett., 1978, 15, 1327-1330.
17. Lewis, K. (1966) Bacteriol.Rev., 30, 101-112.
18. Berezin, I.V., Bogdanovskaya, V.A., Varfolomeev, S.D., Tarasevich, M.R. and Yargopolov, A.I. (1978) Dokl.Akad.Nauk.SSSR, 240 615-618.
19. Ahn, B.K., Wolfson, Jr. S.K., Yao, S.J., Lui, C.C., Todd, R.C. and Wiener, S.B. (1976) J.Biomed.Mater.Res. 10, 283-294.
20. Varfolomeev, S.D., Yaropolov, A.I., Berezin, I.V., Tarasevich, M.R. and Bogdanovskaya, V.A. (1977) Bioelectrochem.Bioenerg. 4, 314-326.
21. Kulis, Y.Y., Malinauskas, A.A. and Kadzyauskene, K.V. (1978) Proc. 3rd Joint US/USSR Enzyme Eng.Sem.US Govt. publication PB283 328 T, pp. 516-528.
22. Goyal, R.N., Nguyen, N.T. and Dryhurst, G. (1982) Bioelectrochem.Bioenerg. 9, 273-285.
23. Neidleman, S.L., Diassi, P.A., Junta, B., Palmere, R.M. and Pan, S.C. (1966) Tetrahedron Lett., 5337-5342.
24. Thomas, J.A., Morris, D.R. and Hager, L.P. (1970) J.Biol.Chem. 245, 3129-3134.
25. Neidleman, S.L., Amon, W.F. and Geigert, J. (1981) US Patent 4 247 641.
26. Geigert, J., Neidleman, S.L. and Dalietos, D.J. (1983) J.Biol.Chem. 258, 2273-2277.
27. Franssen, M.C.R. and Van der Plas, H.C. (1984) Recl.Trav.Chim., Pays-Bas, 103, 99-100.

FLAVIN COFACTORS COVALENTLY ATTACHED TO ELECTRON CONDUCTING SUPPORTS: ELECTROCHEMICAL AND ENZYME ACTIVITY

Lemuel B. Wingard Jr., Krishna Narasimhan and Osato Miyawaki
Department of Pharmacology, School of Medicine, University of Pittsburgh, Pittsburgh, PA 15261, USA

Introduction

The relative spatial arrangement of the individual components that make up flavin cofactor enzyme systems is an important and as yet unresolved aspect of flavin oxidase enzyme chemistry. Although the three dimensional structures of the flavin enzymes glutathione reductase, flavodoxin, and p-hydroxybenzoate hydroxylase are known in considerable detail, very little information is available on the spatial arrangements for oxidase enzymes. For some oxidases there is evidence to suggest that the flavin position-8 may be exposed and thus might be a suitable point for immobilization of FAD on a solid support without loss of enzyme activity. It is not clear if the adenine amino group of FAD also could serve as a point of immobilization with retention of enzyme activity. The adenine amino group is located within the apoenzyme matrix in the flavin enzymes of known three dimensional structure. This suggests that immobilization at the amino group would lead to loss of enzyme activity. In apparent contrast to these predictions is a report of the covalent attachment of 40,000 to 60,000 molecular weight polyethyleneimine to the adenine amino group of FAD with 97% reconstitution of activity on addition of the apoenzyme of glucose oxidase (1).

Flavins and Flavoproteins

The work reported here is part of a longer range project to characterize the electron transfer and enzyme cofactor properties of FAD immobilized on the surface of electron conducting supports. In the present work FAD was coupled covalently to glassy carbon through either the position-8 methyl group or the adenine N^6 amino group (or possibly a ribityl OH group). The presence of the attached flavin was demonstrated electrochemically; and the coenzyme activity of the immobilized flavin was established by incubation of the derivatized glassy carbon with the apoenzyme of glucose oxidase.

Results

Low porosity Tokai glassy carbon electrodes were activated using hot acid to generate surface carboxylic groups. For attachment at the flavin position-8 (2), the glassy carbon carboxylic groups were converted to aldehydes. Riboflavin was acetylated, brominated, and reacted with n-butyllithium. Coupling occurred, presumably to the position-8, as evidenced by differential pulse voltammetry measurements with the derivatized electrode and by homogeneous solution reactions with a model aldehyde and NMR verification of the structures. Control experiments with underivatized glassy carbon showed no signs of coupling taking place. Deacetylation and phosphorylation led to the formation of immobilized FMN. Subsequent reaction with adenosine-5' -phosphoromorpholidate gave immobilized FAD. This again was evidenced by differential pulse voltammetry. Adsorbed FAD could be ruled out, based on earlier work with adsorbed flavins on carbon supports (3). The FAD loading was 2.1×10^{-11} moles/cm^2. The ability of the position-8 immobilized FAD electrode to reconstitute enzyme activity was tested by incubation of the derivatized electrode with the apoenzyme of glucose oxidase.

The apoenxyme was prepared by treatment of _A_. _niger_ glucose oxidase with glycerol/sulfuric acid, as used by Morris et al. (4). The immobilized reconstituted enzyme was assayed by immersing the electrode in buffered glucose solution, taking aliquots at known times, and assaying the aliquots for hydrogen peroxide by the peroxidase/o-dianisidine colormetric method (5). The FAD electrodes showed 0.0021 units of activity/electrode. Several papers have been submitted.

For attachment at the FAD N^6 amino group (or possibly ribityl OH), the carboxylic glassy carbon was activated with a water soluble carbodiimide. This was followed by attachment of either FAD or a spacer molecule. 6-Aminocaproic acid or ethylenediamine were used as spacers. FAD coupling to the aminocaproic acid spacer was accomplished through a second carbodiimide activation and to the ethylenediamine spacer through glutaraldehyde. The presence of the attached FAD was verified by differential pulse and cyclic voltammetry. Adsorbed FAD was removed completely by washing, as demonstrated by rigorous control experiments using differential pulse voltammetry to detect any adsorbed material. Upon incubation of the amino (or ribityl) immobilized FAD with the apoenzyme of glucose oxidase, no reconstituted enzyme activity could be detected with or without the spacer molecules (paper submitted).

From this work we conclude that FAD immobilized at flavin position-8 to glassy carbon can be incubated with the apoenzyme of glucose oxidase to give partial reconstitution of enzyme activity. However, FAD immobilized to glassy carbon through the adenine amino group (or a ribityl OH group) does not lead to reconstitution of enzyme activity, either with or without short spacer molecules between the electrode and the FAD.

This work is continuing with plans to include spacer molecules for the position-8 coupling, immobilization at other positions, repeat of the earlier polyethyleneimine work (1), and the measurement of electron transfer rate constants between the immobilized cofactor and the solid electrode surface.

This work was supported by the National Science Foundation and by the Army Research Office.

References

1. Zappelli, P., Pappa, R., Rossodivita, A., Re, L.: Eur. J. Biochem. 89, 491-499 (1978).
2. Wingard Jr., L.B., Gurecka Jr., J.L.: J. Mol. Cat. 9, 209-217 (1980).
3. Miyawaki, O., Wingard Jr., L.B.: Biotechnol. Bioeng. (in press).
4. Morris, D.L., Ellis, P.B., Carrico, R.J., Yeager, F.M., Schroeder, H.R., Albarella, J.P., Boguslaski, R.C., Hornby, W.E., Rawson, D.: Anal. Chem. 53, 658-665 (1981).
5. Wingard Jr., L.B., Cantin, L.A., Castner, J.F.: Biochim. Biophys. Acta 748, 21-27 (1983).

LUMINOMETRIC DETERMINATION OF IMMUNOADSORBED FLAVOPROTEINS AND OF FLAVIN-ADENINE DINUCLEOTIDE

Ari Hinkkanen, Karl Decker
Biochemisches Institut der Universität Freiburg
D-7800 Freiburg i.Br., F.R.G.

The detection of 6-hydroxy-D-nicotine oxidase and 6-hydroxy-L-nicotine oxidase at the femtomol (μ unit) level was achieved by coupling the oxidation of 6-hydroxy-D- and 6-hydroxy-L-nicotine, respectively, by these flavoenzymes with the peroxidase-catalyzed oxidation of luminol (1). Both oxidases retained their full activity when bound to the respective immobilized antibodies; that allowed the quantitative determination of the enzymes in very dilute solutions. Due to the strict stereospecificity and the absence of immunological cross-reactivity, the flavoproteins could be assayed in the same solution. The method is well suited for the detection of the enzymatic activity of the flavoproteins synthesized in cell-free systems.

The same principle was used for the quantitation of small amounts of FAD (2). Apo-D-amino acid oxidase recombines readily with FAD producing an active holoenzyme. Peroxide formed during the oxidation of D-alanine by the reconstituted D-amino acid oxidase was used for the peroxidase-catalyzed oxidation of luminol. The maximal rate of light emission was proportional to the amount of FAD present in the sample. The method allowed quantitative measurements of 0.1-2 pmol FAD per assay (0.2-4 nmol/l). Other flavin species like riboflavin, FMN and 5-deazaflavin do not interfere with this determination.

Flavins and Flavoproteins

References

1. Hinkkanen, A., Maly, F.E., Decker, K. (1983) Hoppe-Seyler's Z. Physiol. Chem. 364, 407-412.

2. Hinkkanen, A., Decker, K. (1983) Analyt. Biochem. 132, 202-208.

FLAVINS AS LABELS IN IMMUNOASSAYS

I. Design and Synthesis of Flavin Labels

J.P. Albarella, D.L. Morris, and K.F. Yip
Miles Laboratories, Inc., Ames Division

The Apoenzyme Reactivation Immunoassay System (ARIS) is a competitive binding assay in which FAD is used as a label.[1] The concept of the assay is shown below in Figure 1.

FIG. 1 CONCEPT OF THE APOENZYME REACTIVATED IMMUNOASSAY SYSTEM

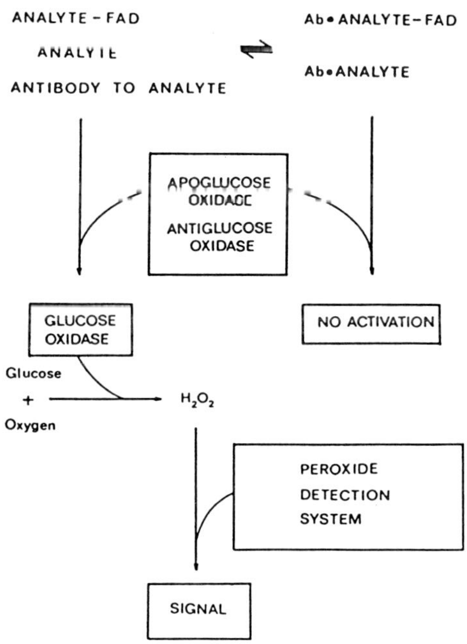

Flavins and Flavoproteins

The FAD label is detected by the measurement of the glucose oxidase activity generated when excess apoglucose oxidase is added to the assay mixture. The assay requires no separation because the free FAD label can be detected in the presence of antibody-bound analyte-FAD conjugate, which does not activate apoenzyme. This homogeneous assay has high sensitivity because of the intrinsic amplification from the conversion of FAD labels into active centers of glucose oxidase. The hydrogen peroxide thus generated in the assay can be detected by a variety of peroxidase catalyzed chromogenic reactions.

The synthetically prepared analyte-FAD conjugate is a key component in the assay. Its structure must be designed to ensure the following certain essential characteristics. First, the FAD moiety must retain its function as a prosthetic group for glucose oxidase. The analyte portion of the conjugate must bind to the analyte-specific antibody such that the FAD activity of the label is quenched. Free analyte must compete effectively with the conjugate for the antibody binding sites. Finally, a general synthetic strategy should be devised to obtain a wide variety of stable, highly purified FAD conjugates.

FAD derivatives for ARIS are prepared from N^6-(6-aminohexyl) AMP according to the synthetic scheme outlined in Figure 2. N^6-(6-aminohexyl) FAD has found wide application in our laboratory for the labeling of numerous drug derivatives, proteins, and hormones of biological interest. Its synthesis begins with the protection of 6-aminohexyl AMP with ethyl trifluoroacetate to give the crystalline N-TFA derivative in 76% yield. This compound is then converted into its FAD derivative by carbonyldiimidazole activation and coupling with triethylammonium 5'-FMN in 1:1 DMF-DMSO. Optimum coupling reaction conditions were found to be with 1.5 molar equivalents of 5'-FMN at 0.1 molar concentration at 30° for 20 h.

The crude conjugate is purified by silicic acid chromatography. The N-trifluoroacetamide group is then removed by treatment at pH 10.5 for 24 h. 6-Aminohexyl FAD is finally isolated in 25% yield by either Bio-Gel P-2 chromatography or semi-preparative reverse phase HPLC.

FIG. 2 SYNTHESIS OF FAD DERIVATIVES

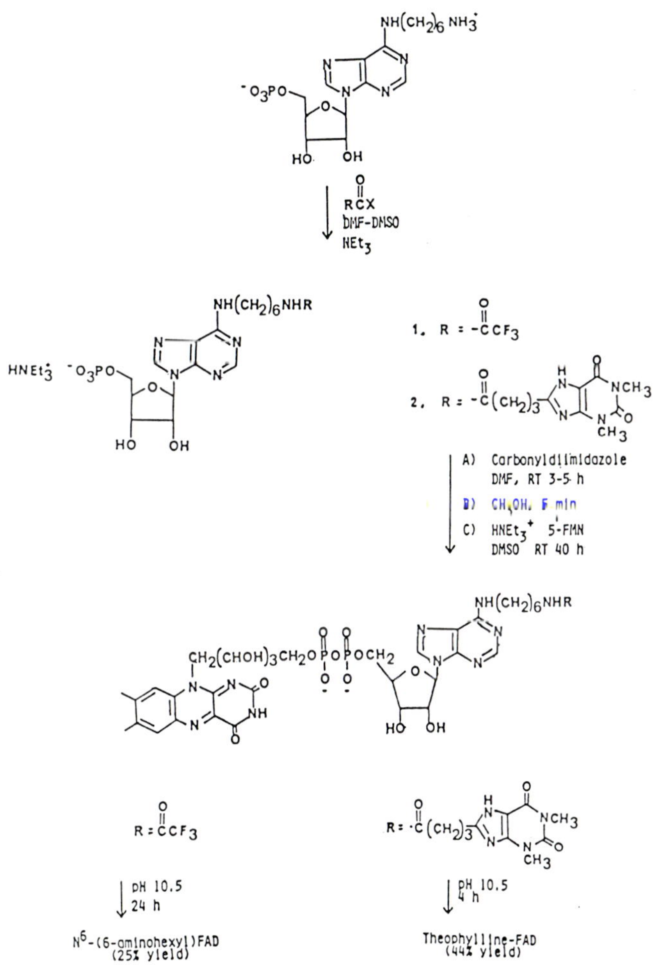

The synthesis of haptenic-FAD conjugates is illustrated in Figure 2 by the method for the preparation of theophylline-FAD. The lactam of 8-(3-carboxypropyl)theophylline is reacted with N^6-(6-aminohexyl) AMP, and is then converted to the FAD conjugate by the carbonyldiimidazole activation method. The crude conjugate and its 2',3'-ribosyl carbonate derivative[2] are isolated by silicic acid chromatography and then hydrolyzed by alkali. Theophylline-FAD is then isolated in 44% yield by either Bio-Gel P-2 or reverse phase chromatography.

A general semi-preparative procedure has been developed in our laboratories for the isolation of 10-300 μmolar quantities of FAD conjugates from crude reaction mixtures.[3] This rapid method, replacing the previous chromatography protocol, is illustrated below for the purification of theophylline-FAD.

FIG. 3 PURIFICATION OF THEOPHYLLINE-FAD BY SEMIPREPARATIVE HPLC

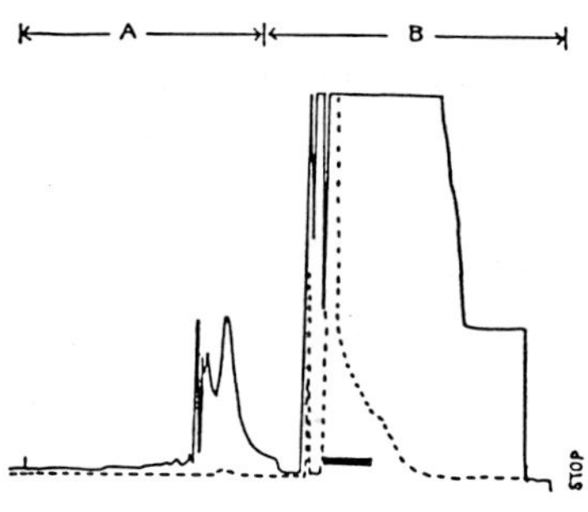

Column: Whatman ODS-3, M-9

Flow Rate: 4.0 mL/min

Mobile Phase: A = H_2O

B = 85:15 (v/v) PPB, 0.02M, pH 5.5: CH_3CN

Detector: Solid line 290 nm
Dashed line 510 nm

Product was eluted from the column in the area designated by the solid bar (—).

REFERENCES

1. D.L. Morris, et. al., Anal. Chem. 53, 658 (1981).

2. M. Malda, A.P. Patel and A. Hampton, Nucleic Acids Res. 4, 2843 (1977).

3. K.F. Yip and J.P. Albarella, Fourth International Symposium on the HPLC of Proteins, Peptides, and Polyncleotides, Dec. 10-12, 1984, Baltimore, MD.

FLAVINS AS LABELS IN IMMUNOASSAYS

II. Properties of Flavin Labels and Their Use in Immunoassay

David L. Morris and James P. Albarella, Immunodiagnostics Department, Ames Division, Miles Laboratories, Inc., Elkhart, IN 46515, USA.

The preceding paper describes how a variety of analyte-FAD conjugates can be synthesized. This is a description, by reference to theophylline-FAD, of how a conjugate can be used to construct an Apoenzyme Reactivation Immunoassay System (ARIS) assay. The goal was to demonstrate a rapid assay for theophylline, using a small sample volume and operating at 37°C.

The other reagents required in the immunoassay are an antibody specific for theophylline, a conjugate detection system which consists of apoglucose oxidase and antiserum against native glucose oxidase, and a glucose oxidase detection system that comprises glucose, peroxidase, 3,5 dicholoro-4-hydroxy benzene sulphonate and 4-aminophenazine. Color is generated and read at 520 nm.

Antiserum to native glucose oxidase is an essential ingredient to make the assay work at 37°C. In Fig. 1 it is shown that whereas theophylline-FAD is active with apoenzyme at 25°C, very little activation occurs at 37°C. FAD, on the other hand, is active at 37°C. Incorporation of antiserum to glucose oxidase induces a dramatic enhancement of the activation of apoenzyme by theophylline-FAD at 37°C, such that activity similar to that shown for theophylline-FAD at 25°C in Fig. 1 is achieved.

ACTIVATION OF CROSS-LINKED APOGLUCOSE OXIDASE BY THEOPHYLLINE -FAD

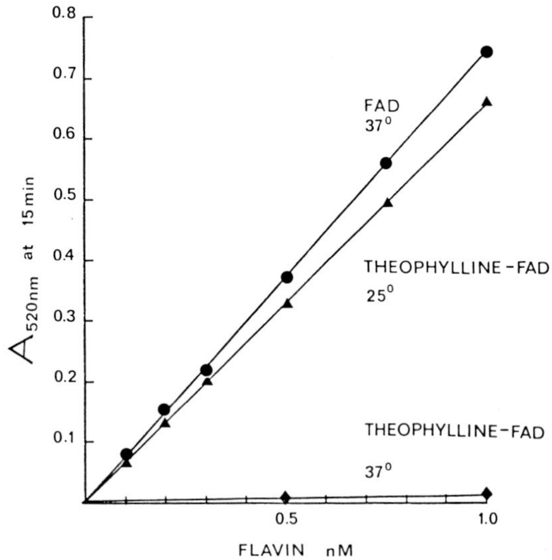

Fig. 1 Assay contains 100 nM cross-linked apoenzyme, 2 mM DHSA, 0.2 mM 4-aminophenazine, 20 μg/ml peroxidase, 0.1 M glucose and 0.1 M phosphate buffer, pH 7.0.

Apoenzyme containing about 1.7 N_{ε} N_{ε}-adipamidino-bis-lysine cross-links per FAD binding site was used. This cross-linked apoenzyme is much more stable than native apoenzyme. Lyophilized cross-linked apoenzyme is indefinitely stable at 4°C, and when stored in a 50% (w/v) glycerol solution retains about 90% of its activity after 10 months at 4°C. This apoenzyme is prepared from glucose oxidase cross-linked with dimethyl adipimidate. The FAD is dissociated at pH 1.7 in the presence of 30% (w/v) glycerol to stabilize the protein, and is removed by gel filtration and absorption with dextran coated charcoal.

Fig. 2 demonstrates that the theophylline-FAD has all the necessary attributes for an ARIS assay. Antibody to theophylline strongly inhibits activity of the conjugate with apoenzyme, and reversal of inhibition is obtained by including 1.0 μl of serum containing 40 μg/ml of theophylline. A suitable standard curve was obtained with 1.5 μl of antiserum to theophylline in the assay, using the conditions of Fig. 2. The performance of the assay is described in Table 2.

EFFECT OF ANTISERUM TO THEOPHYLLINE ON THE ACTIVITY OF THEOPHYLLINE-FAD WITH CROSS-LINKED APOGLUCOSE OXIDASE

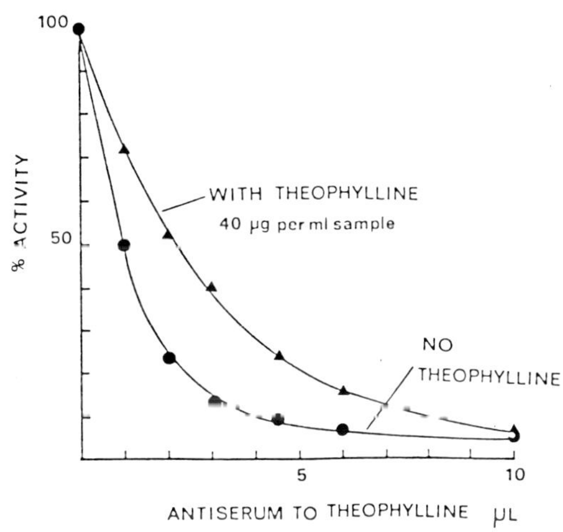

Fig. 2 Assay contains in addition to reagents listed in legend to Fig. 1, 10 nM theophylline-FAD and 5 μl/ml antiserum to glucose oxidase. The theophylline was added as 15 μl of 15-fold prediluted serum (40 μg/ml serum) giving a final assay volume of 2.015 ml. The assay was incubated at 37°C for 4 min.

TABLE I PERFORMANCE OF ARIS ASSAY FOR THEOPHYLLINE AT 37°C

PRECISION

	mean (μg/ml)	standard deviation	(CV(%)
Within assay (n = 20)	14.8	0.7	4.4
Between assay (n = 8, triplicates performed on 8 days)	15.0	1.3	8.7

CORRELATION WITH HPLC*

$$Y = 0.89X + 0.08$$

n = 57

Correlation Coefficient = 0.99

Standard error of estimate = 0.9 μg/ml

*HPLC was performed by South Bend Medical Foundation

ARIS has been found to be applicable to the assay of analytes over a wide range of concentration. An assay for digoxin has been developed with liquid reagents on a totally automated system (Ames Optimate®) which measures digoxin in serum in the range 0 to 5 nM. At the other end of the concentration range a dry reagent manual assay that measures serum theophylline in the range 0 - 200 μM theophylline has been developed that is performed with a reflectance photometer (Ames Seralyzer®).

Author Index

Subject Index